Everyday Mathematics®

The University of Chicago School Mathematics Project

TEACHER'S LESSON GUIDE
VOLUME 1

McGraw Hill

The University of Chicago School Mathematics Project

Max Bell, Director, *Everyday Mathematics* First Edition; James McBride, Director, *Everyday Mathematics* Second Edition; Andy Isaacs, Director, *Everyday Mathematics* Third, CCSS, and Fourth Editions; Amy Dillard, Associate Director, *Everyday Mathematics* Third Edition; Rachel Malpass McCall, Associate Director, *Everyday Mathematics* CCSS and Fourth Editions; Mary Ellen Dairyko, Associate Director, *Everyday Mathematics* Fourth Edition

Authors
Robert Balfanz*, Max Bell, John Bretzlauf, Sarah R. Burns**, William Carroll*, Amy Dillard, Robert Hartfield, Andy Isaacs, James McBride, Kathleen Pitvorec, Denise A. Porter‡, Peter Saecker, Noreen Winningham†

*First Edition only
**Fourth Edition only
†Third Edition only
‡Common Core State Standards Edition only

Fourth Edition Grade 5 Team Leader
Sarah R. Burns

Writers
Melanie S. Arazy, Rosalie A. DeFino, Allison M. Greer, Kathryn M. Rich, Linda M. Sims

Open Response Team
Catherine R. Kelso, Leader, Emily Korzynski

Differentiation Team
Ava Belisle-Chatterjee, Leader, Martin Gartzman, Barbara Molina, Anne Sommers

Digital Development Team
Carla Agard-Strickland, Leader, John Benson, Gregory Berns-Leone, Juan Camilo Acevedo

Virtual Learning Community
Meg Schleppenbach Bates, Cheryl G. Moran, Margaret Sharkey

Technical Art
Diana Barrie, Senior Artist; Cherry Inthalangsy

UCSMP Editorial
Don Reneau, Senior Editor, Rachel Jacobs, Elizabeth Olin, Kristen Pasmore, Loren Santow

Field Test Coordination
Denise A. Porter, Angela Schieffer, Amanda Zimolzak

Field Test Teachers
Diane Bloom, Margaret Condit, Barbara Egofske, Howard Gartzman, Douglas D. Hassett, Aubrey Ignace, Amy Jarrett-Clancy, Heather L. Johnson, Jennifer Kahlenberg, Deborah Laskey, Jennie Magiera, Sara Matson, Stephanie Milzenmacher, Sunmin Park, Justin F. Rees, Toi Smith

Digital Field Test Teachers
Colleen Girard, Michelle Kutanovski, Gina Cipriani, Retonyar Ringold, Catherine Rollings, Julia Schacht, Christine Molina-Rebecca, Monica Diaz de Leon, Tiffany Barnes, Andrea Bonanno-Lersch, Debra Fields, Kellie Johnson, Elyse D'Andrea, Katie Fielden, Jamie Henry, Jill Parisi, Lauren Wolkhamer, Kenecia Moore, Julie Spaite, Sue White, Damaris Miles, Kelly Fitzgerald

Contributors
John Benson, Jeanne Di Domenico, James Flanders, Fran Goldenberg, Lila K. S. Goldstein, Deborah Arron Leslie, Sheila Sconiers, Sandra Vitantonio, Penny Williams

Center for Elementary Mathematics and Science Education Administration
Martin Gartzman, Executive Director, Meri B. Fohran, Jose J. Fragoso, Jr., Regina Littleton, Laurie K. Thrasher

External Reviewers

The *Everyday Mathematics* authors gratefully acknowledge the work of the many scholars and teachers who reviewed plans for this edition. All decisions regarding the content and pedagogy of *Everyday Mathematics* were made by the authors and do not necessarily reflect the views of those listed below.

Elizabeth Babcock, California Academy of Sciences; Arthur J. Baroody, University of Illinois at Urbana-Champaign and University of Denver; Dawn Berk, University of Delaware; Diane J. Briars, Pittsburgh, Pennsylvania; Kathryn B. Chval, University of Missouri–Columbia; Kathleen Cramer, University of Minnesota; Ethan Danahy, Tufts University; Tom de Boor, Grunwald Associates; Louis V. DiBello, University of Illinois at Chicago; Corey Drake, Michigan State University; David Foster, Silicon Valley Mathematics Initiative; Funda Gönülateş, Michigan State University; M. Kathleen Heid, Pennsylvania State University; Natalie Jakucyn, Glenbrook South High School, Glenview, IL; Richard G. Kron, University of Chicago; Richard Lehrer, Vanderbilt University; Susan C. Levine, University of Chicago; Lorraine M. Males, University of Nebraska-Lincoln; Dr. George Mehler, Temple University and Central Bucks School District, Pennsylvania; Kenny Huy Nguyen, North Carolina State University; Mark Oreglia, University of Chicago; Sandra Overcash, Virginia Beach City Public Schools, Virginia; Raedy M. Ping, University of Chicago; Kevin L. Polk, Aveniros LLC; Sarah R. Powell, University of Texas at Austin; Janine T. Remillard, University of Pennsylvania; John P. Smith III, Michigan State University; Mary Kay Stein, University of Pittsburgh; Dale Truding, Arlington Heights District 25, Arlington Heights, Illinois; Judith S. Zawojewski, Illinois Institute of Technology

Note

Many people have contributed to the creation of *Everyday Mathematics*. Visit http://everydaymath.uchicago.edu/authors/ for biographical sketches of *Everyday Mathematics* Fourth Edition staff and copyright pages from earlier editions.

www.everydaymath.com

ISBN: 978-0-07-703845-8
MHID: 0-07-703845-2

Printed in the United States of America.

3 4 5 6 7 8 9 10 11 WEB 27 26 25 24 23 22 21 20

Welcome to *Everyday Mathematics*

The elementary program from the
University of Chicago School Mathematics Project

Dear Teacher,

Everyday Mathematics 4 is designed to help you teach the content required by the Indiana Academic Standards for Mathematics. In fifth grade, that content focuses on procedures, concepts, and applications in three critical areas:

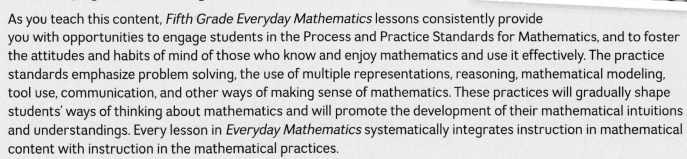

- developing fluency with addition and subtraction of fractions, and an understanding of multiplication of fractions and of division of fractions in limited cases;
- developing fluency with whole-number and decimal operations, extending division to 2-digit divisors, integrating decimals into the place-value system, and developing understanding of operations with decimals to hundredths; and
- developing an understanding of volume.

As you teach this content, *Fifth Grade Everyday Mathematics* lessons consistently provide you with opportunities to engage students in the Process and Practice Standards for Mathematics, and to foster the attitudes and habits of mind of those who know and enjoy mathematics and use it effectively. The practice standards emphasize problem solving, the use of multiple representations, reasoning, mathematical modeling, tool use, communication, and other ways of making sense of mathematics. These practices will gradually shape students' ways of thinking about mathematics and will promote the development of their mathematical intuitions and understandings. Every lesson in *Everyday Mathematics* systematically integrates instruction in mathematical content with instruction in the mathematical practices.

Throughout *Everyday Mathematics*, emphasis is placed on

- problem solving in everyday situations and mathematical contexts;
- an instructional design that revisits topics regularly to ensure depth of knowledge and long-term learning;
- distributed practice through games and other daily activities;
- teaching that supports "productive struggle" and maintains high cognitive demand; and
- lessons and activities that engage *all* students and make mathematics fun.

Your *Teacher's Lesson Guide* and other program features provide support as you implement the program. See Getting Ready to Teach on pages xxxvi–xxxix for some advice about how to begin. As you gain experience with *Fifth Grade Everyday Mathematics*, you will become more comfortable with its content, components, and approaches. By the end of the year, we think you will agree that the rewards are worth the effort.

On a personal note, I would like to acknowledge the efforts of the fifth grade team throughout the development process, but especially during my two maternity leaves. I specifically want to thank Katie Rich, who served as substitute team leader during my leaves. I am indebted to her, and to the entire team, for their support.

Sarah Burns
and the *Fifth Grade Everyday Mathematics* Development Team

An Investment in How Your Children Learn

Behind each student success story is a team of teachers and administrators who set high expectations for themselves and their students. *Everyday Mathematics* is designed to help you achieve those expectations with a research-based approach to teaching mathematics.

The *Everyday Mathematics* Difference

Decades of research show that students who use *Everyday Mathematics* develop deeper conceptual understanding and greater depth of knowledge than students using other programs. They develop powerful, life-long habits of mind such as perseverance, creative thinking, and the ability to express and defend their reasoning.

Daniel LaFlor/Vetta/Getty Images

A Commitment to Educational Equity

Everyday Mathematics was founded on the principle that every student can and should learn challenging, interesting, and useful mathematics. The program is designed to ensure that each of your students develops positive attitudes about math and powerful habits of mind that will carry them through college, career, and beyond.

Provide Multiple Pathways to Learning

Through *Everyday Mathematics*' spiraling structure, your students develop mastery by repeatedly experiencing math concepts in varied contexts, with increasing sophistication, over time. By providing multiple opportunities to access math concepts, you can easily adapt your instruction to better meet the unique learning needs of your children.

Access High Quality Materials

All students deserve strong learning materials especially in early childhood. You can be confident teaching with *Everyday Mathematics* because your instruction is grounded in a century of research in the learning sciences and has been rigorously field tested and proven effective in classrooms for over thirty years.

Use Data to Drive Your Instruction

Using the Quick-Entry Evaluation tool in the ConnectED Teacher Center, you can go beyond tracking progress solely through periodic assessments and easily record evaluations of almost every activity your students engage in every day. The data you collect drives a suite of reports that help you tailor your instruction to meet the needs of every student in your classroom.

Create a System for Differentiation in Your Classroom

Turn your classroom into a rich learning environment that provides multiple avenues for each of your students to master content, make sense of ideas, develop skills, and demonstrate what they know. *Everyday Mathematics* helps you do this by providing the tools you need to effectively address the key components of effective differentiation in your classroom: Content, Process, Product, Classroom Organization, and Learning Environment.*

Build and Maintain Strong Home-School Connections

Research shows that strengthening the link between home and school is integral to your students' success. That's why *Everyday Mathematics* provides a wealth of resources to help you extend what your students learn in your classroom to what they can do at home.

*Tomlinson & Murphy, M (2015). Leading for Differentiation: Growing Teachers Who Grow Kids. ASCD.

Build Mathematical Literacy

Designed for College and Career Readiness, *Everyday Mathematics* builds a solid foundation for success in your mathematics classroom through meaningful practice opportunities, discussion of reasoning and strategies, and engagement in the mathematical practices every day.

Focused Instruction

The instructional design of *Everyday Mathematics* allows you to focus on the critical areas of instruction for each grade.

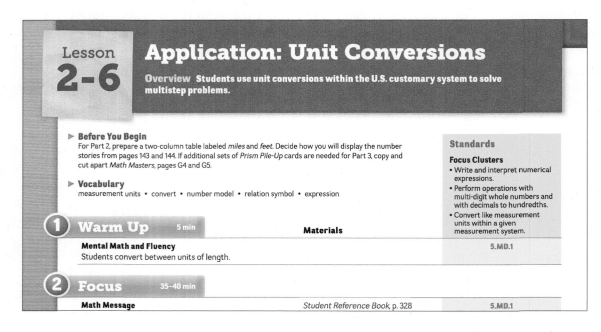

Lesson 2-6

Application: Unit Conversions

Overview Students use unit conversions within the U.S. customary system to solve multistep problems.

▶ **Before You Begin**
For Part 2, prepare a two-column table labeled *miles* and *feet*. Decide how you will display the number stories from pages 143 and 144. If additional sets of *Prism Pile-Up* cards are needed for Part 3, copy and cut apart *Math Masters*, pages G4 and G5.

▶ **Vocabulary**
measurement units • convert • number model • relation symbol • expression

Standards

Focus Clusters
• Write and interpret numerical expressions.
• Perform operations with multi-digit whole numbers and with decimals to hundredths.
• Convert like measurement units within a given measurement system.

1 Warm Up 5 min   **Materials**

Mental Math and Fluency   5.MD.1
Students convert between units of length.

2 Focus 35–40 min

Math Message   *Student Reference Book*, p. 328   5.MD.1

Focus Clusters

Everyday Mathematics identifies the clusters addressed in the Focus part of each lesson to help you understand the content that is being taught in the lesson.

Major Clusters

Each unit focuses on Major Clusters that are clearly identified in the Unit Organizer.

Focus

In this unit, students explore patterns in the base-10 place-value system numbers. Students are also introduced to U.S. traditional multiplication

Major Clusters
5.NBT.A Understand the place value system.
5.NBT.B Perform operations with multi-digit whole numbers with deci

Supporting Cluster
5.MD.A Convert like measurement units within a given measurement

Process and Practice Standards
SMP1 Make sense of problems and persevere in solving them.
SMP6 Attend to precision.

Focus

In Unit 2, students explore patterns in the base-10 place-value system and ways of representing large numbers. Students are also introduced to U.S. traditional multiplication and review partial-quotients division.

Major Clusters
5.NBT.A Understand the place value system.

5.NBT.B Perform operations with multi-digit whole numbers with decimals to hundredths.

Supporting Cluster
5.MD.A Convert like measurement units within a given measurement system.

Coherence Within and Across Grades

Spiral Towards Mastery

Carefully crafted, research-based learning progressions provide opportunities for your students to connect skills, concepts, and applications, while developing deep understanding, long-term learning, and transfer of knowledge and skills to new contexts.

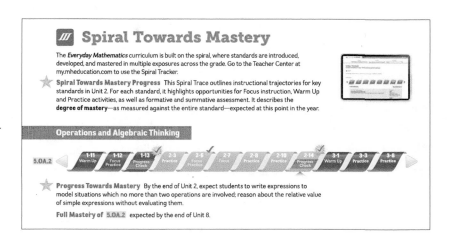

Spiral Towards Mastery

The *Everyday Mathematics* curriculum is built on the spiral, where standards are introduced, developed, and mastered in multiple exposures across the grade. Go to the Teacher Center at my.mheducation.com to use the Spiral Tracker.

⭐ **Spiral Towards Mastery Progress** This Spiral Trace outlines instructional trajectories for key standards in Unit 2. For each standard, it highlights opportunities for Focus instruction, Warm Up and Practice activities, as well as formative and summative assessment. It describes the **degree of mastery**—as measured against the entire standard—expected at this point in the year.

Operations and Algebraic Thinking

5.OA.2 | 1-11 Warm Up | 1-12 Focus Practice | 1-13 Progress Check | 2-3 Practice | 2-6 Focus Practice | 2-7 Focus | 2-8 Practice | 2-10 Practice | 2-14 Progress Check | 3-1 Warm Up | 3-3 Practice | 3-8 Practice

⭐ **Progress Towards Mastery** By the end of Unit 2, expect students to write expressions to model situations which no more than two operations are involved; reason about the relative value of simple expressions without evaluating them.

Full Mastery of 5.OA.2 expected by the end of Unit 8.

Linking Prior and Future Knowledge

Each unit contains information about how the focus standards covered in the unit developed in prior units and grades and how your instruction lays the foundation for future lessons.

Coherence

The table below describes how standards addressed in the Focus parts of the lessons link to the mathematics that students have done in the past and will do in the future.

	Links to the Past	Links to the Future
5.OA.1	In Unit 1, students reviewed how to use grouping symbols in expressions and how to evaluate expressions with grouping symbols. In Grade 3, students inserted parentheses in number sentences to make them true and evaluated number sentences with parentheses.	In Unit 7, students will use grouping symbols in an expression to model how to solve a multistep problem about gauging reaction time. In Grade 6, students will evaluate expressions and perform operations according to the Order of Operations.
5.OA.2	In Unit 1, students represented the volumes of rectangular prisms using expressions. They also wrote expressions to record calculations in the game *Name That Number.* In Grade 4, students represented problems using equations with a letter standing for an unknown quantity.	Throughout Grade 5, students will write expressions to record calculations in a variety of contexts. In Unit 6, they will order and interpret expressions without evaluating them. In Grade 6, students will write expressions in which letters stand for numbers.

Rigorous Content

Everyday Mathematics gives you the tools and resources you need to emphasize conceptual understanding, procedural fluency, and applications with equal intensity.

Planning for Rich Math Instruction

		2-1 Understanding Place Value	2-2 Exponents and Powers of 10	2-3 Applying Powers of 10	2-4 U.S. Traditional Multiplication, Part 1
RIGOR	**Conceptual Understanding**	The relationship between places in multidigit numbers Describing Place-Value Relationships, p. 112 Representing Place Value, p. 113	Exponential notation Introducing Powers of 10, p. 118	Estimation Estimating with Powers of 10, p. 125	Multidigit multiplication Introducing U.S. Traditional Multiplication, p. 130
	Procedural Skill and Fluency	Home Link 2-1, p. 115	Journal p. 44, #1	Math Message, p. 124 Using Powers of 10 to Multiply, p. 124 Readiness, p. 123 Extra Practice, p. 123	Mental Math and Fluency, p. 130 Math Message, p. 130 Introducing U.S. Traditional Multiplication, p. 130 Multiplying 2-Digit Numbers by 1-Digit Numbers, p. 132 Home Link 2-4, p. 133 Readiness, p. 129 Enrichment, p. 129 Extra Practice, p. 129
	Applications		Introducing Powers of 10, p. 118 Solving a Real-World Volume Problem, p. 121 Enrichment, p. 117	Estimating with Powers of 10, p. 125 Writing and Comparing Expressions, p. 127 Home Link 2-3, p. 127	Multiplying 2-Digit Numbers by 1-Digit Numbers, p. 132

Problem-based Instruction

Everyday Mathematics builds problem solving into every lesson. Problem solving is in everything they do.

Warm-up Activity	Daily Routines	Math Message	Focus Activities	Summarize	Practice Activities
Lessons begin with a quick, scaffolded Mental Math and Fluency exercise.	Reinforce and apply concepts and skills with daily activities.	Engage in high cognitive demand problem solving activities that encourage productive struggle	Introduce new content with group problem solving activities and classroom discussion.	Discuss and make connections to the themes of the focus activity.	Lessons end with spiraled review of content from past lessons.

Practice Embedded in Every Lesson

Because *Everyday Mathematics* is a problem-based curriculum, practice opportunities appear naturally in daily instruction, but specific activities in the practice part of lessons help you be confident your students are progressing toward mastery and maintaining and applying knowledge and skills over time.

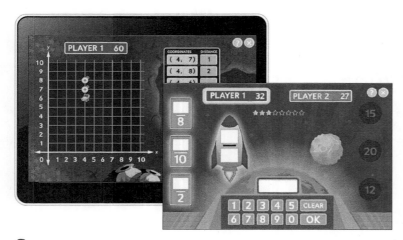

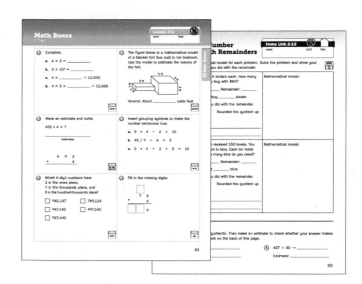

Games

Provide opportunities for fluency practice, along with collaborative learning experiences.

Math Boxes

Provide students with an opportunity to recall previously taught skills and concepts. These are distributed practice activities that include a balance of skills, concepts, and applications.

Home Links

Allow students to practice school mathematics and help family members connect to school.

Mathematical Literacy Sets The Stage for Algebra

Everyday Mathematics encourages students to recognize, analyze, and generalize patterns; represent quantities and relationships symbolically; model problem situations using objects, pictures, words, and symbols; and understand real-world relationships such as direct proportion—which, along with a fluent mastery of basic arithmetic, are the building blocks of algebraic thinking.

GRADE	K	1	2	3	4	5	6	▶

Instruction builds on student curiosity about patterns to explore numbers, shapes, and relationships between them.

Students work with symbolic representations for quantities and relationships, model simple situations, and build arithmetic skills.

Students use symbolic representations to model problem situations, build their understanding of fundamental relations such as direct proportion, and master elementary arithmetic concepts and skills.

McGraw-Hill Education

Be the Teacher They Will Always Remember

An *Everyday Mathematics* classroom has a unique energy that's a result of student engagement and excitement about learning math. This environment builds growth mindset and other positive attitudes about learning that will help your students succeed long after they've left your classroom.

Math Talk

Talking about mathematics is an essential part of learning mathematics. Opportunities for students to share their problem-solving strategies and their reasoning as well as critique others' reasoning are embedded throughout *Everyday Mathematics,* making it easy for you to facilitate math discussions every day.

"I can share my solution!"

Collaboration

Everyday Mathematics was designed to allow your students to share ideas and strategies. They work in small groups and with partners formed according to their needs, helping you create a rich learning environment that supports powerful instruction.

Perseverance and Productive Struggle

Everyday Mathematics helps you create a classroom culture that values and supports productive struggle, that fosters productive dispositions in your students—a belief that mathematics is worthwhile, an inclination to use the mathematics they know to solve problems and confidence in their own mathematical abilities.

"I can do this!"

Hands-on Exploration

Everyday Mathematics includes hands-on activities in every lesson that often involve the use of manipulatives and games to help students make connections to their everyday life. These activities allow students to model mathematics physically, concretely, and visually—deepening their understanding of concepts and skills.

The *Everyday* Mathematics Lesson

Lessons are designed to help teachers facilitate instruction and engineered to accommodate flexible grouping models. The three-part, activity-driven lesson structure helps you easily incorporate research-based instructional methods into your daily instruction.

Embedded Rigor and Spiraled Instruction

Each lesson weaves new content with practice of content introduced in earlier lessons. The structure of the lessons ensures that your instruction includes all elements of rigor in equal measure with problem solving at the heart of everything you do.

Review
Warm Up
FLUENCY

Lessons begin with quick, scaffolded warm up exercises that provide important fluency practice.

Introduction of New Content
Focus
CONCEPTUAL UNDERSTANDING AND APPLICATION

Math Message Students solve a challenging and engaging problem and discuss how they solved it.

Focus Activities Introduce new content, skills, and concepts.

Review
Practice
APPLICATION AND FLUENCY

Spiraled practice that revisits content from earlier lessons.

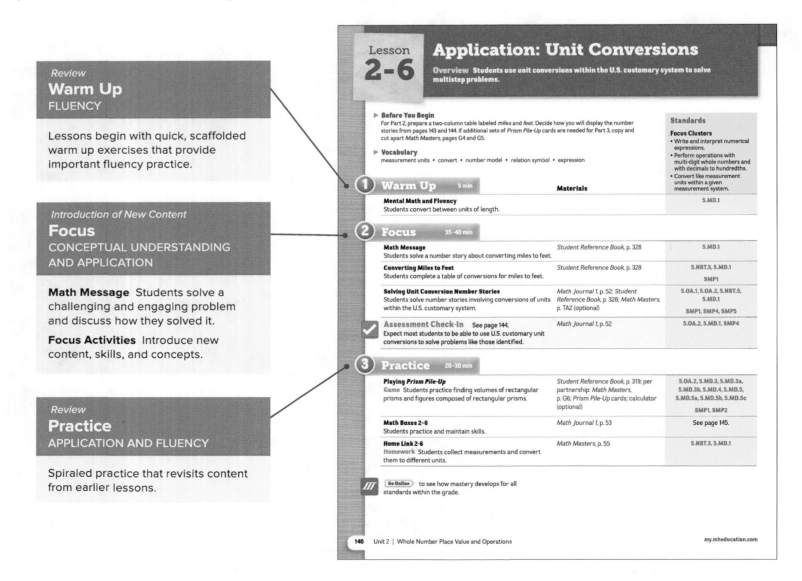

Lesson **2-6**

Application: Unit Conversions

Overview Students use unit conversions within the U.S. customary system to solve multistep problems.

▶ **Before You Begin**
For Part 2, prepare a two-column table labeled *miles* and *feet*. Decide how you will display the number stories from pages 143 and 144. If additional sets of *Prism Pile-Up* cards are needed for Part 3, copy and cut apart *Math Masters*, pages G4 and G5.

▶ **Vocabulary**
measurement units • convert • number model • relation symbol • expression

Standards
Focus Clusters
• Write and interpret numerical expressions.
• Perform operations with multi-digit whole numbers and with decimals to hundredths.
• Convert like measurement units within a given measurement system.

① Warm Up 5 min	Materials	
Mental Math and Fluency Students convert between units of length.		5.MD.1

② Focus 35–40 min		
Math Message Students solve a number story about converting miles to feet.	*Student Reference Book*, p. 328	5.MD.1
Converting Miles to Feet Students complete a table of conversions for miles to feet.	*Student Reference Book*, p. 328	5.NBT.5, 5.MD.1 SMP1
Solving Unit Conversion Number Stories Students solve number stories involving conversions of units within the U.S. customary system.	*Math Journal 1*, p. 52; *Student Reference Book*, p. 328; *Math Masters*, p. TA2 (optional)	5.OA.1, 5.OA.2, 5.NBT.5, 5.MD.1 SMP1, SMP4, SMP5
✓ **Assessment Check-In** See page 144. Expect most students to be able to use U.S. customary unit conversions to solve problems like those identified.	*Math Journal 1*, p. 52	5.OA.2, 5.MD.1, SMP4

③ Practice 20–30 min		
Playing *Prism Pile-Up* **Game** Students practice finding volumes of rectangular prisms and figures composed of rectangular prisms.	*Student Reference Book*, p. 319; per partnership: *Math Masters*, p. G6; *Prism Pile-Up* cards; calculator (optional)	5.OA.2, 5.MD.3, 5.MD.3a, 5.MD.3b, 5.MD.4, 5.MD.5, 5.MD.5a, 5.MD.5b, 5.MD.5c SMP1, SMP2
Math Boxes 2-6 Students practice and maintain skills.	*Math Journal 1*, p. 53	See page 145.
Home Link 2-6 **Homework** Students collect measurements and convert them to different units.	*Math Masters*, p. 55	5.NBT.5, 5.MD.1

Go Online to see how mastery develops for all standards within the grade.

140 Unit 2 | Whole Number Place Value and Operations

my.mheducation.com

Key Components

The *Everyday Mathematics* authors have developed a suite of resources that support your instruction, helping you create a mathematically rich environment every day.

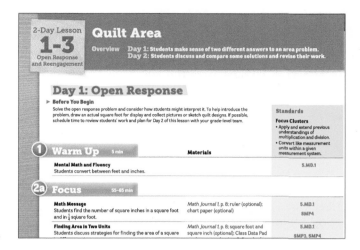

Open Response and Reengagement Lessons

Every unit includes a 2-day lesson that provides your students the opportunity to work with rich tasks and solve complex problems while explicitly engaging in the mathematical practices.

Games

Research shows that games provide a more effective learning experience than tedious drills and worksheets. Games allow for playful, repetitive practice that develops fluency and confidence and helps students learn to strategize.

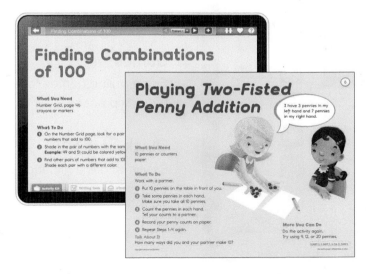

Activity Cards

Activity Cards provide for structured exploration of content tied to the focus of the lesson independently, in partnerships, and in small groups, especially in centers, where students are expected to complete the activity with minimal teacher guidance.

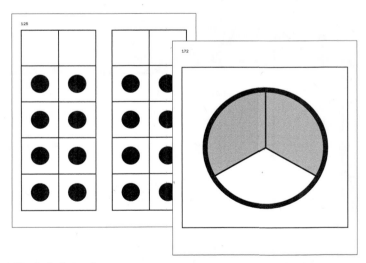

Quick Looks

Quick Look activities are routines that help your students develop the ability to recognize a quantity without counting and to decompose numbers in various ways. As they encounter various combinations of numbers, they also develop strategies for basic facts.

scottdunlap/iStockvectors/Getty Images

Online Resources

Digital tools to help you confidently deliver effective mathematics instruction in your classroom are included with every implementation. Everything you need is included in one easy-to-navigate place and you can customize your lessons by adding resources and notes—and everything is saved and available to you year after year.

The Teacher Center

You'll never waste time looking for resources because everything you need for every lesson is right where you need it, when you need it. When you open the *Everyday Mathematics* Teacher Center, you're automatically taken to the overview of the current lesson.

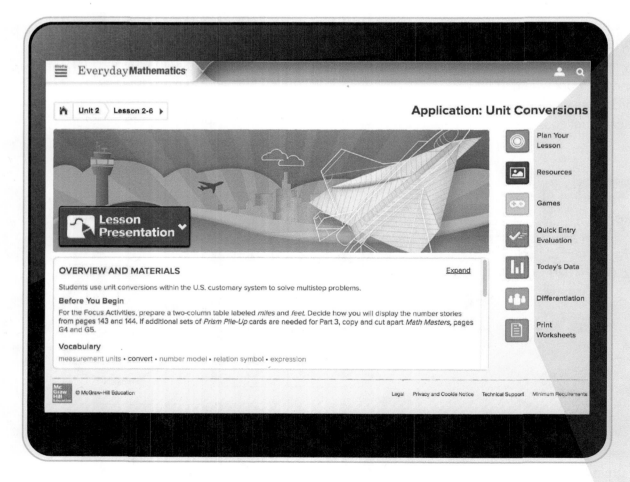

Launch Presentation
Editable versions of digital lessons that help you lead instruction.

Plan Your Lesson
Review all of the activities for the lesson.

Resources
Access lesson resources, additional projects and home-school connections.

Games
Open online games for fluency practice.

Quick Entry
Easily record evaluations of your students' progress.

Today's Data
Easy access to Data Dashboard reports to drive your daily instruction.

Differentiation
Resources to help you adjust the lesson to support all learners.

scottdunlap/iStockvectors/Getty Images

The Student Learning Center

Engineered to help each of your students experience confidence and develop positive feelings about math in a digital environment that keeps them engaged and excited about learning.

Lesson Content

Your students' lessons are synched with your planner so they always have easy access to each day's activities.

My Reference Book

One-click access to the interactive reference book that includes descriptions and examples as well grade-level-appropriate explanations of mathematical content and practices.

eToolkit

eTools and writing tools that enable your students to show their work and explore dynamic extensions.

Geometer's Sketchpad Activities and EM Games Online

Easy to access Fact Practice games and full integration of The Geometer's Sketchpad® activities.

Tutorial Videos

Demonstrations of concepts and skills.

EM at Home

Parents have easy access to resources to help them support their child's learning.

scottdunlap/iStockvectors/Getty Images

Data Driven Instruction

Everyday Mathematics includes a complete set of tools and resources to help teachers evaluate the development of each student's mathematical understanding and skills, while providing actionable data to inform instruction.

Evaluate

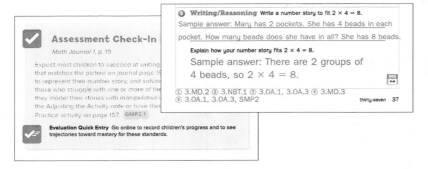

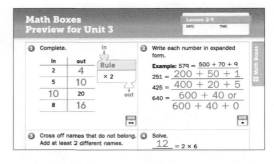

Ongoing Assessments

Assessment Check-In Daily lesson based assessment opportunities.

Writing and Reasoning Prompts Allow students to communicate understanding of concepts and skills and strategies for solving problems.

Pre Unit Assessment

Preview Math Boxes Appear in two lessons toward the end of each unit and help you gauge readiness for upcoming content, plan instruction and choose appropriate differentiation activities.

Data Dashboard Through the reports provided in the ConnectED Teacher Center, data recorded in prior units can provide valuable information to inform instruction in the upcoming unit.

Periodic Assessments

Progress Check lessons at the end of each unit provide formal opportunities to assess students' progress toward mastery of content and process/practice standards.

- **Unit Assessments** Assess students' progress toward mastery of concepts, skills, and applications in the current unit.

- **Self Assessments** Allow students to reflect on their understanding of content and process/practice standards that are the focus of the unit.

- **Challenge Problems** Extend important ideas from the unit, allowing students to demonstrate progress beyond expectations.

- **Cumulative Assessments** Assess students' progress toward mastery of content and process/ practice standards from prior units.

- **Open Response Assessments** Provide information about students' performance on longer, more complex problems and emphasize the process and practice standards for mathematics.

Benchmark Assessments Beginning of Year, Mid-Year, and End of Year benchmarks follow the same format as Unit Assessments.

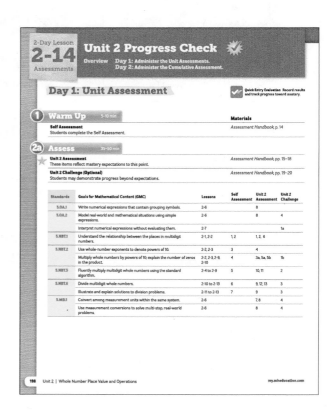

Record

A full suite of tools including rubrics and class checklists are available to help you track your students' progress.

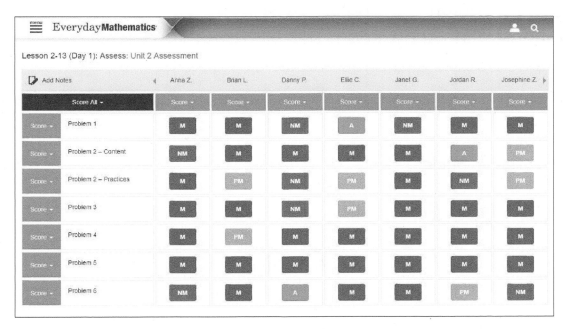

Quick Entry Evaluation Tool

You can quickly and efficiently record evaluations of your students' performance as well as add notes.

Report

The Data Dashboard is a responsive reporting tool that delivers actionable information to help you adapt and personalize your instruction and provide feedback to families and administrators.

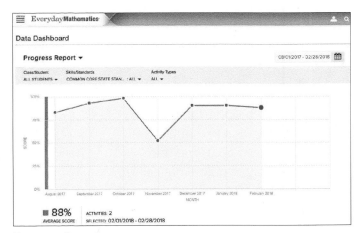

Progress Report

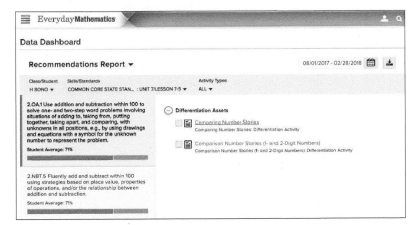

Recommendations Report

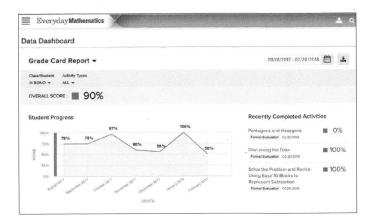

Grade Card Report

Differentiation System

Everyday Mathematics fosters rich learning environments that provide multiple avenues for mastering content, making sense of ideas, developing skills, and demonstrating knowledge. This allows rigorous mathematics content to be accessible and engaging for all students.

Everyday Mathematics Differentiation Model

Content
Clear goals and features that can be readily adapted or scaffolded to adjust the content for individual students.

Process
Engaging activities and point-of-use prompts that help foster rich pedagogical interaction in the classroom.

Product
Multiple opportunities to assess and monitor progress over time and to analyze mathematical strengths and misconceptions.

Classroom Organization
Opportunities for whole-class and small-group instruction built into every lesson, as well as time for students to work in partners, and individually.

Learning Environment
Everyday Mathematics provides multiple opportunities for students to reflect on their own strengths and weaknesses while engaging in productive collaboration.

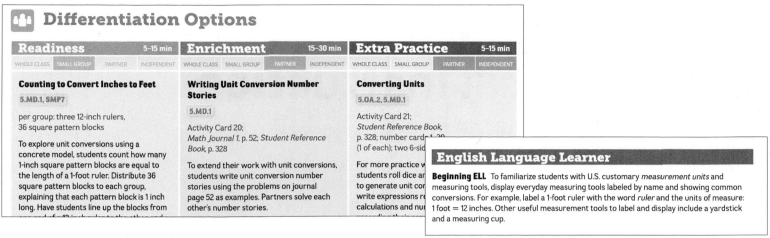

Supplementary Activities

Everyday Mathematics offers specific differentiation options in every lesson for:

- Students who need more scaffolding
- Students who need extra practice
- Advanced Learners
- Beginning English Language Learners
- Intermediate and Advanced English Language Learners

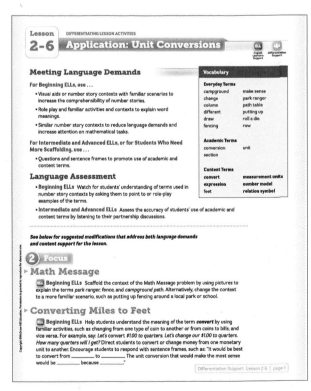

Lesson Supplements

Almost every lesson has Differentiation Support Pages found in the ConnectED Teacher Center that offer extended suggestions for working with diverse learners, including English Language Learners and students who need more scaffolding.

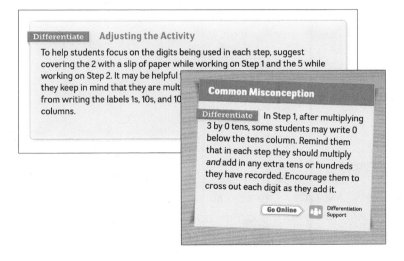

Point-of-Use Differentiation

Assessment Adjustments Suggestions for scaffolding and extending Progress Check assessments.

Game and Activity Adjustments Recommendations for tools, visual aids, and other instructional strategies that provide immediate support.

Adjusting the Activity Suggestions for adapting activities to fit students' needs.

Common Misconceptions Notes that suggest how to use observations of students' work to adapt instruction.

Supporting Rich Mathematical Instruction

Everyday Mathematics includes a wealth of resources to help you deliver effective instruction every day.

Planning

Every Unit Organizer includes a chart that shows where the building-blocks for rich mathematical instruction appear throughout every unit.

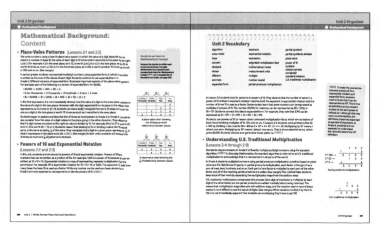

Preparing

Every Unit Organizer also includes important background information on both content and practice standards to help you confidently deliver instruction.

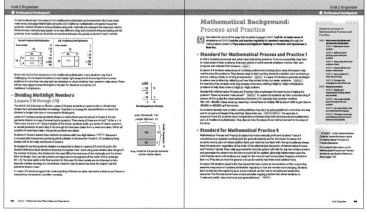

Support

The *Everyday Mathematics* Virtual Learning Community (VLC) at The University of Chicago, provides a free space where you can connect with a network of skilled, passionate educators who are also using the program, and interact with the authors. Resources on the VLC include classroom videos of lessons in action and instructional tools and resources.

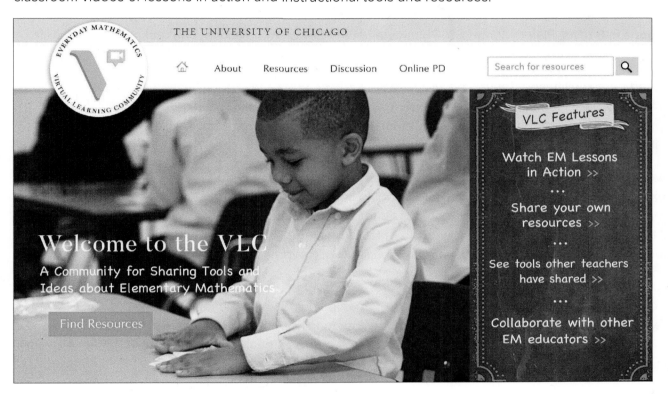

Resources

Everything you need to successfully implement *Everyday Mathematics* is at your fingertips through the ConnectED Resource page of your Teacher Center including videos from the authors, quick start guides for key features, and the Implementation Guide, a comprehensive guide to using the program.

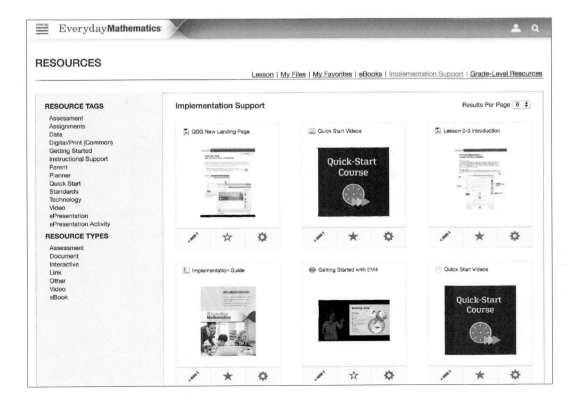

Getting Ready to Teach
Fifth Grade Everyday Mathematics

Welcome to *Fifth Grade Everyday Mathematics*. This guide introduces the organization and pedagogy of Everyday Mathematics and provides tips to help you start planning and teaching right away.

Grade 5 has **113 lessons** in 8 units. Plan to spend 60–75 minutes every day on math so that you complete **3–4 lessons each week** and one **unit every 4–5 weeks**.

This pacing is designed for flexibility and depth. You will have flexibility so you can extend a lesson if discussion has been rich or if students' understandings are incomplete. You can add a day for "journal fix-up" or for differentiation—to provide an Enrichment activity to every student, for example—or for games. There will also be time to accommodate outside mandates, district initiatives, and special projects.

This pacing also gives you time to go deep, to create a classroom culture that values and supports productive struggle. You can expect your students to do their own thinking, to solve problems they have not been shown how to solve, to make connections between concepts and procedures, to explain their thinking, and to understand others' thinking. Creating such a classroom culture takes time, but it's what the Common Core asks you to do in its Standards for Mathematical Practice—and the pacing of *Everyday Mathematics 4* is designed to give you the time you'll need.

The *Teacher's Lesson Guide* is your primary source for information on planning units and teaching lessons. In most lessons, students will complete pages in their *Math Journals* or digitally in the Student Learning Center. Additional pages that require copies are available as *Math Masters*. See the Materials section on pages xxvi-xxvii for information on the teacher and student components.

Preparing for the Beginning of School

- Use the list on pages xxvi-xxvii to check that your **Classroom Resource Package** is complete.
- See page xxix for manipulatives and supplies you will need.
- Read the **Unit 1 Organizer** (pages 2–13) and the **first several lessons in Unit 1** to help you plan for the first week of school.
- Read the *Everyday Mathematics* in Grades 1–6 section of the *Implementation Guide* for more information on getting started.
- Prepare the **Unit 1 Family Letter** on *Math Masters*, pages 2–5 to distribute early in the school year.
- Review the **Beginning-of-Year Assessment** on pages 83–87 in the *Assessment Handbook* and consider when you will administer it.

Go Online ⟩ to join the Virtual Learning Community (VLC) to learn about *Everyday Mathematics* classrooms from other teachers and to find tips for setting up your classroom.

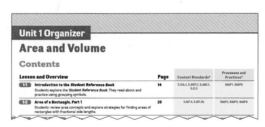

Unit 1 begins on page 2.

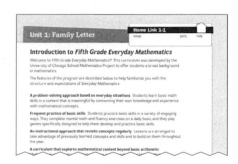

Lesson Types

Fifth Grade Everyday Mathematics includes three types of lessons, which share many of the same features.

Regular Lessons are the most common lesson type. See the tables on the following pages for details about regular lessons.

Open Response and Reengagement Lessons extend over two days and occur in every unit. On Day 1 students solve a challenging problem that involves more than one possible strategy or solution. On Day 2 students reengage in the problem and are asked to defend their reasoning and make sense of the reasoning of other students.

Progress Check Lessons are two-day lessons at the end of every unit. All items on the Progress Check match expectations for progress at that point in the year, and with the exception of the optional challenge assessment, are fair to grade. On Day 1 students complete a self-assessment, a unit assessment, and an optional challenge assessment covering the content and practices that were the focus of the unit. Day 2 includes one of the following types of assessments:

> **Open Response Assessments** are included in odd-numbered units and allow students to think creatively about a problem. They address both content and process/practice standards and are accompanied by task-specific rubrics.

> **Cumulative Assessments** are included in even-numbered units and cover standards from prior units.

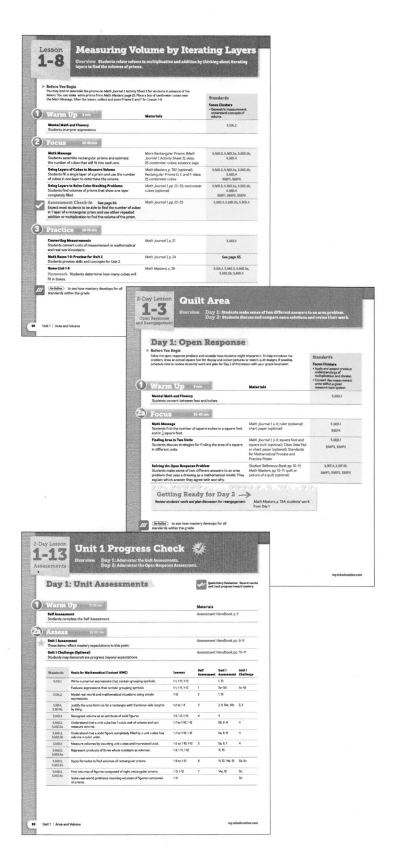

Lesson Parts and Features

Every lesson begins with two planning pages. The remaining pages provide a detailed guide for teaching the three parts of a lesson: Warm Up, Focus, and Practice.

Lesson Parts and Features	Description	Tips
Planning — **Lesson Opener**	An outline of the lesson to assist in your planning that includes information on content and standards, timing suggestions, assessment, and materials.	• See **Before You Begin** for preparation tips. • Follow the time allotments for each part of the lesson.
Differentiation Options	Optional **Readiness**, **Enrichment**, **Extra Practice**, and **English Language Learners (ELL) Support** activities that allow you to differentiate instruction. Additional Differentiation Support pages are available online for each regular lesson.	• Choose to complete Differentiation Options as a whole class, with partners, as a small group, or individually depending on the needs of your students. • Note that some students may benefit from completing the **Readiness** activity prior to the lesson. Go Online to the *Implementation Guide* for information on differentiation.

Part 1: Warm Up	Description	Tips
Instruction — **Mental Math and Fluency**	Quick, leveled warm-up exercises students answer orally, with gestures, or on slates or tablets that provide practice towards fluency.	• Select the levels that make sense for your students and customize for your class. • Spend 5 or fewer minutes on this feature.

Part 2: Focus	Description	Tips
Instruction — **Math Message and Math Message Follow-Up**	An introductory activity to the day's lesson that usually requires students to solve a problem they have not been shown how to solve. The follow-up discussion connects to the focus activities of the lesson and gives students opportunities to discuss their strategies.	• Consider where and how you will display the Math Message and how students will record their answers. • Maintain high cognitive demand by expecting students to work through the problem without your help before the follow-up discussion begins.

Part 2: Focus, con't.		Description	Tips
Instruction	**Focus Activities**	Two to four main instructional activities, including games, in which students explore and engage in new content (skills, concepts, games).	• Encourage students to discuss and work together to solve problems during focus activities. • Remember that many focus skills, concepts, applications, and games will be revisited in later practice. ⟨Go Online⟩ to the Spiral Tracker to see the complete spiral. • Look for Goals for Mathematical Process and Practice icons. GMP1.1 Use these to facilitate discussions about the processes and practices. ⟨Go Online⟩ to the *Implementation Guide* for information on Process and Practice Standards.
	Assessment Check-In ✓	A daily assessment opportunity to assess the focus content standards in the lesson. Assessment Check-Ins provide information on expectations for particular standards at that point in the curriculum.	• Use results to inform instruction. Expectation statements in the Assessment Check-Ins help you decide which students would benefit from differentiation activities. • Consider Assessment Check-Ins as "fair to grade" in most cases. ⟨Go Online⟩ to record students' progress and to see trajectories toward mastery for these and other standards. ⟨Go Online⟩ to the *Implementation Guide* for assessment information.

Part 3: Practice		Description	Tips
Instruction	**Practice Activity**	An opportunity to practice previously taught skills and content through a practice page or a game in many lessons.	• Allow time for practice pages and games because they are critical for students to meet expectations for standards. This is an essential part of the distributed practice in *Everyday Mathematics* • Plan for all students to play *Everyday Mathematics* games at least 60 minutes per week. ⟨Go Online⟩ to the *Implementation Guide* for tips to ensure that all students have ample game time. See also the Virtual Learning Community (VLC) to observe many *Everyday Mathematics* games in action.
	Math Boxes	A daily *Math Journal* page that reviews skills and concepts which students have seen prior to that point in the program. Preview Math Boxes anticipate content in the upcoming unit.	• Aim to have students complete Math Boxes with as little teacher support as possible. • Complete Math Boxes at any point during the day.
	Home Link	A daily homework page that provides practice and informs families about the math from that day's lesson.	Encourage students to do these activities with someone at home, such as a parent, caregiver, or sibling.

Differentiation and Language Features		Description and Purpose
Differentiation	**Adjusting the Activity**	Allows for differentiated instruction by offering modifications to lesson activities.
	Common Misconception	Offers point-of-use intervention tips that address common misconceptions.
	Game Modifications	Provides suggestions online for modifying games to support students who struggle and challenge students who are ready.
	Differentiation Support	Offers two online pages of specific differentiation ideas for each lesson, as well as ELL suggestions and scaffolding for students who need it.
Language Notes	**Academic Language Development**	Suggests how to introduce new academic vocabulary that is relevant to the lesson. These notes benefit all students, not solely English language learners.
	English Language Learners (ELL)	Provides activities and point-of-use ideas for supporting students at different levels of English language proficiency.

Getting to Know Your Classroom Resource Package

Complete access to all digital resources is included in your Classroom Resource Package. To access these resources, log into **my.mheducation.com**.

Planning, Instruction, and Assessment	
Resource	**Description**
Teacher's Lesson Guide **(Volumes 1 and 2)** ☑ digital ☑ print	• Comprehensive guide to the *Everyday Mathematics* lessons and assessments • Standards alignment information: digital version includes online tracking of each content standard • Point-of-use differentiation strategies: Readiness, Enrichment, Extra Practice, English Language Learners support, Academic Language Development, Adjusting the Activity, Game Modifications, Common Misconceptions • Additional Differentiation Support pages available digitally for virtually every lesson • Unit overviews • Planning and calendar tools
eToolkit ☑ digital ☐ print	• Online tools and virtual manipulatives for dynamic instruction • A complete list of Grade 5 eTools on page xxix
ePresentations ☑ digital ☐ print	• Ready-made interactive white board lesson content to support daily instruction
Math Masters ☑ digital ☑ print	• Reproducible masters for lessons, Home Links, Family Letters, and games
Classroom Posters ☑ digital ☑ print	• Posters that display grade-specific mathematical content

Planning, Instruction, and Assessment (con't)

Resource	Description
Assessment Handbook ☑ digital ☑ print	• Assessment masters for unit-based assessments and interim assessments • Record sheets for tracking individual and class progress
Assessment and Reporting Tools ☑ digital ☐ print	• Student, class, school, and district reports • Data available at point-of-use in the planning and teaching materials • Real-time data to inform instruction and differentiation
Spiral Tracker ☑ digital ☐ print	• Online tool that helps you understand how standards develop across the spiral curriculum

Professional Development

Resource	Description
Implementation Guide ☑ digital ☐ print	• Online resource with information on implementing the curriculum
Virtual Learning Community ☑ digital ☐ print	• An online community, sponsored and facilitated by the Center for Elementary Mathematics and Science Education (CEMSE) at the University of Chicago, to network with other educators and share best practices • A collection of resources including videos of teachers implementing lessons in real classrooms, photos, work samples, and planning tools

Family Communications

Resource	Description
Home Connection Handbook ☑ digital ☐ print	• A collections of tips and tools to help you communicate to families about *Everyday Mathematics* • Reproducible masters for home communication for use by both teachers and administrators

Student Materials

Resource	Description
Student Math Journal, (Volumes 1 and 2) ☑ digital ☑ print	• Student work pages that provide daily support for classroom instruction • Provide a long-term record of each student's mathematical development
Student Reference Book ☑ digital ☑ print	• Resource to support student learning in the classroom and at home • Includes explanations of mathematical content and directions for many *Everyday Mathematics* games
Activity Cards ☑ digital ☑ print	• Directions for students for Differentiation Options and other small-group activities
Student Learning Center ☑ digital ☐ print	• Combines *Student Math Journal, Student Reference Book,* eToolkit, and Activity Cards, and other resources for students in one location • Interactive functionality provides access in English and Spanish • Interactive functionality provides immediate feedback on select problems • Animations that can help with skills and concepts and reinforce classroom teaching • Provides access to EM Games Online and Facts Workshop Game
EM Games Online ☑ digital ☐ print	• Digital versions of many of the *Everyday Mathematics* games that provide important practice in a fun and engaging setting

Manipulative Kits and eToolkit

The table below lists the materials that are used on a regular basis throughout *Fifth Grade Everyday Mathematics*. All of the items below are available from McGraw-Hill Education. They may be purchased as a comprehensive classroom manipulatives kit or by individual items. The manipulative kit comes packaged in durable plastic tubs. Note that some lessons call for additional materials, which you or your children can bring in at the appropriate times. The additional materials are listed in the Unit Organizers and in the lessons in which they are used.

Manipulative Kit Contents		eTools
Item	Quantity	Item
Base-10 Big Cube	4 big cubes	✔
Base-10 Flats	3 packs of 10 flats	✔
Base-10 Longs	5 packs of 50 longs	✔
Base-10 Cubes	10 packs of 100 cubes	✔
Counters, Double-Sided	1 pack of 500	✔
Dice, Dot	2 packs of 12	✔
Everything Math Deck	15 decks	✔
Fraction Circle Pieces	25 sets	✔
Metersticks	2 packs of 6	
Number Line, −35 to 180	1 number line (in 3 parts)	✔
Pattern Blocks	1 set of 250	✔
Ruler, 12 in.	5 packs of 5 rulers	
Stopwatch	8 digital stopwatches	✔
Tape Measure, Retractable	15 tape measures	

Clear Pathway to Mastery

You can be confident your students are progressing toward mastery of every standard because *Everyday Mathematics* provides detailed information about the learning trajectories for each standard as well as expectations for mastery at every step of the way.

Unpack

Standards for Mathematical Content

Strand Operations and Algebraic Thinking 5.OA	*Everyday Mathematics* Goals for Mathematical Content

Cluster Write and interpret numerical expressions.

5.OA.1 Use parentheses, brackets, or braces in numerical expressions, and evaluate expressions with these symbols.	**GMC** Write numerical expressions that contain grouping symbols.
	GMC Evaluate expressions that contain grouping symbols.
5.OA.2 Write simple expressions that record calculations with numbers, and interpret numerical expressions without evaluating them. *For example, express the calculation "add 8 and 7, then multiply by 2" as 2 × (8 + 7). Recognize that 3 × (18932 + 921) is three times as large as 18932 + 921, without having to calculate the indicated sum or product.*	**GMC** Model real-world and mathematical situations using simple expressions.
	GMC Interpret numerical expressions without evaluating them.

Cluster Analyze patterns and relationships.

5.OA.3 Generate two numerical patterns using two given rules. Identify apparent relationships between corresponding terms. Form ordered pairs consisting of corresponding terms from the two patterns, and graph the ordered pairs on a coordinate plane. *For example, given the rule "Add 3" and the starting number 0, and given the rule "Add 6" and the starting number 0, generate terms in the resulting sequences, and observe that the terms in one sequence are twice the corresponding terms in the other sequence. Explain informally why this is so.*	**GMC** Generate numerical patterns using given rules.
	GMC Identify relationships between corresponding terms of two patterns.
	GMC Form ordered pairs from corresponding terms of patterns and graph them.

Goals for Mathematical Content

The *Everyday Mathematics* authors developed Goals for Mathematical Content (GMC) that break down each content standard to provide detailed information about the learning trajectories required to meet the full standard. See pages EM3–EM5 for a full view of the content standards and the related GMCs.

Goals for Mathematical Practice

The authors created Goals for Mathematical Practice (GMP) that unpack the practice standards, operationalizing them in ways that are appropriate for elementary students. See pages EM6–EM9 for a full view of the practice standards and the related GMPs.

Standards for Mathematical Process and Practice

Standards for Mathematical Process and Practice	*Everyday Mathematics* Goals for Mathematical Process and Practice

1 Make sense of problems and persevere in solving them.

Mathematically proficient students start by explaining to themselves the meaning of a problem and looking for entry points to its solution. They analyze givens, constraints, relationships, and goals. They make conjectures about the form and meaning of the solution and plan a solution pathway rather than simply jumping into a solution attempt. They consider analogous problems, and try special cases and simpler forms of the original problem in order to gain insight into its solution. They monitor and evaluate their progress and change course if necessary. Older students might, depending on the context of the problem, transform algebraic expressions or change the viewing window on their graphing calculator to get the information they need. Mathematically proficient students can explain correspondences between equations, verbal descriptions, tables, and graphs or draw diagrams of important features and relationships, graph data, and search for regularity or trends. Younger students might rely on using concrete objects or pictures to help conceptualize and solve a problem. Mathematically proficient students check their answers to problems using a different method, and they continually ask themselves, "Does this make sense?" They can understand the approaches of others to solving complex problems and identify correspondences between different approaches.	**GMP1.1** Make sense of your problem.
	GMP1.2 Reflect on your thinking as you solve your problem.
	GMP1.3 Keep trying when your problem is hard.
	GMP1.4 Check whether your answer makes sense.
	GMP1.5 Solve problems in more than one way.
	GMP1.6 Compare the strategies you and others use.

2 Reason abstractly and quantitatively.

Mathematically proficient students make sense of quantities and their relationships in problem situations. They bring two complementary abilities to bear on problems involving quantitative relationships: the ability to	**GMP2.1** Create mathematical representations using numbers, words, pictures, symbols, gestures, tables, graphs, and concrete objects.

Track

Everyday Mathematics provides the tools you need to easily monitor your students' progress toward mastery.

Visible Learning Trajectories

Get a full picture of how each standard develops across a unit—and the entire grade.

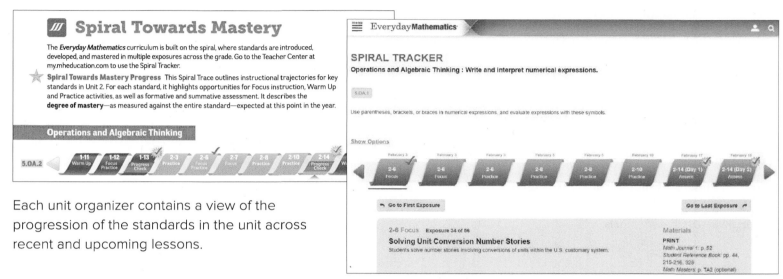

Each unit organizer contains a view of the progression of the standards in the unit across recent and upcoming lessons.

Using the online Spiral Tracker you can see how each standard progresses across the grade.

Master

Unit organizers include mastery expectation statements that provide guidance about what you should expect your students to know by the end of the unit and to help you make decisions about differentiation and groupings.

> **Progress Towards Mastery** By the end of Unit 2, expect students to write expressions to model situations which no more than two operations are involved; reason about the relative value of simple expressions without evaluating them.
>
> **Full Mastery of 5.OA.2** expected by the end of Unit 8.

The Mastery Expectations charts starting on page xl provide a full picture of how every standard develops across the entire grade.

Standards	First Quarter Benchmark Expectations for Units 1 and 2	Second Quarter Benchmark Expectations for Units 3 and 4	Third Quarter Benchmark Expectations for Units 5 and 6	Fourth Quarter Benchmark Expectations for Units 7 and 8
5.OA.1	Use one set of grouping symbols in an expression to model a real-world situation. Evaluate an expression that contains a single set of grouping symbols.	⭐ Use parentheses, brackets, or braces in numerical expressions, and evaluate expressions with these symbols.	Ongoing practice and application.	

Correlation to the Standards for Mathematics

Everyday Mathematics is a standards-based curriculum engineered to focus on specific mathematical content in every lesson and activity. The chart below shows complete coverage of each mathematics standard in the core program throughout the grade level.

*Bold lesson numbers indicate that content from the standard is taught in the Focus part of the lesson. Lesson numbers not in bold indicate that content from the standard is addressed in the Warm Up or Practice part of the lesson. The second set of lesson numbers, which are in parentheses, indicate that content from the standard is being addressed in Home Links or Math Boxes.

Content Standards for Mathematics for Grade 5	*Everyday Mathematics* Grade 5 Lessons*
Operations and Algebraic Thinking 5.OA	
Write and interpret numerical expressions.	
5.OA.1 Use parentheses, brackets, or braces in numerical expressions, and evaluate expressions with these symbols.	**1-1,** 1-5, 1-7, 1-9, **1-11,** 1-12, 2-3, 2-5, **2-6,** 2-8, 2-10, 3-6, **6-13**
	(1-2, 1-3, 1-4, 1-6, 1-8, 1-10, 2-1, 2-2, 2-4, 3-1, 3-2, 3-3, 3-4, 3-9, 3-10, 3-11, 3-12, 3-13, 3-14)
5.OA.2 Write simple expressions that record calculations with numbers, and interpret numerical expressions without evaluating them. *For example, express the calculation "add 8 and 7, then multiply by 2" as 2 × (8 + 7). Recognize that 3 × (18932 + 921) is three times as large as 18932 + 921, without having to calculate the indicated sum or product.*	1-1, 1-5, 1-8, 1-9, 1-11, **1-12,** 2-3, **2-6, 2-7,** 2-8, 2-10, 3-1, 3-3, 3-8, **3-11,** 4-3, 4-10, 4-13, **6-2, 6-8, 7-1**
	(1-2, 1-3, 1-4, 1-6, 1-7, 1-10, 2-2, 2-4, 3-2, 3-4, 3-9, 3-10, 3-13, 3-14)
Analyze patterns and relationships.	
5.OA.3 Generate two numerical patterns using two given rules. Identify apparent relationships between corresponding terms. Form ordered pairs consisting of corresponding terms from the two patterns, and graph the ordered pairs on a coordinate plane. *For example, given the rule "Add 3" and the starting number 0, and given the rule "Add 6" and the starting number 0, generate terms in the resulting sequences, and observe that the terms in one sequence are twice the corresponding terms in the other sequence. Explain informally why this is so.*	**4-9,** 5-6, **7-10, 7-11, 7-12, 7-13,** 8-2, **8-9**
	(6-10, 8-6, 8-10, 8-12)
Number and Operations in Base Ten 5.NBT	
Understand the place value system.	
5.NBT.1 Recognize that in a multi-digit number, a digit in one place represents 10 times as much as it represents in the place to its right and 1/10 of what it represents in the place to its left.	**1-1, 2-1, 2-2,** 2-4, 2-7, 2-10, 2-13, 3-9, 3-10, 3-14, **4-1, 4-2, 4-3, 4-4, 4-5,** 4-8, 4-9, 4-11, 5-4, 5-10, **6-1, 6-2,** 6-6, **6-12**
	(1-2, 1-4, 1-8, 2-3, 2-6, 2-8, 2-9, 2-11, 2-12, 3-1, 3-3, 3-5, 3-6, 3-8, 4-6, 4-13, 4-14, 5-6, 5-8, 6-3, 7-10)
5.NBT.2 Explain patterns in the number of zeros of the product when multiplying a number by powers of 10, and explain patterns in the placement of the decimal point when a decimal is multiplied or divided by a power of 10. Use whole-number exponents to denote powers of 10.	**2-2, 2-3,** 2-4, 2-5, 2-8, **2-9, 2-10,** 2-12, 2-13, 3-2, 3-5, 3-9, 3-10, 3-13, 4-9, **6-1, 6-2, 6-3,** 6-7, **6-9, 6-10,** 6-12, 7-2, 7-3, 7-5, 7-12, 8-1, 8-4, **8-7, 8-8,** 8-10, **8-11, 8-12**
	(1-8, 2-6, 2-7, 3-4, 3-7, 4-5, 4-12, 4-14, 5-10, 6-5, 6-6, 6-8, 6-11, 6-13, 7-1, 7-8, 8-2, 8-9)
5.NBT.3 Read, write, and compare decimals to thousandths.	**4-1, 4-2, 4-3, 4-4, 4-5,** 4-7, 4-8, **4-11,** 4-12, 4-13, 4-14, 5-1, 5-3, 5-4, 5-5, 5-8, 5-10, **6-1, 6-2,** 6-4, 6-6, 6-7, 6-11, **6-13,** 7-3, 8-1
	(3-10, 4-6, 4-9, 5-2, 5-7, 5-11, 5-13, 6-3, 6-5, 6-8)

Content Standards for Mathematics for Grade 5	*Everyday Mathematics* Grade 5 Lessons*
5.NBT.3a Read and write decimals to thousandths using base-ten numerals, number names, and expanded form, e.g., 347.392 = 3 × 100 + 4 × 10 + 7 × 1 + 3 × (1/10) + 9 × (1/100) + 2 × (1/1000).	**4-1, 4-2, 4-3, 4-4, 4-5,** 4-8, **4-11,** 4-12, 4-13, 4-14, 5-4, 5-5, 5-10, **6-1,** 6-2, 6-6, 6-11 (3-10, 4-6, 4-7, 4-9, 5-1, 5-2, 5-3, 5-7, 6-3, 6-8)
5.NBT.3b Compare two decimals to thousandths based on meanings of the digits in each place, using >, =, and < symbols to record the results of comparisons.	**4-4, 4-5,** 4-7, 4-8, 4-13, 5-1, 5-3, 5-4, 5-8, **6-2,** 6-4, 6-6, 6-7, **6-13,** 7-3, 8-1 (3-10, 4-9, 4-12, 4-14, 5-2, 5-5, 5-11, 5-13, 6-5)
5.NBT.4 Use place value understanding to round decimals to any place.	**4-5, 4-12, 4-13,** 4-14, 5-4, 5-6, 5-9, 6-11, **8-5, 8-11, 8-12** (3-10, 4-9, 4-11, 5-5, 5-7, 5-8, 6-2, 6-4)

Perform operations with multi-digit whole numbers and with decimals to hundredths.

5.NBT.5 Fluently multiply multi-digit whole numbers using the standard algorithm.	1-2, 1-7, **2-4, 2-5, 2-6, 2-7, 2-8, 2-9,** 3-1, 3-4, 4-7, 5-8, **6-9, 6-10, 8-1, 8-5, 8-6, 8-7, 8-8, 8-9, 8-10** (1-8, 2-10, 2-11, 2-12, 2-13, 3-3, 3-5, 3-7, 3-8, 3-9, 3-11, 3-12, 3-13, 3-14, 4-1, 4-3, 4-5, 4-11, 4-13, 5-1, 5-2, 5-4, 5-5, 5-7)
5.NBT.6 Find whole-number quotients of whole numbers with up to four-digit dividends and two-digit divisors, using strategies based on place value, the properties of operations, and/or the relationship between multiplication and division. Illustrate and explain the calculation by using equations, rectangular arrays, and/or area models.	1-11, **2-10, 2-11, 2-12, 2-13,** 3-1, 3-2, **3-3, 3-5,** 3-6, 3-9, 3-12, 3-14, 4-4, 5-7, **6-5, 6-11, 6-12, 8-6, 8-7, 8-8, 8-10** (1-8, 3-4, 3-7, 3-8, 3-11, 3-13, 4-2, 4-5, 4-6, 4-7, 4-8, 4-14, 5-1, 5-2, 5-3, 5-5, 5-6, 5-8, 5-14, 6-1, 6-3, 6-8, 8-1)
5.NBT.7 Add, subtract, multiply, and divide decimals to hundredths, using concrete models or drawings and strategies based on place value, properties of operations, and/or the relationship between addition and subtraction; relate the strategy to a written method and explain the reasoning used.	**4-11, 4-12, 4-13, 4-14,** 5-1, 5-3, 5-9, 5-12, 6-4, 6-6, **6-8, 6-9, 6-10, 6-11, 6-12, 6-13,** 7-4, 7-6, 7-7, 7-12, 7-13, **8-1,** 8-2, 8-3, 8-5, **8-6, 8-7, 8-8,** 8-9, **8-10** (5-2, 5-4, 5-5, 5-7, 5-8, 5-10, 5-11, 5-13, 5-14, 6-2, 7-1, 7-2, 7-3, 7-5, 7-8, 7-9, 7-10, 7-11, 8-4, 8-11, 8-12)

Number and Operations—Fractions   5.NF

Use equivalent fractions as a strategy to add and subtract fractions.

5.NF.1 Add and subtract fractions with unlike denominators (including mixed numbers) by replacing given fractions with equivalent fractions in such a way as to produce an equivalent sum or difference of fractions with like denominators. *For example, 2/3 + 5/4 = 8/12 + 15/12 = 23/12. (In general, a/b + c/d = (ad + bc)/bd.)*	1-2, 1-4, **3-10, 3-11, 3-12,** 4-3, **5-1, 5-2, 5-3, 5-4,** 5-11, **6-4,** 6-7, **7-1,** 7-5, 7-6, **7-9, 8-2,** 8-8 (2-10, 3-13, 4-10, 4-11, 4-13, 5-5, 5-6, 5-7, 5-8, 5-9, 5-10, 5-12, 5-13, 5-14, 6-1, 6-2, 6-3, 6-6, 6-8, 6-11, 6-13, 7-2, 7-3, 7-4, 7-8, 7-11, 7-13, 8-4)
5.NF.2 Solve word problems involving addition and subtraction of fractions referring to the same whole, including cases of unlike denominators, e.g., by using visual fraction models or equations to represent the problem. Use benchmark fractions and number sense of fractions to estimate mentally and assess the reasonableness of answers. *For example, recognize an incorrect result 2/5 + 1/2 = 3/7, by observing that 3/7 < 1/2.*	**3-4, 3-6, 3-7, 3-9, 3-10,** 3-11, **3-12,** 4-2, 4-8, **5-1, 5-3, 5-4,** 5-11, **6-4, 6-5,** 7-1, 7-6, **7-9,** 8-8 (2-10, 3-13, 3-14, 4-1, 4-3, 4-4, 4-5, 4-6, 4-7, 4-10, 5-6, 5-8, 5-9, 5-12, 5-14, 6-1, 6-3, 6-11, 7-11, 7-13)

Apply and extend previous understandings of multiplication and division to multiply and divide fractions.

5.NF.3 Interpret a fraction as division of the numerator by the denominator (a/b = a ÷ b). Solve word problems involving division of whole numbers leading to answers in the form of fractions or mixed numbers, e.g., by using visual fraction models or equations to represent the problem. *For example, interpret 3/4 as the result of dividing 3 by 4, noting that 3/4 multiplied by 4 equals 3, and that when 3 wholes are shared equally among 4 people each person has a share of size 3/4. If 9 people want to share a 50-pound sack of rice equally by weight, how many pounds of rice should each person get? Between what two whole numbers does your answer lie?*	**3-1, 3-2, 3-3, 3-4, 3-5,** 3-6, **3-8,** 3-11, **3-12,** 4-2, 4-4, 4-5, 4-9, 5-3, **5-6, 5-11, 5-13,** 6-5, **6-12, 7-2,** 7-4, 7-7 (2-10, 3-7, 4-1, 4-3, 4-6, 4-8, 4-10, 5-1, 5-5, 5-7, 5-10, 6-1, 6-10)
5.NF.4 Apply and extend previous understandings of multiplication to multiply a fraction or whole number by a fraction.	1-2, 1-3, 1-4, 1-6, **3-13, 3-14,** 4-1, 4-6, 4-12, 4-14, **5-5, 5-6, 5-7, 5-8, 5-9, 5-10, 5-11, 5-12,** 5-13, 5-14, 6-5, 6-6, 6-10, 6-13, **7-1, 7-2, 7-3,** 7-6, 7-7, **7-9,** 7-10, 7-12, **8-1, 8-2, 8-3,** 8-6, 8-9, 8-10 (1-5, 1-7, 1-9, 1-10, 1-11, 1-12, 2-1, 2-3, 2-10, 3-2, 3-4, 4-3, 4-9, 4-10, 6-1, 6-3, 6-7, 6-8, 6-9, 6-11, 6-12, 7-4, 7-5, 7-8, 7-11, 7-13, 8-4, 8-5, 8-7, 8-8, 8-11)

Content Standards for Mathematics for Grade 5	*Everyday Mathematics* Grade 5 Lessons*
5.NF.4a Interpret the product $(a/b) \times q$ as a parts of a partition of q into b equal parts; equivalently, as the result of a sequence of operations $a \times q \div b$. *For example, use a visual fraction model to show $(2/3) \times 4 = 8/3$, and create a story context for this equation. Do the same with $(2/3) \times (4/5) = 8/15$. (In general, $(a/b) \times (c/d) = ac/bd$.)*	**3-13, 3-14,** 4-1, 4-6, 4-12, **5-5, 5-6, 5-7, 5-8, 5-10, 5-12,** 5-13, 5-14, 6-5, 6-6, 6-10, 6-13, **7-1, 7-2,** 7-10 (4-3, 4-9, 4-10, 4-14, 5-9, 5-11, 6-1, 6-3, 6-8)
5.NF.4b Find the area of a rectangle with fractional side lengths by tiling it with unit squares of the appropriate unit fraction side lengths, and show that the area is the same as would be found by multiplying the side lengths. Multiply fractional side lengths to find areas of rectangles, and represent fraction products as rectangular areas.	**1-2, 1-3, 1-4,** 1-6, 4-14, **5-7, 5-8, 5-9, 5-10, 5-12,** 6-5, **7-1, 7-2, 7-3,** 7-6, 7-7, **8-1, 8-2, 8-3,** 8-6, 8-9 (1-5, 1-7, 1-9, 1-10, 1-11, 1-12, 2-1, 2-3, 2-10, 3-2, 3-4, 4-10, 5-11, 5-13, 5-14, 6-7, 6-9, 6-10, 6-12, 7-5, 7-10, 7-13, 8-4, 8-5, 8-11)
5.NF.5 Interpret multiplication as scaling (resizing), by:	**4-8, 5-5, 5-6, 5-8, 5-9, 5-11, 5-12,** 5-14, **6-8, 7-1, 7-2, 7-4.** 7-8 (5-10, 5-13, 6-5, 6-7, 7-5, 7-7, 7-9, 7-12, 8-1, 8-3)
5.NF.5a Comparing the size of a product to the size of one factor on the basis of the size of the other factor, without performing the indicated multiplication.	**4-8, 5-5, 5-6, 5-9, 5-11, 5-12,** 5-14, **6-8, 7-1, 7-2,** 7-8 (5-10, 5-13, 6-5, 6-7, 7-5, 7-7, 7-9, 7-12)
5.NF.5b Explaining why multiplying a given number by a fraction greater than 1 results in a product greater than the given number (recognizing multiplication by whole numbers greater than 1 as a familiar case); explaining why multiplying a given number by a fraction less than 1 results in a product smaller than the given number; and relating the principle of fraction equivalence $a/b = (n \times a)/(n \times b)$ to the effect of multiplying a/b by 1.	**5-5, 5-8, 5-9, 5-11, 5-12,** 5-14, **6-8, 7-2, 7-4,** 7-8 (6-5, 7-7, 7-9, 7-12, 8-1, 8-3)
5.NF.6 Solve real world problems involving multiplication of fractions and mixed numbers, e.g., by using visual fraction models or equations to represent the problem.	**3-13, 3-14, 5-5, 5-6, 5-7, 5-9, 5-10, 5-12,** 6-5, 6-6, 6-10, **7-1, 7-2, 7-3,** 7-6, 7-7, 7-10, **8-1, 8-3,** 8-6, 8-9 (5-14, 6-8, 7-4, 7-8, 8-5, 8-7)
5.NF.7 Apply and extend previous understandings of division to divide unit fractions by whole numbers and whole numbers by unit fractions.[1]	**5-13, 5-14,** 6-2, **7-4,** 7-10, 8-7, 8-8 (6-4, 6-5, 6-7, 6-9, 6-10, 6-12, 7-1, 7-2, 7-3, 7-6, 7-7, 7-9, 7-11, 7-12, 7-13, 8-1, 8-2, 8-3, 8-4, 8-5, 8-6, 8-9, 8-11, 8-12)
5.NF.7a Interpret division of a unit fraction by a non-zero whole number, and compute such quotients. *For example, create a story context for $(1/3) \div 4$, and use a visual fraction model to show the quotient. Use the relationship between multiplication and division to explain that $(1/3) \div 4 = 1/12$ because $(1/12) \times 4 = 1/3$.*	**5-13,** 6-2, **7-4,** 7-10, 8-7, 8-8 (6-4, 6-9, 6-12, 7-2, 7-6, 7-13, 8-2, 8-4, 8-5, 8-6, 8-9, 8-11, 8-12)
5.NF.7b Interpret division of a whole number by a unit fraction, and compute such quotients. *For example, create a story context for $4 \div (1/5)$, and use a visual fraction model to show the quotient. Use the relationship between multiplication and division to explain that $4 \div (1/5) = 20$ because $20 \times (1/5) = 4$.*	**5-14,** 6-2, **7-4,** 7-10, 8-7 (6-5, 6-7, 6-10, 7-1, 7-2, 7-3, 7-7, 7-9, 7-11, 7-12, 8-1, 8-3, 8-5, 8-6, 8-9, 8-11, 8-12)
5.NF.7c Solve real world problems involving division of unit fractions by non-zero whole numbers and division of whole numbers by unit fractions, e.g., by using visual fraction models and equations to represent the problem. *For example, how much chocolate will each person get if 3 people share 1/2 lb of chocolate equally? How many 1/3-cup servings are in 2 cups of raisins?*	**5-13, 5-14,** 6-2, **7-4,** 7-10, 8-8 (6-4, 7-1, 7-3, 8-1, 8-2, 8-3, 8-4, 8-9)

Measurement and Data   5.MD

Convert like measurement units within a given measurement system.

5.MD.1 Convert among different-sized standard measurement units within a given measurement system (e.g., convert 5 cm to 0.05 m), and use these conversions in solving multi-step, real world problems.	**1-1, 1-3,** 1-6, 1-8, **1-10, 1-11, 2-6,** 2-10, 4-4, 5-6, 5-13, **6-3, 6-4,** 7-3, **7-11, 8-1, 8-5, 8-6, 8-7, 8-8, 8-9, 8-10** (1-2, 1-4, 1-12, 2-1, 2-3, 2-9, 2-12, 3-1, 3-3, 3-6, 3-8, 4-2, 5-2, 5-4, 7-10)

[1]Students able to multiply fractions in general can develop strategies to divide fractions in general, by reasoning about the relationship between multiplication and division. But division of a fraction by a fraction is not a requirement at this grade.

Content Standards for Mathematics for Grade 5	Everyday Mathematics Grade 5 Lessons*
Represent and interpret data.	
5.MD.2 Make a line plot to display a data set of measurements in fractions of a unit (1/2, 1/4, 1/8). Use operations on fractions for this grade to solve problems involving information presented in line plots. *For example, given different measurements of liquid in identical beakers, find the amount of liquid each beaker would contain if the total amount in all the beakers were redistributed equally.*	**6-4, 6-5, 6-13**, 7-1, **7-9**, 8-8 (6-11, 7-6, 7-8, 8-2, 8-4)
Geometric measurement: Understand concepts of volume and relate volume to multiplication and to addition.	
5.MD.3 Recognize volume as an attribute of solid figures and understand concepts of volume measurement.	**1-5, 1-6, 1-7, 1-8, 1-9, 1-10, 1-12**, 2-1, 2-6, 3-3, 3-13, 4-6, 4-13, **6-6, 6-7, 8-3, 8-4** (1-11, 2-3, 3-1, 7-10)
5.MD.3a A cube with side length 1 unit, called a "unit cube," is said to have "one cubic unit" of volume, and can be used to measure volume.	**1-7, 1-8, 1-9, 1-10, 1-12**, 2-1, 2-6, 3-3, 4-13 (1-11, 2-3)
5.MD.3b A solid figure which can be packed without gaps or overlaps using n unit cubes is said to have a volume of n cubic units.	**1-7, 1-8, 1-9, 1-10, 1-12**, 2-1, 2-6, 3-3, 4-13 (1-11, 2-3, 3-1)
5.MD.4 Measure volumes by counting unit cubes, using cubic cm, cubic in, cubic ft, and improvised units.	**1-5, 1-6, 1-7, 1-8, 1-9, 1-10, 1-12**, 2-1, 2-6, 3-3, 4-13, **6-7** (1-11, 2-3, 3-1)
5.MD.5 Relate volume to the operations of multiplication and addition and solve real world and mathematical problems involving volume.	**1-9, 1-10, 1-11, 1-12**, 2-1, 2-2, 2-6, 3-3, 3-13, 4-6, 4-13, **6-6, 6-7, 8-3, 8-4** (2-4, 2-5, 2-7, 2-8, 2-9, 2-11, 2-12, 2-13, 3-5, 3-7, 3-11, 3-12, 4-2, 4-4, 4-8, 5-10)
5.MD.5a Find the volume of a right rectangular prism with whole-number side lengths by packing it with unit cubes, and show that the volume is the same as would be found by multiplying the edge lengths, equivalently by multiplying the height by the area of the base. Represent threefold whole-number products as volumes, e.g., to represent the associative property of multiplication.	**1-9, 1-11, 1-12**, 2-1, 2-2, 2-6, 3-3, 3-13, 4-6, 4-13, **8-3** (2-4, 2-5, 2-7, 2-9, 2-12, 3-7, 4-2, 6-6)
5.MD.5b Apply the formulas $V = l \times w \times h$ and $V = b \times h$ for rectangular prisms to find volumes of right rectangular prisms with whole-number edge lengths in the context of solving real world and mathematical problems.	**1-9, 1-10, 1-11, 1-12**, 2-1, 2-2, 2-6, 3-3, 3-13, 4-6, 4-13, **6-6, 6-7, 8-3, 8-4** (2-4, 2-5, 2-7, 2-8, 2-9, 2-11, 2-12, 2-13, 3-5, 3-7, 3-11, 3-12, 4-2, 4-4, 5-10)
5.MD.5c Recognize volume as additive. Find volumes of solid figures composed of two non-overlapping right rectangular prisms by adding the volumes of the non-overlapping parts, applying this technique to solve real world problems.	**1-11, 1-12**, 2-2, 2-6, 3-3, 3-13, 4-6, 4-13, **6-6, 8-3** (2-8, 3-11, 3-12, 4-8, 5-10)
Geometry 5.G	
Graph points on the coordinate plane to solve real-world and mathematical problems.	
5.G.1 Use a pair of perpendicular number lines, called axes, to define a coordinate system, with the intersection of the lines (the origin) arranged to coincide with the 0 on each line and a given point in the plane located by using an ordered pair of numbers, called its coordinates. Understand that the first number indicates how far to travel from the origin in the direction of one axis, and the second number indicates how far to travel in the direction of the second axis, with the convention that the names of the two axes and the coordinates correspond (e.g., *x*-axis and *x*-coordinate, *y*-axis and *y*-coordinate).	**4-6, 4-7, 4-8, 4-9, 4-10**, 4-11, 5-2, 5-6, 5-13, 6-1, **7-10, 7-11, 7-12, 7-13**, 8-2, **8-10, 8-11, 8-12** (4-13, 5-1, 5-3, 5-11, 6-2, 6-4, 6-11, 6-13, 7-2, 7-4, 8-6)
5.G.2 Represent real world and mathematical problems by graphing points in the first quadrant of the coordinate plane, and interpret coordinate values of points in the context of the situation.	**4-7, 4-8, 4-9, 4-10**, 5-2, 5-6, 5-13, 6-1, **7-10, 7-11, 7-12, 7-13**, 8-2, **8-10, 8-11, 8-12** (3-10, 5-11, 6-2, 6-4, 6-11, 6-13, 7-2, 7-4, 8-6)
Classify two-dimensional figures into categories based on their properties.	
5.G.3 Understand that attributes belonging to a category of two-dimensional figures also belong to all subcategories of that category. *For example, all rectangles have four right angles and squares are rectangles, so all squares have four right angles.*	**1-1, 7-5, 7-6, 7-7, 7-8**, 7-9, 8-3, 8-8, 8-11, 8-12 (7-12, 8-6, 8-10)
5.G.4 Classify two-dimensional figures in a hierarchy based on properties.	**7-5, 7-6, 7-7, 7-8**, 7-9, 8-3, 8-8, 8-11, 8-12 (6-10, 8-10)

Correlation to the Mathematical Processes and Practices

Everyday Mathematics is a standards-based curriculum engineered to focus on specific mathematical content, processes, and practices in every lesson and activity. The chart below shows complete coverage of each mathematical process and practice in the core program throughout the grade level.

Mathematical Processes and Practices	*Everyday Mathematics* Goals for Mathematical Processes and Practices
1. Make sense of problems and persevere in solving them.	
Mathematically proficient students start by explaining to themselves the meaning of a problem and looking for entry points to its solution. They analyze givens, constraints, relationships, and goals. They make conjectures about the form and meaning of the solution and plan a solution pathway rather than simply jumping into a solution attempt. They consider analogous problems, and try special cases and simpler forms of the original problem in order to gain insight into its solution. They monitor and evaluate their progress and change course if necessary. Older students might, depending on the context of the problem, transform algebraic expressions or change the viewing window on their graphing calculator to get the information they need. Mathematically proficient students can explain correspondences between equations, verbal descriptions, tables, and graphs or draw diagrams of important features and relationships, graph data, and search for regularity or trends. Younger students might rely on using concrete objects or pictures to help conceptualize and solve a problem. Mathematically proficient students check their answers to problems using a different method, and they continually ask themselves, "Does this make sense?" They can understand the approaches of others to solving complex problems and identify correspondences between different approaches.	**Pages** 17, 19, 22, 23, 25, 29, 31, 33, 47, 58, 59, 62, 64, 67, 115, 125, 127, 129, 130, 131, 132, 133, 136, 137, 139, 142, 143, 144, 145, 147, 148, 149, 155, 157, 158, 159, 161, 162, 163, 165, 168, 171, 175, 180, 181, 188, 191, 193, 194, 195, 197, 223, 231, 235, 236, 237, 245, 248, 249, 250, 251, 253, 257, 258, 259, 261, 263, 269, 271, 272, 275, 277, 278, 279, 283, 284, 287, 288, 289, 290, 293, 294, 295, 296, 297, 305, 308, 309, 343, 349, 351, 353, 354, 369, 391, 392, 393, 395, 396, 397, 407, 409, 410, 411, 413, 414, 415, 416, 417, 420, 421, 422, 445, 448, 449, 451, 455, 456, 459, 461, 462, 463, 465, 467, 468, 469, 477, 479, 480, 481, 484, 493, 501, 502, 503, 504, 506, 507, 515, 517, 518, 519, 520, 521, 523, 525, 526, 527, 529, 533, 555, 565, 567, 571, 572, 573, 588, 589, 590, 596, 601, 602, 603, 605, 608, 612, 613, 614, 615, 617, 618, 619, 621, 624, 625, 629, 635, 637, 665, 668, 669, 670, 673, 674, 675, 676, 679, 680, 681, 682, 683, 691, 705, 708, 710, 711, 713, 775, 776, 785, 786, 787, 788, 791, 792, 796, 797, 798, 799, 802, 803, 804, 805, 807, 809, 810, 811, 814, 815, 816, 817, 821, 822, 825, 828
2. Reason abstractly and quantitatively.	
Mathematically proficient students make sense of quantities and their relationships in problem situations. They bring two complementary abilities to bear on problems involving quantitative relationships: the ability to *decontextualize*—to abstract a given situation and represent it symbolically and manipulate the representing symbols as if they have a life of their own, without necessarily attending to their referents—and the ability to *contextualize*, to pause as needed during the manipulation process in order to probe into the referents for the symbols involved. Quantitative reasoning entails habits of creating a coherent representation of the problem at hand; considering the units involved; attending to the meaning of quantities, not just how to compute them; and knowing and flexibly using different properties of operations and objects.	**Pages** 19, 47, 55, 57, 59, 61, 67, 77, 79, 85, 87, 88, 114, 115, 145, 148, 149, 150, 151, 152, 159, 177, 180, 181, 182, 185, 187, 188, 219, 223, 233, 239, 241, 242, 243, 244, 245, 258, 259, 263, 264, 265, 275, 281, 291, 299, 300, 301, 302, 303, 305, 306, 307, 308, 316, 317, 331, 332, 333, 334, 335, 339, 340, 341, 342, 343, 345, 346, 347, 348, 349, 352, 353, 354, 355, 357, 358, 359, 360, 361, 363, 365, 366, 367, 368, 369, 371, 377, 383, 389, 390, 391, 392, 393, 395, 396, 397, 399, 400, 401, 402, 403, 404, 405, 446, 447, 459, 465, 471, 483, 489, 490, 491, 492, 493, 495, 496, 497, 498, 499, 501, 503, 504, 506, 507, 511, 520, 521, 523, 529, 555, 659, 660, 661, 662, 663, 665, 666, 667, 668, 671, 679, 685, 688, 689, 695, 698, 699, 701, 703, 704, 708, 709, 710, 711, 713, 714, 721, 723, 725, 726, 727, 729, 773, 775, 776, 777, 786, 787, 788, 789, 791, 792, 793, 837

Mathematical Processes and Practices	*Everyday Mathematics* Goals for Mathematical Processes and Practices

3. Construct viable arguments and critique the reasoning of others.

Mathematically proficient students understand and use stated assumptions, definitions, and previously established results in constructing arguments. They make conjectures and build a logical progression of statements to explore the truth of their conjectures. They are able to analyze situations by breaking them into cases, and can recognize and use counterexamples. They justify their conclusions, communicate them to others, and respond to the arguments of others. They reason inductively about data, making plausible arguments that take into account the context from which the data arose. Mathematically proficient students are also able to compare the effectiveness of two plausible arguments, distinguish correct logic or reasoning from that which is flawed, and—if there is a flaw in an argument—explain what it is. Elementary students can construct arguments using concrete referents such as objects, drawings, diagrams, and actions. Such arguments can make sense and be correct, even though they are not generalized or made formal until later grades. Later, students learn to determine domains to which an argument applies. Students at all grades can listen or read the arguments of others, decide whether they make sense, and ask useful questions to clarify or improve the arguments.

Pages 21, 23, 28, 29, 30, 31, 33, 34, 35, 44, 45, 46, 50, 52, 64, 67, 69, 70, 129, 220, 221, 222, 223, 225, 239, 257, 258, 259, 260, 261, 264, 265, 266, 267, 270, 271, 272, 273, 281, 299, 307, 308, 375, 379, 380, 413, 472, 474, 475, 477, 479, 480, 481, 497, 517, 530, 531, 532, 533, 600, 602, 621, 623, 624, 659, 679, 813, 825, 831, 832, 837, 838, 839

4. Model with mathematics.

Mathematically proficient students can apply the mathematics they know to solve problems arising in everyday life, society, and the workplace. In early grades, this might be as simple as writing an addition equation to describe a situation. In middle grades, a student might apply proportional reasoning to plan a school event or analyze a problem in the community. By high school, a student might use geometry to solve a design problem or use a function to describe how one quantity of interest depends on another. Mathematically proficient students who can apply what they know are comfortable making assumptions and approximations to simplify a complicated situation, realizing that these may need revision later. They are able to identify important quantities in a practical situation and map their relationships using such tools as diagrams, two-way tables, graphs, flowcharts and formulas. They can analyze those relationships mathematically to draw conclusions. They routinely interpret their mathematical results in the context of the situation and reflect on whether the results make sense, possibly improving the model if it has not served its purpose.

Pages 27, 28, 29, 30, 31, 71, 79, 80, 81, 82, 83, 89, 121, 127, 143, 144, 161, 162, 165, 168, 175, 191, 193, 194, 195, 196, 197, 219, 220, 221, 222, 223, 225, 226, 227, 229, 231, 275, 276, 277, 278, 279, 295, 296, 297, 303, 343, 365, 369, 371, 372, 373, 383, 384, 385, 386, 387, 420, 421, 423, 469, 481, 485, 486, 489, 502, 503, 504, 506, 507, 508, 517, 519, 520, 521, 523, 524, 525, 526, 527, 530, 531, 532, 533, 539, 540, 561, 567, 575, 577, 578, 579, 581, 582, 584, 585, 587, 599, 612, 614, 615, 617, 618, 633, 635, 636, 637, 663, 674, 675, 676, 677, 717, 720, 721, 727, 729, 731, 732, 733, 735, 736, 737, 738, 739, 769, 770, 777, 779, 780, 781, 782, 783, 805, 808, 811, 817, 819, 821, 822, 825, 826, 827, 831, 834, 835, 837, 838, 839, 840, 841

Mathematical Processes and Practices	*Everyday Mathematics* Goals for Mathematical Processes and Practices
5. Use appropriate tools strategically.	
Mathematically proficient students consider the available tools when solving a mathematical problem. These tools might include pencil and paper, concrete models, a ruler, a protractor, a calculator, a spreadsheet, a computer algebra system, a statistical package, or dynamic geometry software. Proficient students are sufficiently familiar with tools appropriate for their grade or course to make sound decisions about when each of these tools might be helpful, recognizing both the insight to be gained and their limitations. For example, mathematically proficient high school students analyze graphs of functions and solutions generated using a graphing calculator. They detect possible errors by strategically using estimation and other mathematical knowledge. When making mathematical models, they know that technology can enable them to visualize the results of varying assumptions, explore consequences, and compare predictions with data. Mathematically proficient students at various grade levels are able to identify relevant external mathematical resources, such as digital content located on a website, and use them to pose or solve problems. They are able to use technological tools to explore and deepen their understanding of concepts.	**Pages** 15, 17, 18, 51, 52, 73, 143, 177, 242, 243, 244, 245, 270, 271, 275, 276, 277, 278, 279, 281, 282, 283, 284, 285, 291, 389, 390, 391, 392, 393, 395, 396, 397, 457, 593, 595, 596, 633, 634, 635, 636, 717
6. Attend to precision.	
Mathematically proficient students try to communicate precisely to others. They try to use clear definitions in discussion with others and in their own reasoning. They state the meaning of the symbols they choose, including using the equal sign consistently and appropriately. They are careful about specifying units of measure, and labeling axes to clarify the correspondence with quantities in a problem. They calculate accurately and efficiently, express numerical answers with a degree of precision appropriate for the problem context. In the elementary grades, students give carefully formulated explanations to each other. By the time they reach high school they have learned to examine claims and make explicit use of definitions.	**Pages** 15, 21, 23, 24, 25, 44, 50, 52, 57, 58, 61, 63, 64, 73, 74, 75, 76, 77, 83, 86, 87, 88, 96, 97, 117, 118, 119, 121, 123, 124, 125, 126, 127, 155, 158, 159, 162, 163, 165, 167, 168, 169, 174, 175, 177, 185, 186, 187, 188, 233, 234, 235, 236, 237, 239, 245, 257, 261, 283, 284, 301, 302, 333, 334, 335, 336, 337, 351, 355, 365, 368, 369, 372, 373, 374, 375, 377, 378, 379, 401, 402, 404, 407, 410, 411, 415, 416, 417, 421, 422, 423, 449, 457, 460, 461, 463, 465, 468, 471, 487, 491, 492, 493, 495, 499, 511, 512, 513, 514, 515, 517, 557, 558, 559, 561, 565, 566, 575, 577, 578, 589, 590, 591, 596, 597, 599, 611, 612, 613, 615, 617, 618, 619, 621, 624, 625, 627, 665, 691, 707, 708, 709, 710, 711, 713, 714, 717, 718, 719, 767, 769, 770, 771, 779, 780, 781, 782, 795, 797, 798, 799, 801, 805, 815, 816, 820, 821, 822, 823, 833, 834

Mathematical Processes and Practices	*Everyday Mathematics* Goals for Mathematical Processes and Practices

7. Look for and make use of structure.

Mathematically proficient students look closely to discern a pattern or structure. Young students, for example, might notice that three and seven more is the same amount as seven and three more, or they may sort a collection of shapes according to how many sides the shapes have. Later, students will see 7×8 equals the well remembered $7 \times 5 + 7 \times 3$, in preparation for learning about the distributive property. In the expression $x^2 + 9x + 14$, older students can see the 14 as 2×7 and the 9 as $2 + 7$. They recognize the significance of an existing line in a geometric figure and can use the strategy of drawing an auxiliary line for solving problems. They also can step back for an overview and shift perspective. They can see complicated things, such as some algebraic expressions, as single objects or as being composed of several objects. For example, they can see $5 - 3(x - y)^2$ as 5 minus a positive number times a square and use that to realize that its value cannot be more than 5 for any real numbers x and y.

Pages 39, 41, 69, 81, 83, 111, 112, 113, 114, 115, 117, 118, 119, 120, 121, 123, 129, 133, 137, 139, 141, 171, 172, 185, 189, 197, 219, 225, 227, 230, 247, 248, 249, 251, 253, 254, 257, 269, 273, 285, 299, 309, 331, 333, 337, 339, 351, 352, 353, 354, 357, 363, 381, 383, 384, 385, 386, 405, 445, 447, 448, 449, 451, 454, 456, 457, 459, 465, 472, 478, 479, 487, 493, 495, 511, 512, 513, 514, 515, 523, 529, 555, 558, 563, 564, 565, 566, 567, 569, 570, 571, 572, 573, 583, 584, 591, 605, 607, 608, 609, 628, 659, 661, 662, 673, 685, 686, 687, 689, 690, 691, 693, 694, 696, 697, 698, 699, 701, 702, 703, 704, 705, 707, 708, 709, 710, 711, 713, 714, 715, 721, 731, 732, 733, 739, 741, 742, 743, 744, 745, 751, 752, 767, 768, 769, 771, 783, 803, 805, 809, 811, 829, 841

8. Look for and express regularity in repeated reasoning.

Mathematically proficient students notice if calculations are repeated, and look both for general methods and for shortcuts. Upper elementary students might notice when dividing 25 by 11 that they are repeating the same calculations over and over again, and conclude they have a repeating decimal. By paying attention to the calculation of slope as they repeatedly check whether points are on the line through (1, 2) with slope 3, middle school students might abstract the equation $(y - 2)/(x - 1) = 3$. Noticing the regularity in the way terms cancel when expanding $(x - 1)(x + 1)$, $(x - 1)(x^2 + x + 1)$, and $(x - 1)(x^3 + x^2 + x + 1)$ might lead them to the general formula for the sum of a geometric series. As they work to solve a problem, mathematically proficient students maintain oversight of the process and practice, while attending to the details. They continually evaluate the reasonableness of their intermediate results.

Pages 39, 69, 70, 135, 155, 227, 230, 231, 236, 247, 248, 249, 250, 251, 253, 254, 255, 257, 269, 273, 285, 339, 361, 362, 377, 380, 381, 387, 391, 393, 395, 396, 413, 472, 473, 495, 498, 511, 513, 514, 523, 558, 559, 560, 570, 571, 572, 573, 583, 601, 659, 681, 682, 703, 723, 724, 725, 726, 727, 729, 737, 738, 741, 742, 743, 744, 745, 777, 811

Mastery Expectations

In Fifth Grade, *Everyday Mathematics* focuses on procedures, concepts, and applications in three critical areas:

- Developing addition/subtraction fluency with fractions, and understanding of multiplication/division of fractions in limited cases.
- Developing fluency with decimal operations, extending division to 2-digit divisors, integrating decimals into the place-value system, and understanding operations with decimals to hundredths.
- Developing an understanding of volume.

Standards	First Quarter Benchmark Expectations for Units 1 and 2	Second Quarter Benchmark Expectations for Units 3 and 4	Third Quarter Benchmark Expectations for Units 5 and 6	Fourth Quarter Benchmark Expectations for Units 7 and 8
5.OA.1	Use one set of grouping symbols in an expression to model a real-world situation. Evaluate an expression that contains a single set of grouping symbols.	☆ Use parentheses, brackets, or braces in numerical expressions, and evaluate expressions with these symbols.	Ongoing practice and application.	
5.OA.2	Write simple expressions to model situations in which no more than two operations are involved. Reason about the relative value of simple expressions without evaluating them.	Write expressions using whole numbers and all four operations to model mathematical and real-world situations. Interpret numerical expressions involving whole numbers without evaluating them.	☆ Write simple expressions that record calculations with numbers, and interpret numerical expressions without evaluating them. *For example, express the calculation "add 8 and 7, then multiply by 2" as 2 × (8 + 7). Recognize that 3 × (18932 + 921) is three times as large as 18932 + 921, without having to calculate the indicated sum or product.*	Ongoing practice and application.

Standards	First Quarter Benchmark Expectations for Units 1 and 2	Second Quarter Benchmark Expectations for Units 3 and 4	Third Quarter Benchmark Expectations for Units 5 and 6	Fourth Quarter Benchmark Expectations for Units 7 and 8
5.OA.3	No expectations for mastery at this point.	Form ordered pairs from data represented in a table with reminders about the conventions of using parentheses to enclose the ordered pairs and commas to separate the numbers in an ordered pair. Graph ordered pairs on a coordinate grid.	Form ordered pairs from data represented in a table and graph them.	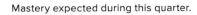 Generate two numerical patterns using two given rules. Identify apparent relationships between corresponding terms. Form ordered pairs consisting of corresponding terms from the two patterns, and graph the ordered pairs on a coordinate plane. *For example, given the rule "Add 3" and the starting number 0, and given the rule "Add 6" and the starting number 0, generate terms in the resulting sequences, and observe that the terms in one sequence are twice the corresponding terms in the other sequence. Explain informally why this is so.*
5.NBT.1	Use place-value understanding to write whole numbers in expanded form. Identify the values of digits in a given whole number. Write whole numbers in which digits represent given values. Recognize that in a multidigit whole number, a digit in one place represents 10 times what it represents in the place to its right.	Recognize that in multidigit whole numbers, a digit in one place represents 10 times what it represents in the place to its right and $\frac{1}{10}$ of what it represents in the place to its left. Recognize that place-value patterns in whole numbers extend to decimals.	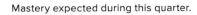 Recognize that in a multi-digit number, a digit in one place represents 10 times as much as it represents in the place to its right and 1/10 of what it represents in the place to its left.	Ongoing practice and application.

Instruction concludes for this standard during this quarter (but the standard may be revisited for review, practice, or application to promote long-term retention, applications, generalization, and transfer).

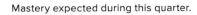 Mastery expected during this quarter.

Standards	First Quarter Benchmark Expectations for Units 1 and 2	Second Quarter Benchmark Expectations for Units 3 and 4	Third Quarter Benchmark Expectations for Units 5 and 6	Fourth Quarter Benchmark Expectations for Units 7 and 8
5.NBT.2	Translate between powers of 10 in exponential notation and standard notation. Correctly multiply a whole number by a power of ten. Notice patterns in the number of zeros in a product when multiplying a whole number by a power of ten.	Use whole-number exponents to denote powers of 10. Correctly multiply whole numbers by powers of 10. Describe patterns in the number of zeros in a product when multiplying a whole number by a power of 10.	Use whole-number exponents to denote powers of 10. Multiply whole numbers by powers of 10 and explain the number of zeros in the product. Multiply or divide a decimal by a power of 10 when no more than one placeholder zero is necessary to write the product or quotient.	☆ Explain patterns in the number of zeros of the product when multiplying a number by powers of 10, and explain patterns in the placement of the decimal point when a decimal is multiplied or divided by a power of 10. Use whole-number exponents to denote powers of 10.
5.NBT.3	No expectations for mastery at this point.	See the mastery expectation statements for the substandards (5.NBT.3a and 5.NBT.3b) for this standard. Students who are meeting expectations for all of the substandards are meeting expectations for this standard.	☆ Read, write, and compare decimals to thousandths.	Ongoing practice and application.
5.NBT.3a	No expectations for mastery at this point.	Represent decimals through thousandths by shading grids. Read and write decimals through thousandths with no placeholder zeros. Read and write decimals in expanded form as sums of decimals (e.g., 0.392 = 0.3 + 0.09 + 0.002).	☆ Read and write decimals to thousandths using base-ten numerals, number names, and expanded form, e.g., $347.392 = 3 \times 100 + 4 \times 10 + 7 \times 1 + 3 \times (1/10) + 9 \times (1/100) + 2 \times (1/1000)$.	Ongoing practice and application.
5.NBT.3b	No expectations for mastery at this point.	Use grids or place-value charts to compare and order decimals through thousandths when the decimals have the same number of digits after the decimal point. Record comparisons using >, =, and < symbols.	☆ Compare two decimals to thousandths based on meanings of the digits in each place, using >, =, and < symbols to record the results of comparisons.	Ongoing practice and application.
5.NBT.4	No expectations for mastery at this point.	Use grids, number lines, or a rounding shortcut to round decimals to the nearest tenth or hundredth in cases when rounding only affects one digit.	☆ Use place value understanding to round decimals to any place.	Ongoing practice and application.

Standards	First Quarter Benchmark Expectations for Units 1 and 2	Second Quarter Benchmark Expectations for Units 3 and 4	Third Quarter Benchmark Expectations for Units 5 and 6	Fourth Quarter Benchmark Expectations for Units 7 and 8
5.NBT.5	Use a strategy to multiply whole numbers. Understand the basic steps of the U.S. traditional multiplication algorithm and successfully apply it to 1-digit by multidigit problems and 2-digit by 2-digit problems in which one factor is less than 20.	Use the U.S. traditional multiplication algorithm to solve 2-digit by 2-digit multiplication problems. Use the U.S. traditional multiplication algorithm to solve multidigit by 2-digit multiplication problems in which only one digit in the second factor requires writing digits above the line. (For example, 636 * 17.)	☆ Fluently multiply multi-digit whole numbers using the standard algorithm.	Ongoing practice and application.
5.NBT.6	Use the partial-quotients algorithm with up to 3-digit dividends and 1-digit or simple 2-digit divisors. Make connections between written partial-quotients work and a given area model representing the same solution.	Use the partial-quotients algorithm with up to 3-digit dividends and 1- or 2-digit divisors. Interpret the remainder of division problems in context, and explain the reasoning. Complete area models to represent solutions to division problems.	☆ Find whole-number quotients of whole numbers with up to four-digit dividends and two-digit divisors, using strategies based on place value, the properties of operations, and/or the relationship between multiplication and division. Illustrate and explain the calculation by using equations, rectangular arrays, and/or area models.	Ongoing practice and application.
5.NBT.7	No expectations for mastery at this point.	Use grids to add and subtract decimals. Use algorithms to add and subtract decimals through tenths with regrouping and through hundredths without regrouping.	Add and subtract decimals to hundredths using models or strategies. Estimate and find products of decimals when both factors are greater than 1. Estimate and find quotients of decimals when the dividend is greater than 1 and the divisor is a whole number.	☆ Add, subtract, multiply, and divide decimals to hundredths, using concrete models or drawings and strategies based on place value, properties of operations, and/or the relationship between addition and subtraction; relate the strategy to a written method and explain the reasoning used.

Instruction concludes for this standard during this quarter (but the standard may be revisited for review, practice, or application to promote long-term retention, applications, generalization, and transfer).

☆ Mastery expected during this quarter.

Standards	First Quarter Benchmark Expectations for Units 1 and 2	Second Quarter Benchmark Expectations for Units 3 and 4	Third Quarter Benchmark Expectations for Units 5 and 6	Fourth Quarter Benchmark Expectations for Units 7 and 8
5.NF.1	No expectations of mastery at this point.	Use tools or visual models to add fractions or mixed numbers with unlike denominators when only one fraction needs to be replaced with an equivalent fraction.	Use tools, visual models, or a strategy to find common denominators. Use tools, visual models, or a strategy to add fractions and mixed numbers with unlike denominators when a common denominator is not difficult to find. Use tools, visual models, or a strategy to subtract fractions and mixed numbers when one of the following is required, but not both: finding a common denominator, or renaming the starting number to have a larger fractional part.	⭐ Add and subtract fractions with unlike denominators (including mixed numbers) by replacing given fractions with equivalent fractions in such a way as to produce an equivalent sum or difference of fractions with like denominators. *For example, 2/3 + 5/4 = 8/12 + 15/12 = 23/12. (In general, a/b + c/d = (ad + bc)/bd.)*
5.NF.2	No expectations of mastery at this point.	Use tools or visual models to solve number stories involving addition and subtraction of fractions and mixed numbers with like denominators.	Use tools, visual models, or equations to solve number stories involving addition and subtraction of fractions and mixed numbers with like denominators. Use tools, visual models, or a strategy to solve number stories involving addition of fractions and mixed numbers with unlike denominators when a common denominator is not difficult to find. Use tools, visual models, or a strategy to solve number stories involving subtraction of fractions and mixed numbers when one of the following is required, but not both: finding a common denominator, or renaming the starting number to have a larger fractional part.	⭐ Solve word problems involving addition and subtraction of fractions referring to the same whole, including cases of unlike denominators, e.g., by using visual fraction models or equations to represent the problem. Use benchmark fractions and number sense of fractions to estimate mentally and assess the reasonableness of answers. *For example, recognize an incorrect result 2/5 + 1/2 = 3/7, by observing that 3/7 < 1/2.*

Standards	First Quarter Benchmark Expectations for Units 1 and 2	Second Quarter Benchmark Expectations for Units 3 and 4	Third Quarter Benchmark Expectations for Units 5 and 6	Fourth Quarter Benchmark Expectations for Units 7 and 8
5.NF.3	No expectations of mastery at this point.	Recognize that a fraction $\frac{a}{b}$ is the result of dividing a by b. Use tools and visual models to solve whole-number division number stories that have fraction or mixed-number answers. Rename mixed numbers and fractions greater than one.	⭐ Interpret a fraction as division of the numerator by the denominator ($a/b = a \div b$). Solve word problems involving division of whole numbers leading to answers in the form of fractions or mixed numbers, e.g., by using visual fraction models or equations to represent the problem. *For example, interpret 3/4 as the result of dividing 3 by 4, noting that 3/4 multiplied by 4 equals 3, and that when 3 wholes are shared equally among 4 people each person has a share of size 3/4. If 9 people want to share a 50-pound sack of rice equally by weight, how many pounds of rice should each person get? Between what two whole numbers does your answer lie?*	Ongoing practice and application.
5.NF.4	No expectations of mastery at this point.	Use tools and visual models to solve fraction-of problems involving a unit fraction and a whole-number.	Understand the relationship between fraction-of problems and fraction multiplication. Use tools and visual models to multiply a fraction by a whole number. Use tools and visual models to multiply a fraction by a fraction.	⭐ Apply and extend previous understandings of multiplication to multiply a fraction or whole number by a fraction.

 Instruction concludes for this standard during this quarter (but the standard may be revisited for review, practice, or application to promote long-term retention, applications, generalization, and transfer).

⭐ Mastery expected during this quarter.

Standards	First Quarter Benchmark Expectations for Units 1 and 2	Second Quarter Benchmark Expectations for Units 3 and 4	Third Quarter Benchmark Expectations for Units 5 and 6	Fourth Quarter Benchmark Expectations for Units 7 and 8
5.NF.4a	No expectations of mastery at this point.	Find a unit fraction of a whole number by partitioning the whole number into the appropriate number of parts and taking one of the parts. Recognize the relationship between the denominator of the unit fraction and the number of parts when partitioning the whole number.	Interpret $\left(\frac{1}{b}\right) \times q$ as 1 part of a partition of q into b equal parts. Find a fraction of a whole number, when the answer is a whole number, by partitioning the whole number into equal parts and taking the appropriate number of parts or by multiplying the whole number by the numerator of the fraction and dividing by the denominator of the fraction. Use paper-folding and other visual representations to partition a fraction into equal parts and find the value of one or more parts. Connect fraction-of problems to fraction multiplication.	⭐ Interpret the product $(a/b) \times q$ as a parts of a partition of q into b equal parts; equivalently, as the result of a sequence of operations $a \times q \div b$. *For example, use a visual fraction model to show $(2/3) \times 4 = 8/3$, and create a story context for this equation. Do the same with $(2/3) \times (4/5) = 8/15$. (In general, $(a/b) \times (c/d) = ac/bd$.)*
5.NF.4b	Find the area of a rectangle with one fractional side length by tiling it with unit squares of side length 1 and counting full and partial squares. Understand that unit squares with fractional side lengths can be used to measure area, but that the count of unit squares with fractional side lengths is different from the measure of area in square units.	Find the area of a rectangle with one fractional side length by tiling it with unit squares of side length 1 and counting full and partial squares, or by using addition. (For example, find the area of a 4 by $2\frac{1}{2}$-unit rectangle by adding $2\frac{1}{2} + 2\frac{1}{2} + 2\frac{1}{2} + 2\frac{1}{2}$.) Understand that unit squares with fractional side lengths can be used to measure area, but that the count of unit squares with fractional side lengths is different from the measure of area in square units.	Find the area of a rectangle with fractional side lengths by counting the number of unit-fraction tiles that cover the rectangle and relating the count to how many tiles cover a unit square. Find the area of rectangles with two fractional side lengths using tools, models, or a fraction multiplication algorithm. Use area models to represent fraction products.	⭐ Find the area of a rectangle with fractional side lengths by tiling it with unit squares of the appropriate unit fraction side lengths, and show that the area is the same as would be found by multiplying the side lengths. Multiply fractional side lengths to find areas of rectangles, and represent fraction products as rectangular areas.
5.NF.5	No expectations of mastery at this point.	No expectations of mastery at this point.	See the mastery expectation statements for the substandards (5.NF.5a and 5.NF.5b) for this standard. Students who are meeting expectations for all of the substandards are meeting expectations for this standard.	⭐ Interpret multiplication as scaling (resizing), by:

Standards	First Quarter Benchmark Expectations for Units 1 and 2	Second Quarter Benchmark Expectations for Units 3 and 4	Third Quarter Benchmark Expectations for Units 5 and 6	Fourth Quarter Benchmark Expectations for Units 7 and 8
5.NF.5a	No expectations of mastery at this point.	No expectations of mastery at this point.	Predict that a product of a whole number and a fraction less than 1 will be less than the whole number, without performing the indicated multiplication. Predict that the product of a whole number or a fraction multiplied by a fraction equal to 1 will be equal to the original whole number or fraction.	⭐ Comparing the size of a product to the size of one factor on the basis of the size of the other factor, without performing the indicated multiplication.
5.NF.5b	No expectations of mastery at this point.	No expectations of mastery at this point.	Explain why multiplying a given number by a fraction less than 1 results in a product smaller than the given number. Understand that multiplying a fraction by another fraction equal to 1 creates an equivalent fraction.	⭐ Explaining why multiplying a given number by a fraction greater than 1 results in a product greater than the given number (recognizing multiplication by whole numbers greater than 1 as a familiar case); explaining why multiplying a given number by a fraction less than 1 results in a product smaller than the given number; and relating the principle of fraction equivalence $a/b = (n \times a)/(n \times b)$ to the effect of multiplying a/b by 1.
5.NF.6	No expectations for mastery at this point.	Use tools and visual models to solve real-world fraction-of problems with unit fractions and whole numbers.	Use tools and models to solve real-world problems involving multiplication of fractions by whole numbers or fractions by fractions. Represent fraction multiplication problems with number sentences.	⭐ Solve real world problems involving multiplication of fractions and mixed numbers, e.g., by using visual fraction models or equations to represent the problem.
5.NF.7	No expectations for mastery at this point.	No expectations for mastery at this point.	See the mastery expectation statements for the substandards (5.NF.7a, 5.NF.7b, and 5.NF.7c) for this standard. Students who are meeting expectations for all of the substandards are meeting expectations for this standard.	⭐ Apply and extend previous understandings of division to divide unit fractions by whole numbers and whole numbers by unit fractions.

Instruction concludes for this standard during this quarter (but the standard may be revisited for review, practice, or application to promote long-term retention, applications, generalization, and transfer).

⭐ Mastery expected during this quarter.

Standards	First Quarter Benchmark Expectations for Units 1 and 2	Second Quarter Benchmark Expectations for Units 3 and 4	Third Quarter Benchmark Expectations for Units 5 and 6	Fourth Quarter Benchmark Expectations for Units 7 and 8
5.NF.7a	No expectations for mastery at this point.	No expectations for mastery at this point.	Use models to solve problems involving division of a unit fraction by a whole number when the problems are in context. Use fraction multiplication to check the quotient of a division problem involving division of a unit fraction by a whole number.	⭐ Interpret division of a unit fraction by a non-zero whole number, and compute such quotients. *For example, create a story context for (1/3) ÷ 4, and use a visual fraction model to show the quotient. Use the relationship between multiplication and division to explain that (1/3) ÷ 4 = 1/12 because (1/12) × 4 = 1/3.*
5.NF.7b	No expectations of mastery at this point.	No expectations for mastery at this point.	Use models to solve problems involving division of a whole number by a unit fraction when the problems are in context. Use fraction multiplication to check the quotient of a division problem involving division of a whole number by a unit fraction.	⭐ Interpret division of a whole number by a unit fraction, and compute such quotients. *For example, create a story context for 4 ÷ (1/5), and use a visual fraction model to show the quotient. Use the relationship between multiplication and division to explain that 4 ÷ (1/5) = 20 because 20 × (1/5) = 4.*
5.NF.7c	No expectation of mastery at this point.	No expectations for mastery at this point.	Use models to solve number stories involving division of a unit fraction by a whole number or division of a whole number by a unit fraction.	⭐ Solve real world problems involving division of unit fractions by non-zero whole numbers and division of whole numbers by unit fractions, e.g., by using visual fraction models and equations to represent the problem. *For example, how much chocolate will each person get if 3 people share 1/2 lb of chocolate equally? How many 1/3-cup servings are in 2 cups of raisins?*
5.MD.1	Perform one-step unit conversions within the same measurement system. Use conversions to solve real-world problems when necessary conversions are identified.	Perform one-step and multi-step unit conversions within the same measurement system, using a resource as necessary to identify difficult measurement equivalents. Use conversions to solve multi-step, real-world problems when necessary conversions are identified.	Perform one-step and multi-step unit conversions within the same measurement system. Use conversions to solve multi-step, real-world problems, using a resource as necessary to identify difficult measurement equivalents.	⭐ Convert among different-sized standard measurement units within a given measurement system (e.g., convert 5 cm to 0.05 m), and use these conversions in solving multi-step, real world problems.

Standards	First Quarter Benchmark Expectations for Units 1 and 2	Second Quarter Benchmark Expectations for Units 3 and 4	Third Quarter Benchmark Expectations for Units 5 and 6	Fourth Quarter Benchmark Expectations for Units 7 and 8
5.MD.2	No expectations for mastery at this point.	No expectations for mastery at this point.	Place fractional data on a line plot when the number line and scale are provided. Use information in line plots to solve single-step problems.	⭐ Make a line plot to display a data set of measurements in fractions of a unit (1/2, 1/4, 1/8). Use operations on fractions for this grade to solve problems involving information presented in line plots. *For example, given different measurements of liquid in identical beakers, find the amount of liquid each beaker would contain if the total amount in all the beakers were redistributed equally.*
5.MD.3	Recognize volume as an attribute of open, three-dimensional figures. (Students may still demonstrate common misconceptions, such as believing that a book does not have volume because it cannot be packed with cubes.)	⭐ Recognize volume as an attribute of solid figures and understand concepts of volume measurement.	Ongoing practice and application.	
5.MD.3a	Understand that cubes are a good unit with which to measure volume because all the edge lengths of a cube are the same.	⭐ A cube with side length 1 unit, called a "unit cube," is said to have "one cubic unit" of volume, and can be used to measure volume.	Ongoing practice and application.	
5.MD.3b	Use unit cubes to pack a solid figure without gaps or overlaps.	⭐ A solid figure which can be packed without gaps or overlaps using *n* unit cubes is said to have a volume of *n* cubic units.	Ongoing practice and application.	
5.MD.4	Find the volume of fully-packed and partially-packed right rectangular prisms by counting unit cubes.	⭐ Measure volumes by counting unit cubes, using cubic cm, cubic in, cubic ft, and improvised units.	Ongoing practice and application.	

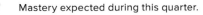 Instruction concludes for this standard during this quarter (but the standard may be revisited for review, practice, or application to promote long-term retention, applications, generalization, and transfer).

⭐ Mastery expected during this quarter.

Standards	First Quarter Benchmark Expectations for Units 1 and 2	Second Quarter Benchmark Expectations for Units 3 and 4	Third Quarter Benchmark Expectations for Units 5 and 6	Fourth Quarter Benchmark Expectations for Units 7 and 8
5.MD.5	See the mastery expectation statements for the substandards (5.MD.5a, 5.MD.5b, and 5.MD.5c) for this standard. Students who are meeting expectations for all of the substandards are meeting expectations for this standard.	⭐ Relate volume to the operations of multiplication and addition and solve real world and mathematical problems involving volume.	Ongoing practice and application.	
5.MD.5a	Understand that packing with unit cubes and multiplying dimensions are two strategies for finding the volume of a right rectangular prism. Use number sentences to represent the volume of a right rectangular prism, when given a formula and whole-number dimensions.	⭐ Find the volume of a right rectangular prism with whole-number side lengths by packing it with unit cubes, and show that the volume is the same as would be found by multiplying the edge lengths, equivalently by multiplying the height by the area of the base. Represent threefold whole-number products as volumes, e.g., to represent the associative property of multiplication.	Ongoing practice and application.	
5.MD.5b	Apply a volume formula to find the volume of a right rectangular prism in mathematical problems when given the formula and the dimensions of the prism.	⭐ Apply the formulas $V = l \times w \times h$ and $V = b \times h$ for rectangular prisms to find volumes of right rectangular prisms with whole number edge lengths in the context of solving real world and mathematical problems.	Ongoing practice and application.	
5.MD.5c	Find volumes of figures composed of right rectangular prisms, when given volume formulas and a clearly labeled representation.	⭐ Recognize volume as additive. Find volumes of solid figures composed of two non-overlapping right rectangular prisms by adding the volumes of the non-overlapping parts, applying this technique to solve real world problems.	Ongoing practice and application.	

Standards	First Quarter Benchmark Expectations for Units 1 and 2	Second Quarter Benchmark Expectations for Units 3 and 4	Third Quarter Benchmark Expectations for Units 5 and 6	Fourth Quarter Benchmark Expectations for Units 7 and 8
5.G.1	No expectation of mastery at this point.	Understand that an ordered pair of numbers identifies an exact location on a coordinate grid. Use coordinates to graph points and to name graphed points in the first quadrant of the coordinate plane.	Make reasonable attempts to explain why an ordered pair of numbers identifies an exact location on a coordinate grid, using terms like origin, x-axis, y-axis, and coordinates. Use coordinates to graph points and to name graphed points in the first quadrant of the coordinate plane.	⭐ Use a pair of perpendicular number lines, called axes, to define a coordinate system, with the intersection of the lines (the origin) arranged to coincide with the 0 on each line and a given point in the plane located by using an ordered pair of numbers, called its coordinates. Understand that the first number indicates how far to travel from the origin in the direction of one axis, and the second number indicates how far to travel in the direction of the second axis, with the convention that the names of the two axes and the coordinates correspond (e.g., x-axis and x-coordinate, y-axis and y-coordinate).
5.G.2	No expectations for mastery at this point.	Understand that information from some real-world and mathematical problems can be represented as ordered pairs and graphed on a coordinate grid. Plot points to represent given information.	Represent real world and mathematical problems by graphing points in the first quadrant of the coordinate plane. Make reasonable attempts to interpret coordinate values of points in context.	⭐ Represent real world and mathematical problems by graphing points in the first quadrant of the coordinate plane, and interpret coordinate values of points in the context of the situation.
5.G.3	No expectations for mastery at this point.	No expectations for mastery at this point.	No expectations for mastery at this point.	⭐ Understand that attributes belonging to a category of two-dimensional figures also belong to all subcategories of that category. *For example, all rectangles have four right angles and squares are rectangles, so all squares have four right angles.*
5.G.4	No expectations for mastery at this point.	No expectations for mastery at this point.	No expectations for mastery at this point.	⭐ Classify two-dimensional figures in a hierarchy based on properties.

Instruction concludes for this standard during this quarter (but the standard may be revisited for review, practice, or application to promote long-term retention, applications, generalization, and transfer).

⭐ Mastery expected during this quarter.

Contents

McGraw-Hill Education

Focus

In **Unit 2,** students explore patterns in the base-10 place-value system and ways of representing large numbers. Students are also introduced to U.S. traditional multiplication and review partial-quotients division.

Major Clusters

5.NBT.A Understand the place value system.

5.NBT.B Perform operations with multi-digit whole numbers with decimals to hundredths.

Supporting Cluster

5.MD.A Convert like measurement units within a given measurement system.

Focus

In **Unit 3,** students build on fractional concepts from previous grades to understand fractions as division. They also use visual models to make estimates, add and subtract fractions and mixed numbers, and check the reasonableness of their answers.

Major Clusters

5.NF.A Use equivalent fractions as a strategy to add and subtract fractions.

5.NF.B Apply and extend previous understandings of multiplication and division to multiply and divide fractions.

Focus

| Unit 6 | Investigations in Measurement; Decimal Multiplication and Division | 542 |

In Unit 6, students multiply and divide decimals by powers of 10. They investigate how patterns can be used to convert measurements in metric units, learn how line plots can be used to organize and analyze data, and explore finding volumes of figures that are not rectangular prisms. Students also multiply and divide decimals.

Major Clusters

5.NBT.A Understand the place value system.

5.NBT.B Perform operations with multi-digit whole numbers and with decimals to hundredths.

Supporting Cluster

5.MD.B Represent and interpret data.

Focus

In Unit 7, students learn two methods for multiplying mixed numbers. They review attributes of 2-dimensional figures and categorize shapes based on their properties. Finally, students graph points on coordinate grids to visualize numerical patterns and represent real-world problems.

Major Cluster

5.NF.B Apply and extend previous understandings of multiplication and division.

Supporting Clusters

5.OA.B Analyze patterns and relationships.

5.G.A Graph points on the coordinate plane to solve real-world and mathematical problems.

5.G.B Classify two-dimensional figures into categories based on their properties.

McGraw-Hill Education

Area and Volume

Contents

*The standards listed here are addressed in the **Focus** of each lesson. For all the standards in a lesson, see the Lesson Opener.

Focus

In this unit, students build on their prior work with area and explore ways to find the areas of rectangles with fractional side lengths. Students also learn about volume.

Major Cluster

5.MD.C Geometric measurement: understand concepts of volume and relate volume to multiplication and to addition.

Supporting Clusters

5.OA.A Write and interpret numerical expressions.

5.MD.A Convert like measurement units within a given measurement system.

Process and Practice Standards

SMP3 Construct viable arguments and critique the reasoning of others.

SMP4 Model with mathematics.

Coherence

The table below describes how standards addressed in the Focus parts of the lessons link to the mathematics that students have done in the past and will do in the future.

	Links to the Past	Links to the Future
5.OA.1	In Grade 3, students inserted parentheses into number sentences and solved number sentences containing parentheses.	Students will model multistep number stories and real-world contexts using expressions with grouping symbols throughout Grade 5. In Grade 6, students will apply Order of Operations to evaluate expressions.
5.NF.4 **5.NF.4b**	In Grade 4, students applied the area formula for rectangles in real-world and mathematical problems.	In Units 5 and 7, students will find areas of rectangles with fractional side lengths and will represent products as rectangular areas as they learn procedures for fraction and mixed-number multiplication. In Grade 6, students will find the areas of triangles and other polygons.
5.MD.3 **5.MD.3a**	In Grade 3, students used unit squares to measure area.	Students will apply their understanding of unit cubes to find the volumes of right rectangular prisms in a variety of contexts in Grade 5. In Grade 6, students will find the volumes of right rectangular prisms with fractional edge lengths.
5.MD.3 **5.MD.3b**	In Grade 3, students found the area of plane figures by tiling them with unit squares.	Students will apply their understanding of packing figures with unit cubes to find the volumes of right rectangular prisms in a variety of contexts in Grade 5. In Grade 6, students will find the volumes of right rectangular prisms with fractional edge lengths by packing them with unit cubes that have unit-fraction edge lengths.
5.MD.4	In Grade 3, students measured area by counting unit squares.	Students will find the volumes of right rectangular prisms in a variety of contexts in Grade 5. In Unit 6, they will measure the volumes of objects in milliliters and will find the relationship between milliliters and cubic centimeters. In Grade 6, students will use appropriate units to measure the volumes of right rectangular prisms with fractional edge lengths.
5.MD.5 **5.MD.5a**	In Grade 4, students developed and applied area formulas for rectangles.	Students will apply their understanding of the relationship between packing prisms with cubes and multiplying dimensions of prisms to find the volumes of rectangular prisms in a variety of contexts in Grade 5. In Grade 6, students will find volumes of right rectangular prisms with fractional edge lengths.

Planning for Rich Math Instruction

	1-1 **Introduction to the** *Student Reference Book*	**1-2** **Area of a Rectangle, Part 1**	**1-3** Open Response **Quilt Area** 2-Day Lesson	**1-4** **Area of a Rectangle, Part 2**
RIGOR				
Conceptual Understanding	Using mathematical tools Examining Student Materials, pp. 16–17	Area Exploring Areas of Rectangles, pp. 22–24	Making sense of others' mathematical thinking in the context of finding areas of rectangles Solving the Open Response Problem, pp. 29–30 Reengaging in the Problem, p. 33	Area Using a Pattern to Find Area, pp. 39–41
Procedural Skill and Fluency	Introducing *Name That Number*, pp. 18–19 Solving Problems Using the *Student Reference Book*, pp. 17–18	Mental Math and Fluency, p. 22 Introducing *Baseball Multiplication*, p. 25	Journal p. 9, #2	Mental Math and Fluency, p. 38
Applications		Journal p. 7, #3, #5	Math Message, p. 27 Solving the Open Response Problem, pp. 29–30	Using a Pattern to Find Area, pp. 39–41 Home Link 1-4, p. 41 Extra Practice, p. 37
Rich Tasks and Mathematical Reasoning	Introducing *Name That Number*, pp. 18–19 Enrichment, p. 15	Exploring Areas of Rectangles, pp. 22–24 Finding Areas of Rectangles, pp. 24–25 Enrichment, p. 21 Extra Practice, p. 21	Math Message, p. 27 Solving the Open Response Problem, pp. 29–30 Reengaging in the Problem, p. 33 Revising Work, pp. 34–35	Using a Pattern to Find Area, pp. 39–41 Enrichment, p. 37
Mathematical Discourse	Examining Student Materials, pp. 16–17 Summarize, p. 18	Exploring Areas of Rectangles, pp. 22–24	Setting Expectations, p. 33 Reengaging in the Problem, p. 33	Discussing Tiling Patterns, pp. 38–39 Using a Pattern to Find Area, pp. 39–41
Distributed Practice	Mental Math and Fluency, p. 16 Introducing *Name That Number*, pp. 18–19 Math Boxes 1-1, p. 19	Mental Math and Fluency, p. 22 Introducing *Baseball Multiplication*, p. 25 Math Boxes 1-2, p. 25	Mental Math and Fluency, p. 27 Math Boxes 1-3, p. 35	Mental Math and Fluency, p. 38 Math Boxes 1-4, p. 41
Differentiation Support	Differentiation Options, p. 15 ELL Support, p. 15 Online Differentiation Support 1-1 Academic Language Development, p. 16 Adjusting the Activity, p. 17	Differentiation Options, p. 21 ELL Support, p. 21 Online Differentiation Support 1-2 Academic Language Development, p. 23 Common Misconception, p. 24	ELL Support, p. 28 Adjusting the Activity, pp. 27, 30 Academic Language Development, p. 29	Differentiation Options, p. 37 ELL Support, p. 37 Online Differentiation Support 1-4 Common Misconception, p. 40

Red text = Game

	1-5 Introduction to Volume	1-6 Exploring Nonstandard Volume Units	1-7 Measuring Volume by Counting Cubes	1-8 Measuring Volume by Iterating Layers	
	Volume Exploring Volume Measurement, pp. 44–45	**Volume measurement** Packing Prisms to Measure Volume, pp. 50–52 Comparing Volume Units, p. 52	**Volume measurement** Measuring Volume with Cubes, pp. 56–57	**Volume measurement** Using Layers of Cubes to Measure Volume, pp. 62–63	**Conceptual Understanding**
	Playing *Name That Number*, p. 47	Home Link 1-6, p. 53	Mental Math and Fluency, p. 56 Playing *Baseball Multiplication*, p. 59	Journal p. 24, #2, #3, #4, #5, #6	**Procedural Skill and Fluency**
	Home Link 1-5, p. 47	Home Link 1-6, p. 53	Journal p. 20, #2	Converting Measurements, p. 64 Home Link 1-8, p. 65	**Applications**
	Exploring Volume Measurement, pp. 44–45 Comparing Volume, p. 46 Enrichment, p. 43 Extra Practice, p. 43	Packing Prisms to Measure Volume, pp. 50–52 Summarize, p. 52 Readiness, p. 49 Enrichment, p. 49	Solving Cube-Stacking Problems, pp. 57–58 Enrichment, p. 55	Summarize, p. 64 Enrichment, p. 61	**Rich Tasks and Mathematical Reasoning**
	Exploring Volume Measurement, pp. 44–45 Comparing Volume, p. 46	Packing Prisms to Measure Volume, pp. 50–52 Comparing Volume Units, p. 52 Readiness, p. 49	Solving Cube-Stacking Problems, pp. 57–58	Using Layers of Cubes to Measure Volume, pp. 62–63	**Mathematical Discourse**
	Mental Math and Fluency, p. 44 Playing *Name That Number*, p. 47 Math Boxes 1-5, p. 47	Mental Math and Fluency, p. 50 Finding Areas of Rectangles, p. 53 Math Boxes 1-6, p. 53	Mental Math and Fluency, p. 56 Playing *Baseball Multiplication*, p. 59 Math Boxes 1-7, p. 59	Mental Math and Fluency, p. 62 Converting Measurements, p. 64	**Distributed Practice**
	Differentiation Options, p. 43 ELL Support, p. 43 Online Differentiation Support 1-5 Academic Language Development, p. 44 Common Misconception, p. 45	Differentiation Options, p. 49 ELL Support, p. 49 Online Differentiation Support 1-6 Adjusting the Activity, p. 51 Academic Language Development, p. 52	Differentiation Options, p. 55 ELL Support, p. 55 Online Differentiation Support 1-7 Academic Language Development, p. 56 Adjusting the Activity, p. 57	Differentiation Options, p. 61 ELL Support, p. 61 Online Differentiation Support 1-8 Academic Language Development, p. 63 Common Misconception, p. 63	**Differentiation Support**

RIGOR

Red text = Game

Planning for Rich Math Instruction

	1-9 **Two Formulas for Volume**	**1-10** **Visualizing Volume Units**	**1-11** **Volume Explorations**	**1-12** **Playing *Prism Pile-Up***
RIGOR				
Conceptual Understanding	Volume measurement Finding a Formula for Volume, pp. 68–70 Finding a Second Volume Formula, pp. 70–71	Volume measurement conversions Visualizing Cubic Units, pp. 75–76	Volumes of figures composed of rectangular prisms Modeling a Suitcase, pp. 80–81	Using clear and precise mathematical explanations to explain volume measurements Sharing and Explaining Strategies, pp. 86–87
Procedural Skill and Fluency	Mental Math and Fluency, p. 68	Journal p. 31, #5	Introducing *Buzz*, p. 83	Understanding Grouping Symbols, p. 89
Applications	Writing and Interpreting Expressions, p. 71	Converting Volume Units, p. 76 Home Link 1-10, p. 77 Extra Practice, p. 73	Math Message, p. 80 Estimating Volumes of Musical Instrument Cases, pp. 81–83 Home Link 1-11, p. 83	Understanding Grouping Symbols, p. 89
Rich Tasks and Mathematical Reasoning	Math Message, p. 68 Writing and Interpreting Expressions, p. 71 Enrichment, p. 67	Converting Volume Units, p. 76 Enrichment, p. 73 Extra Practice, p. 73	Modeling a Suitcase, pp. 80–81 Estimating Volumes of Musical Instrument Cases, pp. 81–83 Journal p. 33: Writing/Reasoning Enrichment, p. 79	Sharing and Explaining Strategies, pp. 86–87 Introducing *Prism Pile-Up*, pp. 87–88 Enrichment, p. 85
Mathematical Discourse	Finding a Formula for Volume, pp. 68–70	Math Message, p. 74		Math Message, p. 86 Sharing and Explaining Strategies, pp. 86–87 Introducing *Prism Pile-Up*, pp. 87–88
Distributed Practice	Mental Math and Fluency, p. 68 Writing and Interpreting Expressions, p. 71 Math Boxes 1-9, p. 71	Mental Math and Fluency, p. 74 More Cube-Stacking Problems, p. 77 Math Boxes 1-10, p. 77	Mental Math and Fluency, p. 80 Introducing *Buzz*, p. 83 Math Boxes 1-11, p. 83	Mental Math and Fluency, p. 86 Understanding Grouping Symbols, p. 89 Math Boxes 1-12, p. 89
Differentiation Support	Differentiation Options, p. 67 ELL Support, p. 67 Online Differentiation Support 1-9 Academic Language Development, p. 68 Adjusting the Activity, p. 69	Differentiation Options, p. 73 ELL Support, p. 73 Online Differentiation Support 1-10 Academic Language Development, p. 74 Common Misconception, p. 75	Differentiation Options, p. 79 ELL Support, p. 79 Online Differentiation Support 1-11 Adjusting the Activity, p. 81	Differentiation Options, p. 85 ELL Support, p. 85 Online Differentiation Support 1-12 Common Misconception, p. 87

Red text = Game

Notes

**1-13 Assessment
Unit 1 Progress Check**

Lesson 1-13 is an assessment lesson. It includes:

- Self Assessment
- Unit Assessment
- Optional Challenge Assessment
- Open Response Assessment
- Suggestions for adjusting the assessments.

Go Online:

Evaluation Quick Entry
Use this tool to record students' performance on assessment tasks.

Data Use the Data Dashboard to view students' progress reports.

Unit 1 Materials

Lesson	*Math Masters*	Activity Cards	Manipulative Kit	Other Materials
1-1	pp. 2–5; TA2 (optional); per partnership: G2	1	per partnership: number cards 0–10 (4 of each) and number cards 11–20 (1 of each) (1 complete Everything Math Deck, if available)	slate; *Student Reference Book;* stick-on notes (optional); per group: nonfiction book with a table of contents, glossary, and index
1-2	pp. 6–9; per partnership: TA3; per group: G3	2	per group: number cards 1–10 (4 of each) (from the Everything Math Deck, if available); 4 counters; per partnership: three 6-sided dice	slate; calculator or multiplication/division facts table
1-3	pp. 10–12; TA4		ruler (optional)	Class Data Pad or chart paper; square foot and square inch (optional); quilt or picture of a quilt (optional); colored pencils (optional); selected samples of students' work; students' work from Day 1
1-4	per partnership: pp. 13; 14–15	3	square pattern blocks	calculator (optional); 3-inch-square stick-on notes; markers; poster paper
1-5	pp. 16; per partnership: G2	4–5	per partnership: number cards 0–10 (4 of each) and number cards 11–20 (1 of each)	slate; variety of empty and full containers, solid objects, and packing materials; small paper plates; 8 index cards; tape; scissors
1-6	pp. 17–20	6–7	pattern blocks	slate; small rectangular prisms; index cards; pennies
1-7	pp. 21 (optional); 22–24; TA3; per group: G3	8	centimeter cubes; square pattern block; per group: number cards 1–10 (4 of each); 4 counters	*Math Journal 1,* Activity Sheet 1 (Rectangular Prism Patterns); scissors; tape; calculator or multiplication/division facts table
1-8	pp. 25 (optional); 26–28; TA2 (optional)	9	centimeter cubes; per group: two 6-sided dice	*Math Journal 1,* Activity Sheet 2 (More Rectangular Prism Patterns); slate; scissors; tape; 1 stick-on note and 1 unused pad of stick-on notes; per group: index cards, paper bag
1-9	pp. 29–30; TA2 (optional)	10	centimeter cubes; ruler	slate; Prisms E and F (from Lesson 1-8 or *Math Masters,* p. 25)
1-10	pp. 31–33	11	centimeter cube; ruler (optional); 3 metersticks; per partnership: tape measure or yardstick, 10 cubes, 10 longs, 10 flats	calculator
1-11	pp. 34; per partnership: 35; 36	12–13	centimeter cubes; per partnership: tape measure, three 6-sided dice	calculator; assembled prism from Lesson 1-8 (or from *Math Masters,* p. TA42, optional)
1-12	pp. 37–38; G4–G5 (optional); per partnership: G6	14	per partnership: centimeter cubes (optional)	*Math Journal 1,* Activity Sheets 3–4 (*Prism Pile-Up* Cards); slate; Class Data Pad; scissors; per partnership: calculator (optional); index cards
1-13	pp. 39–42; *Assessment Handbook,* pp. 5–13		centimeter cubes	

Assessment Check-In

These ongoing assessments offer an opportunity to gauge students' performance on one or more of the standards addressed in that lesson.

 Evaluation Quick Entry Record students' performance online.

 Data View reports online to see students' progress towards mastery.

Lesson	Task Description	Content Standards	Processes and Practices
1-1	Use the *Student Reference Book* to find information to solve problems.		SMP5
1-2	Find the areas of rectangles.	5.NF.4, 5.NF.4b	
1-3	Explain a solution for the area of a quilt.	5.NF.4, 5.NF.4b	SMP3
1-4	Use tiling strategies to find the areas of rectangles with fractional side lengths.	5.NF.4, 5.NF.4b	SMP7
1-5	Understand that volume can be measured by the amount a container holds.	5.MD.3	
1-6	Use pattern blocks to measure the volumes of rectangular prisms.	5.MD.3, 5.MD.4	SMP5
1-7	Count unit cubes to measure the volumes of rectangular prisms.	5.MD.3, 5.MD.3b, 5.MD.4	
1-8	Find the number of unit cubes in one layer and use that number to find the volumes of rectangular prisms.	5.MD.3, 5.MD.3b, 5.MD.4	
1-9	Find the volumes of prisms using formulas.	5.MD.5, 5.MD.5b	
1-10	Reason about the relative size of cubic units.	5.MD.1, 5.MD.3, 5.MD.3a	
1-11	Find the volume of a figure composed of two rectangular prisms.	5.MD.5, 5.MD.5b, 5.MD.5c	
1-12	Apply strategies to find the volumes of figures.	5.MD.3, 5.MD.4, 5.MD.5	

 Virtual Learning Community
vlc.uchicago.edu

While planning your instruction for this unit, visit the *Everyday Mathematics* Virtual Learning Community. You can view videos of lessons in this unit, search for instructional resources shared by teachers, and ask questions of *Everyday Mathematics* authors and other educators. Some of the resources on the VLC related to this unit include:

EM4: Grade 5 Unit 1 Planning Webinar
This webinar provides a preview of the lessons and content in this unit. Watch this video with your grade-level colleagues and plan together under the guidance of an *Everyday Mathematics* author.

Quilt Area: An Open Response and Reengagement Lesson
Watch one classroom work through an Open Response and Reengagement lesson. Explore the introduction and reengagement in practice.

Volume
Watch an EM4 Grade 5 author explain how this edition fulfills the CCSS standards for the teaching of volume.

For more resources, go to the VLC Resources page and search for Grade 5.

Spiral Towards Mastery

The *Everyday Mathematics* curriculum is built on the spiral, where standards are introduced, developed, and mastered in multiple exposures across the grade. Go to the Teacher Center at my.mheducation.com to use the Spiral Tracker.

 Spiral Towards Mastery Progress This Spiral Trace outlines instructional trajectories for key standards in Unit 1. For each standard, it highlights opportunities for Focus instruction, Warm Up and Practice activities, as well as formative and summative assessment. It describes the **degree of mastery**—as measured against the entire standard—expected at this point in the year.

Operations and Algebraic Thinking

5.OA.1

| 1-1 Focus Practice | 1-5 Warm Up Practice | 1-7 Warm Up Practice | 1-9 Warm Up Practice | 1-11 Focus Practice | 1-12 Practice | 1-13 Progress Check | 2-3 Practice | 2-5 Warm Up | 2-6 Focus Practice | 2-8 Practice | 2-10 Practice |

 Progress Towards Mastery By the end of Unit 1, expect students to understand the purpose of grouping symbols and evaluate expressions with one set of symbols; place one set of grouping symbols to make a number sentence true.

Full Mastery of 5.OA.1 expected by the end of Unit 4.

Number and Operations—Fractions

5.NF.4b

| 1-2 through 1-4 Focus Practice | 1-6 Practice | 1-13 Progress Check | 4-14 Practice | 5-7 through 5-10 Focus Practice | 5-12 Focus | 5-15 Progress Check | 6-5 Practice | 7-1 through 7-3 Focus Practice |

 Progress Towards Mastery By the end of Unit 1, expect students to find the area of a rectangle with one fractional side length by tiling it with unit squares of side length 1 and counting full and partial tiles; understand that unit squares with fractional side lengths can be used to measure area, but that the count of unit squares with fractional side lengths is different from the measure of area in square units.

Full Mastery of 5.NF.4b expected by the end of Unit 8.

Measurement and Data

5.MD.3a

| 1-7 Focus Practice | 1-8 Focus Practice | 1-9 Focus Practice | 1-10 Focus Practice | 1-12 Warm Up Focus Practice | 1-13 Progress Check | 2-1 Practice | 2-6 Practice | 2-14 Progress Check | 3-3 Practice | 4-13 Practice | 4-15 Progress Check |

 Progress Towards Mastery By the end of Unit 1, expect students to understand that cubes can be used to measure volume.

Full Mastery of 5.MD.3a expected by the end of Unit 4.

Key  = Assessment Check-In    = Progress Check Lesson   = Current Unit    = Previous or Upcoming Lessons

5.MD.3b

| 1-7 Focus Practice | 1-8 Focus Practice | 1-9 Focus Practice | 1-10 Focus Practice | 1-12 Warm Up Focus Practice | 1-13 Progress Check | 2-1 Practice | 2-6 Practice | 2-14 Progress Check | 3-3 Practice | 4-13 Practice | 4-15 Progress Check |

⭐ **Progress Towards Mastery** By the end of Unit 1, expect students to use unit cubes to pack a solid figure with no gaps and no overlaps.

Full Mastery of 5.MD.3b expected by the end of Unit 4.

5.MD.4

| 1-5 through 1-10 Focus Practice | 1-12 Focus Practice | 1-13 Progress Check | 2-1 Practice | 2-6 Practice | 2-14 Progress Check | 3-1 Practice | 3-3 Practice | 4-13 Practice | 4-15 Progress Check | 6-7 Focus |

⭐ **Progress Towards Mastery** By the end of Unit 1, expect students to find volume by counting the number of unit cubes in a fully-packed prism; find the volume of a partially-packed prism when the dimensions of the prism are clearly shown.

Full Mastery of 5.MD.4 expected by the end of Unit 4.

5.MD.5a

| 1-9 Focus Practice | 1-11 Focus Practice | 1-12 Focus Practice | 1-13 Progress Check | 2-1 Practice | 2-2 Practice | 2-6 Practice | 2-14 Progress Check | 3-3 Practice | 3-13 Practice | 4-6 Practice | 4-13 Practice |

 Progress Towards Mastery By the end of Unit 1, expect students to understand that packing with unit cubes and multiplying dimensions are two strategies for finding the volume of a rectangular prism.

Full Mastery of 5.MD.5a expected by the end of Unit 4.

5.MD.5b

| 1-9 through 1-12 Focus Practice | 1-13 Progress Check | 2-1 Warm Up | 2-2 Practice | 2-6 Practice | 2-14 Progress Check | 3-3 Practice | 3-13 Practice | 4-6 Practice | 4-13 Practice | 4-15 Progress Check |

 Progress Towards Mastery By the end of Unit 1, expect students to apply the appropriate volume formula when given the formulas and a set of whole-number dimensions.

Full Mastery of 5.MD.5b expected by the end of Unit 4.

5.MD.5c

| 1-11 Focus Practice | 1-12 Focus Practice | 1-13 Progress Check | 2-2 Practice | 2-6 Practice | 2-14 Progress Check | 3-3 Practice | 3-13 Practice | 4-6 Practice | 4-13 Practice | 4-15 Progress Check | 6-6 Focus |

⭐ **Progress Towards Mastery** By the end of Unit 1, expect students to understand that the volume of figures composed of two or more right rectangular prisms can be found by adding the volumes of individual parts.

Full Mastery of 5.MD.5c expected by the end of Unit 4.

Key ✓ = Assessment Check-In   🌟 = Progress Check Lesson   ▱ = Current Unit   ▰ = Previous or Upcoming Lessons

Mathematical Background:
Content

▶ **Area of Rectangles with Fractional Side Lengths**
(Lessons 1-2 through 1-4)

Area is a measure of the amount of surface inside a 2-dimensional figure and is typically reported in square units. Students first explore area in Grade 3 by tiling, or covering rectangles with unit squares. They discover that the total number of squares covering a rectangle is equal to the product of the length and the width of the rectangle. In Grade 4 students generalize this pattern and learn to use the area formula $A = l \times w$. Prior to Grade 5 students work solely with rectangles with whole-number side lengths.

In Grade 5 students examine rectangles with fractional side lengths for the first time. In Unit 1 they explore two different tiling strategies for finding the area of rectangles with fractional side lengths:

Strategy 1: Counting Unit Squares and Partial Squares In Lesson 1-2 students build on their work in previous grades by finding area by counting the number of whole and partial unit squares covering a rectangle. A *unit square* is a square that measures 1 unit on each side. For example, to find the area of a $4\frac{1}{2}$-unit by 2-unit rectangle, students count a total of 8 whole squares and 2 half-squares inside the rectangle. (*See margin.*) Reasoning that 2 half-squares make 1 whole square, students find the area of the rectangle to be $8 + 1$, or 9, square units.

Students use this strategy to find areas of tiled and partially tiled rectangles. Note that at this time students are expected to combine partial squares by adding simple fractions, such as $\frac{1}{2} + \frac{1}{2}$, but they are not expected to multiply fractions.

Strategy 2: Tiling with Squares of Unit-Fraction Side Lengths In Lessons 1-3 and 1-4 students consider rectangles tiled with smaller squares of unit-fraction side lengths. **5.NF.4b** For example, the $4\frac{1}{2}$-unit by 2-unit rectangle discussed above could be tiled with $\frac{1}{2}$-unit by $\frac{1}{2}$-unit squares. (*See margin.*) There are 36 smaller squares inside the boundary of the rectangle, but the count of $\frac{1}{2}$-unit by $\frac{1}{2}$-unit squares is *not* equal to the area of the rectangle. In Lesson 1-3 students explore tiling with $\frac{1}{2}$-unit by $\frac{1}{2}$-unit squares and discover that it takes 4 of these smaller squares to cover 1 square unit. They reason that the area of the rectangle can be found by dividing the count of squares by four:
36 small squares ÷ 4 small squares per unit square = 9 square units. In Lesson 1-4 students apply these ideas to squares with other unit-fraction side lengths ($\frac{1}{3}$-unit, $\frac{1}{4}$-unit, and so forth). They use patterns to find a rule for the number of small squares that fit in a square unit and then apply the rule to solve area problems.

By the end of Grade 5 students will be able to show that for rectangles with fractional side lengths, tiling with smaller squares of unit-fraction side lengths (Strategy 2 above) produces the same result as multiplying the side lengths. **5.NF.4b** Unit 1 remains focused on tiling strategies. Students will not be expected to apply the area formula until later in the year, after they have learned methods for multiplying fractions and mixed numbers.

Standards and Goals for Mathematical Content

Because the standards within each strand can be broad, *Everyday Mathematics* has unpacked each standard into Goals for Mathematical Content GMC. For a complete list of Standards and Goals, see page EM1.

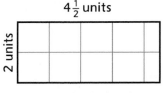

Rectangle tiled with
unit squares

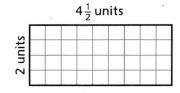

Rectangle tiled with
$\frac{1}{2}$-unit by $\frac{1}{2}$-unit squares

Unit 1 Vocabulary

3-dimensional	cubic unit	rectangular prism
area	expressions	square units
braces	grouping symbols	unit cube
brackets	mathematical model	unit squares
conjecture	nested parentheses	volume

▶ Volume Concepts (Lessons 1-5 through 1-7)

In earlier grades students learned that *length* is a measure of the distance along a path measured in linear units, such as inches or meters. Then students learned that *area* is a measure of the amount of surface inside a 2-dimensional figure, measured in square units. In Grade 5 students explore 3-dimensional objects and aim to answer the question, "How can I measure how much space this object takes up?"

Students begin exploring volume ideas by informally comparing the sizes of cylinders and prisms. They pack cylinders and prisms with improvised units like beans and macaroni. Students discover that the more units it takes to completely fill the object, the more space the object takes up, or the more volume it has. Students understand *volume* as a measure of the amount of 3-dimensional space an object takes up. **5.MD.3**

In everyday life, when a 3-dimensional object is a container, volume is often referred to as *capacity*. However, for mathematical figures such as rectangular prisms, the preferred term is *volume*.

As with measures of length and area, all measures of volume require a unit. After exploring nonstandard units, students recognize that unit cubes make sense as a basic unit of volume, in part because they can be packed without gaps and overlaps. **5.MD.3a** Students fill rectangular prisms and discover that the number of unit cubes it takes to completely fill a prism is the same as that prism's volume. **5.MD.3b** Manipulating unit cubes and making volume estimates facilitates understanding of the 3-dimensional structure of a rectangular prism as unit cubes arranged in rows, columns, and stacks. Students use this structure to visualize and count the number of cubes it would take to fill partially packed prisms. **5.MD.4** (*See margin.*)

> **NOTE** In *Everyday Mathematics*, linear measures are represented in formulas by lowercase letters and areas and volumes are represented by uppercase letters. For example, *b* might be used to represent the length of the base of a rectangle, and *B* to represent the area of the base of a prism.

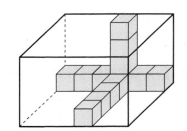

Using rows, columns, and stacks to find volume

▶ Developing Volume Formulas (Lessons 1-8 and 1-9)

Once students understand that counting the unit cubes that fill a figure is a way to find the figure's volume, they can become strategic about finding the total number of unit cubes it takes to fill a figure. One way to find the number of unit cubes in a rectangular prism is to think in terms of layers. (*See margin.*)

In Lesson 1-8 students cover the base of a prism with unit cubes to find the number of cubes in one layer. Because each layer has the same number of cubes, students can find the total number of cubes by iterating layers with repeated addition or multiplication. After applying this strategy to different prisms, students generalize the pattern. The number of unit cubes in the base layer corresponds to the area of the base. The number of layers corresponds to the height. Therefore, the volume (V) can be found by multiplying the area of the base (B) by the height (h): $V = B * h$.

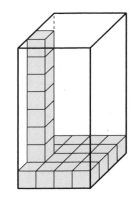

Thinking in layers to find volume

Professional Development

Developing Volume Formulas *Continued*

The area of a rectangle can be found by multiplying the length by the width: $A = l * w$. Using the area formula, the volume formula $V = B * h$ can be rewritten as $V = (l * w) * h$. The resulting volume is the same using either formula. **5.MD.5a**

▶ Volume Applications (Lessons 1-10 through 1-12)

Lessons 1-5 through 1-9 lay the foundation for students to make sense of volume in increasingly complex contexts. Having developed an understanding of what volume is, why unit cubes are the basic unit of volume, and how the total number of unit cubes that fill a rectangular prism can be quickly and reliably calculated, students are able to apply formulas to find the volumes of right rectangular prisms in real-world and mathematical problems. **5.MD.5b** Instead of seeing unit cubes in prisms, students begin to see simple models of rectangular prisms with labeled dimensions.

In real-world situations it is important to choose appropriate units. When measuring the volume of a room, for example, it is necessary to consider the *size* of different cubic units. It is possible to compare cubic units by reasoning about the relative size of related linear units. For example, since an inch is smaller than a foot, a cubic inch is smaller than a cubic foot. Students use concrete models and their familiarity with linear units to visualize and compare volume units. They also relate units by converting among units in the same system. **5.MD.1** This allows students to develop a sense of the relative size of volume units and to choose reasonable units for real-world measures. For example, cubic centimeters may be reasonable for the volume of a small box, but cubic meters may be more reasonable for the volume of a classroom. Students are not expected to fluently convert between units of volume.

Since many real-world objects are irregularly shaped, students use figures made up of rectangular prisms as *mathematical models* of real-world objects to estimate volumes. (*See margin.*) Students also learn that volume is additive, which means that they can divide figures into parts, find the volume of each part, and add to find the volume of the whole. **5.MD.5c** Students practice finding the volume of prisms and figures composed of prisms by playing *Prism Pile-Up*. They will continue to apply volume concepts in practice activities throughout the next few units.

▶ Expressions and Grouping Symbols (All Lessons)

Throughout Unit 1 students gain experience writing and evaluating numerical expressions that contain *grouping symbols*. **5.OA.1** Grouping symbols include parentheses ( ), brackets [ ], and braces { }. They define the order in which operations in an expression are to be carried out. Students are introduced to parentheses and the order of operations in Grade 3 and continue using them in Grade 4. In Grade 5 students are introduced to brackets and braces.

Students review parentheses in Lesson 1-1 with the *Student Reference Book* scavenger hunt and the game *Name That Number.* They write expressions with parentheses to record calculations made during the game and while solving area and volume problems. **5.OA.2** Brackets and braces are introduced in Lesson 1-12. Students continue to explore expressions as representations of real-world and mathematical situations in future units.

A trombone case is an example of an irregularly shaped object.

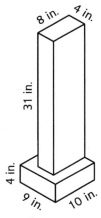

This figure composed of prisms is a mathematical model of the trombone case.

McGraw-Hill Education

Mathematical Background:
Process and Practice

 See below for some of the ways that students engage in **SMP3 Construct viable arguments and critique the reasoning of others** and **SMP4 Model with mathematics** through **Measurement and Data** and the other mathematical content of Unit 1.

▶ Standard for Mathematical Process and Practice 3

When students participate in discussions driven by questions like *How do you know?* and *Why do you think that works?*, they are developing their ability to construct mathematical arguments and to communicate their ideas in a logical manner. **SMP3** As students develop this practice, they often rely on concrete referents, such as drawings, objects, or models, to construct their informal arguments. In Lesson 1-2 students use whole and partial squares to cover rectangles with fractional side lengths. They examine the results of these examples and make a *conjecture*, or mathematical prediction, about what they can expect when working with similar rectangles in the future.
GMP3.1 In Lesson 1-5, when students are asked which of several 3-dimensional objects is largest, they share their conjectures, explain the thinking that led to their choice, and discuss ways to test their ideas. In Lesson 1-9 students reason about *why* multiplying the area of the base times the height of a prism gives the same volume as counting unit cubes. They construct a mathematical *argument* for why $V = B * h$ works for finding the volume of a prism. **GMP3.1**

Standard for Mathematical Process and Practice 3 also emphasizes the importance of making sense of and critically considering others' conjectures and arguments. **GMP3.2** When students respond to questions like *How do you think Ali got her answer? Do you agree or disagree?*, they are making sense of others' mathematical reasoning. In Lesson 1-3 students are asked to make sense of two different answers to a problem and explain which answer they agree with and why.

▶ Standard for Mathematical Process and Practice 4

"Mathematically proficient students who can apply what they know are comfortable making assumptions and approximations to simplify a complicated situation." **SMP4** One way to simplify a real-world situation is to model it with mathematics, using representations such as drawings, diagrams, graphs, symbols, numbers, and tables. **GMP4.1** In Unit 1 students create mathematical models and use the models to solve problems and answer questions. **GMP4.2** In Lesson 1-11 students tackle the problem of finding the volumes of irregularly shaped musical instrument cases. They discover that they can estimate the volume of each case by modeling it with rectangular prisms. Students then use what they already know about volume to arrive at a reasonable estimate of the volume of the case. A problem that was initially daunting is simplified with the help of a model.

Students have used number models to represent problems since the early grades. The use of number models to represent real-world problems continues to be an emphasis in Grade 5. Beginning in Unit 1, students are prompted to record number models to represent real-life problems.

Go Online to the *Implementation Guide* for more information about the Mathematical Process and Practice Standards.

For students' information on the Mathematical Process and Practice Standards, see *Student Reference Book*, pages 1–34.

Introduction to the
Student Reference Book

Overview Students explore the *Student Reference Book*. They read about and practice using grouping symbols.

▶ **Before You Begin**
For the optional Readiness activity, select a few familiar nonfiction books that have a table of contents, glossary, and index.

▶ **Vocabulary**
grouping symbols • expressions

Standards

Focus Clusters
- Write and interpret numerical expressions.
- Understand the place value system.
- Convert like measurement units within a given measurement system.

1 Warm Up   5 min

	Materials	
Mental Math and Fluency Students write numerical expressions.	slate	5.OA.2

2 Focus   35–40 min

Math Message Students read and discuss an introduction to *Fifth Grade Everyday Mathematics*.	*Math Journal 1*, p. 1	
Examining Student Materials Students examine the *Math Journal* and *Student Reference Book* and read about grouping symbols.	*Math Journal 1; Student Reference Book; Math Masters*, p. TA2 (optional); stick-on notes (optional)	5.OA.1 SMP1, SMP5
Solving Problems Using the *Student Reference Book* Students use the *Student Reference Book* to solve problems.	*Math Journal 1*, pp. 2–3; *Student Reference Book*	5.OA.1, 5.NBT.1, 5.MD.1, 5.G.3 SMP1, SMP5
✓ **Assessment Check-In** See page 18. Expect some students who have used the *Student Reference Book* in prior grades to be familiar with the organization of the book and how to find relevant information to solve the scavenger hunt problems.	*Math Journal 1*, pp. 2–3; *Student Reference Book*	SMP5

3 Practice   20–30 min

Introducing *Name That Number* **Game** Students practice writing expressions for calculations and writing expressions with grouping symbols.	*Student Reference Book*, p. 315; per partnership: *Math Masters*, p. G2; number cards 0–10 (4 of each) and 11–20 (1 of each)	5.OA.1, 5.OA.2 SMP1, SMP2
Math Boxes 1-1 Students practice and maintain skills.	*Math Journal 1*, p. 4	See page 19.
Home Link 1-1: Unit 1 Family Letter **Homework** Students take home a Family Letter introducing *Fifth Grade Everyday Mathematics*.	*Math Masters*, pp. 2–5	

 Go Online to see how mastery develops for all standards within the grade.

Differentiation Options

Readiness

Exploring Text Features of the *Student Reference Book*

SMP5

Student Reference Book; per group: nonfiction book with a table of contents, glossary, and index

Language Art Link To build background experience for navigating the *Student Reference Book,* students compare it with another nonfiction book containing a table of contents, a glossary, and an index. Have students identify the features of this book and discuss how each one is used. **GMP5.2** For example, students might suggest using a glossary to look up unfamiliar terms. As each feature is identified, have students find its counterpart in the *Student Reference Book* and examine it. Compare where the features are placed and how they are used in the nonfiction book and the *Student Reference Book.*

Enrichment

Writing a Reference Book Page

SMP5, SMP6

Student Reference Book

To extend their understanding of a mathematics reference book, students design and write a reference book page. Have students brainstorm interesting mathematical topics and then create a reference book page about a topic of their choice. Encourage them to use precise mathematical language and to design their page to be a mathematical tool that others can use. **GMP5.2, GMP6.3**

Extra Practice

Writing and Answering Mathematical Questions

SMP5

Activity Card 1; *Student Reference Book*

For additional practice navigating the *Student Reference Book,* students write mathematical questions and use the book to help them find solutions to a partner's questions. **GMP5.2**

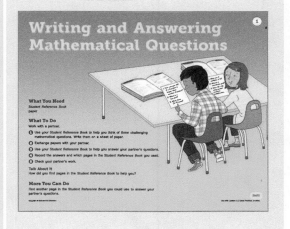

English Language Learner

Beginning ELL To support students' understanding of the Mental Math and Fluency problems, point to numerals on a number line and provide pictorial vocabulary cards illustrating the meaning of *plus, times, sum, double, product, quotient, minus,* and *triple.* For example, to scaffold writing an expression for 5 plus 4, point to the numerals as you display a card showing an addition symbol. Encourage students to repeat the names of the terms.

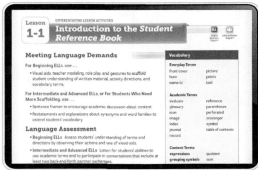

Differentiation Support pages are found in the online Teacher's Center.

Standards and Goals for
Mathematical Process and Practice

SMP1 Make sense of problems and persevere in solving them.

GMP1.3 Keep trying when your problem is hard.

GMP1.5 Solve problems in more than one way.

SMP5 Use appropriate tools strategically.

GMP5.2 Use tools effectively and make sense of your results.

Academic Language Development

Have partnerships select a term from journal page 1 that they feel they know and define it using a 4-Square Graphic Organizer (*Math Masters*, page TA2). Direct students to fill in each of the four quadrants, working in a clockwise direction as follows: drawing a picture, describing an example, describing a non-example, and writing their definition for the term.

1 Warm Up   5 min

▶ Mental Math and Fluency

Dictate the following and have students write an expression for each item on their slates. Students do not need to calculate answers.
Leveled exercises:

● ○ ○ 5 plus 4 $5 + 4$
　　　10 minus 6 $10 - 6$
　　　2 times 4 $2 * 4$

● ● ○ The sum of 60 and 32 $60 + 32$
　　　The product of 100 and 7 $100 * 7$
　　　Double 40 $40 * 2$, or $40 + 40$

● ● ● The sum of 7 and 3 minus 5 $7 + 3 - 5$
　　　The quotient of 25 and 5 $25 / 5$
　　　Triple 10 $10 * 3$, or $10 + 10 + 10$

2 Focus   35–40 min

▶ Math Message

Math Journal 1, p. 1

Read journal page 1. Underline any words or terms that you don't know or you think are interesting. Talk with a partner about one thing you are excited to learn about in math class this year.

▶ Examining Student Materials

Math Journal 1; Student Reference Book

| WHOLE CLASS | SMALL GROUP | PARTNER | INDEPENDENT |

Math Message Follow-Up Ask students to share terms they either found interesting or have questions about. Then have them talk about some of the mathematical ideas they want to explore this year.

Give students several minutes to browse through the *Math Journal* and comment on what they notice. Be sure to cover the following points:

- Most of the journal pages will not be torn out. The journal will serve as a record of the mathematics students have learned over the year. Students will be able to refer back to previous pages as they learn new concepts.
- Many of the pages include familiar routines, such as Math Boxes.
- The final pages in the journal are perforated Activity Sheets that will be torn out at the appropriate time.
- The inside front and back covers contain resources for students to use while they work.

Next, ask students to look through the *Student Reference Book.* After a few minutes, bring the class together to discuss what students noticed. Point out the index, the glossary, and the color-coding scheme for the different topics, and guide a discussion of how they might be useful. Read *Student Reference Book,* page xi as a class and talk about how the book can be useful as a mathematical tool. Explain that, as with any tool, it is important to know how to use it effectively. GMP5.2

Work through *Student Reference Book,* page 42 to model how the book might be used. Use the activity as an opportunity to review how to use parentheses as **grouping symbols** in mathematical **expressions** and how to evaluate expressions with grouping symbols. Have students try Check Your Understanding Problems 1-3 and check their answers. Encourage them to analyze mistakes by finding what they did incorrectly. GMP1.3 Discuss problems that the class finds difficult. If necessary, provide a few more problems where students must either insert parentheses or evaluate expressions with parentheses. *Suggestions:*

- Evaluate the expression $6 * (4 + 3) - 1$. 41
- Insert parentheses to make the following number sentence true: $21 - 8 + 2 + 3 = 14$. $21 - (8 + 2) + 3 = 14$

▶ Solving Problems Using the *Student Reference Book*

Math Journal 1, pp. 2–3; *Student Reference Book*

| WHOLE CLASS | SMALL GROUP | PARTNER | INDEPENDENT |

Explain to students that they will go on a scavenger hunt through the *Student Reference Book* to find answers for the problems on journal pages 2 and 3. In addition to solving the problems, they will record the *Student Reference Book* page number where they found the information that helped them. Explain that for the scavenger hunt, they will score 3 points for each correct answer and 5 points for each correct page number. This keeps the focus on locating helpful pages in the *Student Reference Book,* rather than solely obtaining the correct answer. GMP5.2

Have small groups or partners complete the journal pages. Circulate and assist, pointing out that helpful information can often be found in several places in the *Student Reference Book.* Explain that students can increase their point total by recording multiple page numbers. GMP1.5

Math Journal 1, p. 1

Welcome to *Fifth Grade Everyday Mathematics*

Lesson 1-1

DATE   TIME

This year in math class you will continue to build on the mathematical skills and ideas you have learned in previous years. You will learn new mathematics and think about the importance of mathematics in your life now and how math will be useful to you in the future. Many of the new ideas you learn this year will be ones that your parents, or even your older brothers and sisters, may not have learned until much later than fifth grade. The authors of *Everyday Mathematics* believe that today's fifth graders are able to learn more and do more than fifth graders in the past. They think that mathematics is fun and they think you will find it enjoyable too.

Here are some of the things you will do in *Fifth Grade Everyday Mathematics:*

- Extend your understanding of place value to decimals and use what you learn to explain how our place-value system works.
- Review and extend your skills doing arithmetic, using a calculator, and thinking about problems and their solutions. You will add, subtract, multiply, and divide whole numbers and decimals.
- Use your knowledge of fractions and operations to compute with fractions. You will think about how adding, subtracting, multiplying, and dividing fractions is similar to and different from doing the same computations with whole numbers and decimals.
- Explore the concept of volume. You will learn how volume differs from other measurements you have studied. You will find the volume of 3-dimensional figures in multiple ways, and you will develop strategies for finding the volume of rectangular prisms. Look at journal page 2. Without telling anyone, write the number one hundred twelve in the top right-hand corner of the page.
- Learn about coordinate grids and find out how graphing can help you solve mathematical and real-world problems.
- Deepen your understanding of 2-dimensional figures, their attributes, and how different 2-dimensional figures are related to each other.

We want you to become better at using mathematics so you can better understand your world. We hope you enjoy the activities in *Fifth Grade Everyday Mathematics* and that they help you appreciate the beauty and usefulness of mathematics in your daily life.

1

Adjusting the Activity

Differentiate If students struggle locating information in the *Student Reference Book,* have them create tabs for each section of the book using stick-on notes. The tabs will help them quickly locate relevant sections.

 Go Online Differentiation Support

Math Journal 1, p. 2

Student Reference Book
Scavenger Hunt

Lesson 1-1
DATE   TIME

Solve the problems on this page and page 3. Use your *Student Reference Book* to find information about each problem. Record the page numbers.

112

	Problem Points	Page Points

① 5 meters = __500__ centimeters   _____ _____

 page __213, 328__

② 300 mm = __30__ cm   _____ _____

 page __213, 328__

③ Solve.   _____ _____
(15 − 4) • 3 = __33__
25 + (47 − 18) = __54__

 page __42__

④ Write the value of the 5 in each of the following numbers.   _____ _____
9,652 __50__
15,690 __5,000__
1,052,903 __50,000__

 page __66–67__

⑤ Name two fractions equivalent to 4/6. Sample answers: _____ _____
__2/3__ and __12/18__

 page __166, 168–170__

⑥ 460 ÷ 5 = __92__   _____ _____

 page __108__

2   5.OA.1, 5.NBT.1, 5.MD.1, SMP5

Math Journal 1, p. 3

Student Reference Book
Scavenger Hunt (continued)

Lesson 1-1
DATE   TIME

	Problem Points	Page Points

⑦ a. What is the definition of a trapezoid?   _____ _____
A quadrilateral that has at
least one pair of parallel sides.
b. Draw two different trapezoids. Sample answers:

 page __268__

⑧ What materials do you need to play *Name That Number*?   _____ _____
1 complete deck of
number cards

 page __315__

Record your scavenger hunt scores in the table below. Then calculate the totals.

Problem Number	Problem Points	Page Points	Total Points
1			
2			
3			
4			
5			
6			
7			
8			
Total Points			

5.G.3, SMP5   3

When most students have completed the journal pages, review the answers and page references. Then have students calculate their scores. Finally, to reinforce the importance of always reading carefully, refer students to journal page 2 and ask how many wrote the number 112 in the top right-hand corner. Point out that they were directed to do so by the bullet on journal page 1. Tell students who followed this direction to add 10 extra points to their scores.

✓ ## Assessment Check-In

Math Journal 1, pp. 2–3; Student Reference Book

Observe as students use the *Student Reference Book* to find information while solving the problems on the journal pages. **GMP5.2** Some will have used the *Student Reference Book* in prior grades, so expect them to be familiar with the organization of the book and how to find relevant information. If students struggle locating relevant pages, consider reviewing reference skills, such as using a table of contents, glossary, and index.

✓ **Evaluation Quick Entry** Go online to record students' progress and to see trajectories toward mastery for these standards.

Summarize Direct students to journal page 4, drawing their attention to the SRB icons that appear with the Math Box exercises. Explain that they will see this icon on many pages they work on. It is there to indicate where information on the topic can be found in the *Student Reference Book*. Have partners discuss how the icon and the *Student Reference Book* will be helpful to them as a mathematical tool.

③ Practice   20–30 min

▶ Introducing *Name That Number*

Student Reference Book, p. 315; *Math Masters*, p. G2

WHOLE CLASS	SMALL GROUP	PARTNER	INDEPENDENT

Have students locate the *Student Reference Book* page giving game directions for *Name That Number*. Read the directions with the class and play a few sample rounds. Model how to complete the game record sheet on *Math Masters*, page G2. As students write expressions recording their calculations, remind them that each one should be equivalent to the target number. Encourage students to use grouping symbols in the expressions they record.

Have students play the game in small groups or partnerships.

Observe

- Which students easily write simple expressions that are equivalent to the target number? Which students struggle writing simple expressions?
- Which students correctly use grouping symbols in their expressions? **GMP2.1**

Discuss

- *How did you decide which operations to use in your expressions?* **GMP1.1**
- *Can you make the target number a different way? How?* **GMP1.5**

Differentiate **Game Modifications** **Go Online** Differentiation Support

▶ Math Boxes 1-1

Math Journal 1, p. 4

WHOLE CLASS | SMALL GROUP | PARTNER | INDEPENDENT

Mixed Practice Math Boxes 1-1 are paired with Math Boxes 1-3.

▶ Home Link 1-1: Unit 1 Family Letter

Math Masters, pp. 2–5

Homework The Family Letter introduces the content of *Fifth Grade Everyday Mathematics* and the topics to be covered in Unit 1. Consider distributing this letter to parents during your introductory meeting or curriculum night.

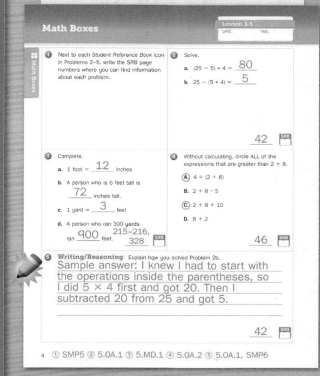

Math Journal 1, p. 4

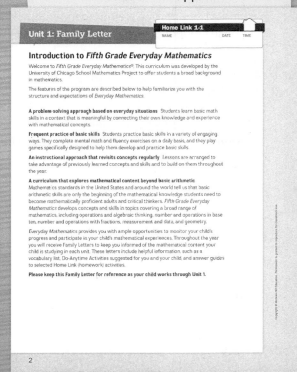

Math Masters, pp. 2–5

Lesson 1-2

Area of a Rectangle, Part 1

Overview Students review area concepts and explore strategies for finding areas of rectangles with fractional side lengths.

▶ **Before You Begin**

For Part 2, decide how you will display the rectangles on *Math Masters,* page 6 so that students do not see them all at once. Copy and cut apart *Math Masters,* page 7 so that each partnership has one copy of Rectangle D.

▶ **Vocabulary**

area • unit squares • square units

Standards

Focus Cluster
• Apply and extend previous understandings of multiplication and division.

	Materials	Standards
① Warm Up 5 min		
Mental Math and Fluency Students add simple fractions.	slate	5.NF.1
② Focus 35–40 min		
Math Message Students read about area and find the area of rectangles.	*Math Journal 1,* p. 5; *Student Reference Book,* p. 221	5.NF.4, 5.NF.4b SMP1
Exploring Areas of Rectangles Students review area and discuss strategies for finding areas of rectangles with and without fractional side lengths.	*Math Journal 1,* p. 5; *Student Reference Book,* p. 221; *Math Masters,* pp. 6–7	5.NF.4, 5.NF.4b SMP1, SMP3, SMP6
Finding Areas of Rectangles Students find areas of rectangles with fractional side lengths.	*Math Journal 1,* p. 6; *Student Reference Book,* p. 225	5.NF.4, 5.NF.4b SMP6
✓ **Assessment Check-In** See page 25. Expect most students to be able to find the correct areas of rectangles with whole-number and fractional side lengths and write appropriate number sentences in problems like those identified.	*Math Journal 1,* p. 6	5.NF.4, 5.NF.4b
③ Practice 20–30 min		
Introducing *Baseball Multiplication* **Game** Students practice multiplication facts to prepare for multidigit multiplication in Unit 2.	*Student Reference Book,* p. 292; per group: *Math Masters,* p. G3; number cards 1–10 (4 of each) (from the Everything Math Deck, if available); 4 counters; calculator or multiplication/division facts table	5.NBT.5 SMP1, SMP6
Math Boxes 1-2 Students practice and maintain skills.	*Math Journal 1,* p. 7	See page 25.
Home Link 1-2 **Homework** Students find the area of rectangles with fractional side lengths.	*Math Masters,* p. 9	5.OA.1, 5.NF.4, 5.NF.4b

/// **Go Online** to see how mastery develops for all standards within the grade.

Differentiation Options

Readiness — 5–15 min

WHOLE CLASS | **SMALL GROUP** | PARTNER | INDEPENDENT

Reviewing Equal Groups

5.NF.4, 5.NF.4b

To prepare for thinking about using a row or column of units to fill a rectangle, students find the total number of objects in equal groups of both whole-number and fractional sizes. Encourage students to draw pictures to help them solve the problems, and have them share strategies. Ask:

- *Three groups with 2 in each group. How many in all?* 6; Sample strategies: I drew 3 groups of 2 Xs and there were 6 in all. I multiplied 3 and 2 to find out there are 6 in all.

- *Three groups with $\frac{1}{2}$ in each group. How many in all?* $\frac{3}{2}$, or $1\frac{1}{2}$; Sample strategy: I drew 3 squares and shaded half of each one. I can put 2 halves together to make 1 and then there is 1 more half left. That's $1\frac{1}{2}$ in all.

- *Three groups with $2\frac{1}{2}$ in each group. How many in all?* $7\frac{1}{2}$; Sample strategy: I put 2 groups together first. $2 + 2 = 4$ and $\frac{1}{2} + \frac{1}{2} = 1$, so that's 5 in all. Adding in the last group of $2\frac{1}{2}$ makes $7\frac{1}{2}$ in all.

Repeat the activity with other examples as needed.

Enrichment — 15–30 min

WHOLE CLASS | SMALL GROUP | **PARTNER** | INDEPENDENT

Finding the Area of Figures with Fractional Side Lengths

5.NF.4, 5.NF.4b, SMP6

Math Masters, p. 8

To extend their work finding areas of rectangles with fractional side lengths, students solve problems on *Math Masters,* page 8 involving rectilinear figures with fractional side lengths. They use precise language to explain their strategies. **GMP6.1, GMP6.3**

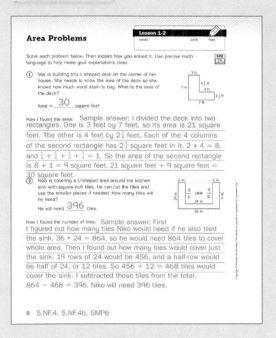

Extra Practice — 5–15 min

WHOLE CLASS | SMALL GROUP | **PARTNER** | INDEPENDENT

Finding Areas of Rectangles

5.NF.4, 5.NF.4b, SMP3

Activity Card 2; per partnership: *Math Masters,* p. TA3; three 6-sided dice

To practice finding areas of rectangles with fractional side lengths, students generate dimensions of rectangles by rolling dice. They draw and find the area of the rectangles they generate. Then they explain their strategies to a partner and try to apply their partners' strategies. **GMP3.2**

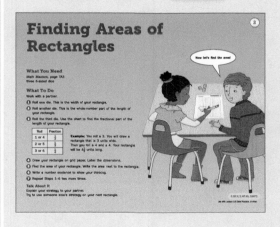

English Language Learner

Beginning ELL Provide a visual of a rectangle with some whole and partial square units on the inside. Label the key words students will hear and see during the lesson, including the following terms: *boundary, closed boundary, unit squares, whole, partial, area, region, overlap,* and *gap.* Ask volunteers to share what they wrote about area for the Math Message, using the visual to illustrate the terms by pointing to them as they are used.

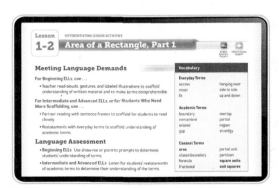

Differentiation Support pages are found in the online Teacher's Center.

Standards and Goals for
Mathematical Process and Practice

SMP1 **Make sense of problems and persevere in solving them.**
 GMP1.6 Compare the strategies you and others use.

SMP3 **Construct viable arguments and critique the reasoning of others.**
 GMP3.1 Make mathematical conjectures and arguments.

SMP6 **Attend to precision.**
 GMP6.3 Use clear labels, units, and mathematical language.

---

1 Warm Up 5 min

▶ Mental Math and Fluency

Display fraction addition problems for students to solve on their slates.
Leveled exercises:

◑○○ $\frac{1}{8} + \frac{1}{8} + \frac{1}{8}$ $\frac{3}{8}$
$\frac{1}{9} + \frac{1}{9} + \frac{1}{9} + \frac{1}{9} + \frac{1}{9}$ $\frac{5}{9}$
$\frac{2}{6} + \frac{2}{6} + \frac{1}{6}$ $\frac{5}{6}$

◑◑○ $\frac{1}{2} + \frac{1}{2} + \frac{1}{2} + \frac{1}{2}$ $\frac{4}{2}$, or 2
$\frac{6}{8} + \frac{6}{8} + \frac{6}{8} + \frac{6}{8}$ $\frac{24}{8}$, or 3
$\frac{3}{7} + \frac{4}{7} + \frac{2}{7} + \frac{5}{7}$ $\frac{14}{7}$, or 2

◑◑◑ $\frac{1}{3} + \frac{1}{3} + \frac{1}{3} + \frac{1}{3}$ $\frac{4}{3}$, or $1\frac{1}{3}$
$\frac{2}{3} + \frac{2}{3} + \frac{2}{3} + \frac{2}{3} + \frac{2}{3}$ $\frac{10}{3}$, or $3\frac{1}{3}$
$\frac{3}{12} + \frac{7}{12} + \frac{5}{12} + \frac{10}{12}$ $\frac{25}{12}$, or $2\frac{1}{12}$

---

2 Focus 35–40 min

▶ Math Message

Math Journal 1, p. 5; *Student Reference Book,* p. 221

Read page 221 in the Student Reference Book. *Then complete journal page 5.* GMP1.6

▶ Exploring Areas of Rectangles

Math Journal 1, p. 5; *Student Reference Book,* p. 221; *Math Masters,* pp. 6–7

| WHOLE CLASS | SMALL GROUP | PARTNER · | INDEPENDENT |

Math Message Follow-Up Invite students to share what they wrote about area. Guide a discussion covering the following points:

• **Area** is a measure of the surface, or region, inside a closed boundary. It is the number of whole or partial **unit squares** that fit inside the boundary.
• The whole or partial unit squares should cover the region completely, without extending outside the boundary.
• The whole unit squares should all be the same size.
• The unit squares should not overlap.
• There should be no gaps between the unit squares.

Ask: *Why are squares a convenient unit to use when you measure the area of a rectangle?* Sample answers: They fit together easily. They fit nicely in the corners of rectangles.

Math Journal 1, p. 5

Areas of Rectangles

Lesson 1-2
DATE   TIME

Write two important facts that you learned about area after reading
Student Reference Book, page 221.

① Answers vary.

②

In Problems 3 and 4 each grid square is 1 square unit.
Find the area of each rectangle. Don't forget to include a unit.

③ 3 units / 4 units
Area: 12 square units

④ 4½ units / 2 units
Area: 9 square units

⑤ Think about how you found the area of the rectangle in Problem 3 and how you found the area of the rectangle in Problem 4. What was the same? What was different? Record your thoughts. Be prepared to share them with the class.
Sample answer: In both problems, I thought about counting squares. In Problem 3 I just had to count whole squares. In Problem 4 there were some squares that weren't whole, so I had to think about how to add those into the area.

5.NF.4, 5.NF.4b, SMP1   5

Invite several students to share their strategies for solving Problem 3 on journal page 5. Expect that most will have counted the squares inside the rectangle to find the area of 12 **square units.** Some may have applied the area formula and multiplied the dimensions: 3 units * 4 units = 12 square units.

Have students share their strategies for solving Problem 4. Expect most to have used reasoning similar to the following: "I counted 8 whole squares. Then there were 2 half-squares on the end, so I put those together to make 1 more whole square. That's 9 squares in all, so the area is 9 square units."

Ask: *How do you know that the partial squares are both exactly $\frac{1}{2}$ square unit?* Sample answer: The length of the rectangle is $4\frac{1}{2}$ units. That means the partial squares are $\frac{1}{2}$ unit wide.

Ask: *How was finding the area of the rectangle in Problem 4 different from finding the area of the rectangle in Problem 3?* GMP1.6 Sample answer: I had to think about partial squares and putting partial squares together. I couldn't just count whole squares. *Why can't whole unit squares completely cover the rectangle in Problem 4?* One of the side lengths is $4\frac{1}{2}$ units. The last square in each row doesn't fit. *Will you always need to think about partial squares when a rectangle has a fractional side length? How do you know?* GMP3.1 Yes. Sample explanation: There will always be some squares that hang off the edge because a whole square will not fit along the fractional part of the length or width.

Acknowledge that finding the area of rectangles with fractional side lengths is challenging because it requires thinking about partial squares. Tell students that today they will review strategies for finding the area of rectangles with whole-number side lengths and see if they can use the same strategies for finding the areas of rectangles with fractional side lengths. GMP1.6

Display Rectangle A from *Math Masters,* page 6. Have students find the area and share their strategies. 36 square units Encourage students to use words like *row, column,* and *square unit* in their explanations, pointing out how using precise mathematical language helps others understand our thinking. GMP6.3 When appropriate, ask students to suggest a number sentence that summarizes their strategy. *Sample strategies:*

- I counted the squares. There are 36, so the area is 36 square units.
- Each of the 4 rows has 9 squares, so there are 36 squares in all. The area is 36 square units. Number sentence: $9 + 9 + 9 + 9 = 36$
- The formula for area of a rectangle is $A = l * w$. The length is 9 and the width is 4, so the area is 36 square units. Number sentence: $9 * 4 = 36$

Next display Rectangles B and C from *Math Masters,* page 6. Have partnerships find the area of each rectangle and write a number sentence. Rectangle B: 15 square units; Rectangle C: 24 square units Invite students to share their strategies and number sentences.

Academic Language Development

Help students understand the terms *boundary* and *closed boundary* using more familiar terms like *lines on the outside* that close in a space and an *outline* that encloses a space or region. For further language development, refer to maps and real-world examples of boundaries, such as state boundaries or playground fences.

Professional Development

In Grade 5 students are expected to find the areas of rectangles with fractional side lengths using two types of reasoning: tiling and applying the formula $A = l \times w$. In Unit 1 students explore two tiling strategies: counting whole and partial unit squares in this lesson and tiling with squares with fractional side lengths in the next two lessons. Students are not expected to apply the area formula or multiply mixed numbers until later in the year. For more information, see the Mathematical Background section of the Unit 1 Organizer.

Math Journal 1, p. 6

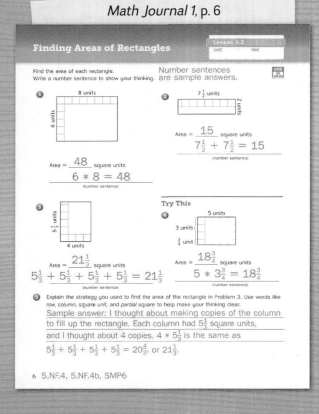

Expect that many students will have thought about partitioning the whole rectangle into unit squares and then counting the squares or applying the area formula. If no one suggests it, point out that rows and columns can be used to define units for filling a rectangle with squares. Discuss the following strategies:

- Rectangle B has 3 squares in 1 row. Imagine making copies of that row and using the copies to fill in the other rows. There would be 3 squares in all 5 rows, or an area of 15 square units. Possible number sentence: $3 + 3 + 3 + 3 + 3 = 15$
- Rectangle C has 4 squares in 1 column. That column can be used to fill in the other columns. There would be 4 squares in each of the 6 columns and 24 unit squares in all. Possible number sentence: $4 * 6 = 24$

Display Rectangle D from *Math Masters*, page 7 and give each partnership a copy. Remind students of the three strategies they have used for finding area: counting squares, thinking about a row or column, and using the area formula. Have partners try to apply these strategies to find the area of Rectangle D and write number sentences to represent their thinking. $13\frac{1}{3}$ square units Then ask students to share their thinking. *Sample strategies and number sentences:*

- I partitioned the whole rectangle into squares and partial squares. There are 12 whole squares and 4 partial squares. Each partial square is $\frac{1}{3}$ square unit, so I can put 3 of those partial squares together to make a whole square with $\frac{1}{3}$ square left over. I have 13 whole squares plus $\frac{1}{3}$ square. The area is $13\frac{1}{3}$ square units. Number sentence: $12 + 1 + \frac{1}{3} = 13\frac{1}{3}$
- I can think of making copies of the filled-in row. Each row has $3\frac{1}{3}$ squares, and there are 4 rows in all. Four groups of $3\frac{1}{3}$ is $13\frac{1}{3}$ squares in all. Number sentence: $3\frac{1}{3} + 3\frac{1}{3} + 3\frac{1}{3} + 3\frac{1}{3} = 13\frac{1}{3}$

Ask: *Did anyone use the area formula?* Sample answer: No. I would have to find $4 * 3\frac{1}{3}$, and I don't know how to multiply mixed numbers. Explain that students will learn how to multiply mixed numbers later in the year. Until then, they can use other strategies to find areas of rectangles with fractional side lengths. These strategies will still be useful later on to check their work with the formula.

▶ Finding Areas of Rectangles

Math Journal 1, p. 6; *Student Reference Book*, p. 225

| WHOLE CLASS | SMALL GROUP | **PARTNER** | INDEPENDENT |

Have partners complete journal page 6 to practice finding the area of rectangles with fractional side lengths. Remind students to use precise mathematical language when explaining their strategies. **GMP6.3**

Assessment Check-In 5.NF.4, 5.NF.4b

Math Journal 1, p. 6

Expect most students to find the correct areas and write appropriate number sentences for Problems 1 and 2 on journal page 6. Some may be able to find the correct areas in Problems 3 and 4. If students struggle, suggest that they start by partitioning each rectangle into squares.

 Evaluation Quick Entry Go online to record students' progress and to see trajectories toward mastery for these standards.

Summarize Read *Student Reference Book*, page 225 as a class.

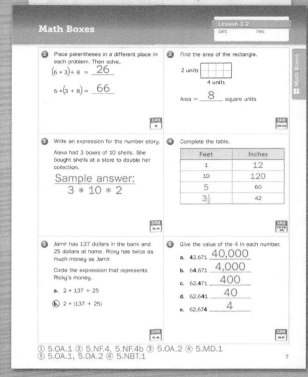

3 Practice 20–30 min

▶ Introducing *Baseball Multiplication*

Student Reference Book, p. 292; *Math Masters*, p. G3

| WHOLE CLASS | SMALL GROUP | PARTNER | INDEPENDENT |

Read the game directions on *Student Reference Book*, page 292 as a class. Distribute copies of the game mat and demonstrate how to play an inning. Then have students play in small groups. This game refreshes students' fact knowledge and provides an opportunity to informally assess their recall of basic facts. Knowing multiplication facts is an important prerequisite for multiplying multidigit whole numbers, a focus of Unit 2.

Observe
- What strategies do students use to find products?
- Which students show fluency in basic multiplication facts?

Discuss
- *How did you find the product of the two numbers?* GMP6.4
- *Which facts do you know? Which do you still need to learn?* GMP1.2

| Differentiate | Game Modifications | Go Online | Differentiation Support |

▶ Math Boxes 1-2

Math Journal 1, p. 7

| WHOLE CLASS | SMALL GROUP | PARTNER | INDEPENDENT |

Mixed Practice Math Boxes 1-2 are paired with Math Boxes 1-4.

▶ Home Link 1-2

Math Masters, p. 9

Homework Students find areas of rectangles with fractional side lengths.

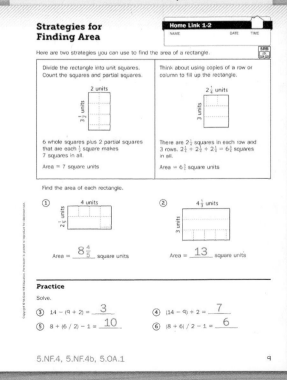

Quilt Area

Overview   **Day 1:** Students make sense of two different answers to an area problem.
Day 2: Students discuss and compare some solutions and revise their work.

Day 1: Open Response

▶ Before You Begin

Solve the open response problem and consider how students might interpret it. To help introduce the problem, draw an actual square foot for display and collect pictures or sketch quilt designs. If possible, schedule time to review students' work and plan for Day 2 of this lesson with your grade-level team.

Standards

Focus Clusters
- Apply and extend previous understandings of multiplication and division.
- Convert like measurement units within a given measurement system.

① Warm Up 5 min	**Materials**	
Mental Math and Fluency Students convert between feet and inches.		5.MD.1

②a Focus 55–65 min		
Math Message Students find the number of square inches in a square foot and in $\frac{1}{4}$ square foot.	*Math Journal 1*, p. 8; ruler (optional); chart paper (optional)	5.MD.1 SMP4
Finding Area in Two Units Students discuss strategies for finding the area of a square in different units.	*Math Journal 1*, p. 8; square foot and square inch (optional); Class Data Pad or chart paper (optional); Standards for Mathematical Process and Practice Poster	5.MD.1 SMP3, SMP4
Solving the Open Response Problem Students make sense of two different answers to an area problem that uses a drawing as a mathematical model. They explain which answer they agree with and why.	*Student Reference Book*, pp. 10–11; *Math Masters*, pp. 10–11; quilt or picture of a quilt (optional)	5.NF.4, 5.NF.4b SMP1, SMP3, SMP4

Getting Ready for Day 2 →

Review students' work and plan discussion for reengagement.   *Math Masters*, p. TA4; students' work from Day 1

 Go Online to see how mastery develops for all standards within the grade.

1 Warm Up   `5 min`

▶ Mental Math and Fluency

Have students convert between feet and inches. *Leveled exercises:*

◒○○ 12 inches equals how many feet? 1
    36 inches equals how many feet? 3
    5 feet equals how many inches? 60

◒◒○ 6 inches equals how many feet? $\frac{1}{2}$
    $1\frac{1}{2}$ feet equals how many inches? 18
    $3\frac{1}{2}$ feet equals how many inches? 42

◒◒◒ 8 inches equals how many feet? $\frac{2}{3}$
    3 inches equals how many feet? $\frac{1}{4}$
    27 inches equals how many feet? $2\frac{1}{4}$

2a Focus   `55–65 min`

▶ Math Message

Math Journal 1, p. 8

Complete journal page 8. GMP4.1

> **Differentiate** **Adjusting the Activity**
>
> For students who may be unsure how to determine the number of square inches in the square because the dimensions are given in feet, pose the following questions: *How many inches are in 1 foot?* 12 inches *Does the picture on the journal page show that there are 12 inches in a foot?* GMP4.1 Sample answer: Yes, there are lines that mark each of the 12 inches on each 1-foot side of the square. Help students use a ruler to draw an actual square foot on chart paper, mark 12 inches on each side, and then begin to partition the larger square into square inches. This will help students find the total number of square inches in a square foot. It will also help them visualize the relative sizes of an actual square foot and square inch. That is, a square foot is 144 times as large as a square inch.

Professional Development

Each Open Response and Reengagement lesson focuses on one Goal for Mathematical Process and Practice Standard. Through the focus GMPs, each of the Mathematical Process and Practice Standards will be addressed in the Open Response and Reengagement lessons in Grade 5. The focus GMP in this lesson is GMP3.2. Students examine both a correct and an incorrect answer to an area problem, consider the thinking behind each answer, and decide which one they agree with. For information on Mathematical Process and Practice Standards in this unit, see the Mathematical Background in the Unit Organizer.

Go Online for information about SMP3 in the *Implementation Guide.*

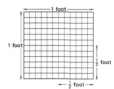
▶ Finding Area in Two Units

Math Journal 1, p. 8

| WHOLE CLASS | SMALL GROUP | PARTNER | INDEPENDENT |

Math Message Follow-Up Display the picture of the large square from journal page 8. Consider showing an actual square foot and square inch to help develop students' understanding of their relative sizes. Have partners discuss how they solved Problem 1 before sharing with the class. Strategies might include counting squares, using the array model of multiplication (multiply 12 rows by 12 columns to get 144 squares), or using the area formula ($A = 12$ in. $* 12$ in. $= 144$ in.^{2}). GMP4.1 Consider listing students' strategies on the Class Data Pad or chart paper as they share. GMP3.2

Remind students that square inches and square feet are both standard units of measure for area. Students should pay close attention to the size of the squares in order to label the area with the correct unit. For example, in Problem 1 the area of Noah's design (the large square) could be described as either 1 square foot or 144 square inches.

Continue to display the large square from the journal page. Have partners discuss how they solved Problem 2, and then have them share their strategies with the class. Expect that students used strategies similar to Problem 1. GMP4.1   36 square inches. Sample answer: $A = 6$ in. $* 6$ in. $= 36$ in.^{2} Consider listing students' strategies on the Class Data Pad as they share. GMP3.2

Students may think that the area of the square with a side length of $\frac{1}{2}$ foot is half the area of the large square. On the large square, shade in a square with a side length of $\frac{1}{2}$ foot. Ask: *What fraction of the large square is shaded?* $\frac{1}{4}$ of the large square *How many squares of this size would fit in the large square?* 4 squares *If you knew the area in square inches of $\frac{1}{4}$ of the large square, how could you find the area in square inches of the large square?* Sample answer: I could multiply the area by 4 to get the area of the large square. Consider drawing pictures like the following:

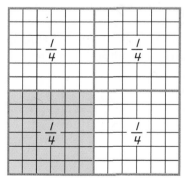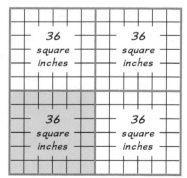

If the area of $\frac{1}{4}$ of the large square is 36 square inches, then the large square has an area of 144 square inches.

Explain that when students discussed the problems with their partner, they had to make sense of each other's thinking. Making sense of others' mathematical thinking is a skill that mathematicians often use. Although it may sometimes be hard to do, especially if the thinking contains errors, understanding the solutions and reasoning of others strengthens students' own understanding. Tell students they will practice this skill in today's problem. They will look at two possible answers for an area problem and explain how the students might have arrived at each answer. Refer students to **GMP3.2** on the Standards for Mathematical Process and Practice Poster.

> **NOTE** This is the first Open Response and Reengagement lesson in Grade 5. These two-day lessons appear in each unit to provide students consistent opportunities to engage in the Mathematical Process and Practice Standards as they solve problems. In these lessons students solve an open response problem on Day 1 and reengage with the same problem on Day 2 to deepen their understanding of the content and process/practice standards.

 for information about Open Response and Reengagement lessons in the *Implementation Guide.*

▶ Solving the Open Response Problem

Student Reference Book, pp. 10–11; *Math Masters,* pp. 10–11

| WHOLE CLASS | SMALL GROUP | PARTNER | INDEPENDENT |

Distribute *Math Masters,* pages 10 and 11. Read the problem as a class and review the directions. Consider displaying a quilt or pictures of a quilt to demonstrate how one is made by piecing together squares of fabric. Explain that people who make quilts often create a special design or pattern. Have partners or groups work together, making sure they understand the problem. Remind students that it is important to explain how they think Justin and Allyson might have arrived at their answers, even if they do not agree with them. Students should work together, but they should each record a solution.

Academic Language Development Explain to students that they need to include whether they agree or disagree with an answer, as well as the reasons why. Provide sentence frames like this one: "I agree/disagree with _____'s answer because _____."

Monitor students as they work. Ask: *How do you think Allyson and Justin got their answers?* **GMP3.2** *How can you use the picture to help you understand what each of them was thinking? How does the picture help you think about the problem?* **GMP1.2, GMP4.1** Answers vary.

Math Masters, p. 10

Quilt Area   Lesson 1-3
NAME   DATE   TIME

Allyson and Justin are working together to sew a quilt. Justin wrote down the length and width of the quilt and started to sketch a plan for the design. He showed Allyson his sketch and told her they will use 54 square feet of fabric. Allyson disagrees and says they will only use $13\frac{1}{2}$ square feet of fabric.

$4\frac{1}{2}$ ft

3 ft

① Why might Justin think they will use 54 square feet of fabric? Do you agree or disagree with Justin's answer? Why?
Sample answers:
- Justin might think there are 54 square feet of fabric because the drawing shows 6 rows and 9 columns. When you multiply 6 * 9, you get 54.
- If he finished drawing this plan, there would be a total of 54 squares.
- I disagree with Justin because even though there are 54 squares, each square does not represent a square foot. Those lines are not marked at each foot, but are marked at every $\frac{1}{2}$ foot. The area in square feet will be less than 54.

10   5.NF.4, 5.NF.4b, SMP1, SMP3, SMP4

Math Masters, p. 11

Quilt Area (continued)   Lesson 1-3
NAME   DATE   TIME

② Why might Allyson think they will use $13\frac{1}{2}$ square feet of fabric? Do you agree or disagree with Allyson's answer? Why?
See sample students' work on page 35 of the *Teacher's Lesson Guide.*

5.NF.4, 5.NF.4b, SMP1, SMP3, SMP4   11

Make Conjectures and Arguments

A **conjecture** is a statement that might be true. In mathematics, conjectures are not simply guesses. They are claims based on information or mathematical thinking. **Arguments** in mathematics use mathematical reasoning to show whether a conjecture is true or false. Mathematical arguments can use words, pictures, symbols, or other representations.

Think about this discussion between two students:

Josh makes a conjecture about multiplication.

My conjecture is that when you multiply two numbers, the answer is always greater than both numbers.

He defends his conjecture with this argument:

$9 \cdot 2 = 18$

18 is more than 9 and 2.

$3 \cdot 5 = 15$

15 is more than 3 and 5.

$6 \cdot 9 = 54$

54 is more than 6 and 9

Josh explains, "Each time I multiply two numbers, the answer I get is larger than both numbers, so my conjecture is true."

Josh, think about this. $\frac{1}{3} \cdot 9$ means $\frac{1}{3}$ of 9, which is less than 9.

Emily disagrees with Josh's conjecture and explains why.

Here is Emily's argument for why she thinks Josh's conjecture is false:

$\frac{1}{3} \cdot 9 = 3$

Since 3 is less than 9, Josh's conjecture is false.

Emily

SRB 10 ten

NOTE Use this information to help you prepare for the reengagement activity on Day 2 of this lesson.

If students struggle identifying why Allyson thinks there are $13\frac{1}{2}$ ft² of fabric in Problem 2, suggest that they fill in the remainder of the squares on the quilt. Then have students mark each foot on the sides of the quilt.

GMP3.2, GMP4.1

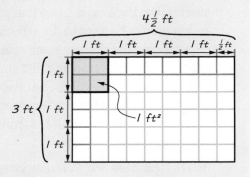

Summarize Have students turn to *Student Reference Book,* pages 10 and 11 and read it together to learn more about **SMP3** and **GMP3.2**. Ask: *Why is it important to make sense of others' mathematical thinking?* **GMP3.2** Sample answers: You can learn new ideas and make mathematical connections when you make sense of someone else's answer. It is difficult to think about how others might have solved a problem when you do not see their work, but it challenges you to think about the types of mistakes others might make or the understandings others may have about the problem.

Tell students that they will have a more in depth conversation about the problem during the reengagement discussion.

Collect students' work so that you can evaluate it and prepare for Day 2.

Getting Ready for Day 2

Math Masters, p. TA4

Planning a Follow-Up Discussion

Review students' work. Use the Reengagement Planning Form (*Math Masters,* page TA4) and the rubric on page 32 to plan ways to help students meet expectations for both the content and process/practice standards. Look for common misconceptions, such as students thinking that both Justin's and Allyson's solutions are correct, as well as clear drawings and explanations of the solutions.

Organize the discussion in one of the following ways or in another way you choose. If students' work is unclear or if you prefer to show work anonymously, recopy the work for display.

Go Online> for sample students' work that you can use in your discussion.

1. Show students' solutions for Problems 1 and 2 in which they used a picture to show why Justin answered 54 square feet and Allyson answered $13\frac{1}{2}$ square feet, such as in Student A's and B's work. Ask:
   • *How did this student make sense of Justin's or Allyson's mathematical thinking using the picture?* GMP3.2, GMP4.1 Sample answers: For Problem 1: It looks like Student A filled in the rest of the squares, marked them with Xs to keep track of the small squares, and counted a total of 54 squares. For Problem 2: Student B might have counted the 54 squares but realized that it takes 4 small squares to make 1 ft², so the student marked groups of 4 squares and counted $13\frac{1}{2}$ ft².
   • *Who is correct?* Allyson *Why?* Sample answer: Each small square is $\frac{1}{4}$ ft², not 1 ft², so Justin should have divided the number of little squares by 4. Student B showed this by marking groups of 4 small squares and counting them as 1 ft².

2. Show a solution in which a misconception is evident, such as in Student C's work for Problem 1. Ask:
   • *Does this student agree with Justin's solution of 54 square feet of fabric?* GMP3.2 No. *How much fabric does Student C say there is in the quilt?* 27 ft *How do you think this student arrived at that answer?* Sample answer: This student says, "2 squares are 1 foot," and the picture shows that the student made one long mark across every two squares. If you count the marks, there are 27.
   • *Do you agree that the area is 27 feet?* GMP1.2 No. *What would you say to help this student understand why the answer is not 27 feet?* Sample answer: If you are thinking about the distance around the rectangle, then the side of each small square is $\frac{1}{2}$ ft long, so 2 **sides** of the small squares together are 1 ft. But we are finding the area of the quilt, and it takes 4 small squares that each have a side length of $\frac{1}{2}$ ft to fill in 1 ft². So if you mark off 4 small squares to make a larger square with an area of 1 ft², you can count the larger squares and you get $13\frac{1}{2}$ ft².

Planning for Revisions

Have copies of *Math Masters,* pages 10–11 or extra paper available for students to use in revisions. You might want to ask students to use colored pencils so you can see what they revised.

Sample student's work, Student A

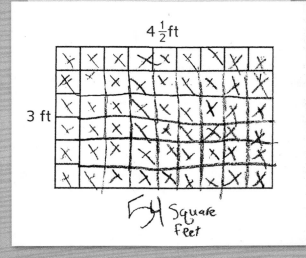

Sample student's work, Student B

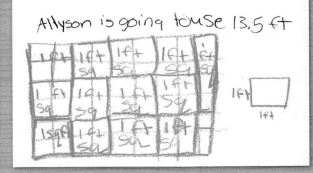

Sample student's work, Student C

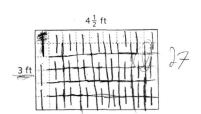

Quilt Area

Day 2: Reengagement

▶ **Before You Begin**

Have extra copies available of *Math Masters*, pages 10 and 11 for students to revise their work.

Standards

Focus Cluster
• Apply and extend previous understandings of multiplication and division.

2b Focus   50–55 min

	Materials	
Setting Expectations Students review the open response problem and discuss what a good response might include. They also discuss how to talk about others' work respectfully.	Class Data Pad or chart paper	SMP1, SMP3
Reengaging in the Problem Students discuss and compare various strategies and explanations.	selected samples of students' work	5.NF.4, 5.NF.4b SMP1, SMP3, SMP4
Revising Work Students revise their work.	*Math Masters*, pp. 10–11 (optional); students' work from Day 1; colored pencils (optional)	5.NF.4, 5.NF.4b SMP3
✓ **Assessment Check-In** See page 34 and the rubric below. Expect most students to be able to make sense of others' mathematical thinking and find the area of a rectangle with fractional side lengths.		5.NF.4, 5.NF.4b SMP3

Goal for Mathematical Process and Practice Standards GMP3.2 Make sense of others' mathematical thinking.	Not Meeting Expectations	Partially Meeting Expectations	Meeting Expectations	Exceeding Expectations
	Does not provide a clear explanation for how Allyson found an answer of $13\frac{1}{2}$ square feet, or the explanation does not refer to putting groups of 4 small squares together or multiplying dimensions.	Uses words or pictures to show that Allyson either: • used a strategy to multiply 3 by $4\frac{1}{2}$ to get $13\frac{1}{2}$, without referring to units, or • counted sets of 4 small squares, without referring to the idea that 4 small squares equal 1 square foot.	Uses words or pictures to show that Allyson either: • used a strategy to multiply the dimensions in feet to get the area in square feet, or • counted sets of 4 small squares because 4 small squares equal 1 square foot.	Uses words or pictures to show both explanations of Allyson's thinking described under Meeting Expectations.

3 Practice   10–15 min

Math Boxes 1-3 Students practice and maintain skills.	*Math Journal 1*, p. 9	See page 34.
Home Link 1-3 **Homework** Students find the areas of rectangles and explain their strategies.	*Math Masters*, p. 12	5.NF.4, 5.NF.4b, 5.MD.1

/// **Go Online** to see how mastery develops for all standards within the grade.

 **Focus** 50–55 min

▶ Setting Expectations

| WHOLE CLASS | SMALL GROUP | PARTNER | INDEPENDENT |

Establishing Guidelines for Reengagement

A significant part of Day 2 is a class discussion about students' strategies and explanations. To promote a cooperative environment, develop class guidelines for discussion and record them on the Class Data Pad or chart paper. Elicit suggestions from the class and include items you feel are important. Use and expand the list throughout the year during any group discussion. (*See sample poster in the margin.*)

Consider modeling or having students role-play situations based on one or more of the guidelines on your poster. For example, use a simple math problem like 10 * 20 = 20 to model how you can learn from a mistake in your own work. Use the following sentence frames, encouraging students to use the frames and other appropriate language when discussing other students' work:

- I noticed _____.
- Could you explain _____?
- I agree because _____.
- I don't understand _____.
- I disagree because _____.
- I'd like to add _____.

Reviewing the Problem

Briefly review the open response problem from Day 1. Ask: *What were you asked to do?* **GMP1.2** Sample answer: We had to figure out how Justin and Allyson got their answers to the area problem and explain whose answer we agreed with. *What do you think a good response should include?* **GMP3.2** Sample answer: It should explain or show how Justin and Allyson might have gotten their answers. It should also explain who we agree with and why. The explanations should be clear so that someone else can understand them.

▶ Reengaging in the Problem

| WHOLE CLASS | SMALL GROUP | PARTNER | INDEPENDENT |

Students reengage in the problem by analyzing and critiquing other students' work in pairs and in a whole-group discussion. Have students discuss with partners before sharing with the whole group. Guide this discussion based on the decisions you made in Getting Ready for Day 2.
GMP1.2, GMP3.2, GMP4.1

> **NOTE** These Day 2 activities will ideally take place within a few days of Day 1. Prior to beginning Day 2, see Planning a Follow-Up Discussion from Day 1.

> ### Guidelines for Discussion
>
> During our class discussions, we can:
> - ✓ Make mistakes and learn from them
> - ✓ Share ideas and strategies respectfully
> - ✓ Change our minds about how to solve a problem
> - ✓ Ask questions of our teacher and classmates

▶ Revising Work

| WHOLE CLASS | SMALL GROUP | PARTNER | INDEPENDENT |

Pass back students' work from Day 1. Before students revise anything, ask them to examine their responses and decide how they can be improved. Ask the following questions one at a time. Have partners discuss their responses and give a thumbs-up or thumbs-down based on their own work.

- *Did you include a reasonable explanation for how Justin and Allyson could have gotten their answers?* GMP3.2
- *Did you clearly explain why you agree or disagree with each student's answer?*

Tell students they now have a chance to revise their work. Students who wrote complete and correct explanations on Day 1 can try to find a different strategy to check or show their work. Help students see that the explanations presented during the reengagement discussion are not the only correct ones. Tell them to add to their earlier work using colored pencils or to use another sheet of paper, instead of erasing their original work.

Summarize Have students reflect on their work and revisions. Ask: *What did you do to improve your explanations?* Answers vary.

✓ Assessment Check-In 5.NF.4, 5.NF.4b

Collect and review students' revised work. Expect students to improve their work based on the class discussion. For the content standards, expect most students to agree with Allyson's answer and show how to find the area of $13\frac{1}{2}$ square feet by partitioning the square and counting groups of 4 smaller squares. You can use the rubric on page 32 to evaluate students' revised work for **GMP3.2**.

Evaluation Quick Entry Go online to record students' progress and to see trajectories toward mastery for these standards.

(Go Online) for optional generic rubrics in the *Assessment Handbook* that can be used to assess any additional GMPs addressed in this lesson.

Math Journal 1, p. 9

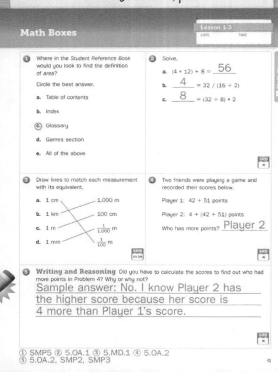

Math Boxes
Lesson 1-3

① Where in the *Student Reference Book* would you look to find the definition of area?

Circle the best answer.

a. Table of contents
b. Index
c. Glossary
d. Games section
e. All of the above

② Solve.

a. $(4 * 12) + 8 =$ __56__
b. __4__ $= 32 / (16 ÷ 2)$
c. __8__ $= (32 ÷ 8) * 2$

③ Draw lines to match each measurement with its equivalent.

a. 1 cm — 1,000 m
b. 1 km — 100 cm
c. 1 m — $\frac{1}{1,000}$ m
d. 1 mm — $\frac{1}{100}$ m

④ Two friends were playing a game and recorded their scores below.

Player 1: $42 + 51$ points
Player 2: $4 + (42 + 51)$ points
Who has more points? Player 2

⑤ **Writing and Reasoning** Did you have to calculate the scores to find out who had more points in Problem 4? Why or why not?

Sample answer: No. I know Player 2 has the higher score because her score is 4 more than Player 1's score.

① SMP5 ② 5.OA.1 ③ 5.MD.1 ④ 5.OA.2
⑤ 5.OA.2, SMP2, SMP3

9

Sample Students' Work—Evaluated

See the sample in the margin. This work meets expectations for the content standards because it shows that the student agreed with Allyson and correctly tiled a rectangle with fractional side lengths in order to determine that the area is $13\frac{1}{2}$ square feet. The work meets expectations for the mathematical process and practice standard because the student made sense of Allyson's mathematical thinking by showing that Allyson put groups of 4 small squares together to make 1 square foot and counted to get $13\frac{1}{2}$ square feet. This student also labeled the dimensions and each group of 4 small squares as 1 square foot and the group of 2 squares as half a square foot. **GMP3.2**

Go Online for other samples of evaluated students' work.

3 Practice   10–15 min

▶ Math Boxes 1-3 ✏️

Math Journal 1, p. 9

WHOLE CLASS | **SMALL GROUP** | PARTNER | INDEPENDENT

Mixed Practice Math Boxes 1-3 are paired with Math Boxes 1-1.

▶ Home Link 1-3

Math Masters, p. 12

Homework Students find the area of rectangles and explain their strategies.

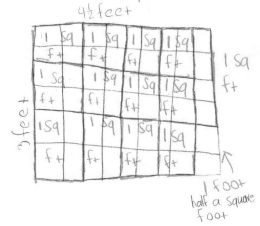

2. Why might Allyson think they will use $13\frac{1}{2}$ square feet of fabric? Do you agree or disagree with Allyson's answer? Why?

I agree with Allyson because we made a grid showing how she got $13\frac{1}{2}$. So she made 13 plus two square's that equals a half. So $13 + .5 = 13\frac{1}{2}$ square feet.

Math Masters, p. 12

Finding the Area of Rectangles

Home Link 1-3
NAME   DATE   TIME

Find the area of the rectangles below. Write a number sentence for each problem and explain how you found the area.

|← 1 cm →| □ 1 cm²

Sample number sentences given.

① $1\frac{1}{2}$ cm

cm $\frac{1}{2}$ ⟩ 3

Area = ___3 cm²___
Number sentence: $1\frac{1}{2} + 1\frac{1}{2} = 3$
Explanation:
Sample answer: I counted the 2 whole square centimeters. I put the 2 half square centimeters together to make another whole square centimeter. So all together there are 3 cm².

② 3 cm

2$\frac{1}{2}$ cm

1	2	3
4	5	6
7	$\frac{1}{2}$	

Area: ___$7\frac{1}{2}$ cm²___
Number sentence: $2\frac{1}{2} + 2\frac{1}{2} + 2\frac{1}{2} = 7\frac{1}{2}$
Explanation:
Sample answer: I drew lines to make rows and columns. I counted 6 whole square centimeters. I put 2 of the half square centimeters together to make another whole square centimeter. So there are 7 whole square centimeters. $7 + \frac{1}{2} = 7\frac{1}{2}$, so the area is $7\frac{1}{2}$ cm².

Practice

Solve.

③ 36 inches = ___3___ feet

④ ___60___ inches = 5 feet

⑤ 18 inches = ___$1\frac{1}{2}$___ feet

⑥ $\frac{1}{2}$ foot = ___6___ inches

12   5.NF.4, 5.NF.4b, 5.MD.1

Area of a Rectangle, Part 2

Overview Students find areas of rectangles with fractional side lengths by tiling them with squares of the appropriate unit-fraction side length.

▶ **Before You Begin**
For the optional Readiness activity, gather square pattern blocks.

Standards

Focus Cluster
• Apply and extend previous understandings of multiplication and division.

---

① Warm Up 5 min

	Materials	
Mental Math and Fluency Students add mixed numbers.		5.NF.1

---

② Focus 35–40 min

Math Message Students determine how many squares with a side length of $\frac{1}{3}$ foot fit in 1 square foot.	*Math Journal 1*, p. 10	5.NF.4, 5.NF.4b
Discussing Tiling Patterns Students find and justify a pattern about how many squares with fractional side lengths fit in 1 square foot.	*Math Journal 1*, p. 10	5.NF.4, 5.NF.4b SMP7, SMP8
Using a Pattern to Find Area Students solve area problems involving tiling with squares with fractional side lengths.	*Math Journal 1*, pp. 10–11; calculator (optional)	5.NF.4, 5.NF.4b SMP7
✓ **Assessment Check-In** See page 41. Expect most students to use the tick marks to tile the rectangles, count to find the number of tiles, and realize that the count of squares is not the area when finding the areas of rectangles in problems like those identified.	*Math Journal 1*, p. 10 (optional), p. 11	5.NF.4, 5.NF.4b, SMP7

---

③ Practice 20–30 min

Math Boxes 1-4 Students practice and maintain skills.	*Math Journal 1*, p. 12	See page 41.
Home Link 1-4 **Homework** Students divide a square mile into squares with fractional side lengths.	*Math Masters*, p. 15	5.NF.4, 5.NF.4b, 5.MD.1

 Go Online to see how mastery develops for all standards within the grade.

Differentiation Options

Readiness	5–15 min	Enrichment	15–30 min	Extra Practice	5–15 min
WHOLE CLASS SMALL GROUP **PARTNER** INDEPENDENT		WHOLE CLASS **SMALL GROUP** PARTNER INDEPENDENT		WHOLE CLASS **SMALL GROUP** PARTNER INDEPENDENT	

Tiling with Squares with Fractional Side Lengths

5.NF.4, 4.NF.4b

per partnership: *Math Masters,* p. 13; 3-inch square stick-on notes; square pattern blocks

For concrete experience with tiling using squares with fractional side lengths, students use stick-on notes and square pattern blocks to tile a larger square. Give each partnership one copy of *Math Masters,* page 13 as well as 5 or 6 stick-on notes and a handful of square pattern blocks. Explain that the page shows one square unit, so each side of the square is 1 unit long. Ask: *What is the length of a side of one stick-on note?* $\frac{1}{2}$ unit *How do you know?* Two stick-on notes fit along the side of the square. Have students tile the square with stick-on notes. Ask: *How many squares with $\frac{1}{2}$ unit side lengths fit in this unit square?* 4 Repeat the activity with the square pattern blocks. Students should determine that the sides of the blocks are $\frac{1}{6}$ unit long and that 36 blocks fit in the unit square. Depending on how many square pattern blocks you have, partnerships may need to take turns tiling the square.

Showing Area Unit Conversions

5.MD.1

Activity Card 3; *Student Reference Book,* pp. 218–219; per partnership: poster paper, markers

To apply their understanding of tiling with squares with fractional side lengths, students create a poster showing area unit conversions. Consider inviting groups to present their posters to the class.

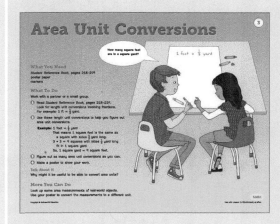

Finding Areas of Rectangles

5.NF.4, 5.NF.4b

Math Masters, p. 14

To practice finding areas of rectangles with fractional side lengths, students solve the problems on *Math Masters,* page 14.

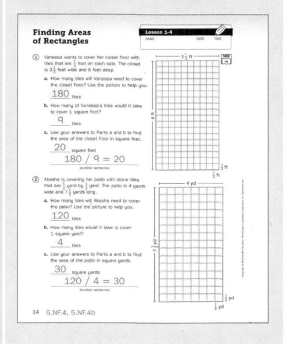

English Language Learner

Beginning ELL Provide scaffolding for students when introducing the term *side length* by explaining the individual words. Use a think-aloud to describe the number of sides of various shapes. Ask students one-word response questions like: *How many sides does the triangle have?* Demonstrate measuring the side lengths of various shapes and report the measurements using paired sentences like these: *The length of the side is ____ inches. The side length is ____ inches.* Have students measure the sides of various shapes, using the sentence frames to report their measurements.

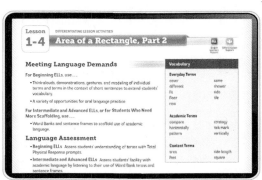

Differentiation Support pages are found in the online Teacher's Center.

SMP7 Look for and make use of structure.

  GMP7.1 Look for mathematical structures such as categories, patterns, and properties.

  GMP7.2 Use structures to solve problems and answer questions.

SMP8 Look for and express regularity in repeated reasoning.

  GMP8.1 Create and justify rules, shortcuts, and generalizations.

Professional Development

The Grade 5 standards require students to find the area of a rectangle with fractional side lengths by tiling it with squares that have unit fraction side lengths.

To use this strategy students must understand that:

- the count of the squares is not equal to the area in square units, because each square covers only a fractional part of a square unit.

- the number of smaller squares needed to cover one square unit is equal to the square of the denominator of the unit fraction. For example, 9 squares with a side length of $\frac{1}{3}$ unit fit in one square unit.

In Lesson 1-3, students explored these ideas in the special case of squares with $\frac{1}{2}$-unit side lengths. In this lesson students use patterns to generalize the idea to other unit fraction side lengths.

1 Warm Up 5 min

▶ Mental Math and Fluency

Display addition problems with mixed numbers for students to solve. *Leveled exercises:*

◉○○ $1\frac{1}{2} + 1\frac{1}{2}$ 3
    $4\frac{1}{3} + 4\frac{1}{3}$ $8\frac{2}{3}$
    $6\frac{1}{6} + 6\frac{1}{6}$ $12\frac{2}{6}$

◉◉○ $4\frac{1}{4} + 4\frac{1}{4} + 4\frac{1}{4}$ $12\frac{3}{4}$

    $3\frac{1}{5} + 4\frac{2}{5}$ $7\frac{3}{5}$
    $2\frac{1}{8} + 3\frac{1}{8} + 1\frac{3}{8}$ $6\frac{5}{8}$

◉◉◉ $2\frac{1}{3} + 2\frac{1}{3} + 2\frac{1}{3}$ 7
    $3\frac{4}{7} + 2\frac{5}{7}$ $5\frac{9}{7}$, or $6\frac{2}{7}$
    $3\frac{5}{8} + 2\frac{3}{8} + 5\frac{7}{8}$ $10\frac{15}{8}$, or $11\frac{7}{8}$

2 Focus 35–40 min

▶ Math Message

Math Journal 1, p. 10

Complete Problem 1 on journal page 10.

▶ Discussing Tiling Patterns

Math Journal 1, p. 10

| WHOLE CLASS | SMALL GROUP | PARTNER | INDEPENDENT |

Math Message Follow-Up Invite volunteers to share their strategies and answers to the Math Message problem. Guide a discussion to cover the following ideas:

- Because the smaller square has a side length of $\frac{1}{3}$ foot, 3 of the smaller squares will fit horizontally along the side of a square foot.
- Similarly, 3 squares with a side length of $\frac{1}{3}$ foot will fit vertically along the side of a square foot.
- This means that 3 rows of 3 squares with a side length of $\frac{1}{3}$ foot will fit inside a square foot. Since $3 * 3 = 9$, there are 9 squares in all.

Consider showing how the large square in the Math Message can be partitioned to show 9 smaller squares fitting inside it. Ask: *How many squares with a side length of $\frac{1}{4}$ foot do you think would fit in 1 square foot?* 16 squares *Why?* Sample answer: 16 squares with a side length of $\frac{1}{4}$ foot will fit in a square foot because there will be 4 rows of 4 smaller squares.

Ask partners to discuss and share their thinking. Encourage students to make a drawing to help them answer the question.

Sketch a two-column table with these headings: Side Length of Smaller Square and Number of Squares That Fit in a Square Foot. Fill in rows for $\frac{1}{2}$ foot, $\frac{1}{3}$ foot, and $\frac{1}{4}$ foot. (*See margin.*) Ask: *Do you see any patterns?* **GMP7.1** Sample answer: The number of smaller squares needed to cover the larger square is the denominator of the smaller square side length multiplied by itself. *Do you think this pattern would continue if we used squares with other side lengths, such as $\frac{1}{5}$ foot or $\frac{1}{6}$ foot?* Yes. *Why does the pattern always work?* **GMP8.1** Sample answer: The denominator of the fraction shows how many small squares fit along the side of a square foot, so it tells you how many rows of small squares fit in a square foot and how many are in each row. Have students use the pattern to help you fill in a few more rows of the table.

Explain that rectangular objects in the real world often have fractional side lengths. Tell students that they will learn how to apply the pattern they just discovered using squares with fractional side lengths to find areas of rectangles.

▶ Using a Pattern to Find Area

Math Journal 1, pp. 10–11

| WHOLE CLASS | SMALL GROUP | PARTNER | INDEPENDENT |

Read Problem 2 on journal page 10 as a class and have students solve Problem 2a. Circulate and observe, making sure students understand how to use tick marks to help them draw and count the tiles. Have students share answers. 72 tiles Ask: *It takes 72 tiles to cover the shower floor. Does that mean the area of the shower floor is 72 square feet?* No. *Why not?* Sample answer: The tiles are only $\frac{1}{3}$ foot on each side, so they are not square feet.

Have students read Problem 2b. Refer to the two-column table from the previous activity. Ask: *How could we use the pattern we noticed in the table to figure out how many tiles would cover 1 square foot?* Sample answer: Since the tiles are $\frac{1}{3}$ foot on a side, I multiplied $3 * 3$ and found that 9 tiles would cover 1 square foot. Have students record the answer to Problem 2b.

Ask: *How could we use the information in Problems 2a and 2b to find the area of the shower floor in square feet?* **GMP7.2** Let small groups discuss the question, and then bring the class together to share ideas. Help students understand that they must divide 72 (the total number of tiles) by 9 (the number of tiles needed to cover 1 square foot) to find an area of 8 square feet. Discuss several different ways of thinking through the problem, using students' ideas.

Side Length of Smaller Square	Number of Squares That Fit in a Square Foot
$\frac{1}{2}$ foot	4
$\frac{1}{3}$ foot	9
$\frac{1}{4}$ foot	16
$\frac{1}{5}$ foot	25
$\frac{1}{10}$ foot	100

Math Journal 1, p. 10

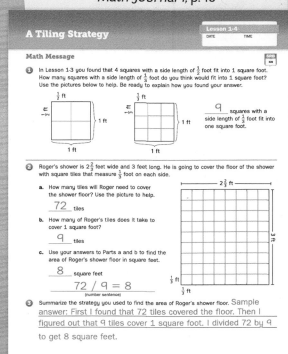

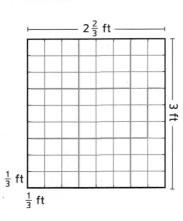

Tiles	Square Feet
9	1
18	2
27	3
36	4
45	5
72	?

Table Relating Tiles and Square Feet

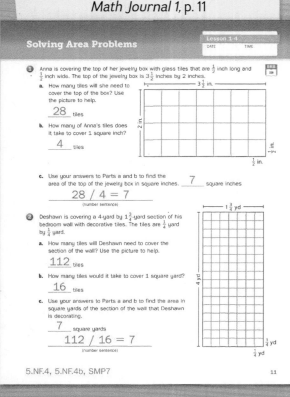

Groups of 9 Tiles

Math Journal 1, p. 11

Sample explanations:

- I know that 9 tiles cover 1 square foot, so that means 18 tiles cover 2 square feet, 27 tiles cover 3 square feet, and so on. I made a table to help me think about it. (*See margin.*) I noticed a pattern in the table: The number of tiles is 9 times the number of square feet. $9 * 8 = 72$, so the area is 8 square feet.

- Because it takes 9 tiles to cover 1 square foot, I can think of making groups of 9 tiles. (*See margin.*) There are 8 groups of 9 tiles, so the area is 8 square feet. This makes sense because $72 \div 9 = 8$.

Differentiate   **Common Misconception**

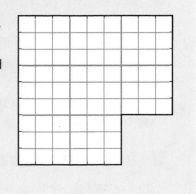

Some students may believe that the two groups of 9 tiles shown on the right side of the diagram in the margin do not cover 1 square foot because they are not arranged in the shape of a square. Encourage these students to think about how these tiles could be rearranged to cover eight 1-foot by 1-foot squares, as shown at the right.

Emphasize the idea that area is a measure of how much 2-dimensional space is covered, and the tiles cover the same amount of space no matter how they are arranged.

Go Online Differentiation Support

After the class has agreed that the shower has an area of 8 square feet and discussed how division can be used to solve the problem, have students record the answer and a division number sentence for Problem 2c.

Have students summarize the strategy they used to find the area of Roger's shower floor and record it in their own words for Problem 3. Students' descriptions will vary, but should include some version of these three steps:

1. Cover the floor completely with tiles and count the number of tiles.

2. Determine how many tiles cover a square foot.

3. Divide the total number of tiles by the number of tiles that cover a square foot.

Ask: *Do you think you could use a similar strategy to solve other problems using different measurement units? For example, if there were a $5\frac{1}{4}$-yard by 4-yard floor that was covered with $\frac{1}{4}$-yard by $\frac{1}{4}$-yard tiles, could you use a similar strategy to find the area of the floor in square yards?* GMP7.2 Yes. *What would be the same?* Sample answer: I would still have to think about how many tiles I need and how many fit on a square unit. *What would be different?* Sample answers: Since the tiles are $\frac{1}{4}$ yard by $\frac{1}{4}$ yard, I'd have to divide by 16 instead of 9. I would be finding the area in square yards, not square feet.

Have partnerships complete journal page 11. Allow students to use a calculator, if needed, to help with the multiplication and division.

 Assessment Check-In   5.NF.4, 5.NF.4b

Math Journal 1, p. 11

Expect most students to use the tick marks to tile the rectangles and count to find the number of tiles on journal page 11. Most should also realize that the count of squares is not the area measurement. Some students may struggle trying to apply the pattern from the Math Message Follow-Up to find the correct area. GMP7.2 If students have trouble tiling the rectangles, refer them to the squares at the top of journal page 10 for help visualizing the row-and-column structure.

 Evaluation Quick Entry Go online to record students' progress and to see trajectories toward mastery for these standards.

Summarize Have students share with a partner their favorite method for finding the area of a rectangle with a fractional side length.

3 Practice   20–30 min

▶ Math Boxes 1-4

Math Journal 1, p. 12

| WHOLE CLASS | SMALL GROUP | PARTNER | INDEPENDENT |

Mixed Practice Math Boxes 1-4 are paired with Math Boxes 1-2.

▶ Home Link 1-4

Math Masters, p. 15

Homework Students divide a square mile into squares with fractional side lengths.

Math Journal 1, p. 12

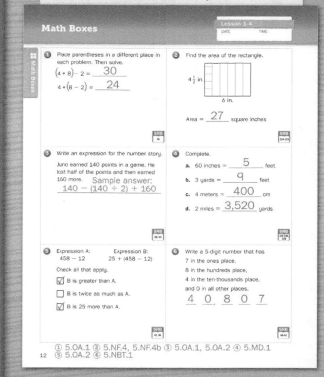

Math Masters, p. 15

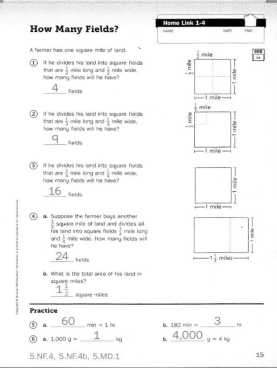

Introduction to Volume

Overview Students explore the concept of volume as they informally compare volumes of 3-dimensional objects.

▶ **Before You Begin**

Place three empty, open, similar-size containers, such as an oatmeal container, a tissue box, and a cereal box, near the Math Message. Gather enough packing material, such as dry macaroni, dry beans, or centimeter cubes, to fill one of the containers. For Part 2, gather two identical containers, one full and one empty, and a few solid objects. Each small group will need 2 half-sheets of paper, tape, a small paper plate, and about $1\frac{1}{2}$ cups of packing material. For the optional Readiness activity, gather four small 3-dimensional objects. For the optional Extra Practice activity, gather 8–10 small containers.

▶ **Vocabulary**

3-dimensional • volume • conjecture

Standards

Focus Cluster
• Geometric measurement: understand concepts of volume.

	**Materials**	
Mental Math and Fluency Students write numerical expressions.	slate	**5.OA.1, 5.OA.2**

② **Focus** 35–40 min		
Math Message Students explain how they know which object is largest.	3 open, empty containers	**5.MD.3**
Exploring Volume Measurement Students explore the attribute of volume.	*Student Reference Book*, p. 270; materials listed in Before You Begin	**5.MD.3, 5.MD.4** **SMP3, SMP6**
Comparing Volume Students predict which container has a greater volume and test their predictions.	*Math Journal 1*, p. 13; per group: materials listed in Before You Begin	**5.MD.3, 5.MD.4** **SMP3**
✓ **Assessment Check-In** See page 46. Expect most students to be able to report that they are comparing volume or comparing the amount that each cylinder holds after pouring items into two cylinders to compare the cylinders' volumes.	*Math Journal 1*, p. 13	**5.MD.3**

③ **Practice** 20–30 min		
Playing *Name That Number* **Game** Students practice writing expressions for calculations and writing expressions with grouping symbols.	*Student Reference Book*, p. 315; per partnership: *Math Masters*, p. G2; number cards 0–10 (4 of each) and 11–20 (1 of each)	**5.OA.1, 5.OA.2** **SMP1, SMP2**
Math Boxes 1-5 Students practice and maintain skills.	*Math Journal 1*, p. 14	See page 47.
Home Link 1-5 **Homework** Students compare volumes of everyday objects.	*Math Masters*, p. 16	**5.NF.4, 5.NF.4b, 5.MD.3, 5.MD.4**

 ⟨Go Online⟩ to see how mastery develops for all standards within the grade.

Differentiation Options

Readiness	10–15 min
WHOLE CLASS	**SMALL GROUP**	PARTNER	INDEPENDENT

Enrichment	20–30 min
WHOLE CLASS	SMALL GROUP	**PARTNER**	INDEPENDENT

Extra Practice	10–15 min
WHOLE CLASS	**SMALL GROUP**	PARTNER	INDEPENDENT

Identifying Measurable Attributes

`5.MD.3`

per group: four common 3-dimensional objects

To prepare for exploring volume as a measurable attribute, students review the idea that objects have many measurable attributes. Display or list four familiar 3-dimensional objects, including at least one that is a container. For each object, ask students to list on a blank piece of paper any attributes that can be measured. *For example:* book Sample answers: height; thickness; weight; number of pages; **soccer ball** Sample answers: weight; distance around; number of white/black segments; how high it can bounce; **trash can** Sample answers: height; weight; how much trash it can hold; thickness of the plastic; **desk** Sample answers: weight; height; number of legs; area of the top; how many books could fit inside As students work, encourage them to consider attributes that lead them toward the idea of volume, such as the number of books that can fit inside a desk. Have students share their responses and incorporate other students' ideas by adding them to their lists.

Creating Prisms: The Volume Challenge

`5.MD.3, 5.MD.4`

Activity Card 4; 8 index cards; tape; scissors; packing material (beans, macaroni, or cubes)

To further explore volume, students use index cards to create prisms with the greatest and least possible volumes.

Detecting Volume by Touch

`5.MD.3, 5.MD.4`

Activity Card 5; per group: 8–10 empty containers (small gift bags, yogurt containers, small boxes, paper cups or bowls, small milk cartons); packing material (beans, macaroni, or cubes)

For more practice with volume comparisons, students order containers by volume, using only touch to make their predictions.

English Language Learner

Beginning ELL Students may have heard the term *hold* used in everyday contexts, such as *holding hands* or *holding your breath*. Demonstrate other everyday uses of the term, using contexts with which you think your students are familiar. Then demonstrate the meaning of the term as related to volume in the sense of containing. Use statements like these: *This mug holds a lot of water. This small cup does not hold much water.* Ask one-word response questions like this one: *Which holds more water, the mug or the cup?*

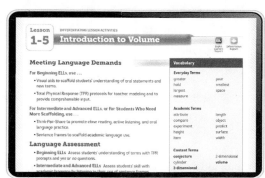

Differentiation Support pages are found in the online Teacher's Center.

Standards and Goals for
Mathematical Process and Practice

SMP3 **Construct viable arguments and critique the reasoning of others.**

  GMP3.1 Make mathematical conjectures and arguments.

  GMP3.2 Make sense of others' mathematical thinking.

SMP6 **Attend to precision.**

  GMP6.1 Explain your mathematical thinking clearly and precisely.

Academic Language Development

Extend students' vocabulary knowledge with a discussion of the term *volume*. Students may be familiar with the use of *volume* to talk about sound and degrees of loudness. They may also have heard the term used to describe a book. Point out the different meanings of the term, noting the fact that the same spelling is used in each instance, as you explain the mathematical meaning.

---

1 Warm Up   5 min

▶ Mental Math and Fluency

Dictate the following and have students write an expression for each item on their slates. Students do not need to calculate answers.
Leveled exercises:

● ○ ○ 9 less than 18 $18 - 9$
          The product of 5 and 2 5×2
          Double 8 2×8, or $8 + 8$

● ● ○ 3 times the product of 2 and 4 $3 \times (2 \times 4)$
          Double the sum of 5 and 3 $2 \times (5 + 3)$
          7 times the sum of 3 and 2 $7 \times (3 + 2)$

● ● ● Triple the product of 4 and 2 $3 \times (4 \times 2)$
          5 less than the product of 7 and 6 $(7 \times 6) - 5$
          3 less than double the sum of 9 and 1 $2 \times (9 + 1) - 3$

2 Focus   35–40 min

▶ Math Message

Look at these 3-dimensional objects. Which is the largest? Be prepared to explain your answer.

▶ Exploring Volume Measurement

Student Reference Book, p. 270

| WHOLE CLASS | SMALL GROUP | PARTNER | INDEPENDENT |

Math Message Follow-Up Remind students that in previous lessons they explored the 2-dimensional attribute of *area.* Invite volunteers to explain what makes the objects displayed for the Math Message **3-dimensional** objects. Sample answer: They have length, width, and thickness or height. Have students read *Student Reference Book,* page 270 to review the difference between 2- and 3-dimensional objects.

Ask students to explain their reasoning for choosing one of the 3-dimensional objects as the largest. GMP3.1, GMP6.1 Point out that asking which is *largest* allows for a variety of interpretations because the attribute being compared is unclear. Pose questions that focus on specific attributes. Ask: *How could we find out which object:*

• *is tallest?* Sample answer: Measure the height of each object.

• *has a side with the largest surface?* Sample answer: Calculate the area of the largest side of each object.

• *takes up the most space?* Answers vary.

Students may suggest measuring the height of each object or the area each object covers on the table to determine which object takes up the most space. Point out that both attributes—height and area—affect how much space is taken up by an object. Guide the discussion to help students see that one way to think about how much space something takes up is to see how much it holds. Explain that **volume** is the attribute that describes how much space is taken up by an object.

Display two identical containers, such as soup cans. Make sure one container is open and empty and one is closed and full. Ask students which object has more volume. Explain that both containers take up the same amount of space, even though one is filled and the other is empty. Since they take up the same amount of space, they have the same volume. Next display several solid objects like a book, an eraser, or a block. Ask: *Do these objects have volume? How do you know?* Sample answer: Yes. They are all 3-dimensional and take up space. Emphasize that *all* 3-dimensional objects have volume—whether they are solid objects like a book or a block, or containers like a box or an empty soup can.

> **Differentiate** | **Common Misconception**
>
> Some students may think of volume as the amount of a substance *contained in* an object rather than the amount of space defined by the boundaries of the object. They might disagree with the idea that an unopened soup can and an empty soup can have the same volume. To help students grasp this distinction, be explicit in discussions about containers. Share ideas like these:
>
> - When we talk about volume, we are talking about the amount of space an object takes up. Imagine you have two identical boxes of cereal. One is empty; the other is full. The cereal boxes have the same volume because each box would take up the same amount of space in a cabinet.
>
> - When comparing the volumes of several 3-dimensional objects, it helps to think: *If each object were hollow inside, or empty, which would hold the most?*
>
>  Go Online | Differentiation Support

Return to the three containers from the Math Message. Ask students to predict which one has the greatest volume. Explain that a mathematical prediction is sometimes called a **conjecture.** Conjectures, like predictions, are based on careful thinking. Provide an example like this one: *When meteorologists predict the weather, they base their predictions on what they notice and what they know about clouds, temperature, air pressure, and wind. They can explain the thinking that led them to make a certain prediction. When we make mathematical conjectures, we should be able to use mathematical thinking to explain* why *we made the conjecture we did.* Encourage students to share their conjectures along with the thinking behind their ideas. GMP3.1

Ask students how they might test their conjectures. If no one suggests it, fill one of the objects with packing material. Ask students how they could compare the volumes of the three objects using the packing material. Have volunteers pour the contents of the filled container into each of the other containers in turn. Discuss how the results of the pouring experiment show which container has the greatest volume, which has the least volume, and which is in the middle.

**Cylinder A
and Cylinder B**

Math Journal 1, p. 13

Comparing Volume   Lesson 1-5
                         DATE   TIME

1   Create Cylinders A and B from half-sheets of paper.
What could you measure about these cylinders?
Sample answers: height; surface area;
distance around the cylinders; weight; volume

2   Which cylinder do you think has a greater volume? Explain your answer.
Answers vary.

3   Test your prediction. Which cylinder has a greater volume? Explain your answer.
Sample answer: Cylinder A has a greater volume than Cylinder
B. I filled A with beans and then poured the beans into B. Not
all of the beans from A fit into B, which means that Cylinder A
holds more and has a greater volume.

4   Think about all the attributes you listed in Problem 1.

a. How are the cylinders different?
Sample answer: They differ in height, in the
distance around them, and in volume.

b. How are the cylinders the same?
Sample answer: The cylinders are both made from the
same-size paper, so they must have the same area on the
outside and the same weight.

5.MD.3, 5.MD.4, SMP3   13

▶ Comparing Volume

Math Journal 1, p. 13

| WHOLE CLASS | SMALL GROUP | PARTNER | INDEPENDENT |

Distribute 2 half-sheets of paper to each small group of students. Demonstrate how to create two different cylinders by taping one half-sheet of paper together along the long edges and the other together along the short edges. Label the short cylinder A. Label the tall cylinder B. (*See margin.*)

After they create their cylinders, have students complete Problems 1 and 2 on journal page 13 and discuss their predictions for Problem 2.   GMP3.1   Encourage them to ask questions, making sure they understand the thinking of the other students in their group. Invite a few volunteers to summarize another student's explanation for the whole class.   GMP3.2

Distribute about $1\frac{1}{2}$ cups of packing material and a small paper plate to each group. Instruct students to test their predictions by doing a pouring experiment and to then complete the rest of the journal page. Because the cylinders have no bottoms, have students place them on the paper plate before conducting the pouring experiment.

Discuss the results of the experiment as a class. Have students share anything that surprised them as they performed the experiment.

 Assessment Check-In   5.MD.3

Math Journal 1, p. 13

Circulate and observe as students conduct the pouring experiment and complete the journal page. Ask: *What are you comparing when you pour the items into the cylinders?* Expect most to report that they are comparing volume or comparing the amount that each cylinder holds. If anyone describes the activity in vague terms, such as "We're trying to see which is bigger," follow up with questions such as: *In what way is one bigger than the other?* If necessary, focus on the term *volume* to help students incorporate it into their vocabulary.

Evaluation Quick Entry   Go online to record students' progress and to see trajectories toward mastery for these standards.

Summarize   Ask a volunteer to summarize the meaning of volume in the activity they just finished. Then give students 30 seconds to list as many items as they can for which the attribute of volume could be measured. Share ideas as time permits.

3 Practice 20–30 min

▶ Playing *Name That Number*

Student Reference Book, p. 315; *Math Masters,* p. G2

| WHOLE CLASS | SMALL GROUP | PARTNER | INDEPENDENT |

Students practice writing expressions for calculations and writing expressions with grouping symbols. See *Student Reference Book,* page 315 for detailed directions. Have students record the expressions they use on *Math Masters,* page G2.

Observe

• Which operations do students consistently use in their solutions?

• What strategies do students use when carrying out operations?

Discuss

• *How did you decide which numbers to use? How did you decide which operation to use?* GMP1.1

• *Do parentheses help you in this game? If so, how?* GMP2.2

| Differentiate | Game Modifications | Go Online | Differentiation Support |

▶ Math Boxes 1-5

Math Journal 1, p. 14

| WHOLE CLASS | SMALL GROUP | PARTNER | INDEPENDENT |

Mixed Practice Math Boxes 1-5 are paired with Math Boxes 1-7.

▶ Home Link 1-5

Math Masters, p. 16

Homework Students compare the volumes of household objects.

Planning Ahead

To ensure adequate time for instruction in Lessons 1-7 and 1-8, consider assembling prisms in advance. See the Before You Begin notes in Lessons 1-7 and 1-8.

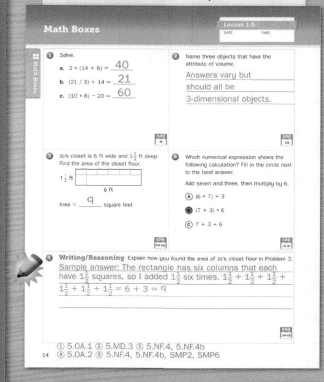

Math Journal 1, p. 14

Math Boxes
Lesson 1-5

① Solve.
a. $2 * (14 + 6) =$ __40__
b. $(21 / 3) + 14 =$ __21__
c. $(10 * 8) - 20 =$ __60__

② Name three objects that have the attribute of volume.
Answers vary but should all be 3-dimensional objects.

③ Jo's closet is 6 ft wide and $1\frac{1}{2}$ ft deep. Find the area of the closet floor.
Area = __9__ square feet

④ Which numerical expression shows the following calculation? Fill in the circle next to the best answer.
Add seven and three, then multiply by 6.
(A) $(6 * 7) + 3$
(B) $(7 + 3) * 6$
(C) $7 + 3 + 6$

⑤ **Writing/Reasoning** Explain how you found the area of Jo's closet floor in Problem 3.
Sample answer: The rectangle has six columns that each have $1\frac{1}{2}$ squares, so I added $1\frac{1}{2}$ six times. $1\frac{1}{2} + 1\frac{1}{2} + 1\frac{1}{2} + 1\frac{1}{2} + 1\frac{1}{2} + 1\frac{1}{2} = 6 + 3 = 9$

① 5.OA.1 ② 5.MD.3 ③ 5.NF.4, 5.NF.4b
14 ④ 5.OA.2 ⑤ 5.NF.4, 5.NF.4b, SMP2, SMP6

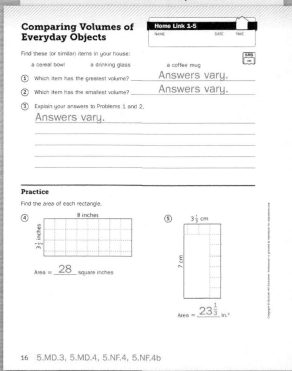

Math Masters, p. 16

Comparing Volumes of Everyday Objects
Home Link 1-5

Find these (or similar) items in your house:
a cereal bowl   a drinking glass   a coffee mug

① Which item has the greatest volume?   Answers vary.
② Which item has the smallest volume?   Answers vary.
③ Explain your answers to Problems 1 and 2.
Answers vary.

Practice
Find the area of each rectangle.

④ 8 inches, $3\frac{1}{2}$ inches
Area = __28__ square inches

⑤ $3\frac{1}{3}$ cm, 7 cm
Area = __$23\frac{1}{3}$__ in.²

16   5.MD.3, 5.MD.4, 5.NF.4, 5.NF.4b

Lesson 1-6

Exploring Nonstandard Volume Units

Overview Students use nonstandard units to measure volumes of rectangular prisms. They discuss packing units without gaps or overlaps to obtain an accurate volume measurement.

▶ **Before You Begin**
For Part 2, each small group will need an empty, open rectangular prism, such as a small box for paper clips, crayons, or tea bags. If your pattern block supply is limited, consider having groups rotate through stations—one with squares, one with triangles, and one with hexagons—with enough blocks to completely pack the prisms. For the optional Extra Practice activity, each student or partnership will need two small empty rectangular prisms.

▶ **Vocabulary**
rectangular prism

Standards

Focus Cluster
• Geometric measurement: understand concepts of volume.

① **Warm Up** 5 min	**Materials**	
Mental Math and Fluency Students convert among units of length.	slate	5.MD.1

② **Focus** 35–40 min		
Math Message Students discuss whether a rectangle has volume.		5.MD.3 SMP3, SMP6
Packing Prisms to Measure Volume Students use three different kinds of pattern blocks to measure the volume of a rectangular prism.	*Math Journal 1*, p. 15; pattern blocks (triangles, squares, hexagons); small rectangular prisms	5.MD.3, 5.MD.4 SMP3, SMP5, SMP6
✓ **Assessment Check-In** See page 52. Expect most students to be able to minimize gaps and overlaps when packing rectangular prisms with pattern blocks and report the volume measurements of the prisms by counting the individual blocks.	*Math Journal 1*, p. 15	5.MD.3, 5.MD.4, SMP5
Comparing Volume Units Students compare nonstandard units used to measure the volume of a rectangular prism.	*Math Journal 1*, p. 15	5.MD.3, 5.MD.4 SMP3, SMP5, SMP6

③ **Practice** 20–30 min		
Finding Areas of Rectangles Students find areas of rectangles with fractional side lengths.	*Math Journal 1*, p. 16	5.NF.4, 5.NF.4b
Math Boxes 1-6 Students practice and maintain skills.	*Math Journal 1*, p. 17	See page 53.
Home Link 1-6 **Homework** Students identify objects that have volume and describe how to measure the volume of one of the objects.	*Math Masters*, p. 20	5.OA.1, 5.MD.3, 5.MD.4

 Go Online to see how mastery develops for all standards within the grade.

my.mheducation.com

Differentiation Options

Readiness 10–15 min	**Enrichment** 20–30 min	**Extra Practice** 10–15 min
WHOLE CLASS · **SMALL GROUP** · PARTNER · INDEPENDENT	WHOLE CLASS · SMALL GROUP · **PARTNER** · INDEPENDENT	WHOLE CLASS · SMALL GROUP · **PARTNER** · INDEPENDENT

Measuring with Nonstandard Units

`5.MD.4`

Math Masters, p. 17; index card; pennies

To review the idea that accurate measurement depends on placing the units without gaps and overlaps, students examine the measurement techniques illustrated on *Math Masters*, page 17. Have students complete Problems 1–3 and then discuss their responses. For Problem 4, provide each student with an index card. Have students use pennies to measure the area with as few gaps and overlaps as possible.

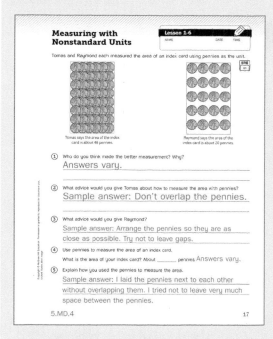

Building and Measuring the Volume of a Polyhedron

`5.MD.3, 5.MD.4`

Activity Card 6; *Student Reference Book*, pp. 272–274; *Math Masters*, p. 18; pattern blocks

To extend their work with nonstandard volume units, students build polyhedrons with one kind of pattern block, recording the volumes in pattern-block units. They figure out how to express that same volume using other pattern-block units. For example, a polyhedron made from 10 hexagons would have a volume of 60 triangles because there are 6 triangles in 1 hexagon.

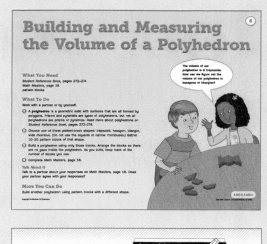

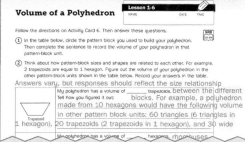

Estimating Volume in Nonstandard Units

`5.MD.3, 5.MD.4`

Activity Card 7; *Math Masters*, p. 19; pattern blocks; 2 small empty rectangular prisms

For more practice with nonstandard volume units, students estimate the number of pattern blocks required to fill 3-dimensional objects.

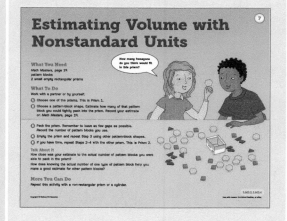

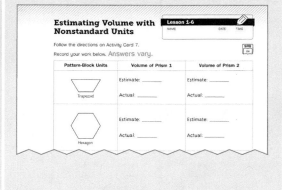

English Language Learner

Beginning ELL Students may have seen and heard the consonant cluster *sm* in everyday terms such as *small*. Extend their knowledge of English spelling conventions by having them repeat the term *prism* several times, pointing out the pronunciation of the *s* as a *z* sound and explaining that the *sm* blend has a different pronunciation in this context.

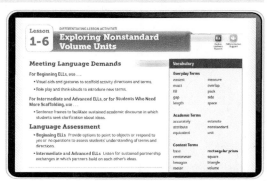

Differentiation Support pages are found in the online Teacher's Center.

SMP3 **Construct viable arguments and critique the reasoning of others.**
GMP3.2 Make sense of others' mathematical thinking.

SMP5 **Use appropriate tools strategically.**
GMP5.2 Use tools effectively and make sense of your results.

SMP6 **Attend to precision.**
GMP6.1 Explain your mathematical thinking clearly and precisely.

---

1 Warm Up 5 min

▶ Mental Math and Fluency

Have students record on their slates measurement equivalencies for units of length. *Leveled exercises*:

⬤○○ 1 meter equals how many centimeters? 100
2 meters equals how many centimeters? 200
3 meters equals how many centimeters? 300

⬤⬤○ 14 meters equals how many centimeters? 1,400
900 centimeters equals how many meters? 9
2,400 centimeters equals how many meters? 24

⬤⬤⬤ 50 centimeters equals how many meters? $\frac{1}{2}$
850 centimeters equals how many meters? $8\frac{1}{2}$
$\frac{1}{4}$ meter equals how many centimeters? 25

---

2 Focus 35–40 min

▶ Math Message

Draw a rectangle on a piece of paper. Think about what you would do if someone asked you to find the volume of your rectangle. Record your ideas and then talk about them with a partner. GMP3.2, GMP6.1

▶ Packing Prisms to Measure Volume

Math Journal 1, p. 15

| WHOLE CLASS | SMALL GROUP | PARTNER | INDEPENDENT |

Math Message Follow-Up Invite students to share the ideas they discussed with their partners, encouraging them to explain their own thinking clearly and to ask questions to be sure they understand each other's thinking. GMP3.2, GMP6.1 Guide the discussion to emphasize the idea that volume is an attribute of 3-dimensional objects only. Because a drawing of a rectangle is 2-dimensional, it does not have volume. Ask: *What objects in our room do you see that have the attribute of volume?* Sample answers: desk; tissue box; pencil case; garbage can *What attributes does the 2-dimensional rectangle have that we can measure?* Sample answers: length; width; perimeter; area Explain that in today's lesson, students will review how to accurately measure attributes of 2-dimensional objects and then apply similar techniques when measuring the volume of 3-dimensional objects.

Math Journal 1, p. 15

Packing Prisms to Measure Volume   Lesson 1-6
DATE   TIME

① Use pattern blocks to measure the volume of your rectangular prism. Record your results below. Answers vary.

Our prism has a volume of about _____ **square** pattern blocks.
Our prism has a volume of about _____ **triangle** pattern blocks.
Our prism has a volume of about _____ **hexagon** pattern blocks.

② What was important to remember as you packed the prism with pattern blocks so you could measure volume as accurately as possible?
Sample answer: We had to pack the blocks as tightly as possible and try to avoid gaps and overlaps.

③ Which pattern block did you need the **most** of to fill your prism? Why?
Sample answer: We used the greatest number of triangles. They are the smallest unit, so we needed the most of them.

④ Which pattern block did you need the **least** of to fill your prism? Why?
Sample answer: We used the least number of hexagons. They are also the largest unit.

⑤ What other objects could you use to fill the prism?
Sample answers: other pattern blocks, marbles, cubes, dried beans, popcorn, small pebbles

⑥ What 3-dimensional shape do you think would be easiest to pack tightly into a rectangular prism without gaps or overlaps? Why do you think so?
Sample answer: A cube would pack tightly because each face would fit exactly against the face of another cube. It would also fit in the corners of the prism.

5.MD.3, 5.MD.4, SMP5   15

Provide each student with several square pattern blocks. Instruct students to use the blocks to measure the length of the rectangle they drew for the Math Message. Remind them that their measurement unit is the *side length* of the pattern block square, not the entire square. If students drew a rectangle that is smaller than the pattern block, ask them to draw a larger one. Ask: *What do you have to keep in mind in order to accurately measure length with a pattern block—or any unit?* GMP5.2 Sample answer: Align or repeat the unit with no gaps or overlaps. Cover the whole distance. Now ask students to use the square pattern blocks to measure the area of their rectangle. Remind them that because they are measuring area, their unit is the entire square. Ask: *What do you have to keep in mind in order to accurately measure area with a pattern block—or any unit?* GMP5.2 Sample answer: Align or repeat the unit with no gaps or overlaps. Fill all the space. Tell students that these same measurement techniques—no gaps, no overlaps, fill all the space—will also be important when they measure volume.

Divide the class into small groups. Provide each group with a rectangular prism and a collection of square, triangle, and hexagon pattern blocks, or have groups rotate through pattern block stations. (*See Before You Begin.*) Briefly review with students that a **rectangular prism** is a prism that has rectangular bases. Have students find the bases of the prisms you distributed and confirm that the prisms are rectangular prisms. Remind students that in the previous lesson they compared the volume of two containers by filling each with packing material. Today they will measure the volume of one rectangular prism three different times by filling it with three different units—triangle pattern blocks, square pattern blocks, and hexagon pattern blocks.

Ask: *What should you keep in mind as you pack your prisms?* Sample answers: The blocks should be close together so there are no gaps. We should fill as much of the prism as we can. *After you pack your prism with one type of pattern block, how will you report the volume?* Sample answer: Remove the pattern blocks and count them. The number of blocks is the volume measurement.

Have students take turns packing their group's prism with each type of pattern block. They record their findings on journal page 15 and then answer the remaining questions.

Packing a rectangular prism with pattern blocks

NOTE It is likely that students' rectangles will not be an exact number of pattern blocks in length or area. Encourage students to use the pattern blocks to find the approximate length and area. Keep the focus of this activity on reviewing the proper techniques for measuring length and area.

Adjusting the Activity

Differentiate If students have difficulty tightly packing the pattern blocks into the prism, suggest that they draw an outline of the base of the prism. Encourage them to experiment with different configurations to determine the best way to arrange the blocks in just one layer, minimizing gaps as much as possible. Students will likely find it easier to tightly pack the prism if they approach the task as a layering exercise.

Go Online Differentiation Support

Assessment Check-In   5.MD.3, 5.MD.4

Math Journal 1, p. 15

Observe students as they pack their prisms and complete Problems 1 and 2 on journal page 15. Expect that most will be able to minimize gaps and overlaps when packing their prisms and report the volume measurement by counting the individual blocks.   GMP5.2   To support students who struggle, refer to the Adjusting the Activity note. Encourage students who want a challenge to complete the Enrichment activity.

 Evaluation Quick Entry Go online to record students' progress and to see trajectories toward mastery for these standards.

▶ Comparing Volume Units

Math Journal 1, p. 15

| WHOLE GROUP | SMALL GROUP | PARTNER | INDEPENDENT |

As a class, discuss students' responses to the questions on the journal page. Encourage them to explain their thinking about why different quantities of pattern blocks were needed to fill the same prism.   GMP6.1   During the discussion, emphasize the importance of packing units tightly without gaps or overlaps, acknowledging that gaps are impossible to avoid with some pattern-block shapes.   GMP5.2   Ask students which pattern block was the easiest to pack into the prism to measure the volume. Expect most to reply that the square was the easiest unit to use because it packed neatly without gaps and fit easily into the square corners of the prism.

Ask: *Do any of your measurements give an exact volume of the prism? Why or why not?* Sample answer: No. There were always some gaps on the edges, and sometimes the blocks did not make it all the way to the top. So I could not fill all the space with the blocks. *How could you get a measurement that is closer to the exact volume?* Sample answer: I could use a unit that leaves fewer gaps.

Summarize Ask partners to explain to each other their thinking about the final question on the journal page: *What 3-dimensional shape do you think would be easiest to pack tightly into a rectangular prism without gaps or overlaps? Why do you think so?* Invite volunteers to summarize their partner's thinking.   GMP3.2   Tell students that in the next lesson they will use another unit—cubes—to measure the volume of different rectangular prisms.

3 Practice   20–30 min

▶ Finding Areas of Rectangles

Math Journal 1, p. 16

| WHOLE CLASS | SMALL GROUP | PARTNER | INDEPENDENT |

Students complete journal page 16 to practice finding areas of rectangles with fractional side lengths.

▶ Math Boxes 1-6

Math Journal 1, p. 17

| WHOLE CLASS | SMALL GROUP | PARTNER | INDEPENDENT |

Mixed Practice Math Boxes 1-6 are grouped with Math Boxes 1-10 and 1-12.

▶ Home Link 1-6

Math Masters, p. 20

Homework Students identify objects that have volume and describe how the volume of one of the objects could be measured.

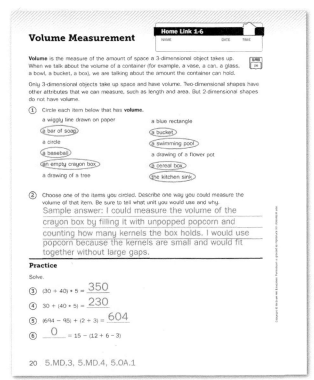

Math Masters, p. 20

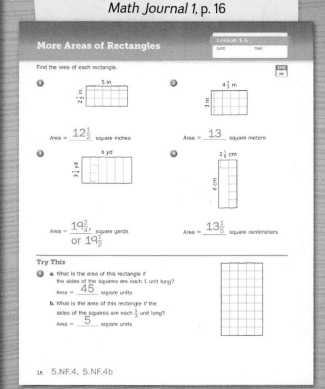

Math Journal 1, p. 16

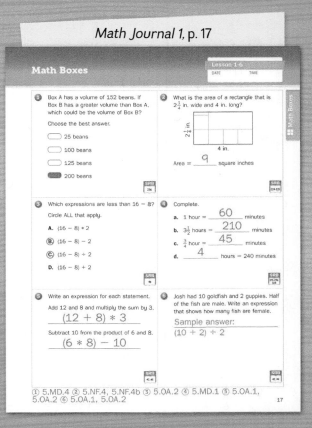

Math Journal 1, p. 17

Measuring Volume by Counting Cubes

Overview Students discuss the benefits of using unit cubes to measure volume. They measure volume by counting the number of cubes it takes to fill a rectangular prism.

▶ **Before You Begin**

Consider assembling the prisms on *Math Journal 1*, Activity Sheet 1 for students in advance. You can make extra prisms from *Math Masters*, page 21. Place centimeter cubes near the Math Message.

▶ **Vocabulary**

unit cube • cubic unit

Standards

Focus Cluster
• Geometric measurement: understand concepts of volume.

Warm Up   5 min

	Materials	
Mental Math and Fluency Students evaluate expressions with grouping symbols.		5.OA.1

Focus   35–40 min

Math Message Students assemble rectangular prisms and estimate how many cubes will fit in each one.	*Math Journal 1*, p. 18; Rectangular Prisms (*Math Journal 1*, Activity Sheet 1); 25 centimeter cubes; scissors; tape	5.MD.3, 5.MD.3a, 5.MD.3b, 5.MD.4
Measuring Volume with Cubes Students determine the volume of rectangular prisms by counting the number of cubes it takes to fill them.	*Math Journal 1*, p. 18; Rectangular Prisms A, B, and C; square pattern block; 25 centimeter cubes	5.MD.3, 5.MD.3a, 5.MD.3b, 5.MD.4 SMP6
Solving Cube-Stacking Problems Students examine partially filled rectangular prisms and determine the number of cubes needed to fill each one.	*Math Journal 1*, pp. 18–19; *Math Masters*, p. TA3 (optional); centimeter cubes	5.MD.3, 5.MD.3a, 5.MD.3b, 5.MD.4 SMP1, SMP2, SMP6
✓ **Assessment Check-In** See page 58. Expect most students to be able to look at a picture of a rectangular prism or build the prism with centimeter cubes to determine the number of cubes needed to fill the prism.	*Math Journal 1*, p. 19	5.MD.3, 5.MD.3b, 5.MD.4

Practice   20–30 min

Playing *Baseball Multiplication* **Game** Students practice multiplication facts to prepare for multidigit multiplication in Unit 2.	*Student Reference Book*, p. 292; per group: *Math Masters*, p. G3; number cards 1–10 (4 of each); 4 counters; calculator or multiplication/division facts table	5.NBT.5 SMP1
Math Boxes 1-7 Students practice and maintain skills.	*Math Journal 1*, p. 20	See page 59.
Home Link 1-7 **Homework** Students find volumes of partially filled prisms.	*Math Masters*, p. 24	5.OA.1, 5.MD.3, 5.MD.3a, 5.MD.3b, 5.MD.4, SMP2

 Go Online to see how mastery develops for all standards within the grade.

Differentiation Options

Solving 1-Layer Prism Problems

5.MD.3, 5.MD.3a, 5.MD.3b, 5.MD.4, SMP2

Math Masters, p. 22; centimeter cubes (optional)

To prepare for more difficult cube-stacking problems, students practice finding the number of cubes needed to fill prisms that are only one layer tall. Ask students to determine how many cubes would fill each prism on *Math Masters*, page 22.

GMP2.2 Be sure they understand that they need to find the total number of cubes that would fit in the prism, not the number of additional cubes they would need to fill it. Provide centimeter cubes for those who would benefit from modeling the problem. Have students share their strategies for finding the total number of cubes.

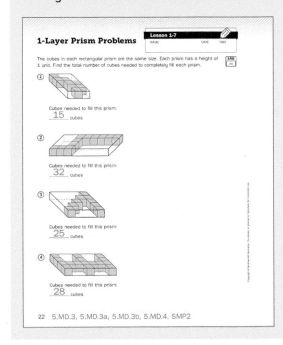

Exploring Penticubes

5.MD.3, 5.MD.3a, 5.MD.3b, 5.MD.4

Activity Card 8; *Math Masters*, p. 23; centimeter cubes

To investigate volume and surface area, students use centimeter cubes to build penticubes.

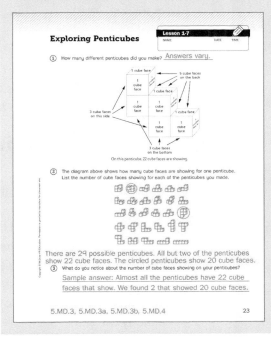

Creating Prism Patterns and Finding Volume with Cubes

5.MD.3, 5.MD.3a , 5.MD.3b, 5.MD.4

grid paper (*Math Masters*, p. TA3); centimeter cubes; scissors; tape

For additional experience using cubes to find volume, students create prism patterns on grid paper. Demonstrate how to draw a pattern for an open box on grid paper. Instruct students to create a box pattern of any dimension they choose with a height of at least 2 units. Have students exchange patterns with a partner, cut out the pattern, fold up the sides, and tape it together. Students use centimeter cubes to find the volume.

English Language Learner

Beginning ELL Help students understand the meaning of the terms *partial* and *partially* by displaying them with the base word *part* underlined. Display visuals like a complete and a partially eaten apple and use a think-aloud to describe them, pointing to the displays and saying: *A part of this apple is gone. This is not the whole apple. It is a partial apple. It has been partially eaten.* Point out how the sound of the base word *part* changes in the two derivatives *partial* and *partially*.

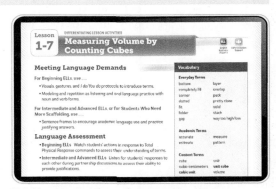

Differentiation Support pages are found in the online Teacher's Center.

Standards and Goals for
Mathematical Process and Practice

SMP1 Make sense of problems and persevere in solving them.
GMP1.3 Keep trying when your problem is hard.

SMP2 Reason abstractly and quantitatively.
GMP2.2 Make sense of the representations you and others use.

SMP6 Attend to precision.
GMP6.4 Think about accuracy and efficiency when you count, measure, and calculate.

1 Warm Up 5 min

▶ **Mental Math and Fluency**

Display the following expressions for students to evaluate.
Leveled exercises:

◉○○ $2 \times (2 + 3)$ 10
    $(2 \times 2) + 3$ 7

◉◉○ $(24 \div 4) + 8$ 14
    $5 + (18 \div 3)$ 11
    $10 \times (56 \div 8)$ 70

◉◉◉ $(3 \times 2) + (4 \times 3)$ 18
    $(42 \div 6) \times (27 \div 9)$ 21
    $12 + (48 \div 8) - 7$ 11

2 Focus 35–40 min

▶ **Math Message**

Math Journal 1, p. 18 and Activity Sheet 1

Cut out and assemble Rectangular Prisms A, B, and C. Take 25 cubes. Estimate how many cubes will fit in each prism. Record your estimates in the second column of the table on journal page 18.

▶ **Measuring Volume with Cubes**

Math Journal 1, p. 18

| WHOLE CLASS | SMALL GROUP | PARTNER | INDEPENDENT |

Math Message Follow-Up Show students a square pattern block and a cube. Ask: *Could both shapes be used to pack a prism without gaps or overlaps?* Yes. *Could both shapes fit into the corners of a prism?* Yes. *Do you think a cube or a square pattern block is better for measuring volume? Why?* Sample answer: A cube is better. Since each edge length is exactly the same, the same number of cubes will fit in a row in the prism no matter which way you turn it. The edges on a square pattern block are not all the same length. Because the height is smaller than the length and the width, it fits differently if it is on its side than if it has the square part on the bottom. Tell students that today they will use cubes to measure the volume of rectangular prisms. Remind them that when they packed prisms with pattern blocks, they reported their results as a number of *pattern-block units*. Explain that when cubes are used to measure, each cube has a side length of 1 unit and is called a **unit cube.** Its volume is 1 **cubic unit.** Explain that cubic units can also be written as units^3.

Academic Language Development

Help students construct their own understanding of the rationale for using the number 2 for squared and 3 for cubed or cubic. Remind them that when we use squares to measure area, we are measuring two dimensions (length and width). Have students work in pairs to come up with explanations for why it makes sense to use the number 3 when thinking about a cube and why units^3 is read as "units cubed." To help students get started, ask: *How many dimensions are involved when measuring volume?* Practice reading units written as units^2 and units^3, using the terms *squared* and *cubed*.

Ask students to share how many unit cubes they estimated would fit into each prism and to explain how they arrived at their estimate. Display a few of the students' estimates to demonstrate the notation. For example: *Rhonda estimated that the volume of the prism is about 20 cubic units, or 20 units^3.*

> **NOTE** Some students may point out that the length of each edge of a unit cube is 1 centimeter, and so the volumes of the prisms could be reported in *cubic centimeters*, or *cm^3*. Acknowledge that this is correct, but keep the focus of this lesson on the more general cubic units.

Instruct students to pack Prism A with cubes, reminding them of the importance of packing without gaps or overlaps. **GMP6.4** Have them count the total number of cubes they used, and record the number in the table on journal page 18. Ask: *How many cubes did you need to completely fill Prism A?* 24 *What is the volume of Prism A?* 24 cubic units Ask: *Why is the number of cubes you used to fill the prism the same as the volume of the prism?* Sample answer: One cube has a volume of 1 cubic unit. If I count the total number of cubes, it gives the number of cubic units that fill the prism. That is the volume.

Have students pack Prisms B and C and record their results. Prism B: 20 cubic units; Prism C: 18 cubic units Invite volunteers to share their results and discuss how the actual number of cubes needed to pack the prisms compared to their estimates. Ask: *Of the different materials we have used to measure volume—packing material, pattern blocks, cubes—which unit do you think results in the most accurate measurement? Why?* **GMP6.4** Sample answers: Cubes provide the most accurate measurement because they can be packed tightly without gaps or overlaps. They are small, so it is easy to fill the container almost exactly.

▶ Solving Cube-Stacking Problems

Math Journal 1, pp. 18–19

| WHOLE CLASS | SMALL GROUP | PARTNER | INDEPENDENT |

Display Prism 1 from journal page 18 as students look at it in their journals. Explain that in these problems they must examine a drawing of a prism partially filled with cubes and determine the prism's volume, or the number of cubes that would be needed to completely fill the prism. **GMP2.2** Ask students to share what they notice about the drawing of Prism 1. Be sure to discuss how the drawing shows the number of cubes that fit across the length and the width of the bottom of the prism and how many cubes tall the prism is. Have students look at the other prisms on journal pages 18 and 19. Point out that each prism has at least one stack of cubes that goes to the top.

 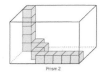
Adjusting the Activity

Differentiate To help students who struggle solving cube-stacking problems, provide grid paper (*Math Masters,* page TA3) and centimeter cubes. Have them go through the following steps:

- Refer to the picture and reproduce the configuration of cubes of *just the bottom layer* on grid paper.
- Draw the outline of the entire base on the grid paper. Remind students that the base of a rectangular prism is always a rectangle.
- Fill in the base with cubes.
- Refer to the picture and determine the height of the prism.
- Reproduce a column with cubes, connecting it to the base.
- Fill in with cubes, remembering not to go beyond the base or higher than the column.
- Count the total number of cubes.

(**Go Online**) Differentiation Support

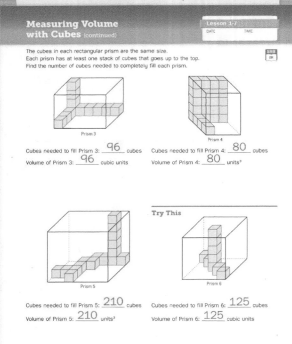

Measuring Volume with Cubes (continued)

Lesson 1-7

DATE   TIME

The cubes in each rectangular prism are the same size.
Each prism has at least one stack of cubes that goes up to the top.
Find the number of cubes needed to completely fill each prism.

Prism 3

Prism 4

Cubes needed to fill Prism 3: __96__ cubes

Volume of Prism 3: __96__ cubic units

Cubes needed to fill Prism 4: __80__ cubes

Volume of Prism 4: __80__ units³

Try This

Prism 5

Prism 6

Cubes needed to fill Prism 5: __210__ cubes

Volume of Prism 5: __210__ units³

Cubes needed to fill Prism 6: __125__ cubes

Volume of Prism 6: __125__ cubic units

5.MD.3, 5.MD.3a, 5.MD.3b, 5.MD.4, SMP2   19

Games

Baseball Multiplication (1 to 10 Facts)

Materials	☐ 1 *Baseball Multiplication* Game Mat (*Math Masters*, p. G3)
	☐ number cards 1–10 (4 of each)
	☐ 4 counters
	☐ 1 calculator or 1 multiplication/division table
Players	2 teams of one or more players each
Skill	Multiplying with automaticity
Object of the Game	To score more runs in a 3-inning game.

Directions

1. Shuffle the cards and place the deck number-side down on the table.

2. Teams take turns being the pitcher and the batter. The rules are similar to the rules of baseball, but this game lasts only 3 innings.

3. The batter puts a counter on home plate. The pitcher draws 2 cards. The batter multiplies the numbers on the cards and gives the answer. The pitcher checks the answer and may use a calculator to do so.

4. If the answer is correct, the batter looks up the product in the Hitting Table to the right. If it is a hit, the batter moves all counters on the field the number of bases shown in the table. The pitcher tallies each out on the scoreboard.

5. An incorrect answer is a strike and another pitch is thrown (2 more cards are drawn). Three strikes make an out.

6. A run is scored each time a counter crosses home plate. The batter tallies each run scored on the scoreboard.

7. After each hit or out, the batter puts a counter on home plate. The batting and pitching teams switch roles after the batting team has made 3 outs. The inning is over when both teams have made 3 outs. Shuffle all cards and replace the deck between innings or when all cards have been used.

The team with more runs at the end of 3 innings wins the game. If the game is tied at the end of 3 innings, play continues in extra innings until one team wins.

Scoreboard

Inning	1	2	3	Total
Team 1	outs			
	runs			
Team 2	outs			
	runs			

Hitting Table 1 to 10 Facts

1 to 21	Out
24 to 45	Single (1 base)
48 to 70	Double (2 bases)
72 to 81	Triple (3 bases)
90 to 100	Home Run (4 bases)

Make sure cubes are available for students to use, and have partners work together to figure out the number of cubes needed to fill Prism 1. Be sure students understand that they need to find the total number of cubes that fit in the prism, not the number of additional cubes they would add to fill it.

When most of the class has finished, invite volunteers to share some of the strategies they used. *Sample strategies:*

- I used cubes to build what was shown in the picture. Then I could see how many cubes were hidden and how many were missing from the prism.

- I saw that each of the 6 rows along the bottom had 5 cubes. I counted by 5s to figure out how many would fit on the bottom of the prism: 5, 10, 15, 20, 25, 30. Then I saw that I would need 4 sets of 30 to fill up the prism. $30 * 4 = 120$.

After volunteers share their strategies, have partners work together to find the volume of Prism 2. When most have finished, discuss how Prism 2 differed from Prism 1. Sample answer: The drawing of Prism 1 showed all the dimensions of the prism. The drawing of Prism 2 did not. I had to fill in some of the spaces with cubes to figure out how to find the total number of cubes. Ask students to share difficulties they encountered, how they overcame them, and how they made sure they were accurately counting the hidden cubes. **GMP1.3, GMP6.4**

When most partnerships seem comfortable with a strategy for the cube-stacking problems, have students complete journal page 19.

 Assessment Check-In   5.MD.3, 5.MD.3b, 5.MD.4

Math Journal 1, p. 19

Expect most students to be able to determine the number of cubes needed to fill Prism 3 on journal page 19 by looking at the picture or by building the prism with centimeter cubes. Most will also understand that the number of cubes needed to pack the prism is the volume of the prism in cubic units. Some students may be able to find the number of cubes that would fill Prisms 4–6. To support students who struggle, refer to the Adjusting the Activity note.

 Evaluation Quick Entry Go online to record students' progress and to see trajectories toward mastery for these standards.

Summarize Ask partners to share their thoughts about which cube-stacking problem was the easiest to solve and which was the most difficult. **GMP1.3** Summarize by recalling how in today's lesson they found volume by counting cubes. In the next lesson they will continue to measure volume with cubes, but they will explore a new method for doing so.

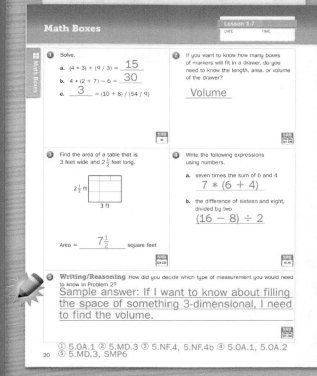

③ Practice 20-30 min

▶ Playing *Baseball Multiplication*

Student Reference Book, p. 292; *Math Masters*, p. G3

| WHOLE CLASS | **SMALL GROUP** | PARTNER | INDEPENDENT |

Have students play *Baseball Multiplication*. See *Student Reference Book*, page 292 for detailed directions. Playing this game refreshes students' fact knowledge and provides an opportunity to informally assess their recall of basic facts. Knowing multiplication facts is an important prerequisite for multiplying multidigit whole numbers, a focus of Unit 2.

Observe

• What strategies do students use to determine the products?

• Which students show fluency with basic multiplication facts?

Discuss

• *How did you find the product of the two numbers?* **GMP1.2**

• *Which facts do you know automatically? Which do you still need to learn?* **GMP1.2**

| **Differentiate** | **Game Modifications** | **Go Online** 📖 Differentiation Support |

▶ Math Boxes 1-7 ✏️

Math Journal 1, p. 20

| WHOLE CLASS | SMALL GROUP | PARTNER | **INDEPENDENT** |

Mixed Practice Math Boxes 1–7 are paired with Math Boxes 1–5.

▶ Home Link 1-7

Math Masters, p. 24

Homework Students examine drawings of rectangular prisms that are partially filled with cubes. They determine the total number of cubes needed to fill each prism and the volume of each prism. **GMP2.2**

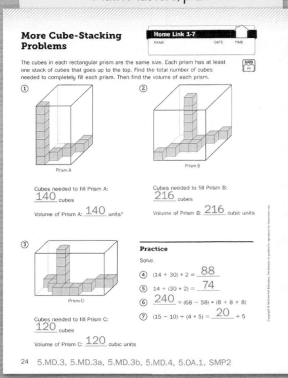

Lesson 1-8

Measuring Volume by Iterating Layers

Overview Students relate volume to multiplication and addition by thinking about iterating layers to find the volumes of prisms.

▶ **Before You Begin**

You may wish to assemble the prisms on *Math Journal 1*, Activity Sheet 2 for students in advance of the lesson. You can make extra prisms from *Math Masters*, page 25. Place a box of centimeter cubes near the Math Message. After the lesson, collect and store Prisms E and F for Lesson 1-9.

Standards

Focus Cluster
• Geometric measurement: understand concepts of volume.

	Materials	
① Warm Up 5 min		
Mental Math and Fluency Students interpret expressions.		5.OA.2
② Focus 35–40 min		
Math Message Students assemble rectangular prisms and estimate the number of cubes that will fit into each one.	More Rectangular Prisms (*Math Journal 1*, Activity Sheet 2); slate; 25 centimeter cubes; scissors; tape	5.MD.3, 5.MD.3a, 5.MD.3b, 5.MD.4
Using Layers of Cubes to Measure Volume Students fill a single layer of a prism and use the number of cubes in one layer to determine the volume.	*Math Masters*, p. TA2 (optional); Rectangular Prisms D, E, and F; slate; 25 centimeter cubes	5.MD.3, 5.MD.3a, 5.MD.3b, 5.MD.4 SMP1, SMP6
Using Layers to Solve Cube-Stacking Problems Students find volumes of prisms that show one layer completely filled.	*Math Journal 1*, pp. 22–23; centimeter cubes (optional)	5.MD.3, 5.MD.3a, 5.MD.3b, 5.MD.4 SMP1, SMP3, SMP6
✓ **Assessment Check-In** See page 64. Expect most students to be able to find the number of cubes in 1 layer of a rectangular prism and use either repeated addition or multiplication to find the volume of the prism.	*Math Journal 1*, pp. 22–23	5.MD.3, 5.MD.3b, 5.MD.4
③ Practice 20–30 min		
Converting Measurements Students convert units of measurement in mathematical and real-world contexts.	*Math Journal 1*, p. 21	5.MD.1
Math Boxes 1-8: Preview for Unit 2 Students preview skills and concepts for Unit 2.	*Math Journal 1*, p. 24	See page 65.
Home Link 1-8 **Homework** Students determine how many cubes will fit in boxes.	*Math Masters*, p. 28	5.OA.1, 5.MD.3, 5.MD.3a, 5.MD.3b, 5.MD.4

  **Go Online** to see how mastery develops for all standards within the grade.

Differentiation Options

Readiness 10–15 min	**Enrichment** 20–30 min	**Extra Practice** 20–30 min
WHOLE CLASS · **SMALL GROUP** · PARTNER · INDEPENDENT	WHOLE CLASS · SMALL GROUP · **PARTNER** · INDEPENDENT	WHOLE CLASS · **SMALL GROUP** · PARTNER · INDEPENDENT

Laying Down Layers

5.MD.4, SMP6

per group: index cards, paper bag

For experience working with layers, students find the total number of items drawn on a set of stacked cards. Provide each student with 2 or 3 index cards and have them sketch 12 items, like stars or smiley faces, on each one. Put the cards in a bag and choose a student to remove cards one at a time and stack them on a table. After each card is placed on the stack, students call out the running total of items. After the last card is stacked, have students share strategies for finding the total. Repeat the activity, but have students draw a different number of items on each card. After a few rounds, tell the group to wait until all the cards are stacked before giving the total number. Have students discuss their strategies for finding the total number of items. Encourage them to explain how to use both multiplication and repeated addition to find the total. **GMP6.1**

Finding the Volume of One Stick-On Note

5.MD.3, 5.MD.3a, 5.MD.3b, 5.MD.4, SMP6

Math Masters, p. 26; 1 stick-on note; 1 unused pad of stick-on notes; centimeter cubes

To apply their understanding of finding the volume of a rectangular prism, students determine the volume of a single stick-on note. Have them record their work on *Math Masters*, page 26. If time permits, invite volunteers to share their answers and solution strategies. **GMP6.1** For an additional challenge, have students find the volume of one sheet of notebook paper.

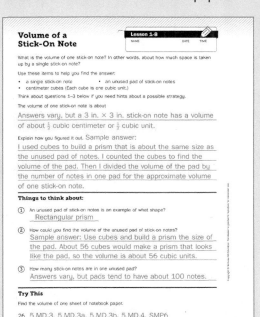

Rolling for Prisms

5.MD.3, 5.MD.3a, 5.MD.3b, 5.MD.4, SMP2

Activity Card 9; *Math Masters*, p. 27; per group: centimeter cubes; two 6-sided dice

For additional experience measuring volume by iterating layers of cubes, students roll dice to determine the dimensions of prisms. They use cubes to partially build the prisms and then calculate and compare volumes. **GMP2.1**

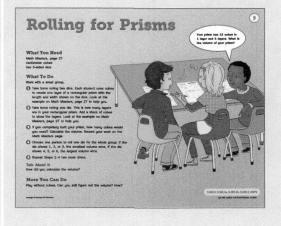

English Language Learner

Beginning ELL Use visual aids and demonstrations to explain the meaning of the term *layer*. Display pictures or segments of a cooking video showing layers in food items, like casseroles or desserts. Speak in simple sentences like the following: *First, there is a layer of pasta. Then there is a layer of sauce. Next there is a layer of cheese.* Ask questions with one-word answers like: *What is in the next layer? What is in the top layer?* Have students complete and repeat simple sentences like this one: "The next layer will have _____."

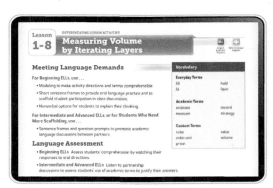

Differentiation Support pages are found in the online Teacher's Center.

Standards and Goals for
Mathematical Process and Practice

SMP1 **Make sense of problems and persevere in solving them.**
GMP1.2 Reflect on your thinking as you solve your problem.

SMP3 **Construct viable arguments and critique the reasoning of others.**
GMP3.2 Make sense of others' mathematical thinking.

SMP6 **Attend to precision.**
GMP6.1 Explain your mathematical thinking clearly and precisely.

Math Journal 1, Activity Sheet 2

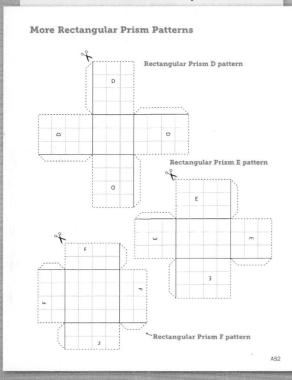

More Rectangular Prism Patterns

Rectangular Prism D pattern

Rectangular Prism E pattern

Rectangular Prism F pattern

AS2

---

① Warm Up   5 min

► Mental Math and Fluency

Display expressions and ask students to show thumbs-up for yes or thumbs-down for no. Encourage them to share how they know. *Leveled exercises:*

◐○○ Is the value of the expression greater than 4×5?
$2 + (4 \times 5)$ Thumbs-up
$2 \times (4 \times 5)$ Thumbs-up
$(4 \times 5) - 2$ Thumbs-down

◐◐○ Is the value of the expression twice as large as $8 + 17$?
$2 \times (8 + 17)$ Thumbs-up
$(8 + 17) \times 2$ Thumbs-up
$2 / (8 + 17)$ Thumbs-down

◐◐◐ Is the value of the expression less than 12×4?
$1 \div (12 \times 4)$ Thumbs-up
$(12 \times 4) \div 4$ Thumbs-up
$(12 \times 4) \times (4 \div 2)$ Thumbs-down

② Focus   35–40 min

► Math Message

Math Journal 1, Activity Sheet 2

Cut out and assemble Rectangular Prisms D, E, and F. Take 25 cubes. Estimate how many cubes will fit in each prism. Record your estimates on your slate.

► Using Layers of Cubes to Measure Volume

| WHOLE CLASS | SMALL GROUP | PARTNER | INDEPENDENT |

Math Message Follow-Up Invite students to share their estimation strategies. Explain that when mathematicians have a new problem to solve, they think about how they have solved similar problems in the past. GMP1.2 Ask: *Did you find it easier to make estimates today than in the previous lesson? Why or why not?* Sample answer: It was easier today because we had a better idea of how many cubes fit in a prism.

Ask students to check their volume estimate for Prism D by filling it with cubes. When they realize that there are not enough cubes to fill the prism, ask how they could use the cubes to find the volume of the prism, even without enough cubes to completely fill it. If necessary, prompt students to think about the layers inside the prism and how they might be used. Tell students to arrange cubes in a single layer on the bottom of Prism D. Ask: *How many cubes fit in a single layer?* 9 cubes *If we put another layer of cubes on top of the first layer, how many cubes would that be all together?* 18 cubes *How many layers fit in this prism?* 4 layers *How do you know?* The prism is 4 cubes tall, so 4 layers of cubes would fit. *How many cubes would fill the prism?* 36 cubes *Explain your strategy.* Sample answers: I thought about $9 + 9 + 9 + 9 = 36$. I multiplied the number of cubes in one layer by the number of layers. $9 * 4 = 36$

Emphasize that the number of cubes, 36, is equal to the volume of the prism in cubic units. Remind the class that in the previous lesson they counted individual cubes to find volume. In this lesson they will think about layers of cubes as they measure volume.

Display a table similar to the one below. Complete the row for Prism D as a class.

Prism	Number of Cubes in 1 Layer	Number of Layers	Total Number of Cubes That Fill the Prism
D	9	4	36
E	12	3	36
F	16	2	32

Have partnerships use the layering strategy with Prism E. They should find the number of cubes that will fit in one layer and then think about how many layers of cubes would fill the prism. Have students record their results on slates. Invite a few students to explain their thinking and share their results. **GMP6.1** Discuss the solution and strategies until everyone is in agreement. Then have the class repeat the process with Prism F. Ask: *What is the volume of Prism F?* The volume of Prism F is 32 cubic units. *Does Prism E or Prism F have the greater volume? How much greater?* The volume of Prism E is 4 cubic units greater than the volume of Prism F.

Academic Language Development

Promote students' understanding and use of mathematical terms by having them define terms after they have had experiences with them. In this lesson, have partnerships collaborate to define the term *cubic unit*, using the 4-Square Graphic Organizer (*Math Masters*, page TA2) to draw a picture, describe an example, describe a non-example, and write a definition for the term.

Common Misconception

Differentiate When determining how many layers a prism contains, some students may count only the cubes that appear above the base layer, neglecting to include the bottom cube. Suggest thinking about layers as different floors in an apartment building. Point to the top cube in the vertical stack and ask the student to describe it in terms of which floor it represents. Suggest the phrase "Don't forget the ground floor!" as a reminder for students to count all the layers. Go Online Differentiation Support

Math Journal 1, p. 22

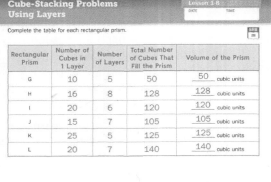

Cube-Stacking Problems Using Layers

Lesson 1-8

DATE   TIME

Complete the table for each rectangular prism.

Rectangular Prism	Number of Cubes in 1 Layer	Number of Layers	Total Number of Cubes That Fill the Prism	Volume of the Prism
G	10	5	50	50 cubic units
H	16	8	128	128 cubic units
I	20	6	120	120 cubic units
J	15	7	105	105 cubic units
K	25	5	125	125 cubic units
L	20	7	140	140 cubic units

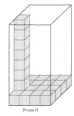

Prism G   Prism H

22 5.MD.3, 5.MD.3a, 5.MD.3b, 5.MD.4

Math Journal 1, p. 23

Cube-Stacking Problems Using Layers (continued)

Lesson 1-8

DATE   TIME

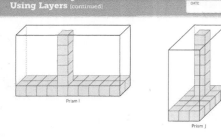

Prism I   Prism J

Try This

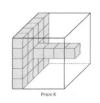
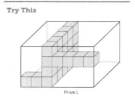

Prism K   Prism L

5.MD.3, 5.MD.3a, 5.MD.3b, 5.MD.4   23

▶ Using Layers to Solve Cube-Stacking Problems

Math Journal 1, pp. 22–23

| WHOLE CLASS | SMALL GROUP | PARTNER | INDEPENDENT |

As a class, discuss the drawing of Prism G on journal page 22. Have students complete the first row of the table and then reflect on their thinking with a partner. **GMP1.2** Invite several partnerships to share their solutions and strategies. In the discussion, emphasize how calculating volume is related to both multiplication and addition. Some students may think about finding the volume of Prism G by adding: 10 + 10 + 10 + 10 + 10. Others may find the volume by multiplying 10 * 5.

Have students complete journal pages 22 and 23. Remind them that each prism has at least one stack of cubes that goes up to the top. Provide centimeter cubes for students who may wish to use them.

 Assessment Check-In 5.MD.3, 5.MD.3b, 5.MD.4

Math Journal 1, pp. 22–23

Expect most students to find the number of cubes in 1 layer and use either repeated addition or multiplication to find the volumes of Prisms H, I, and J on journal pages 22 and 23. Some students may be able to find the volumes of Prisms K and L. For students who struggle, translate the problem into a real-world scenario. For example: *Pretend you are a builder. Prism J is your project and the first floor is completed. Each cube is a room. How many rooms are on the first floor? You build a second floor. How many rooms are in the building now?* Continue until all 7 floors are built, concluding with a statement like this: *The building has _____ rooms. Its volume is _____ room units.* Encourage students who excel to try the Enrichment activity.

 Evaluation Quick Entry Go online to record students' progress and to see trajectories toward mastery for these standards.

Summarize Ask students to choose one problem from the journal pages and explain to a partner how they solved it. **GMP6.1** Invite a few volunteers to explain their partner's strategy. **GMP3.2** Remind students that whether they count individual cubes or add or multiply the number of cubes in a single layer, the result is the same: they are calculating volume in cubic units.

③ Practice 20–30 min

▶ Converting Measurements

Math Journal 1, p. 21

| WHOLE CLASS | SMALL GROUP | PARTNER | INDEPENDENT |

Students convert units of measurement in mathematical and real-world contexts.

▶ Math Boxes 1-8: Preview for Unit 2

Math Journal 1, p. 24

WHOLE CLASS	SMALL GROUP	PARTNER	INDEPENDENT

Mixed Practice Math Boxes 1-8 are paired with Math Boxes 1-13. These problems focus on skills and understandings that are prerequisite for Unit 2. You may want to use information from these Math Boxes to plan instruction and grouping in Unit 2.

▶ Home Link 1-8

Math Masters, p. 28

Homework Students determine whether partially filled boxes are big enough to hold a specified number of cubes.

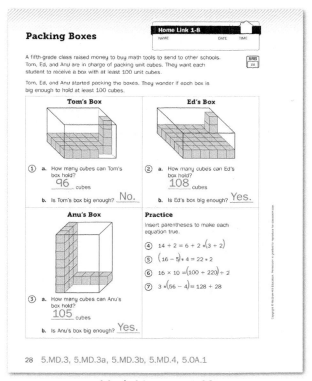

Math Masters, p. 28

Planning Ahead

Collect Prisms E and F from students. They will be reused in Lesson 1-9.

Math Journal 1, p. 21

Converting Measurements

Lesson 1-8
DATE   TIME

Solve. If necessary, look up measurement equivalents in the *Student Reference Book.*

① Record measurement equivalents in the 2-column tables below.

Minutes	Seconds
5	300
10	600
15	900
20	1,200
25	1,500

Kiloliters	Liters
5	5,000
$1\frac{1}{2}$	1,500
$2\frac{1}{2}$	2,500
4	4,000
$3\frac{1}{2}$	3,500

Miles	Yards
1	1,760
2	3,520
3	5,280
4	7,040
5	8,800

② Students recorded their running distances over the weekend using different units. Complete the table to convert them to the same units.

	Kilometers	Meters
Jason	3	3,000
Kayla	$4\frac{1}{2}$	4,500
Lohan	$2\frac{1}{2}$	2,500
Malik	5	5,000
Jada	$3\frac{1}{2}$	3,500

③ Jordan needed to convert measurements for his recipes. Fill in the blanks.

2 quarts milk = __8__ cups   __4__ cups pasta = 2 pints

32 oz flour = __2__ lb $2\frac{1}{2}$ cups water = __20__ fl oz

8 cups rice = __4__ pints $\frac{1}{2}$ cup oil = __8__ tbs

④ Mahalia is making cloth napkins. She bought fabric that is 12 inches wide and 6 yards long.

How many napkins can Mahalia make if each napkin is 1 foot by 1 foot? __18__ napkins

5.MD.1   21

Math Journal 1, p. 24

Math Boxes: Preview for Unit 2

Lesson 1-8
DATE   TIME

① What is the value of the 8 in the following numbers?

a. 1,384 __80__
b. 8,294 __8,000__
c. 418 __8__
d. 6,897 __800__

② Solve.

a. 3 * 10 = __30__
b. 3 * 100 = __300__
c. 3 * 1,000 = __3,000__
d. 30 * 10 = __300__

③ Write four multiples of 4.

Sample answers:
__24__ __16__ __40__ __28__

④ Solve.

a.   2 3   b.   2 4 2
    *   3   *   2
    ———   ———
      6 9   4 8 4

⑤ Solve.

How many 10s in 30? __3__
How many 10s in 300? __30__
How many 7s in 21? __3__
How many 7s in 210? __30__

⑥ Solve.

50 + 6 = __56__
300 + 20 = __320__
200 + 50 + 6 = __256__

① 5.NBT.1 ② 5.NBT.2 ③ 5.NBT.6 ④ 5.NBT.5 ⑤ 5.NBT.6
⑥ 5.NBT.1

24

Lesson 1-9

Two Formulas for Volume

Overview Students explain and apply two different formulas for finding the volume of a rectangular prism.

► **Before You Begin**

For Part 2 students will need Prisms E and F from Lesson 1–8. If students do not have them, make one copy of *Math Masters,* page 25 for each student and have them cut out and assemble the prisms before the lesson begins. Place a box of centimeter cubes near the Math Message.

Standards

Focus Cluster
• Geometric measurement: understand concepts of volume.

Warm Up 5 min	**Materials**	
Mental Math and Fluency Students evaluate expressions with grouping symbols.	slate	5.OA.1

② Focus 35–40 min		
Math Message Students use cubes to find the volume of a rectangular prism and compare strategies with a partner.	Prisms E and F (*See Before You Begin*.), 40 centimeter cubes	5.MD.3, 5.MD.3a, 5.MD.3b, 5.MD.4
Finding a Formula for Volume Students discuss how to find the height of a rectangular prism and the area of its base without using cubes. They use this information to generalize the formula $V = B \times h$.	*Math Masters,* p. TA2 (optional); Prisms E and F; 40 centimeter cubes; ruler	5.MD.3, 5.MD.3a, 5.MD.3b, 5.MD.4, 5.MD.5, 5.MD.5a, 5.MD.5b SMP3, SMP7, SMP8
Finding a Second Volume Formula Students discuss the relationship between $V = B \times h$ and $V = l \times w \times h$. They apply both formulas to find volumes of rectangular prisms.	*Math Journal 1,* p. 25; Prisms E and F, ruler	5.MD.5, 5.MD.5a, 5.MD.5b SMP3, SMP8
✓ **Assessment Check-In** See page 71. Expect most students to be able to find the volumes of rectangular prisms by referring to the formulas displayed in the lesson and multiplying the dimensions shown.	*Math Journal 1,* p. 25	5.MD.5, 5.MD.5b

③ Practice 20–30 min		
Writing and Interpreting Expressions Students write expressions that model real-world and mathematical contexts.	*Math Journal 1,* p. 26	5.OA.1, 5.OA.2 SMP4
Math Boxes 1-9 Students practice and maintain skills.	*Math Journal 1,* p. 27	See page 71.
Home Link 1-9 **Homework** Students use formulas to find volumes of prisms.	*Math Masters,* p. 30	5.NF.4, 5.NF.4b, 5.MD.5, 5.MD.5b

 Go Online to see how mastery develops for all standards within the grade.

Differentiation Options

Readiness 5–15 min

WHOLE CLASS | SMALL GROUP | PARTNER | INDEPENDENT *

Reviewing an Area Formula

5.MD.5, 5.MD.5b, SMP1, SMP2

To prepare for using volume formulas, students review how to use the formula for the area of a rectangle. Display the formula $A = l \times w$ and ask students to explain what each letter means. Draw and label several different rectangles, asking students to identify where the labeled information belongs in the formula. Be sure to include an example in which the area of the rectangle is known, but one side length is not. Have students solve for the missing information. *For example:* Draw a rectangle labeled with an area of 36 cm² and a width of 9 cm. Ask students to rewrite the formula with the known information filled in. 36 cm² = $l \times 9$ cm Then ask students to solve for the missing length. 4 cm As you work through various examples, invite students to share how they knew where the information from the drawing belonged in the formula. **GMP1.1, GMP2.2**

Enrichment 5–15 min

WHOLE CLASS | SMALL GROUP | PARTNER | INDEPENDENT

Finding Dimensions for a Given Volume

5.MD.5, 5.MD.5a, 5.MD.5b, SMP1, SMP3

Activity Card 10; 24 centimeter cubes

To apply their understanding of volume, students find all possible combinations of base area and height for a rectangular prism with a volume of 24 cm³. **GMP1.5**
$B = 24$ cm², $h = 1$ cm; $B = 12$ cm², $h = 2$ cm; $B = 8$ cm², $h = 3$ cm; $B = 6$ cm², $h = 4$ cm; $B = 4$ cm², $h = 6$ cm; $B = 3$ cm², $h = 8$ cm; $B = 2$ cm², $h = 12$ cm; $B = 1$ cm², $h = 24$ cm
They consider whether prisms with the same base area but different lengths and widths are still the same prism. **GMP3.1**

Extra Practice 5–15 min

WHOLE CLASS | SMALL GROUP | PARTNER | INDEPENDENT

Using Volume Formulas

5.MD.5, 5.MD.5a, 5.MD.5b

Math Masters, p. 29

For additional practice using volume formulas, students find the volumes of rectangular prisms and list possible dimensions of rectangular prisms with a given volume.

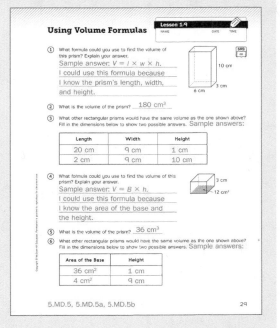

English Language Learner

Beginning ELL Extend students' comprehension of the word *find* beyond the everyday association with something being lost. Use the phrase *solve for the answer* as you point to the blank space or to the *V* (volume) before or after the equal sign so that students will understand the directions *finding the volume* as meaning *to solve for the answer.*

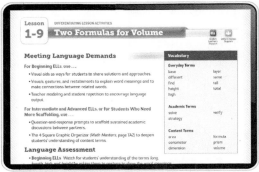

Differentiation Support pages are found in the online Teacher's Center.

Standards and Goals for
Mathematical Process and Practice

SMP3 **Construct viable arguments and critique the reasoning of others.**

GMP3.1 Make mathematical conjectures and arguments.

SMP7 **Look for and make use of structure.**

GMP7.1 Look for mathematical structures such as categories, patterns, and properties.

SMP8 **Look for and express regularity in repeated reasoning.**

GMP8.1 Create and justify rules, shortcuts, and generalizations.

Academic Language Development

Have partnerships use the 4-Square Graphic Organizer (*Math Masters*, page TA2) to define the term *formula*. Tell them to use the headings: My Definition, Number Example, Visual Representation, and Reminds Me Of. Encourage partnerships to compare their definitions with one from another partnership and then compare their definitions with the one in the *Student Reference Book*. To scaffold for academic conversation, provide a sentence frame, such as: "Our definition is like/unlike the *Student Reference Book*'s because _____."

---

1 Warm Up   5 min

▶ Mental Math and Fluency

Display the following expressions for students to evaluate. Have them record their answers on slates. *Leveled exercises:*

● ○ ○ $4 \times (6 - 2)$ 16
$(4 \times 6) - 2$ 22
$(4 - 2) \times 6$ 12

● ● ○ $(10 \times 4) \times 2$ 80
$10 \times (4 \times 2)$ 80
$48 + (12 \times 2)$ 72

● ● ● $(6 \times 5 \times 2) \div 5$ 12
$(15 \times 2) \div (5 \times 6)$ 1
$(21 \div 7) + (36 \div 6)$ 9

2 Focus   35–40 min

▶ Math Message

Take 40 cubes and Prisms E and F from Lesson 1-8. Find the volume of Prism E. Talk with a partner about the strategy you used to find the volume. Did you and your partner use the same strategy?

▶ Finding a Formula for Volume

| WHOLE CLASS | SMALL GROUP | PARTNER | INDEPENDENT |

Math Message Follow-Up Display one of the cubes you placed near the Math Message. Tell students that it is called a centimeter cube because all of its edges are 1 centimeter in length. Have students use a ruler to confirm that every edge is 1 centimeter long. Ask: *What is the volume of this cube? How do you know?* 1 cm³. It is a unit cube with a length of 1 cm on each edge, so its volume is 1 cubic centimeter, or 1 cm³.

Verify the volume of Prism E. 36 cubic units Point out that because 1 cube is 1 cubic centimeter, we can express this volume as 36 cm³. Ask several students to share their strategies for finding the volume of the prism. Make sure the strategies shared include counting the total number of cubes and counting the number of cubes in one layer and multiplying by the number of layers. Have students confirm that both strategies result in the same measure of volume. Tell them that in today's lesson they will learn another way to find the volumes of rectangular prisms.

Remind the class that one of the strategies they used in previous lessons to determine the volume of a prism involved finding the number of cubes in a single layer and multiplying that number by the total number of layers in the prism. Display the following:

Volume = number of cubes in 1 layer × number of layers

Ask students if they can think of a way they could find the number of cubes in one layer of Prism E without using cubes. Suggest examining the inside of Prism E. Ask: *What do you notice about the squares on the base?* GMP7.1 Sample answer: Each square is the same size as one face of a centimeter cube. The squares are 1 centimeter on each side, so they are each 1 square centimeter. *How might counting the number of squares help us find the volume of the prism without packing it?* Sample answer: One centimeter cube fits on each square, so we can count the squares to figure out how many cubes would be in one layer. Remind the class that the number of centimeter squares covering the base of the prism is the area of the base of the prism. Ask: *Will finding the area of the base always tell us how many cubes are in one layer? How do you know?* Yes. The area is the number of square centimeters on the base, and 1 cube fits on each square centimeter. Write "area of base" below "number of cubes in 1 layer." (*See below.*)

Ask: *What else do we need to know to find the volume?* The number of layers *How could we find the number of layers in Prism E without using cubes?* Measure the height of the prism *Why does this work?* GMP7.1 One cube is 1 centimeter tall, so each layer is 1 centimeter in height. The total height of Prism E in centimeters is the same as the number of layers that would fit in the prism. Write "height" below "number of layers" so that your display looks like the following:

Volume = number of cubes in 1 layer × number of layers

Volume =   area of base   ×   height

Ask students to multiply the area of the base by the height of Prism E and verify that the product is the same as what they found for the volume of Prism E in the Math Message. $12 \text{ cm}^2 \times 3 \text{ cm} = 36 \text{ cm}^3$. The volume of the prism is 36 cm^3. Summarize by asking: *Why does multiplying the area of the base by the height of a prism give the volume?* GMP3.1, GMP8.1 Sample answer: The area of the base gives the number of cubes that fit in a single layer of the prism. Because the height is the number of layers, multiplying the area of the base by the height gives the total number of cubes that can pack the prism.

Display the formula $V = B \times h$ for students, making sure they understand what each letter represents. Have them examine Prism F to find the area of the base and measure its height. After they have applied the formula to find the volume, have them confirm their findings by packing Prism F with centimeter cubes. Area of the base: 16 cm^2; height: 2 cm; volume: 32 cm^3

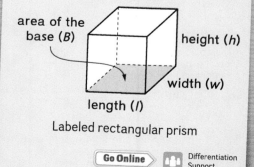

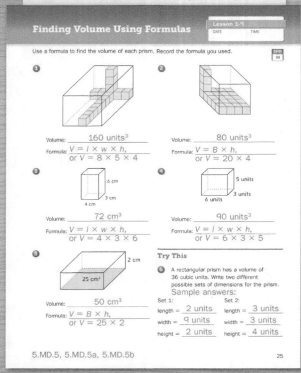

Finding Volume Using Formulas

Lesson 1-9
DATE   TIME

Use a formula to find the volume of each prism. Record the formula you used.

① Volume: 160 units³
Formula: $V = l \times w \times h$,
or $V = 8 \times 5 \times 4$

② Volume: 80 units³
Formula: $V = B \times h$,
or $V = 20 \times 4$

③ 6 cm, 3 cm, 4 cm
Volume: 72 cm³
Formula: $V = l \times w \times h$,
or $V = 4 \times 3 \times 6$

④ 5 units, 3 units, 6 units
Volume: 90 units³
Formula: $V = l \times w \times h$,
or $V = 6 \times 3 \times 5$

⑤ 2 cm, 25 cm²
Volume: 50 cm³
Formula: $V = B \times h$,
or $V = 25 \times 2$

Try This

⑥ A rectangular prism has a volume of 36 cubic units. Write two different possible sets of dimensions for the prism.
Sample answers:

Set 1:
length = 2 units
width = 9 units
height = 2 units

Set 2:
length = 3 units
width = 3 units
height = 4 units

5.MD.5, 5.MD.5a, 5.MD.5b   25

Give students a variety of dimensions of rectangular prisms and have them find the volume using the formula $V = B \times h$. Record the dimensions and the volume.

Suggestions:

Area of the Base (B)	Height (h)	Volume (V)
10 units²	8 units	80 units³
4 units²	5 units	20 units³
30 cm²	7 cm	210 cm³

▶ Finding a Second Volume Formula

Math Journal 1, p. 25

| WHOLE CLASS | SMALL GROUP | PARTNER | INDEPENDENT |

Once most students seem comfortable using the formula $V = B \times h$, ask: *What shape is the base of the prism?* Rectangle *What is the formula for finding the area of a rectangle?* Area = $l \times w$ Display the formula $V = B \times h$ and the formula $V = l \times w \times h$. Ask: *What is the relationship between these two formulas?* GMP3.1, GMP8.1 Sample answer: Since $l \times w$ is the same as the area of the base (B), the two formulas mean the same thing. They both give you the volume of a prism.

Have students use rulers to measure the length, width, and height of Prisms E and F. List the dimensions and direct students to apply the formula $V = l \times w \times h$. Confirm that the volume they calculate this time is the same as the volume they calculated using the formula $V = B \times h$.

Prism E: length = 4 cm; width = 3 cm; height = 3 cm; $4 \times 3 \times 3 = 36$ cm³
Prism F: length = 4 cm; width = 4 cm; height = 2 cm; $4 \times 4 \times 2 = 32$ cm³

Give the dimensions of several prisms and have students use the dimensions to find the volume. Invite volunteers to share which formula they used. *Suggestions:*

- length = 8 cm; width = 3 cm; height = 5 cm 120 cm³; $V = l \times w \times h$
- height = 7 units; area of the base = 10 square units 70 cubic units; $V = B \times h$
- width = 4 units; height = 10 units; length = 7 units 280 units³; $V = l \times w \times h$

When most students seem comfortable using the volume formulas, have them work in partnerships or small groups to complete journal page 25, where they use formulas to find the volume of a variety of rectangular prisms. When most students have finished, review their strategies.

Writing and Interpreting Expressions

Lesson 1-9
DATE   TIME

Write an expression that models the calculation described in words.

① The sum of 13 and 12, which is then multiplied by 2.
$(13 + 12) * 2$

② Divide 16 by 4 and add the sum of 3 and 8 to the quotient.
$(16 / 4) + (3 + 8)$

③ Multiply 12 and 6 and divide the product by 9.
$12 * 6 / 9$, or $(12 * 6) / 9$

Without calculating, circle the expression with the greater value.

④ $3 * (126 + 12)$ $6 * (126 + 12)$

⑤ $(18 - 8) / 2$ $(18 - 8) / 5$

⑥ Explain how you knew which expression had a greater value in Problem 5.
Sample answer: If I divide by 5, that creates more equal parts than if I divide by 2, so I know the parts will be smaller. Dividing by 2 will give a larger quotient.

⑦ Ivan was playing a video game. He had 1,300 points and on the next level earned 120 more. Then he lost 12 points. When his turn ended, his score doubled. Write an expression that shows the number of points Ivan has at the end of his turn.
$(1,300 + 120 - 12) * 2$

⑧ Write a situation that can be modeled by the expression 6 * (24 − 5).
Sample answer: Sam earns 24 dollars each week at his job. He keeps 5 dollars and puts the rest in his savings account. After 6 weeks, how much money does Sam have in his savings account?

26   5.OA.1, 5.OA.2, SMP4

Assessment Check-In   5.MD.5, 5.MD.5b

Math Journal 1, p. 25

Expect most students to be able to find the volume of the prisms in Problems 1–5 on journal page 25 by referring to the formulas displayed in the lesson and multiplying the dimensions shown. If students have difficulty deciding which formula to use, suggest that they list the information they know and match their lists to the two formulas. For a challenge, have students list as many dimensions as they can for Problem 6.

Evaluation Quick Entry Go online to record students' progress and to see trajectories toward mastery for these standards.

Summarize Ask partners to discuss how the two volume formulas they learned are similar and different.

③ Practice   20–30 min

▶ Writing and Interpreting Expressions

Math Journal 1, p. 26

| WHOLE CLASS | SMALL GROUP | PARTNER | INDEPENDENT |

Students write expressions to model real-world and mathematical situations. They describe situations that could be modeled by an expression.   GMP4.1

▶ Math Boxes 1-9

Math Journal 1, p. 27

| WHOLE CLASS | SMALL GROUP | PARTNER | INDEPENDENT |

Mixed Practice Math Boxes 1-9 are paired with Math Boxes 1-11.

▶ Home Link 1-9

Math Masters, p. 30

Homework Students use formulas to find the volume of rectangular prisms.

Math Journal 1, p. 27

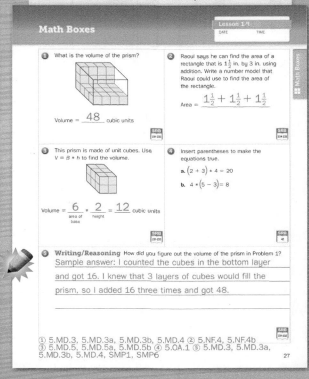

Math Masters, p. 30

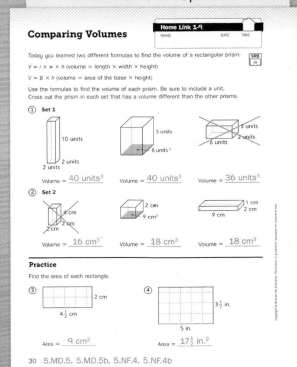

Visualizing Volume Units

Overview Students explore units of volume and convert between them.

▶ **Before You Begin**

For Part 2, assemble a partial frame of a cubic meter by placing two metersticks on a flat surface at right angles to each other and using packing tape to hold them in place. Tape a third meterstick perpendicular to the other two so that all three metersticks meet in one corner. See the illustration on page 76.

Standards

Focus Clusters
- Convert like measurement units within a given measurement system.
- Geometric measurement: understand concepts of volume.

1 Warm Up 5 min

Materials

Mental Math and Fluency Students convert units of length.		5.MD.1

2 Focus 35–40 min

Math Message Students compare units of length, area, and volume.		5.MD.3, 5.MD.3a
Understanding Cubic Units Students generate a list of length, area, and volume units and discuss the relationships between them.	centimeter cube	5.MD.3, 5.MD.3a SMP6
Visualizing Cubic Units Students determine the number of cubic centimeters in a cubic decimeter and the number of cubic decimeters in a cubic meter.	centimeter cube; 3 metersticks; per partnership: 10 cubes, 10 longs, 10 flats	5.MD.1, 5.MD.3, 5.MD.3a, 5.MD.3b, 5.MD.4 SMP6
Converting Volume Units Students convert between cubic units.	*Math Journal 1*, pp. 28–29	5.MD.1, 5.MD.3, 5.MD.3a, 5.MD.3b, 5.MD.4, 5.MD.5, 5.MD.5b, SMP6
✓ **Assessment Check-In** See page 76. Expect most students to understand the relative size of the units of volume given in the identified problems and to name objects with volumes they might reasonably measure with each unit.	*Math Journal 1*, pp. 28–29; rulers (optional)	5.MD.1, 5.MD.3, 5.MD.3a

3 Practice 20–30 min

More Cube-Stacking Problems Students find the volume of partially packed prisms.	*Math Journal 1*, pp. 30	5.MD.3, 5.MD.3a, 5.MD.3b, 5.MD.4, SMP2
Math Boxes 1-10 Students practice and maintain skills.	*Math Journal 1*, pp. 31	See page 77.
Home Link 1-10 **Homework** Students find objects at home that they might measure in a given unit of volume.	*Math Masters*, p. 33	5.MD.3, 5.MD.3a, 5.MD.5, 5.MD.5b SMP6

⟋⟋⟋ [**Go Online** ⟩ to see how mastery develops for all standards within the grade.

Differentiation Options

Readiness | 5–15 min

WHOLE CLASS | SMALL GROUP | PARTNER | INDEPENDENT

Converting Linear Measurements

5.MD.1, SMP5, SMP6

Math Masters, p. 31;
measuring tools (optional)

To review the relative size of measurement units, students complete a chart of measurement equivalents. They list objects they would measure with a given unit of length. **GMP6.2** Have students work in small groups or partnerships to complete the tables of equivalents. Consider having them use measuring tools like rulers, tape measures, metersticks, or yardsticks to help them find objects to measure with a given unit of length. **GMP5.1**

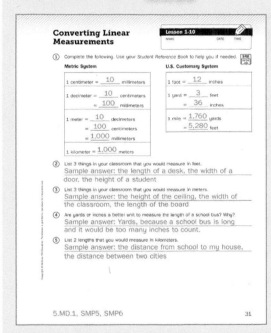

Enrichment | 15–30 min

WHOLE CLASS | SMALL GROUP | PARTNER | INDEPENDENT

Packing Cubes in a Box

5.NBT.7, 5.MD.1, 5.MD.5, 5.MD.5b, SMP6

Math Masters, p. 32; calculator

To extend their work converting between units of volume, students solve a problem involving converting between cubic inches and cubic centimeters. They explain their thinking using clear labels, units, and mathematical language. **GMP6.3**

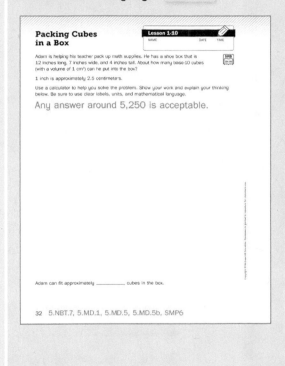

Extra Practice | 15–30 min

WHOLE CLASS | SMALL GROUP | PARTNER | INDEPENDENT

Estimating the Volume of Your Classroom

5.MD.1, 5.MD.5, 5.MD.5b, SMP6

Activity Card 11, per partnership: tape measure or yardstick

To practice choosing volume units, finding volume, and converting volume units, students choose a unit of volume and estimate the volume of the classroom in that unit. **GMP6.2** They convert the volume measurement to a different unit.

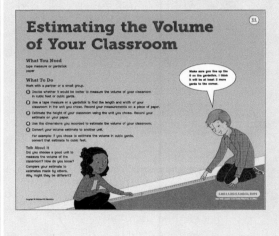

English Language Learner

Beginning ELL For the Math Message, aid students by showing a labeled display. For example, use a ruler with a centimeter highlighted and labeled with the words *centimeter* and *length*. Similarly, show centimeter grid paper with a square centimeter highlighted and labeled with the words *square centimeter* and *area*. Finally, show a centimeter cube labeled with the words *cubic centimeter* and *volume*. Point to the different displays, model naming them aloud, and have students chorally repeat the terms.

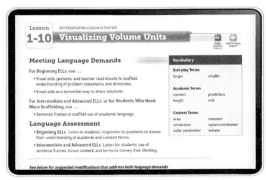

Differentiation Support pages are found in the online Teacher's Center.

Academic Language Development

Highlight or underline the base word *meter* in the word *centimeter* and circle the prefix *centi-*. Ask students what part of the prefix *centi-* is familiar. Cent Then ask them to think about other words in which they *see* cent or *centi*. Sample answers: centipede, century Ask: *How many cents make a dollar?* 100 *How many years are in a century?* 100 Build on this knowledge to explain the meaning of the prefix *centi-* as $\frac{1}{100}$, making a centimeter a hundredth of a meter. You may wish to create a prefix chart and add terms as they come up in future lessons. Include columns for the prefix, the meaning of the prefix, and examples of words where it appears.

1 Warm Up · 5 min

▶ Mental Math and Fluency

Have students give measurement equivalencies for length.
Leveled exercises:

How many inches are in . . .

● ○ ○ 1 foot? 12

● ● ○ $\frac{1}{2}$ foot? 6

● ● ● $2\frac{1}{2}$ feet? 30

How many feet are in . . .

● ○ ○ 1 yard? 3

● ● ○ $\frac{1}{2}$ yard? $1\frac{1}{2}$

● ● ● $6\frac{1}{2}$ yards? $19\frac{1}{2}$

How many meters are in . . .

● ○ ○ 1 kilometer? 1,000

● ● ○ $\frac{1}{2}$ kilometer? 500

● ● ● $3\frac{1}{2}$ kilometers? 3,500

How many centimeters are in . . .

● ○ ○ 1 decimeter? 10

● ● ○ $\frac{1}{2}$ decimeter? 5

● ● ● $4\frac{1}{2}$ decimeters? 45

2 Focus · 35–40 min

▶ Math Message

A unit you can use to measure length is a centimeter. A unit you can use to measure area is a square centimeter. A unit you can use to measure volume is a cubic centimeter. Talk with a partner about how these units are similar and how they are different. Record your ideas to share.

▶ Understanding Cubic Units

| WHOLE CLASS | SMALL GROUP | PARTNER | INDEPENDENT |

Math Message Follow-Up Ask several students to share their responses to the Math Message. As they discuss the units, hold up a centimeter cube and use it to show the length 1 centimeter (the length of an edge), the area 1 square centimeter (the area of a face), and the volume 1 cubic centimeter (the volume of the cube). Display the notation for each unit (cm, cm², cm³) and make sure students are comfortable with each one. **GMP6.3**

Create a table like the one below. Ask students to name other units they use to measure length. Sample answers: inch, meter, foot, kilometer, yard Record each unit in the table. After the class has generated a list of length units, have partners talk about which area and volume units they think correspond to each length unit. Record the units in the table.

Length	Area	Volume
inch	square inch	cubic inch
meter	square meter	cubic meter

Ask questions to make sure students understand the relationship between units of length, area, and volume. *Suggestions:*

- *What is the length of each side of a square meter?* 1 meter
- *If you have a cubic kilometer, what is the length of each edge of the cube?* 1 kilometer *What is the area of each face of the cube?* 1 square kilometer

Tell students that today they are going to explore the sizes of various volume units, think about the most appropriate units for measuring volume, and discuss the relationship between units of volume.

▶ Visualizing Cubic Units

WHOLE CLASS   SMALL GROUP   PARTNER   INDEPENDENT

Display a centimeter cube and ask: *What objects have volumes that it makes sense to measure in cubic centimeters?* GMP6.2 Sample answers: pencil box, granola bar, food storage container, juice box Next help students think about the relative size of a cubic decimeter. Ask: *How many centimeters are in a decimeter?* 10 *Is a cubic decimeter larger or smaller than a cubic centimeter?* Larger *How many cubic centimeters do you think are in a cubic decimeter?* Answers vary. Tell students that now they will work together to figure out the answer.

Have partners or small groups line up centimeter cubes in a row to show a decimeter. Confirm that students have used 10 centimeter cubes and demonstrate the equivalence in length of 10 cubes and 1 long. Ask: *How many centimeter cubes do you think it will take to cover a square decimeter?* Answers vary. Have students test their predictions by using base-10 longs to show a square decimeter. Confirm that there are 100 square centimeters in a square decimeter.

Ask: *How many centimeter cubes do you think it would take to fill a cubic decimeter?* Answers vary. Have students test their predictions by using base-10 flats to show a cubic decimeter. Confirm that there are 1,000 cubic centimeters in a cubic decimeter. Ask: *What objects have volumes that it makes sense to measure in cubic decimeters?* GMP6.2 Answers vary, but objects should be larger than those listed for cubic centimeters, such as a suitcase, a drawer, or a shipping box.

NOTE This is the first lesson where students encounter unit conversions for volume. Do not expect most students to master this concept at this time. Support students' work with conversions, but be sure to keep the emphasis in the lesson on understanding the relative sizes of volume units. Also, help students think about real-world situations that might call for volume to be reported in a particular unit. Students will have additional practice with unit conversions in future lessons.

Common Misconception

Differentiate Some students may think that because there are 10 centimeters in a decimeter, there are just 10 square centimeters in a square decimeter or 10 cubic centimeters in a cubic decimeter. Point out that 10 square centimeters is the number of square centimeters in only 1 row of a square decimeter. To get the total number of square centimeters in all 10 rows of a square decimeter, students must multiply by 100. Similarly, they must multiply by 1,000 to get all of the rows and layers of a cubic decimeter. Continue using a variety of models to help students visualize the differences between units.

Go Online Differentiation Support

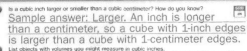

Converting Cubic Units   Lesson 1-10
                                          DATE   TIME

① Is a cubic inch larger or smaller than a cubic centimeter? How do you know?
Sample answer: Larger. An inch is longer
than a centimeter, so a cube with 1-inch edges
is larger than a cube with 1-centimeter edges.

② List objects with volumes you might measure in cubic inches.
Sample answers: a shoe box, a desk drawer

③ a. How many cubic inches do you think are in a cubic foot? Answers vary.
   b. How many inches are in a foot? 12
   c. How many square inches are in a square foot?
      144 square inches
      How did you find your answer?
      Since 12 in. = 1 ft, a square foot would be
      12 in. long and 12 in. high. 12 × 12 = 144
   d. How many cubic inches are in a cubic foot?
      1,728 cubic inches
      How did you find your answer? There are 144 square inches
      in 1 square foot, so it takes 144 cubic
      inches to cover a square foot. A cubic foot is
      12 inches high, so there would be 12 layers.
      144 × 12 = 1,728

④ List objects with volumes you might measure in cubic feet.
Answers vary, but objects should be larger
than those listed for cubic inches, such as a
closet, a locker, or the trunk of a car.

28   5.MD.1, 5.MD.3, 5.MD.3a, 5.MD.3b, 5.MD.4, SMP6

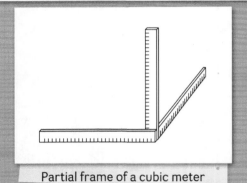

Partial frame of a cubic meter

Converting Cubic Units (continued)   Lesson 1-10
                                               DATE   TIME

⑤ How many cubic feet are in a cubic yard?
   27 cubic feet
   How did you find your answer? Sample answer: I know that
   there are 3 feet in a yard. That means that there
   are 3 × 3, or 9 square feet, in a square yard
   and 9 × 3, or 27 cubic feet, in a cubic yard.

⑥ List objects with volumes you might measure in cubic yards. Answers vary,
   but objects should be larger than those listed
   for cubic feet, such as a warehouse, a
   hallway, or a cargo ship.

⑦ Deena's family has a freezer that is 2 yards in width, 1 yard in length, and 1 yard in height.
   a. What is the volume of the freezer?
      2 cubic yards
   b. How many cubic feet of food will fit in the freezer?
      54 cubic feet
      How did you find your answer? Sample answer: I converted
      the dimensions to feet. 2 yards = 6 feet,
      and 1 yard = 3 feet. 6 × 3 × 3 = 54
      cubic feet
   c. Do you think cubic yards or cubic feet are better units to measure the volume
      of the freezer? Why?
      Answers vary.

5.MD.1, 5.MD.3, 5.MD.3a, 5.MD.3b, 5.MD.4, 5.MD.5,
5.MD.5b, SMP6   29

Ask: *How many decimeters are in 1 meter?* 10 *How many cubic decimeters do you think are in a cubic meter?* Answers vary. Guide the discussion so that students understand that there are 10 ∗ 10 ∗ 10 or 1,000 cubic decimeters in a cubic meter. Help them visualize the size of a cubic meter by displaying the partial frame you assembled. (*See margin and Before You Begin.*) Ask: *What objects have volumes that it makes sense to measure in cubic meters?* **GMP6.2** Answers vary, but objects should be larger than those listed for cubic decimeters, such as a room, a truck, or a ship.

▶ Converting Volume Units

Math Journal 1, pp. 28–29

| WHOLE CLASS | SMALL GROUP | PARTNER | INDEPENDENT |

Have students work in small groups or partnerships to complete the journal pages, where they determine the number of cubic inches in a cubic foot and the number of cubic feet in a cubic yard. They also think about what volumes it makes sense to measure using different cubic units.

When most students have finished the journal pages, invite volunteers to share some of the objects they listed for Problems 2, 4, and 6. Point to several classroom objects and ask which units students would use to measure the volume of each object. **GMP6.2** *Suggestions:*

- tissue box Cubic inches or cubic centimeters
- file cabinet drawer Cubic feet or cubic decimeters
- storage closet Cubic yards or cubic meters

 Assessment Check-In 5.MD.1, 5.MD.3, 5.MD.3a

Math Journal 1, pp. 28–29

Expect most students to understand the relative size of units of volume given on journal pages 28 and 29 and to name objects with volumes they might reasonably measure with each unit. For those who struggle, offer as many visual supports as possible. For example, consider making a frame of a cubic foot using rulers, similar to the frame of a cubic meter. Do not expect students to be fluent in converting between cubic units of measure. This is the first time they have encountered unit conversions for volume, and they will continue to practice unit conversions in upcoming lessons.

Evaluation Quick Entry Go online to record students' progress and to see trajectories toward mastery for these standards.

Summarize Ask partners to tell each other two things they learned about cubic units and measuring with cubic units.

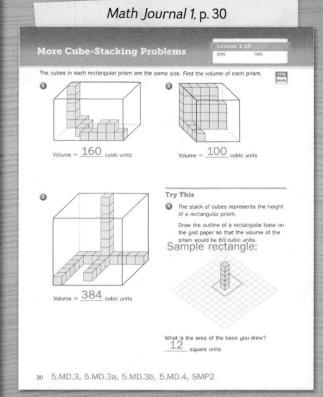

More Cube-Stacking Problems

Math Journal 1, p. 30

| WHOLE CLASS | SMALL GROUP | **PARTNER** | **INDEPENDENT** |

Students examine prisms partially filled with cubes and determine their volume. **GMP2.2**

Math Boxes 1-10

Math Journal 1, p. 31

| WHOLE CLASS | **SMALL GROUP** | PARTNER | INDEPENDENT |

Mixed Practice Math Boxes 1-10 are grouped with Math Boxes 1-6 and 1-12.

Home Link 1-10

Math Masters, p. 33

Homework Students compare units of volume. They find objects at home that they might measure with a given unit of volume. **GMP6.2**

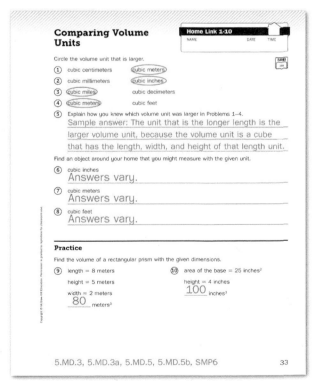

Math Masters, p. 33

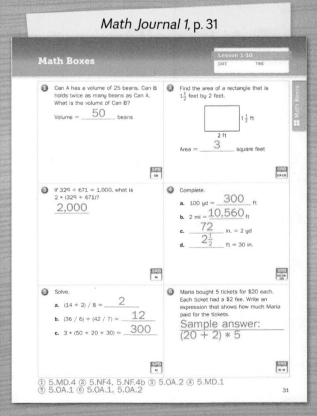

Lesson 1-11

Volume Explorations

Overview Students find volumes of figures composed of rectangular prisms and solve real-world problems involving volume.

▶ **Before You Begin**

The context of this lesson is musical instrument cases. If you think your students may not be familiar with the musical instruments referenced, consider finding photos that show each instrument. Decide how you will display *Math Masters*, page 34. Be sure calculators are available. For the optional Enrichment and Extra Practice activities, you may wish to demonstrate how to make sketches of 3-dimensional figures.

▶ **Vocabulary**

mathematical model

Standards

Focus Clusters

- Write and interpret numerical expressions.
- Convert like measurement units within a given measurement system.
- Geometric measurement: understand concepts of volume.

1 Warm Up 5 min

	Materials	
Mental Math and Fluency Students interpret numerical expressions.		5.OA.2

2 Focus 35–40 min

	Materials	
Math Message Students find the volume of a suitcase.	calculator	5.MD.5, 5.MD.5b
Modeling a Suitcase Students use a rectangular prism to model the suitcase and show that the volume of the suitcase stays the same if the suitcase is turned.	calculator; assembled prisms from Lesson 1-8 (or from *Math Masters*, p. 25, optional); centimeter cubes (optional)	5.OA.1, 5.MD.5, 5.MD.5a, 5.MD.5b SMP4, SMP7
Estimating Volumes of Musical Instrument Cases Students find volumes of figures composed of rectangular prisms by adding the volumes of the prisms.	*Math Journal 1*, p. 32; *Math Masters*, p. 34 (for display); calculator	5.MD.1, 5.MD.5, 5.MD.5b, 5.MD.5c SMP4
✔ **Assessment Check-In** See page 82. Expect most students to be able to use a strategy to break apart the model and apply a volume formula to the prisms to estimate the volume of a real-world object that can be decomposed into prisms.	*Math Journal 1*, p. 32	5.MD.5, 5.MD.5b, 5.MD.5c

3 Practice 20–30 min

	Materials	
Introducing Buzz Game Students practice finding multiples of numbers to prepare for multidigit division in Unit 2.	*Student Reference Book*, p. 294	5.NBT.6 SMP6, SMP7
Math Boxes 1-11 Students practice and maintain skills.	*Math Journal 1*, p. 33	See page 83.
Home Link 1-11 Homework Students find volumes of figures composed of rectangular prisms.	*Math Masters*, p. 36	5.MD.5, 5.MD.5b, 5.MD.5c SMP4

  **Go Online** to see how mastery develops for all standards within the grade.

Differentiation Options

Readiness	5–15 min	Enrichment	15–30 min	Extra Practice	5–15 min
WHOLE CLASS **SMALL GROUP** PARTNER INDEPENDENT		WHOLE CLASS **SMALL GROUP** PARTNER INDEPENDENT		WHOLE CLASS **SMALL GROUP** PARTNER INDEPENDENT	

Using Cubes to Find Volumes

`5.MD.3, 5.MD.3a-b, 5.MD.4, 5.MD.5, 5.MD.5b-c`

centimeter cubes

To explore adding volumes using a concrete model, students build figures with centimeter cubes. Distribute a handful of centimeter cubes to each student. Have them build structures from the cubes and then separate the structures into rectangular prisms. (*See example below.*) Ask students to find the volume of each prism by counting cubes or applying a formula and then adding the volumes to find the total volume of the structure. Then have them count the total number of cubes they used. Ask: *Is the volume of the structure the same as the sum of the volumes of your prisms?* Yes. *Can you add volumes of parts of a figure to find the total volume?* Yes.

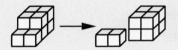

Estimating the Volume of a Classroom Object

`5.MD.1, 5.MD.5, 5.MD.5b, 5.MD.5c, SMP4`

Activity Card 12; per partnership: *Math Masters*, p. 35; tape measure; calculator

To extend their work with modeling real-world objects and finding volumes of figures composed of rectangular prisms, students model objects in the classroom using prisms and use the models to estimate the volume of the objects. `GMP4.1, GMP4.2`

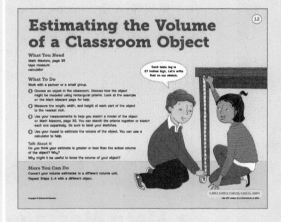

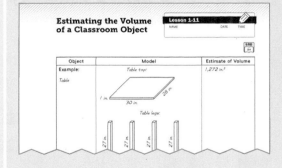

Adding to Find Volumes

`5.MD.5, 5.MD.5a, 5.MD.5b, 5.MD.5c, SMP2`

Activity Card 13; per partnership: three 6-sided dice

For additional practice finding volumes of figures composed of rectangular prisms, students roll dice to determine dimensions of figures and then find volumes. `GMP2.1`

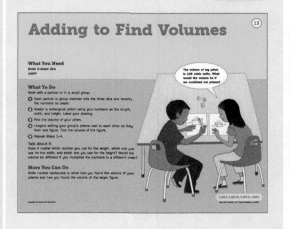

English Language Learner

Beginning ELL For the Math Message, display a picture of a suitcase with the three dimensions from the problem labeled and indicated by arrows. Build on students' understanding of suitcases as used for transporting clothes to explain the use of *cases* to carry other items, such as musical instruments.

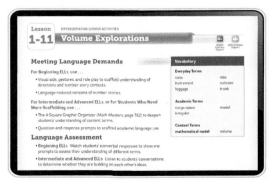

Differentiation Support pages are found in the online Teacher's Center.

1 Warm Up   5 min

▶ Mental Math and Fluency

Display expressions and have students show thumbs-up for yes or thumbs-down for no. Encourage students to share how they know. *Leveled exercises:*

●○○ Is the value of the expression greater than 27×10?
$(27 \times 10) - 50$ Thumbs-down
$2 + (27 \times 10)$ Thumbs-up
$2 \times (27 \times 10)$ Thumbs-up

●●○ Is the value of the expression less than $498 + 672$?
$(498 + 672) - 125$ Thumbs-up
$(498 + 672) - (498 + 672)$ Thumbs-up
$(498 + 672) / 2$ Thumbs-up

●●● Is the value of the expression half of $1{,}315 \times 6$?
$(1{,}315 \times 6) \div 2$ Thumbs-up
$(1{,}315 \times 6) \times \frac{1}{2}$ Thumbs-up
$2 \times (1{,}315 \times 6)$ Thumbs-down

2 Focus   35–40 min

▶ Math Message

You have a suitcase that is 14 inches long, 8 inches wide, and 20 inches tall. What is the volume of the suitcase? 2,240 cubic inches *If you lay the suitcase on its side, does it still have the same volume?* Yes. *You may use a calculator. Be ready to explain your answer.*

▶ Modeling a Suitcase

| WHOLE CLASS | SMALL GROUP | PARTNER | INDEPENDENT |

Math Message Follow-Up Before students share their Math Message responses, ask: *Why might you want to know the volume of a suitcase?* Sample answer: To figure out how much you can pack Have students confirm the volume and share a formula they could use to find it.
$V = l \times w \times h$

Sketch a rectangular prism and label it with the suitcase dimensions. (*See margin.*) Point out that the sketch is a **mathematical model** of the suitcase. The model does not look exactly like a suitcase because it does not include details like pockets or handles. Explain that when an object is about the same shape as a geometric figure, mathematicians often adopt a simplified picture as a model and use the model to help them solve problems. GMP4.1

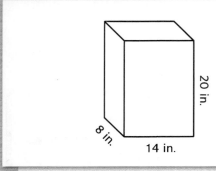

$V = 14 \text{ in.} \times 8 \text{ in.} \times 20 \text{ in.} = 2{,}240 \text{ in.}^3$

Ask: *How could this model help you solve the Math Message?* GMP4.2
Sample answer: If I label the model with the dimensions of the suitcase, it helps me see which formula to use to find the volume of the suitcase. Write the equation $V = 14$ in. $\times$ 8 in. $\times$ 20 in. $= 2{,}240$ in.³ under the mathematical model. (*See margin on page 80.*)

Ask: *Does the suitcase have the same volume if you lay it on its side?* Yes. *How do you know?* Sample answer: Even if you turn it, the dimensions do not change. It takes up the same amount of space.

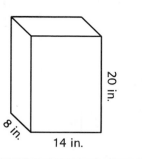

$V = (14 \times 8) \times 20$
$V = 112 \times 20 = 2{,}240$ in.³

| Differentiate | **Adjusting the Activity** |

If students struggle trying to visualize turning the suitcase, give them one of the assembled prisms from Lesson 1-8. Have them fill it with cubes to find the volume, and then empty it, turn it on its side, and find the volume again to verify that it is the same.

Go Online 〉 Differentiation Support

Invite volunteers to sketch a few ways the suitcase could be turned on its side. (*See margin.*) Challenge them to show mathematically why the volume stays the same. Allow students to use calculators as needed. If no one suggests it, point out that no matter which way the suitcase is turned, we still multiply the same three numbers to find the volume. GMP7.2 Display number sentences that match the sketch of each orientation to illustrate that the order in which the dimensions are multiplied does not affect the calculated volume. (*See margin.*)

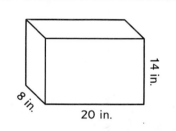

$V = (20 \times 8) \times 14$
$V = 160 \times 14 = 2{,}240$ in.³

> **Professional Development** The Associative Property of Multiplication states that three or more numbers can be multiplied in any order without changing the product. The suitcase activity is an illustration of the Associative Property of Multiplication. You do not need to use the name of the property with students.

Tell students that in today's lesson they will use more mathematical models to estimate volumes.

▶ Estimating Volumes of Musical Instrument Cases

Math Journal 1, p. 32; *Math Masters,* p. 34

| WHOLE CLASS | SMALL GROUP | PARTNER | INDEPENDENT |

Ask students whether they have ever seen a piece of luggage not shaped like a rectangular prism. Some may say they have seen oddly shaped pet carriers or bags for sports equipment like skis or surfboards.

Explain that when musicians travel they usually put their instruments in cases. Because musical instruments have irregular shapes, the cases are often not shaped like simple rectangular prisms. Nevertheless, we can use figures made of rectangular prisms to model the cases and then use the models to estimate their volume. GMP4.1, GMP4.2

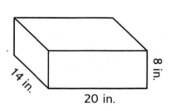

$V = (20 \times 14) \times 8$
$V = 280 \times 8 = 2{,}240$ in.³

Display the top of *Math Masters,* page 34. Explain that on the left side is a picture of a guitar case and on the right is a model of the case. Have partnerships use the model to estimate the volume of the guitar case, using calculators as needed. **GMP4.2** 2,025 in.³ Then have them share strategies. Be sure the following two strategies are shared:

- Divide the model into two rectangular prisms and use the formula $V = l \times w \times h$. The wide part of the case is a 20-inch by 5-inch by 15-inch prism, so the volume of that part is $20 * 5 * 15 = 1{,}500$ in.³. The narrow part of the case is a 21-inch by 5-inch by 5-inch prism, so the volume of that part is $21 * 5 * 5 = 525$ in.³. Add the volumes of the two prisms to get the total volume: 1,500 in.³ + 525 in.³ = 2,025 in.³.

- Think about turning the model so that the widest part lies flat on the ground. Find the area of the whole base. One part of the base is a 20-inch by 15-inch rectangle with an area of $20 * 15 = 300$ in.², and one part of the base is a 21-inch by 5-inch rectangle, with an area of $21 * 5 = 105$ in.². The total area of the base is 300 in.² + 105 in.² = 405 in.². The height of the case is 5 inches. Use $V = B \times h$ to find that the volume of the case is 405 in.² × 5 in. = 2,025 in.³.

Point out that both strategies involve thinking about the model in separate pieces. The pieces are different in each strategy, but both strategies give the same volume. Explain that breaking apart measurements and adding them together at the end can be a helpful strategy for solving measurement problems.

Display the baritone case and model shown on the bottom of *Math Masters,* page 34. Have partnerships use the model to estimate the volume of the case and then share strategies. 3,563 in.³ Expect most students to break the model into three rectangular prisms and add the volumes to find the total volume.

Have partners complete journal page 32. Note that Problem 4 provides practice with volume unit conversions.

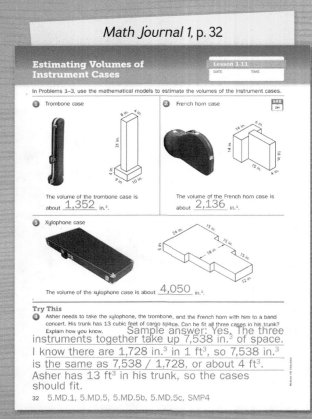
 Assessment Check-In   5.MD.5, 5.MD.5b, 5.MD.5c

Math Journal 1, p. 32

Expect most students to be able to use a strategy to break apart the model and apply a volume formula to the prisms to estimate the volume of the trombone case in Problem 1 on journal page 32. Some may be able to estimate the volumes of the more complex cases in Problems 2 and 3. If students struggle finding the volumes of the cases, encourage them to think about breaking the models into rectangular prisms and to sketch the prisms separately before finding the volumes.

 Evaluation Quick Entry Go online to record students' progress and to see trajectories toward mastery for these standards.

Summarize Ask: *When solving real-world volume problems, why is it useful to model objects with rectangular prisms?* Sample answer: I know two formulas for the volume of a rectangular prism, so it is easy to use those to estimate real-world volumes.

Math Journal 1, p. 33

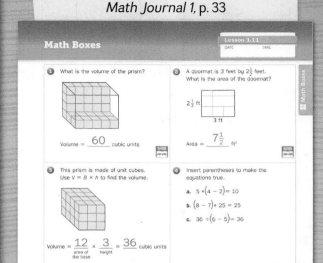

③ Practice   20–30 min

▶ Introducing *Buzz*

Student Reference Book, p. 294

| WHOLE CLASS | SMALL GROUP | PARTNER | INDEPENDENT |

Read the directions for *Buzz* on *Student Reference Book,* page 294 as a class. Then have students play the game. *Buzz* gives students practice finding multiples and prepares them for dividing multidigit whole numbers, which will be a focus of Unit 2.

Observe
• Which students can easily recognize multiples of the *BUZZ* number?
• Which *BUZZ* numbers are the most or least challenging?

Discuss
• *How do you keep track of which numbers to* BUZZ? **GMP6.1, GMP7.2**
• *What patterns do you notice in the* BUZZ *numbers?* **GMP7.1**

| Differentiate | **Game Modifications** | Go Online | Differentiation Support |

▶ Math Boxes 1-11

Math Journal 1, p. 33

| WHOLE CLASS | SMALL GROUP | PARTNER | INDEPENDENT |

Mixed Practice Math Boxes 1-11 are paired with Math Boxes 1-9.

▶ Home Link 1-11

Math Masters, p. 36

Homework Students find volumes of figures composed of rectangular prisms and name real-world objects the figures could model. **GMP4.1, GMP4.2**

Math Masters, p. 36

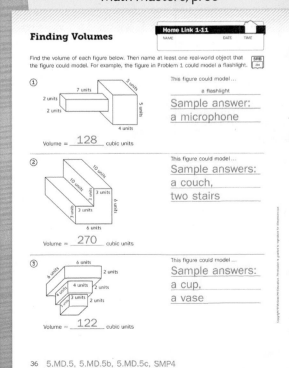

Playing *Prism Pile-Up*

Overview Students play a game to practice finding volumes of rectangular prisms, and they write number models for the volumes.

▶ **Before You Begin**

Near the Math Message display the rectangular prism shown on page 86. For Part 2, students will cut out the *Prism Pile-Up* cards from Activity Sheets 3 and 4 at the back of their journals. The cards are also used in the optional Readiness and Enrichment activities. If extra cards are needed, make copies of *Math Masters*, pages G4–G5. After the lesson, collect and store the cards for future use.

▶ **Vocabulary**

nested parentheses • brackets • braces

Standards

Focus Clusters
• Write and interpret numerical expressions.
• Geometric measurement: understand concepts of volume.

Warm Up   5 min

	Materials	
Mental Math and Fluency Students answer true/false questions about volume.	slate	5.MD.3, 5.MD.3a, 5.MD.3b

Focus   30–35 min

Math Message Students list strategies for finding volume.		5.MD.3, 5.MD.3a, 5.MD.3b, 5.MD.4, 5.MD.5, 5.MD.5b
Sharing and Explaining Strategies Students share their strategies for finding volume and clearly and precisely explain how to apply them.	Class Data Pad	5.MD.3, 5.MD.3a, 5.MD.3b, 5.MD.4, 5.MD.5, 5.MD.5a, 5.MD.5b SMP6
Introducing *Prism Pile-Up* **Game** Students practice finding volumes of rectangular prisms and figures composed of rectangular prisms.	*Student Reference Book*, p. 319; per partnership: *Prism Pile-Up* cards (*See Before You Begin*.); *Math Masters*, p. G6; scissors; calculator and centimeter cubes (optional)	5.OA.2, 5.MD.3, 5.MD.3a, 5.MD.3b, 5.MD.4, 5.MD.5, 5.MD.5a, 5.MD.5b, 5.MD.5c SMP2, SMP6
✓ **Assessment Check-In** See page 88. Expect most students to be able to apply a strategy and to use a displayed formula to find the volumes of the figures on the cards identified while playing the game *Prism Pile-Up*.	*Math Masters*, p. G6; centimeter cubes (optional)	5.MD.3, 5.MD.4, 5.MD.5

Practice   25–35 min

Understanding Grouping Symbols Students evaluate expressions with nested grouping symbols	*Math Journal 1*, p. 34; slate	5.OA.1, 5.OA.2 SMP4
Math Boxes 1-12 Students practice and maintain skills.	*Math Journal 1*, p. 35	See page 89.
Home Link 1-12 **Homework** Students circle the winning card in rounds of *Prism Pile-Up*.	*Math Masters*, p. 38	5.OA.2, 5.MD.3, 5.MD.3a, 5.MD.3b, 5.MD.4, 5.MD.5, 5.MD.5a, 5.MD.5b, 5.MD.5c

  Go Online ▷ to see how mastery develops for all standards within the grade.

Differentiation Options

Readiness — 5–15 min

WHOLE CLASS · **SMALL GROUP** · PARTNER · INDEPENDENT

Choosing Volume Strategies

5.MD.3, 5.MD.3a-b, 5.MD.4, 5.MD.5, 5.MD.5a-c

Prism Pile-Up cards
(*See Before You Begin.*)

For experience choosing an appropriate strategy for finding the volume of a prism, students examine the *Prism Pile-Up* cards and discuss strategies that could be used to find the volume of each figure. Ask questions like these: Sample answers given.

- *For which card(s) could you use the formula V = l * w * h?* Card 1; Card 10 *How do you know you can use that strategy?* The length, width, and height are labeled. I can count cubes to figure out the length, width, and height. *Could you use that formula for Card 14? Why or why not?* No. I can't tell from the figure what the length and width are.

- *What strategy could you use to find the volume of the figure on Card 16?* Divide the figure into two pieces and use V = l * w * h for each piece. *Why did you choose that strategy?* I have to use the formula more than once because there is more than one height.

Enrichment — 15–30 min

WHOLE CLASS · SMALL GROUP · **PARTNER** · INDEPENDENT

Creating *Prism Pile-Up* Cards

5.MD.3, 5.MD.3a-b, 5.MD.4, 5.MD.5, 5.MD.5a-c, SMP2

Activity Card 14; Prism Pile-Up cards (*see Before You Begin*);
index cards

To extend their work finding volumes of rectangular prisms and figures composed of rectangular prisms, students create additional *Prism Pile-Up* cards and find the volumes of the figures on their cards. If time permits, have students play the game with their new cards. **GMP2.1, GMP2.2**

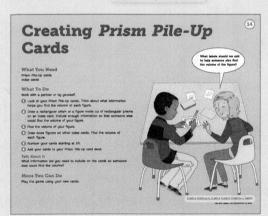

Extra Practice — 5–15 min

WHOLE CLASS · SMALL GROUP · **PARTNER** · INDEPENDENT

Solving Volume Problems

5.MD.3, 5.MD.3a-b, 5.MD.4, 5.MD.5, 5.MD.5a-c

Math Masters, p. 37

For further practice finding volumes of rectangular prisms and writing number models, students complete *Math Masters*, page 37.

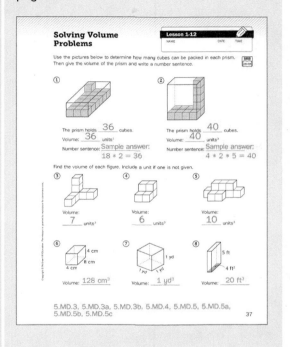

English Language Learner

Beginning ELL To scaffold student participation in the Math Message discussion, encourage students at the pre-production stage of English acquisition to use drawings or unit cubes to demonstrate their strategies for finding the volume of the prism.

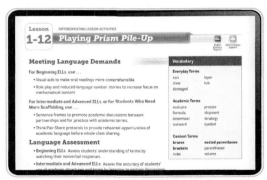

Differentiation Support pages are found in the online Teacher's Center.

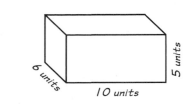

Prism for the Math Message

---

1 Warm Up   5 min

▶ Mental Math and Fluency

Read statements about volume to the class. Have students write *true* or *false* on their slates. *Leveled exercises:*

●○○ Prisms are 3-dimensional shapes. True.
      Volume measures the surface of a 2-dimensional shape. False.

●●○ The amount of space enclosed by a 3-dimensional shape can be measured in cubic units. True.
      If two containers have the same volume, they must have the same length, width, and height. False.

●●● The volume of a prism is the number of unit cubes that fit in its base. False.
      A rectangular prism completely packed with 24 unit cubes has a volume of 24 cubic units. True.

2 Focus   30–35 min

▶ Math Message

Display the prism shown in the margin.
Imagine you had this prism in front of you. Talk to a partner about how you could find the volume of the prism. List as many strategies as you can. Use one of your strategies to find the volume of the prism.

▶ Sharing and Explaining Strategies

| WHOLE CLASS | SMALL GROUP | PARTNER | INDEPENDENT |

Math Message Follow-Up Have students share the strategies they listed. Record them on the Class Data Pad. Strategies may include the following:

• Pack the prism with unit cubes and count the cubes.

• Pack one layer with unit cubes. Multiply the number of cubes that fit in one layer by the number of layers.

• Use the formula $V = l * w * h$.

• Find the area of the base and then use the formula $V = B * h$.

Ask a volunteer to explain the strategy he or she used to find the volume of the prism. Sample answer: To use the formula $V = l * w * h$, I can multiply 10 and 6 and 5 to get 300.

Discuss how explanations are most useful to others when they are both clear and precise. An explanation is clear when it makes sense to the listener or reader. An explanation is precise when it uses specific mathematical language. GMP6.1

Discuss several ways students could make their explanations clearer or more precise. *For example:*

- Saying that you multiplied the area of the base times the height is clearer than saying you multiplied *B* times *h*, since a listener might not know what *B* and *h* stand for.
- Saying that 10 units * 6 units * 5 units = 300 cubic units is more precise than saying 10 * 6 * 5 = 300 because it includes the units.

Ask students to suggest ways the explanation given by the first volunteer could be made clearer or more precise. Sample answers: It would be clearer to say what *l*, *w*, and *h* stand for. It would be more precise to include units. Have students explain how to use the other strategies on the list, and encourage the class to suggest ways the explanations could be made clearer or more precise. GMP6.1

Tell students that today they will play a game to practice finding volumes of rectangular prisms and practice giving clear and precise explanations.

▶ Introducing *Prism Pile-Up*

| WHOLE CLASS | SMALL GROUP | PARTNER | INDEPENDENT |

Math Journal 1, Activity Sheets 3–4; *Student Reference Book,* p. 319; *Math Masters,* p. G6

Distribute one copy of *Math Masters,* page G6 to each partnership and have students cut it in half. Each partnership will need one set of *Prism Pile-Up* cards (*Math Journal 1,* Activity Sheets 3–4). Have partners share the work of cutting out one set, or have each student cut out the set from his or her journal and collect and store one set from each partnership for future use.

Read as a class the directions for *Prism Pile-Up* on *Student Reference Book,* page 319. Explain that the cubes on Cards 1–9 are meant to represent centimeter cubes, so the volume of those prisms can be reported in cubic centimeters. Ask students to discuss with a partner whether they agree with the volumes the players found in the example on the *Student Reference Book* page.

Differentiate | **Common Misconception**

Some students may count only the cubes they see in the pictures. For example, they may report the volume of the prism on Card 1 as 23 cubic centimeters because 23 cubes are visible. Others may count only the squares on the picture. They may report the volume of the prism on Card 2 to be 47 cm³. Have these students build the prisms with cubes to help them find the volumes.

Go Online ▸ | Differentiation Support

Discuss the number sentences in the example. Ask: *Why did Libby write 4 * 2 * 4 = 32 as her number sentence?* GMP2.2 Sample answers: She saw that there were 4 cubes along the front, 2 cubes along the side, and 4 cubes in a stack. She thought about $V = l * w * h$.

Student Reference Book, p. 319

Games

Prism Pile-Up

Prism Pile-Up Cards 1–9

Materials	☐ 1 set of *Prism Pile-Up* Cards (Math Journal 1, Activity Sheets 3–4)
	☐ 1 Prism Pile-Up Record Sheet for each player (Math Masters, p. G6)
	☐ calculator (optional)
Players	2
Skill	Finding volumes of rectangular prisms
Object of the Game	To collect more cards.

Directions

① Shuffle the cards and place them facedown in a pile.

② Each player draws a card from the pile and finds the volume of the figure. The cubes on cards 1–9 represent centimeter cubes, so their volume can be reported in cm³. Players may use a calculator.

③ Players record the card number and volume of their figure on the *Prism Pile-Up* Record Sheet. They also record one or more number sentences for the volume.

④ Players check each other's work. The player whose figure has the greater volume takes both cards.

⑤ The game ends when there are no cards remaining. The player with more cards wins. If players have the same number of cards, they look through their cards to find the figure with the greatest volume. The player whose figure has the greatest volume wins.

Prism Pile-Up Cards 10–18

Example
Libby draws Card 1. She finds the volume to be 32 cubic centimeters and writes 4 * 2 * 4 = 32.

Christopher draws Card 16. He calculates the volume to be 76 cubic centimeters and writes these number sentences: 2 * 4 * 2 = 16; 3 * 4 * 5 = 60; 16 + 60 = 76.

Christopher's figure has the greater volume, so he takes both cards.

Libby | Christopher

three hundred nineteen | SRB 319

Prism Pile-Up Card Deck			
Card	Volume (cubic centimeters)	Card	Volume (cubic centimeters)
1	32	10	45
2	60	11	48
3	25	12	20
4	50	13	54
5	64	14	125
6	68	15	40
7	30	16	76
8	42	17	52
9	36	18	232

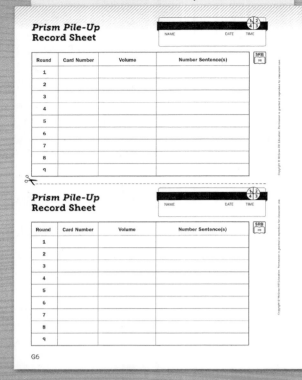

Understanding Grouping Symbols

Lesson 1-12

Evaluate the following expressions.

① $10 * [13 + (12 - 7)] = $ __180__

② __8__ $= \{(5 * 6) + 2\} / 4$

③ __45__ $= \{13 + (2 * 1)\} * 3$

④ $64 / [20 - (4 * 3)] = $ __8__

Insert grouping symbols to make the following number sentences true.

⑤ $4 = 4 * [6 - (2 + 3)]$

⑥ $300 \div \{(6 + 4) * (2 + 8)\} = 3$

⑦ $70 / [13 - (2 + 1)] = 7$

⑧ $160 = 8 * [16 + \{(12 - 4) \div 2\}]$

Brackets, braces, and parentheses are all acceptable grouping symbols.

Write an expression that models the story. Then evaluate the expression.

⑨ Tommy had a bag of 100 balloons. He took out 2 red balloons and 1 blue balloon for each party favor. He created 12 party favors. How many balloons did Tommy have left?

Expression: $100 - [12 * (2 + 1)]$

Answer: __64__ balloons

⑩ A grocery store received a shipment of 100 cases of apple juice. Each case contained four 6-packs of cans. After inspection, the store found that 9 cans were damaged. How many cans were undamaged?

Expression: $\{100 * (4 * 6)\} - 9$

Answer: __2,391__ cans

34   5.OA.1, 5.OA.2, SMP4

Ask students to suggest other number sentences Libby could write for Card 1. **GMP2.1** *For example:*

- If she counted the cubes in the top layer and then multiplied by the number of layers, she could write $8 * 4 = 32$.
- If Libby saw that there were 16 cubes in the front layer and 2 layers, she could write $16 + 16 = 32$ or $16 * 2 = 32$.

Ask: *Why did Christopher write three different number sentences?*
Sample answer: He split the figure on Card 16 into 2 rectangular prisms. The first two number sentences show the volume of each smaller prism. The third number sentence shows how he added the volumes together to find the total volume.

Tell students they can record any number sentence that makes sense to them. The number sentence could show how they found the volume, or they could use the formulas $V = l * w * h$ or $V = B * h$ to help them write number sentences. Demonstrate a sample round. Then have students play the game in partnerships. Be sure the volume formulas are displayed.

Observe

- Which students are counting to find the volumes of the figures? Which students are applying formulas?
- Which students need additional support to play the game?

Discuss

- *Explain how you found the volume of one of the figures.* **GMP6.1**
- *How could you explain your thinking more clearly?* **GMP6.1**

| **Differentiate** | **Game Modifications** | **Go Online** | Differentiation Support |

Assessment Check-In 5.MD.3, 5.MD.4, 5.MD.5

Math Journal 1, p. G6

Expect most students to be able to apply a strategy to correctly find the volumes of the figures on *Prism Pile-Up* Cards 1–3 and 7–9. Most should also be able to use a displayed formula to find the volumes of the figures on Cards 10–15. Some students may be able to find the volumes of the figures composed of prisms on Cards 4–6 and 16–18. (*See the margin on page 87 for volumes.*) Some students may be able to write number sentences for the volumes of the figures. If students struggle trying to find volumes, provide cubes so they can build the figures and count the cubes.

Evaluation Quick Entry Go online to record students' progress and to see trajectories toward mastery for these standards.

Summarize Have volunteers explain how they found the volume of one of the figures on the game cards. Encourage other students to give suggestions about how to make the explanation clearer or more precise. **GMP6.1**

3 Practice 25–35 min

▶ Understanding Grouping Symbols

Math Journal 1, p. 34

| WHOLE CLASS | SMALL GROUP | PARTNER | INDEPENDENT |

Review the use of parentheses by having students write an expression on their slates to represent this situation: *Ben bought three concert tickets. They were $6 each and he had to pay a $2 service charge for each ticket. How much money did he spend?* 3 * (6 + 2) **GMP4.1** Remind students that operations inside parentheses are carried out first. Tell students that there are some situations in which more than one set of parentheses is needed, and sometimes one set of parentheses appears inside another. We call these **nested parentheses,** and the operations in the innermost parentheses are carried out first.

Extend the previous example to illustrate nested parentheses: *Ben paid $50 for the concert tickets. How much change did he receive?* Write the expression 50 − (3 * (6 + 2)) and discuss how it represents the story. Tell students to first evaluate the innermost parentheses, (6 + 2), and then work outward to evaluate the expression. 26

Explain that when parentheses are nested, the outer parentheses are sometimes replaced with different symbols. Display and label **brackets [ ]** and **braces { }.** Rewrite the expression from the story above using both brackets and braces. Explain that parentheses, brackets, and braces all mean the same thing in an expression.

Display expressions with nested grouping symbols for students to evaluate. *Suggestions:*

- 6 * [12 − (3 + 4)] 30
- {32 + (29 − 21)} / 8 5
- 16 + {13 + (21 − 8)} 42

Have students work in partnerships or small groups to complete journal page 34. Discuss the solutions as a class if time permits.

▶ Math Boxes 1-12

Math Journal 1, p. 35

| WHOLE CLASS | SMALL GROUP | PARTNER | INDEPENDENT |

Mixed Practice Math Boxes 1-12 are grouped with Math Boxes 1-6 and 1-10.

▶ Home Link 1-12

Math Masters, p. 38

Homework Students circle the winning card in rounds of *Prism Pile-Up.*

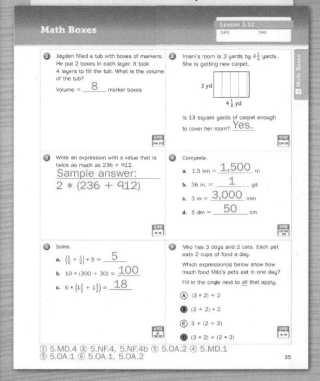

Math Masters, p. 38

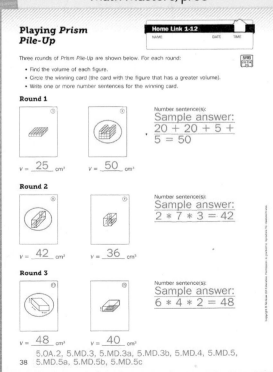

Unit 1 Progress Check

Overview   Day 1: Administer the Unit Assessments.
Day 2: Administer the Open Response Assessment.

Day 1: Unit Assessment

 **Quick Entry Evaluation** Record results and track progress toward mastery.

1 Warm Up   5–10 min

Materials

Self Assessment
Students complete the Self Assessment.

Assessment Handbook, p. 5

2a Assess   35–50 min

★ **Unit 1 Assessment**
These items reflect mastery expectations to this point.

Assessment Handbook, pp. 6–9

Unit 1 Challenge (Optional)
Students may demonstrate progress beyond expectations.

Assessment Handbook, pp. 10–11

Standards	Goals for Mathematical Content (GMC)	Lessons	Self Assessment	Unit 1 Assessment	Unit 1 Challenge
5.OA.1	Write numerical expressions that contain grouping symbols.	1-1, 1-11, 1-12		1, 13	
	Evaluate expressions that contain grouping symbols.	1-1, 1-11, 1-12	1	3a–3d	1a–1d
5.OA.2	Model real-world and mathematical situations using simple expressions.	1-12	2	1, 13	
5.NF.4, 5.NF.4b	Justify the area formula for a rectangle with fractional side lengths by tiling.	1-2 to 1-4	3	2, 9, 10a, 10b	2, 3
5.MD.3	Recognize volume as an attribute of solid figures.	1-5, 1-6, 1-10	4	4	
5.MD.3, 5.MD.3a	Understand that a unit cube has 1 cubic unit of volume and can measure volume.	1-7 to 1-10, 1-12		5b, 6–8	4
5.MD.3, 5.MD.3b	Understand that a solid figure completely filled by *n* unit cubes has volume *n* cubic units.	1-7 to 1-10, 1-12		5a, 6–8	4
5.MD.4	Measure volumes by counting unit cubes and improvised units.	1-5 to 1-10, 1-12	5	5a, 6, 7	4
5.MD.5, 5.MD.5a	Represent products of three whole numbers as volumes.	1-9, 1-11, 1-12		11, 15	
5.MD.5, 5.MD.5b	Apply formulas to find volumes of rectangular prisms.	1-9 to 1-12	6	11, 12, 14a, 15	5b, 5c
5.MD.5, 5.MD.5c	Find volumes of figures composed of right rectangular prisms.	1-11, 1-12	7	14a, 15	5b
	Solve real-world problems involving volumes of figures composed of prisms.	1-11			5b

Standards	Goals for Mathematical Process and Practice (GMP)	Lessons	Self Assessment	Unit 1 Assessment	Unit 1 Challenge
SMP1	Compare the strategies you and others use. GMP1.6	1-2		8	
SMP4	Model real-world situations using graphs, drawings, tables, symbols, numbers, diagrams, and other representations. GMP4.1	1-3, 1-11		13	5a
	Use mathematical models to solve problems and answer questions. GMP4.2	1-11			5b–5d
SMP6	Explain your mathematical thinking clearly and precisely. GMP6.1	1-5, 1-6, 1-8, 1-12		14b	3
	Use clear labels, units, and mathematical language. GMP6.3	1-2, 1-10		5a	
	Think about accuracy and efficiency when you count, measure, and calculate. GMP6.4	1-7		5b	
SMP7	Look for mathematical structures such as categories, patterns, and properties. GMP7.1	1-4, 1-9		10a	
	Use structures to solve problems and answer questions. GMP7.2	1-4, 1-11		10b	3

 Go Online to see how mastery develops for all standards within the grade.

1 Warm Up   5–10 min

▶ Self Assessment

Assessment Handbook, p. 5

| WHOLE CLASS | SMALL GROUP | PARTNER | **INDEPENDENT** |

Students complete the Self Assessment to reflect on their progress in Unit 1.

Some students may benefit from recalling the types of problems cited in each row on the Self Assessment. Show students where these pointers appear on their Self Assessments. In Unit 2 students begin using these pointers themselves.

Assessment Handbook, p. 5

NAME   DATE   TIME   Lesson 1-13 ✓

Unit 1 Self Assessment

Think about each skill listed below. Assess your own progress by checking the most appropriate box.

Skills	I can do this on my own and explain how to do it.	I can do this on my own.	I can do this if I get help or look at an example.
① Evaluate expressions with grouping symbols. MJ1 20 [Problem 1]			
② Write expressions to model situations. MJ1 26			
③ Find the area of a rectangle with one fractional side length. MJ1 6			
④ Identify objects with volume. MJ1 14 [Problem 2]			
⑤ Use cubes to find volume. MJ1 24			
⑥ Use formulas to find volume. MJ1 28			
⑦ Find the volume of a figure made of rectangular prisms. MJ1 32			

Assessment Masters   5

NAME   DATE   TIME   **Lesson 1-13** ✓

Unit 1 Assessment

① Kayla was playing *Name That Number*. She had the cards shown below.
Write two different expressions that show how Kayla could play her cards.
Use grouping symbols in at least one of the expressions.

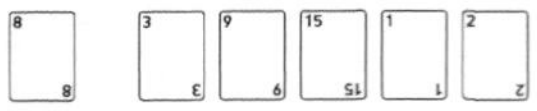

| 8 | 3 | 9 | 15 | 1 | 2 |

Target Number

Sample answers: $2 * (3 + 1)$;
$15 - 9 + 2 * 1$; $(15 + 1) \div 2$

② Find the area of the rectangle. Write a number sentence to show your thinking.

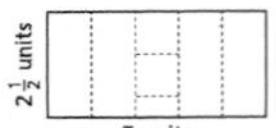

$2\frac{1}{2}$ units

5 units

Area = $12\frac{1}{2}$ square units
Sample answer: $5 + 5 + \frac{1}{2} + \frac{1}{2} + \frac{1}{2} + \frac{1}{2} + \frac{1}{2}$
$= 12\frac{1}{2}$
(number sentence)

③ Solve.

a. $12 \cdot (6 + 4) = $ __120__   b. $(12 \cdot 6) + 4 = $ __76__

c. __30__ $= (48 \div 2) + 6$ d. __6__ $= 48 \div (2 + 6)$

④ Circle the items that have volume.

a wiggly line   (a trash can)   a drawing of a truck

the top of your desk   (a coffee mug)   (a cereal box)

6   Assessment Handbook

NAME   DATE   TIME   **Lesson 1-13** ✓

Unit 1 Assessment (continued)

⑤ a. Jonah filled a box and said its volume was 25 balls.
Shandra filled the same box and said its volume was 36 cubes.

Explain how Jonah and Shandra could get different volumes for the same box.
Sample answer: They used different units. The cubes were
smaller than the balls, so it took more to fill the same box.

b. Are balls or cubes better for measuring the volume of a rectangular prism? Why?
Sample answer: Cubes are better. They fit together tightly
without gaps. They fit into the corners of rectangular prisms.

⑥ 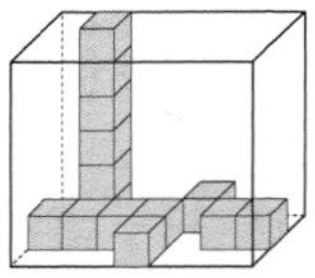 How many cubes would it take to fill this prism?
__126__ cubes

What is the volume of this prism?
__126__ cubic units

⑦ 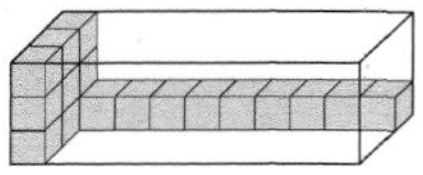 How many cubes would it take to fill this prism?
__90__ cubes

What is the volume of this prism?
__90__ cubic units

⑧ Compare the strategies you used to find the volume in Problem 6 and in Problem 7.
How were they the same? How were they different?
Sample answer: For both problems, I found how many
cubes fill the prism. In Problem 6, I counted the cubes to
find the length and width and multiplied these. Then I
multiplied by the height. In Problem 7, I saw there were
9 cubes in a layer and multiplied 9 by 10 layers.

Assessment Masters   7

2a Assess   35–50 min

▶ Unit 1 Assessment

Assessment Handbook, pp. 6–9

| WHOLE CLASS | SMALL GROUP | PARTNER | **INDEPENDENT** |

Students complete the Unit 1 Assessment to demonstrate their progress on the standards covered in this unit.

The online assessment and reporting tools provide additional resources for monitoring student progress. You may use differentiation options to intervene as indicated.

Generic rubrics in the *Assessment Handbook* can be used to evaluate student progress on the Mathematical Process and Practice Standards.

Written assessments are one way students can demonstrate what they know. The table below shows adjustments you can make to the Unit 1 Assessment to maximize opportunities for individual students or for your entire class.

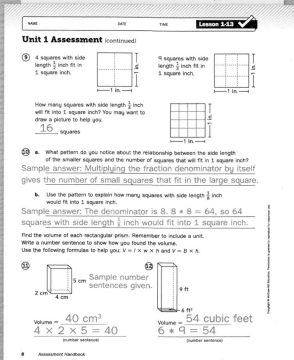

Differentiate — Adjusting the Assessment

Item(s)	Adjustments
1	To extend Item 1, have students write an expression that uses nested grouping symbols.
2	To scaffold Item 2, have students partition the entire rectangle into squares and partial squares.
3	To scaffold Item 3, have students first perform the operation in parentheses and then rewrite each expression without the parentheses.
4	To scaffold Item 4, provide access to each of the objects listed.
5	To extend Item 5, ask students to list other objects they could use to measure volume and write about why they would or would not be good units for measuring volume.
6–8	To scaffold Items 6–8, provide access to centimeter cubes.
9, 10	To extend Items 9 and 10, ask students to write about whether or not the patterns work for other units, such as square centimeters or square feet.
11, 12	To scaffold Items 11 and 12, have students list the meaning of each variable that appears in the equations.
13	To extend Item 13, have students describe another situation that could be modeled by the expression they write.
14	To scaffold Item 14, have students sketch and label two different rectangular prisms that make up the figure in the problem.
15	To scaffold Item 15, choose two different *Prism Pile-Up* cards for students to compare, such as two cards that do not show composite figures or two cards showing figures that are easy to compare visually..

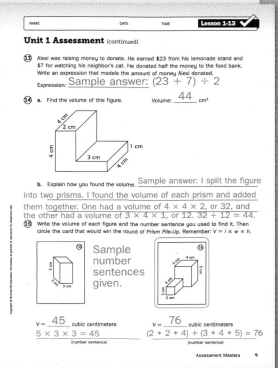

NAME   DATE   TIME   **Lesson 1-13** ✔

Unit 1 Challenge

① Solve.

  a. $8 * [(15 - 9) \div 3)] = $ __16__   **b.** __6__ $= (160 \div (4 * 20)) \cdot 3$

  c. __50__ $= [(6 + 2) * (9 + 16)] \div 4$ **d.** $100 \div ((2 + 3) * (6 - 4)) = $ __10__

② Find the area of the rectangle. Remember to include a unit.

  Area $= $ __$19\frac{2}{4}$ in.², or $19\frac{1}{2}$ in.²__

  [rectangle: $3\frac{1}{4}$ in. by 6 in.]

  Sample answer: $6 * 3 + \frac{1}{4} + \frac{1}{4} + \frac{1}{4} + \frac{1}{4} + \frac{1}{4} + \frac{1}{4} = 19\frac{1}{2}$ in.²
  (number sentence)

③ Annika is making a quilt with squares that are $\frac{1}{2}$ foot in length on each side. The finished quilt will be 4 feet long and $3\frac{1}{2}$ feet wide.

  How many quilt squares will Annika need? You may draw a picture to help you.

  Answer: __56__ quilt squares

  What is the area of the quilt? Explain how you got your answer.
  14 square feet; Sample answer: Four $\frac{1}{2}$-foot squares are in 1 square foot. Annika is using 56 quilt squares. $56 \div 4 = 14$, so the area is 14 square feet.

10   Assessment Handbook

NAME   DATE   TIME   **Lesson 1-13** ✔

Unit 1 Challenge (continued)

④ Draw a figure that would beat this card in a game of *Prism Pile-Up*.

  ⑧

  Sample answer: [3 cm × 3 cm × 7 cm prism]

⑤ **a.** Sketch a mathematical model of the coffee mug using rectangular prisms. Use your model to answer the following questions about the mug.

  Sample answers given.

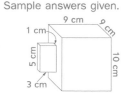

  b. The volume of the entire coffee mug is about __825 cm³__

  c. The volume of coffee that the mug would hold is about __810 cm³__

  d. Why might you want to know the volume of the entire coffee mug?
  If I were trying to fit a certain number of mugs in a box, I would want to know the volume to see if all the mugs would fit.

Assessment Masters   11

Advice for Differentiation

Because this is the beginning of the school year, all of the content included on the Unit 1 Assessment was recently introduced and will be revisited in subsequent units.

Go Online:

Quick Entry Evaluation Record students' progress and see trajectories toward mastery for these standards.

Data Review your students' progress reports. Differentiation materials are available online to help you address students' needs.

> **NOTE** See the Unit Organizer on pages 8–9 or the online Spiral Tracker for details on Unit 1 focus topics and the spiral.

(Go Online) for additional information in the *Implementation Guide* about assessment in *Everyday Mathematics,* including grading and differentiation.

▶ Unit 1 Challenge (Optional)

Assessment Handbook, pp. 10–11

| WHOLE CLASS | SMALL GROUP | PARTNER | **INDEPENDENT** |

Students can complete the Unit 1 Challenge after they complete the Unit 1 Assessment. The Unit 1 Challenge offers students an opportunity to demonstrate a deeper understanding of the content and process/practice standards addressed so far this year. Do not expect all students to succeed at the Challenge problems. However, student responses to these problems may help you choose appropriate interventions, including Enrichment activities.

Unit 1 Progress Check

Day 2: Open Response Assessment

2b Assess   50–55 min

Materials

Solving the Open Response Problem
After a brief introduction, students solve a volume problem and explain their thinking.

Assessment Handbook, pp. 12–13; centimeter cubes

Discussing the Problem
After completing the problem, students share strategies.

Assessment Handbook, pp. 12–13

Standards	Goals for Mathematical Content (GMC)	Lessons
5.MD.3, 5.MD.3b	Understand that a solid figure completely filled by *n* unit cubes has volume *n* cubic units.	1-7 to 1-10, 1-12
5.MD.4	Measure volumes by counting unit cubes and improvised units.	1-5 to 1-10, 1-12
	Goal for Mathematical Process and Practice (GMP)	
SMP6	Explain your mathematical thinking clearly and precisely. GMP6.1	1-5, 1-6, 1-8, 1-12

▶ **Evaluating Students' Responses**

Evaluate students' abilities to measure volume by counting cubic units. Use the rubric below to evaluate their work based on **GMP6.1**.

Goal for Mathematical Process and Practice GMP6.1 Explain your mathematical thinking clearly and precisely.	Not Meeting Expectations	Partially Meeting Expectations	Meeting Expectations	Exceeding Expectations
	Does not attempt to explain or provides an inconsistent or vague explanation for Problems 1 and 2.	Clearly explains thinking for Problem 1 or Problem 2 (see Meeting Expectations).	Clearly explains thinking for Problems 1 and 2 using words or drawings. For Problem 1, explains that because there are 30 balls in boxes with a volume of 1 cubic foot, the minimum size box is 30 cubic feet. For Problem 2, explains that the large box will hold fewer than 30 balls using one of the these methods: Explains that the length × width × height is less than 30 cubic feet. Indicates that 2 layers of 12 boxes is not enough room for 30 boxes. Adds to the drawing or uses cubes to show that 30 unit cubes will not fit.	Meets expectations and uses correct and formal mathematical terminology such as the word dimensions to collectively describe the volume, length, width, and height, or uses correct units or formulas such as $V = l \times w \times h$ and then substitutes in values and calculates for Problems 1 and 2.

3 Look Ahead   10–15 min

Materials

Math Boxes 1-13: Preview for Unit 2
Students preview skills and concepts for Unit 2.

Math Journal 1, p. 36

Home Link 1-13
Students take home the Family Letter that introduces Unit 2.

Math Masters, pp. 39–42

  Go Online to see how mastery develops for all standards within the grade.

Assessment Handbook, p. 12

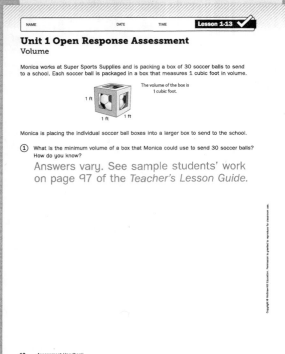

Assessment Handbook, p. 13

2b Assess   50–55 min

▶ Solving the Open Response Problem

Assessment Handbook, pp. 12–13

| WHOLE CLASS | SMALL GROUP | PARTNER | **INDEPENDENT** |

The open response problem requires students to apply skills and concepts from Unit 1 to a real-world problem involving volume. The focus of this task is **GMP6.1:** Explain your mathematical thinking clearly and precisely. Tell students that today they will solve a problem about volume and explain their mathematical thinking.

Distribute *Assessment Handbook,* pages 12 and 13. Have students read the pages and discuss any questions about the directions. Make centimeter cubes available to the class. Tell students that they can solve the problem in any way that makes sense to them, but they need to clearly explain their solutions. GMP6.1

Students should complete the problem independently.

> **Differentiate   Adjusting the Activity**
>
> If students struggle to determine the dimensions of the box, have them recreate the picture with centimeter cubes.

▶ Discussing the Problem

Assessment Handbook, pp. 12–13

| **WHOLE CLASS** | SMALL GROUP | **PARTNER** | INDEPENDENT |

After students complete their work, have them share their solutions and explanations with a partner before inviting a few students to share with the class. Be sure to discuss both correct and incorrect solutions as well as clear and unclear explanations. It may be helpful to start with an unclear explanation and let students suggest how to improve it. Ask: *What would you keep in this explanation? What would you do to improve the explanation?* GMP6.1   Answers vary.

Evaluating Student's Responses

5.MD.3, 5.MD.3b, 5.MD.4

Collect students' work. For the content standards, expect most students to apply their understanding of volume to determine that the minimum volume of a box that Monica can use is 30 cubic feet and that 30 boxes of soccer balls will not fit into the large box. You can use the rubric on page 95 to evaluate students' work for **GMP6.1**.

The sample in the margin shows an example of work that meets expectations for the content standards. For Problem 1 the student determined that the minimum volume of the box is 30 cubic feet. In Problem 2 this student counted cubic units and found that only 24 soccer ball boxes fit into the large box, not 30. The work meets expectations for the mathematical process and practice standard because the student clearly explained the thinking in words. For Problem 1 the student stated that the volume of 1 soccer ball box is 1 cubic foot, so for 30 balls, "the box has to be 30 cubic ft." The student used correct units throughout. For Problem 2 the student explained that he or she counted actual cubes to find that 30 cubic units will not fit in the box. **GMP6.1**

 Evaluation Quick Entry Go online to record students' progress and to see trajectories toward mastery for these standards.

NOTE Additional samples of evaluated students' work can be found in the *Assessment Handbook* appendix.

3 Look Ahead 10–15 min

▶ **Math Boxes 1-13:** Preview for Unit 2

Math Journal 1, p. 36

| WHOLE CLASS | SMALL GROUP | PARTNER | INDEPENDENT |

Mixed Practice Math Boxes 1-13 are paired with Math Boxes 1-8. These problems focus on skills and understandings that are prerequisite for Unit 2. You may want to use information from these Math Boxes to plan instruction and grouping in Unit 2.

▶ **Home Link 1-13**

Math Masters, pp. 39–42

Home Connection The Unit 2 Family Letter provides information and activities related to Unit 2 content.

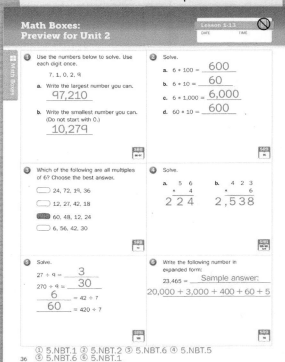

Math Journal 1, p. 36

Whole Number Place Value and Operations

Contents

*The standards listed here are addressed in the **Focus** of each lesson. For all the standards in a lesson, see the Lesson Opener.

Focus

In this unit, students explore patterns in the base-10 place-value system and ways of representing large numbers. Students are also introduced to U.S. traditional multiplication and review partial-quotients division.

Major Clusters

5.NBT.A Understand the place value system.

5.NBT.B Perform operations with multi-digit whole numbers with decimals to hundredths.

Supporting Cluster

5.MD.A Convert like measurement units within a given measurement system.

Process and Practice Standards

SMP1 Make sense of problems and persevere in solving them.

SMP6 Attend to precision.

Coherence

The table below describes how standards addressed in the Focus parts of the lessons link to the mathematics that students have done in the past and will do in the future.

	Links to the Past	Links to the Future
5.OA.1	In Unit 1, students reviewed how to use grouping symbols in expressions and how to evaluate expressions with grouping symbols. In Grade 3, students inserted parentheses into number sentences to make them true and evaluated number sentences with parentheses.	In Unit 7, students will use grouping symbols in an expression to model how to solve a multistep problem about gauging reaction time. In Grade 6, students will evaluate expressions and perform operations according to the Order of Operations.
5.OA.2	In Unit 1, students represented the volumes of rectangular prisms using expressions. They also wrote expressions to record calculations in the game *Name That Number.* In Grade 4, students represented problems using equations with a letter standing for an unknown quantity.	Throughout Grade 5, students will write expressions to record calculations in a variety of contexts. In Unit 6, they will order and interpret expressions without evaluating them. In Grade 6, students will write expressions in which letters stand for numbers.
5.NBT.1	In Grade 4, students worked with place-value concepts in whole numbers through 1,000,000.	In Unit 4, students will extend place-value concepts and patterns to decimals through thousandths. In Grade 6, students will extend their understanding of place value by applying their reasoning to make sense of decimal computation.
5.NBT.2	In Grade 4, students developed a rule for solving multiplication problems involving multiples of 10.	After students gain more experience with using exponents to denote powers of 10, they will multiply and divide decimals by powers of 10 and develop rules for doing so. In Unit 8, students will apply their knowledge of powers of 10 to solve rich, real-world problems. In Grade 6, students will write and evaluate numerical expressions with whole-number exponents.
5.NBT.5	In Grade 4, students used partial-products multiplication and lattice multiplication to solve multidigit multiplication problems.	Throughout Grade 5, students will use U.S. traditional multiplication to solve multiplication problems in mathematical and rich, real-world contexts. In Grade 6, students will use U.S. traditional multiplication to solve multidigit decimal multiplication problems.
5.NBT.6	In Grade 4, students used partial-quotients division to solve division problems with 4-digit dividends and 1-digit divisors.	Throughout Grade 5, students will use partial-quotients division to solve division problems in mathematical and rich, real-world contexts. In Grade 6, students will use the U.S. traditional division algorithm to solve division problems.
5.MD.1	In Unit 1, students converted square units and cubic units. In Grade 4, students expressed measurement quantities in a larger unit in terms of a smaller unit.	In Unit 6, students will convert between metric units. In Units 7 and 8, students will use unit conversions to help them solve rich, real-world problems. In Grade 6, students will use ratio reasoning to convert measurement units.

Planning for Rich Math Instruction

	2-1 **Understanding Place Value**	**2-2** **Exponents and Powers of 10**	**2-3** **Applying Powers of 10**	**2-4** **U.S. Traditional Multiplication, Part 1**
RIGOR — **Conceptual Understanding**	**The relationship between places in multidigit numbers** Describing Place-Value Relationships, p. 112 Representing Place Value, p. 113	**Exponential notation** Introducing Powers of 10, p. 118	**Estimation** Estimating with Powers of 10, p. 125	**Multidigit multiplication** Introducing U.S. Traditional Multiplication, p. 130
Procedural Skill and Fluency	Home Link 2-1, p. 115	Journal p. 44, #1	Math Message, p. 124 Using Powers of 10 to Multiply, p. 124 Readiness, p. 123 Extra Practice, p. 123	Mental Math and Fluency, p. 130 Math Message, p. 130 Introducing U.S. Traditional Multiplication, p. 130 Multiplying 2-Digit Numbers by 1-Digit Numbers, p. 132 Home Link 2-4, p. 133 Readiness, p. 129 Enrichment, p. 129 Extra Practice, p. 129
Applications		Introducing Powers of 10, p. 118 Solving a Real-World Volume Problem, p. 121 Enrichment, p. 117	Estimating with Powers of 10, p. 125 Writing and Comparing Expressions, p. 127 Home Link 2-3, p. 127 Enrichment, p. 123	Multiplying 2-Digit Numbers by 1-Digit Numbers, p. 132
Rich Tasks and Mathematical Reasoning	Journal p. 40: Writing/Reasoning Enrichment, p. 111	Math Message, p. 118 Introducing *High-Number Toss*, p. 120	Estimating with Powers of 10, p. 125 Enrichment, p. 123	Math Message, p. 130
Mathematical Discourse	Representing Place Value, p. 113 Introducing *Number Top-It*, p. 114	Introducing Powers of 10, p. 118 Connecting Expanded Form and Exponential Notation, p. 120 Introducing *High-Number Toss*, p. 120 Extra Practice, p. 117	Estimating with Powers of 10, p. 125	Playing *Number Top-It*, p. 133
Distributed Practice	Mental Math and Fluency, p. 112 Finding Volumes of Rectangular Prisms, p. 115 Math Boxes 2-1, p. 115	Mental Math and Fluency, p. 118 Solving a Real-World Volume Problem, p. 121 Math Boxes 2-2, p. 121	Mental Math and Fluency, p. 124 Writing and Comparing Expressions, p. 127 Math Boxes 2-3, p. 127	Mental Math and Fluency, p. 130 Playing *Number Top-It*, p. 133 Math Boxes 2-4, p. 133
Differentiation Support	Differentiation Options, p. 111 ELL Support, p. 111 Online Differentiation Support 2-1 Adjusting the Activity, p. 113 Academic Language Development, p. 113	Differentiation Options, p. 117 ELL Support, p. 117 Online Differentiation Support 2-2 Common Misconception, p. 119	Differentiation Options, p. 123 ELL Support, p. 123 Online Differentiation Support 2-3 Common Misconception, pp. 124, 125 Academic Language Development, p. 125	Differentiation Options, p. 129 ELL Support, p. 129 Online Differentiation Support 2-4 Common Misconception, p. 132 Adjusting the Activity, p. 132

Red text = Game

	2-5 U.S. Traditional Multiplication, Part 2	**2-6** Application: Unit Conversions	**2-7** U.S. Traditional Multiplication, Part 3	**2-8** U.S. Traditional Multiplication, Part 4	
Conceptual Understanding	**Multidigit multiplication** Extending U.S. Traditional Multiplication, p. 136	**Measurement conversions** Converting Miles to Feet, p. 142	**Multidigit multiplication** Introducing U.S. Traditional Multiplication with 2-Digit Factors, p. 148 Comparing Multiplication Methods, p. 151	**Multidigit multiplication** Extending U.S. Traditional Multiplication to Larger Numbers, p. 156 Choosing Multiplication Strategies, p. 158	R
Procedural Skill and Fluency	Mental Math and Fluency, p. 136 Extending U.S. Traditional Multiplication, p. 136 *Introducing Multiplication Top-It: Larger Numbers,* p. 138 Home Link 2-5, p. 139 *Readiness, p. 135* Enrichment, p. 135 Extra Practice, p. 135	Solving Unit Conversion Number Stories, p. 143 Home Link 2-6, p. 145 Enrichment, p. 141 Extra Practice, p. 141	Estimating and Multiplying, p. 152 *Introducing Multiplication Bull's Eye,* p. 153 Home Link 2-7, p. 153 *Readiness, p. 147* Extra Practice, p. 147	Mental Math and Fluency, p. 156 Math Message, p. 156 Choosing Multiplication Strategies, p. 158 *Playing Name That Number,* p. 159 Home Link 2-8, p. 159 *Extra Practice, p. 155*	I G O R
Applications	Practicing with Powers of 10, p. 139 Enrichment, p. 135	Math Message, p. 142 Solving Unit Conversion Number Stories, p. 143 Home Link 2-6, p. 145		Math Message, p. 156 Choosing Multiplication Strategies, p. 158 Home Link 2-8, p. 159	
Rich Tasks and Mathematical Reasoning	Math Message, p. 136 Journal p. 51: Writing/Reasoning Enrichment, p. 135	Solving Unit Conversion Number Stories, p. 143	Math Message, p. 148 Estimating and Multiplying, p. 152 Enrichment, p. 147	Summarize, p. 159 Enrichment, p. 155	
Mathematical Discourse	Math Message, p. 136 Extending U.S. Traditional Multiplication, p. 136 *Introducing Multiplication Top-It: Larger Numbers,* p. 138	Solving Unit Conversion Number Stories, p. 143 *Playing Prism Pile-Up,* p. 145	Comparing Multiplication Methods, p. 151 *Introducing Multiplication Bull's Eye,* p. 153 *Readiness, p. 147*	Choosing Multiplication Strategies, p. 158 *Extra Practice, p. 155*	
Distributed Practice	Mental Math and Fluency, p. 136 Practicing with Powers of 10, p. 139 Math Boxes 2-5, p. 139	Mental Math and Fluency, p. 142 *Playing Prism Pile-Up,* p. 145 Math Boxes 2-6, p. 145	Mental Math and Fluency, p. 148 *Introducing Multiplication Bull's Eye,* p. 153 Math Boxes 2-7, p. 153	Mental Math and Fluency, p. 156 *Playing Name That Number,* p. 159 Math Boxes 2-8, p. 159	
Differentiation Support	Differentiation Options, p. 135 ELL Support, p. 135 Online Differentiation Support 2-5 Academic Language Development, p. 136 Adjusting the Activity, p. 137 Common Misconception, p. 138	Differentiation Options, p. 141 ELL Support, p. 141 Online Differentiation Support 2-6 Academic Language Development, p. 143 Adjusting the Activity, p. 144	Differentiation Options, p. 147 ELL Support, p. 147 Online Differentiation Support 2-7 Adjusting the Activity, p. 149	Differentiation Options, p. 155 ELL Support, p. 155 Online Differentiation Support 2-8 Common Misconception, p. 157	

Red text = Game

Planning for Rich Math Instruction

		2-9 Open Response **One Million Taps** 2-Day Lesson	2-10 **A Mental Division Strategy**	2-11 **Reviewing Partial-Quotients Division**	2-12 **Strategies for Choosing Partial Quotients**
RIGOR	**Conceptual Understanding**	**Multiplying by powers of 10 in the context of calculating with efficiency** Discussing Efficient Strategies, p. 161	**Multidigit division** Using Multiples to Divide Mentally, p. 173	**Multidigit division** Reviewing Partial-Quotients Division, p. 178 Making Area Models, p. 180	**Multidigit division** Using Partial-Quotients Division with Lists of Multiples, p. 187
RIGOR	**Procedural Skill and Fluency**	Journal p. 59, #1, #2	Solving Extended Division Facts, p. 172 Using Multiples to Divide Mentally, p. 173 Introducing *Division Dash*, p. 174 Home Link 2-10, p. 175 Readiness, p. 171 Enrichment, p. 171 Extra Practice, p. 171	Mental Math and Fluency, p. 178 Estimating and Dividing, p. 181 Home Link 2-11, p. 183 Extra Practice, p. 177	Choosing Partial Quotients, p. 186 Using Partial-Quotients Division with Lists of Multiples, p. 187 Home Link 2-12, p. 189 Readiness, p. 185 Enrichment, p. 185 Extra Practice, p. 185
RIGOR	**Applications**	Math Message, p. 161 Solving the Open Response Problem, p. 163 Home Link 2-9, p. 169	Practicing Unit Conversions, p. 175	Math Message, p. 178	Journal p. 66, #1 Enrichment, p. 185
	Rich Tasks and Mathematical Reasoning	Solving the Open Response Problem, p. 163 Reengaging in the Problem, p. 168 Revising Work, p. 168	Using Multiples to Divide Mentally, p. 173 Enrichment, p. 171	Estimating and Dividing, p. 181	Enrichment, p. 185
	Mathematical Discourse	Discussing Efficient Strategies, p. 161 Setting Expectations, p. 167 Reengaging in the Problem, p. 168	Introducing *Division Dash*, p. 174 Summarize, p. 175 Enrichment, p. 171 Extra Practice, p. 171	Estimating and Dividing, p. 181 Enrichment, p. 177	Math Message, p. 186 Summarize, p. 189 Introducing *Power Up*, p. 189 Readiness, p. 185 Extra Practice, p. 185
	Distributed Practice	Mental Math and Fluency, p. 161 Math Boxes 2-9, p. 169	Mental Math and Fluency, p. 172 Practicing Unit Conversions, p. 175	Mental Math and Fluency, p. 178 Math Boxes 2-11, p. 183	Mental Math and Fluency, p. 186 Introducing *Power Up*, p. 189 Math Boxes 2-12, p. 189
	Differentiation Support	ELL Support, p. 162 Adjusting the Activity, pp. 163, 168	Differentiation Support, p. 171 ELL Support, p. 171 Online Differentiation Support 2-10 Adjusting the Activity, pp. 172, 174	Differentiation Support, p. 177 ELL Support, p. 177 Online Differentiation Support 2-11 Adjusting the Activity, p. 179	Differentiation Support, p. 185 ELL Support, p. 185 Online Differentiation Support 2-12 Adjusting the Activity, p. 186

Red text = Game

Notes

2-13

Interpreting the Remainder

Interpreting division contexts

Modeling a Division Problem,
pp. 192–194

Mental Math and Fluency, p. 192

Interpreting Remainders,
pp. 194–196

Home Link 2-13, p. 197

Enrichment, p. 191

Extra Practice, p. 191

Math Message, p. 192

Modeling a Division Problem,
pp. 192–194

Interpreting Remainders,
pp. 194–196

Home Link 2-13, p. 197

Differentiation Options, p. 191

Interpreting Remainders,
pp. 194–196

Enrichment, p. 191

Modeling a Division Problem,
pp. 192–194

Interpreting Remainders,
pp. 194–196

Mental Math and Fluency, p. 192

Playing *High-Number Toss*, p. 197

Math Boxes 2-13, p. 197

Differentiation Options, p. 191

ELL Support, p. 191

Online Differentiation
Support 2-13

Common Misconception, p. 193

Academic Language
Development, p. 193

**2-14 Assessment
Unit 2 Progress Check**

**Lesson 2-14 is an assessment
lesson. It includes:**

• **Self Assessment**

• **Unit Assessment**

• **Optional Challenge
Assessment**

• **Cumulative Assessment**

• **Suggestions for adjusting
the assessments.**

Go Online:

Evaluation Quick Entry
Use this tool to record students'
performance on assessment tasks.

Data Use the Data Dashboard to
view students' progress reports.

Unit 2 Materials

Lesson	Math Masters	Activity Cards	Manipulative Kit	Other Materials
2-1	pp. 43–44; TA5; per partnership: G7–G9	15	base-10 blocks; number cards 1–9 (1 of each); per partnership: number cards 0–9 (4 of each)	calculator
2-2	pp. per partnership: 45–46; 47; per partnership: G10; G11	16	per partnership: two 6-sided dice	slate; scissors; per group: calculator (optional)
2-3	pp. 48–49; TA5 (optional); TA6	17	number cards 1–10 (1 of each); per partnership: number cards 0–9 (4 of each)	slate; per partnership: poster paper
2-4	pp. 50–51; TA7 (optional); per partnership: G7–G9	18	per partnership: number cards 0–9 (4 of each)	slate; calculator
2-5	pp. 52–54; TA6; per group: G3	19	per partnership: number cards 0–9 (4 of each); per group: number cards 1–10 (4 of each); 4 counters	per group: calculator or multiplication/division facts table
2-6	pp. 55; TA2 (optional); per partnership: G4–G5 (optional), G6	20–21	number cards 1–20 (1 of each); two 6-sided dice; per group: three 12-inch rulers, 36 square pattern blocks	per partnership: *Prism Pile-Up* cards, calculator (optional)
2-7	pp. 56–59; G12		per partnership: number cards 0–9 (4 of each), 6-sided die	slate
2-8	pp. 60; per partnership: G2	22	per partnership: number cards 0–10 (4 of each) and number cards 11–20 (1 of each)	per group: poster paper, crayons or markers
2-9	pp. 61–64; TA4		per partnership: stopwatch (optional)	slate; Guidelines for Discussion Poster; colored pencils (optional); selected samples of students' work; students' work from Day 1
2-10	pp. 65; per partnership: TA8; G13	23–24	per partnership: number cards 1–9 (4 of each), number cards 10–20 (1 of each), two 6-sided dice, 40 counters	slate
2-11	pp. 66; TA7 (optional); TA9	25	per partnership: number cards 0–9 (4 of each), tape measure	calculator (optional)
2-12	pp. 67–69; TA7 (optional); TA9–TA10; per partnership: G11	26	number cards 10–20 (1 of each); per partnership: two 6-sided dice	slate; calculator (optional)
2-13	pp. 70–72; TA7 (optional); TA10 (optional); TA11; per partnership: G10	27	6-sided die	
2-14	pp. 73–76; *Assessment Handbook*, pp. 14–22			

Literature Link   Optional Books: **2-3** *Two of Everything: A Chinese Folktale*   **2-9** *One Odd Day; My Even Day*

Go Online for a complete literature list for Grade 2 and to download all Quick Look Cards.

Assessment Check-In

These ongoing assessments offer an opportunity to gauge students' performance on one or more of the standards addressed in that lesson.

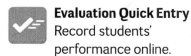 **Evaluation Quick Entry** Record students' performance online.

Data View reports online to see students' progress towards mastery.

Lesson	Task Description	Content Standards	Processes and Practices
2-1	Write numbers in expanded form and identify values of digits.	5.NBT.1	SMP7
2-2	Multiply whole numbers by powers of ten and write the product in standard notation.	5.NBT.2	
2-3	Use powers of 10 to estimate products and explain reasoning.	5.NBT.2	SMP6
2-4	Multiply 2-digit numbers by 1-digit numbers using U.S. traditional multiplication and other strategies.	5.NBT.5	SMP1
2-5	Multiply multidigit numbers by 1-digit numbers using U.S. traditional multiplication.	5.NBT.5	
2-6	Solve number stories involving U.S. customary unit conversions and write expressions to model problems.	5.OA.2, 5.MD.1	SMP4
2-7	Multiply two 2-digit numbers using U.S. traditional multiplication.	5.NBT.5	
2-8	Multiply multidigit numbers using U.S. traditional multiplication.	5.NBT.5	
2-9	Use patterns of powers of 10 to calculate an estimate.	5.NBT.2	SMP6
2-10	Divide multidigit numbers using informal strategies.	5.NBT.6	SMP6
2-11	Use partial-quotients division to solve problems with 3-digit and 4-digit dividends.	5.NBT.6	SMP2
2-12	Use partial-quotients division to solve problems with 4-digit dividends.	5.NBT.6	
2-13	Create mathematical models to solve division problems and interpret remainders.	5.NBT.6	SMP4

Virtual Learning Community
vlc.uchicago.edu

While planning your instruction for this unit, visit the *Everyday Mathematics* Virtual Learning Community. You can view videos of lessons in this unit, search for instructional resources shared by teachers, and ask questions of *Everyday Mathematics* authors and other educators. Some of the resources on the VLC related to this unit include:

EM4: Grade 5 Unit 2 Planning Webinar
This webinar provides a preview of the lessons and content in this unit. Watch this video with your grade-level colleagues and plan together under the guidance of an *Everyday Mathematics* author.

Choosing Multiplication Strategies
Watch students solve a multiplication problem in two ways and discuss what they like and dislike about each method. The teacher concludes the discussion by pointing out a third method that also works.

ACI Booklet: Grade 5, Unit 2
This booklet is a collection of PDFs of all the student pages in the unit that include concepts leading up to the ACI. Each page shows you where to find the ACI information in the Teacher's Lesson Guide.

For more resources, go to the VLC Resources page and search for Grade 5.

Spiral Towards Mastery

The *Everyday Mathematics* curriculum is built on the spiral, where standards are introduced, developed, and mastered in multiple exposures across the grade. Go to the Teacher Center at my.mheducation.com to use the Spiral Tracker.

Spiral Towards Mastery Progress This Spiral Trace outlines instructional trajectories for key standards in Unit 2. For each standard, it highlights opportunities for Focus instruction, Warm Up and Practice activities, as well as formative and summative assessment. It describes the **degree of mastery**—as measured against the entire standard—expected at this point in the year.

Operations and Algebraic Thinking

5.OA.2

Progress Towards Mastery By the end of Unit 2, expect students to write expressions to model situations in which no more than two operations are involved; reason about the relative value of simple expressions without evaluating them.

Full Mastery of 5.OA.2 expected by the end of Unit 8.

Number and Operations in Base Ten

5.NBT.1

Progress Towards Mastery By the end of Unit 2, expect students to use place-value understanding to write whole numbers in expanded form; identify the values of digits in a given whole number; write whole numbers in which digits represent given values; recognize that in a multidigit whole number, a digit in one place represents 10 time what it represents in the place to its right.

Full Mastery of 5.NBT.1 expected by the end of Unit 8.

5.NBT.2

Progress Towards Mastery By the end of Unit 2, expect students to translate between powers of 10 in exponential notation and standard notation; correctly multiply a whole number by a power of 10; notice patterns in the number of zeros in a product when multiplying a whole number by a power of 10.

Full Mastery of 5.NBT.2 expected by the end of Unit 4.

Key ✓ = Assessment Check-In   = Progress Check Lesson   = Current Unit   = Previous or Upcoming Lessons

McGraw-Hill Education

5.NBT.5

2-4 through 2-9 Focus Practice | 2-11 through 2-13 Practice | 2-14 Progress Check | 3-1 Practice | 3-3 through 3-5 Practice | 3-7 Practice | 3-9 Practice | 3-11 through 3-14 Practice

Progress Towards Mastery By the end of Unit 2, expect students to use a strategy to multiply whole numbers; understand the basic steps of the U.S. traditional multiplication algorithm and successfully apply it to 1-digit by multidigit problems and 2-digit by 2-digit problems in which one factor is less than 20.

Full Mastery of 5.NBT.5 expected by the end of Unit 7.

5.NBT.6

2-10 through 2-13 Warm Up Focus Practice | 2-14 Progress Check | 3-1 Practice | 3-2 Warm Up Practice | 3-3 Warm Up Focus Practice | 3-5 Focus Practice | 3-6 Practice | 3-9 Practice | 3-12 Practice | 3-14 Warm Up | 3-15 Progress Check

Progress Towards Mastery By the end of Unit 2, expect students to use partial-quotient algorithm with up to 3-digit dividends and 1-digit or simple 2-digit divisors; make connections between written partial-quotients work and a given area model representing the same solution.

Full Mastery of 5.NBT.6 expected by the end of Unit 5.

Measurement and Data

5.MD.1

1-3 Warm Up Focus Practice | 1-6 Warm Up Practice | 1-8 Practice | 1-10 Warm Up Focus Practice | 1-11 Focus | 2-6 Warm Up Focus Practice | 2-9 Practice | 2-10 Practice | 2-12 Practice | 2-14 Progress Check | 3-1 Practice | 3-3 Practice

Progress Towards Mastery By the end of Unit 2, expect students to perform one-step unit conversions within the same measurement system; use conversions to solve real-world problems when necessary conversions are identified.

Full Mastery of 5.MD.1 expected by the end of Unit 6.

Key  = Assessment Check-In    = Progress Check Lesson   = Current Unit   = Previous or Upcoming Lessons

🍎 **Professional Development**

Mathematical Background:
Content

▶ **Place-Value Patterns** (Lessons 2-1 and 2-2)

We write numbers using a base-10 place-value system in which the value of a digit depends on its place in a number. In base 10, the value of each digit is 10 times what it would be in the place to its right. **5.NBT.1** For example, a 2 in the ones place, as in **72**, is worth just 2, but a 2 in the tens place, as in **23**, is worth 10 times as much, or 20. A 2 in the hundreds place, as in **2**30, is worth another 10 times as much, or 200, and so on. *(See margin.)*

In earlier grades, students represented multidigit numbers using *expanded form,* in which a number is written as the sum of the values of each digit. Students continue to use expanded form in Grade 5. Different versions of expanded form illuminate important aspects of the place-value system. For example, each of the following is a version of expanded form for 65,682:

- 60,000 + 5,000 + 600 + 80 + 2
- 6 ten thousands + 5 thousands + 6 hundreds + 8 tens + 2 ones
- (6 * 10,000) + (5 * 1,000) + (6 * 100) + (8 * 10) + (2 * 1)

In the first expression, it is not immediately obvious how the value of a digit in the ones place relates to the value of a digit in the tens place. However, with the digit separated from its place in the other two expressions, as in 2 ones or (2 * 1), students can more easily recognize that *ten* is 10 times as much as *one*, so a digit in the tens place is worth 10 times as much as it would be in the ones place.

Students began to explore and describe this *10 times as much* pattern in Grade 4. In Grade 5, students also consider how the value of a digit relates to the place going in the other direction. They observe that if a digit moves one place to the *right*, its value is divided by 10. For example, the 2 in **27** is worth 20, but 2 in **72** is worth $20 \div 10$, or 2. Students reason that dividing by 10, or dividing a value into 10 equal parts, is the same as taking $\frac{1}{10}$ of the value. They recognize that a digit in a given place represents $\frac{1}{10}$ of what it represents in the place to its left. **5.NBT.1** *(See margin.)* In later units, students will extend the *10 times as much* and $\frac{1}{10}$ *of* patterns to decimals.

▶ **Powers of 10 and Exponential Notation**
(Lessons 2-2 and 2-3)

In this unit, students are introduced to powers of 10 and exponential notation. *Powers of 10* are numbers that can be written as a product of 10s. For example, 1,000 is a power of 10 because it can be written as 10 * 10 * 10. *Exponential notation* is a way of representing repeated multiplication by the same factor. For example, 10^3 is exponential notation for 10 * 10 * 10, or 1,000. The *exponent,* 3, tells how many times the *base,* 10, is used as a factor. While any number can be used as a base, students in Grade 5 are only expected to use exponents to denote powers of 10. **5.NBT.2**

Standards and Goals for Mathematical Content

Because the standards within each strand can be broad, *Everyday Mathematics* has unpacked each standard into Goals for Mathematical Content GMC. For a complete list of Standards and Goals, see page EM1.

	10 *		10 *		10 *	

1,000s Thousands	100s Hundreds	10s Tens	1s Ones
			2
		2	3
	2	3	0
2	3	0	0

A place-value chart showing the *10 times as much* relationship between places

1,000s Thousands	100s Hundreds	10s Tens	1s Ones
6	5	0	0
	6	5	0

A place-value chart showing the $\frac{1}{10}$ *of* relationship between places

Unit 2 Vocabulary

algorithm	exponent	partial quotient
area model	exponential notation	partial-quotients division
base	expression	place value
convert	extended multiplication fact	power of 10
dividend	mathematical model	quotient
divisor	measurement units	relation symbol
efficient	multiple	remainder
estimate	number model	standard notation
expanded form	partial-products multiplication	U.S. traditional multiplication

In Lesson 2-2 students look for patterns in powers of 10. They observe that the number of zeros in a power of 10 written in standard notation matches both the exponent in exponential notation and the number of times 10 is used as a factor. Students also learn that some numbers can be expressed as multiples of powers of 10. The number 65,000, for instance, can be represented as $65 * 1,000$, or $65 * 10^3$. Students connect this idea to expanded form. For example, they note that 3,745 can be expressed as $(3 * 10^3) + (7 * 10^2) + (4 * 10^1) + (5 * 10^0)$.

Students use powers of 10 to reason about *extended multiplication facts*, which are variations of basic facts involving multiples of 10, 100, and so on. In Lesson 2-3 students solve problems like $50 * 400$ by thinking: *I can rewrite $50 * 400$ as $5 * 10^1 * 4 * 10^2$. $5 * 4 = 20$. Multiplying by 10^1 means I attach one zero. Multiplying by 10^2 means I attach two zeros. That is three attached zeros, which gives 20,000.* Students discuss and generalize these patterns. **5.NBT.2**

> ## Understanding U.S. Traditional Multiplication
> ### (Lessons 2-4 through 2-9)

Standards require students in Grade 5 to fluently multiply multidigit numbers using the standard algorithm. **5.NBT.5** In *Everyday Mathematics,* the standard algorithm is referred to as *U.S. traditional multiplication* to acknowledge that it is not standard in all parts of the world.

In Grade 4, students multiplied numbers using *partial-products multiplication,* a method based on place value and the Distributive Property. In partial-products multiplication, each factor is thought of as a sum of ones, tens, hundreds, and so on. Each part of one factor is multiplied by each part of the other factor, and all of the resulting *partial products* are added. (*See margin.*) This method helps students keep track of their work by separating the multiplication steps from the addition steps.

U.S. traditional multiplication compresses this process. Each digit of one factor is multiplied by each digit of the other factor, but the partial products are added mentally before being recorded. This means that multiplication steps alternate with addition steps, and the notation used to record steps makes it more difficult to see the values of digits. (*See margin.*) When students multiply 3 by the 5 in 752, it is not immediately apparent that students are multiplying 3 by 5 *tens* to get 150.

NOTE To make the connection between powers of 10 in exponential notation and expanded form, students need to know that 1 can be represented as 10^0. This is true by definition, as any nonzero number to the zero power is defined as 1. Although this may seem counterintuitive, the definition preserves important properties of exponents. In Grade 5, students are not expected to understand the rationale behind the 10^0 definition.

$$
\begin{array}{rrrr}
 & 7 & 5 & 2 \\
* & & & 3 \\
\hline
\end{array}
$$

$3 * 700 \rightarrow \quad 2 \quad 1 \quad 0 \quad 0$
$3 * 50 \rightarrow \quad\quad 1 \quad 5 \quad 0$
$3 * 2 \rightarrow \quad + \quad\quad\quad 6$

$$
\begin{array}{rrrr}
2, & 2 & 5 & 6 \\
\end{array}
$$

Partial-products multiplication

$$
\begin{array}{rrrr}
 & 1 & & \\
 & 7 & 5 & 2 \\
* & & & 3 \\
\hline
2, & 2 & 5 & 6 \\
\end{array}
$$

U.S. traditional multiplication

Professional Development

To help students learn the steps of U.S. traditional multiplication and understand why those steps make sense, *Everyday Mathematics* presents U.S. traditional multiplication alongside the partial-products method. Students solve problems using both methods and compare the steps and results. While the two methods may appear to be very different, they both involve finding and adding partial products. Area models can illustrate connections between the partial products in each method.

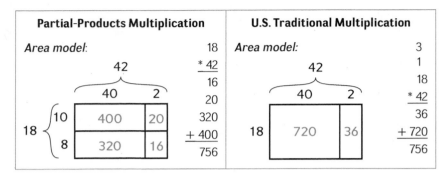

Since Unit 2 is the first exposure to U.S. traditional multiplication, many students may find it challenging. Do not expect students to use it easily right away, but do encourage them to solve problems in more than one way and use estimates to check whether their answers make sense. There will be many opportunities throughout the year for students to practice U.S. traditional multiplication.

▶ Dividing Multidigit Numbers
(Lessons 2-10 through 2-13)

The end of Unit 2 focuses on division. Lesson 2-10 gives students an opportunity to refresh their division-fact and extended-fact knowledge. They learn a strategy for mental division in which the dividend is broken into two or more easy-to-divide parts.

Lesson 2-11 reviews *partial-quotients division,* a method that was introduced in Grade 4. Partial-quotients division is a way of answering the question, "How many of these are in that?" Or for $a \div b$, "How many b's are in a?" Using multiples of the divisor, students build up a series of interim answers, or *partial quotients.* At each step, if not enough b's have been taken from a, more are taken. When all possible b's have been taken, the partial quotients are added.

Students in Grade 5 extend this method to problems with two-digit divisors. (*See margin.*) **5.NBT.6** Because it is conceptually transparent, partial-quotients division is the focus for Grade 5. U.S. traditional long division will be formally introduced in Grade 6.

Strategies for partial-quotients division are described in detail in Lessons 2-11 and 2-12 and in the *Student Reference Book.* Students illustrate and explain their work using area models. (*See margin.*) In the context of division, the dividend (in this case 156) is the total area of the rectangle, and the divisor (12) is the length. Each partial quotient corresponds to one segment of the width of the rectangle (10 + 3). The total width is the final quotient (in this case 13). Area models are not intended to be a separate solution strategy but are instead meant to help students see what the steps in partial-quotients division mean.

In Lesson 2-13 students apply their understanding of division to solve real-world problems and focus on interpreting remainders in problem contexts.

Partial-quotients division

Area (Dividend): 156
Length (Divisor): 12

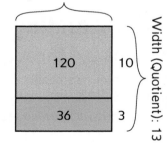

Area model for the partial-quotients solution shown above

Mathematical Background:
Process and Practice

 See below for some of the ways that students engage in **SMP1 Make sense of problems and persevere in solving them** and **SMP6 Attend to precision** through the mathematical content of **Number and Operations in Base-Ten Measurement and Data** and **Measurement and Data.**

▶ Standard for Mathematical Process and Practice 1

In Unit 2, students encounter and solve many interesting problems. To do so successfully, they have to make sense of their problems, find entry points to work towards solutions, monitor their own progress, and evaluate their answers. **SMP1**

In Lesson 2-6 students make sense of multistep problems by thinking about what information they need to solve the problems. They discuss ways to start working towards a solution, such as drawing a picture, making a table, or writing an expression. **GMP1.1** In Lesson 2-5 students generate strategies to solve a new problem by reflecting on how they solved similar, but easier problems. **GMP1.2** For example, they consider what they already know about solving 2-digit by 1-digit multiplication problems to help them solve a 3-digit by 1-digit problem.

Standard for Mathematical Process and Practice 1 also emphasizes the importance of asking the question: "Does my answer make sense?" **GMP1.4** In Lesson 2-3 students use their understanding of powers of 10 to judge the reasonableness of answers. For example, they consider whether $492 * 63 = 480{,}992$ makes sense by reasoning: *I would have to multiply 492 by about 1,000 to get close to 480,992, so 480,992 can't be correct.*

As students develop their problem-solving abilities, they learn to solve problems in more than one way and to compare strategies they and their classmates use. **GMP1.5, GMP1.6** For example, in Lessons 2-4 and 2-8, students solve multiplication problems using both partial-products multiplication and U.S. traditional multiplication. They discuss how the steps of one method connect to the steps of the other.

▶ Standard for Mathematical Practice 6

When students "attend to precision," they pay attention to accuracy and efficiency as they count, measure, and calculate. **GMP6.4** In Lesson 2-3, for example, students use estimation, rounding, and their knowledge of powers of 10 to arrive at answers more efficiently. As students develop division strategies in Lessons 2-10 through 2-12, they use extended facts to make division easier. They also discover that while all sets of partial quotients will produce the same answer to a problem, some partial quotients allow for completing the division in fewer steps.

Standard for Mathematical Practice 6 also emphasizes the importance of clear and precise mathematical explanations. **GMP6.1** In Lesson 2-2 students are encouraged to explain their thinking clearly as they talk to a partner about what they think a raised number, or exponent, means. In Lesson 2-3 students clearly explain how they solved problems mentally. Throughout *Fifth Grade Everyday Mathematics,* question prompts and sample explanations are provided to encourage both teachers and students to clearly communicate their mathematical ideas.

Standards and Goals for Mathematical Process and Practice

SMP1   Make sense of problems and persevere in solving them.

GMP1.1 Make sense of your problem.

GMP1.2 Reflect on your thinking as you solve your problem.

GMP1.3 Keep trying when your problem is hard.

GMP1.4 Check whether your answer makes sense.

GMP1.5 Solve problems in more than one way.

GMP1.6 Compare the strategies you and others use.

SMP6   Attend to precision.

GMP6.1 Explain your mathematical thinking clearly and precisely.

GMP6.2 Use an appropriate level of precision for your problem.

GMP6.3 Use clear labels, units, and mathematical language.

GMP6.4 Think about accuracy and efficiency when you count, measure, and calculate.

Go Online to the *Implementation Guide* for more information about the Mathematical Process and Practice Standards.

For students' information on the Mathematical Process and Practice Standards, see *Student Reference Book*, pages 1–34.

Lesson 2-1

Understanding Place Value

Overview Students explore the multiplicative relationships between places in multidigit numbers.

▶ **Before You Begin**

For Part 2, decide how you will display the place-value chart. Consider assembling the *Number Top-It* gameboards from *Math Masters*, pages G7 and G8.

▶ **Vocabulary**

place value • standard notation • expanded form

Standards

Focus Cluster
• Understand the place value system.

① Warm Up 5 min	**Materials**	
Mental Math and Fluency Students find volumes of rectangular prisms.		5.MD.5, 5.MD.5b

② Focus 30–40 min		
Math Message Students record the value of digits in numbers.	*Math Journal 1*, p. 37	5.NBT.1 SMP7
Describing Place-Value Relationships Students describe relationships in the place-value system.	*Math Journal 1*, p. 37; *Math Masters*, p. TA5; base-10 blocks (optional)	5.NBT.1 SMP7
Representing Place Value Students use standard notation and expanded form.	*Math Journal 1*, pp. 37–38	5.NBT.1 SMP2, SMP7
✓ **Assessment Check-In** See page 114. Expect that most students will be able to write numbers that have up to 8 digits in expanded form and identify values of digits in numbers that contain up to 9 digits.	*Math Journal 1*, pp. 37–38; *Math Masters*, p. TA5 (optional); base-10 blocks (optional)	5.NBT.1 SMP7
Introducing *Number Top-It* **Game** Students use place-value understanding to build and compare multidigit numbers.	*Student Reference Book*, p. 316; per partnership: *Math Masters*, pp. G7–G9; number cards 0–9 (4 of each)	5.NBT.1 SMP7

③ Practice 20–30 min		
Finding Volumes of Rectangular Prisms Students find volumes of rectangular prisms that are partially packed with cubes.	*Math Journal 1*, p. 39	5.MD.3, 5.MD.3a, 5.MD.3b, 5.MD.4, 5.MD.5, 5.MD.5a SMP2
Math Boxes 2-1 Students practice and maintain skills.	*Math Journal 1*, p. 40	See page 115.
Home Link 2-1 **Homework** Students use place-value clues to solve riddles.	*Math Masters*, p. 44	5.OA.1, 5.NBT.1 SMP1, SMP7

 Go Online to see how mastery develops for all standards within the grade.

Differentiation Options

Readiness	5–15 min
WHOLE CLASS	**SMALL GROUP**	PARTNER	INDEPENDENT

Modeling Numbers with Base-10 Blocks

5.NBT.1, SMP7

base-10 blocks

To explore place-value relationships using concrete models, students identify relationships between the values of base-10 blocks. Display a unit cube, a long, a flat, and a thousands cube. Ask: *Which block is worth 10 times as much as the unit cube?* One long If needed, demonstrate how 10 unit cubes can be lined up to make 1 long, so 1 long is worth *10 times as much as* a cube. Repeat with other base-10 blocks. When students seem comfortable, ask them to identify the block that is worth $\frac{1}{10}$ of the one you display. If necessary, demonstrate how to find $\frac{1}{10}$ of something by dividing it into 10 equal-size pieces and taking 1 of the pieces. A unit cube is $\frac{1}{10}$ of a long because a long can be divided into 10 cubes. Continue with $\frac{1}{10}$ of a flat One long and $\frac{1}{10}$ of a thousands cube. One flat Finally, ask students to model the numbers 2; 20; 200; and 2,000. As they make each new number, ask: *What number is 10 times as much as this number? What number is $\frac{1}{10}$ of this number?* **GMP7.1**

Enrichment	5–15 min
WHOLE CLASS	**SMALL GROUP**	PARTNER	INDEPENDENT

Exploring Base-5 Place Value

5.NBT.1, SMP7

Math Masters, p. 43

To extend their understanding of base-10 place-value relationships, students explore the base-5 number system. They study a table of numbers in base-5 and then practice converting between base-10 and base-5. They represent base-5 numbers in expanded form. **GMP7.1, GMP7.2**

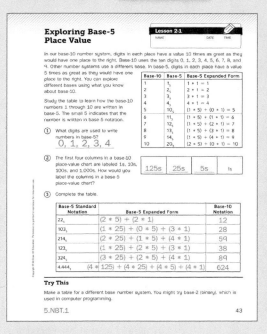

Extra Practice	5–15 min
WHOLE CLASS	**SMALL GROUP**	**PARTNER**	INDEPENDENT

Calculating to Explore Place-Value Relationships

5.NBT.1, SMP7

Activity Card 15;
Math Masters, p. TA5; number cards 1–9 (1 of each); calculator

For more practice describing place-value relationships, students use calculators to multiply and divide numbers by 10 and record the results. **GMP7.1**

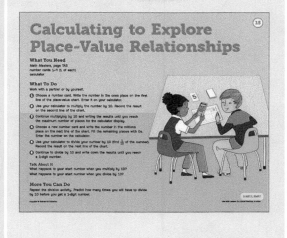

English Language Learner

Beginning ELL Use everyday experiences and terms to help students actively understand and make connections between the terms *value* and *worth*. Show examples of car advertisements featuring new and used cars from several years. Point to the prices, making think-aloud statements like these: *This 2014 car costs the most. It has the highest value. It is worth the most money. This 2009 model is worth the least. It has the least value. It costs the least amount of money.*

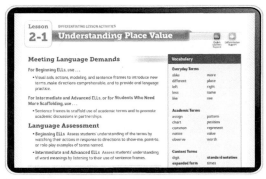

Differentiation Support pages are found in the online Teacher's Center.

SMP2 **Reason abstractly and quantitatively.**
GMP2.3 Make connections between representations.

SMP7 **Look for and make use of structure.**
GMP7.1 Look for mathematical structures such as categories, patterns, and properties.
GMP7.2 Use structures to solve problems and answer questions.

1 Warm Up 5 min

▶ Mental Math and Fluency

Display the formulas $V = l * w * h$ and $V = B * h$. Give dimensions of rectangular prisms and have students find the volumes. *Leveled exercises:*

● ○ ○ 5 unit cubes by 2 unit cubes by 3 unit cubes 30 cubic units
base area 20 square units and height 4 units 80 cubic units

● ● ○ 3 inches by 4 inches by 5 inches 60 cubic inches
base area 25 square inches and height 5 inches 125 cubic inches

● ● ● 7 feet by 5 feet by 4 feet 140 cubic feet
base area 100 square feet and height 12 feet 1,200 cubic feet

2 Focus 30–40 min

▶ Math Message

Math Journal 1, p. 37

Complete Problems 1 and 2 on journal page 37. Talk with a partner about the patterns you notice in the values of the 2 and 6 in each number. GMP7.1

▶ Describing Place-Value Relationships

Math Journal 1, p. 37; *Math Masters*, p. TA5

| WHOLE CLASS | SMALL GROUP | PARTNER | INDEPENDENT |

Math Message Follow-Up Invite several students to share what they noticed about the position, or place, of the 2 in each number in Problem 1. It moves one place to the left in each number. Then discuss the value of the 2 in each number, using students' responses to highlight how each 2 has a value that is 10 times the value it would have one place to the right. For example, the 2 in 23 is worth 20 because it is in the tens place, but the 2 in 230 is worth 200 because it is in the hundreds place. GMP7.1

Ask students to share their observations about the position and the value of the 6 in each number in Problem 2. The 6 moves one place to the right in each number. Its value is less each time it moves to the right. Discuss how our number system assigns every digit a value based on its place in a number. This system is a **place-value** system. Tell students that in today's lesson they will learn more about our place-value system and how to represent the values of digits in numbers.

Math Journal 1, p. 37

Place-Value Relationships Lesson 2-1 DATE TIME

Math Message

① What is the value of the 2 in the following numbers?
2 _2_
23 _20_
230 _200_
2,300 _2,000_
23,000 _20,000_

② What is the value of the 6 in the following numbers?
65,000 _60,000_
6,500 _6,000_
650 _600_
65 _60_
6 _6_

③ Write these numbers in expanded form. Sample answers:
a. 2,387,926 = (2 * 1,000,000) + (3 * 100,000) + (8 * 10,000) + (7 * 1,000) + (9 * 100) + (2 * 10) + (6 * 1)
b. 92,409,224 = 90,000,000 + 2,000,000 + 400,000 + 9,000 + 200 + 20 + 4

④ Write these numbers in standard notation.
a. 4 (100,000s) + 5 (10,000s) + 0 (1,000s) + 3 (100s) + 6 (10s) + 2 (1s) = _450,362_
b. (9 * 10,000) + (3 * 1,000) + (4 * 100) + (9 * 10) + (1 * 1) = _93,491_
c. 3 ten-thousands + 2 thousands + 5 hundreds + 7 tens + 9 ones = _32,579_

⑤ How does expanded form help you see the patterns in our place-value system?
Sample answer: Expanded form shows how much each digit is worth. It shows how the value of each digits depends on its place in the number.

5.NBT.1, SMP7 37

Display the place-value chart on *Math Masters*, page TA5 and have students help you complete it using the numbers from Problem 1.

Add arrows above the chart as shown below to illustrate the relationship between places. Have volunteers read the following numbers and add them to the chart: 23,000; 230,000; and 2,300,000.

	10 *		10 *		10 *	

1,000s Thousands	100s Hundreds	10s Tens	1s Ones
			2
		2	3
	2	3	0
2	3	0	0

A partially filled place-value chart showing
the *10 times* relationship between places

Repeat the process for Problem 2. Highlight how each 6 has a value that is $\frac{1}{10}$ of the value it has in the place to its left. For example, the 6 in 6,500 is worth 6,000 because it is in the thousands place, but the 6 in 650 is worth $\frac{1}{10}$ of 6,000, or 600, because it is in the hundreds place. Finding $\frac{1}{10}$ of a number is the same as dividing that number by 10. Add arrows above the place-value chart as shown below to illustrate this relationship.

	$\frac{1}{10}$ *		$\frac{1}{10}$ *		$\frac{1}{10}$ *	

1,000s Thousands	100s Hundreds	10s Tens	1s Ones
6	5	0	0
	6	5	0

A partially filled place-value chart showing
the $\frac{1}{10}$ *of* relationship between places

Explain that a pattern is a mathematical structure. Observing and describing mathematical structures, like place-value relationships, help us understand our number system and solve problems. `GMP7.1`

▶ Representing Place Value

Math Journal 1, pp. 37–38

WHOLE CLASS	SMALL GROUP	PARTNER	INDEPENDENT

Display the number 65,582 and explain that it is written in **standard notation.** Standard notation is the most common way of representing numbers. Another way to represent a number is by writing it in **expanded form.** Expanded form represents a number as the sum of the values of each digit.

Adjusting the Activity

Differentiate   For students who confuse changes in magnitude to the left and right, suggest that they act out a place-value machine using base-10 blocks. Line up four students and label them 1,000s; 100s; 10s; and 1s. The student labeled 1s chooses a small number of cubes, states the value, and passes them to the "10s" student, saying, "times 10." The "10s" student trades the cubes for longs, states the value, and passes them to the "100s" student, saying, "times 10," and so on. Repeat the process beginning with the student labeled 1,000s and the largest place-value block, saying, "$\frac{1}{10}$ of" or "divided by 10" when passing them back to the "100s" student.

Go Online Differentiation Support

Academic Language Development

Contrast numbers written in standard notation and expanded form to help students actively construct the meaning of *standard* as "that which is normally used." Ask students which of the two forms—standard or expanded—they are more likely to see in everyday use.

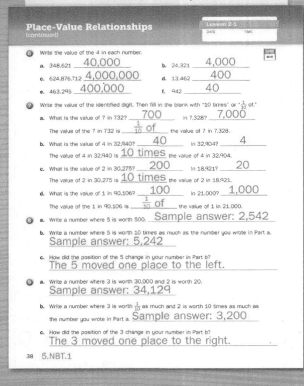

Math Journal 1, p. 39

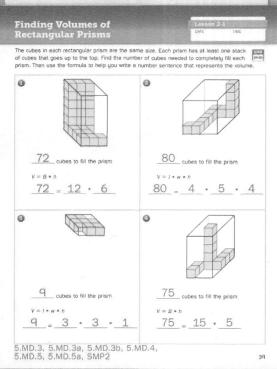

Show students several ways to represent the number 65,582 in expanded form.

- 6 ten thousands + 5 thousands + 5 hundreds + 8 tens + 2 ones
- 6 [10,000s] + 5 [1,000s] + 5 [100s] + 8 [10s] + 2 [1s]
- 60,000 + 5,000 + 500 + 80 + 2
- (6 * 10,000) + (5 * 1,000) + (5 * 100) + (8 * 10) + (2 * 1)

Compare and contrast different versions of expanded form, highlighting how the value of each digit is represented. **GMP2.3** Guide the discussion to include these points:

- The first example uses words to show the place names, and the second example uses numbers in brackets to show the place names.
- The third example shows the value of each digit in standard notation.
- The parentheses in the fourth example are not necessary, but they make it easier to see the place-value patterns.
- All of the different versions represent the same number.

Ask: *What do you notice about the values of the 5s in the number?*
Sample answer: One 5 is worth 500 and the other is worth 10 times as much, or 5,000.

Have students complete journal page 37. Invite volunteers to share their answers. Discuss how expanded form helps them identify place-value patterns. Then have students complete journal page 38. **GMP7.2**

 Assessment Check-In 5.NBT.1

Math Journal 1, pp. 37–38

Expect that most students will be able to write numbers in expanded form in Problem 3 and identify values of digits in Problem 6 on journal pages 37 and 38. Some will be able to apply the *10 times* and $\frac{1}{10}$ *of* relationships to solve Parts b and c of Problems 8 and 9. If students struggle, suggest that they write the numbers on a place-value chart (*Math Masters,* page TA5) or use base-10 blocks to visualize the value of the numbers in each place. **GMP7.2**

 Evaluation Quick Entry Go online to record students' progress and to see trajectories toward mastery for these standards.

▶ **Introducing *Number Top-It***

Student Reference Book, p. 316; *Math Masters,* pp. G7–G9

| WHOLE CLASS | SMALL GROUP | PARTNER | INDEPENDENT |

Students apply their understanding of place-value relationships and expanded form by playing *Number Top-It*.

Review the game directions on *Student Reference Book,* page 316. Play a sample round with a volunteer. Demonstrate how players should record their final numbers in both standard notation and expanded form on *Math Masters,* page G9. **GMP7.2**

Observe

• Which students seem to have a strategy for placing the cards?

• Which students need assistance playing the game?

Discuss

• *How did you decide where to place your cards?* GMP7.2

• *What would happen to the value of the number on the card if you placed it one place to the left (or the right)?* GMP7.1

Differentiate **Game Modifications** (Go Online) Differentiation Support

Summarize After one round of the game, have partners share which number card they could move to a new place on their board to make a larger number and explain why they chose it.

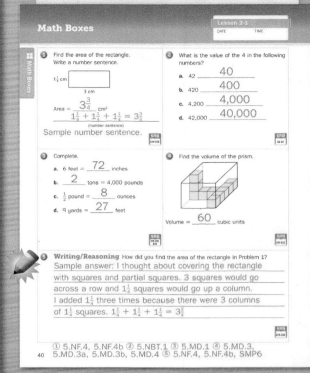
Math Journal 1, p. 40

③ Practice 20–30 min

▶ Finding Volumes of Rectangular Prisms

Math Journal 1, p. 39

| WHOLE CLASS | SMALL GROUP | **PARTNER** | **INDEPENDENT** |

Students find volumes of partially packed rectangular prisms. GMP2.2

▶ Math Boxes 2-1

Math Journal 1, p. 40

| WHOLE CLASS | **SMALL GROUP** | PARTNER | INDEPENDENT |

Mixed Practice Math Boxes 2-1 are paired with Math Boxes 2-3.

▶ Home Link 2-1

Math Masters, p. 44

Homework Students solve number riddles involving place-value clues.
GMP1.1, GMP7.2

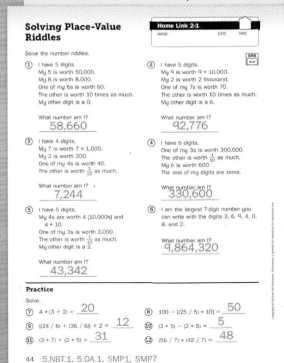
Math Masters, p. 44

Lesson 2-2

Exponents and Powers of 10

Overview Students explain patterns in the number of zeros when multiplying by powers of 10. They use whole-number exponents to denote powers of 10.

▶ **Before You Begin**

For the optional Enrichment activity, photocopy one set of cards from *Math Masters,* page 46 on cardstock for each small group.

▶ **Vocabulary**

exponent • base • power of 10 • exponential notation

Standards

Focus Cluster
• Understand the place value system.

① Warm Up 5 min

	Materials	
Mental Math and Fluency Students use place-value relationships to write numbers.	slate	5.NBT.1

② Focus 35–40 min

		Standards
Math Message Students study numbers written in exponential notation.		5.NBT.2 SMP6, SMP7
Introducing Powers of 10 Students discuss patterns in powers of 10 and write numbers in standard and exponential notation.	*Math Journal 1,* p. 41; *Student Reference Book,* p. 68	5.NBT.1, 5.NBT.2 SMP6, SMP7
Connecting Expanded Form and Exponential Notation Students write numbers in expanded form in different ways.	*Math Journal 1,* p. 42	5.NBT.1, 5.NBT.2 SMP7
Introducing *High-Number Toss* **Game** Students practice reading, writing, and comparing numbers in standard and exponential notation.	*Student Reference Book,* p. 312; per partnership: *Math Masters,* p. G10; 6-sided die	5.NBT.1, 5.NBT.2 SMP6, SMP7
✓ **Assessment Check-In** See page 121. Expect most students to be able to write the numbers they create while playing the game *High-Number Toss* in standard notation.	*Student Reference Book,* p. 68 (optional); *Math Masters,* p. G10	5.NBT.2

③ Practice 20–30 min

Solving a Real-World Volume Problem Students solve a real-world problem by finding the volume of figures composed of rectangular prisms.	*Math Journal 1,* p. 43	5.MD.5, 5.MD.5a, 5.MD.5b, 5.MD.5c SMP4
Math Boxes 2-2 Students practice and maintain skills.	*Math Journal 1,* p. 44	See page 121.
Home Link 2-2 **Homework** Students evaluate and compare expressions with exponential notation.	*Math Masters,* p. 47	5.NBT.1, 5.NBT.2, 5.MD.5, 5.MD.5b, 5.MD.5c

 Go Online to see how mastery develops for all standards within the grade.

Differentiation Options

Readiness 5–10 min	**Enrichment** 15–25 min	**Extra Practice** 10–20 min
WHOLE CLASS · **SMALL GROUP** · PARTNER · INDEPENDENT	WHOLE CLASS · **SMALL GROUP** · **PARTNER** · INDEPENDENT	WHOLE CLASS · SMALL GROUP · **PARTNER** · INDEPENDENT

Exploring Multiplication by Powers of 10

`5.NBT.2`

per group: 6-sided die,
slate or blank paper,
calculator (optional)

For experience with the magnitude of numbers represented in exponential notation, students compare addition and multiplication expressions. Divide students into two or more teams. Choose a number between 2 and 12 and have each team write that number of 10s on a slate or piece of paper, leaving space between each 10. For example, if you say *3*, students write: 10   10   10. Have a volunteer from each team roll a die. For a roll of 1, 2, or 3 the team writes an addition symbol between each 10. For a roll of 4, 5, or 6 the team writes multiplication symbols. Teams evaluate the resulting expression, using a calculator as needed. The value of the evaluated expression is the team's score for that round. Play several rounds. As students calculate, contrast the effect of adding 10s with multiplying by 10. For example: *Adding five 10s is 50. Multiplying by 10 five times is 100,000.*

Solar System Sightseeing

`5.NBT.2`

Activity Card 16; per partnership: *Math Masters*, pp. 45–46; scissors

To extend their understanding of exponential notation and powers of 10, students create multiplication expressions representing distances in the solar system.

Playing *Power Up*

`5.NBT.2, SMP6, SMP7`

Student Reference Book,
p. 318; per partnership: *Math Masters*, p. G11;
two 6-sided dice

For more practice using exponents to denote powers of 10 and multiplying whole numbers by powers of 10, students play *Power Up*. You may wish to review the directions and play a sample round. Students should keep a record of each round on *Math Masters*, page G11.

Observe

- Which students can accurately convert from exponential notation to standard notation? `GMP6.4`

Discuss

- *What patterns did you notice that helped you multiply?* `GMP7.1, GMP7.2`

English Language Learner

Beginning ELL Have students prepare a vocabulary card for each of the new terms introduced in the lesson. They should write the term on one side of the card and an example on the other. Have students use their cards to respond to Total Physical Response commands like these: *Show me the word* exponent. *Show your partner an example of a number written in exponential notation.*

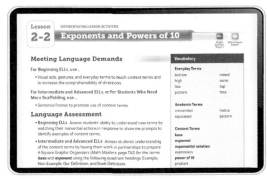

Differentiation Support pages are found in the online Teacher's Center.

Standards and Goals for
Mathematical Process and Practice

SMP6 **Attend to precision.**
    GMP6.1 Explain your mathematical thinking clearly and precisely.

SMP7 **Look for and make use of structure.**
    GMP7.1 Look for mathematical structures such as categories, patterns, and properties.

    GMP7.2 Use structures to solve problems and answer questions.

Math Journal 1, p. 41

Powers of 10 and Exponential Notation
    Lesson 2-2
    DATE   TIME

Follow your teacher's instructions to complete this table.

Exponential Notation	Product of 10s	Standard Notation
10^6	10 * 10 * 10 * 10 * 10 * 10	1,000,000
10^5	10 * 10 * 10 * 10 * 10	100,000
10^8	10 * 10 * 10 * 10 * 10 * 10 * 10 * 10	100,000,000
10^3	10 * 10 * 10	1,000

Complete the number sentences.

1. $3 * 10^2 = 3 * 100 = $ __300__
2. $7 * 10^9 = 7 * $ __1,000,000,000__ $ = $ __7,000,000,000__
3. $3 * 10^4 = $ __3__ $ * $ __10,000__ $ = $ __30,000__
4. $25 * 10^3 = $ __25__ $ * $ __1,000__ $ = $ __25,000__
5. __93__ $* 10^6 = 93 * $ __1,000,000__ $ = $ __93,000,000__

6. The numbers in Problems 1–5 are the answers to the following questions. Fill in each blank with your best guess. Write your answer in exponential or standard notation. $93 * 10^6$,

  a. The distance from the sun to Earth is about __or 93,000,000__ miles.
  b. The Statue of Liberty is about __3 * 10², or 300__ feet tall.
  c. In 2014, the population of Earth was about __7 * 10⁹,__ people. __or 7,000,000,000__
  d. If you walked around Earth's equator, you would walk about __25 * 10³, or 25,000__ miles.
  e. Mount Everest, the highest mountain on Earth, is about __3 * 10⁴, or 30,000__ feet high.

5.NBT.2, SMP7
    41

---

1 Warm Up   5 min

▶ Mental Math and Fluency

Have students write numbers on slates based on place-value clues.
Leveled exercises: Sample answers given.

- ●○○ Write a number in which 5 is worth 50.   53
    Write a number in which 5 is worth 10 times as much.   530

- ●●○ Write a number in which 8 is worth 800.   832
    Write a number in which 8 is worth $\frac{1}{10}$ as much.   82

- ●●● Write a number in which 7 is worth 7,000.   7,241
    Write a number in which 7 is worth 100 times as much.   700,241

2 Focus   35–40 min

▶ Math Message

Study these number sentences: $10^2 = 10 * 10 = 100$ and $10^5 = 10 * 10 * 10 * 10 * 10 = 100,000$. Talk with a partner about what you think the small raised number means.   **GMP6.1**   *Then try to write a similar number sentence starting with 10^6.*   **GMP7.1, GMP7.2**

▶ Introducing Powers of 10

Math Journal 1, p. 41; Student Reference Book, p. 68

WHOLE CLASS	SMALL GROUP	PARTNER	INDEPENDENT

Math Message Follow-Up Have students share their answers to the Math Message, pointing out instances in which they used patterns to solve the problem.   **GMP7.2** $10^6 = 10 * 10 * 10 * 10 * 10 * 10 = 1,000,000$ Explain that the small raised number is called an **exponent** and that the number before the exponent is the **base.** The exponent tells how many times the base is multiplied. When the base is 10, the number is called a **power of 10.**

Remind students about their recent work with standard notation and expanded form. Explain that today they will use powers of 10 to write numbers in **exponential notation.**

Read and discuss *Student Reference Book,* page 68. Have students examine the Powers of 10 chart and share patterns they notice in the chart.   **GMP7.1**

Guide the discussion to bring out the following points:

- Each number is 10 times the number above it and $\frac{1}{10}$ of the number below it.

- The number of zeros in a power of 10 written in standard notation is the same as the number of 10s that are multiplied to give that product.
- The number of zeros in a power of 10 written in standard notation is equal to the exponent of that number when it is written in exponential notation. For example, 1,000,000 has 6 zeros, so the exponent is 6; $1{,}000{,}000 = 10^6$.

Ask: *How do the patterns in the chart help you know how to write 10 billion in exponential notation?* GMP7.2 Sample answer: I noticed that for every additional zero in a number, the number of 10s in the middle column increased by 1, and the exponent increased by 1. The chart shows that 1 billion is 10^9. 10 billion is $10 * 1$ billion, so it has 1 additional zero, and the exponent will increase by 1. 10 billion $= 10^{10}$

Work as a class to add a few more rows to the bottom of the chart. Have students continue to describe the patterns they notice. GMP7.1

NOTE The next three place-value periods to the left of billions are trillions, quadrillions, and quintillions.

Call attention to the patterns as you move *up* each column. For example, remind students that 100 is $\frac{1}{10}$ of the value of 1,000 and 10 is $\frac{1}{10}$ of the value of 100. Ask: *If we added a row before the 10, what would we write in the standard notation column? Explain your thinking clearly.* GMP6.1, GMP7.1, GMP7.2 Sample answer: We would write 1. If we continue the pattern, the number will be $\frac{1}{10}$ of 10, which is 1. Point out that the exponents also follow a pattern: they decrease by 1 as you move up the column. Following this pattern, 1 can be written as 10^0. Explain that it is impossible to follow the pattern in the middle column to fill in a row for 1 because 10 cannot be used as a product 0 times. Point out that since they cannot multiply 10s to help them write 10^0 in standard notation, they should just remember that $10^0 = 1$. This is a convention that mathematicians agree upon because it follows the patterns in the other two columns.

Have students check their understanding of powers of 10 by completing the table on journal page 41. Discuss the first row as a class. Then dictate the following, instructing students to fill in a row for each number: *10 to the fifth power; 10 to the eighth power; 10 cubed.* Prompt students to discuss their answers.

Explain that numbers are sometimes represented as the product of a power of 10 and another number. Display the expression $3 * 10^2$ and invite students to discuss with a partner how they would write this number in standard notation. Ask volunteers to share their answers. Sample answer: $3 * 10^2 = 3 * 100 = 300$

Display the number 65,000,000. Have students discuss with a partner how they could write the number in exponential notation as a product of a whole number and a power of 10. Invite volunteers to share. Sample answer: 65 million is $65 * 1$ million. 1 million is 10^6, so 65,000,000 can be written as $65 * 10^6$.

Common Misconception

Differentiate Some students may misunderstand the meaning of the *exponent*, erroneously interpreting 10^5 as $10 * 5$ rather than $10 * 10 * 10 * 10 * 10$. Be sure to address this common misconception explicitly. Write both a correct and an incorrect interpretation of exponential notation, asking students to explain in their own words which interpretation is correct and how they know. It may also be helpful to have students evaluate both expressions ($10 * 5 = 50$; $10 * 10 * 10 * 10 * 10 = 100{,}000$) to demonstrate that they represent very different numbers.

Go Online Differentiation Support

Professional Development

The Grade 5 standards require students to use whole-number exponents to denote powers of 10. 5.NBT.2 In *Fifth Grade Everyday Mathematics* the use of the whole number *zero* as an exponent is limited to situations in which powers of 10 are used to represent numbers in expanded form. Students only need to know that $10^0 = 1$. They are not expected to thoroughly explain why this is so. For more information about this convention, see the Mathematical Background of the Unit 2 Organizer.

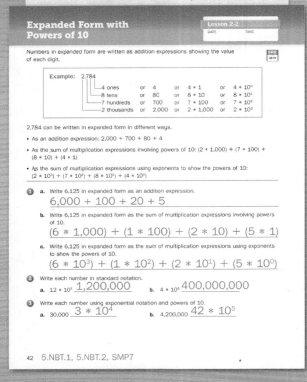

Expanded Form with Powers of 10

Lesson 2-2

Numbers in expanded form are written as addition expressions showing the value of each digit.

Example: 2,784

4 ones	or	4	or	4 * 1	or	4 * 10^0
8 tens	or	80	or	8 * 10	or	8 * 10^1
7 hundreds	or	700	or	7 * 100	or	7 * 10^2
2 thousands	or	2,000	or	2 * 1,000	or	2 * 10^3

2,784 can be written in expanded form in different ways.

• As an addition expression: 2,000 + 700 + 80 + 4
• As the sum of multiplication expressions involving powers of 10: (2 * 1,000) + (7 * 100) + (8 * 10) + (4 * 1)
• As the sum of multiplication expressions using exponents to show the powers of 10: (2 * 10^3) + (7 * 10^2) + (8 * 10^1) + (4 * 10^0)

1. **a.** Write 6,125 in expanded form as an addition expression.
   6,000 + 100 + 20 + 5
   b. Write 6,125 in expanded form as the sum of multiplication expressions involving powers of 10.
   (6 * 1,000) + (1 * 100) + (2 * 10) + (5 * 1)
   c. Write 6,125 in expanded form as the sum of multiplication expressions using exponents to show the powers of 10.
   (6 * 10^3) + (1 * 10^2) + (2 * 10^1) + (5 * 10^0)

2. Write each number in standard notation.
   a. 12 * 10^5 1,200,000   **b.** 4 * 10^8 400,000,000

3. Write each number using exponential notation and powers of 10.
   a. 30,000 3 * 10^4 **b.** 4,200,000 42 * 10^5

42   5.NBT.1, 5.NBT.2, SMP7

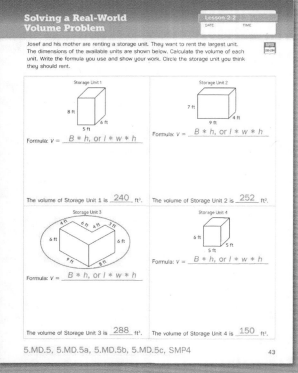

Solving a Real-World Volume Problem

Lesson 2-2

Josef and his mother are renting a storage unit. They want to rent the largest unit. The dimensions of the available units are shown below. Calculate the volume of each unit. Write the formula you use and show your work. Circle the storage unit you think they should rent.

Storage Unit 1
8 ft, 6 ft, 5 ft
Formula: V = _B * h, or l * w * h_

Storage Unit 2
7 ft, 4 ft, 9 ft
Formula: V = _B * h, or l * w * h_

The volume of Storage Unit 1 is 240 ft³.   The volume of Storage Unit 2 is 252 ft².

Storage Unit 3
6 ft, 6 ft
Formula: V = _B * h, or l * w * h_

Storage Unit 4
6 ft, 5 ft, 5 ft
Formula: V = _B * h, or l * w * h_

The volume of Storage Unit 3 is 288 ft³.   The volume of Storage Unit 4 is 150 ft³.

5.MD.5, 5.MD.5a, 5.MD.5b, 5.MD.5c, SMP4   43

Have partnerships complete Problems 1–5 on journal page 41. Tell them to use their answers to Problems 1–5 to fill in the blanks in Problem 6, which will allow them to discover interesting facts. Take a few minutes to discuss the journal page.

▶ ## Connecting Expanded Form and Exponential Notation

Math Journal 1, p. 42

| WHOLE CLASS | SMALL GROUP | PARTNER | INDEPENDENT |

Remind students that in the previous lesson they learned about writing numbers in expanded form. Have them read the top of journal page 42 and share what they notice about the ways of writing numbers in expanded form. Guide the discussion to emphasize the following points:

• The first bullet shows the value of each digit in standard notation.
• The second bullet shows the value of each digit as the product of two factors, one of them a power of 10 written in standard notation.
• The third bullet is similar to the second one, but it shows the power of 10 written in exponential notation.

Have partnerships complete the journal page. Encourage them to discuss how to use the patterns they notice to write numbers in expanded form using exponential notation. GMP7.2

▶ ## Introducing *High-Number Toss*

Student Reference Book, p. 312; *Math Masters*, p. G10

| WHOLE CLASS | SMALL GROUP | PARTNER | INDEPENDENT |

Review the directions for *High-Number Toss* on *Student Reference Book*, page 312. Play a sample round with a volunteer and demonstrate how to complete the record sheet. Then have partnerships play the game.

Observe
• Which students can represent their number in standard notation?
• Which students have a strategy for placing cards to create larger numbers?

Discuss
• *If you roll a 6, where should you place it to increase your chance of winning? Why?* GMP7.2
• *Is the player with the highest exponent always the winner? Why or why not?* GMP7.1

| Differentiate | **Game Modifications** | Go Online | Differentiation Support |

Assessment Check-In 5.NBT.2

Math Journal 1, p. G10

Expect most students to be able to correctly write the numbers they create while playing *High-Number Toss* in standard notation. Refer those who struggle to the Powers of 10 chart on *Student Reference Book*, page 68. Encourage them to write out the standard notation for the power of 10 above the exponential notation before multiplying to find the total.

Evaluation Quick Entry Go online to record students' progress and to see trajectories toward mastery for these standards.

Summarize Have students talk with a partner about any patterns they notice on their completed *High-Number Toss* record sheets. GMP7.1
Encourage them to clearly explain how they determined the standard notation of their number. GMP6.1

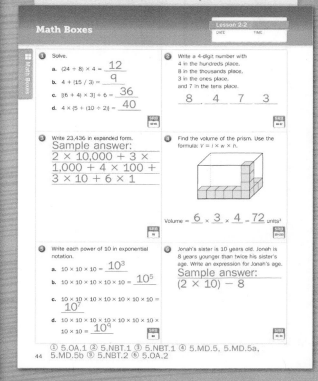

Math Boxes

Lesson 2-2

1. Solve.
   a. $(24 \div 8) \times 4 =$ __12__
   b. $4 + (15 / 3) =$ __9__
   c. $[(6 + 4) \times 3] + 6 =$ __36__
   d. $4 \times (5 + (10 \div 2)) =$ __40__

2. Write a 4-digit number with 4 in the hundreds place, 8 in the thousands place, 3 in the ones place, and 7 in the tens place.

   __8__ __4__ __7__ __3__

3. Write 23,436 in expanded form.
   Sample answer:
   $2 \times 10,000 + 3 \times 1,000 + 4 \times 100 + 3 \times 10 + 6 \times 1$

4. Find the volume of the prism. Use the formula: $V = l \times w \times h$.

   Volume = __6__ × __3__ × __4__ = __72__ units³

5. Write each power of 10 in exponential notation.
   a. $10 \times 10 \times 10 =$ __10^3__
   b. $10 \times 10 \times 10 \times 10 =$ __10^5__
   c. $10 \times 10 \times 10 \times 10 \times 10 \times 10 =$ __10^7__
   d. $10 \times 10 \times 10 \times 10 \times 10 \times 10 \times 10 \times 10 \times 10 =$ __10^9__

6. Jonah's sister is 10 years old. Jonah is 8 years younger than twice his sister's age. Write an expression for Jonah's age.
   Sample answer:
   $(2 \times 10) - 8$

① 5.OA.1 ② 5.NBT.1 ③ 5.NBT.1 ④ 5.MD.5, 5.MD.5a, 5.MD.5b ⑤ 5.NBT.2 ⑥ 5.OA.2

44

3 Practice 20–30 min

▶ Solving a Real-World Volume Problem

Math Journal 1, p. 43

| WHOLE CLASS | SMALL GROUP | **PARTNER** | **INDEPENDENT** |

Students practice finding the volume of rectangular prisms and figures composed of rectangular prisms. They use models of storage units to solve a real-world problem. GMP4.2

▶ Math Boxes 2-2

Math Journal 1, p. 44

| WHOLE CLASS | **SMALL GROUP** | PARTNER | INDEPENDENT |

Mixed Practice Math Boxes 2-2 are paired with Math Boxes 2-4.

▶ Home Link 2-2

Math Masters, p. 47

Homework Students write and compare numbers using standard and exponential notation.

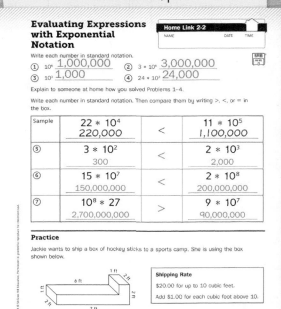

Evaluating Expressions with Exponential Notation Home Link 2-2

Write each number in standard notation.
① 10^6 __1,000,000__   ② $3 * 10^6$ __3,000,000__
③ 10^3 __1,000__   ④ $24 * 10^3$ __24,000__

Explain to someone at home how you solved Problems 1–4.

Write each number in standard notation. Then compare them by writing >, <, or = in the box.

Sample	$22 * 10^4$ 220,000	<	$11 * 10^5$ 1,100,000
⑤	$3 * 10^2$ 300	<	$2 * 10^3$ 2,000
⑥	$15 * 10^7$ 150,000,000	<	$2 * 10^8$ 200,000,000
⑦	$10^8 * 27$ 2,700,000,000	>	$9 * 10^7$ 90,000,000

Practice

Jackie wants to ship a box of hockey sticks to a sports camp. She is using the box shown below.

Shipping Rate
$20.00 for up to 10 cubic feet.
Add $1.00 for each cubic foot above 10.

⑧ What is the volume of the box?
   About __16__ cubic feet

⑨ How much will Jackie pay for shipping? $ __26.00__

5.NBT.1, 5.NBT.2, 5.MD.5, 5.MD.5b, 5.MD.5c   47

Applying Powers of 10

Overview Students estimate with powers of 10 to solve multiplication problems and check the reasonableness of products.

▶ **Before You Begin**
For Part 2, copy and cut apart enough Exit Slips (*Math Masters*, page TA6) for each student to have one. For the optional Enrichment activity, each partnership will need a sheet of poster paper.

▶ **Vocabulary**
extended multiplication fact • estimate

Standards

Focus Cluster
• Understand the place value system.

① Warm Up 5 min	**Materials**	
Mental Math and Fluency Students convert between standard and exponential notation for powers of 10.	slate	5.NBT.2

② Focus 35–40 min		
Math Message Students use powers of 10 to mentally calculate extended multiplication facts.		5.NBT.2 SMP6
Using Powers of 10 to Multiply Students use powers of 10 to solve multiplication fact extensions.	*Math Masters*, p. TA5 (optional)	5.NBT.2 SMP6
Estimating with Powers of 10 Students use powers of 10 and estimation to solve number stories and check answers.	*Math Journal 1*, p. 45	5.NBT.2 SMP1, SMP6
✓ **Assessment Check-In** See page 127. Expect most students to be able to round factors to the nearest multiple of a power of 10 and attempt to calculate the estimated product.	*Math Masters*, p. TA6	5.NBT.2 SMP6

③ Practice 20–30 min		
Writing and Comparing Expressions Students write and compare the values of simple expressions in the context of real-world and mathematical situations.	*Math Journal 1*, p. 46	5.OA.1, 5.OA.2 SMP1, SMP4
Math Boxes 2-3 Students practice and maintain skills.	*Math Journal 1*, p. 47	See page 127.
Home Link 2-3 Homework Students use powers of 10 and estimation to solve number stories.	*Math Masters*, p. 49	5.NBT.2 SMP6

Go Online to see how mastery develops for all standards within the grade.

Differentiation Options

Readiness 5–10 min	**Enrichment** 15–25 min	**Extra Practice** 10–20 min
WHOLE CLASS · SMALL GROUP · **PARTNER** · INDEPENDENT	WHOLE CLASS · SMALL GROUP · **PARTNER** · INDEPENDENT	WHOLE CLASS · **SMALL GROUP** · PARTNER · INDEPENDENT

Practicing Extended Multiplication Facts

5.NBT.2, SMP7

Student Reference Book, pp. 97–98; *Math Masters,* p. 48; number cards 1–10 (1 of each)

For experience with extended multiplication facts, students use number cards to generate and solve extended-fact problems. They describe or write about patterns they see. **GMP7.1** Students may refer to *Student Reference Book,* pages 97 and 98.

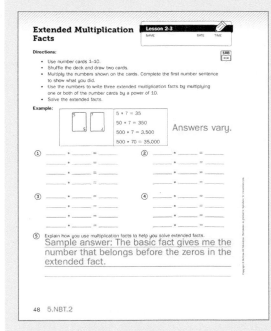

Freight Train Wrap-Around

5.NBT.2, 5.MD.1, SMP6

Activity Card 17, per partnership: poster paper

To extend their understanding of multiplication by powers of 10, students estimate whether all of the freight train cars in the United States, placed end to end, would wrap all the way around Earth. They make a poster to explain their thinking. **GMP6.1** Sample answer: No. There are about 1,283,000 freight train cars, each about 50 feet long. Estimate: 1,300,000 * 50 = 65,000,000 feet of train cars. Earth is about 24,900 miles around at the equator. 1 mile = 5,280 feet. Estimate: 25,000 * 5,000 = 125,000,000 feet around Earth. The train cars would extend about half the distance around Earth.

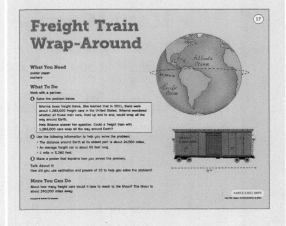

Playing *Multiplication Top-It: Extended Facts*

5.NBT.2, SMP7

Student Reference Book, p. 325; per partnership: number cards 0–9 (4 of each)

For more practice with extended multiplication facts, students play *Multiplication Top-It: Extended Facts.* See the rules on *Student Reference Book,* page 325. Students draw two number cards, attach a zero to the first card, and multiply by the second card. For example, with 5 as the first card and 9 as the second, students compute 50 * 9 = 450.

Observe
- Which students can accurately name the product of each extended fact?

Discuss
- If you attached a zero to the number on both cards, how would you calculate the answer? **GMP7.2**

English Language Learner

Beginning ELL Use concrete objects such as a Slinky® or an extension cord to teach the terms *extend* and *extension.* Extend the object as you define contexts by making think-aloud statements like these: *I extend the Slinky®, so it's longer. I can't reach the outlet. I need an extension cord.*

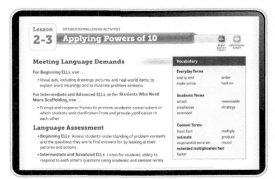

Differentiation Support pages are found in the online Teacher's Center.

Standards and Goals for
Mathematical Process and Practice

SMP1 **Make sense of problems and persevere in solving them.**
  GMP1.4 Check whether your answer makes sense.

SMP6 **Attend to precision.**
  GMP6.1 Explain your mathematical thinking clearly and precisely.

  GMP6.4 Think about accuracy and efficiency when you count, measure, and calculate.

Common Misconception

Differentiate Pay attention to students' word choice when they describe multiplying by a power of 10 by appending zeros. Students are likely to use the term *add*, as in: "I added a zero." Take time to point out that, mathematically, they are not *adding* zero, as that would leave the value of the number unchanged. Encourage students to use terms like *attach, tack on,* and *append* to more accurately convey their strategy. Emphasize how with every additional zero the number shifts one place to the left, which is the same as multiplying by 10.

Go Online Differentiation Support

---

1 Warm Up 5 min

▶ Mental Math and Fluency

Display the following numbers. Have students convert between standard and exponential notation, recording their answers on slates. *Leveled exercises:*

●○○ $100 = 10^2$
      $10^3 = 1,000$
      $10 = 10^1$

●●○ $10,000 = 10^4$
      $100,000 = 10^5$
      $10^7 = 10,000,000$

●●● $10^8 = 100,000,000$
      $1,000,000,000 = 10^9$
      $1 = 10^0$

2 Focus 35–40 min

▶ Math Message

Explain to your partner how you could do the following problems in your head: $5 * 40$; $50 * 40$; $50 * 400$. GMP6.1

▶ Using Powers of 10 to Multiply

| WHOLE CLASS | SMALL GROUP | PARTNER | INDEPENDENT |

Math Message Follow-Up Discuss the strategies partners shared during the Math Message. Possible strategies include:

• For $5 * 40$, I know that 5 [20s] = 100. Then I doubled 100 because 40 is twice as much as 20. So the answer is 200.

• For $50 * 40$, I know that $5 * 4 = 20$, so I just attached 2 more zeros to the end to get 2,000.

• For $50 * 400$, I know that $5 * 400$ is 2,000, so I attached an extra zero because 50 is 5 tens, not just 5. So the answer is 20,000.

Explain to the class that knowing basic multiplication facts makes it easier to solve problems like $5 * 40$ and $50 * 400$. Ask: *Why do you think problems like $5 * 40$ are called **extended multiplication facts?*** Sample answer: We are building on—or extending—facts we already know. We can solve the basic fact, and then attach more zeros, depending on whether we are multiplying by 10s, 100s, or 1,000s. *What basic fact can you use to help you solve $5 * 40$?* $5 * 4 = 20$ *How many zeros do you need to attach?* One *How do you know?* Sample answer: Instead of multiplying by 4, we are multiplying by 4 tens. So we need to attach 1 zero to move the digits one place to the left.

Rewrite the first problem from the Math Message as 5 * 4 * 10 and 5 * 4 * 10^1. Ask: *How does rewriting this problem as a product involving a power of 10 help you think about how to use a basic fact to solve the problem?* Sample answer: I can see the basic fact separately. I can figure out how many zeros to attach to the answer by counting the zeros on the power of 10 or by using the exponent.

Invite volunteers to explain how to rewrite and solve the other Math Message problems using powers of 10. *Sample explanations:*

- I can rewrite 50 * 40 as 5 * 10 * 4 * 10. This is the same as 5 * 4 * 10 * 10, because the order of the numbers does not change the product. I know that 5 * 4 = 20, but I need to multiply by 10 twice, which means attaching two zeros. So the answer is 2,000.

- I can rewrite 50 * 400 as 5 * 10^1 * 4 * 10^2. I know that 5 * 4 = 20. Multiplying by 10^1 means I attach one zero, and multiplying by 10^2 means I attach two zeros. That is three attached zeros, which gives 20,000.

Give the class a few more problems to solve mentally. For some problems, ask students to write out the factors in exponential notation before calculating mentally. *Examples:* 300 * 90 27,000 [3 * 10^2 * 9 * 10^1]; 8,000 * 50 400,000 [8 * 10^3 * 5 * 10^1]; 7 * 9,000 63,000 [7 * 9 * 10^3]; 400 * 400 160,000 [4 * 10^2 * 4 * 10^2] Emphasize how using extended multiplication facts is an efficient way to calculate. GMP6.4

Tell students that today they will use powers of 10 and extended multiplication facts to check whether their answers are reasonable and to solve more complicated multiplication number stories.

▶ **Estimating with Powers of 10**

Math Journal 1, p. 45

| WHOLE CLASS | SMALL GROUP | PARTNER | INDEPENDENT |

Remind students that when solving problems they should always think about whether their answer makes sense. One way to do this is to make an **estimate,** or an answer that is close to the exact answer. This is an important step in determining whether their thinking is on the right track. GMP1.4 Pose this problem: *Miles solved 492 * 63 and got 480,992. Is Miles's answer reasonable?* Encourage students to use their knowledge of powers of 10 to think through the reasonableness of the answer. Guide them to pay attention to the factors. Ask: *What power of 10 would we have to multiply 492 by to come close to 480,992?* About 1,000 *What power of 10 would we have to multiply 63 by?* About 10,000 *What does that tell us about the reasonableness of Miles's answer?* The answer is not reasonable. Neither factor multiplied by the other results in a product that is close to 480,992.

Academic Language Development

Build on students' knowledge of the term *base*, as in *base word*, to help them actively construct understanding of the meanings of *basic* and *extended*. Display examples of several words with the same base word. For example, write the words disappear, reappear, and appearance. Ask students to identify the base word and say what was added to extend the base. Have partners work together to identify the basic facts on which given extended multiplication facts are based. Encourage students to use the terms *basic* and *extended* in their discussions.

Common Misconception

Differentiate When multiplying numbers that are multiples of powers of 10, some students may misinterpret the number of zeros in the product, particularly when the basic fact ends in a zero. For example, when students evaluate 50 * 600, they may erroneously include the zero in the 30 as one of the additional three zeros, writing 3,000 instead of 30,000. Students who struggle keeping track of the zeros may benefit from using a place-value chart (*Math Masters,* page TA5). Suggest that they begin by writing the number of zeros they need to attach (beginning with the 1s place and moving left) and then insert the product of the basic fact.

Estimating with Powers of 10

Lesson 2-3

DATE   TIME

Use estimation to solve.

1. A hardware store sells ladders that extend up to 12 feet. The store's advertising says:

   Largest inventory in the country! If you put all our ladders end to end, you could climb to the top of the Empire State Building!

   The company has 295 ladders in stock. The Empire State Building is 1,453 feet tall.

   Is it true that the ladders would reach the top of the building? **Yes.**

   Explain how you solved the problem. **Sample answer: To find out exactly how high all the ladders would reach, I would multiply 12 * 295. To estimate, I multiplied 10 * 300. That's 3,000 feet, enough to get to the top of a 1,453-foot building.**

2. The school library received a donation of 42 boxes of books. Each box contains 15–18 books. The library has 10 empty bookshelves. Each bookshelf can hold up to 60 books.

   Does the library have enough shelf space for all the new books? **No.**

   Explain how you solved the problem. **Sample answer: I rounded 42 to 40 and 15–18 to 20 and multiplied 40 * 20. About 800 books were donated. I multiplied 10 * 60 to find that there's only room for 600 books. There is not enough space for the new books.**

3. Nishant is in charge of collecting cereal box tops to trade in for technology items for his school. He keeps the box tops in 38 folders. Each folder contains 80 box tops.

   Does Nishant have enough box tops to trade in for a printer? **Yes.**

   For a digital camera? **No.**

   For a tablet computer? **No.**

Item	Number of Box Tops Needed
Printer	1,500
Digital camera	3,500
Tablet computer	5,000

   Explain how you solved the problem. **Sample answer: Nishant has 38 * 80 box tops. To estimate, I rounded 38 to 40, then multiplied 40 * 80. Nishant has about 3,200 box tops. That's enough to get a printer, but not a camera or a computer.**

   5.NBT.2, SMP6   45

Writing and Comparing Expressions

Lesson 2-3

DATE   TIME

1. Write each statement as an expression using grouping symbols. Do not evaluate the expressions.

   a. Find the sum of 13 and 7, then subtract 5. $(13 + 7) - 5$

   b. Multiply 3 and 4 and divide the product by 6. $(3 \times 4) \div 6$

   c. Add 138 and 127 and multiply the sum by 5. $(138 + 127) \times 5$

   d. Divide the sum of 45 and 35 by 10. $(45 + 35) \div 10$

   e. Add 300 to the difference of 926 and 452. $(926 - 452) + 300$

2. Compare the two expressions. Do not evaluate them. How are their values different?

   a. 2 * (489 + 126) and 489 + 126
      The value of 2 * (489 + 126) is twice as large as the value of 489 + 126.

   b. (367 × 42) − 328 and 367 × 42
      The value of (367 × 42) − 328 is 328 less than the value of 367 × 42.

3. Below are advertisements for two stores having sales on T-shirts.   Sample answers:

   a. Write an expression that represents the cost of two shirts at each store.

Shirts-R-Us	T-Shirt Mart
Half off T-shirts! Regular price: $14.00.	T-shirt Sale Price: $8.00.
Expression: $\left(\frac{1}{2} * \$14.00\right) + \left(\frac{1}{2} * \$14.00\right)$	Expression: $\$8.00 + \$8.00$

   b. Which store has the better deal? How do you know?
      Shirts-R-Us; $\frac{1}{2}$ of $14.00 is less than $8.00, so the shirts will cost less at Shirts-R-Us.

   46   5.OA.1, 5.OA.2, SMP1, SMP4

Another way to use an estimate to determine whether an answer makes sense is to round the numbers in a problem to the nearest multiple of a power of 10 (the nearest 10, 100, or 1,000). Have students use this strategy with another example. Ask: *Jenna multiplied 72 * 689 and got 490,608. Is Jenna's answer reasonable? Use an estimate to explain how you know.* Sample answer: No, Jenna's answer is not reasonable. I rounded 72 to 70 and 689 to 700. I multiplied to get 49,000. Jenna's answer is more than 10 times my estimate of 49,000. Emphasize that rounding to the nearest 10 or 100 creates an extended multiplication fact. Since students know that $7 * 7 = 49$, they can easily calculate $70 * 700 = 49,000$.

Explain that powers of 10 and estimation can also be used to solve number stories. Pose this problem: *A grocery store manager likes to have at least 4,000 cans of vegetables in his storeroom. He has 67 pallets of 36 cans. Does he need to order more? Explain how you figured it out.*

Take a few minutes to discuss what the problem is asking. Emphasize that it is not asking "How many more . . . ," but simply whether or not the grocer needs to place another order. Have students share how they used estimation and powers of 10 to quickly figure out the answer. **GMP6.4** Guide the discussion to emphasize the following points:

- The number of cans currently in the storeroom is the product of 67 and 36.
- It is not necessary to find an exact answer. Each factor can be rounded to the nearest 10 to estimate the product: $70 * 40 = 2,800$.
- The grocer has about 2,800 cans in stock. If he wants to have at least 4,000 cans available, he needs to order more.

Ask: *If the problem had asked us to find the number of cans the grocer needed to order, would we solve it differently? If so, how?* **GMP6.1** Yes. We would multiply 67 * 36 to find the exact answer, and then subtract that number from 4,000.

Provide another example for students to discuss and solve. *A restaurant is hosting a dinner for 365 people. The restaurant owner has 48 boxes of pasta. Each box makes 12 servings of pasta. Does the owner have enough pasta? Explain how you figured it out.*

Have students share their solution strategies. Possible strategies:

- Round both factors to the nearest 10 to get $10 * 50$, which is 500. The restaurant has more than enough servings of pasta for 365 people.
- Round the 12 to 10 and multiply $10 * 48$ to get 480. There is enough pasta for 365 people.

Point out that students' estimates may not always be the same. In the pasta example some students may have rounded both the 48 and the 12, while others may have rounded only the 12. Stress the point that the goal of rounding numbers is to make mental calculations easier and more efficient. **GMP6.4** Students may have different preferences about which numbers are easier to use. Have students work in partnerships or small groups to solve the problems on journal page 45.

Assessment Check-In 5.NBT.2

Math Masters, p. TA6

Distribute an Exit Slip (*Math Masters,* page TA6) to each student. Ask: *Is the product of 284 * 79 closer to 2,400; 24,000; or 240,000?* Have students answer the question and explain their thinking on the Exit Slip. GMP6.1

Sample answer: It is closer to 24,000. I can round 284 to 300 and 79 to 80. I can estimate the answer as 300 * 80. I multiply 3 * 8 to get 24. Then I attach 3 zeros: 24,000. Students should be familiar with estimating from previous grades, but this lesson is the first time students have been asked to think about powers of 10 when estimating. Expect most to be able to successfully round the factors and attempt to calculate the estimated product. Some may struggle determining the correct number of zeros to attach to the product. Students will improve their skills as they estimate products throughout the year.

Evaluation Quick Entry Go online to record students' progress and to see trajectories toward mastery for these standards.

Summarize Invite volunteers to share their solutions and strategies from journal page 45.

3 Practice   20–30 min

▶ Writing and Comparing Expressions

Math Journal 1, p. 46

| WHOLE CLASS | SMALL GROUP | PARTNER | INDEPENDENT |

Students practice writing expressions to model real-world and mathematical situations. GMP4.1 They interpret expressions to solve a real-world problem. GMP 1.1, GMP4.2

▶ Math Boxes 2-3

Math Journal 1, p. 47

| WHOLE CLASS | SMALL GROUP | PARTNER | INDEPENDENT |

Mixed Practice Math Boxes 2-3 are paired with Math Boxes 2-1.

▶ Home Link 2-3

Math Masters, p. 49

Homework Students use powers of 10 and estimation to solve number stories. They explain their solutions. GMP6.1

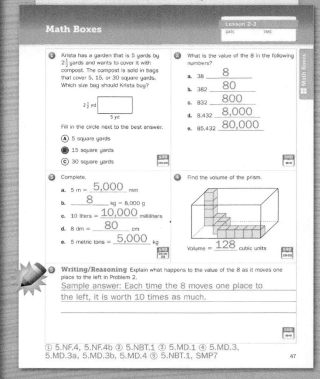

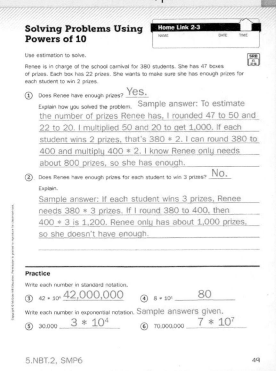

Lesson 2-4

U.S. Traditional Multiplication, Part 1

Overview Students use U.S. traditional multiplication to multiply 2-digit numbers by 1-digit numbers.

▶ **Before You Begin**
If your students are not comfortable using partial-products multiplication, have them complete the Readiness activity.

▶ **Vocabulary**
partial-products multiplication • U.S. traditional multiplication • area model

Standards

Focus Cluster
• Perform operations with multi-digit whole numbers and with decimals to hundredths.

1 Warm Up 5 min

	Materials	
Mental Math and Fluency Students use powers of 10 to solve extended multiplication facts.	slate	5.NBT.2

2 Focus 35–40 min

Math Message Students multiply a 2-digit number by a 1-digit number.		SMP1
Introducing U.S. Traditional Multiplication Students share strategies and use U.S. traditional multiplication to multiply 2-digit numbers by 1-digit numbers.		5.NBT.5 SMP1
Multiplying 2-Digit Numbers by 1-Digit Numbers Students practice using U.S. traditional multiplication.	*Math Journal 1*, p. 48; *Math Masters*, p. TA7 (optional)	5.NBT.5 SMP1
✓ **Assessment Check-In** See page 132. This is the first exposure to U.S. traditional multiplication. Expect some students to be able to multiply a 2-digit number by a 1-digit number when they choose their own strategy.	*Math Journal 1*, p. 48	5.NBT.5, SMP1

3 Practice 20–30 min

Playing *Number Top-It* **Game** Students use place-value understanding to practice building and comparing multidigit numbers.	*Student Reference Book*, p. 316; per partnership: *Math Masters*, pp. G7–G9; number cards 0–9 (4 of each)	5.NBT.1 SMP1, SMP7
Math Boxes 2-4 Students practice and maintain skills.	*Math Journal 1*, p. 49	See page 133.
Home Link 2-4 **Homework** Students multiply using U.S. traditional multiplication.	*Math Masters*, p. 51	5.NBT.1, 5.NBT.5

Go Online to see how mastery develops for all standards within the grade.

Differentiation Options

Readiness 5–15 min	**Enrichment** 5–15 min	**Extra Practice** 5–15 min
WHOLE CLASS · SMALL GROUP · PARTNER · INDEPENDENT	WHOLE CLASS · SMALL GROUP · PARTNER · INDEPENDENT	WHOLE CLASS · SMALL GROUP · PARTNER · INDEPENDENT

Reviewing Partial-Products Multiplication

`5.NBT.5, SMP3`

Student Reference Book, p. 100

To build readiness for understanding U.S. traditional multiplication, students review partial-products multiplication. Guide students through the steps for partial-products multiplication on *Student Reference Book,* page 100. Have them use partial-products multiplication to solve 24 ∗ 8 192 and review the steps together. Then ask students to work with a partner to solve 36 ∗ 7 twice using partial-products multiplication: once multiplying the tens first (30 ∗ 7) and once multiplying the ones first (6 ∗ 7). 252 Ask: *Did you get the same answer when you multiplied the ones first as when you multiplied the tens first?* Yes. *Will you always get the same answer, no matter which numbers you multiply first? Why?* `GMP3.1` Yes. No matter what you multiply first, you still get the same partial products to add together, so the final answer will be the same.

Using Place Value to Multiply

`5.NBT.5, SMP7`

Math Masters, p. 50; calculator

To extend their understanding of U.S. traditional multiplication, students multiply 2-digit numbers by 1-digit numbers using U.S. traditional multiplication. Then they use place-value patterns to solve related problems. `GMP7.1, GMP7.2`

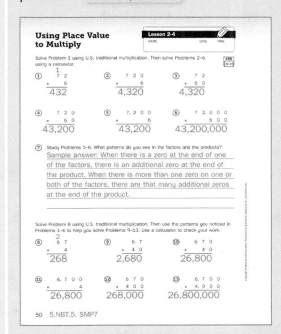

Practicing Multiplication Strategies

`5.NBT.5, SMP1`

Activity Card 18; *Student Reference Book,* pp. 100 and 102 (optional); per partnership: number cards 1–9 (4 of each)

For additional practice with U.S. traditional multiplication, students generate problems using number cards. They solve the problems using both U.S. traditional multiplication and partial-products multiplication, and then compare the strategies. `GMP1.5, GMP1.6`

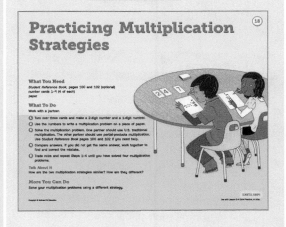

English Language Learner

Beginning ELL Introduce the term *partial* by building on concrete examples of parts of a whole. Display the written term and pronounce it for students to hear. Then have them chorally repeat the term. Underline the base word *part* and show a picture of a whole pizza divided into parts. Ask: *How many parts are there in this pizza?* Show a picture of part of a pizza, saying: *This is not the whole pizza. It only has _____ parts. It is a partial pizza.*

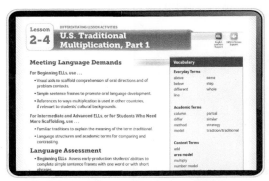

Differentiation Support pages are found in the online Teacher's Center.

Mathematical Process and Practice

Professional Development

Everyday Mathematics calls the algorithm introduced in this lesson U.S. traditional multiplication because it is not the standard algorithm in other parts of the world. The Grade 5 standards require students to fluently multiply multidigit whole numbers using U.S. traditional multiplication, so it is the focus of *Fifth Grade Everyday Mathematics* lessons. Students should learn this method, but if they prefer a different multiplication strategy, they should be allowed to use it to solve multiplication problems.

1 Warm Up   5 min

▶ Mental Math and Fluency

Have students multiply and write the answers to the following problems on slates. *Leveled exercises:*

● ○ ○ $7 * 6$ 42
        $7 * (6 * 10^1)$ 420
        $7 * 60$ 420

● ● ○ $4 * 7$ 28
        $7 * 10^2$ 700
        $4 * (7 * 10^2)$ 2,800
        $4 * 700$ 2,800

● ● ● $9 * 7$ 63
        $9 * 10^2$ 900
        $7 * 10^2$ 700
        $900 * 700$ 630,000

2 Focus   35–40 min

▶ Math Message

*Solve: $34 * 9 = $ _____*

Compare your strategy to a partner's strategy.   GMP1.6

▶ Introducing U.S. Traditional Multiplication

| WHOLE CLASS | SMALL GROUP | PARTNER | INDEPENDENT |

Math Message Follow-Up Invite several students to share their strategies for finding the product. 306 If no one mentions partial-products multiplication, be sure to demonstrate it yourself. *Sample strategies:*

• Use **partial-products multiplication.** First write 34 in expanded form: $30 + 4$. Then multiply each part of 34 by 9: $30 * 9 = 270$ and $4 * 9 = 36$. Finally, add the two partial products: $270 + 36 = 306$.

• First find $34 * 10 = 340$. I know that $34 * 10$ is 34 more than $34 * 9$, so I have to subtract 34. I subtracted to find $340 - 34 = 306$.

• Double 34, and then double the result two more times. $34 * 2 = 68$; $68 * 2 = 136$; and $136 * 2 = 272$. That's 8 [34s] so far. Add one more [34] to find the answer: $272 + 34 = 306$.

Ask: How are these strategies different? Sample answer: With partial products, you take apart 34 and find parts of the product separately. With other strategies, you think about 34 as one chunk. *How are they similar?*   GMP1.6   Sample answers: They all give the right answer. I can do some of the steps in my head for all the strategies.

Tell students that today they will explore a new strategy for solving multiplication problems called **U.S. traditional multiplication** and learn how this strategy relates to partial-products multiplication.

Have students solve 56 * 7 using partial-products multiplication. Record the steps for this problem and sketch the **area model** shown in the margin. Briefly review how the area model shows both the partial products and the answer to the multiplication problem. The area of the right part is equal to the first partial product; the area of the left part is equal to the second partial product; and the area of the whole model is equal to the answer.

Demonstrate how to multiply 56 * 7 using U.S. traditional multiplication. Discuss how each step of U.S. traditional multiplication is similar to one or two steps in partial-products multiplication, and relate the steps to the area model as shown below. GMP1.6

U.S. Traditional Multiplication	
Step 1: Multiply the ones. 7 * 6 ones = 42 ones, or 4 tens and 2 ones Write 2 below the line in the 1s column. Write 4 above the 10s column. **Compare:** This is like finding the partial product 7 * 6 → 42, or the area of the right part of the area model.	4 5 6 * 7 —— 2
Step 2: Multiply the tens. 7 * 5 tens = 35 tens Remember the 4 tens from Step 1. 35 tens + 4 tens = 39 tens in all 39 tens = 3 hundreds and 9 tens Write 9 below the line in the 10s column. Write 3 below the line in the 100s column. **Compare:** This is like finding the partial product 7 * 50 → 350, or the area of the left part of the area model, and then adding the partial products to get 392.	4 5 6 * 7 —— 3 9 2

Point out that the final answer (392) is the sum of the two partial products shown in the partial-products multiplication work and the sum of the two areas shown in the area model. Explain that with U.S. traditional multiplication, students record the partial products so that they can do the addition as they multiply, instead of at the end. However, the final answer is the same.

Solve other 2-digit by 1-digit multiplication problems using U.S. traditional multiplication. Invite volunteers to explain each step and point out connections to partial-products multiplication. GMP1.6

Suggestions:

- 26 * 4 104
- 35 * 6 210

- 41 * 8 328
- 72 * 7 504


```
            5   6
    *           7
           —————————
7 * 6  →        4   2
7 * 50 →    3   5   0
           —————————
            3   9   2
```

Partial-products
multiplication for 56 * 7

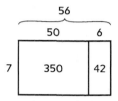

Area model showing the
partial products for 56 * 7

NOTE Area models are not intended to be viewed as a separate strategy for solving multiplication problems. The models can help students make sense of a variety of multiplication methods and explain where each partial product comes from. Some students may find it useful to sketch area models, but they are not required to draw them.

Remind students that when solving problems it is important to check whether their answers make sense, especially when using a new strategy. **GMP1.4** Ask students to suggest how they can check whether their answers make sense. Guide the discussion to cover the following points:

- Students can solve the problem again using a different method and check that they get the same answer. For example, it makes sense that the class got the same answer·using U.S. traditional multiplication that they got using partial-products multiplication when solving the problem 56 * 7. **GMP1.5**

- Students can use an estimate to check the reasonableness of an answer. For example, they know that 26 * 4 should be close to 25 * 4, or 100. Because 104 is close to 100, the answer seems reasonable. **GMP1.4**

▶ Multiplying 2-Digit Numbers by 1-Digit Numbers

Math Journal 1, p. 48

| WHOLE CLASS | SMALL GROUP | **PARTNER** | **INDEPENDENT** |

Have students complete journal page 48 individually or in partnerships. Remind them to check that their answers make sense. **GMP1.4** For Problems 4–7, students write number models using a symbol for the unknown. They have done this in Grades 2, 3, and 4. An example is provided on the journal page. If students struggle writing number models, you may want to write the number model for Problem 5 together as a class.

Differentiate **Adjusting the Activity**

If students struggle lining up the digits when they are multiplying, provide a computation grid (*Math Masters*, page TA7) and demonstrate how to use it to keep the digits in the numbers aligned.

Go Online ⊞ Differentiation Support

 ## Assessment Check-In 5.NBT.5

Math Journal 1, p. 48

Expect most students to correctly solve the multiplication problems on journal page 48 when they choose their own strategy. Most will also be able to describe how they know their answers make sense in Problem 3. **GMP1.4** This is the first exposure to U.S. traditional multiplication, so some students may struggle using it. They will have plenty of opportunities throughout the year to practice U.S. traditional multiplication.

 Evaluation Quick Entry Go online to record students' progress and to see trajectories toward mastery for these standards.

Math Journal 1, p. 48

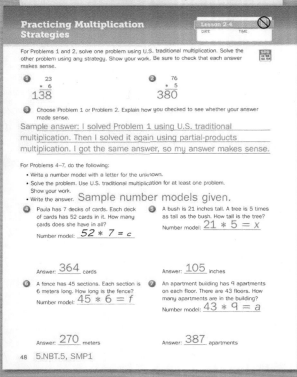

Summarize Have students tell a partner one similarity they noticed between U.S. traditional multiplication and partial-products multiplication. **GMP1.6**

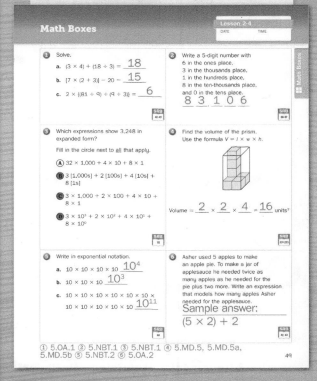

③ Practice 20–30 min

▸ **Playing *Number Top-It***

Student Reference Book, p. 316; *Math Masters,* pp. G7–G9

| WHOLE CLASS | SMALL GROUP | PARTNER | INDEPENDENT |

Students practice building and comparing numbers based on their understanding of the place-value system.

Observe
- Which students have a strategy for placing their cards on the gameboard?
- Which students correctly report the values of the digits when they place them?

Discuss
- *How do you decide where to place the numbers that you draw?* **GMP1.2, GMP7.2**
- *What happens to your number if you move one of your cards one place to the right? To the left?* **GMP7.2**

| Differentiate | Game Modifications | Go Online | Differentiation Support |

▸ **Math Boxes 2-4**

Math Journal 1, p. 49

| WHOLE CLASS | SMALL GROUP | PARTNER | INDEPENDENT |

Mixed Practice Math Boxes 2-4 are paired with Math Boxes 2-2.

▸ **Home Link 2-4**

Math Masters, p. 51

Homework Students use U.S. traditional multiplication to multiply.

Lesson 2-5

U.S. Traditional Multiplication, Part 2

Overview Students use U.S. traditional multiplication to multiply multidigit numbers by 1-digit numbers.

➤ **Before You Begin**
Copy and cut apart enough Exit Slips (*Math Masters*, page TA6) for each student to have one.

Standards

Focus Cluster
• Perform operations with multi-digit whole numbers and with decimals to hundredths.

	Materials	Standards
① Warm Up 5 min		
Mental Math and Fluency Students evaluate expressions with grouping symbols.		5.OA.1
② Focus 35–40 min		
Math Message Students study a problem solved using U.S. traditional multiplication.		5.NBT.5
Extending U.S. Traditional Multiplication Students discuss the worked example and use U.S. traditional multiplication to multiply 1-digit numbers by multidigit numbers.	*Student Reference Book*, p. 102; *Math Masters*, p. TA6	5.NBT.5 SMP1, SMP7
✓ **Assessment Check-In** See page 138. Expect most students to be able to find the product of a multidigit number and 1-digit number, which does not involve writing any digits above the line, using U.S. traditional multiplication.	*Math Masters*, p. TA6	5.NBT.5
Introducing *Multiplication Top-It: Larger Numbers* **Game** Students practice multiplying multidigit numbers by 1-digit numbers.	*Student Reference Book*, p. 325; per partnership: number cards 0–9 (4 of each)	5.NBT.5 SMP1
③ Practice 20–30 min		
Practicing with Powers of 10 Students practice writing and interpreting powers of 10 in exponential notation.	*Math Journal 1*, p. 50	5.NBT.2 SMP7
Math Boxes 2-5 Students practice and maintain skills.	*Math Journal 1*, p. 51	See page 139.
Home Link 2-5 **Homework** Students play *Multiplication Top-It: Larger Numbers* at home and record their work for two rounds.	*Math Masters*, p. 54	5.NBT.2, 5.NBT.5

 Go Online to see how mastery develops for all standards within the grade.

Differentiation Options

Playing *Baseball Multiplication*

5.NBT.5

Student Reference Book, p. 292; per group: *Math Masters,* p. G3; number cards 1–10 (4 of each); 4 counters; calculator or multiplication/division facts table

To practice basic multiplication facts, students play *Baseball Multiplication.* Knowing basic multiplication facts is an important prerequisite skill for fluently multiplying multidigit numbers using U.S. traditional multiplication.

Observe

- Which students are fluent with basic multiplication facts?

Discuss

- *Which facts do you know? Which facts do you still need to learn?*

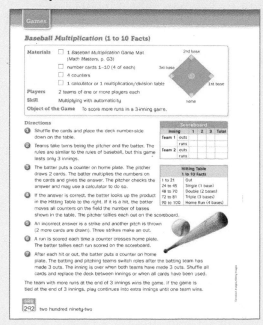

Multiplying Larger Numbers

5.NBT.5, SMP8

Math Masters, p. 52

To extend their work with U.S. traditional multiplication, students describe how to use the strategy to multiply a 1-digit number by a number of any size. **GMP8.1** They use U.S. traditional multiplication to solve number stories involving large numbers.

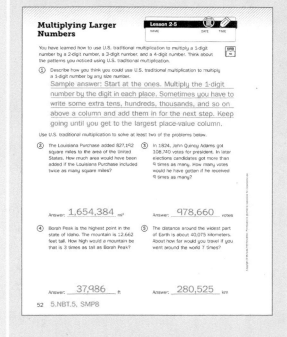

Solving Silly Multiplication Number Stories

5.NBT.5

Activity Card 19; *Math Masters,* p. 53

For additional practice with multiplication, students work in partnerships to create and solve silly number stories.

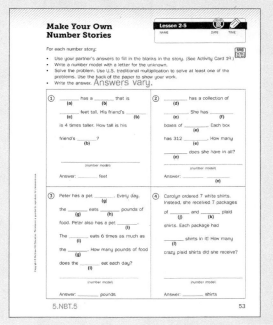

English Language Learner

Beginning ELL Build students' mathematical vocabulary by explicitly pointing out terms that belong to the same word family. For this lesson, display the terms *multiply* and *multiplication.* Pronounce the words and then have students repeat them chorally. Use Total Physical Response prompts to provide practice using the terms. For example, say: *Multiply 352 by 7. Show me the multiplication facts for the numbers 9 and 7.*

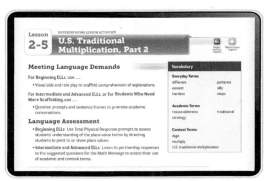

Differentiation Support pages are found in the online Teacher's Center.

Mathematical Process and Practice

SMP1 **Make sense of problems and persevere in solving them.**

GMP1.2 Reflect on your thinking as you solve your problem.

GMP1.6 Compare the strategies you and others use.

SMP7 **Look for and make use of structure.**

GMP7.1 Look for mathematical structure such as categories, patterns, and properties.

Academic Language Development

Explain the term *traditional* by showing examples of traditions. Use visual aids for familiar practices or role-play familiar and unfamiliar practices to show how traditions describe established ways of doing things in various parts of the world. Discuss different ways of doing the same thing. For example, celebrating Thanksgiving is a tradition in both the United States and Canada, although the traditional dates for celebrating the holiday are different. Extend the discussion to the diverse ways people in different parts of the world multiply large numbers. As a way of validating other multiplication methods students might know, encourage the class to share other ways of multiplying that they have seen being used.

---

1 Warm Up — 5 min

▶ Mental Math and Fluency

Display the following expressions for students to evaluate.
Leveled exercises:

● ○ ○ $5 * (3 + 2)$ 25
    $(5 * 3) + 2$ 17
    $(10 \div 2) * 6$ 30

● ● ○ $(32 / 8) * (16 / 4)$ 16
    $(27 / 9) + (100 / 10)$ 13
    $(25 + 10) + (49 / 7)$ 42

● ● ● $9 * \{24 \div (6 * 2)\}$ 18
    $5 * [3 + (15 / 3)]$ 40
    $3 * [(14 \div 2) - (4 - 1)]$ 12

2 Focus — 35–40 min

▶ Math Message

This multiplication problem was solved using U.S. traditional multiplication. Talk to a partner about how you think the strategy works.

$$
\begin{array}{r}
{\scriptstyle 4\ \ 2} \\
2\ 6\ 4 \\
*\ \ \ \ \ 7 \\
\hline
1,\ 8\ 4\ 8
\end{array}
$$

▶ Extending U.S. Traditional Multiplication

Student Reference Book, p. 102; *Math Masters*, p. TA6

| WHOLE CLASS | SMALL GROUP | PARTNER | INDEPENDENT |

Math Message Follow-Up Remind students that in the previous lesson they multiplied 2-digit numbers by 1-digit numbers using U.S. traditional multiplication. Encourage them to think about how they can use what they learned to solve the Math Message problem. Discuss how to use U.S. traditional multiplication, recording the steps as follows.

Step 1: Multiply the ones.	
7 * 4 ones = 28 ones, or 2 tens and 8 ones	$\begin{array}{r}{\scriptstyle 2}\\ 2\ 6\ \mathbf{4}\\ *\ \ \ \ 7\\ \hline 8\end{array}$
Write 8 in the 1s column below the line.	
Write 2 above the 10s column.	

Point out that so far this is the same as using the algorithm to multiply a 1-digit number by a 2-digit number. GMP1.2

<table>
<tr><td>

Step 2: Multiply the tens.

7 * 6 tens = 42 tens

Remember the 2 tens from Step 1.

42 tens + 2 tens = 44 tens, or 4 hundreds and 4 tens

Write 4 in the 10s columns below the line.

Write 4 above the 100s column.

</td><td>

```
  4  2
  2  6  4
*        7
─────────
        4  8
```

</td></tr>
</table>

Ask: *How is this step similar to multiplying a 1-digit number by a 2-digit number?* Sample answer: We are still multiplying the tens and adding in the tens from the first step. *How is it different?* **GMP1.2** Sample answer: We have to write the hundreds above the hundreds column instead of below the line because we still have more to multiply.

<table>
<tr><td>

Step 3: Multiply the hundreds.

7 * 2 hundreds = 14 hundreds

Remember the 4 hundreds from Step 2.

14 hundreds + 4 hundreds = 18 hundreds, or 1 thousand and 8 hundreds

Write 8 in the 100s column below the line.

Write 1 in the 1,000s column below the line.

</td><td>

```
  4  2
  2  6  4
*        7
──────────
1, 8  4  8
```

</td></tr>
</table>

Ask: *Do you notice any patterns in the steps of U.S. traditional multiplication?* **GMP7.1** Sample answer: You multiply by each place in the number and then split up the product above and below the line if you still have more to multiply. If you are done multiplying, you can write the whole product below the line.

Tell students they can use these patterns to help when multiplying larger numbers using U.S. traditional multiplication. Solve 2,192 * 3 as a class, using U.S. traditional multiplication, as shown below.

<table>
<tr><td>

Step 1: Multiply the ones.

3 * 2 ones = 6 ones

Write 6 in the 1s column below the line.

Since there are no tens, do not write anything above the 10s column.

</td><td>

```
2, 1  9  2
*         3
──────────
          6
```

</td></tr>
</table>

Be sure to highlight the idea that sometimes there is no need to write an extra digit above one of the columns. In those cases there are no extra tens (or hundreds, or thousands) to add in the next step.

<table>
<tr><td>

Step 2: Multiply the tens.

3 * 9 tens = 27 tens

There are no tens to add from Step 1.

27 tens = 2 hundreds and 7 tens

Write 7 in the 10s columns below the line.

Write 2 above the 100s column.

</td><td>

```
       2
2, 1  9  2
*         3
──────────
       7  6
```

</td></tr>
</table>

Lesson 2-5   **137**

Differentiate Some students may benefit from solving these problems using partial-products multiplication so they can compare their work with partial products to their work with U.S. traditional multiplication, as shown in Lesson 2-4.

Here is how 264 * 7 can be solved using partial-products multiplication:

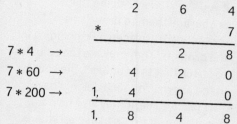

		2	6	4
	*			7
7 * 4 →			2	8
7 * 60 →		4	2	0
7 * 200 →	1,	4	0	0
	1,	8	4	8

Students may also benefit from sketching or looking at an area model for the problem, as shown below.

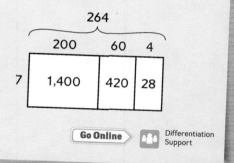

264
200 | 60 | 4

| 7 | 1,400 | 420 | 28 |

Go Online ⬛ Differentiation Support

Math Journal 1, p. 50

Step 3: Multiply the hundreds.	
3 * 1 hundred = 3 hundreds	²
Remember the 2 hundreds from Step 2.	2, **1** 9 2
3 hundreds + 2 hundreds = 5 hundreds	* 3
Write 5 in the 100s columns below the line.	‾‾‾‾‾‾‾
Since there are no thousands, do not write anything above the 1,000s column.	5 7 6

Step 4: Multiply the thousands.	
3 * 2 thousands = 6 thousands	²
	2, 1 9 2
There are no thousands to add from Step 3.	* 3
Write 6 in the 1,000s column below the line.	‾‾‾‾‾‾‾
	6, 5 7 6

Distribute an Exit Slip (*Math Masters*, page TA6) to each student. Have students work in partnerships to solve the two multiplication problems below using U.S. traditional multiplication. Refer students to *Student Reference Book*, page 102 as needed.

- 423 * 3 1,269
- 2,681 * 5 13,405

Assessment Check-In 5.NBT.5

Math Masters, p. TA6

Expect most students to be able to solve 423 * 3, which does not involve writing any digits above the line, using U.S. traditional multiplication. Some may also be able to solve 2,681 * 5, which does involve writing digits above the line and remembering to add them.

✓ᐧ **Evaluation Quick Entry** Go online to record students' progress and to see trajectories toward mastery for these standards.

▶ **Introducing *Multiplication Top-It: Larger Numbers***

Student Reference Book, p. 325

| WHOLE CLASS | SMALL GROUP | PARTNER | INDEPENDENT |

Review the directions for *Multiplication Top-It: Larger Numbers* on *Student Reference Book*, page 325. Students are familiar with *Top-It* games from previous years of *Everyday Mathematics*. In this variation of *Multiplication Top-It*, partners draw 4 cards and use them to make a 3-digit number and a 1-digit number. They multiply the numbers and compare products. The student with the larger product wins the round. Demonstrate a sample round, and then have partnerships or small groups play the game.

Students may use whatever multiplication strategy they wish during this game. However, to provide additional practice with U.S. traditional multiplication, you may want to suggest that students use this strategy for at least one round.

Observe

• Which students have an efficient strategy for multiplication?

• Which strategies do students seem to prefer?

Discuss

• *Compare a strategy you or your partner used to U.S. traditional multiplication. How are they alike? How are they different?* GMP1.6

• *Which strategy is the easiest to use? Which is the hardest? Why?* GMP1.6

Differentiate **Game Modifications** Go Online Differentiation Support

Summarize Have students share whether they prefer to use U.S. traditional multiplication, partial-products multiplication, or a different multiplication strategy while playing *Multiplication Top-It: Larger Numbers*. Ask students to explain their choice.

③ Practice 20–30 min

▶ Practicing with Powers of 10

Math Journal 1, p. 50

WHOLE CLASS | SMALL GROUP | PARTNER | INDEPENDENT

Students practice writing and interpreting powers of 10 in exponential notation. They look for and describe a pattern in multiplication problems involving powers of 10. GMP7.1

▶ Math Boxes 2-5

Math Journal 1, p. 51

WHOLE CLASS | SMALL GROUP | PARTNER | INDEPENDENT

Mixed Practice Math Boxes 2-5 are paired with Math Boxes 2-7.

▶ Home Link 2-5

Math Masters, p. 54

Homework Students play *Multiplication Top-It: Larger Numbers* at home and record their work for two rounds.

Math Journal 1, p. 51

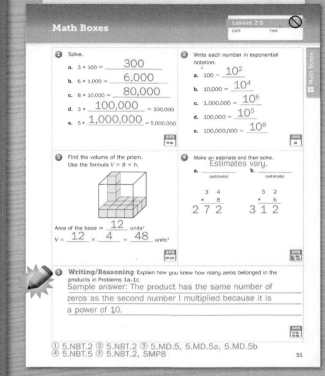

Math Masters, p. 54

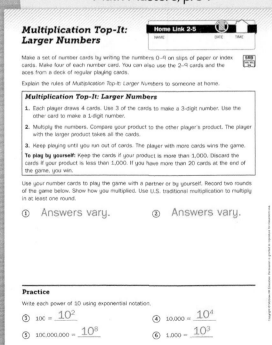

Application: Unit Conversions

Overview Students use unit conversions within the U.S. customary system to solve multistep problems.

▶ **Before You Begin**
For Part 2, prepare a two-column table labeled *miles* and *feet*. Decide how you will display the number stories from pages 143 and 144. If additional sets of *Prism Pile-Up* cards are needed for Part 3, copy and cut apart *Math Masters*, pages G4 and G5.

▶ **Vocabulary**
measurement units • convert • number model • relation symbol • expression

Standards

Focus Clusters
- Write and interpret numerical expressions.
- Perform operations with multi-digit whole numbers and with decimals to hundredths.
- Convert like measurement units within a given measurement system.

 Warm Up 5 min

Materials

	Materials	Standards
Mental Math and Fluency Students convert between units of length.		5.MD.1

 Focus 35–40 min

	Materials	Standards
Math Message Students solve a number story about converting miles to feet.	*Student Reference Book*, p. 328	5.MD.1
Converting Miles to Feet Students complete a table of conversions for miles to feet.	*Student Reference Book*, p. 328	5.NBT.5, 5.MD.1 SMP1
Solving Unit Conversion Number Stories Students solve number stories involving conversions of units within the U.S. customary system.	*Math Journal 1*, p. 52; *Student Reference Book*, p. 328; *Math Masters*, p. TA2 (optional)	5.OA.1, 5.OA.2, 5.NBT.5, 5.MD.1 SMP1, SMP4, SMP5
✓ **Assessment Check-In** See page 144. Expect most students to be able to use U.S. customary unit conversions to solve problems like those identified.	*Math Journal 1*, p. 52	5.OA.2, 5.MD.1, SMP4

 Practice 20–30 min

	Materials	Standards
Playing *Prism Pile-Up* **Game** Students practice finding volumes of rectangular prisms and figures composed of rectangular prisms.	*Student Reference Book*, p. 319; per partnership: *Math Masters*, p. G6; *Prism Pile-Up* cards; calculator (optional)	5.OA.2, 5.MD.3, 5.MD.3a, 5.MD.3b, 5.MD.4, 5.MD.5, 5.MD.5a, 5.MD.5b, 5.MD.5c SMP1, SMP2
Math Boxes 2–6 Students practice and maintain skills.	*Math Journal 1*, p. 53	See page 145.
Home Link 2–6 **Homework** Students collect measurements and convert them to different units.	*Math Masters*, p. 55	5.NBT.5, 5.MD.1

Go Online to see how mastery develops for all standards within the grade.

Differentiation Options

Readiness	5–15 min	Enrichment	15–30 min	Extra Practice	5–15 min

WHOLE CLASS	SMALL GROUP	PARTNER	INDEPENDENT

Counting to Convert Inches to Feet

5.MD.1, SMP7

per group: three 12-inch rulers,
36 square pattern blocks

To explore unit conversions using a concrete model, students count how many 1-inch square pattern blocks are equal to the length of a 1-foot ruler. Distribute 36 square pattern blocks to each group, explaining that each pattern block is 1 inch long. Have students line up the blocks from one end of a 12-inch ruler to the other and then count them. Ask: *How many inches do you need to make a foot?* 12 inches Repeat with two 12-inch rulers and three 12-inch rulers. 24 inches; 36 inches Record the information in a two-column table. Ask: *What patterns do you see?*

GMP7.1 Sample answer: There are 12 more inches every time you add a foot.

Writing Unit Conversion Number Stories

5.MD.1

Activity Card 20;
Math Journal 1, p. 52; *Student Reference Book*, p. 328

To extend their work with unit conversions, students write unit conversion number stories using the problems on journal page 52 as examples. Partners solve each other's number stories.

Converting Units

5.OA.2, 5.MD.1

Activity Card 21;
Student Reference Book, p. 328; number cards 1–20 (1 of each); two 6-sided dice

For more practice with unit conversions, students roll dice and draw number cards to generate unit conversion problems. They write expressions recording their calculations and number sentences recording their conversions.

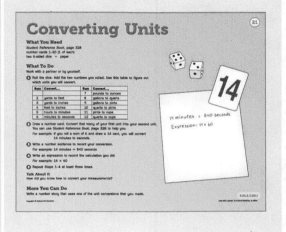

English Language Learner

Beginning ELL To familiarize students with U.S. customary *measurement units* and measuring tools, display everyday measuring tools labeled by name and showing common conversions. For example, label a 1-foot ruler with the word *ruler* and the units of measure: 1 foot = 12 inches. Other useful measurement tools to label and display include a yardstick and a measuring cup.

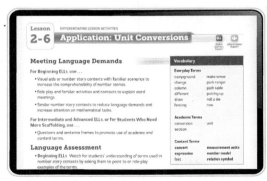

Differentiation Support pages are found in the online Teacher's Center.

Standards and Goals for
Mathematical Process and Practice

SMP1 **Make sense of problems and persevere in solving them.**
  GMP1.1 Make sense of your problem.

SMP4 **Model with mathematics.**
  GMP4.1 Model real-world situations using graphs, drawings, tables, symbols, numbers, diagrams, and other representations.

SMP5 **Use appropriate tools strategically.**
  GMP5.2 Use tools effectively and make sense of your results.

Miles	Feet
1	5,280
2	10,560
3	15,840
4	21,120
5	26,400

Conversions between miles and feet

1 Warm Up 5 min

▶ Mental Math and Fluency

Have students convert between units of length. *Leveled exercises:*

● ○ ○ 1 foot equals how many inches? 12
       1 yard equals how many feet? 3
       2 yards equals how many feet? 6

● ● ○ 4 feet equals how many inches? 48
       5 yards equals how many feet? 15
       36 inches equals how many feet? 3

● ● ● $1\frac{1}{2}$ feet equals how many inches? 18
       $5\frac{1}{2}$ feet equals how many inches? 66
       54 inches equals how many feet? $4\frac{1}{2}$

2 Focus 35–40 min

▶ Math Message

Student Reference Book, p. 328

Park rangers are putting up a fence along a 2-mile section of campground path. How many feet of fencing will they need? Use Student Reference Book, *page 328 to help you.*

▶ Converting Miles to Feet

Student Reference Book, p. 328

| WHOLE CLASS | SMALL GROUP | PARTNER | INDEPENDENT |

Math Message Follow-Up Have students share answers. 10,560 feet Ask: *What information did you need to know before you could solve the problem?* GMP1.1 The number of feet in 1 mile Discuss how students found the number of feet in 1 mile, and then ask: *Which is longer, a 1-mile section of path or a 5,280-foot section of path?* They are the same length. Explain that 1 mile and 5,280 feet are the same distance expressed in different **measurement units.** Explain to students that when they change the unit in which a measurement is expressed, they are **converting** a measurement to a different unit.

Display a two-column table labeled Miles and Feet. (*See margin.*) Fill in the numbers 1–5 in the Miles column and complete the first two rows of the Feet column. Ask: *How many feet of fencing would the rangers need for a 3-mile section of path?* 15,840 *How do you know?* There are 5,280 feet in each mile, and 3 ∗ 5,280 = 15,840. *How many feet would they need for a 4-mile section?* 21,120 *A 5-mile section?* 26,400 Record the conversions in the table.

Tell students that for many problems it is necessary to convert units before the problem can be solved. In today's lesson students will use multiplication to solve problems involving unit conversions.

▶ Solving Unit Conversion Number Stories

Math Journal 1, p. 52; *Student Reference Book,* p. 328

WHOLE CLASS   SMALL GROUP   PARTNER ·   INDEPENDENT

Remind students that when solving problems, they should start by *making sense* of the problem, or thinking about what the problem asks and what information they need to solve it.   GMP1.1   Techniques for making sense of a problem might include making a table or drawing a picture in addition to determining what information they need. Read or display the following number story. Have students solve it in partnerships or small groups. Tell them to refer to *Student Reference Book,* page 328 as needed.   GMP5.2

> *An art teacher has 5 pounds of clay. Each student needs 1 ounce of clay to complete an art project. How many students can complete the art project?*

After students have worked on the problem, invite them to share strategies. Strategies students may use to help them make sense of the problem include drawing pictures like the one below or creating a conversion table for pounds and ounces, similar to the one shown for the Math Message Follow-Up.   GMP1.1

Sample picture:

1 pound	1 pound	1 pound	1 pound	1 pound
16 ounces	16 ounces	16 ounces	16 ounces	16 ounces

Ask:

- *What information did you need to solve this problem?*   GMP1.1
  1 pound = 16 ounces
- *Where did you find that information?*   GMP5.2   In the *Student Reference Book*
- *How can we find the number of ounces in 5 pounds?* Multiply 5 by 16

Remind students that **number models** represent real-world problems using only numbers and mathematical symbols.   GMP4.1   Ask: *What number model can we use to show how we found the number of ounces in 5 pounds?* 5 * 16 Tell students that a number model that has no **relation symbol** (=, >, <, ≤, ≥, or ≠) is called an **expression.** Expressions are often useful models because they can be evaluated to solve problems. The expression 5 * 16 can be evaluated to find the number of ounces in 5 pounds.

Have students use U.S. traditional multiplication to multiply 5 by 16 and then check whether they get the same answer using other methods. Ask again: *How many students can complete the art project?* 80 students

Professional Development

This lesson focuses on conversions within the U.S. customary system. Because the number of smaller units in a larger unit varies greatly in the U.S. customary system (for example, there are 12 inches in a foot but 3 feet in a yard), converting between units is a good application of whole-number multiplication and division. To keep the focus on multiplication, this lesson emphasizes conversions from a larger unit to a smaller unit. Converting from smaller to larger units will be covered in ongoing practice following the division lessons later in Unit 2. In the metric system there are usually 10 smaller units in each next-larger unit (for example, a centimeter is equal to 10 millimeters, a decimeter to 10 centimeters, and so on). Conversions within the metric system are a good application of multiplying and dividing whole numbers and decimals by powers of 10. Conversions among metric units are a focus in Unit 4.

Academic Language Development

Students may be familiar with the term *expression* in the sense of an idiomatic or cultural phrase, as in: "It's just an expression." To extend students' understanding to the mathematical meaning of expression, have them work in groups to complete a 4-Square Graphic Organizer (*Math Masters,* page TA2), showing an example, a non-example, a student definition, and a description of a real-life scenario in which a mathematical expression might be used.

Read or display the following number story and have partnerships or groups work together to solve it. Encourage students to draw pictures to help them make sense of the problem, and tell them to write an expression to record the calculations. **GMP1.1, GMP4.1**

> *A camp counselor is building a bench to put near the fire pit. She has one piece of wood that is 3 yards long and one piece of wood that is 7 feet long. If she places these pieces of wood end to end to make the bench seat, what will the length of the bench be in feet?*

After students have had time to work on the problem, invite them to share strategies. Some students may have drawn pictures like the one below.

Sample picture:

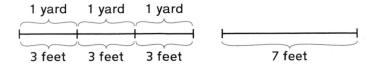

1 yard 1 yard 1 yard

3 feet 3 feet 3 feet   7 feet

Ask:

- *What unit conversion do you need to know to solve this problem?* The number of feet in 1 yard *What expression shows how to find the number of feet in 3 yards?* $3 * 3$ Record the expression $3 * 3$.
- *What would you do next to solve the problem?* Add the length of the other piece of wood *What could we add to this expression to show that step?* $+ 7$ Add to the expression to show $3 * 3 + 7$.
- *How could we show that the multiplication happens first?* Add parentheses, brackets, or braces around $3 * 3$ Add grouping symbols to show $(3 * 3) + 7$.
- *Evaluate this expression. How long will the bench be?* 16 feet

Have partnerships complete journal page 52, where they model and solve problems involving unit conversions. **GMP1.1, GMP4.1**

✓ **Assessment Check-In** 5.OA.2, 5.MD.1

Math Journal 1, p. 52

Expect most students to be able to use U.S. customary unit conversions to solve Problems 1 and 2 on journal page 52. Some may be able to solve Problems 3 and 4, which do not identify the necessary conversions. Some students may also be able to write expressions to model the problems. **GMP4.1** For students who struggle to solve the problems, suggest that they make a two-column table relating the units in the problem, similar to the table of mile and feet equivalencies for the Math Message Follow-Up..

✓= **Evaluation Quick Entry** Go online to record students' progress and to see trajectories toward mastery for these standards.

Summarize Invite students to share and explain the number models they wrote for the problems on journal page 52.

---

3 Practice · 20–30 min

▶ Playing *Prism Pile-Up*

Student Reference Book, p. 319; *Math Masters,* p. G6

| WHOLE CLASS | **SMALL GROUP** | **PARTNER** | INDEPENDENT |

Students practice calculating the volumes of rectangular prisms and figures composed of rectangular prisms. Have them record the volume of each figure and the number sentences they used for their calculations on *Math Masters,* page G6.

Observe

- Which students are counting to find the volumes of the figures? Which students are applying formulas?
- Which students can write a number sentence to represent their strategy? **GMP2.1**

Discuss

- *Did you use a formula to find the volume of the figure? If so, how did you decide which formula to use?* **GMP2.2**
- *Could you find the volume of the figure in a different way? How?* **GMP1.5**

| **Differentiate** | **Game Modifications** | **Go Online** | Differentiation Support |

▶ Math Boxes 2-6

Math Journal 1, p. 53

| WHOLE CLASS | **SMALL GROUP** | **PARTNER** | **INDEPENDENT** |

Mixed Practice Math Boxes 2-6 are paired with Math Boxes 2-8.

▶ Home Link 2-6

Math Masters, p. 55

Homework Students collect measurements and convert them to different units.

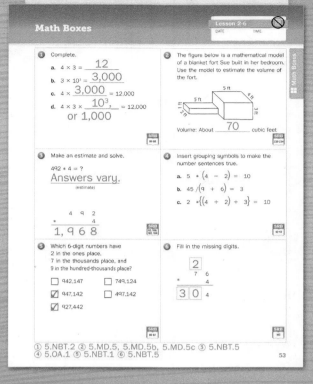

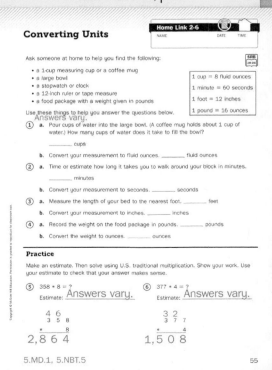

U.S. Traditional Multiplication, Part 3

Overview Students use U.S. traditional multiplication to multiply 2-digit numbers by 2-digit numbers.

▶ **Before You Begin**
If your students are unfamiliar with partial-products multiplication, you may want to review *Student Reference Book*, page 100 as a class before starting this lesson. Some students may benefit from completing the Readiness activity before reviewing the *Student Reference Book* page.

▶ **Vocabulary**
area model

Standards

Focus Clusters
- Write and interpret numerical expressions.
- Perform operations with multi-digit whole numbers and with decimals to hundredths.

1 Warm Up 5 min	**Materials**	
Mental Math and Fluency Students write numbers in expanded form.	slate	**5.NBT.1**

2 Focus 35–40 min		
Math Message Students discuss how to solve multiplication problems.		**5.NBT.5**
Introducing U.S. Traditional Multiplication with 2-Digit Factors Students multiply 2-digit numbers by 2-digit numbers.		**5.NBT.5** **SMP1, SMP2**
Comparing Multiplication Methods Students compare area models for multiplication methods.		**5.NBT.5** **SMP2**
Estimating and Multiplying Students estimate to check whether products make sense.	*Math Journal 1*, p. 54	**5.OA.2, 5.NBT.5** **SMP2**
✓ **Assessment Check-In** See page 152. This is the first exposure to U.S. traditional multiplication with two 2-digit factors. Expect some students to be able to multiply 2-digit numbers by 2-digit numbers using this method.	*Math Journal 1*, p. 54	**5.NBT.5**

3 Practice 20–30 min		
Introducing *Multiplication Bull's Eye* **Game** Students practice making estimates and solving multiplication problems.	per partnership: number cards 0–9 (4 of each), 6-sided die; *Student Reference Book*, p. 313	**5.NBT.5**
Math Boxes 2-7 Students practice and maintain skills.	*Math Journal 1*, p. 55	See page 153.
Home Link 2-7 **Homework** Students make estimates and multiply.	*Math Masters*, p. 59	**5.NBT.2, 5.NBT.5**

/// (Go Online) to see how mastery develops for all standards within the grade.

Differentiation Options

Readiness 5–15 min	**Enrichment** 5–15 min	**Extra Practice** 5–15 min
WHOLE CLASS · SMALL GROUP · **PARTNER** · INDEPENDENT	WHOLE CLASS · SMALL GROUP · **PARTNER** · INDEPENDENT	WHOLE CLASS · SMALL GROUP · **PARTNER** · INDEPENDENT

Readiness

Playing *Multiplication Wrestling*

5.NBT.5

Student Reference Book, p. 314;
Math Masters, p. G12; per partnership:
number cards 0–9 (4 of each)

For experience finding partial products, students play *Multiplication Wrestling.* While many will be familiar with the game from *Fourth Grade Everyday Mathematics,* you may want to review the directions before partners play on their own. Note that students' ability to find partial products easily will help them understand and use U.S. traditional multiplication.

Observe
• Which students have a strategy for finding all of the partial products?

Discuss
• *How does this game help you understand multiplication of 2-digit numbers?*

Enrichment

Using an Ancient Multiplication Strategy

5.NBT.5, SMP1

Math Masters, pp. 56–57

To extend their work with multiplication, students use an ancient multiplication strategy and compare it to U.S. traditional multiplication. **GMP1.5, GMP1.6**

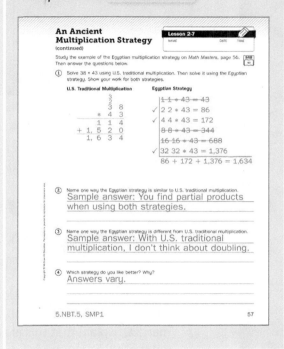

Extra Practice

Practicing U.S. Traditional Multiplication

5.NBT.5

Math Masters, p. 58

For additional practice with U.S. traditional multiplication, students use the algorithm to fill in missing numbers in multiplication problems on *Math Masters,* page 58.

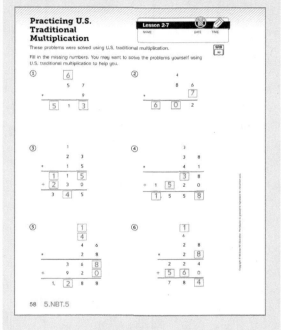

English Language Learner

Beginning ELL Display models (such as toy shoes, toy hats, or small items from a board game) and corresponding real-life objects. Explain that a *model* is sometimes used to represent a real object. Show a real-life object and say: *Pick up the model of the _____.* Extend the idea to an understanding of using *area models* for multiplication by using base-10 blocks to model the numbers being multiplied. Use the blocks to cover the area of different parts of a rectangle as you multiply numbers.

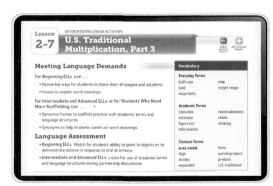

Differentiation Support pages are found in the online Teacher's Center.

Standards and Goals for
Mathematical Process and Practice

SMP1 **Make sense of problems and persevere in solving them.**
GMP1.2 Reflect on your thinking as you solve your problem.

SMP2 **Reason abstractly and quantitatively.**
GMP2.1 Create mathematical representations using numbers, words, pictures, symbols, gestures, tables, graphs, and concrete objects.
GMP2.2 Make sense of the representations you and others use.

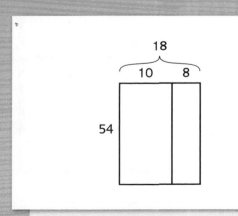

An area model representing
54 * 18

1 Warm Up 5 min

▶ Mental Math and Fluency

Have students write numbers in expanded form on slates. *Leveled exercises:*

● ○ ○ **4,723** Sample answer: $4 * 1{,}000 + 7 * 100 + 2 * 10 + 3 * 1$

● ● ○ **28,954** Sample answer: $2 * 10^4 + 8 * 10^3 + 9 * 10^2 + 5 * 10^1 + 4 * 10^0$

● ● ● **892,395** Sample answer: 8 [100,000s] + 9 [10,000s] + 2 [1,000s] + 3 [100s] + 9 [10s] + 5 [1s]

2 Focus 35–40 min

▶ Math Message

*Suppose you already figured out that 54 * 8 = 432. How could you use that information to help you solve 54 * 18? Talk about it with a partner.*

▶ Introducing U.S. Traditional Multiplication with 2-Digit Factors

| WHOLE CLASS | SMALL GROUP | PARTNER | INDEPENDENT |

Math Message Follow-Up Ask partnerships to share their responses to the Math Message. Then display a divided rectangle as shown in the margin. Explain that this is an **area model** for the problem 54 * 18, and point out that the model divides the product into two parts. Ask: *Where do you see 54 * 8 in this area model?* **GMP2.2** The rectangle on the right *What is the area of that rectangle?* 432 Write 432 in the right part of the area model. Ask: *What do we still need to figure out to solve 54 * 18?* **GMP1.2** The area of the left rectangle *What is the area of the left rectangle?* 54 * 10, or 540 Write 540 in the left part of the area model. Ask: *How can we use this area model to find 54 * 18?* Add the two areas: 540 + 432 = 972

Discuss the idea that 54 * 18 is the same as (54 * 10) + (54 * 8), which means that the two products can be calculated separately and then added to determine the final answer. Explain that students will apply this idea when they use U.S. traditional multiplication to multiply by 2-digit numbers. Tell students that in today's lesson they will learn to use U.S. traditional multiplication to multiply two 2-digit numbers, and they will use area models to help explain why it works.

Display the problem 37 ∗ 25 vertically and sketch a corresponding area model. (*See margin.*) Ask: *How does this area model show 37 ∗ 25?* GMP2.2 Sample answer: The model is a rectangle that is 37 by 25. It divides the total area, or product, into two pieces. One is 37 ∗ 20 and one is 37 ∗ 5. Explain that you are going to show how to use U.S. traditional multiplication to multiply 37 ∗ 25. Tell students that after each step you are going to ask them how the step relates to the area model. GMP1.2, GMP2.2

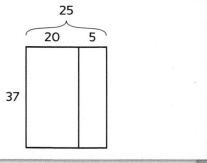

Step 1: Multiply 37 by the 5 in 25, as if the problem were 37 ∗ 5. 5 ∗ 7 ones = 35 ones, or 3 tens and 5 ones Write 5 in the 1s column and 3 above the 10s column. 5 ∗ 3 tens = 15 tens 15 tens + 3 tens = 18 tens, or 1 hundred and 8 tens. Write these numbers in the appropriate columns.	$$\begin{array}{cccc} & & \overset{3}{3} & 7 \\ * & & 2 & 5 \\ \hline & 1 & 8 & 5 \end{array}$$

Ask: *How does this step relate to the area model?* Sample answer: 185 is the area of the rectangle that is 37 by 5. Write 185 in the appropriate part of the area model. (*See below.*)

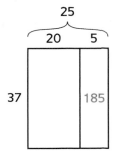

NOTE The area models in this lesson are not meant to be seen as a separate strategy for solving multiplication problems. They are intended to help students make sense of the multiplication methods. Some students may find it useful to sketch area models as they solve multiplication problems, but students should not be required to do so. Note that area models will be used later in the year for mixed number multiplication.

Differentiate   Adjusting the Activity

To help students focus on the digits being used in each step, suggest covering the 2 with a slip of paper while working on Step 1 and the 5 while working on Step 2. It may be helpful for some to cover the 5 with a 0 so that they keep in mind that they are multiplying by 20. Students may also benefit from writing the labels 1s, 10s, and 100s over the appropriate place-value columns.

Go Online 📖 Differentiation Support

NOTE In Step 2 students may question why they write the 1 that represents 100 above the tens column rather than the hundreds column. Writing the digit here helps students remember when to add in the digit. In this case students add in the extra 1 hundred after multiplying 20 by 3 tens, so the 1 is written above the digit representing the 3 tens. This reminder will become important when students use U.S. traditional multiplication to multiply numbers with more than 2 digits.

Professional Development

Many students learning U.S. traditional multiplication are taught to put in a zero or zeros as placeholders in the partial products. In this example, they would start Step 2 by writing a 0 in the ones column, and then they would multiply 37 by 2 to get 74. This strategy works because students are really multiplying by a multiple of 10 (20 in this case), so the zero is needed to show that the partial product is actually 740, not 74. The *Everyday Mathematics* approach preserves place-value understanding by encouraging students to think about the values of the numbers they are multiplying. In Step 2 of this example students think about multiplying 37 by 20, not 2, so explaining zero as a placeholder is not necessary.

Step 2: Think: *What is the value of the 2 in 25?* It's 20, so multiply 37 by the 2 in 25, as if the problem were 37 * 20.

20 * 7 ones = 140 ones, or 1 hundred, 4 tens, and 0 ones. Write 0 in the ones column and 4 in the tens column. Write 1 above the 3. Remember that it represents 1 hundred.

20 * 3 tens = 60 tens, or 6 hundreds

6 hundreds + 1 hundred = 7 hundreds

Write 7 in the 100s column.

```
        1
        3
      3 7
*     2 5
    1 8 5
    7 4 0
```

Ask: *How does this step relate to the area model?* Sample answer: 740 is the area of the rectangle that is 37 by 20. Write 740 in the appropriate part of the area model.

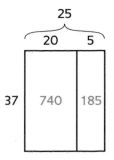

Step 3: Add the products from Step 1 and Step 2.

```
        1
        3
      3 7
*     2 5
    1 8 5
+   7 4 0
    9 2 5
```

Ask: *How does this step relate to the area model?* Sample answer: Adding 185 and 740 gives the total area of the rectangle.

Work through a few more examples as a class, asking students to draw area models and explain how the models connect to the steps of U.S. traditional multiplication. **GMP2.1, GMP2.2** Leave the area model and work for each example displayed. *Suggestions:*

• 73 * 42 3,066   • 11 * 89 979

After completing several problems, ask students to examine their work and see if they notice any patterns or similarities in the problems. If no one mentions it, ask: *What do you notice about the 1s digit in the second partial product for each problem?* The 1s digit is always zero. *Why is the 1s digit always 0?* Sample answer: When we multiply by the 10s digit in the second factor, we are multiplying by a multiple of 10. I know from fact extensions that the 1s digit is always 0 when you multiply by 10, 20, 30, and so on. Point out that understanding patterns such as this one can help students catch mistakes and check their work when they use U.S. traditional multiplication.

▶ Comparing Multiplication Methods

WHOLE CLASS | SMALL GROUP | PARTNER | INDEPENDENT

Have students work in small groups to solve the problem 18 * 42 in two different ways: using partial-products multiplication and using U.S. traditional multiplication. Invite volunteers to display their group's work.

Partial-Products Multiplication	U.S. Traditional Multiplication
``` 1 8 * 4 2 ───── 1 6 2 0 3 2 0 + 4 0 0 ───── 7 5 6 ```	``` 3 1 1 8 * 4 2 ───── 3 6 + 7 2 0 ───── 7 5 6 ```

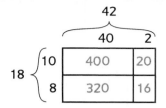

Area model for partial-products multiplication of 18 * 42

Sketch area models illustrating both methods, like the ones shown in the margin (without the partial products). Ask students to help you fill in the partial products in each area model using their displayed work to help. **GMP2.1** Ask them to compare the two area models. **GMP2.2** Guide the discussion to cover the following points:

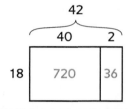

Area model for U.S. traditional multiplication of 18 * 42

- Both area models show the partial products from each method. The partial-products area model shows the four partial products that appear in the written method. The U.S. traditional area model shows the two partial products that appear in the written method.

- The sum of the partial products on the left side of the partial-products area model is the same as the number on the left side of the U.S. traditional area model.

- The sum of the partial products on the right side of the partial-products area model is the same as the number on the right side of the U.S. traditional area model.

- The area models show why both methods give the same answer. When using partial-products multiplication, one step is to find 8 * 2 and 10 * 2. The sum of these two partial products is the same as 18 * 2, which is found in the first step of U.S. traditional multiplication. Similarly, the sum of 8 * 40 and 10 * 40, which are two of the partial products in partial-products multiplication, is the same as 18 * 40, which is found in Step 2 of U.S. traditional multiplication.

▶ Estimating and Multiplying

Math Journal 1, p. 54

| WHOLE CLASS | SMALL GROUP | PARTNER | INDEPENDENT |

Display the work shown at the right. Ask: *Do you think 384 is the correct answer for the problem 64 * 15?* No. *How do you know?* Guide the discussion to cover the following strategies:

$$
\begin{array}{r}
\overset{2}{6}\;4 \\
*\quad 1\;5 \\
\hline
3\;2\;0 \\
+\quad 6\;4 \\
\hline
3\;8\;4
\end{array}
$$

- Examine the work and look for a mistake. The work does not have a zero in the 1s place of the second partial product. This shows that this student multiplied 64 by 1 in the second step of U.S. traditional multiplication instead of multiplying 64 by 10. So the answer must be wrong.

- Interpret the expression 64 * 15 without evaluating it. 64 * 15 is equivalent to a number that is 15 times as large 64. Ten times as much as 64 is 640, so that means 64 * 15 is greater than 640. Since 384 is less than 640, it cannot be the correct answer.

- Round the factors and make an estimate. For example, 64 is close to 60, and 15 is close to 20. 60 * 20 = 1,200. 384 is not very close to 1,200, so it does not seem like a reasonable answer.

Tell students that making an estimate is often the easiest and most reliable way to check whether their answers make sense. Explain that, because U.S. traditional multiplication is new to them, they might make mistakes when using this algorithm. That is why they should always use estimates to check the reasonableness of their answers. Ask: *What are some strategies you can use to make estimates for multiplication problems?* Sample answers: Round the factors to the nearest ten or hundred and then multiply. Use close-but-easier numbers.

Have partnerships complete journal page 54. Problem 5 asks them to make sense of an area model.   GMP2.2

✓ Assessment Check-In   5.NBT.5

Math Journal 1, p. 54

This is the first exposure to U.S. traditional multiplication with two 2-digit factors. Some students may successfully solve the problems on journal page 54, but expect that many will struggle. Plenty of additional practice with U.S. traditional multiplication will be provided throughout the year.

Evaluation Quick Entry Go online to record students' progress and to see trajectories toward mastery for these standards.

Summarize Have partners discuss why it is important to make estimates when solving multiplication problems.

3 Practice 20–30 min

▶ Introducing *Multiplication Bull's Eye*

Student Reference Book, p. 313

| WHOLE CLASS | SMALL GROUP | **PARTNER** | INDEPENDENT |

Read the directions for *Multiplication Bull's Eye* on *Student Reference Book*, page 313 as a class. Demonstrate a sample turn and then have students play the game in partnerships.

Observe

- What strategies are students using to make estimates?
- Which students can successfully use U.S. traditional multiplication?

Discuss

- *How did you decide which numbers to make with your number cards?*
- *What do you like about U.S. traditional multiplication? What is difficult about U.S. traditional multiplication?*

| **Differentiate** | **Game Modifications** | **Go Online** | Differentiation Support |

▶ Math Boxes 2-7

Math Journal 1, p. 55

| WHOLE CLASS | SMALL GROUP | PARTNER | INDEPENDENT |

Mixed Practice Math Boxes 2-7 are paired with Math Boxes 2-5.

▶ Home Link 2-7

Math Masters, p. 59

Homework Students make estimates and solve multiplication problems.

Math Boxes

Lesson 2-7
DATE   TIME

① Complete. Use exponential notation for Parts d and e.

a. $8 \times 10^2 =$ ___800___

b. $3 \times 10^3 =$ ___3,000___

c. $5 \times 10^4 =$ ___50,000___

d. $2 \times$ ___10^6___ $= 2,000,000$

e. $7 \times$ ___10^5___ $= 700,000$

② Complete the table.

Standard Notation	Exponential Notation
10,000	10^4
100,000	10^5
10,000,000	10^7
1,000,000	10^6

③ Find the volume of the prism. Use the formula $V = B \times h$.

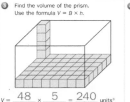

$V =$ ___48___ $\times$ ___5___ $=$ ___240___ units³

④ Make an estimate and solve.
Estimates vary.

a. (estimate) ___   b. (estimate) ___

```
  6 4       8 9
* 5         * 3
3 2 0       2 6 7
```

⑤ **Writing/Reasoning** Explain how you know the formula $V = B \times h$ tells you how many unit cubes could be packed into the prism in Problem 3. Sample answer: The area of the base tells you how many cubes fit in one layer of the prism. The height tells you how many layers there are. When you multiply the cubes in one layer by the number of layers, you find out how many cubes could be packed in the whole prism.

① 5.NBT.2 ② 5.NBT.2 ③ 5.MD.5, 5.MD.5a, 5.MD.5b
④ 5.NBT.5 ⑤ 5.MD.5, 5.MD.5a, SMP2, SMP8   55

Estimating and Multiplying

Home Link 2-7
NAME   DATE   TIME

Make an estimate for each multiplication problem. Write a number sentence to show how you estimated.

Then solve ONLY the problems that have answers that are *more than 1,000*. Use your estimates to help you decide which problems to solve.

Use U.S. traditional multiplication to solve at least one of the problems. Show your work.
Estimates are sample answers.

① $23 * 41 = ?$
$20 * 40 = 800$
(estimate)

```
  2 3
* 4 1
```

② $72 * 56 = ?$
$70 * 60 = 4,200$
(estimate)

```
  7 2
* 5 6
4,032
```

③ $32 * 15 = ?$
$30 * 15 = 450$
(estimate)

```
  3 2
* 1 5
```

④ $82 * 11 = ?$
$80 * 10 = 800$
(estimate)

```
  8 2
* 1 1
```

⑤ $63 * 39 = ?$
$60 * 40 = 2,400$
(estimate)

```
  6 3
* 3 9
2,457
```

⑥ $91 * 46 = ?$
$100 * 46 = 4,600$
(estimate)

```
  9 1
* 4 6
4,186
```

Practice

Solve.

⑦ a. $7 * 10,000 =$ ___70,000___
b. $7 * 10^4 =$ ___70,000___

⑧ a. $2 * 400 =$ ___800___
b. $2 * 4 * 10^2 =$ ___800___

⑨ a. $6,000 * 300 =$ ___1,800,000___
b. $6 * 10^3 * 3 * 10^2 =$ ___1,800,000___

5.NBT.5, 5.NBT.2   59

U.S. Traditional Multiplication, Part 4

Overview Students use U.S. traditional multiplication to multiply multidigit numbers.

▶ **Before You Begin**
For the optional Enrichment activity, have poster paper available.

▶ **Vocabulary**
algorithm

Standards

Focus Cluster
• Perform operations with multi-digit whole numbers and with decimals to hundredths.

1 Warm Up   5 min

	Materials	
Mental Math and Fluency Students solve extended multiplication facts with exponents.		5.NBT.2

2 Focus   35–40 min

Math Message Students solve a multiplication number story.		5.NBT.5
Extending U.S. Traditional Multiplication to Larger Numbers Students learn to multiply multidigit numbers using U.S. traditional multiplication.		5.NBT.5 SMP1
Choosing Multiplication Strategies Students choose strategies for solving multiplication problems and explain their choices.	*Math Journal 1*, p. 56	5.NBT.5 SMP1, SMP6
Assessment Check-In See page 159. Expect most students to have a sense of the steps involved in the U.S. traditional multiplication algorithm and to make reasonable efforts to use it to multiply a 3-digit number by a 2-digit number.	*Math Journal 1*, p. 56	5.NBT.5

3 Practice   20–30 min

Playing *Name That Number* **Game** Students practice writing expressions for calculations and writing expressions with grouping symbols.	*Student Reference Book*, p. 315; per partnership: *Math Masters*, p. G2; number cards 0–10 (4 of each) and 11–20 (1 of each)	5.OA.1, 5.OA.2 SMP1, SMP2
Math Boxes 2-8 Students practice and maintain skills.	*Math Journal 1*, p. 57	See page 159.
Home Link 2-8 **Homework** Students choose multiplication strategies and explain their choices.	*Math Masters*, p. 60	5.NBT.2, 5.NBT.5 SMP6

 Go Online to see how mastery develops for all standards within the grade.

Differentiation Options

Readiness	5–15 min	Enrichment	15–30 min	Extra Practice	5–15 min
WHOLE CLASS · **SMALL GROUP** · PARTNER · INDEPENDENT		WHOLE CLASS · **SMALL GROUP** · 'PARTNER · INDEPENDENT		WHOLE CLASS · SMALL GROUP · **PARTNER** · INDEPENDENT	

Readiness

Using U.S. Traditional Multiplication with Large Numbers

5.NBT.5, SMP8

To prepare for using U.S. traditional multiplication to multiply larger numbers, students solve a series of problems using that strategy. First, have students solve 261 * 3. 783 Next, have them solve 3,261 * 3. 9,783 Discuss how the steps for the two problems are similar and different. Sample answer: I started the same way, but I had to multiply an additional number for the second problem. Finally, have them solve 23,261 * 3. 69,783 Discuss how solving it was like and unlike the previous problems. Sample answer: I used the same steps but just multiplied more times. Ask: *If I gave you a 10-digit number to multiply by 3, could you use U.S. traditional multiplication to solve the problem? How?* **GMP8.1** Yes. I would just keep using the same steps until I multiplied all the numbers.

Enrichment

Comparing Multiplication Strategies

5.NBT.5, SMP1, SMP6

Activity Card 22, *Student Reference Book;* per group: poster paper, crayons or markers

To extend their work with multiplication, students make a poster comparing different multiplication strategies. **GMP1.6** They discuss why different strategies might be more efficient for different problems. **GMP6.4**

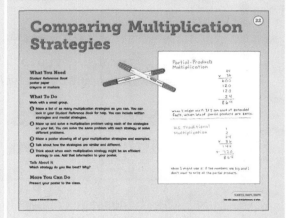

Extra Practice

Playing *Multiplication Top-It: Larger Numbers*

5.NBT.5

Student Reference Book, p. 325; per partnership: number cards 0–9 (4 of each)

For more practice with multiplication, students play *Multiplication Top-It: Larger Numbers.* Depending on their skill level, have them play the game as in the standard version or use one of these variations:

• Draw 3 cards and use them to create one 2-digit number and one 1-digit number.

• Draw 5 cards and use them to create one 3-digit number and one 2-digit number.

Students should choose their own multiplication strategies, but encourage them to use U.S. traditional multiplication for at least one of the rounds.

Observe
• What strategies are students using to multiply?

Discuss
• *Which strategy do you prefer to use to multiply? Why?*

English Language Learner

Beginning ELL To scaffold students' understanding of an algorithm as a set of steps used to solve a problem, display an example of illustrated step-by-step instructions, such as directions for putting together a model car or setting up a video game system. Extend by going through the steps of a familiar math algorithm like addition or subtraction with regrouping. Name the algorithm and number the steps as you demonstrate. Display the term *algorithm*, directing students to say and repeat the word.

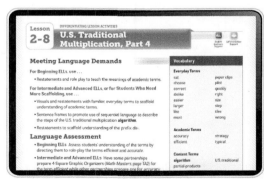

Differentiation Support pages are found in the online Teacher's Center.

SMP1 **Make sense of problems and persevere in solving them.**
  GMP1.5 Solve problems in more than one way.
  GMP1.6 Compare the strategies you and others use.

SMP6 **Attend to precision.**
  GMP6.4 Think about accuracy and efficiency when you count, measure, and calculate.

1 Warm Up 5 min

▶ Mental Math and Fluency

Have students solve extended multiplication facts. *Leveled exercises:*

●○○		●●○		●●●	
$6 * 8$	48	$3 * 9$	27	$5 * 7 = 35$	
$8 * 10^1$	80	$3 * 10^1$	30	$5 * 10^2 = 500$	
$6 * 80$	480	$9 * 10^1$	90	$7 * 10^2 = 700$	
		$30 * 90$	2,700	$500 * 700 = 350,000$	

2 Focus 35–40 min

▶ Math Message

A typical cat sleeps up to 18 hours per day. There are 365 days in 1 year. How many hours does a typical cat sleep in 1 year? 6,570 hours

▶ Extending U.S. Traditional Multiplication to Larger Numbers

| WHOLE CLASS | SMALL GROUP | PARTNER | INDEPENDENT |

Math Message Follow-Up Have students share their strategies for solving the Math Message problem. Expect that many will have chosen partial-products multiplication. If a student used U.S. traditional multiplication, invite him or her to explain how to use it to multiply a 3-digit number by a 2-digit number. If no one used U.S. traditional multiplication, have partners talk briefly about how it might be used to multiply 365 and 18.

Explain that U.S. traditional multiplication is an example of an **algorithm,** or a set of steps that can be used to solve a certain kind of problem. Once students understand the patterns in an algorithm, they can use it to solve problems involving numbers of any size. Ask students to name some other algorithms they have used and the kinds of problems they can solve with each one. Sample answers: Partial-sums addition can be used to add any whole numbers. U.S. traditional subtraction can be used to subtract any whole numbers.

Tell students that in today's lesson they will learn to extend U.S. traditional multiplication to the multiplication of larger numbers.

Display the problem 506 ∗ 43 vertically. Have students work with a partner to solve it using partial-products multiplication and record their work on paper. 21,758 Then demonstrate how to solve the problem using U.S. traditional multiplication. GMP1.5 Pause after each step, inviting partnerships to check what was done against how they solved the problem using partial-products multiplication. This is the first example involving a zero in one of the factors, so pay particular attention to multiplying by zero. (*See also the* Common Misconception *note in the margin.*)

Common Misconception

Differentiate In Step 1, after multiplying 3 by 0 tens, some students may write 0 below the tens column. Remind them that in each step they should multiply *and* add in any extra tens or hundreds they have recorded. Encourage them to cross out each digit as they add it.

Go Online Differentiation Support

Step 1: Multiply 506 by the 3 in 43, as if the problem were 506 ∗ 3.
 	
Check: The product for Step 1 is 1,518 and should be equal to the sum of the partial products involving 3:
 	
3 ∗ 500 = 1,500   3 ∗ 0 = 0   3 ∗ 6 = 18
 	
1,500 + 0 + 18 = 1,518
	1
	    5 0 6
	 * **4 3**
	  1, 5 1 8

Step 2: Think: *What is the value of the 4 in 43?* It's 40, so multiply 506 by the 4 in 43 as if the problem were 506 ∗ 40.
 	
Check: The product for Step 2 is 20,240 and should be equal to the sum of the partial products involving 40:
 	
40 ∗ 500 = 20,000   40 ∗ 0 = 0   40 ∗ 6 = 240
 	
20,000 + 0 + 240 = 20,240
	2
	     1
	    5 0 6
	 * **4 3**
	  1, 5 1 8
	2 0, 2 4 0

Step 3: Add the partial products.
 	
Check: The final answer should be equal to the sum of all the partial products.
 	
3 ∗ 500 = 1,500   3 ∗ 0 = 0   3 ∗ 6 = 18
 	
40 ∗ 500 = 20,000   40 ∗ 0 = 0   40 ∗ 6 = 240
 	
1,500 + 0 + 18 + 20,000 + 0 + 240 = 21,758
	2
	     1
	    5 0 6
	 * 4 3
	   1, 5 1 8
	+ 2 0, 2 4 0
	  2 1, 7 5 8

Next display the problem 457 ∗ 309 vertically and have partnerships solve it using partial-products multiplication. Then solve the problem as a class using U.S. traditional multiplication, once again inviting volunteers to explain each step and show how to check it against what they did using partial-products multiplication. GMP1.5 (*See margin.*)

```
            4 5 7
   *        3 0 9
            ─────
              6 3
            4 5 0
          3, 6 0 0
          2, 1 0 0
        1 5, 0 0 0
      1 2 0, 0 0 0
      ───────────
      1 4 1, 2 1 3
```

Partial-products multiplication

```
              1   2
              5   6
            4 5 7
   *        3 0 9
          ─────────
          4, 1 1 3
          0 0 0 0
    + 1 3 7, 1 0 0
    ─────────────
    1 4 1, 2 1 3
```

U.S. traditional multiplication

Math Journal 1, p. 56

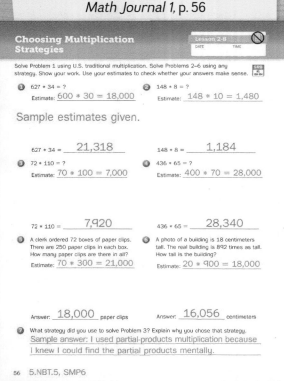

Discuss the following points:

- When the second factor is a 3-digit number, there will be three partial products to add.
- When there is a 0 in the second factor, one of the partial products will be 0.

▶ Choosing Multiplication Strategies

Math Journal 1, p. 56

| WHOLE CLASS | SMALL GROUP | PARTNER | INDEPENDENT |

Display the problem 417 ∗ 99. 41,283 Have students work independently or in partnerships to solve it in two ways: using partial-products multiplication and using U.S. traditional multiplication. **GMP1.5** Have volunteers demonstrate each strategy. (*See margin.*) Then have students compare the strategies. **GMP1.6** Ask:

- *What do you like about partial-products multiplication?* Sample answer: I can see all the partial products separately.
- *What do you dislike about partial-products multiplication?* Sample answers: I have to write down a lot of numbers. There are a lot of numbers to add at the end.
- *What do you like about U.S. traditional multiplication?* Sample answer: I only have to add two partial products.
- *What do you dislike about U.S. traditional multiplication?* Sample answers: I have to think about adding and multiplying at the same time. It's hard to keep track of everything.

Explain to students that when they are given a multiplication problem to solve, they will usually be able to choose the strategy they want to use. Two things to think about when choosing a strategy are *accuracy* and *efficiency.* **GMP6.4** *Accuracy* means getting the correct answer. For example, if students almost always get the right answer using partial-products multiplication, that strategy might be a good choice for them. *Efficiency* means getting the answer quickly or in just a few steps. For example, for students who can use U.S. traditional multiplication successfully, it might be a more efficient strategy than partial-products multiplication.

Point out that for some problems other strategies can be even more efficient than U.S. traditional multiplication. Ask: *Can anyone think of an efficient strategy for multiplying 417 by 99?* Sample answer: Multiply 417 by 100 to get 41,700. Then just subtract 417 to get 41,283. Have students complete journal page 56 independently.

Assessment Check-In 5.NBT.5

Math Journal 1, p. 56

Expect most students to have a sense of the steps involved in the U.S. traditional multiplication algorithm and to make reasonable efforts using it to solve Problem 1 on journal page 56. However, many may still make minor computation errors. Ongoing practice should allow all students to develop fluency with U.S. traditional multiplication by the end of the year.

Evaluation Quick Entry Go online to record students' progress and to see trajectories toward mastery for these standards.

Summarize Have students share the strategies they chose on journal page 56 and their reasons for choosing the ones they did.

3 Practice 20–30 min

▶ Playing *Name That Number*

Student Reference Book, p. 315; *Math Masters,* p. G2

| WHOLE CLASS | **SMALL GROUP** | **PARTNER** | INDEPENDENT |

Students practice writing expressions for calculations and using grouping symbols in expressions. Have students record the expressions they use on *Math Masters,* page G2.

Observe
- Which operations do students use in their solutions?
- Which students are correctly using parentheses? GMP2.1

Discuss
- *How did you decide which operation to use?* GMP1.1
- *Can you write an expression that names the number with (or without) parentheses?* GMP1.5

▶ Math Boxes 2-8

Math Journal 1, p. 57

| WHOLE CLASS | SMALL GROUP | PARTNER | **INDEPENDENT** |

Mixed Practice Math Boxes 2-8 are paired with Math Boxes 2-6.

▶ Home Link 2-8

Math Masters, p. 60

Homework Students choose multiplication strategies and explain their choices. GMP6.4

Math Journal 1, p. 57

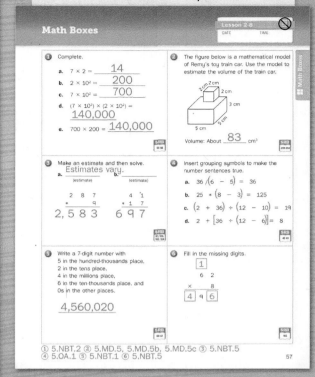

Math Masters, p. 60

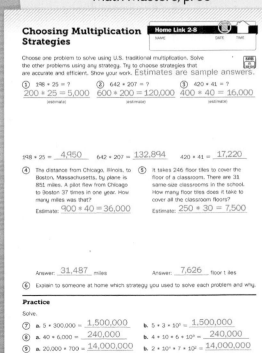

2-Day Lesson
2-9
Open Response and Reengagement

One Million Taps

Overview **Day 1:** Students estimate how much time it would take to tap their desks one million times. **Day 2:** Students examine others' solutions using a rubric or in a class discussion, and they revise their work.

Day 1: Open Response

▶ Before You Begin

Solve the open response problem in as many ways as you can. If possible, schedule time to review students' work and plan for Day 2 of this lesson with your grade-level team.

▶ Vocabulary

efficient

Standards

Focus Clusters
- Understand the place value system.
- Perform operations with multi-digit whole numbers and with decimals to hundredths.

1 Warm Up 5 min

	Materials	
Mental Math and Fluency Students write numbers in exponential notation.	slate	5.NBT.2

2a Focus 55–65 min

Math Message Students estimate the amount of time it takes to address 10 and 100 envelopes based on the amount of time it takes to address 1 envelope.	*Math Journal 1*, p. 58	5.NBT.2, 5.NBT.5
Discussing Efficient Strategies Students discuss strategies for solving the Math Message and consider which ones are more efficient.	*Math Journal 1*, p. 58; Standards for Mathematical Process and Practice Poster	5.NBT.2 SMP1, SMP4, SMP6
Solving the Open Response Problem Students find the time it takes to tap their desks 100 times and estimate how much time it would take to tap their desks 1,000,000 times.	*Math Masters*, pp. 61–62; per partnership: stopwatch (optional)	5.NBT.2 SMP1, SMP6

Getting Ready for Day 2 →

Review students' work and plan discussion for reengagement. *Math Masters*, p. TA4, p. 63 (optional); students' work from Day 1

Go Online to see how mastery develops for all standards within the grade.

1 Warm Up 5 min

▶ Mental Math and Fluency

Display the following. Have students write the number or product as a power of 10 with exponents on slates. *Leveled exercises:*

● ○ ○ $10 * 1 = 10^1$
$10 * 10 = 10^2$
$10 * 10 * 10 = 10^3$

● ● ○ $1,000 = 10^3$
$10,000 = 10^4$
$1,000,000 = 10^6$

● ● ● $10 * 100 = 10^3$
$100 * 100 = 10^4$
$1,000 * 100 = 10^5$

2a Focus 55–65 min

▶ Math Message

Math Journal 1, p. 58

Work with a partner to complete journal page 58.

> **Differentiate** **Adjusting the Activity**
>
> For students who have trouble getting started, suggest that they draw a picture to represent the amount of time it took to address each envelope. For example, they might draw 10 envelopes and label each 30 seconds for the time it takes to label one. Ask: *How can you find the total amount of time it takes to address 10 envelopes?* Sample answers: I can multiply 30 by 10. I can add up all of the seconds. *How did your picture help you solve the problem?* **GMP4.2** Sample answer: It helped me see that I needed to add up all the seconds it would take to address all 10 envelopes.

▶ Discussing Efficient Strategies

Math Journal 1, p. 58

| WHOLE CLASS | SMALL GROUP | PARTNER | INDEPENDENT |

Math Message Follow-Up Have partners discuss how they solved Problem 1 on the journal page and then share strategies with the class. Strategies might include drawing a picture of the 10 envelopes, using repeated addition, or using multiplication. **GMP1.6, GMP4.2**

Standards and Goals for
Mathematical Process and Practice

SMP1 Make sense of problems and persevere in solving them.
GMP1.6 Compare the strategies you and others use.

SMP4 Model with mathematics.
GMP4.2 Use mathematical models to solve problems and answer questions.

SMP6 Attend to precision.
GMP6.4 Think about accuracy and efficiency when you count, measure, and calculate.

Professional Development

The focus for this lesson is **GMP6.4**. While multiple strategies can be used to solve the open response problem, the emphasis here is on efficiency. *Efficiency* in this context means solving a problem in a way that minimizes time and effort. As students compare strategies, they will discuss which are most efficient. For more information on **GMP6.4**, see the Mathematical Background in the Unit Organizer.

Go Online for information about **SMP6** in the *Implementation Guide*.

Math Journal 1, p. 58

Invitations Lesson 2-9

Zoey is mailing invitations for a fifth-grade party. It takes her about 30 seconds to address 1 envelope.

① About how many seconds would it take Zoey to address 10 envelopes? Show your work.
About __300__ seconds
Sample answers:
• 10 envelopes * 30 seconds per envelope = 300 seconds
• 10 * 30 = 300 seconds
•

| 30 | 30 | 30 | 30 | 30 |
| 30 | 30 | 30 | 30 | 30 |

② About how many seconds would it take Zoey to address 100 envelopes? Show your work.
About __3,000__ seconds
Sample answers:
• In Problem 1 I found that it takes 300 seconds to address 10 envelopes. There are 10 times as many envelopes to address in Problem 2 as in Problem 1.
10 * 300 = 3,000 seconds
• 30 seconds per envelope * 100 envelopes = 3,000 seconds

58 5.NBT.2, 5.NBT.5

Prior to the lesson, use role-play activities to introduce students to the contexts of addressing an envelope and tapping a desk. Point out various uses of the word *address*. Once students understand the basic vocabulary, use simple problems to familiarize them with the concept of scaling, such as: *If I can write 2 addresses in 1 minute, how many addresses can I write in 3 minutes? How long will it take me to address 6 envelopes? Explain how you found your answer.* Ask similar questions about tapping a desk. During this discussion, introduce other vocabulary that may be new to students, such as *interruption, estimate, strategy,* or *guess.*

Time (seconds)	30	300	3,000
Number of Envelopes	1	10	100

Table modeling the Math Message problem

Have partners discuss how they solved Problem 2, and then have them share their thinking. Students may have used strategies similar to those in Problem 1, or they may have used their solution from Problem 1 to solve Problem 2. **GMP1.6, GMP4.2** Ask: *Of the strategies we used in Problem 1, which could we also use in Problem 2?* Sample answer: We could multiply the number of seconds it took to address 1 envelope by the number of envelopes we need to address. *Were there any strategies from Problem 1 that you would not use in Problem 2?* Sample answer: Drawing a picture of the exact number of envelopes would not make sense because it would take a lot of time to draw 100 envelopes.

Display the table shown in the margin. Ask: *How does the table model the problems?* **GMP4.2** Sample answer: The first column shows that it takes 30 seconds to address 1 envelope. The bottom row shows the number of envelopes to address. We can complete the top row to answer how long it takes to address 10 and 100 envelopes.

Ask: *What patterns do you notice in the row for the number of envelopes?* Sample answer: As you move to the right, the number of envelopes is 10 times the number in the column to the left. $1 * 10 = 10$ and $10 * 10 = 100$

Have partners discuss how they think they could use this table to solve the problem. Sample answers: If you know the time it takes to address 1 envelope, you can find the amount of time it takes to address 10 or 100 envelopes. You can multiply $30 * 10$ to find the number of seconds it takes to address 10 envelopes. You can multiply $30 * 100$ to get the amount of time it takes to address 100 envelopes.

Ask: *Does using the table give you the same answer as the strategies we discussed earlier?* Yes. *How does the table help you?* **GMP4.2, GMP6.4** Sample answers: It models the problem; organizes the information; helps you see patterns; and helps you think efficiently.

Tell students that even though there are often multiple ways to solve a problem, mathematicians try to solve problems in the most **efficient** way. Efficiency refers to solving a problem in a way that minimizes time and effort. Refer students to the Standards for Mathematical Process and Practice Poster for **GMP6.4**. Ask: *Of the strategies we discussed for this problem, which are most efficient? Why?* Sample answer: Using a table or number sentence is more efficient than drawing a picture of each envelope because it takes a long time to draw and label each envelope. It takes less time to write out a number sentence. Tell students that they should think about efficiency when solving the open response problem. **GMP6.4**

▶ Solving the Open Response Problem

Math Masters, pp. 61–62

| WHOLE CLASS | SMALL GROUP | **PARTNER** | INDEPENDENT |

Distribute *Math Masters,* pages 61 and 62. Read Problems 1–3 as a class and review the directions. Partners should work together to ensure that they understand the problems. For Problem 2, tell students that they can tap their desks at any speed as long as they are able to count each tap. One partner should keep time with a stopwatch or a clock with a second hand while the other taps to 100. Then they switch roles. Remind students that for Problem 3 they do not need to write anything, but they should discuss their thinking with a partner. GMP1.6

When students have completed Problem 3, read Problems 4 and 5 as a class and answer any questions about them. Point out that the task in Problem 4 is to make sense of Maya's strategy and explain whether they think it is efficient. GMP6.4 Remind students to use their answers to Problem 2 to make an estimate for Problem 5. Have students write their answers to Problems 4 and 5 on a separate sheet of paper.

While students work, circulate and ask questions such as:

- *In Problem 4, how did Maya start? What was her next step?* GMP1.6
  Answers vary.

- *Why did you decide to make your estimate for Problem 5 this way? Is there a more efficient way to solve the problem?* GMP6.4
  Answers vary.

Differentiate **Adjusting the Activity**

If students have trouble developing a plan that is more efficient than Maya's, ask: *Do you notice any patterns in the number of zeros? Can you use patterns to solve the problem more efficiently?* GMP6.4 Answers vary. Remind students of the table discussed in the Math Message Follow-Up.

Summarize Ask: *How does your guess for Problem 1 compare to the calculated estimate for Problem 5?* Answers vary. *Did you calculate the exact time it would take to make 1,000,000 taps? Why or why not?* GMP6.4 No, the estimate I calculated is not the exact time, but since we used the number of taps we counted, it is more accurate than the first guess.

Remind students that they will continue to discuss how to solve the problem more efficiently during the reengagement discussion. Collect students' work so that you can evaluate it and prepare for Day 2.

Math Masters, p. 61

One Million Taps

Lesson 2-9
NAME   DATE   TIME

① How many seconds do you think it would take to tap your desk 1 million times without any interruptions? Be prepared to tell your partner how you made your guess.

About _____ seconds Answers vary.

② Work with a partner to time how many seconds it takes to tap your desk 100 times.
_____ seconds Sample answer: 26

③ Discuss with a partner how you could use your time from Problem 2 to estimate the number of seconds it would take to tap your desk 1 million times without any interruptions.

④ Look at Maya's work on *Math Masters,* page 62.
Did Maya use an efficient strategy? Explain your thinking on another sheet of paper.
See Problem 5 below.

⑤ Estimate the time it would take you to tap your desk 1 million times without any interruptions. Use the time it took you to make 100 taps in your estimate. Use a strategy that is more efficient than Maya's strategy. Show your strategy on another sheet of paper.

About _____ Answers vary.

See sample students' work on page 169 of the *Teacher's Lesson Guide.*

5.NBT.2, SMP1, SMP4, SMP6   61

Math Masters, p. 62

One Million Taps (continued)

Lesson 2-9
NAME   DATE   TIME

Maya solved the open response problem. She explained her thinking to a partner.

"It took me 22 seconds to tap my desk 100 times. I wondered how that information would help me figure out how long it would take to tap my desk 1,000,000 times."

"If I multiply 100 taps times 10, it gives me 1,000 taps. How can I get to 1,000,000 taps?"

"I kept multiplying my answer by 10 until the product was 1,000,000."

100 taps	1,000 taps	10,000 taps	100,000 taps
* 10	* 10	* 10	* 10
000	0,000	00,000	000,000
+ 1,000	+ 10,000	+ 100,000	+ 1,000,000
1,000 taps	10,000 taps	100,000 taps	1,000,000 taps

"I looked back at my work and saw that I multiplied 100 by 10 * 10 * 10 * 10."

"I reorganized that into 100 * 100, which is 10,000. That meant that if I multiply 100 taps by 10,000, it gives me 1,000,000 taps."

(10 * 10) * (10 * 10)
  10   10   100
* 10   * 10   * 100
  00   00   000
+ 100   + 100   + 10,000
  100   100   10,000

"It took 22 seconds for 100 taps. So 1,000,000 taps would take 22 seconds * 10,000. My answer was 220,000 seconds."

22 seconds * 10,000
   10,000
 *   22
   20,000
 + 200,000
   220,000 seconds to make 1,000,000 taps

62   5.NBT.2, SMP1, SMP4, SMP6

Student Rubric for One Million Taps

Lesson 2-9

NAME   DATE   TIME

Goal: Think about accuracy and efficiency when you count, measure, and calculate.

Meets Expectations	Student 1	Student 2	Student 3	
For Problem 4, explains ways in which Maya's solution strategy is efficient or not and why.				
For Problem 5, shows a more efficient solution strategy than Maya's.				
Exceeds Expectations				
Correctly explains how the strategy used for Problem 5 is more efficient than Maya's.				

63

Sample student's work, Student A

4. Yes, it was efficient, it makes sence and gets her a good estimate. I do belive that she could've done it faster though. She could've used division or done it all in 1 problem. Her way of finding how long it would take her is good though, she did in the end get a pretty good estimate.

5. 100 × N = 1,000,000   1,000,000/100 = N
N = 10,000
10000
× 29
290000.

I tapped 100 times in 29 secs
290000   I used fact extensions

Getting Ready for Day 2

Math Masters, p. TA4

Planning a Follow-Up Discussion

Review students' work. Use the Reengagement Planning Form (*Math Masters,* page TA4) and the rubric on page 166 to plan ways to help students meet expectations for both the content and practice standards.

This lesson introduces the use of a student-friendly rubric. Organize a peer discussion using a student-friendly rubric as described in Option 1 below. Or, facilitate a class discussion as described in Options 2 through 4 or in another way you choose. If students' work is unclear or if you prefer to show work anonymously, rewrite the work for peer review or display.

Go Online for sample students' work that you can use in your discussion.

1. Have partners review and discuss student work using the student-friendly rubric on *Math Masters,* page 63. Choose work from three students showing a range of explanations for Problem 4. Be sure to choose work with mathematically reasonable estimates for Problem 5 so that the peer review can focus on the efficiency of the strategy instead of calculation errors. Choose at least one sample that meets expectations because the student met criteria in the student rubric for both Problems 4 and 5, as in Student A's work. Choose a second sample that partially meets expectations because it met only one of the criteria. The third sample can meet expectations by showing thinking in a different way or exceed expectations. Label the three samples Student 1, Student 2, and Student 3 (or print them on different colored paper), and make enough copies so that students can review all three samples in partnerships. Plan to model how to use the rubric with Student 1's work and for partners to work together to review the work of Students 2 and 3. See the section on Reengaging in the Problem on Day 2 for more information.

For a whole-class discussion, use questions similar to those below.

2. Display a response for Problem 5, such as Student B's, that shows a different strategy than Maya's. Ask: *Which strategy is more efficient and why?* GMP1.6, GMP4.2, GMP6.4 Sample answer: This strategy is more efficient than Maya's because this student used powers of 10, so it was much faster. The student figured out that 100 * 10,000 = 1,000,000 by counting the difference in zeros between 1,000,000 and 100. Maya multiplied 100 by 10 again and again until she got to 1,000,000 and then had to multiply all of those 10s together to get 10,000.

3. Display a response to Problem 5 that is incomplete or incorrect, as in Student C's work. Ask: *Do you agree or disagree with this solution? Explain.* GMP1.6 Sample answer: I disagree because multiplying 100 taps by 10,000 gives you 1,000,000 taps. We already knew we were looking for 1,000,000 taps. The student needed to multiply the amount of time it took to make 100 taps by 10,000 to find how many seconds it would take to make 1,000,000 taps.

4. Display samples of student work containing different computation errors. Ask: *What was this student trying to do? What would you say to this student to explain how to correct the errors?* Answers vary.

Planning for Revisions

Have copies of *Math Masters,* pages 61 and 62 or extra paper available for students to use in revisions. You might want to ask students to use colored pencils so that you can see what they revised.

Sample student's work, Student B

Sample student's work, Student C

One Million Taps

Day 2: Reengagement

▶ Before You Begin

Have extra copies of *Math Masters*, pages 61 and 62 available for students to revise their work. See Option 1 in Getting Ready for Day 2 for information on preparing for a peer review using a student-friendly rubric.

(2b) Focus   50–55 min

Standards

Focus Cluster
- Understand the place value system.

	Materials	
Setting Expectations Students review how to discuss other students' work respectfully. They also review the open response problem and discuss what a good response might include.	Guidelines for Discussions Poster, Standards for Mathematical Process and Practice Poster	SMP6
Reengaging in the Problem Students examine other students' work using a rubric as a guide or in a class discussion.	*Math Masters*, p. 63 (optional); selected samples of students' work	5.NBT.2 SMP1, SMP4 , SMP6
Revising Work Students revise their work from Day 1.	*Math Masters*, pp. 61–62 (optional), p. 63; students' work from Day 1; colored pencils (optional)	5.NBT.2 SMP1, SMP6

 Assessment Check-In   See page 169 and the rubric below. Expect that most students will be able to calculate a reasonable estimate of the time it takes to make one million taps using patterns of powers of 10.   **5.NBT.2, SMP6**

Goal for Mathematical Process and Practice **GMP6.4** Think about accuracy and efficiency when you count, measure, and calculate.	Not Meeting Expectations	Partially Meeting Expectations	Meeting Expectations	Exceeding Expectations
	For Problem 4, does not address the efficiency of Maya's strategy, and for Problem 5, does not use a more efficient strategy.	For Problem 4, addresses an aspect of the efficiency of Maya's strategy (see Meeting Expectations), or for Problem 5, uses a more efficient strategy than Maya's (see Meeting Expectations).	For Problem 4, addresses an aspect of the efficiency of Maya's strategy (such as saying it is inefficient because of too many steps or because the steps are tedious; or it is efficient because she timed just 100 taps), and for Problem 5, uses a more efficient strategy than Maya's (such as applying powers of 10).	Meets expectations and correctly explains how the strategy used for Problem 5 is more efficient than Maya's.

(3) Practice   10–15 min

Math Boxes 2-9 Students practice and maintain skills.	*Math Journal 1*, p. 59	See page 168.
Home Link 2-9 **Homework** Students multiply by multiples of 10 to make estimates.	*Math Masters*, p. 64	5.NBT.2, 5.NBT.5

 Go Online to see how mastery develops for all standards within the grade.

▶ Setting Expectations

| WHOLE CLASS | SMALL GROUP | PARTNER | INDEPENDENT |

Revisiting Guidelines for Reengagement

To promote a cooperative environment, consider revisiting the class guidelines for discussion that you developed in Unit 1. After reviewing the guidelines, have students reflect on how well they are following them. Solicit additional guidelines from the class. Your revised list might look like the one in the margin.

Revisit some of the sentence frames from Unit 1 to model using appropriate language and encourage students to do the same when discussing others' work. Add more frames to the list, such as the following:

- I like how _____.
- I wonder why _____.

Reviewing the Problem

Briefly review the open response problem from Day 1. Ask: *What were you asked to do?* GMP6.4 Sample answer: We had to find the time it took to tap our desks 100 times and use that information to estimate how much time it would take to tap our desks 1,000,000 times. We had to decide whether Maya's solution strategy was efficient or not and try to solve the problem in a more efficient way. *What do you think a good response would include?* It should have an explanation of whether Maya's solution was efficient and show how it was possible to calculate an estimate of 1,000,000 taps using a strategy that is more efficient than Maya's. It also might explain why the solution strategy is more efficient than Maya's.

After this brief discussion, tell students that they are going to look at other students' work and see whether they thought about the problem in the same way. Refer to GMP6.4 on the Standards for Mathematical Process and Practice Poster. Explain to students that they will figure out how other students decided whether Maya's solution was efficient. They will also look at how other students tried to solve the problem in a more efficient way than Maya.

NOTE These Day 2 activities will ideally take place within a few days of Day 1. Prior to beginning Day 2, see Planning a Follow-Up Discussion from Day 1.

Guidelines for Discussion

During our discussions, we can:

✓ Make mistakes and learn from them.

✓ Share ideas and strategies respectfully.

✓ Change our minds about how to solve a problem.

✓ Ask questions of our teacher and classmates.

✓ Feel confused.

✓ Listen closely to others' ideas.

✓ Be patient.

▶ **Reengaging in the Problem**

WHOLE CLASS	SMALL GROUP	PARTNER	INDEPENDENT

Students reengage in the problem by analyzing and critiquing other students' work through a peer review or class discussion. Guide this discussion based on the decisions you made in Getting Ready for Day 2. **GMP1.6, GMP4.2, GMP6.4**

If you planned to facilitate a peer review using a student-friendly rubric as described in Option 1 on page 164, use *Math Masters,* page 63 to structure students' analysis of sample work. Distribute copies of the samples you chose for Students 1, 2, and 3 and student-friendly rubrics to each partnership. Briefly discuss **GMP6.4**, which is written at the top of the student rubric. Model reviewing Student 1's work with the class. Point out that to meet expectations the work must clearly meet the criteria listed under Meets Expectations for both Problems 4 and 5. Ask students to explain how the work meets or does not meet each of the criteria and write "Yes" or "No" in the appropriate boxes. Ask: *What would a paper look like that exceeds or goes beyond expectations?* Sample answer: The student would correctly explain how his or her strategy is more efficient than Maya's.

Have partners review the problem together and come to a decision on how they would evaluate work from Students 2 and 3 using the rubric. Conclude by discussing partners' choices for each work sample. Ask students to support their choices by showing how each piece of work met or did not meet each of the criteria. **GMP1.6, GMP6.4**

▶ **Revising Work**

Math Masters, p. 63

WHOLE CLASS	SMALL GROUP	PARTNER	INDEPENDENT

Pass back students' work from Day 1. Before students revise anything, ask them to examine their own work. Whether you chose to conduct a peer review or a class discussion, have students use the student-friendly rubric to decide whether their work meets expectations for Problems 4 and 5. Have students add their names to the last column of the rubric and write "Yes" or "No" in the boxes for their own work. **GMP1.6, GMP6.4**

Tell students they now have a chance to revise their work. Those who wrote complete explanations for Maya's strategy and found an efficient estimate on Day 1 can explain how their strategy is more efficient than Maya's. Help students see that the explanations presented during the reengagement discussion are not the only correct ones. Tell them to add to their earlier work using colored pencils or another sheet of paper, instead of erasing their original work. **GMP1.6, GMP6.4**

Math Journal 1, p. 59

Math Journal 1, p. 59

Summarize Ask students to reflect on their work and revisions. Ask: *What did you do to improve your explanation or estimate more efficiently?* Answers vary.

 Assessment Check-In 5.NBT.2

Collect and review students' revised work. Expect students to improve their work based on the class discussion. For the content standard, expect most students to calculate a reasonable estimate of the time it takes to make one million taps using patterns of powers of 10. You can use the rubric on page 166 to evaluate students' revised work for **GMP6.4**.

 Evaluation Quick Entry Go online to record students' progress and to see trajectories toward mastery for these standards.

(Go Online) for optional generic rubrics in the *Assessment Handbook* that can be used to assess any additional GMPs addressed in this lesson.

Sample Students' Work—Evaluated

See the sample in the margin. This work meets expectations for the content standard because the student used patterns of powers of 10 to figure out "100 * ? = 1,000,000." The work meets expectations for the mathematical process and practice standard because for Problem 4 the student showed how to use "division" (by finding the missing factor) and extended facts to improve the efficiency of Maya's solution. Although the student used lattice multiplication for Problem 5, which is less efficient than using powers of 10 and extended facts, the student's strategy is more efficient than Maya's because it required fewer steps. GMP6.4

(Go Online) for other samples of evaluated students' work.

3 Practice 10–15 min

▶ Math Boxes 2-9

Math Journal 1, p. 59

| WHOLE CLASS | SMALL GROUP | PARTNER | INDEPENDENT |

Mixed Practice Math Boxes 2-9 are paired with Math Boxes 2-12.

▶ Home Link 2-9

Math Masters, p. 64

Homework Students multiply by multiples of 10 to make estimates.

Sample student's work, "Meeting Expectations"

4. No, because she could've done a differently. To get to 1,000,000 she could have done 100 x ? = 1,000,000. To figure out the ? she should have thought, since I know there are 6 0's in 1,000,000 and I already have 2, how much more would I need? Once she figured that out it would be 100 x 10,000 = 1,000,000. I do think the way she figured out 229,000 seconds is efficient because, thats what I would have done.

5.   100
   x 10,000
   ─────────
   1,000,000

   533
   [lattice multiplication grid]
   530,000 seconds

Math Masters, p. 64

Using Multiples of 10 to Estimate Home Link 2-9

NAME   DATE   TIME

① Estimate about how many meters Martin swims in June if he swims about 200 meters per day. There are 30 days in June. Show how you made your estimate.
About **6,000** meters
Sample answer: 20 meters * 30 days = 600 meters

② Estimate how many days it would take Martin to swim 60,000 meters. Show how you made your estimate.
About **300** days
Sample answers:
• 200 meters per day * ? = 60,000 meters
  60,000 / 200 = 300 days

Days	30	300
Meters	6,000	60,000

Practice

Make an estimate and solve. Estimates vary.

③ 107 * 19 = ?   ④ 86 * 975 = ?
Estimate: _____   Estimate: _____

      1 0 7   9 7 5
   ×   1 9   ×   8 6
   ─────────   ──────────
     2,033   83,850

64   5.NBT.2, 5.NBT.5

Lesson 2-10

A Mental Division Strategy

Overview Students use the relationship between multiplication and division to mentally divide multidigit numbers.

▶ **Before You Begin**

For Part 2, decide how you will display a name-collection box to demonstrate using multiples to rename numbers. Prepare a deck of number cards (4 each of numbers 1–9) for demonstration. Copy and cut apart *Math Masters*, page G13 so that each student has one half-sheet.

▶ **Vocabulary**

dividend • divisor • quotient • multiple • remainder

Standards

Focus Clusters

• Understand the place value system.
• Perform operations with multi-digit whole numbers and with decimals to hundredths.

1 **Warm Up** 5 min	**Materials**	
Mental Math and Fluency Students convert from expanded form to standard notation.	slate	**5.NBT.1**

2 **Focus** 40–45 min		
Math Message Students read about extended division facts and use patterns they notice to solve division problems.	*Student Reference Book*, p. 106	**5.NBT.2, 5.NBT.6** SMP7
Solving Extended Division Facts Students describe and use patterns when dividing multiples of 10.	*Student Reference Book*, p. 106	**5.NBT.2, 5.NBT.6** SMP7
Using Multiples to Divide Mentally Students generate equivalent names for dividends using multiples of the divisor.	*Math Journal 1*, p. 60; per partnership: number cards 1–9 (4 of each)	**5.NBT.2, 5.NBT.6** SMP6
Introducing *Division Dash* **Game** Students practice solving division problems.	*Student Reference Book*, p. 301; per partnership: *Math Masters*, p. G13; number cards 1–9 (4 of each)	**5.NBT.2, 5.NBT.6** SMP6
✓ **Assessment Check-In** See page 175. Expect most students to use lists of multiples and written strategies to find quotients while playing the game *Division Dash*.	*Math Masters*, p. G13, p. TA8 (optional)	**5.NBT.6, SMP6**

3 **Practice** 15–25 min		
Practicing Unit Conversions Students practice converting units.	*Math Journal 1*, p. 61; *Student Reference Book*, p. 328	**5.OA.1, 5.OA.2, 5.MD.1** SMP1, SMP4
Math Boxes 2-10: Preview for Unit 3 Students preview skills and concepts for Unit 3.	*Math Journal 1*, p. 62	See page 175.
Home Link 2-10 Students use multiplication and division facts to mentally solve division problems.	*Math Masters*, p. 65	**5.NBT.2, 5.NBT.5, 5.NBT.6**

 Go Online to see how mastery develops for all standards within the grade.

Differentiation Options

Readiness 5–15 min	**Enrichment** 5–15 min	**Extra Practice** 5–15 min
WHOLE CLASS · SMALL GROUP · PARTNER · INDEPENDENT	WHOLE CLASS · SMALL GROUP · PARTNER · INDEPENDENT	WHOLE CLASS · SMALL GROUP · PARTNER · INDEPENDENT

Playing *Division Arrays*

5.NBT.6, SMP1

Student Reference Book,
p. 300; per partnership: number cards 6–20
(1 of each), 6-sided die, 40 counters

To review the relationship between
division and multiplication, students play
Division Arrays. Read the directions on
Student Reference Book, page 300 and play
a sample round with students. Ask
questions showing the connection between
the number of rows or columns and the
quotient. For example: *You had 26 counters
and made an array with 4 columns. How
many 4s are in 26?* 6 *How do you know?* I
used 26 counters to make 6 rows of 4. There
were 2 counters left over. Have students
write division number sentences for each
array as they play.

Observe

• Which students accurately find
quotients?

Discuss

• *What multiplication fact can you use
to check your quotient?* GMP1.4

A New Division Strategy

5.NBT.2, 5.NBT.6, SMP1

Activity Card 23, number cards 1–9 (4 of
each), two 6-sided dice

To extend their work with mental division,
students think of division problems as
missing-factor multiplication problems.
They draw cards and roll dice to generate
division problems and then apply the
missing-factor strategy. They compare it to
the renaming strategy taught in the lesson.
GMP1.5, GMP1.6

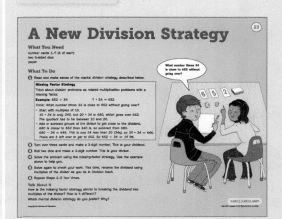

Renaming Dividends to Divide Mentally

5.NBT.2, 5.NBT.6, SMP7

Activity Card 24;
per partnership: *Math Masters*, p. TA8;
number cards 1–9 (4 of each)

To practice applying extended facts in a
mental division strategy, students multiply
numbers by powers of 10 to generate
dividends and divisors. They break up
dividends into multiples of the divisor to
help them solve the problems. They explain
the patterns they see in the number of
zeros in the problems. GMP7.1

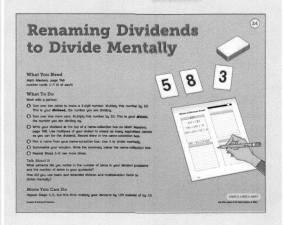

English Language Learner

Beginning ELL Use think-alouds and objects to help students understand the terms *divide*,
dividend, *divisor*, and *quotient* by connecting them to real-world contexts. For example, use
15 small objects (like crayons) and 3 cups to demonstrate the meaning of division. Say: *I will
divide these crayons among you, me, and another friend. Let's divide the crayons: 1 for you,
1 for me, 1 for my friend. 1 for you, 1 for me, 1 for my friend . . .* Have students divide a group of
objects with 2 or 3 of their classmates, repeating the language: *1 for you, 1 for you, . . . , and
1 for me.* Post a division problem chart with the dividend, divisor, and quotient labeled for
reference. Include a sketch of 15 crayons with the caption *dividend*, 3 cups with the caption
divisor, and 5 crayons with the caption *quotient*.

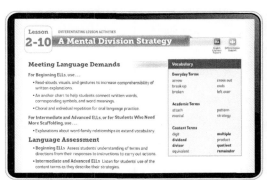

Differentiation Support pages are
found in the online Teacher's Center.

Standards and Goals for
Mathematical Process and Practice

SMP6 **Attend to precision.**
GMP6.4 Think about accuracy and efficiency when you count, measure, and calculate.

SMP7 **Look for and make use of structure.**
GMP7.1 Look for mathematical structures such as categories, patterns, and properties.
GMP7.2 Use structures to solve problems and answer questions.

Adjusting the Activity

Differentiate To help students who struggle identifying patterns, ask questions like the following:

- *What basic fact would help solve 32,000 / 4?* 32 / 4 = 8

- *How is the basic fact related to the extended fact?* Sample answer: 32,000 is 1,000 times as much as 32, so the quotient of the extended fact will be 1,000 times as large as the basic fact. Since 32 / 4 = 8, then 32,000 / 4 = 1,000 * 8 = 8,000.

- *How can we check that our quotient has digits in the right place-value positions?* Sample answer: Multiply the quotient (8,000) by the divisor (4) to check that the product is our original dividend (32,000).

 Go Online | **Differentiation Support**

1 Warm Up 5 min

▶ Mental Math and Fluency

Display numbers in expanded form. Have students write the numbers in standard notation on slates. *Leveled exercises:*

- ●○○ 300 + 20 + 5 325
  5 * 100 + 6 * 10 + 8 * 1 568
  9 [100s] + 3 [10s] + 2 [1s] 932

- ●●○ 4 * 1,000 + 6 * 100 + 4 * 10 + 3 * 1 4,643
  8 thousands + 2 hundreds + 9 tens + 5 ones 8,295
  6 [1,000s] + 3 [100s] + 8 [10s] + 1 [1] 6,381

- ●●● 2 * 10,000 + 8 * 1,000 + 4 * 100 + 1 * 10 + 5 * 1 28,415
  90,000 + 4,000 + 60 + 3 94,063
  9 [100,000s] + 4 [10,000s] + 6 [100s] + 1 [1] 940,601

2 Focus 40–45 min

▶ Math Message

Student Reference Book, p. 106

Read page 106 in your Student Reference Book. *Use the patterns you notice to solve the Check Your Understanding problems.* GMP7.1, GMP7.2

▶ Solving Extended Division Facts

Student Reference Book, p. 106

| WHOLE CLASS | SMALL GROUP | PARTNER | INDEPENDENT |

Math Message Follow-Up Remind students that in a division problem the number being divided is called the **dividend;** the number that divides the dividend is called the **divisor;** and the answer is called the **quotient.** Encourage students to use this language when sharing the patterns they used to solve the Math Message. GMP7.1, GMP7.2

Make sure that students discuss both of the strategies shown on the *Student Reference Book* page. Throughout the discussion, emphasize how multiplication can be used to reason through and check division problems.

Pose additional extended division facts. *Suggestions:*

- What is 630 divided by 7? 90
- How many 10s are in 1,000? 100
- 20 times what number equals 200? 10
- 5,000 is how many times as great as 5? 1,000

Tell students that today they will learn a mental division strategy in which they use basic and extended facts to solve division problems.

▶ Using Multiples to Divide Mentally

Math Journal 1, p. 60

WHOLE CLASS | SMALL GROUP | PARTNER | INDEPENDENT

Remind students that **multiples** are products of a given number and a counting number (1, 2, 3, and so on). Ask: *What are some multiples of 3?* Sample answer: 3, 30, 15, 6, 33 *What are some multiples of 30?* Sample answer: 30, 300, 90, 600 Remind students that multiples can be found for any number, including numbers with two or more digits.

Display a name-collection box. Draw two cards from a set of number cards. (*See Before You Begin.*) Form a 2-digit number and write it at the top of the name-collection box. Tell students that this number is the dividend. Draw another card and say that it is the divisor. Explain to students that in this activity they will look for equivalent names for the dividend that contain multiples of the divisor. Proceed as follows:

- Ask students to give multiples of the divisor, and display the multiples near the name-collection box. (*See margin for an example.*)
- Prompt students to think of equivalent names for the dividend that contain multiples of the divisor. For example, if 68 is the dividend and 5 is the divisor, ask:
  - *What is a multiple of 5 that is less than 68?* Sample answer: 60
  - *What would we need to add to that multiple to make 68?* 8 because $60 + 8 = 68$
  - *Are there other multiples of 5 we could use to break up what is left of the dividend?* 5 is a multiple of 5 that is less than 8. Use $5 + 3$ in place of 8.
- Record the equivalent name in the name-collection box (in this example, $60 + 5 + 3$). (*See margin.*) Ask students to identify other multiples they could use to rename the dividend. Help them see that the larger multiples can be broken into smaller multiples to make other equivalent names. For example, in the name $60 + 5 + 3$, 60 can be broken into $40 + 20$ to make a new equivalent name for 68: $40 + 20 + 5 + 3$.
- After you have recorded several names, pick one and ask: *How many times does the divisor go into each part of this name?* Sample answer for this example: There are 12 [5s] in 60. There is 1 [5] in 5. There are 0 [5s] in 3. Record number sentences for each part, emphasizing the way basic and extended division facts are being used. (*See margin.*) Remind students that when a number cannot be evenly divided by another number, the quotient will include a leftover part, or **remainder.** Ask: *In total, how many times does the divisor go into the dividend?* Sample answer: 12 [5s] + 1 [5] = 13 [5s] in 68. 13 is the quotient, and 3 is left over, so the remainder is 3. Display 68 / 5 → 13 R3 to summarize the solution.

6	8		5
Dividend			Divisor

Multiples of 5: 5, 10, 15, 20, 25, 30, 35, 40, 45, 50, 55, 60, 65, 70, . . .

68
$60 + 5 + 3$
$40 + 20 + 5 + 3$
$35 + 30 + 3$
$55 + 10 + 3$
$50 + 15 + 3$

Name-collection box for 68 (the dividend) including names that contain multiples of 5 (the divisor)

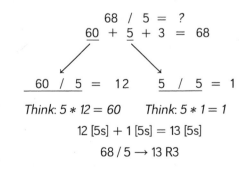

$68 / 5 = ?$

$\underline{60} + \underline{5} + 3 = 68$

$\underline{60} / 5 = 12 \qquad \underline{5} / 5 = 1$

*Think: 5 * 12 = 60*   *Think: 5 * 1 = 1*

12 [5s] + 1 [5s] = 13 [5s]

68 / 5 → 13 R3

Summarizing a mental strategy for dividing 68 by 5

Professional Development When the result of division is expressed as a quotient and a nonzero remainder, *Everyday Mathematics* uses an arrow rather than an equal sign, as in 127 ÷ 7 → 18 R1. We use this notation because 127 ÷ 7 = 18 R1 is not a proper number sentence. The arrow is read as "is," "yields," or "results in." Point out this notation for students in the division examples. Consider labeling and displaying the arrow for student reference.

• Confirm the quotient by selecting another equivalent name and repeating the reasoning above. Ask: *Did the quotient change when we used a different name for the dividend? Why or why not?* No, because the names were equivalent.

Repeat this process with another example, this time multiplying both the dividend and the divisor by 10 to create a 3-digit dividend and a 2-digit divisor. For example, drawing the cards 4 and 9 for the dividend and 6 for the divisor creates the division problem 490 / 60. Point out that dividing each multiple by the divisor is like solving an extended division fact.

Distribute number cards to partnerships. Have partners draw cards to create division problems. On journal page 60 have students record the problem, list multiples of each divisor, record equivalent names for each dividend in a name-collection box, and solve the division problems mentally. Encourage partnerships to discuss how using multiples to rename the dividend makes it easier to divide. **GMP6.4** If appropriate, suggest that students create a problem with a 3-digit dividend and 2-digit divisor by multiplying the dividend and divisor by 10.

▶ Introducing *Division Dash*

Student Reference Book, p. 301; *Math Masters*, p. G13

| WHOLE CLASS | SMALL GROUP | PARTNER | INDEPENDENT |

Division Dash uses randomly generated numbers to create dividends and divisors. Read the directions on *Student Reference Book*, page 301 aloud and play a round together. After students understand the rules, have them play the game in partnerships. Encourage students to divide mentally, but do not restrict the use of paper and pencil. **GMP6.4** Circulate and assist.

Observe
• Which students are using multiples of the divisor to divide?
• Which students need support to break apart the dividend?

Discuss
• *What strategy did you use to divide?*
• *How did you decide where to use each number card?*

Differentiate **Game Modifications** **Go Online** Differentiation Support

---

Assessment Check-In 5.NBT.6

Math Masters, p. G13

Expect most students playing *Division Dash* to use lists of multiples and written strategies to find quotients. GMP6.4 Some might be able to use a mental strategy. If students struggle determining quotients accurately, suggest writing out multiples of the divisor and using them to rename the dividend before dividing. Consider providing blank Name-Collection Boxes (*Math Masters*, page TA8) as a scaffold.

Evaluation Quick Entry Go online to record students' progress and to see trajectories toward mastery for these standards.

Summarize Have students explain how they used multiples and extended facts while playing *Division Dash*.

3 Practice 15–25 min

▶ Practicing Unit Conversions

Math Journal 1, p. 61; *Student Reference Book*, p. 328

| WHOLE CLASS | SMALL GROUP | **PARTNER** | INDEPENDENT |

Students practice converting units of measure. They complete tables of conversions and model and solve unit conversion number stories. GMP1.1, GMP4.1 Remind them to refer to *Student Reference Book*, page 328 as needed.

▶ Math Boxes 2-10: Preview for Unit 3

Math Journal 1, p. 62

| WHOLE CLASS | SMALL GROUP | PARTNER | INDEPENDENT |

Mixed Practice Math Boxes 2-10 are paired with Math Boxes 2-14. These problems focus on skills and understandings that are prerequisite for Unit 3. You may want to use information from these Math Boxes to plan instruction and grouping in Unit 3.

▶ Home Link 2-10

Math Masters, p. 65

Homework Students use multiplication and division facts to mentally solve division problems.

Math Journal 1, p. 62

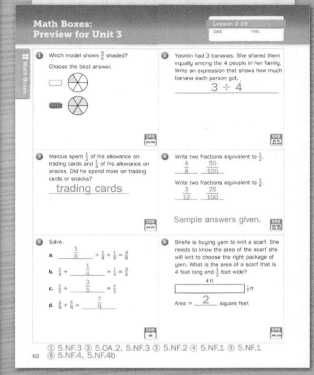

Math Masters, p. 65

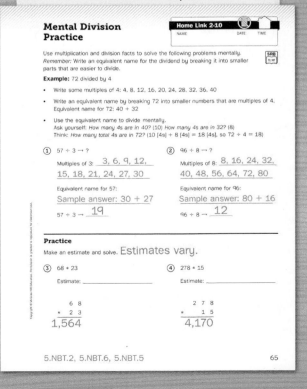

Reviewing Partial-Quotients Division

Overview Students review and practice strategies for using partial-quotients division to divide whole numbers.

▶ **Before You Begin**

Display the rectangle from the margin of page 178 near the Math Message. Have plenty of computation grids (*Math Masters*, page TA7) and area model templates (*Math Masters*, page TA9) available to students.

▶ **Vocabulary**

partial-quotients division • partial quotient • area model

Standards

Focus Cluster
- Perform operations with multi-digit whole numbers and with decimals to hundredths.

① Warm Up 5 min	Materials	
Mental Math and Fluency Students solve extended division facts.		5.NBT.6

② Focus 40–50 min		
Math Message Students solve a division number story.		5.NBT.6
Reviewing Partial-Quotients Division Students review partial-quotients division and use it to solve the Math Message problem.	*Math Masters*, p. TA7 (optional)	5.NBT.6 SMP1
Making Area Models Students create area models to represent solutions to the Math Message problem.	*Math Masters*, p. TA9	5.NBT.6 SMP1, SMP2
Estimating and Dividing Students make estimates and solve division problems.	*Math Journal 1*, p. 63	5.NBT.6 SMP1, SMP2
✓ **Assessment Check-In** See page 182. Expect most students to be able to solve division problems involving 3-digit dividends and 2-digit divisors with and without remainders in the quotients.	*Math Journal 1*, p. 63	5.NBT.6 SMP2

③ Practice 15–20 min		
Math Boxes 2-11 Students practice and maintain skills.	*Math Journal 1*, p. 64	See page 183.
Home Link 2-11 **Homework** Students make estimates and solve problems using partial-quotients division.	*Math Masters*, p. 66	5.NBT.5, 5.NBT.6

　(Go Online) to see how mastery develops for all standards within the grade.

Differentiation Options

Readiness · 5–15 min
WHOLE CLASS · **SMALL GROUP** · PARTNER · INDEPENDENT

Drawing Area Models for Division

5.NBT.6, SMP2

To explore how area models represent division, students draw area models for multiplication facts. Have students draw an area model for
7 ∗ 5. **GMP2.1** Ask: *What do you know that you can label on this model?* The length and width of the rectangle *What do you need to find out?* The area of the rectangle Have students calculate the area and label the rectangle. Next help them draw an area model for the division problem 48 ÷ 6.
GMP2.1 Encourage them to think of the division problem as 6 ∗ ? = 48. Ask: *What do you know that you can label on this model?* The length of one side and the area Have students label the model. Ask: *What do you need to find out?* The missing side length Have students determine the missing side length of the rectangle and label the area model. 8 Repeat with other division problems, asking: *What do you notice about area models for division?* **GMP2.2** Sample answers: We're always trying to find a missing side length. It's like finding the missing factor.

Enrichment · 15–30 min
WHOLE CLASS · SMALL GROUP · **PARTNER** · INDEPENDENT

Dividing to Convert Units of Length

5.NBT.6, 5.MD.1, SMP5

Activity Card 25;
Math Masters, p. TA7 (optional); per partnership: tape measure

To apply their division skills, students measure lengths in inches and convert their measurements to feet and yards.
GMP5.2 Provide copies of *Math Masters*, page TA7 for students who might benefit from using a computation grid.

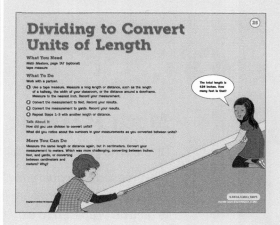

Extra Practice · 5–15 min
WHOLE CLASS · **SMALL GROUP** · PARTNER · INDEPENDENT

Playing *Division Top-It: Larger Numbers*

5.NBT.6, SMP6

Student Reference Book,
p. 325; per partnership: number cards 0–9 (4 of each)

Have students read the directions for *Division Top-It: Larger Numbers* on *Student Reference Book,* page 325 and play the game. Students divide 3-digit numbers by 1-digit numbers and compare quotients.

Observe
• Which students are dividing accurately and efficiently? **GMP6.4**

Discuss
• *Do you always need to find the exact quotient to know whether yours is largest?* **GMP6.2**

English Language Learner

Beginning ELL Review the division terms on the chart created in Lesson 2-10. Build background knowledge for understanding the meaning of *partial* by using jigsaw pieces to demonstrate the meaning of *part* and *partial*. Show a puzzle piece, saying: *This is a part of the puzzle.* Show several pieces, saying: *These are parts of the puzzle. This is a partial puzzle.* Provide practice with the terms using Total Physical Response commands like these: *Show me one part. Count the parts. Use the parts to make the whole. Show me a whole puzzle. Point to a partial puzzle.*

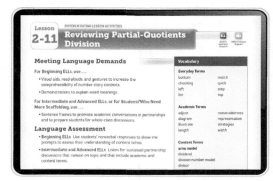

Differentiation Support pages are found in the online Teacher's Center.

Length: 12 ft

156 ft²

Width: ?

Picture for Math Message

1 Warm Up   5 min

▶ Mental Math and Fluency

Display extended division facts and have students solve them.
Leveled exercises:

● ○ ○   6 / 3   2
60 / 3   20
600 / 3   200

● ● ○   25 / 5   5
250 / 50   5
2,500 / 50   50

● ● ●   560 / 80   7
5,600 / 800   7
56,000 / 80   700

2 Focus   40–50 min

▶ Math Message

Display the picture shown in the margin.

Steven used 156 square feet of carpet to cover the floor of his bedroom. The length of his bedroom is 12 feet. What is the width of his bedroom? 13 feet *Be prepared to explain how the picture can help you solve the problem and how you got your answer.*

▶ Reviewing Partial-Quotients Division

| WHOLE CLASS | SMALL GROUP | PARTNER | INDEPENDENT |

Math Message Follow-Up Ask students to suggest a number model for the Math Message problem. Sample answers: $12 * w = 156$; $156 / 12 = w$ Ask: *What operation would you use to solve this problem?* Division *Why?* Sample answer: I know that Area = length * width. We know the area but not the width, so we can divide the area by the length to find the width.

Invite students to share strategies for carrying out the division. Expect that some will suggest renaming 156 using multiples of 12. Others may reason that $12 * 12 = 144$, which is 12 less than 156. That would mean $12 * 13 = 156$, and $156 / 12 = 13$.

Tell students that today they will review how to use **partial-quotients division** to solve the Math Message problem.

> **Professional Development** Partial-quotients division was introduced in *Fourth Grade Everyday Mathematics.* The focus in Grade 5 is to develop efficient strategies for choosing which partial quotients to work with, allowing many of the intermediate calculations to be done mentally and the division to be completed in fewer steps. Having efficient strategies for choosing partial quotients will prepare students to learn and use the U.S. traditional long division algorithm in Grade 6.

Remind students that in this problem 156 is the dividend and 12 is the divisor. Demonstrate how to divide 156 by 12 using partial-quotients division as students follow along on blank paper or computation grids (*Math Masters*, page TA7). (*See margin.*)

1. Write the problem in traditional form. Draw a vertical line on the right. Remind students that with this notation they list **partial quotients** on the right of the vertical line and then do the related subtraction on the left.

2. Remind students that they can think of this problem as "How many 12s are in 156?" Explain that because it is easy to multiply by 10, an easy strategy to start with is checking whether there are at least 10 [12s] in the dividend. Ask: *Are there at least 10 [12s] in 156?* Yes, because 12 * 10 = 120, and 120 is less than 156. Write 120 under 156.

3. Ask: *How many 12s are in 120?* 10, because 10 * 12 = 120 Write 10 to the right of the vertical line, and say: *10 [12s] are in 120.* Explain that 10 is the first partial quotient.

4. Subtract 120 from 156, saying: *156 minus 120 is 36.* Write 36 below 120, explaining that 36 is what remains to be divided.

5. Ask: *How many 12s are in 36?* 3, because 3 * 12 = 36 Write 36 under 36 and 3 next to it on the right side of the vertical line and say: *3 [12s] are in 36.* Subtract, saying: *36 minus 36 equals 0.* Explain that 3 is the second partial quotient and 0 is what remains to be divided.

6. The division is complete when the subtraction leaves a number less than the divisor. That number (0 in this example) is the remainder.

7. Add the partial quotients. Write 13 below the 3, saying: *10 [12s] plus 3 [12s] equals 13 [12s].* Write 13 above the dividend. This is the quotient. Write "R0" next to the quotient to represent the remainder.

Write a number model summarizing the problem: 156 / 12 = 13. Return to the number story and interpret the answer. Ask: *What is the width of Steven's bedroom?* 13 feet Keep the partial-quotients work displayed for reference in the next activity.

Steps 1-4:

```
    12)156
     − 120   | 10     10 [12s] are in 120.
    ──────   |
       36    |
```

Steps 5-7:

```
        13   | R0
    12)156
     − 120   | 10     10 [12s] are in 120.
    ──────
       36
     − 36    | 3      3 [12s] are in 36.
    ──────   |
        0    | 13
```

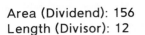
Area (Dividend): 156
Length (Divisor): 12

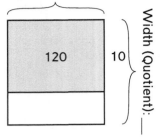
120 | 10
Width (Quotient): ____

Area (Dividend): 156
Length (Divisor): 12

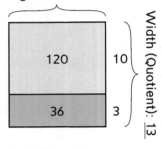
120 | 10
36 | 3
Width (Quotient): 13

Two steps for drawing an area
model of 156 / 12

NOTE Some students may find it useful to include units in the labels. For example, they might write "156 ft²" on the area line and "12 ft" on the length line. It is fine for students to do this if it helps them make sense of the model. However, in the next lesson these models will also be used to represent problems that are not contextualized. If students are able to understand the models without units, leaving the units off may help them transition to creating the models for other problems.

Remind students that when using the partial-quotients algorithm the same problem can be solved in more than one way. GMP1.5 For example, in Step 5 some students may not have recognized immediately that there are 3 [12s] in 36, so they might have taken out 1 [12] at a time. Demonstrate this approach using the partial quotients 10, 1, 1, and 1, emphasizing the point that adding these partial quotients produces the same final quotient: $10 + 1 + 1 + 1 = 13$. Similarly, instead of starting by taking out 10 [12s], or 120, students could start by taking out a smaller number, such as 5 [12s], or 60. Give students a few minutes to experiment with different partial quotients to see that they always get the same final answer. GMP1.5

▶ Making Area Models

Math Masters, p. TA9

| WHOLE CLASS | SMALL GROUP | PARTNER | INDEPENDENT |

Display a copy of *Math Masters,* page TA9. Tell students that you are going to create a representation of how they solved the Math Message problem. GMP2.1 The representation is called an **area model** for a division problem. Explain that just as area models can help students make sense of what they are doing when they multiply, they can also help make sense of the steps of partial-quotients division.

Refer to the picture from the Math Message, reminding students that they knew the area and length of the bedroom. Fill in the area and length on the appropriate lines of the area model. Ask: *What were we trying to find?* Width Tell students that you are going to leave the width line blank for now.

Refer back to the partial-quotients work from earlier in the lesson. Ask: *What was the first step in solving this problem?* We used 10 as our first partial quotient and subtracted 120. Tell students that this was like thinking: *A part of the room that has an area of 120 square feet would have a width of 10 feet.* Divide the rectangle into two pieces. Shade the top part and write 120 inside it and 10 alongside it on the right. (*See the first model in the margin.*)

> **NOTE** The area model in the margin is not proportional. It is meant as a tool to help students understand division and does not need to be proportional. Don't expect students to draw models that are proportional.

Ask: *In the partial-quotients work, what was left after we subtracted 120?* 36 *Where can we show 36 in this diagram?* It is the area of the bottom part of the rectangle, or the rest of the bedroom. Shade the rest of the rectangle with a different color and label the bottom section 36. Ask: *The area of this part of the bedroom is 36 and the length is 12. What is the width?* 3 Write 3 next to the bottom part of the model and connect this to the second partial quotient. Ask: *What is the total width of the bedroom?* 10 + 3, or 13 Record this on the width line and point out that this is the quotient. (*See the second model in the margin.*)

Tell students that they can create area models for division problems even when the problems are not about area. For example, if the Math Message problem had simply asked students to solve 156 / 12, they could still use this area model to illustrate their solutions. They simply have to think of the dividend as an area and the divisor as a side length. The partial quotients, when added together, will give them the width. Explain to students that in the next lesson they will create area models for other division problems.

If time permits, distribute copies of *Math Masters,* page TA9 and have students create area models for 156 / 12 to match solutions using other partial quotients. For example, an area model matching a solution using the partial quotients 10, 1, 1, and 1 is shown in the margin. GMP1.5, GMP2.1

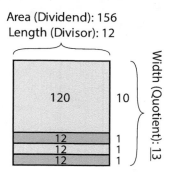

Another area model for 156 / 12

▶ Estimating and Dividing

Math Journal 1, p. 63

WHOLE CLASS | SMALL GROUP | PARTNER | INDEPENDENT

Tell students that making estimates for division problems is one way to check the reasonableness of their answers. GMP1.4 Remind them that one method they have used to estimate answers to multiplication problems is to find numbers that are close to the factors but easier to multiply, and then use those numbers to make an estimate. Explain that a similar strategy can be used to make estimates for division problems.

Display the problem 190 / 27. Ask: *What close-but-easier numbers could we use to help us make an estimate?* Some students may suggest estimates that can be calculated mentally, such as 200 / 25 = 8. However, expect that some may arrive at a division problem they cannot solve mentally, such as 200 / 30. These students may try to solve their "easier" problems using partial-quotients division or another paper-and-pencil strategy. Explain that when making estimates the goal is to make a quick, mental calculation. Students should understand that if the close-but-easier numbers they choose first do not produce a problem that is quick and easy to solve, they should make adjustments to the numbers.

Ask students to suggest some strategies they could use to adjust 200 / 30 to a similar problem that is easy to solve mentally. Be sure the following strategies are discussed:

• Keep the dividend the same, but adjust the divisor to a close number that divides the dividend evenly. Ask: *What is a number close to 30 that divides 200 evenly?* Sample answers: 20; 25 *What would be our adjusted estimate?* Sample answers: 200 / 20 = 10; 200 / 25 = 8

• Keep the divisor the same, but adjust the dividend to a close number that the divisor goes into evenly. One way to do this is to think of multiples of the divisor. For example, think: *The multiples of 30 are 30, 60, 90, 120, 150, 180, 210, 240, and so on.* Ask: *What multiple of 30 is close to 200?* Sample answers: 180; 210 *What would be our adjusted estimate?* Sample answers: 180 / 30 = 6; 210 / 30 = 7

• Think of multiplication fact extensions. For example, think: *30 times what number would be about 200? 30 * 7 = 210, so 210 / 30 = 7. My estimate is 7.*

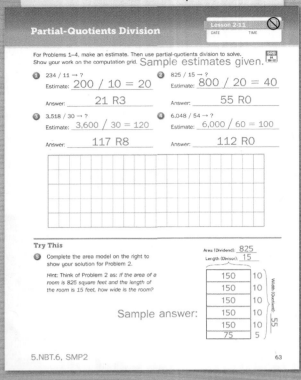

Partial-Quotients Division

Lesson 2-11

For Problems 1–4, make an estimate. Then use partial-quotients division to solve. Show your work on the computation grid. *Sample estimates given.*

① 234 / 11 → ?
Estimate: 200 / 10 = 20

Answer: 21 R3

② 825 / 15 → ?
Estimate: 800 / 20 = 40

Answer: 55 R0

③ 3,518 / 30 → ?
Estimate: 3,600 / 30 = 120

Answer: 117 R8

④ 6,048 / 54 → ?
Estimate: 6,000 / 60 = 100

Answer: 112 R0

Try This

⑤ Complete the area model on the right to show your solution for Problem 2.

Hint: Think of Problem 2 as: If the area of a room is 825 square feet and the length of the room is 15 feet, how wide is the room?

Sample answer:

Area (Dividend): 825
Length (Divisor): 15

150	10
150	10
150	10
150	10
150	10
75	5

Width (Quotient): 55

5.NBT.6, SMP2   63

Pose a few more division problems, prompting students to suggest estimates and explain how they estimated. *Suggestions:*

- **414 / 61** Sample answer: 400 / 50 = 8; First I rounded the numbers to get 400 / 60, but I didn't know the answer to that. So I thought, "What number is close to 60 and can divide 400 evenly?" I knew that 8 ∗ 5 = 40, so 8 ∗ 50 = 400. My estimate is 8.

- **591 / 73** Sample answer: 630 / 70 = 9; I started out with 600 / 70. I couldn't solve that in my head. So I thought about fact extensions with 70 that have an answer close to 600. I know 70 ∗ 9 = 630, so I can use 630 / 70 = 9. My estimate is 9.

Have students work in partnerships to complete journal page 63. Students create an area model in Problem 5. GMP2.1

✓ **Assessment Check-In** 5.NBT.6

Math Journal 1, p. 63

Expect most students to be able to solve Problems 1 and 2 involving 3-digit dividends on journal page 63. Some may be able to solve Problems 3 and 4, which involve 4-digit dividends and divisors greater than 20. Students will learn strategies for efficiently solving division problems with larger numbers in the next lesson. For students who struggle solving Problems 1 and 2, suggest that they begin by using 100s, 10s, or 1s as their partial quotients. Some students may be able to create an area model in Problem 5. GMP2.1 Expect many students to struggle when making quick estimates. With ongoing practice, they will become more fluent in choosing appropriate numbers to make division estimates.

✓ **Evaluation Quick Entry** Go online to record students' progress and to see trajectories toward mastery for these standards.

Summarize Have students share the partial quotients they used for each problem.

③ Practice   15–20 min

▶ Math Boxes 2-11 ✎

Math Journal 1, p. 64

| WHOLE CLASS | SMALL GROUP | PARTNER | INDEPENDENT |

Mixed Practice Math Boxes 2-11 are paired with Math Boxes 2-13.

▶ Home Link 2-11

Math Masters, p. 66

Homework Students make estimates and use partial-quotients division to solve division problems.

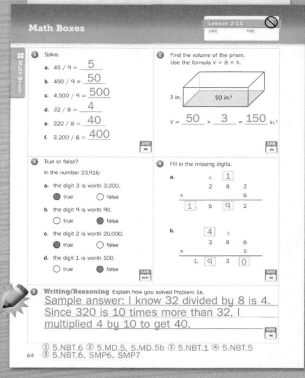

Math Boxes

① Solve.
a. $45 / 9 =$ __5__
b. $450 / 9 =$ __50__
c. $4,500 / 9 =$ __500__
d. $32 / 8 =$ __4__
e. $320 / 8 =$ __40__
f. $3,200 / 8 =$ __400__

② Find the volume of the prism.
Use the formula $V = B \times h$.

3 in.   50 in.²

$V =$ __50__ $\times$ __3__ $=$ __150__ in.³

③ True or false?
In the number 23,916:
a. the digit 3 is worth 3,000.
● true   ○ false
b. the digit 9 is worth 90.
○ true   ● false
c. the digit 2 is worth 20,000.
● true   ○ false
d. the digit 1 is worth 100.
○ true   ● false

④ Fill in the missing digits.
a.   1
         2 8 2
      ×   6
      1 6 9 2

b.   4 3
         3 8 6
      ×   5
      1 9 3 0

⑤ **Writing/Reasoning** Explain how you solved Problem 1e.
Sample answer: I know 32 divided by 8 is 4.
Since 320 is 10 times more than 32, I
multiplied 4 by 10 to get 40.

64 ① 5.NBT.6 ② 5.MD.5, 5.MD.5b ③ 5.NBT.1 ④ 5.NBT.5
⑤ 5.NBT.6, SMP6, SMP7

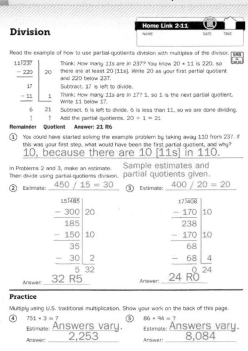

Division   Home Link 2-11

Read the example of how to use partial-quotients division with multiples of the divisor.

11)237
 −220 | 20   Think: *How many 11s are in 237?* You know 20 * 11 is 220, so there are at least 20 [11s]. Write 20 as your first partial quotient and 220 below 237.
  17   Subtract. 17 is left to divide.
 −11 | 1   Think: *How many 11s are in 17?* 1, so 1 is the next partial quotient. Write 11 below 17.
   6 | 21   Subtract. 6 is left to divide. 6 is less than 11, so we are done dividing.
   ↑ | ↑   Add the partial quotients. 20 + 1 = 21
Remainder Quotient   **Answer: 21 R6**

① You could have started solving the example problem by taking away 110 from 237. If this was your first step, what would have been the first partial quotient, and why?
10, because there are 10 [11s] in 110.

In Problems 2 and 3, make an estimate. Then divide using partial-quotients division.   *Sample estimates and partial quotients given.*

② Estimate: **450 / 15 = 30**
15)485
 − 300 | 20
   185
 − 150 | 10
    35
 − 30 | 2
     5 | 32
Answer: **32 R5**

③ Estimate: **400 / 20 = 20**
17)408
 − 170 | 10
   238
 − 170 | 10
    68
 − 68 | 4
     0 | 24
Answer: **24 R0**

Practice

Multiply using U.S. traditional multiplication. Show your work on the back of this page.
④ $751 * 3 = ?$
Estimate: **Answers vary.**
Answer: **2,253**

⑤ $86 * 94 = ?$
Estimate: **Answers vary.**
Answer: **8,084**

66 5.NBT.6, 5.NBT.5

Strategies for Choosing Partial Quotients

Overview Students use lists of multiples to find and choose partial quotients.

▶ **Before You Begin**

Throughout the lesson, have plenty of computation grids (*Math Masters*, page TA7), area-model templates (*Math Masters*, page TA9), and blank lists of multiples (*Math Masters*, page TA10) available to students.

▶ **Vocabulary**

multiple

Standards

Focus Cluster
• Perform operations with multi-digit whole numbers and with decimals to hundredths.

	Materials	
① Warm Up 5 min		
Mental Math and Fluency Students write division problems related to multiplication problems.	slate	**5.NBT.6**
② Focus 35–40 min		
Math Message Students decide whether 10 or 100 can be used as the first partial quotient in a division problem.		**5.NBT.6**
Choosing Partial Quotients Students discuss why lists of multiples might be useful when using partial-quotients division.	*Math Masters*, p. TA10; calculator (optional)	**5.NBT.6** **SMP6**
Using Partial-Quotients Division with Lists of Multiples Students use lists of multiples to help them find partial quotients. They represent their solutions with area models.	*Math Journal 1*, p. 65; *Math Masters*, pp. TA9 and TA10; calculator (optional)	**5.NBT.6** **SMP1, SMP2, SMP6**
✓ **Assessment Check-In** See page 189. Expect most students to be able to divide 4-digit dividends by 2-digit divisors with the help of a list of multiples, although they may not be strategic in choosing the most efficient partial quotients.	*Math Journal 1*, p. 65; *Math Masters*, p. TA10 (optional)	**5.NBT.6**
③ Practice 20–30 min		
Introducing *Power Up* **Game** Students practice interpreting exponential notation and multiplying by powers of 10.	*Student Reference Book*, p. 318; per partnership: *Math Masters*, p. G11; two 6-sided dice	**5.NBT.2** **SMP7**
Math Boxes 2-12 Students practice and maintain skills.	*Math Journal 1*, p. 66	See page 189.
Home Link 2-12 **Homework** Students use a list of multiples to solve a division problem.	*Math Masters*, p. 69	**5.NBT.6**

 Go Online to see how mastery develops for all standards within the grade.

Differentiation Options

Readiness 5–15 min	**Enrichment** 10–20 min	**Extra Practice** 5–15 min
WHOLE CLASS · SMALL GROUP · PARTNER · INDEPENDENT	WHOLE CLASS · SMALL GROUP · PARTNER · INDEPENDENT	WHOLE CLASS · SMALL GROUP · PARTNER · INDEPENDENT

Playing *Buzz*

5.NBT.6, SMP7

Student Reference Book, p. 294

To review finding multiples, students play *Buzz*. Read the directions on *Student Reference Book*, page 294 with students and play a sample round before having students play in small groups or as a class.

Observe

• Which students can quickly identify multiples of the *BUZZ* number?

Discuss

• What strategies did you use to figure out multiples of the BUZZ number?

GMP7.1

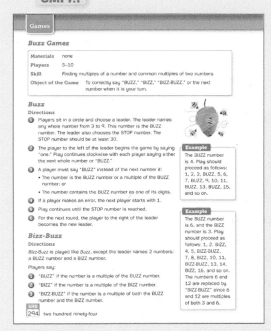

Exploring Life Spans

5.NBT.6, 5.MD.1

Math Masters, p. 67

To practice division in context, students convert ages of the world's oldest people from days to years. They also convert days to weeks and months.

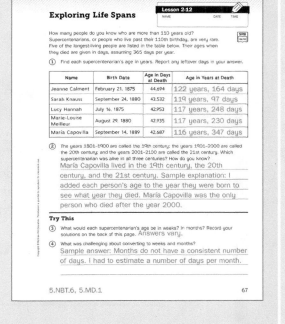

Dividing with Lists of Multiples

5.NBT.6, SMP2, SMP6

Activity Card 26; *Math Masters*, p. 68, pp. TA7 and TA9 (optional);
number cards 10–20 (1 of each);
6-sided die

For additional practice using lists of multiples with partial-quotients division, students generate division problems with 4-digit dividends and 2-digit divisors. They make estimates; list multiples on *Math Masters*, page 68; and solve the problems using partial-quotients division. They discuss which partial quotients make dividing easier. **GMP6.4** Students draw area models as an extension. **GMP2.1**

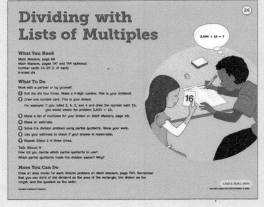

English Language Learner

Beginning ELL To prepare students for the word *multiple* as used in the lesson, skip count on a number grid, highlighting the multiples of a given number. Have students choose a number from the grid using the following sentence frame: "This is _____. A multiple of _____ is _____." Have students take turns repeating the activity.

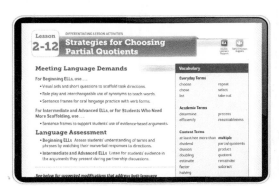

Differentiation Support pages are found in the online Teacher's Center.

Standards and Goals for
Mathematical Process and Practice

SMP1 **Make sense of problems and persevere in solving them.**
  GMP1.5 Solve problems in more than one way.

SMP2 **Reason abstractly and quantitatively.**
  GMP2.1 Create mathematical representations using numbers, words, pictures, symbols, gestures, tables, graphs, and concrete objects.

SMP6 **Attend to precision.**
  GMP6.4 Think about accuracy and efficiency when you count, measure, and calculate.

Adjusting the Activity

Differentiate Throughout the lesson, allow students to use calculators as needed to complete the lists of multiples.

Go Online 　 Differentiation Support

$$27 \overline{)\ 1{,}847}$$

− 270	10
1,577	
− 270	10
1,307	
− 270	10
1,037	

Starting to solve 1,847 / 27 using 10 as a partial quotient

---

① Warm Up — 5 min

▶ Mental Math and Fluency

For each multiplication problem, have students write two related division problems on slates. *Leveled exercises:*

● ○ ○ $2 * 3 = 6$ 　 $6 / 3 = 2; 6 / 2 = 3$
　　　 $4 * 2 = 8$ 　 $8 / 2 = 4; 8 / 4 = 2$

● ● ○ $60 * 4 = 240$ 　 $240 / 4 = 60; 240 / 60 = 4$
　　　 $70 * 3 = 210$ 　 $210 / 70 = 3; 210 / 3 = 70$

● ● ● $2{,}200 * 4 = 8{,}800$ 　 $8{,}800 / 4 = 2{,}200; 8{,}800 / 2{,}200 = 4$
　　　 $1{,}250 * 3 = 3{,}750$ 　 $3{,}750 / 3 = 1{,}250; 3{,}750 / 1{,}250 = 3$

② Focus — 35–40 min

▶ Math Message

Imagine you are solving 1,847 / 27 using partial-quotients division. Can you use 100 as your first partial quotient? Can you use 10 as your first partial quotient? Explain to a partner how you know.

▶ Choosing Partial Quotients

Math Masters, p. TA10

WHOLE CLASS	SMALL GROUP	PARTNER	INDEPENDENT

Math Message Follow-Up Ask volunteers to share their answers to the Math Message problem. Sample answer: I can't use 100 as my first partial quotient because 100 [27s] is 2,700 and that's more than 1,847. I can use 10 because 10 [27s] is 270 and that's less than 1,847.

Display the problem 1,847 / 27 and begin dividing, using 10 as the first partial quotient. After subtracting the first time, ask: *Are there still 10 [27s] left?* Yes. *How do you know?* 1,577 is left, and that's more than 270. Repeat this process a few more times, taking out 10 [27s] at a time. (*See margin.*)

Explain to students that they could continue taking out 10 [27s], but that it seems to be taking a long time. Suggest using a different first partial quotient to make the division go faster. **GMP6.4** Ask: *100 [27s] was too much to take out of 1,847, and 10 [27s] is too little. Could we use 50 [27s]?* Sample answer: I'm not sure because I can't multiply 27 * 50 in my head. Explain that knowing some **multiples** of the divisor, 27, could help students determine whether 50 or some other number could be used as the first partial quotient.

Distribute *Math Masters,* page TA10. Have students write 27 as the second factor in each of the number sentences in the first box and compute the products. Encourage them to use doubling and halving strategies to find the products mentally. 27,000; 13,500; 2,700; 1,350; 540; 270; 135; and 54

Ask: *How can this list of multiples help you divide 1,847 by 27?* Sample answer: I can see that 50 * 27 = 1,350, which is less than 1,847, so I can use 50 as my first partial quotient. Tell students that today they will learn how to use the partial-quotients algorithm more efficiently by using lists of multiples to choose partial quotients. GMP6.4

▶ Using Partial-Quotients Division with Lists of Multiples

Math Journal 1, p. 65; Math Masters, pp. TA9 and TA10

WHOLE CLASS | SMALL GROUP | PARTNER | INDEPENDENT

Have students refer to their list of multiples for 27 as you guide them to use partial-quotients division to solve 1,847 / 27. (*See margin.*)

1. Have students make an estimate. Sample answer: 1,800 / 30 = 60

2. Help students use an "at least/not more than" strategy to choose a partial quotient from the list by asking questions. GMP6.4
   Suggestions:
   • *Are there at least 10 [27s] in 1,847?* Yes. 27 * 10 = 270
   • *Are there at least 50 [27s] in 1,847?* Yes. 27 * 50 = 1,350
   • *Are there at least 100 [27s] in 1,847?* No. 27 * 100 = 2,700
   50 will be a good partial quotient because there are at least 50 [27s] but not 100 [27s] in 1,827.

3. Use 50 as the first partial quotient. Subtract. That leaves 497 to divide.

4. Ask: *How many 27s are in 497?* Tell students to use the same "at least/not more than" strategy, referring to their list of multiples. Ask:
   • *Are there at least 10 [27s] in 497?* Yes. 27 * 10 = 270
   • *Are there at least 20 [27s] in 497?* No. 27 * 20 = 540
   Use 10 as the next partial product because there are at least 10 [27s] but not 20 [27s] in 497. Subtract. That leaves 227 to divide.

5. Ask: *How many 27s are in 227?* Use the same strategy to find that there are at least 5 [27s], but not 10 [27s], in 227. Use 5 as the next partial quotient and subtract, resulting in 92 left to divide.

6. Repeat the process, using 2 and 1 as the next two partial quotients. Note that when 11 is left, students can stop dividing because 11 is less than 27.

7. Add the partial quotients. Record the quotient and the remainder above the problem. 68 R11

8. Have students check the reasonableness of the answer by comparing it to the estimate from Step 1. Display 1,847 / 27 → 68 R11 to summarize.

Display *Math Masters,* page TA9 and have students help you create an area model for this solution as shown in the main column on the next page. GMP2.1 Remind them that they can think of the dividend as the area of a rectangle and the divisor as the length. The partial quotients will add up to the width.

Step 3:

```
27) 1,847
  − 1,350   50    At least 50 [27s]
  ───────          are in 1,847.
      497
```

Step 4:

```
27) 1,847
  − 1,350   50    At least 50 [27s]
  ───────          are in 1,847.
      497
  −   270   10    At least 10 [27s]
  ───────          are in 497.
      227
```

Steps 5–7:

```
         68   R11
27) 1,847
  − 1,350   50    At least 50 [27s]
  ───────          are in 1,847.
      497
  −   270   10    At least 10 [27s]
  ───────          are in 497.
      227
  −   135    5    At least 5 [27s]
  ───────          are in 227.
       92
  −    54    2    At least 2 [27s]
  ───────          are in 92.
       38
  −    27    1    At least 1 [27]
  ───────          is in 38.
       11   68
```

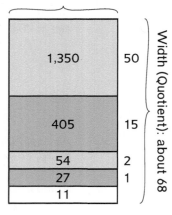

Partial Quotients with Multiples

Lesson 2-12
DATE   TIME

For Problems 1–4, make an estimate. Then use partial-quotients division to solve. Show your work. You can make lists of multiples on *Math Masters*, page TA10 to help you.

Sample estimates given.

1 1,647 / 28 → ?
Estimate: 1,500 / 30 = 50
Answer: 58 R23

2 4,319 / 42 → ?
Estimate: 4,200 / 42 = 100
Answer: 102 R35

3 2,628 / 36 → ?
Estimate: 2,800 / 40 = 70
Answer: 73 R0

4 9,236 / 41 → ?
Estimate: 8,000 / 40 = 200
Answer: 225 R11

Try This

5 Paul drew the area model at the right for his solution to Problem 1. What partial quotients did he use to solve the problem?
50, 5, 2, and 1

5.NBT.6   65

Area (Dividend): 1,847
Length (Divisor): 27

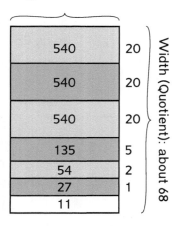

1,350	50
405	15
54	2
27	1
11	

Width (Quotient): about 68

Area (Dividend): 1,847
Length (Divisor): 27

540	20
540	20
540	20
135	5
54	2
27	1
11	

Width (Quotient): about 68

Area models for two other solutions to 1,847 / 27

List each partial quotient along the right side of the area model and mark off a rectangle with each partial quotient as the width. Write the area of each rectangle inside and shade each one in a different color. Point out that shading part of the area is like subtracting that number in the partial-quotients work. When you have drawn a rectangle for each partial quotient, point out that there are still 11 units of area left in the area model.

Ask: *Can we write a 1 next to this part with area 11?* No. *Why?* Sample answer: The area of a rectangle that is 1 by 27 equals 27, and the area of this piece is less than 27. Explain that the width of this section is less than 1 because there is less than 1 [27] in 11. Tell students that they do not have to write a number next to this part of the model, but when they add up the numbers along the side and record the width, they should write "about 68" as a reminder that a small part of the width is not included in the quotient.

Show students at least one other solution to the problem, emphasizing how there is not one "correct" set of partial quotients to use. Two other solutions are shown below. **GMP1.5** Remind the class that all sets of partial quotients produce the same answer, 68 R11, but with some partial quotients the division is completed in fewer steps. **GMP6.4** Challenge students to create area models for at least one solution on their own copies of *Math Masters,* page TA9. (*See margin.*) **GMP2.1**

Area (Dividend): 1,847
Length (Divisor): 27

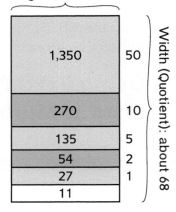

1,350	50
270	10
135	5
54	2
27	1
11	

Width (Quotient): about 68

An area model for 1,847 / 27

```
      68   R11
27) 1,847
  − 1,350    50    At least 50 [27s]
    497            are in 1,847.
  −  405    15    At least 15 [27s]
     92            are in 497.
  −   54     2    At least 2 [27s]
     38            are in 92.
  −   27     1    At least 1 [27]
     11    68     is in 38.
```

```
      68   R11
27) 1,847
  −  540    20    At least 20 [27s]
   1,307           are in 1,847.
  −  540    20    At least 20 [27s]
    767            are in 1,307.
  −  540    20    At least 20 [27s]
    227            are in 767.
  −  135     5    At least 5 [27s]
     92            are in 227.
  −   54     2    At least 2 [27s]
     38            are in 92.
  −   27     1    At least 1 [27]
     11    68     is in 38.
```

If time permits, have students solve additional division problems using lists of multiples. Discuss their solutions, pointing out the different ways each problem can be solved. Then have them complete journal page 65 independently or in partnerships.

Assessment Check-In 5.NBT.6

Math Journal 1, p. 65

Expect most students to be able to solve Problems 1–4 on journal page 65 with the help of a list of multiples, although they may not yet be strategic in choosing the most efficient partial quotients. Some students may be able to interpret the area model in Problem 5. For students who struggle solving the problems, suggest thinking of them in real-world contexts. For example, for Problem 1 they might think: *How many boxes of 28 pencils can I make from 1,647 pencils?*

Evaluation Quick Entry Go online to record students' progress and to see trajectories toward mastery for these standards.

Summarize Have students discuss with a partner how lists of multiples can be helpful when solving division problems.

3 Practice 20–30 min

▶ Introducing *Power Up*

Student Reference Book, p. 318; *Math Masters*, p. G11

| WHOLE CLASS | SMALL GROUP | PARTNER | INDEPENDENT |

Have students read the directions for *Power Up* on *Student Reference Book*, page 318. Play a sample round and then have students play with a partner. Students should keep a record of each round on *Math Masters*, page G11.

Observe
- How do students interpret exponential notation?
- Which students are using patterns in the number of zeros when multiplying by powers of 10? **GMP7.2**

Discuss
- *How did you evaluate expressions in exponential notation?*
- *What patterns did you notice that helped you multiply?* **GMP7.1**

▶ Math Boxes 2-12

Math Journal 1, p. 66

| WHOLE CLASS | SMALL GROUP | PARTNER | INDEPENDENT |

Mixed Practice Math Boxes 2-12 are paired with Math Boxes 2-9.

▶ Home Link 2-12

Math Masters, p. 69

Homework Students use multiples to a solve division problem.

Math Journal 1, p. 66

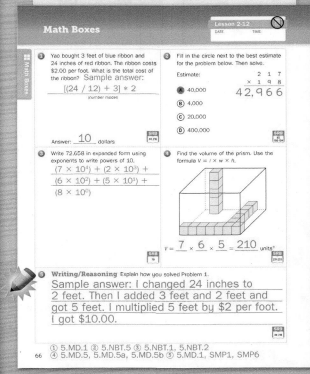

Game Modifications

Differentiate Go online for game modifications.

Go Online Differentiation Support

Math Masters, p. 69

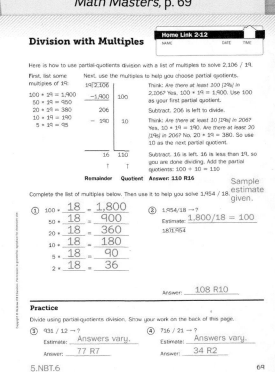

Interpreting the Remainder

Overview Students solve division number stories and practice interpreting remainders.

▶ **Before You Begin**

For Part 2, decide how you will display the Problem-Solving Diagram from *Student Reference Book*, page 30 and students' models for the Math Message problem. Have computation grids (*Math Masters*, page TA7) and blank lists of multiples (*Math Masters*, page TA10) available for student use.

▶ **Vocabulary**

mathematical model • remainder • quotient • dividend • divisor

Standards

Focus Cluster
• Perform operations with multi-digit whole numbers and with decimals to hundredths.

① **Warm Up** 5 min	**Materials**	
Mental Math and Fluency Students practice extended division facts.		5.NBT.6

② **Focus** 35–40 min		
Math Message Students solve a division number story.		5.NBT.6
Modeling a Division Problem Students read about mathematical models and use them to reason through the Math Message problem.	*Student Reference Book*, pp. 12-14 and 30	5.NBT.6 SMP1, SMP4
Interpreting Remainders Students solve division number stories and practice interpreting remainders.	*Math Journal 1*, pp. 68–69; *Math Masters*, pp. TA7 and TA10 (optional)	5.NBT.6 SMP1, SMP4
✓ **Assessment Check-In** See page 196. Expect most students to create models to help them solve division number stories. Students should be able to distinguish situations in which it makes sense to ignore the remainder or round the quotient up.	*Math Journal 1*, pp. 68–69	5.NBT.6 SMP4

③ **Practice** 20–30 min		
Playing *High-Number Toss* **Game** Students practice reading, writing, and comparing numbers in standard and exponential notations.	*Student Reference Book*, p. 312; per partnership: *Math Masters*, p. G10; 6-sided die	5.NBT.1, 5.NBT.2 SMP1, SMP7
Math Boxes 2-13 Students practice and maintain skills.	*Math Journal 1*, p. 67	See page 197.
Home Link 2-13 **Homework** Students solve division number stories and interpret remainders.	*Math Masters*, p. 72	5.NBT.6 SMP4

[Go Online] to see how mastery develops for all standards within the grade.

Differentiation Options

Readiness 10–15 min	**Enrichment** 5–15 min	**Extra Practice** 5–15 min
WHOLE CLASS · SMALL GROUP · PARTNER · INDEPENDENT	WHOLE CLASS · SMALL GROUP · PARTNER · INDEPENDENT	WHOLE CLASS · SMALL GROUP · PARTNER · INDEPENDENT

Thinking About Remainders in Context

5.NBT.6, SMP1

Math Masters, p. 70

To prepare for working with remainders, students interpret remainders in real-world problems. Distribute *Math Masters*, page 70 and solve the first problem together. Identify both the remainder and the quotient in the problem and ask questions like these: *What would it mean if we ignored the remainder in this story? What if we rounded our quotient up? How would our answer change? Which option makes more sense for this story?* GMP1.2, GMP1.4 Guide students through the remaining problems, emphasizing that the treatment of the remainder depends on what is happening in the number story.

Writing Division Number Stories

5.NBT.6, SMP4

Activity Card 27;
Math Masters, p. TA11;
6-sided die

To extend their work interpreting remainders, students write their own number stories to match division problems. They roll a die to generate a 3-digit dividend and a 2-digit divisor. After finding the quotient and remainder, they write two different stories that match the division problem, one for which the remainder can be ignored and one for which the quotient should be rounded up. GMP4.2

Interpreting Remainders in Division Number Stories

5.NBT.6, SMP4

Math Masters, p. 71

To practice interpreting remainders in different contexts, students model and solve division number stories. GMP4.1, GMP4.2 They write explanations for why they rounded up the quotient or ignored the remainder.

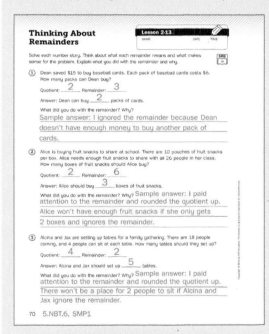

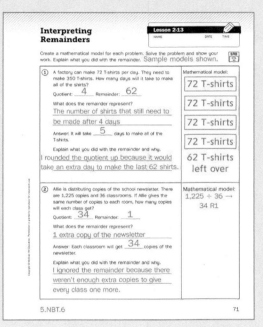

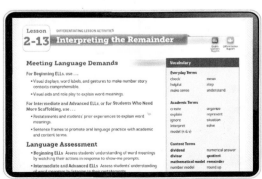

Differentiation Support pages are found in the online Teacher's Center.

Mathematical Process and Practice

SMP1 **Make sense of problems and persevere in solving them.**
GMP1.4 Check whether your answer makes sense.

SMP4 **Model with mathematics.**
GMP4.1 Model real-world situations using graphs, drawings, tables, symbols, numbers, diagrams, and other representations.
GMP4.2 Use mathematical models to solve problems and answer questions.

---

① Warm Up `5 min`

▶ Mental Math and Fluency

Display extended division facts and have students solve them.
Leveled exercises:

● ○ ○ 9 / 3 = 3
       90 / 3 = 30
       900 / 3 = 300

● ● ○ 36 / 6 = 6
       360 / 60 = 6
       3,600 / 60 = 60

● ● ● 8,100 / 90 = 90
       81,000 / 90 = 900
       81,000 / 900 = 90

---

② Focus `35–40 min`

▶ Math Message

A roller coaster holds 30 people. There are 252 people waiting in line. How many times will the roller coaster need to run for all 252 people to get a ride? 9 times

▶ Modeling a Division Problem

Student Reference Book, pp. 12–14 and 30

| WHOLE CLASS | SMALL GROUP | PARTNER | INDEPENDENT |

Math Message Follow-Up Have volunteers share how they solved the Math Message problem, and list the strategies used. Strategies might include drawing a picture or using partial-quotients division. Emphasize the idea that there are many valid ways to approach the problem.

Point out that this Math Message problem required students not only to find a numerical answer but also to interpret the answer in the context of the problem. Tell students that in today's lesson they will use **mathematical models** to help them make sense of real-world problems.

Have partners read *Student Reference Book*, pages 12–14. When they have finished, ask them to list mathematical models they use to solve problems. Sample answers: number sentences; drawings; graphs; situation diagrams
Ask: *How are models helpful when solving problems?* Sample answers: They help us understand the problem better. They summarize how we solve a problem.

Student Reference Book, p. 12

Create and Use Mathematical Models

You can use mathematics to represent situations or objects in the real world. When you do this, you create a **mathematical model**. Your model might use graphs, drawings, tables, symbols, numbers, diagrams, or words.

When you solve a problem using a model, you analyze the real-world situation, create a model for it, and use the model to answer the question. Your model should help you find an answer that makes sense. If it doesn't, you should revise your model or create a new one that better fits the real-world situation or helps you answer the question.

A fifth-grade class is helping their principal solve this problem:

Principal Pippen plans to paint one wall of a school hallway purple. The school engineer sent Mrs. Pippen this note and information:

*Dear Mrs. Pippen,
I measured the area of the wall. It is 187 square feet.
Make sure you buy enough paint.
Mr. Price, School Engineer*

Show how you would solve this problem.

Mrs. Pippen asks the class, "How many pints of paint should I buy?" The students use models to solve the problem and share their thinking.

Aiden creates a model using numbers:

$187 \div 25 \rightarrow 7\ R\ 12$

Aiden, I don't understand what your model represents or how it can help answer the question.

I know that division should work, but I'm not sure how my solution answers the question either. My model is not really helping me.

Ava | Aiden

SRB 12 twelve

PAINT
Color: Purple
Contains: 1 pint
Covers: 25 ft²

Stress the point that there is no one correct model for a problem and that many models can be used to represent the same problem.

Refer students back to the Math Message and ask partners to create a mathematical model to help them solve the problem. **GMP4.1** Invite several partnerships to share their models. As they share, guide a discussion about how the models are useful for solving problems.

Students might use a picture to model the problem:

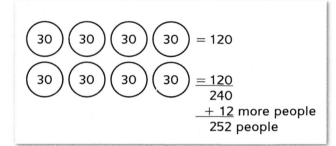

Ask:

- *How does this model the problem?* Sample answer: It shows the groups of 30 people that can ride at one time.
- *How does this model help you solve the problem?* **GMP4.2** Sample answer: It helps me see that 8 groups of 30 with 12 people left over equals 252 people.
- *Does this model help you think about the **remainder**, or what is left over, after you divide? How?* Sample answer: Yes. It shows that there are 12 more people who need to ride the roller coaster after 8 groups of 30 have ridden.
- *Does this model help you check your answer? How?* **GMP1.4** Sample answer: Yes. I can count the groups of 30 and add 12 to make sure that all 252 people got a ride.

Other students might use a number model, such as $252 \div 30 = t$, to model the problem. Ask:

- *How does this model the problem?* Sample answer: It shows the total of 252 people who need to be split into groups of 30 to ride the roller coaster.
- *How does this model help you solve the problem?* **GMP4.2** Sample answer: It shows what operation I need to do to find an answer.
- *Does this model help you think about the remainder? How?* Sample answer: No. I will divide and then I will have to think about what to do with the remainder.
- *Does this model help you check your answer? How?* **GMP1.4** Sample answer: Yes, but I have to be careful with the remainder. Since the model shows the operation of division, I can use multiplication to check my work. I will need to add the remainder after I multiply.

Discuss how different models serve different purposes. Some models can help students organize the information in a problem. Some help students decide what operations they should use. Others help them interpret the solution in context.

Ask partnerships to use one or more of the mathematical models to interpret the quotient and the remainder in the context of the roller coaster problem. Remind the class that the **quotient** is the result of the division, the **dividend** is the number being divided, and the **divisor** is the number that divides the dividend. Ask: *What does the quotient 8 mean?* The roller coaster would need to run 8 times. *What does the remainder 12 mean?* There are 12 people who still have not gotten a ride. *Should the 12 remaining people be ignored? Why or why not?* No. In this case everyone wants a ride, so the remainder should be included as part of the answer. *How can we revise our solution so that everyone gets a ride?* Have the roller coaster run one more time.

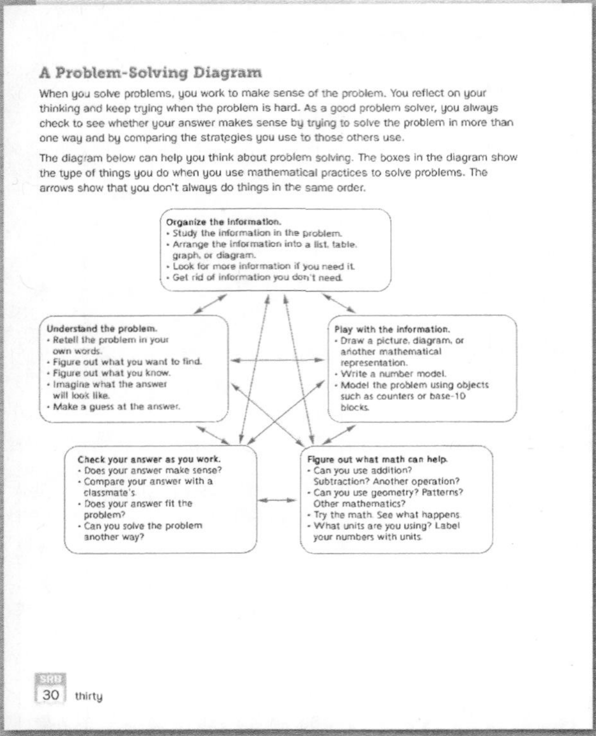

Summarize the solution, concluding that the roller coaster will need to run 9 times to give all 252 people a ride. GMP1.4

Explain how problem solving can be a messy process, and creating a mathematical model is often an important step in that process. Display the Problem-Solving Diagram from *Student Reference Book,* page 30. Explain that this diagram shows one way to think about problem solving. Ask: *Where do mathematical models fit into this process?* Sample answer: We can use models throughout the process to understand the problem, organize the information, play with the information, figure out what math can help, and check our answers. *Why do you think there are arrows connecting all of the boxes?* Sample answers: We can go back and forth between different steps. There's no "right" order. Invite students to share how this diagram could help them solve real-world problems.

NOTE Students may be familiar with a more step-by-step problem-solving process from the Guide for Solving Number Stories in *Third Grade* and *Fourth Grade Everyday Mathematics.* The problem-solving diagram introduced here is designed to encourage them to think more flexibly about real-world contexts. Allow students to continue to use the Guide for Solving Number Stories if they wish.

▶ Interpreting Remainders

Math Journal 1, pp. 68–69

| WHOLE CLASS | SMALL GROUP | PARTNER | INDEPENDENT |

Provide problems in which students must interpret remainders like the ones suggested on the next page. Have students create a model for each number story and solve.

As partnerships work, circulate and pose questions like these to facilitate discussion:

• *What is going on in this story? What model(s) are you using to make sense of the situation?* GMP4.1

• *How does your model help you organize the information in the problem?*

- *How does your model help you understand how to solve the problem?* GMP4.2
- *What did you do with the remainder? How did your model help you interpret the remainder?*
- *How can you use your model to check your work?* GMP1.4

Problem 1: It costs $3 for a 24-hour digital movie rental. Beatrice has $14. How many movies can she rent?

Students might write a number model like $14 \div 3 = v$ or draw a picture showing $14 split into 3s (see below). GMP4.2

Students might use a drawing, mental division, counting by 3s, or the multiplication fact $3 * 4 = 12$ to find that $14 \div 3 \rightarrow 4$ R2. Ask: *What do the quotient 4 and the remainder 2 mean?* The quotient 4 means that Beatrice can rent 4 movies. The remainder 2 means that Beatrice will have $2 left if she rents 4 movies. Ask: *What should be done with the remainder?* Beatrice does not have enough money to rent another movie, so it does not make sense to round the quotient up. The remainder should be ignored.

Problem 2: Shawn is helping the music teacher set up for a concert. There are 76 chairs. Shawn can fit 8 chairs in each row. How many rows can Shawn set up?

Students may write a number model like $76 \div 8 = r$ or draw a picture showing rows of 8 chairs each. GMP4.2

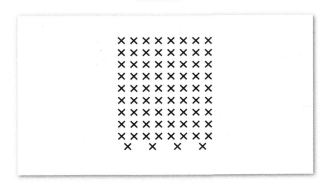

Students might use a drawing, mental division, counting by 8s, or the multiplication fact $8 * 9 = 72$ to find that $76 \div 8 \rightarrow 9$ R4. They can check their work by multiplying $8 * 9$ and adding the remainder of 4 as follows:

76 chairs $\div$ 8 chairs per row $\rightarrow$ 9 rows R4

8 chairs per row $*$ 9 full rows $= 72$ chairs

$72 + 4$ extra chairs $= 76$ chairs

Ask: *What do the quotient 9 and the remainder 4 mean?* The quotient 9 means that Shawn can set up 9 full rows of chairs. The remainder 4 means that there will be 4 chairs left that do not make a full row. Ask: *What should be done with the remainder?* Shawn can set up a tenth row for the 4 extra chairs. It just won't be a full row. We round the quotient up to 10.

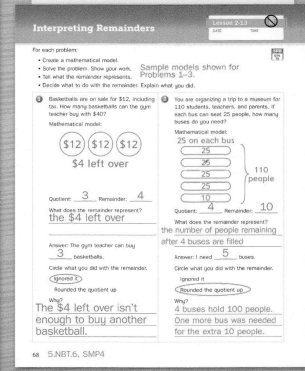

Math Journal 1, p. 69

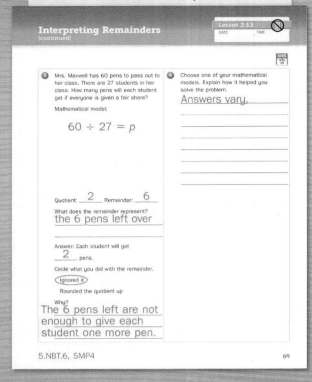

Student Reference Book, p. 312

When partnerships have finished solving the problems, invite volunteers to share their models. Ask them to explain what the quotient and remainder mean in context, as suggested on the previous page. Guide a discussion about how there is more than one way to interpret a remainder. Include the following points:

- In some problems the quotient is rounded up to account for the remainder. For example, in the Math Message, the roller coaster runs were rounded up from 8 runs to 9 runs so that the 12 leftover people would get a ride. In Problem 2, the rows of chairs were rounded up from 9 rows to 10 rows to allow for the 4 extra chairs.
- In other problems the remainder is ignored. For example, in Problem 1 the remainder is ignored because the remaining money is not enough to rent another movie.
- To decide whether to round up the quotient or ignore the remainder, students must consider the situation and understand what the problem is asking.

NOTE There are many real-world contexts in which it makes sense to report the remainder as a fraction or a decimal. Students will learn to report a remainder as a fraction in Unit 3. In this unit problems are limited to contexts in which the remainder can be ignored or the quotient can be rounded up.

Have students complete journal pages 68 and 69. Provide computation grids (*Math Masters,* page TA7) and lists of multiples (*Math Masters,* page TA10) as needed. If students confuse the term *quotient* with the terms *dividend* or *divisor,* scaffold by displaying a labeled division number sentence. As students decide how to interpret remainders, remind them to think about what each remainder means in the context of the problem.

 Assessment Check-In 5.NBT.6

Math Journal 1, pp. 68–69

Expect most students to create models to help them solve Problems 1–3 on journal pages 68 and 69. GMP4.1 Students should be able to distinguish situations in which it makes sense to ignore the remainder or round the quotient up. For those who struggle, ask questions like these: *Does the model help you organize the information in the problem? If not, what other model could you use? How does the model help you understand what the quotient means?* GMP4.2 Have students who are ready for a challenge write some number stories of their own.

 Evaluation Quick Entry Go online to record students' progress and to see trajectories toward mastery for these standards.

Summarize Have students explain to a partner how they decided what to do with each remainder.

3 Practice 20–30 min

▶ Playing *High-Number Toss*

Student Reference Book, p. 312; *Math Masters*, p. G10

| WHOLE CLASS | SMALL GROUP | PARTNER | INDEPENDENT |

Students practice reading, writing, and comparing numbers in standard and exponential notations.

Observe

• Which students have a strategy for building the largest number?
• Which students are comfortable applying powers of 10?

Discuss

• *How did you decide where to place the numbers you rolled?* GMP1.2, GMP7.2
• *What patterns do you notice between the power of 10 used and the number that wins?* GMP7.1

| Differentiate | Game Modifications | Go Online | Differentiation Support |

▶ Math Boxes 2-13

Math Journal 1, p. 67

| WHOLE CLASS | SMALL GROUP | PARTNER | INDEPENDENT |

Mixed Practice Math Boxes 2-13 are paired with Math Boxes 2-11.

▶ Home Link 2-13

Math Masters, p. 72

Homework Students use mathematical models to solve division number stories and interpret remainders. GMP4.1, GMP4.2

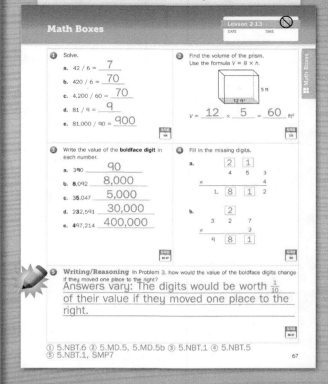

Math Boxes

Lesson 2-13

① Solve.
a. $42 / 6 = \underline{7}$
b. $420 / 6 = \underline{70}$
c. $4,200 / 60 = \underline{70}$
d. $81 / 9 = \underline{9}$
e. $81,000 / 90 = \underline{900}$

② Find the volume of the prism. Use the formula $V = B \times h$.

$V = \underline{12} \times \underline{5} = \underline{60}$ ft³

③ Write the value of the **boldface digit** in each number.
a. 3**9**0 _____ 90
b. **8**,092 _____ 8,000
c. 3**5**,047 _____ 5,000
d. 2**3**2,591 _____ 30,000
e. **4**97,214 _____ 400,000

④ Fill in the missing digits.

⑤ **Writing/Reasoning** In Problem 3, how would the value of the boldface digits change if they moved one place to the right?
Answers vary: The digits would be worth $\frac{1}{10}$ of their value if they moved one place to the right.

① 5.NBT.6 ② 5.MD.5, 5.MD.5b ③ 5.NBT.1 ④ 5.NBT.5
⑤ 5.NBT.1, SMP7

67

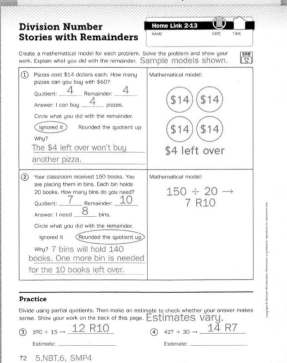

Division Number Stories with Remainders

Home Link 2-13

Create a mathematical model for each problem. Solve the problem and show your work. Explain what you did with the remainder. Sample models shown.

① Pizzas cost $14 dollars each. How many pizzas can you buy with $60?
Quotient: **4** Remainder: **4**
Answer: I can buy **4** pizzas.
Circle what you did with the remainder.
(Ignored it) Rounded the quotient up
Why? The $4 left over won't buy another pizza.

Mathematical model:
($14) ($14)
($14) ($14)
$4 left over

② Your classroom received 150 books. You are placing them in bins. Each bin holds 20 books. How many bins do you need?
Quotient: **7** Remainder: **10**
Answer: I need **8** bins.
Circle what you did with the remainder.
Ignored it (Rounded the quotient up)
Why? 7 bins will hold 140 books. One more bin is needed for the 10 books left over.

Mathematical model:
$150 \div 20 \rightarrow$
7 R10

Practice

Divide using partial quotients. Then make an estimate to check whether your answer makes sense. Show your work on the back of this page. Estimates vary.

③ $190 \div 15 \rightarrow$ **12 R10**
Estimate: _____

④ $427 \div 30 \rightarrow$ **14 R7**
Estimate: _____

72 5.NBT.6, SMP4

Unit 2 Progress Check

Overview   Day 1: Administer the Unit Assessments.
Day 2: Administer the Cumulative Assessment.

Day 1: Unit Assessment

Quick Entry Evaluation Record results and track progress toward mastery.

1 Warm Up   5–10 min

Materials

Self Assessment
Students complete the Self Assessment.

Assessment Handbook, p. 14

2a Assess   35–50 min

☆ **Unit 2 Assessment**
These items reflect mastery expectations to this point.

Assessment Handbook, pp. 15–18

Unit 2 Challenge (Optional)
Students may demonstrate progress beyond expectations.

Assessment Handbook, pp. 19–20

Standards	Goals for Mathematical Content (GMC)	Lessons	Self Assessment	Unit 2 Assessment	Unit 2 Challenge
5.OA.1	Write numerical expressions that contain grouping symbols.	2-6		8	
5.OA.2	Model real-world and mathematical situations using simple expressions.	2-6		8	4
	Interpret numerical expressions without evaluating them.	2-7			1a
5.NBT.1	Understand the relationship between the places in multidigit numbers.	2-1, 2-2	1, 2	1, 2, 6	
5.NBT.2	Use whole-number exponents to denote powers of 10.	2-2, 2-3	3	4	
	Multiply whole numbers by powers of 10; explain the number of zeros in the product.	2-2, 2-3, 2-9, 2-10	4	3a, 5a, 5b	1b
5.NBT.5	Fluently multiply multidigit whole numbers using the standard algorithm.	2-4 to 2-9	5	10, 11	2
5.NBT.6	Divide multidigit whole numbers.	2-10 to 2-13	6	9, 12, 13	3
	Illustrate and explain solutions to division problems.	2-11 to 2-13	7	9	3
5.MD.1	Convert among measurement units within the same system.	2-6		7, 8	4
	Use measurement conversions to solve multi-step, real-world problems.	2-6		8	4

Standards	Goals for Mathematical Process and Practice (GMP)	Lessons	Self Assessment	Unit 2 Assessment	Unit 2 Challenge
SMP1	Make sense of your problem. GMP1.1	2-6		8, 9	1c, 4
SMP2	Create mathematical representations using numbers, words, pictures, symbols, gestures, tables, graphs, and concrete objects. GMP2.1	2-7, 2-11, 2-12			3
SMP4	Model real-world situations using graphs, drawings, tables, symbols, numbers, diagrams, and other representations. GMP4.1	2-6, 2-13		8, 9	4
	Use mathematical models to solve problems and answer questions. GMP4.2	2-9, 2-13		9	
SMP6	Explain your mathematical thinking clearly and precisely. GMP6.1	2-2, 2-3		3b, 5b	1c
	Think about accuracy and efficiency when you count, measure, and calculate. GMP6.4	2-3, 2-8 to 2-10, 2-12		3b	2
SMP7	Use structures to solve problems and answer questions. GMP7.2	2-1, 2-2, 2-10		5b	

 Go Online to see how mastery develops for all standards within the grade.

1 Warm Up   5–10 min

▶ Self Assessment

Assessment Handbook, p. 14

| WHOLE CLASS | SMALL GROUP | PARTNER | **INDEPENDENT** |

Students complete the Self Assessment to reflect on their progress in Unit 2.

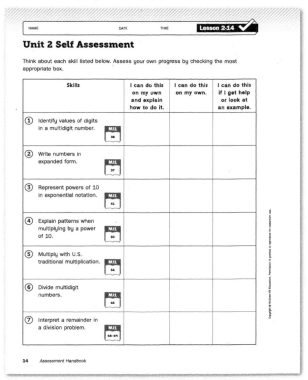

Assessment Handbook, p. 14

NAME   DATE   TIME   **Lesson 2-14** ✓

Unit 2 Assessment

Solve the following number riddles.

① I am a 5-digit number.
My 3 is worth 3 • 10,000.
My 4 is worth 4,000.
One of my 2s is worth 20. The other 2 is worth 10 times as much.
My other digit is 7.
What number am I? __34,227__

② I am a 6-digit number.
One of my 7s is worth 700,000. The other 7 is worth $\frac{1}{10}$ as much.
My 8 is worth 8 [100s].
My 9 is worth 90.
My other digits are 0.
What number am I? __770,890__

③ a. Jesse collects cans for recycling. When he has 1,500 cans, the recycling center will pick them up from his house. Jesse has 120 bags with about 35 cans in each bag. Should he call the recycling center to arrange a pick-up? Explain how you know.

Sample answer: Yes, Jesse should call the recycling center. I estimated 100 * 35 = 3,500 cans. That is more than 1,500 cans.

b. Did you have to find an exact answer to solve Problem 3a? Explain why or why not.
Sample answer: No. When I estimated it gave me enough information to answer the question.

Assessment Masters   **15**

---

② a **Assess**   35–50 min

▶ Unit 2 Assessment

Assessment Handbook, pp. 15–18

| WHOLE CLASS | SMALL GROUP | PARTNER | **INDEPENDENT** |

Students complete the Unit 2 Assessment to demonstrate their progress on the standards covered in this unit.

Generic rubrics in the *Assessment Handbook* can be used to evaluate student progress on the Mathematical Process and Practice Standards.

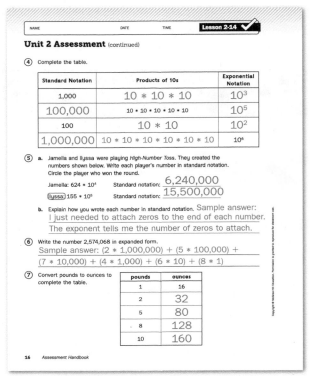

NAME   DATE   TIME   **Lesson 2-14** ✓

Unit 2 Assessment (continued)

④ Complete the table.

Standard Notation	Products of 10s	Exponential Notation
1,000	10 * 10 * 10	10^3
100,000	10 * 10 * 10 * 10 * 10	10^5
100	10 * 10	10^2
1,000,000	10 * 10 * 10 * 10 * 10 * 10	10^6

⑤ a. Jamella and Ilyssa were playing *High-Number Toss.* They created the numbers shown below. Write each player's number in standard notation. Circle the player who won the round.

Jamella: 624 * 10^4 Standard notation: 6,240,000
(Ilyssa:) 155 * 10^5 Standard notation: 15,500,000

b. Explain how you wrote each number in standard notation. Sample answer: I just needed to attach zeros to the end of each number. The exponent tells me the number of zeros to attach.

⑥ Write the number 2,574,068 in expanded form.
Sample answer: (2 * 1,000,000) + (5 * 100,000) + (7 * 10,000) + (4 * 1,000) + (6 * 10) + (8 * 1)

⑦ Convert pounds to ounces to complete the table.

pounds	ounces
1	16
2	32
5	80
8	128
10	160

16   Assessment Handbook

Assessment Handbook, p. 16

Differentiate   Adjusting the Assessment

Item(s)	Adjustments
1, 2	To scaffold Items 1 and 2, have students use a place-value chart.
3	To extend Item 3, have students explain whether they overestimated or underestimated the actual number of cans Jesse has.
4	To scaffold Item 4, have students use calculators to check that the product of 10s is correct.
5	To extend Item 5, have students write numbers that would beat both Jamella and Ilyssa in *High-Number Toss*.
6	To extend Item 6, have students write the number in expanded form in a different way.
7	To scaffold Item 7, have students use words to describe the relationship between pounds and ounces. Record the relationship with an expression and have students evaluate the expression to fill in the remaining rows.
8	To scaffold Item 8, have students use the table in Item 7 to figure out the number of ounces in 4 pounds. Then have them find the total weight of the package.
9	To extend Item 9, have students write and solve another number story in which they have to interpret the remainder.
10, 11	To extend Items 10 and 11, have students solve the problems using both partial-products multiplication and U.S. traditional multiplication and compare the methods.
12, 13	To scaffold Items 12 and 13, provide copies of *Math Masters,* page TA10, and have students write lists of multiples for the divisors.

Advice for Differentiation

Because this is the beginning of the school year, all of the content included on the Unit 2 Assessment was recently introduced and will be revisited in subsequent units.

Go Online:

 Quick Entry Evaluation Record students' progress and see trajectories toward mastery for these standards.

 Data Review your students' progress reports. Differentiation materials are available online to help you address students' needs.

NOTE See the Unit Organizer on pages 104–105 or the online Spiral Tracker for details on Unit 2 focus topics and the spiral.

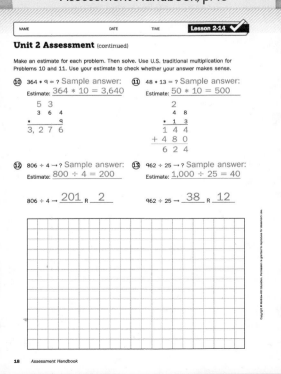

Assessment Handbook, p. 17

Assessment Handbook, p. 18

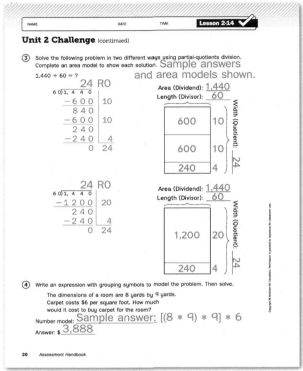

NAME   DATE   TIME   Lesson 2-14 ✓

Unit 2 Challenge

① a. Use the following expressions to complete the statements below.

$6 * 10^8$ $68 * 10^3$ $16 * 10^5$ $5 * 10^7$

In 1 year Earth travels about ___$6 * 10^8$___ miles in its orbit around the Sun.

In 1 month Earth travels about ___$5 * 10^7$___ miles.

In 1 day Earth travels about ___$16 * 10^5$___ miles.

In 1 hour Earth travels about ___$68 * 10^3$___ miles.

b. Evaluate each of the expressions.

$6 * 10^8 =$ ___600,000,000___

$68 * 10^3 =$ ___68,000___

$16 * 10^5 =$ ___1,600,000___

$5 * 10^7 =$ ___50,000,000___

c. Do your answers to Part a make sense? How do you know? Sample answer: Yes. Earth travels fewer miles in less time. When I evaluated the expressions, I saw that I had put them in order from longest to shortest in Part a.

② Sally and Paul solved the same multiplication problem. Sally used U.S. traditional multiplication. Paul used a different strategy.

Sally
$101 * 26 = ?$

```
  1 0 1
*   2 6
  6 0 6
+2,0 2 0
2,6 2 6
```

Paul
$101 * 26 = ?$

$100 * 26 = 2,600$
$1 * 26 = 26$
$2,600 + 26 = 2,626$

Which strategy seems more efficient? Why?
Sample answer: Paul's strategy seems more efficient because I can do all his steps in my head.

Assessment Masters   19

▶ Unit 2 Challenge (Optional)

Assessment Handbook, pp. 19–20

| WHOLE CLASS | SMALL GROUP | PARTNER | **INDEPENDENT** |

Students can complete the Unit 2 Challenge after they complete the Unit 2 Assessment.

NAME   DATE   TIME   Lesson 2-14 ✓

Unit 2 Challenge (continued)

③ Solve the following problem in two different ways using partial-quotients division. Complete an area model to show each solution. Sample answers and area models shown.

$1,440 ÷ 60 = ?$

```
        24 R0
60)1,4 4 0
  - 6 0 0   10
    8 4 0
  - 6 0 0   10
    2 4 0
  - 2 4 0    4
        0   24
```

Area (Dividend): 1,440
Length (Divisor): 60

600	10
600	10
240	4

Width (Quotient): 24

```
        24 R0
60)1,4 4 0
  -1 2 0 0   20
    2 4 0
  - 2 4 0    4
        0   24
```

Area (Dividend): 1,440
Length (Divisor): 60

| 1,200 | 20 |
| 240 | 4 |

Width (Quotient): 24

④ Write an expression with grouping symbols to model the problem. Then solve.

The dimensions of a room are 8 yards by 9 yards. Carpet costs $6 per square foot. How much would it cost to buy carpet for the room?

Number model: Sample answer: $[(8 * 9) * 9] * 6$

Answer: $___3,888___

20   Assessment Handbook

Assessment Handbook, p. 20

Unit 2 Progress Check

Overview   Day 2: Administer the Cumulative Assessment.

Day 2: Cumulative Assessment

2b Assess   35–45 min

Materials

Cumulative Assessment
These items reflect mastery expectations to this point.

Assessment Handbook, pp. 21–22

Standards	Goals for Mathematical Content (GMC)	Cumulative Assessment
5.OA.1	Write numerical expressions that contain grouping symbols.	1–4
5.OA.2	Interpret numerical expressions without evaluating them.	5a, 5b
5.MD.1	Convert among measurement units within the same system.	9b
	Use measurement conversions to solve multi-step, real-world problems.	9b
5.MD.3, 5.MD.3a	Understand that a unit cube has 1 cubic unit of volume and can measure volume.	7
5.MD.3, 5.MD.3b	Understand that a solid figure completely filled by *n* unit cubes has volume *n* cubic units.	6a, 6b, 7
5.MD.4	Measure volumes by counting unit cubes and improvised units.	6a, 6b
5.MD.5, 5.MD.5a	Represent products of three whole numbers as volumes.	8
5.MD.5, 5.MD.5b	Apply formulas to find volumes of rectangular prisms.	8, 9a
5.MD.5, 5.MD.5c	Find volumes of figures composed of right rectangular prisms.	9a
	Solve real-world problems involving volumes of figures composed of prisms.	9a
	Goals for Mathematical Process and Practice (GMP)	
SMP1	Make sense of your problem. GMP1.1	5b, 8, 9c
SMP2	Make sense of the representations you and others use. GMP2.2	6a, 6b
SMP6	Explain your mathematical thinking clearly and precisely. GMP6.1	5b, 6b, 7
	Think about accuracy and efficiency when you count, measure, and calculate. GMP6.4	5b

3 Look Ahead   10–15 min

Materials

Math Boxes 2-14: Preview for Unit 3
Students preview skills and concepts for Unit 3.

Math Journal 1, p. 70

Home Link 2-14
Students take home the Family Letter that introduces Unit 3.

Math Masters, pp. 73–76

    Go Online to see how mastery develops for all
standards within the grade.

NAME   DATE   TIME   **Lesson 2-14** ✔

Unit 2 Cumulative Assessment

For Problems 1–4, insert grouping symbols to make the number sentences true.

① $160 \div (16 + 4) = 8$ ② $(4 + 6) \cdot 8 \div 2 = 40$

③ $3 = 120 \div (64 - 24)$ ④ $135 = (42 + 3) \cdot 3$

⑤ a. Write each expression in the correct column.

$2 * (12 + 4)$ $12 + 4 - 4$ $(12 + 4) \div 8$

$6 + 12 + 4$ $(12 + 4) * 10$ $(12 + 4) - 10$

> (12 + 4)	< (12 + 4)
$2 * (12 + 4)$	$12 + 4 - 4$
$6 + 12 + 4$	$(12 + 4) \div 8$
$(12 + 4) * 10$	$(12 + 4) - 10$

b. Did you need to evaluate the expressions to solve Problem 5a? Why or why not?
Sample answer: No. Every expression had 12 + 4 in it. I just had to decide whether (12 + 4) was getting larger or smaller in each expression.

⑥ a. Find the volume of the prism.

Volume = __72__ cubic units

b. Explain how you found the volume of the prism.
Sample answer: I filled in the rest of the base with cubes and saw that it would be 12. The height is 6. 12 * 6 = 72 cubes.

Assessment Masters   21

NAME   DATE   TIME   **Lesson 2-14** ✔

Unit 2 Cumulative Assessment (continued)

⑦ Why is a unit cube a good unit for measuring volume?
Sample answers: Cubes fit into corners and pack neatly. A unit cube has a volume of 1 cubic unit, so you count the cubes that fit into a prism to find the volume.

⑧ Damien is buying boxes from a moving company. He wants to buy the box that will fit the most of his belongings. He can choose between the following three options:

Box 1: 3 ft, 2 ft, 3 ft
Box 2: 3 ft, 9 ft²,
Box 3: 2 ft, 4 ft, 3 ft

What is the volume of each box? Write a number sentence that shows how you found the volume. Remember: $V = l \times w \times h$ and $V = B \times h$

Volume of Box 1: __18 ft³__ Number sentence: $3 * 2 * 3 = 18$

Volume of Box 2: __27 ft³__ Number sentence: $9 * 3 = 27$

Volume of Box 3: __24 ft³__ Number sentence: $4 * 3 * 2 = 24$

Which box should Damien buy? Explain your answer. Sample answer: He should buy Box 2 because it has the largest volume.

⑨ Josh's family is renting a storage unit. The storage facility gave Josh's family this sketch of a storage unit.

a. What is the volume of the storage unit? __180__ ft³

b. Josh's family estimated that they need a unit with a volume of 10 cubic yards. How many cubic feet of storage space do they need? Hint: There are 3 feet in 1 yard.

Josh's family needs __270__ ft³ of storage space.

c. Is the storage unit large enough for Josh's family? Explain why or why not.
No. Sample explanation: Josh's family needs a unit with about 270 cubic feet. This unit has only 180 cubic feet of volume.

22   Assessment Handbook

▶ Cumulative Assessment

Assessment Handbook, pp. 21–22

| WHOLE CLASS | SMALL GROUP | PARTNER | **INDEPENDENT** |

Students complete the Cumulative Assessment. The problems in the Cumulative Assessment address content from Unit 1. It can help you monitor learning and retention of some (but not all) of the content and process/practice standards that were the focus of that unit, as detailed in the Cumulative Assessment table on page 203. Successful responses to these problems indicate adequate progress at this point in the year.

Monitor student progress on the standards using the online assessment and reporting tools.

Generic rubrics in the *Assessment Handbook* can be used to evaluate student progress on the Mathematical Process and Practice Standards.

Written assessments are one way students can demonstrate what they know. The table below shows adjustments you can make to the Cumulative Assessment to maximize opportunities for individual students or for your entire class.

Differentiate	Adjusting the Assessment
Item(s)	**Adjusting the Assessment**
1–4	To scaffold Items 1–4, provide students with several examples of where grouping symbols could be placed and have them choose the correct answer from the examples.
5	To extend Item 5, have students write additional expressions that could be placed in each column of the table.
6	To scaffold Item 6, provide students with unit cubes and allow them to build the rectangular prism shown.
7	To scaffold Item 7, give students pattern blocks and a prism. Ask them to pack the prism with different pattern blocks and compare them to cubes.
8	To scaffold Item 8, remind students what each of the variables represents in the formulas $V = l * w * h$ and $V = B * h$.
9	To extend Item 9, ask students to sketch a storage unit that would be large enough to fit the family's belongings.

Advice for Differentiation

Because this is the beginning of the school year, all of the content included on the Cumulative Assessment was recently introduced and will be revisited in subsequent units.

Go Online:

 Quick Entry Evaluation Record students' progress and see trajectories toward mastery for these standards.

 Data Review your students' progress reports. Differentiation materials are available online to help you address students' needs.

③ Look Ahead 10–15 min

▶ **Math Boxes 2-14:** Preview for Unit 3

Math Journal 1, p. 70

WHOLE CLASS	SMALL GROUP	PARTNER	INDEPENDENT

Mixed Practice Math Boxes 2-14 are paired with Math Boxes 2-10. These problems focus on skills and understandings that are prerequisite for Unit 3. You may want to use information from these Math Boxes to plan instruction and grouping in Unit 3.

▶ **Home Link 2-14:** Unit 3 Family Letter

Math Masters, pp. 73–76

Home Connection The Unit 3 Family Letter provides information and activities related to Unit 3 content.

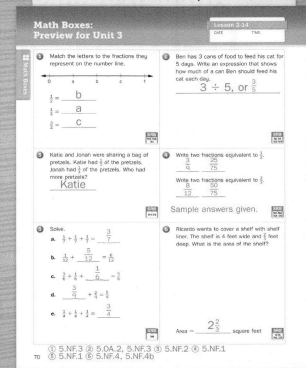

Math Journal 1, p. 70

Math Masters, pp. 73–76

Unit 3 Organizer

Fraction Concepts, Addition, and Subtraction
Contents

*The standards listed here are addressed in the **Focus** of each lesson. For all the standards in a lesson, see the Lesson Opener.

Focus

In this unit, students build on fraction concepts from previous grades to understand fractions as division. They also use visual models to make estimates, add and subtract fractions and mixed numbers, and check the reasonableness of their answers.

Major Clusters

5.NF.A Use equivalent fractions as a strategy to add and subtract fractions.

5.NF.B Apply and extend previous understandings of multiplication and division to multiply and divide fractions.

Process and Practice Standards

SMP5 Use appropriate tools strategically.

SMP8 Look for and express regularity in repeated reasoning.

Coherence

The table below describes how standards addressed in the Focus parts of the lessons link to the mathematics that students have done in the past and will do in the future.

	Links to the Past	**Links to the Future**
5.NF.1	In Grade 4, students used fraction circles, number lines, fraction strips, and strategies to add and subtract fractions and mixed numbers with like denominators.	In Unit 5, students will develop formal strategies, including using equivalent fractions with common denominators, to add and subtract fractions and mixed numbers. They will apply these strategies to solve real-world and mathematical problems. In Grade 7, students will add and subtract rational numbers.
5.NF.2	In Grade 4, students used fraction circles, number lines, fraction strips, and strategies to solve real-world problems involving the addition and subtraction of fractions and mixed numbers with like denominators.	Throughout Grade 5, students will use estimates to help them assess the reasonableness of their answers to fraction addition and subtraction problems. In Unit 5, they will apply formal strategies to add and subtract fractions in real-world contexts. In Grade 7, students will solve real-world problems involving all operations with rational numbers.
5.NF.3	In Grade 3, students interpreted whole-number quotients of whole numbers.	Throughout Grade 5, students will apply their understanding of fractions as division to solve number stories and interpret those solutions in context. In Unit 5, they will use this knowledge to make sense of fraction multiplication strategies. In Grade 6, students will apply this knowledge to help them make sense of fraction division.
5.NF.4 **5.NF.4a**	In Grade 4, students used a variety of strategies, including repeated addition and fraction representations, to multiply a fraction by a whole number.	In Unit 5, students will develop formal strategies for multiplying fractions and in Unit 7, they will apply those strategies to multiply mixed numbers. In Grade 6, students will multiply fractions in the context of solving equations.
5.NF.5	In Grade 4, students used visual fraction models and other strategies to solve number stories involving multiplication of a fraction by a whole number.	In Unit 5, students will apply formal fraction multiplication strategies to solve real-world problems. In Unit 7, students will solve real-world problems involving multiplication of mixed numbers. In Grade 6, students will solve real-world fraction multiplication problems in the context of solving equations.

Planning for Rich Math Instruction

	3-1 Connecting Fractions and Division, Part 1	3-2 Connecting Fractions and Division, Part 2	3-3 Application: Interpreting Remainders	3-4 Fractions on a Number Line
RIGOR — Conceptual Understanding	Modeling and solving division problems with fractional answers Modeling with Fraction Circle Pieces, pp. 220–221 Creating Other Models for Fair Share Stories, pp. 222–223	Understanding fractions as division Writing Division Models for Fair Share Stories, pp. 226–227 Interpreting Fractions as Division, pp. 228–230	Understanding fractions as division Reporting Remainders as Fractions, pp. 234–236	Fraction number sense Representing Fractions on Number Lines, pp. 240–243
RIGOR — Procedural Skill and Fluency	Multiplying and Dividing, p. 223	Mental Math and Fluency, p. 226	Mental Math and Fluency, p. 234	Playing *Multiplication Top-It: Larger Numbers,* p. 245
RIGOR — Applications	Mental Math and Fluency, p. 220 Math Message, p. 220 Modeling with Fraction Circle Pieces, pp. 220–221 Creating Other Models for Fair Share Stories, pp. 222–223 Home Link 3-1, p. 223 Enrichment, p. 219 Extra Practice, p. 219	Math Message, p. 226 Writing Division Models for Fair Share Stories, pp. 226–227 Interpreting Fractions as Division, pp. 228–230 Home Link 3-2, p. 231 Differentiation Options, p. 225	Math Message, p. 234 Reporting Remainders as Fractions, pp. 234–236 Interpreting Remainders in Division Number Stories, pp. 236–237 Home Link 3-3, p. 237 Enrichment, p. 233 Extra Practice, p. 233	Mental Math and Fluency, p. 240 Math Message, p. 240 Using Number Lines to Solve Problems, pp. 243–245
Rich Tasks and Mathematical Reasoning	Modeling with Fraction Circle Pieces, pp. 220–221 Creating Other Models for Fair Share Stories, pp. 222–223 Enrichment, p. 219	Interpreting Fractions as Division, pp. 228–230 Journal p. 76: Writing/Reasoning Enrichment, p. 225	Math Message, p. 234 Reporting Remainders as Fractions, pp. 234–236 Interpreting Remainders in Division Number Stories, pp. 236–237 Playing *Prism Pile-Up,* p. 237 Home Link 3-3, p. 237 Enrichment, p. 233	Using Number Lines to Solve Problems, pp. 243–245 Journal p. 82: Writing/Reasoning Enrichment, p. 239
Mathematical Discourse	Modeling with Fraction Circle Pieces, pp. 220–221 Creating Other Models for Fair Share Stories, pp. 222–223 Extra Practice, p. 219	Interpreting Fractions as Division, pp. 228–230	Reporting Remainders as Fractions, pp. 234–236 Extra Practice, p. 233	Representing Fractions on Number Lines, pp. 240–243 Using Number Lines to Solve Problems, pp. 243–245
Distributed Practice	Mental Math and Fluency, p. 220 Multiplying and Dividing, p. 223 Math Boxes 3-1, p. 223	Mental Math and Fluency, p. 226 Playing *Power Up,* p. 231 Math Boxes 3-2, p. 231	Mental Math and Fluency, p. 234 Playing *Prism Pile-Up,* p. 237 Math Boxes 3-3, p. 237	Mental Math and Fluency, p. 240 Playing *Multiplication Top-It: Larger Numbers,* p. 245 Math Boxes 3-4, p. 245
Differentiation Support	Differentiation Options, p. 219 ELL Support, p. 219 Online Differentiation Support 3-1 Adjusting the Activity, p. 221	Differentiation Options, p. 225 ELL Support, p. 225 Online Differentiation Support 3-2 Common Misconception, p. 227 Academic Language Development, p. 229	Differentiation Options, p. 233 ELL Support, p. 233 Online Differentiation Support 3-3 Adjusting the Activity, p. 235	Differentiation Options, p. 239 ELL Support, p. 239 Online Differentiation Support 3-4 Common Misconception, p. 241 Adjusting the Activity, p. 243

Red text = Game

	3-5 Open Response **Game Strategies** (2-Day Lesson)	**3-6** **Fraction Estimation with Number Sense**	**3-7** **Fraction Estimation with Benchmarks**	**3-8** **Renaming Fractions and Mixed Numbers**	
	Understanding fractions as division Solving the Open Response Problem, p. 249 Reengaging in the Problem, p. 253	**Fraction number sense** Estimating Fraction Sums and Differences, pp. 258–260	**Fraction number sense** Using Benchmarks to Estimate Sums and Differences of Fractions, pp. 264–266	**Equivalent names for fractions** Renaming by Making and Breaking Apart Wholes, pp. 270–272 Introducing *Rename That Mixed Number,* pp. 272–273	**Conceptual Understanding**
	Mental Math and Fluency, p. 247	Interpreting Remainders, p. 261	Journal p. 89, #2, #3, #4	Journal p. 92, #2 Home Link 3-8, p. 273	**Procedural Skill and Fluency**
	Math Message, p. 247 Sharing Game Strategies, p. 248 Solving the Open Response Problem, p. 249 Home Link 3-5, p. 255	Estimating Fraction Sums and Differences, pp. 258–260 Interpreting Remainders, p. 261	Using Benchmarks to Estimate Sums and Differences of Fractions, pp. 264–266 Home Link 3-7, p. 267 Extra Practice, p. 263	Connecting Fractions and Division, p. 273	**Applications**
	Math Message, p. 247 Solving the Open Response Problem, p. 249 Reengaging in the Problem, p. 253 Revising Work, pp. 254–255 Home Link 3-5, p. 255	Math Message, p. 258 Estimating Fraction Sums and Differences, pp. 258–260 Introducing *Build-It,* pp. 260–261 Home Link 3-6, p. 261 Enrichment, p. 257 Extra Practice, p. 257	Math Message, p. 264 Using Benchmarks to Estimate Sums and Differences of Fractions, pp. 264–266 Introducing *Fraction Spin,* p. 267	Renaming by Making and Breaking Apart Wholes, pp. 270–272 Introducing *Rename That Mixed Number,* pp. 272–273 Connecting Fractions and Division, p. 273 Journal p. 92: Writing/Reasoning Enrichment, p. 269 Extra Practice, p. 269	**Rich Tasks and Mathematical Reasoning**
	Sharing Game Strategies, p. 248 Setting Expectations, p. 253 Reengaging in the Problem, p. 253	Estimating Fraction Sums and Differences, pp. 258–260	Using Benchmarks to Estimate Sums and Differences of Fractions, pp. 264–266	Renaming by Making and Breaking Apart Wholes, pp. 270–272 Extra Practice, p. 269	**Mathematical Discourse**
	Mental Math and Fluency, p. 247 Math Boxes 3-5, p. 255	Mental Math and Fluency, p. 258 Interpreting Remainders, p. 261 Math Boxes 3-6, p. 261	Mental Math and Fluency, p. 264 Math Boxes 3-7, p. 267	Mental Math and Fluency, p. 270 Connecting Fractions and Division, p. 273 Math Boxes 3-8, p. 273	**Distributed Practice**
	ELL Support, p. 247 Adjusting the Activity, p. 249	Differentiation Options, p. 257 ELL Support, p. 257 Online Differentiation Support 3-6 Academic Language Development, p. 258 Adjusting the Activity, p. 259	Differentiation Options, p. 263 ELL Support, p. 263 Online Differentiation Support 3-7 Adjusting the Activity, p. 265 Academic Language Development, p. 265	Differentiation Options, p. 269 ELL Support, p. 269 Online Differentiation Support 3-8 Common Misconception, p. 271	**Differentiation Support**

RIGOR

Red text = Game

Planning for Rich Math Instruction

RIGOR		**3-9** **Introduction to Adding and Subtracting Fractions and Mixed Numbers**	**3-10** **Exploring Addition of Fractions with Unlike Denominators**	**3-11** **Playing *Fraction Capture***	**3-12** **Solving Fraction Number Stories**
	Conceptual Understanding	Addition and subtraction of fractions and mixed numbers Exploring Addition and Subtraction of Fractions and Mixed Numbers, pp. 276–278	Addition of fractions Finding Fraction Sums, pp. 283–284	Addition of fractions Breaking Apart Fractions, pp. 288–289 Introducing *Fraction Capture*, pp. 289–291	Applying problem-solving strategies to fraction problems Math Message, p. 294 Identifying Problem-Solving Strategies, pp. 294–296
	Procedural Skill and Fluency	Solving More Fraction Addition and Subtraction Number Stories, pp. 278–279 Playing *Division Dash,* p. 279	Practicing Fraction Addition, pp. 284–285	Mental Math and Fluency, p. 288 Introducing *Fraction Capture*, pp. 289–291 Home Link 3-11, p. 291	Solving Fraction Number Stories, p. 296 Practicing Division, p. 297 Home Link 3-12, p. 297
	Applications	Math Message, p. 276 Exploring Addition and Subtraction of Fractions and Mixed Numbers, pp. 276–278 Solving More Fraction Addition and Subtraction Number Stories, pp. 278–279 Home Link 3-9, p. 279 Enrichment, p. 275 Extra Practice, p. 275	Home Link 3-10, p. 285	Mental Math and Fluency, p. 288	Identifying Problem-Solving Strategies, pp. 294–296 Solving Fraction Number Stories, p. 296 Practicing Division, p. 297 Home Link 3-12, p. 297 Enrichment, p. 293 Extra Practice, p. 293
	Rich Tasks and Mathematical Reasoning	Exploring Addition and Subtraction of Fractions and Mixed Numbers, pp. 276–278 Solving More Fraction Addition and Subtraction Number Stories, pp. 278–279 Playing *Division Dash,* p. 279 Journal p. 94: Writing/Reasoning	Math Message, p. 282 Finding Fraction Sums, pp. 283–284 Explaining Place-Value Patterns, p. 285 Home Link 3-10, p. 285 Enrichment, p. 281 Extra Practice, p. 281	Breaking Apart Fractions, pp. 288–289 Introducing *Fraction Capture*, pp. 289–291 Enrichment, p. 287	Identifying Problem-Solving Strategies, pp. 294–296 Solving Fraction Number Stories, p. 296 Practicing Division, p. 297 Journal p. 104: Writing/Reasoning
	Mathematical Discourse	Exploring Addition and Subtraction of Fractions and Mixed Numbers, pp. 276–278	Math Message, p. 282 Summarize, p. 285	Breaking Apart Fractions, pp. 288–289	Identifying Problem-Solving Strategies, pp. 294–296
	Distributed Practice	Mental Math and Fluency, p. 276 Playing *Division Dash,* p. 279 Math Boxes 3-9, p. 279	Mental Math and Fluency, p. 282 Explaining Place-Value Patterns, p. 285	Mental Math and Fluency, p. 288 Renaming Fractions and Mixed Numbers, p. 291 Math Boxes 3-11, p. 291	Mental Math and Fluency, p. 294 Practicing Division, p. 297 Math Boxes 3-12, p. 297
	Differentiation Support	Differentiation Options, p. 275 ELL Support, p. 275 Online Differentiation Support 3-9 Academic Language Development, p. 277 Common Misconception, p. 277	Differentiation Options, p. 281 ELL Support, p. 281 Online Differentiation Support 3-10 Common Misconception, p. 283 Academic Language Development, p. 284	Differentiation Options, p. 287 ELL Support, p. 287 Online Differentiation Support 3-11 Adjusting the Activity, p. 289 Academic Language Development, p. 289	Differentiation Options, p. 293 ELL Support, p. 293 Online Differentiation Support 3-12 Adjusting the Activity, p. 295 Academic Language Development, p. 295

　　　　Red text = Game

Notes

3-13 Fraction-Of Problems, Part 1	**3-14** Fraction-Of Problems, Part 2	**3-15 Assessment** Unit 3 Progress Check
Multiplication of fractions and whole numbers Solving Fraction-Of Problems, pp. 300–302	**Multiplication of fractions and whole numbers** Solving Fraction-Of Problems with Answers Greater Than 1, pp. 306–307 Solving Fraction-Of Problems with Answers Less Than 1, pp. 307–308	Lesson 3-15 is an assessment lesson. It includes: • **Self Assessment** • **Unit Assessment** • **Optional Challenge Assessment** • **Open Response Assessment** • **Suggestions for adjusting the assessments.**
Mental Math and Fluency, p. 300	Mental Math and Fluency, p. 306	
Math Message, p. 300 Using Models to Estimate Volume, p. 303 Home Link 3-13, p. 303	Home Link 3-14, p. 309 Enrichment, p. 305	
Solving Fraction-Of Problems, pp. 300–302 Introducing *Fraction Of,* pp. 302–303 Enrichment, p. 299	Solving Fraction-Of Problems with Answers Greater Than 1, pp. 306–307 Solving Fraction-Of Problems with Answers Less Than 1, pp. 307–308 Enrichment, p. 305	
Introducing *Fraction Of,* pp. 302–303 Extra Practice, p. 299	Extra Practice, p. 305	
Mental Math and Fluency, p. 300 Using Models to Estimate Volume, p. 303 Math Boxes 3-13, p. 303	Mental Math and Fluency, p. 306 Playing *Number Top-It,* p. 309 Math Boxes 3-14, p. 309	
Differentiation Options, p. 299 ELL Support, p. 299 Online Differentiation Support 3-13 Common Misconception, p. 300 Academic Language Development, p. 301 Adjusting the Activity, p. 302	Differentiation Options, p. 305 ELL Support, p. 305 Online Differentiation Support 3-14 Academic Language Development, p. 306 Adjusting the Activity, p. 307	

Go Online:

Evaluation Quick Entry Use this tool to record students' performance on assessment tasks.

Data Use the Data Dashboard to view students' progress reports.

Red text = Game

Unit 3 Materials

Lesson	*Math Masters*	Activity Cards	Manipulative Kit	Other Materials
3-1	pp. 77–80; TA12–TA14 (optional)	28	fraction circles; 6-sided die	slate; Fraction Circles Poster
3-2	pp. 81–82; TA11; per partnership: G11	29	fraction circles; counters; number cards 1–12 (1 of each); per partnership: two 6-sided dice	slate
3-3	pp. per partnership: 83–84; 85–86; per group: TA12 (14 copies of red fraction circle); per partnership: G6	30	fraction circles; counters (2 different colors)	*Prism Pile-Up* cards; calculator (optional); slate; index cards (optional)
3-4	pp. 87–89; TA15	31	fraction circles; number cards 1–20 (1 of each); 6-sided die; per partnership: number cards 0–9 (4 of each)	slate; Fraction Number Lines Poster; large paper (at least 8.5" × 14"); scissors; tape
3-5	pp. 90–92; TA4; TA31 (optional); G14		fraction circles (optional); per partnership: number cards 1–9 (4 of each)	Guidelines for Discussion Poster; colored pencils (optional); selected samples of students' work; students' work from Day 1
3-6	pp. 93; per partnership: 94; 95–96; TA12–TA14 (optional); TA16–TA19 (optional); G15	32	fraction circles	slate; Fraction Number Lines Poster; scissors; per partnership: *Math Journal 1,* Activity Sheets 1–6 (fraction cards); per group: 12 index cards, 25 paper clips, about 2 feet of string
3-7	pp. 97–99; per partnership: G14, G16–G17	33	fraction circles	per partnership: 30 index cards, large paper clip
3-8	pp. 100; per partnership: 101; 102; per partnership: G18	34	fraction circles; base-10 blocks; per partnership: number cards 1–9 (4 of each); stopwatch (optional)	per partnership: scissors; timer (optional)
3-9	pp. 103–104; TA11; TA20; per partnership: G13	35	fraction circles; 6-sided die; per partnership: number cards 1–9 (4 of each)	Fraction Number Lines Poster
3-10	pp. 105–107	36	fraction circles	slate
3-11	pp. 108; TA2 (optional); TA8; TA20 (optional); per partnership: G19–G20	37–38	per partnership: fraction circles, two 6-sided dice, number cards 1–10 (1 of each)	
3-12	pp. 109–110; TA11; TA20 (optional)	39	fraction circles (optional); number cards 1–16 (1 of each); 6-sided die	
3-13	pp. 111–112; TA20 (optional); G24	40	fraction circles; counters	slate; per partnership: scissors, *Math Journal 1,* Activity Sheets 1–2 *(Fraction Of cards)*
3-14	pp. 113–114; per partnership: G7–G9	41	fraction circles (optional); number cards 1–20 (1 of each); 6-sided die; per partnership: number cards 0–9 (4 of each)	slate; scissors; stick-on notes; index cards (optional)
3-15	pp. 115–118; *Assessment Handbook,* pp. 23–30		fraction circles	Fraction Number Lines Poster

Assessment Check-In

These ongoing assessments offer an opportunity to gauge students' performance on one or more of the standards addressed in that lesson.

 Evaluation Quick Entry Record students' performance online.

 Data View reports online to see students' progress towards mastery.

Lesson	Task Description	Content Standards	Processes and Practices
3-1	Use fraction circles or drawings to model and solve fair-share number stories.	5.NF.3	SMP4
3-2	Solve fair-share number stories and use an understanding of fractions as division to write a division number story.	5.NF.3	
3-3	Divide whole numbers, interpret remainders, and express remainders as fractions when appropriate.	5.NBT.6, 5.NF.3	SMP1
3-4	Solve comparison and renaming problems using the Fraction Number Lines Poster.	5.NF.2, 5.NF.3	SMP5
3-5	Interpret fractions as division of the numerator by the denominator to compare fractions and develop a rule for creating the largest possible fraction.	5.NF.3	SMP8
3-6	Write arguments about whether fraction sums and differences make sense by reasoning about the size of the numbers in each problem.	5.NF.2	SMP3
3-7	Use benchmarks to estimate fraction sums and differences.	5.NF.2	
3-8	Rename fractions and mixed numbers by making and breaking apart wholes.	5.NF.3	SMP3
3-9	Make estimates and solve fraction addition and subtraction number stories using fraction circles, drawings, or number lines.	5.NF.2	SMP4, SMP5
3-10	Add fractions with unlike denominators using fraction circle pieces.	5.NF.1	SMP5
3-11	Find fractions that add to a given sum and write expressions to record calculations.	5.OA.2, 5.NF.1	
3-12	Solve a variety of fraction number stories.	5.NF.1, 5.NF.2, 5.NF.3	
3-13	Use manipulatives or drawings to solve fraction-of problems.	5.NF.4, 5.NF.4a	SMP2
3-14	Solve fraction-of problems.	5.NF.4, 5.NF.4a, 5.NF.6	

Virtual Learning Community
vlc.uchicago.edu

While planning your instruction for this unit, visit the *Everyday Mathematics* Virtual Learning Community. You can view videos of lessons in this unit, search for instructional resources shared by teachers, and ask questions of *Everyday Mathematics* authors and other educators. Some of the resources on the VLC related to this unit include:

EM4: Grade 5 Unit 3 Planning Webinar
This webinar provides a preview of the lessons and content in this unit. Watch this video with your grade-level colleagues and plan together under the guidance of an *Everyday Mathematics* author.

ACI Booklet: Grade 5 Unit 3
This booklet is a collection of PDFs of all the student pages in the unit that include concepts leading up to the ACI. Each page shows you where to find the ACI information in the Teacher's Lesson Guide.

For more resources, go to the VLC Resources page and search for Grade 5.

University of Chicago

Spiral Towards Mastery

The *Everyday Mathematics* curriculum is built on the spiral, where standards are introduced, developed, and mastered in multiple exposures across the grade. Go to the Teacher Center at my.mheducation.com to use the Spiral Tracker.

 Spiral Towards Mastery Progress This Spiral Trace outlines instructional trajectories for key standards in Unit 3. For each standard, it highlights opportunities for Focus instruction, Warm Up and Practice activities, as well as formative and summative assessment. It describes the **degree of mastery**—as measured against the entire standard—expected at this point in the year.

Number and Operations—Fractions

5.NF.1

 Progress Towards Mastery By the end of Unit 3, expect students to use tools and visual models to generate equivalent fractions and to add fractions with unlike denominators when only one fraction needs to be replaced with an equivalent fraction.

Full Mastery of 5.NF.1 expected by the end of Unit 7.

5.NF.2

 Progress Towards Mastery By the end of Unit 3, expect students to use tools and visual models to solve number stories involving addition and subtraction of fractions and mixed numbers with like denominators; identify benchmarks close to fractions less than or equal to 2 and use them to make reasonable estimates for fraction sums and differences.

Full Mastery of 5.NF.2 expected by the end of Unit 7.

5.NF.3

 Progress Towards Mastery By the end of Unit 3, expect students to recognize that a fraction $\frac{a}{b}$ is the result of dividing a by b; use tools and visual models to solve whole-number division problems that have fraction or mixed-number answers; use tools and visual models to rename mixed numbers and fractions greater than one.

Full Mastery of 5.NF.3 expected by the end of Unit 6.

Key  = Assessment Check-In    = Progress Check Lesson   = Current Unit   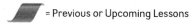 = Previous or Upcoming Lessons

McGraw-Hill Education

5.NF.4

| 1-2 through 1-4 Focus Practice | 1-6 Practice | 1-13 Progress Check | 3-13 Focus Practice | 3-14 Focus Practice | 3-15 Progress Check | 4-1 Practice | 4-6 Warm Up | 4-12 Practice | 4-14 Practice | 4-15 Progress Check |

⭐ **Progress Towards Mastery** By the end of Unit 3, expect students to solve fraction-of problems to build a conceptual foundation for multiplication of fractions by whole numbers.

Full Mastery of 5.NF.4 expected by the end of Unit 8.

5.NF.4a

| 3-13 Focus Practice | 3-14 Focus Practice | 3-15 Progress Check | 4-1 Practice | 4-6 Warm Up | 4-12 Practice | 4-15 Progress Check | 5-5 through 5-8 Warm Up Focus Practice | 5-10 Focus Practice | 5-12 Focus Practice | 5-13 Warm Up Practice |

⭐ **Progress Towards Mastery** By the end of Unit 3, expect students to find a unit fraction of a whole number by partitioning the whole number into the appropriate number of parts and taking one of the parts.

Full Mastery of 5.NF.4a expected by the end of Unit 7.

5.NF.6

| 3-13 Focus Practice | 3-14 Focus Practice | 3-15 Progress Check | 5-5 through 5-7 Focus Practice | 5-9 Focus Practice | 5-10 Focus | 5-12 Focus | 5-15 Progress Check | 6-5 Practice | 6-6 Warm Up Practice | 6-10 Warm Up |

⭐ **Progress Towards Mastery** By the end of Unit 3, expect students to use tools and visual models to solve real-world fraction-of problems with unit fractions and whole numbers.

Full Mastery of 5.NF.6 expected by the end of Unit 8.

Key ✓ = Assessment Check-In    = Progress Check Lesson   = Current Unit   = Previous or Upcoming Lessons

Mathematical Background:
Content

▶ ## Connecting Fractions and Division
(Lessons 3-1 through 3-3, 3-5)

In early grades, students partition shapes into equal parts and use fraction language to name the parts. For example, students might divide a rectangle into 5 equal parts and name 3 of the 5 equal parts as 3 fifths. In Grade 3, students are introduced to conventional fraction notation, learning that a fraction $\frac{a}{b}$ represents a whole divided into b equal parts, with a of those equal parts under consideration. They also place fractions on a number line and learn that fractions can name numbers between whole numbers.

Students in Grade 4 continue to develop their understanding of the numerator as a count, or multiple, of equal shares and the denominator as the number of equal parts in 1 whole. For example, Grade 4 students would interpret $\frac{5}{4}$ as 5 copies of the equal share created by dividing a whole into 4 equal parts, or $5 * \frac{1}{4}$.

In Grade 5, students build on the concept that a fraction represents the result of equal-sharing and connect fractions to whole-number division. Having deepened their understanding of division as an operation at the end of Unit 2, students explore situations in which the result of whole-number division is a fraction in Lessons 3-1 and 3-2. For example: *3 people share 2 pancakes equally. How many pancakes does each person get?* Students use fraction circle pieces and draw pictures to help them solve: *Each person gets $\frac{2}{3}$ pancake.*

Solving a series of fair-share problems leads students to discover patterns relating fractions and division. In the number story above, they notice $2 \div 3 = \frac{2}{3}$. Students generalize this pattern to interpret fractions as division of the numerator by the denominator. In other words, for any fraction $\frac{a}{b}$, $\frac{a}{b} = a \div b$. **5.NF.3** In Lesson 3-3 students apply this understanding to report the remainders in whole-number division problems as fractions. Students further explore the connections between fractions and division in the Open Response task in Lesson 3-5.

▶ ## Estimation and Building Fraction Number Sense
(Lessons 3-4, 3-6 through 3-8)

One way that students strengthen their estimation skills and develop their fraction number sense in *Everyday Mathematics* is by using a consistent set of fraction representations across multiple grades. Beginning in Grade 3, students use *fraction circle pieces* (also called *fraction circles*) to construct mental images of specific fractions and develop a sense of the relative size of fractions. Students also use *fraction strips* (rectangular strips partitioned to show fractions) and number lines to represent fractions. In Grade 5, students make frequent use of fraction circles, number lines, and their own drawings as they develop fraction concepts.

Standards and Goals for Mathematical Content

Because the standards within each strand can be broad, *Everyday Mathematics* has unpacked each standard into Goals for Mathematical Content **GMC**. For a complete list of Standards and Goals, see page EM1.

Note Fraction circle pieces (or fraction circles) are circles divided into equal-size pieces. Each size represents a different unit fraction, and all the same-size pieces are the same color.

NOTE When solving fair-share number stories, students will encounter problems in which there is more than one way to express the quotient. For example, if 4 bags of popcorn are equally shared among 6 people, each person's fair share could be expressed as $\frac{4}{6}$ or $\frac{2}{3}$ bag of popcorn. While some names for fractions are clearer illustrations of the relationship between fractions and division $\left(\text{for example, } 4 \div 6 = \frac{4}{6} \text{ is more transparent than } 4 \div 6 = \frac{2}{3} \right)$, it is essential that students recognize that fractional amounts have many equivalent names. Encourage students to check equivalence by representing each fraction with fraction circle pieces. It is not necessary or important for students to use the simplest form of a fraction.

Unit 3 Vocabulary

argument	divisor	model
benchmark	equivalent fraction	numerator
compare	fraction greater than 1	quotient
conjecture	fraction-of problem	representation
denominator	interval	
dividend	mixed number	

▶ # Estimation and Building Fraction Number Sense *Continued*

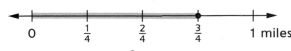

Representing $\frac{3}{4}$ mile on a number line

Number lines are reintroduced in Lesson 3-4, when students reason about how a number line could be used to show distances, such as $\frac{3}{4}$ mile. If 1 mile is represented by the distance between 0 and 1, then students can imagine dividing this space into four equal parts and traveling over three of them to $\frac{3}{4}$. (*See margin.*) This reasoning reinforces the idea that the denominator represents how the whole is divided and the numerator represents the number of equal parts under consideration.

Students also use the number line to make sense of the relationship between fractions greater than 1 and mixed numbers when they recognize that they can divide the interval between any two consecutive whole numbers into equal parts according to the denominator of a fraction. For example, they understand that they can travel 1 whole unit and 1 half unit from 0 to locate $1\frac{1}{2}$ or travel 3 half-units from 0 to locate $\frac{3}{2}$.

In Lessons 3-6 and 3-7, students use number lines and fraction circles as tools for estimation. Students are presented with addition and subtraction scenarios and asked to assess the reasonableness of answers. **5.NF.2** For example, when asked to determine the reasonableness of an answer such as $\frac{1}{2} + \frac{2}{3} = \frac{3}{5}$, students represent $\frac{1}{2}$ and $\frac{2}{3}$ with fraction circle pieces or on a number line. They reason that, when combined, the pieces make up more than 1 whole, so the sum $\frac{3}{5}$ is incorrect.

Lesson 3-7 introduces the use of *benchmarks*, such as $0, \frac{1}{2}, 1, 1\frac{1}{2}, 2,$ and $2\frac{1}{2}$, to estimate sums and differences. For example, to estimate a solution to $\frac{3}{5} - \frac{1}{8}$, students may reason that $\frac{3}{5}$ is close to $\frac{1}{2}$, and $\frac{1}{8}$ is close to 0. Since they know that $\frac{1}{2} - 0 = \frac{1}{2}$, students reason that the difference should be close to $\frac{1}{2}$. Students need ample time to practice estimation skills to develop their fraction number sense. Having students justify their reasoning with visual models also deepens their sense of the relative sizes of fractions.

In Lesson 3-8 students use number lines and fraction circles to rename mixed numbers and fractions greater than 1. They think about groups of fractions that make wholes and connect this thinking to their understanding of fractions as division. For example, a student might find that $\frac{9}{4}$ is the same as $2\frac{1}{4}$ by thinking: "There are two groups of 4 fourths, or 2 wholes, in 9 fourths, with 1 fourth left over." **5.NF.3** Students also rename mixed numbers by breaking apart wholes. For example, they might break one of the wholes in $2\frac{1}{4}$ into fourths to rename the number as $1\frac{5}{4}$. Being able to think flexibly about names for fractions and mixed numbers is an important prerequisite for adding and subtracting mixed numbers. **5.NF.1**

Professional Development

► Exploring Addition and Subtraction with Fractions and Mixed Numbers
(Lessons 3-9 through 3-12)

Students often struggle to compute with fractions, likely because the rules for computing with fractions can seem very different from the rules for whole numbers. Having a solid sense of what fractions mean and how fractions relate to each other helps students make sense of fraction computation procedures and allows them to judge the reasonableness of their answers.

To support the development of fraction number sense, *Fifth Grade Everyday Mathematics* delays the introduction of formal procedures for finding common denominators and for adding and subtracting fractions until later units. In Lessons 3-9 through 3-11, students explore addition and subtraction with fractions and mixed numbers using visual models and their understanding of equivalent names for numbers.

In Lesson 3-9 students develop informal strategies for reasoning through addition and subtraction problems with like denominators. For example, to add $\frac{2}{3}$ and $\frac{2}{3}$, a student could "hop" 2 thirds from 0 on a number line, and then hop 2 thirds more, landing at $\frac{4}{3}$. The student could reason that 3 thirds $\left(\frac{3}{3}\right)$ makes 1 whole, and the sum, 4 thirds, is 1 third past 1, so $\frac{2}{3} + \frac{2}{3} = 1\frac{1}{3}$. (*See margin.*) Similarly, to subtract $1\frac{4}{6}$ from $2\frac{1}{6}$, a student might break apart 1 whole, trading 1 whole fraction circle for 6 sixths. After taking away 1 whole and 4 sixths, the student would find that 3 sixths are left and conclude that $2\frac{1}{6} - 1\frac{4}{6} = \frac{3}{6}$. (*See margin.*)

Students continue to use visual models and informal reasoning to add fractions and mixed numbers with unlike denominators in Lesson 3-10. For example, a student could solve $\frac{1}{4} + \frac{1}{8}$ by reasoning that $\frac{1}{4} = \frac{2}{8}$, so $\frac{1}{4} + \frac{1}{8}$ is the same as $\frac{2}{8} + \frac{1}{8}$, which is $\frac{3}{8}$. **5.NF.4** To support students' use of informal strategies, these lessons limit problems to numbers that can be easily represented with fraction circle pieces and number lines. **5.NF.1**

► Fraction-Of Problems (Lessons 3-13 and 3-14)

In *Everyday Mathematics* a "fraction-of" problem is a problem that requires finding a fractional part of a whole. In some cases, the whole is a set or collection of objects, such as 8 counters. In others, the whole is a fractional amount, such as $\frac{3}{4}$ pan of lasagna. Finding $\frac{1}{2}$ of 8, $\frac{3}{4}$ of 12, and $\frac{2}{3}$ of $\frac{3}{4}$ are all fraction-of problems. Problems in Unit 3 are limited to finding unit fractions of whole numbers in order to deepen students' understanding of what it means to find a fraction of a whole, as well as to develop students' strategies for partitioning various wholes into equal parts. This informal fraction-of reasoning lays a conceptual foundation for fraction multiplication, which is a focus of Unit 5. **5.NF.4**

Using a number line to solve $\frac{2}{3} + \frac{2}{3}$

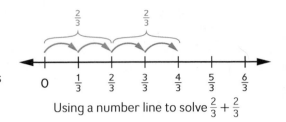

Show $2\frac{1}{6}$.

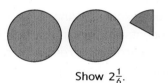

Trade 1 whole for 6 sixths.

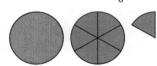

Take away $1\frac{4}{6}$. $\frac{3}{6}$ is left.

Using fraction circle pieces to solve $2\frac{1}{6} - 1\frac{4}{6}$

Mathematical Background:
Process and Practice

 See below for some of the ways that students engage in **SMP5 Use appropriate tools strategically** and **SMP8 Look for and express regularity in repeated reasoning** through **Number and Operations—Fractions** and the other mathematical content of Unit 3.

▶ Standard for Mathematical Process and Practice 5

Proficient problem solvers "consider the available tools" and "make sound decisions about when each of these tools might be helpful." **SMP5** By emphasizing the power of tools and helping students learn how to choose the proper tools, *Everyday Mathematics* makes problem solving at school resemble how mathematics is done in everyday life.

In Unit 3 students use fraction circle pieces and number lines as tools to help them understand and solve problems with fractions. For example, in Lesson 3-4 students visually compare fractions with unlike denominators by using number lines and fraction circle pieces to represent each fraction. In Lesson 3-8 they use fraction circle pieces to rename mixed numbers and fractions greater than 1. For example, to rename $\frac{13}{4}$ as a mixed number, a student might use fraction circle pieces to show 13 one-fourth pieces. Trading groups of 4 fourths for wholes results in 3 wholes with 1 fourth remaining, confirming that $\frac{13}{4} = 3\frac{1}{4}$. In Lessons 3-9 and 3-10, students use fraction circle pieces and number lines to solve addition and subtraction problems before they develop formal computation strategies in Unit 5.

▶ Standard for Mathematical Process and Practice 8

Standard for Mathematical Process and Practice 8 is emphasized each time students are encouraged to use patterns and relationships they notice to make generalizations or create rules.

In Lesson 3-2 students solve fair-share stories and examine the resulting number models. For example, they notice that problems asking them to split 3 cartons of juice among 8 students or share 5 yards of tape among 6 tables yield the number models $3 \div 8 = \frac{3}{8}$ and $5 \div 6 = \frac{5}{6}$, respectively. Looking for patterns helps students arrive at the generalization that a fraction $\left(\frac{a}{b}\right)$ can be thought of as a division problem $(a \div b)$. In Lesson 3-3 students examine the remainders in several division problems. They notice that a whole-number remainder is the same as the numerator of a fractional remainder and that the divisor is the same as the denominator of the fractional remainder. They create a rule about how to report the remainder as a fraction for *any* division problem.

As students develop their ability to "look both for general methods and shortcuts," they learn how rules can be used to help them solve problems more efficiently and see connections among concepts. **SMP8**

Standards and Goals for Mathematical Process and Practice

SMP5 Use appropriate tools strategically.
 GMP5.1 Choose appropriate tools.
 GMP5.2 Use tools effectively and make sense of your results.

SMP8 Look for and express regularity in repeated reasoning.
 GMP8.1 Create and justify rules, shortcuts, and generalizations.

Go Online to the *Implementation Guide* for more information about the Mathematical Process and Practice Standards.

For students' information on the Mathematical Process and Practice Standards, see *Student Reference Book*, pages 1–34.

Connecting Fractions and Division, Part 1

Overview Students solve division number stories that lead to fractional answers.

▶ **Before You Begin**

Fraction circle pieces will be used throughout Unit 3 and should be readily available to students. Decide how you will store and distribute them. You may want to assign a letter to each set and label each piece so that loose fraction pieces can be easily identified. If fraction circle manipulatives are not available, have students cut out fraction circles from Activity Sheets 5–7. Be sure to allow a few minutes for students to explore the pieces, or do the optional Readiness activity as a class. When students need to model multiple wholes, consider having them work in partnerships or create spare sets of fraction pieces from *Math Masters*, pages TA12–TA14. Display the Fraction Circles Poster for reference throughout Unit 3.

▶ **Vocabulary**

model

Standards

Focus Cluster
• Apply and extend previous understandings of multiplication and division.

① Warm Up 5 min

	Materials	
Mental Math and Fluency Students write expressions that model number stories.	slate	5.OA.2

② Focus 35–40 min

Math Message Students use fraction circle pieces to solve a division number story with a fractional answer.	fraction circles	5.NF.3 SMP4
Modeling with Fraction Circle Pieces Students use fraction circle pieces to model and solve division number stories.	*Math Masters*, pp. TA12–TA14 (optional); fraction circles	5.NF.3 SMP3, SMP4
Creating Other Models for Fair Share Stories Students draw pictures to solve division number stories with noncircular wholes.	*Math Journal 1*, p. 71; fraction circles; slate (optional)	5.NF.3 SMP3, SMP4
✓ **Assessment Check-In** See page 223. Expect most students to be able to use tools or drawings to solve division number stories, which can be directly modeled with fraction circle pieces and result in a fractional answer less than 1.	*Math Journal 1*, p. 71	5.NF.3, SMP4

③ Practice 20–30 min

Multiplying and Dividing Students multiply and divide whole numbers.	*Math Journal 1*, p. 72	5.NBT.5, 5.NBT.6 SMP1, SMP2
Math Boxes 3-1 Students practice and maintain skills.	*Math Journal 1*, p. 73	See page 223.
Home Link 3-1 **Homework** Students solve fair share number stories.	*Math Masters*, p. 80	5.NBT.5, 5.NF.3 SMP4

  Go Online ▷ to see how mastery develops for all standards within the grade.

Differentiation Options

Readiness — 5–15 min

WHOLE CLASS · **SMALL GROUP** · PARTNER · INDEPENDENT

Revisiting Fraction Circle Pieces

5.NF.3, SMP2

Math Masters, p. 79; fraction circles

To prepare for using fraction circles to model division, students review the meaning of the pieces. Explain that in this activity the red circle represents 1 whole. Have students create 1 whole circle with pieces of each color. Ask questions like these about each piece: *How many pink pieces are in 1 whole circle?* 2 *What is 1 pink piece worth?* 1 half, or $\frac{1}{2}$ *How could we rename 1 red circle using pink pieces?* 2 halves, or $\frac{2}{2}$ Have students complete *Math Masters*, page 79 to keep as a reference. **GMP2.2**

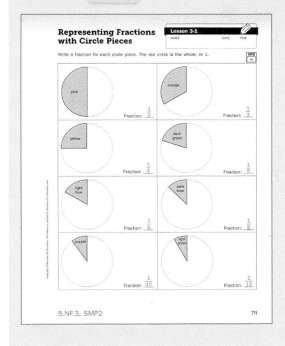

Enrichment — 10–15 min

WHOLE CLASS · SMALL GROUP · **PARTNER** · INDEPENDENT

Looking for Patterns in Fair Share Number Stories

5.NF.3, SMP7

Math Masters, p. 78; fraction circles (optional)

To extend their work with fair share number stories, students solve number stories and look for patterns in the number of objects being shared, the number of fair shares, and the solution. **GMP7.1**

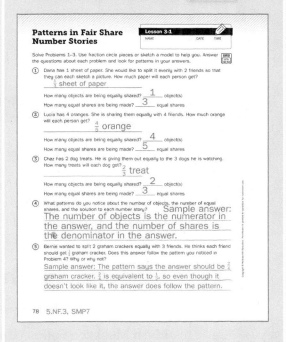

Extra Practice — 10–15 min

WHOLE CLASS · SMALL GROUP · **PARTNER** · INDEPENDENT

Solving More Fair Share Number Stories

5.NF.3, SMP4

Activity Card 28;
Math Masters, p. 77; 6-sided die; fraction circles

For additional practice modeling division problems with fractional answers, students generate and solve fair share number stories. **GMP4.1, GMP4.2**

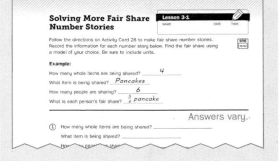

English Language Learner

Beginning ELL To support students' understanding of the term *fair share,* provide collections of items like counters, pencils, or stickers and have students role-play while you think aloud: *I have so many _____. I want to share some of them with you. I want to make fair shares. Then we will all have the same amount.* Model this sharing by saying *One for you, one for me . . .* as you work with the students who are role-playing. Provide additional practice by giving students items and directing them to make sure everyone in the group has a *fair share.* Ask questions like: *How many does everyone have?*

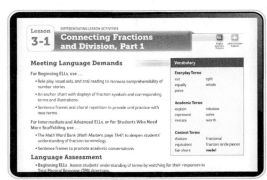

Differentiation Support pages are found in the online Teacher's Center.

Pose several division number stories. Have students work in partnerships to model the stories with fraction circle pieces. **GMP4.1** *Suggestions:*

Problem 1: Darian and Ashley made an apple pie for their family. They cut the pie into 8 equal slices. If 4 people split the pie equally, how much pie will each person get? $\frac{2}{8}$, or $\frac{1}{4}$, pie

Ask partnerships to share their models and solution strategies. **GMP3.2** Expect most students to use the red circle to represent 1 pie. Some will use dark blue pieces to divide the pie into eighths, giving each person 2 eighths, while others will use yellow pieces to divide the pie into fourths and give each person 1 fourth. **GMP4.2** Ask: *Do these two solutions give the same answer? How do you know?* Yes. $\frac{2}{8}$ is equivalent to $\frac{1}{4}$ because 2 dark blue pieces cover 1 yellow piece.

Problem 2: Amisha's family made pancakes for breakfast. There were two pancakes left, and Amisha and her two sisters all wanted more. If the three girls split the two pancakes equally, what portion of a pancake will each girl get? $\frac{2}{3}$ pancake

Ask several partnerships to share how they modeled and solved the problem. **GMP4.1** Expect most students to use the red circle to represent 1 pancake. Some will divide each pancake into thirds and distribute the thirds, finding that each girl gets $\frac{1}{3} + \frac{1}{3} = \frac{2}{3}$ pancake. Others might split each pancake in half, give each person $\frac{1}{2}$, and then divide the last half into 3 equal parts (sixths of a pancake) to distribute to each person, resulting in each person getting $\frac{1}{2} + \frac{1}{6}$ pancake. Have students show how to use their fraction circle pieces to name the amount made up by $\frac{1}{2}$ and $\frac{1}{6}$. If students need help, ask questions like these: *How would we show $\frac{1}{2}$ pancake with sixth pieces?* 3 sixth pieces cover 1 half piece, so $\frac{1}{2} = \frac{3}{6}$. *If we exchange $\frac{1}{2}$ for $\frac{3}{6}$ and then add the $\frac{1}{6}$ piece, what is the total amount?* $\frac{4}{6}$ Have students show that $\frac{4}{6} = \frac{2}{3}$ using their fraction circle pieces.

Ask the class to compare the strategies used to solve Problems 1 and 2. Be sure to discuss the following points:

- One amount can be named in different but equivalent ways. In Problem 1, $\frac{2}{8}$ and $\frac{1}{4}$ are different names for the same amount, so both are acceptable answers.
- Different models can lead to the same answer. In Problem 2, using thirds and using a combination of halves and sixths resulted in equivalent answers, so both were effective models. **GMP4.2**

Math Journal 1, p. 71

Solving Fair Share Number Stories

Lesson 3-1

DATE   TIME

Use fraction circle pieces or a drawing to model each number story. Then solve.

1. Mary and her two friends were working on a science project. They shared 1 pizza equally as a snack. How much pizza did each person get?

   Solution: _____ $\frac{1}{3}$ pizza

   Models: Sample models shown.

2. Jose is taking care of a neighbor's cat. The neighbor will be gone for 5 days and left 3 cans of cat food. The cat is supposed to eat the same amount each day. How much food should Jose give the cat each day?

   Solution: _____ $\frac{3}{5}$ can

3. A school received a shipment of 4 boxes of paper. The school wants to split the paper equally among its 3 printers. How much paper should go to each printer?

   Solution: $\frac{4}{3}$, or $1\frac{1}{3}$, boxes of paper

4. Adrian brought 2 loaves of olive bread to school for a class celebration. There were 12 people who wanted to try the bread. They decided to split the loaves evenly. How much bread did each person receive?

   Solution: _____ $\frac{2}{12}$, or $\frac{1}{6}$, loaf

5.NF.3, SMP4   71

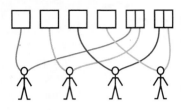

Sample model for Problem 3

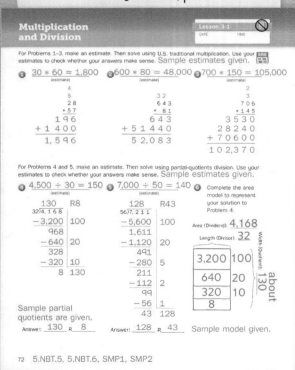

Math Journal 1, p. 72

▶ Creating Other Models for Fair Share Stories

Math Journal 1, p. 71

| WHOLE CLASS | SMALL GROUP | PARTNER | INDEPENDENT |

Explain that the number stories students just solved are called fair share stories because they involve sharing something equally, or fairly. Although fraction circle pieces are useful models, they are not the only models that can be used to represent fair shares. For example, drawing a picture could be a clearer way to represent whole items that are not circular. Pose other fair share problems. Have students create a model for each problem and discuss their models and solutions. **GMP3.2, GMP4.1, GMP4.2** If students choose to draw models, consider having them draw on slates for ease of sharing.

Problem 3: Leila brought 6 graham cracker squares for a snack. She wants to share them with 3 friends. If they share the squares equally, how much will each person get? $\frac{6}{4}$, $1\frac{2}{4}$, or $1\frac{1}{2}$ graham cracker squares *Sample strategies:*

- Draw 6 squares to represent the 6 graham crackers. Break each square into fourths. Distribute $\frac{1}{4}$ from each square to each person. Each person gets $\frac{1}{4} + \frac{1}{4} + \frac{1}{4} + \frac{1}{4} + \frac{1}{4} + \frac{1}{4}$, or $6 * \frac{1}{4}$, which equals $\frac{6}{4}$ crackers.
- Draw 6 squares to represent the 6 graham crackers. Distribute 1 whole square to each person. Break the remaining 2 squares into fourths. Distribute $\frac{1}{4}$ from each square to each person. Each person gets $1 + \frac{1}{4} + \frac{1}{4}$, or $1\frac{2}{4}$, crackers.
- Draw 6 squares to represent the 6 graham crackers. Distribute 1 whole square to each person. Break the remaining 2 squares in half. Give $\frac{1}{2}$ to each person. Each person gets $1 + \frac{1}{2}$, or $1\frac{1}{2}$, crackers. (*See margin.*)

Problem 4: Jamie went on a field trip with his class. His group had 4 large sub sandwiches to share for lunch. There were 6 people in his group. What part of a whole sandwich did each person get? $\frac{4}{6}$, or $\frac{2}{3}$, sub *Sample strategies:*

- Draw 4 rectangles to represent the 4 subs. Divide each sub into sixths. Distribute the pieces to each person. Each person gets $\frac{1}{6} + \frac{1}{6} + \frac{1}{6} + \frac{1}{6}$, or $4 * \frac{1}{6}$, which equals $\frac{4}{6}$ sub.
- Draw 4 rectangles to represent the 4 subs. Cut each sub in half. Distribute $\frac{1}{2}$ to each person. Cut the remaining 2 halves to create sixths. Distribute $\frac{1}{6}$ to each person. Each person gets $\frac{1}{2} + \frac{1}{6}$, or $\frac{4}{6}$, sub.

Point out that there are multiple ways to model each problem. Encourage students to use the models that make the most sense to them as they solve the number stories on journal page 71. **GMP4.1, GMP4.2** Circulate and assist.

Assessment Check-In 5.NF.3

Math Journal 1, p. 71

Expect most students to be able to use tools or drawings to solve Problems 1 and 2 on journal page 71, which can be directly modeled with fraction circle pieces and result in a fractional answer less than 1.

GMP4.1, GMP4.2   Some may be able to solve Problems 3 and 4. If students struggle making sense of the number stories, suggest that they first represent the starting amount and then ask: *Into how many equal parts are we splitting this amount?* Guide them to divide each whole into the given number of shares and then add the unit fractions in each share.

Evaluation Quick Entry Go online to record students' progress and to see trajectories toward mastery for these standards.

Summarize Have partnerships compare the models they used to solve the number stories on journal page 71.   GMP3.2, GMP4.2

③ Practice   20–30 min

▶ Multiplying and Dividing

Math Journal 1, p. 72

| WHOLE CLASS | SMALL GROUP | **PARTNER** | INDEPENDENT |

Students multiply whole numbers using U.S. traditional multiplication and divide whole numbers using partial-quotients division. They use estimates to check whether their answers make sense.   GMP1.4   They draw an area model to represent their solution to one division problem.   GMP2.1

▶ Math Boxes 3-1

Math Journal 1, p. 73

| WHOLE CLASS | **SMALL GROUP** | PARTNER | INDEPENDENT |

Mixed Practice Math Boxes 3-1 are paired with Math Boxes 3-3.

▶ Home Link 3-1

Math Masters, p. 80

Homework Students practice modeling and solving fair share number stories.   GMP4.1, GMP4.2

Math Journal 1, p. 73

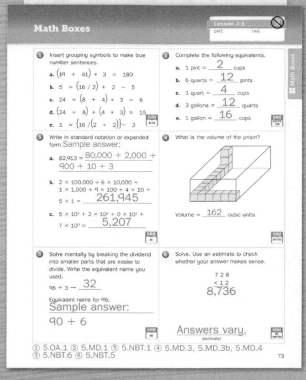

Math Masters, p. 80

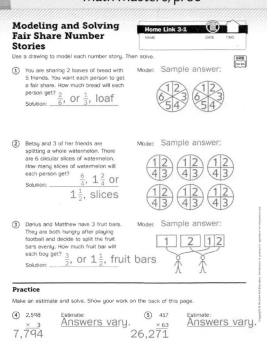

Connecting Fractions and Division, Part 2

Overview Students solve division number stories and write number models to build an understanding of fractions as division.

▶ **Before You Begin**

For Part 2, consider having students work in partnerships so they can combine their fraction circles as needed. Alternatively, make copies of *Math Masters,* pages TA12–TA14 so students have enough pieces to model multiple wholes.

▶ **Vocabulary**

dividend • numerator • quotient • divisor • denominator • mixed number

	Materials	Standards

Standards

Focus Cluster
• Apply and extend previous understandings of multiplication and division.

① **Warm Up** 5 min

	Materials	
Mental Math and Fluency Students solve extended multiplication facts and write related division problems.	slate	**5.NBT.2, 5.NBT.6**

② **Focus** 35–40 min

Math Message Students solve and model a fair share number story.	fraction circles	**5.NF.3** SMP4
Writing Division Models for Fair Share Stories Students solve fair share number stories, write number models, and discuss patterns that they notice.	fraction circles	**5.NF.3** SMP4, SMP7, SMP8
Interpreting Fractions as Division Students write division number stories by working backward from a fractional answer.	*Math Journal 1,* pp. 74–75; fraction circles	**5.NF.3** SMP4, SMP7, SMP8
✓ **Assessment Check-In** See page 230. Expect most students to be able to find correct solutions and write matching division number models with answers less than 1 when solving division number stories.	*Math Journal 1,* p. 75	**5.NF.3**

③ **Practice** 20–30 min

Playing *Power Up* **Game** Students practice converting from exponential notation to standard notation.	*Student Reference Book,* p. 318; per partnership: *Math Masters,* p. G11; two 6-sided dice	**5.NBT.2** SMP1, SMP8
Math Boxes 3-2 Students practice and maintain skills.	*Math Journal 1,* p. 76	See page 231.
Home Link 3-2 **Homework** Students solve fair share number stories and write number models summarizing their solutions.	*Math Masters,* p. 82	**5.NBT.6, 5.NF.3** SMP4

  ⟨Go Online⟩ to see how mastery develops for all standards within the grade.

Differentiation Options

Readiness 10–15 min	**Enrichment** 5–15 min	**Extra Practice** 10–15 min
WHOLE CLASS · SMALL GROUP · PARTNER · INDEPENDENT	WHOLE CLASS · SMALL GROUP · PARTNER · INDEPENDENT	WHOLE CLASS · SMALL GROUP · PARTNER · INDEPENDENT

Reviewing the Meaning of Operations

5.NF.3, SMP4

counters

To prepare for writing number models to solve fair share number stories, students model and discuss contexts in which different operations apply. **GMP4.1** Have students take 3 counters in one hand and 2 counters in the other and then put them together. Ask: *When in real life might we add 3 and 2?* Sample answer: I checked out 3 books from the library and then checked out 2 more. Emphasize the meaning of addition as putting together. Model and brainstorm contexts for subtraction, multiplication, and division. Use easy numbers to keep the focus on the operation. *Suggestions:*

- *Take away 3 counters from 10.* Sample context: There were 10 carrots, and then 3 were eaten. Emphasize the meaning of subtraction as taking away or finding the difference.
- *Make 5 groups of 4 counters.* Sample context: There are 5 tables with 4 students at each. Emphasize the meaning of multiplication as repeating equal groups.
- *Distribute 18 counters into groups of 3.* Sample context: 18 students work in groups of 3. Emphasize the meaning of division as splitting into equal groups.

Consider displaying operation symbols for reference, including reminders about the meaning of each operation.

Exploring Relationships in Number Stories

5.NF.3, SMP3, SMP7

Math Masters, p. 81

To explore how number relationships can be used to solve number stories, students use the Data Bank on *Math Masters,* page 81 to help them determine missing numbers in a number story. They explain how they found their answers. Then they decide whether there is more than one way to assign numbers in the problem and defend their reasoning. **GMP3.1, GMP7.2**

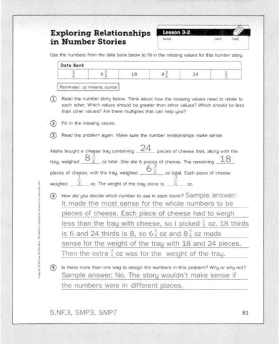

Writing Number Stories from Fractional Answers

5.NF.3, SMP7

Activity Card 29;
Math Journal 1, p. 75;
Math Masters, p. TA11;
number cards 1–12 (1 of each)

For additional practice interpreting fractions as division, students form fractions by drawing number cards. They write division number stories using the generated fractions as the answers. **GMP7.2**

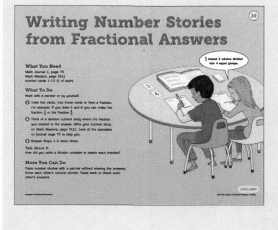

English Language Learner

Beginning ELL Introduce the terms *numerator* and *denominator* by preparing an anchor chart titled *Fractions* that shows several examples of sets of objects divided into equal subsets. Also include corresponding fraction numerical representations with the terms *numerator* and *denominator* clearly labeled. Have students repeat the terms as you point to them. Follow with show-me prompts, asking students to point to the numerator or denominator. For students who can respond in English, ask: *Where is the numerator? The denominator?*

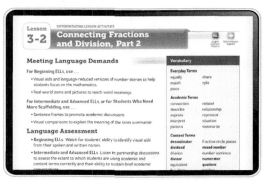

Differentiation Support pages are found in the online Teacher's Center.

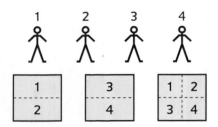

Each friend gets $\frac{1}{2}$ of one sandwich and $\frac{1}{4}$ of one sandwich. So each friend gets $\frac{1}{2} + \frac{1}{4}$, or $\frac{3}{4}$, of a whole sandwich.

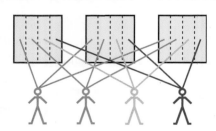

Each sandwich is cut into fourths because there are 4 friends. Each friend is given $\frac{1}{4}$ of each sandwich, or $\frac{1}{4} + \frac{1}{4} + \frac{1}{4} = \frac{3}{4}$ in all.

① Warm Up   5 min

▶ Mental Math and Fluency

Display multiplication problems for students to solve. Then have them write two related division problems on slates. *Leveled exercises:*

●○○ **3 * 50** 150; 150 / 3 = 50; 150 / 50 = 3

●●○ **40 * 20** 800; 800 / 20 = 40; 800 / 40 = 20

●●● **3,000 * 25** 75,000; 75,000 / 3,000 = 25; 75,000 / 25 = 3,000

② Focus   35–40 min

▶ Math Message

Julia came home from school to find that her mom had made 3 sandwiches. She had 3 friends with her. The 4 of them decided to share the 3 sandwiches equally. How much did each friend get? Use your fraction circle pieces to help you. After you solve the problem, write a number sentence that models the problem. **GMP4.1** $\frac{3}{4}$ sandwich; $3 \div 4 = \frac{3}{4}$

▶ Writing Division Models for Fair Share Stories

| WHOLE CLASS | SMALL GROUP | PARTNER | INDEPENDENT |

Math Message Follow-Up Invite several volunteers to share their strategies for the Math Message. (*See margin for sample strategies.*) Be sure to discuss several alternatives. Point out that the object being shared in this problem is not circular. Ask: *Could we model this problem with fraction circle pieces? How?* Sample answer: Yes. I could use 3 red circles to represent the 3 sandwiches. I can divide each circle into 4 yellow pieces, or fourths, and give 1 fourth from each circle to each friend. Each friend would get three $\frac{1}{4}$ pieces, or $\frac{3}{4}$ in all.

Ask students to share their number models. Some may have written a number model representing the fractional parts of a sandwich each friend received (for example, each friend received $\frac{1}{4} + \frac{1}{4} + \frac{1}{4} = \frac{3}{4}$ sandwich). Accept such a number model as valid, but point out that it represents the action of *totaling* the pieces each friend received. Ask: *How did we get those pieces in the first place?* We split up the sandwiches.

What operation matches splitting the sandwiches? Why? Division, because we are dividing the sandwiches into equal parts. *What division number model matches this problem?* $3 \div 4 = \frac{3}{4}$ Point to the corresponding parts of the number model as you say: *Three sandwiches divided by 4 friends means that each friend gets $\frac{3}{4}$ sandwich.*

Emphasize the idea that writing a number model is one way to clearly describe the relationships between numbers in a real-world situation. Tell students that today they will continue to explore fair share number stories, representing each story by writing number models. Encourage students to look for patterns as they model and solve each problem. GMP7.1

Pose fair share problems, making sure to vary the contexts. Allow students to use fraction circle pieces, but encourage them to draw pictures for some of the problems. Have them write number models for each problem, and guide them to relate the number model back to the context of the problem. GMP4.1 *Suggestions:*

- Angela's family is having lasagna for dinner. There are 2 small pans of lasagna and 5 people in her family. If each person gets an equal portion, how much lasagna will each person get? $\frac{2}{5}$ pan of lasagna; Number model: $2 \div 5 = \frac{2}{5}$; This means that 2 pans divided by 5 people is $\frac{2}{5}$ pan for each person.
- Coach Dean brought 3 cartons of orange juice to split among the 8 students on the basketball team. If Coach Dean pours the same amount for each student, what portion of a carton of orange juice will each student receive? $\frac{3}{8}$ carton of orange juice; Number model: $3 \div 8 = \frac{3}{8}$; This means that 3 cartons divided by 8 students is $\frac{3}{8}$ carton for each student.
- Ms. Martinez has to distribute yarn equally to 6 table groups for an art project. She has 5 yards of yarn. How many yards of yarn will each table group get? $\frac{5}{6}$ yard of yarn; Number model $5 \div 6 = \frac{5}{6}$; This means that 5 yards divided by 6 groups equals $\frac{5}{6}$ yard for each group.

After discussing several problems, display the division number models and ask: *What patterns do you notice in these number models?* GMP7.1 Highlight the following points:

- The **dividend** is the same as the **numerator** of the **quotient.**
- The **divisor** is the same as the **denominator** of the quotient.

Ask: *What does this tell you about how fractions and division are related?* GMP8.1 Sample answer: A fraction can be thought of as a division problem. For example, the fraction $\frac{2}{5}$ can be thought of as the division problem $2 \div 5$, and the fraction $\frac{10}{6}$ can be thought of as the division problem $10 \div 6$. Explain to students that by creating their own number stories they will be practicing interpreting fractions as division.

Common Misconception

Differentiate Watch for students who confuse the order of numbers in division number models. For example, a student might write $4 \div 3$ for the Math Message problem, thinking, "4 people are dividing 3 sandwiches." If you see this mistake, display a number model for a fair share story with the dividend, divisor, and quotient labeled. Emphasize the meaning of each number:

- The *dividend* comes first. It is the amount that is being divided.
- The *divisor* comes after the division symbol. It tells how the dividend is being split up.
- The *quotient* is the result of division. It is one fair share.

Help students write number models for specific problems by asking: *What is being split up? How is the start amount being divided? What is one fair share?*

  Differentiation Support

Common Misconception

Differentiate Look for students who confuse the terms *dividend* and *divisor* or *numerator* and *denominator.* Consider creating a poster or other visual reference of a worked division problem with the dividend and divisor labeled, alongside the corresponding fraction with labels for the numerator and denominator.

  Differentiation Support

Writing Division Number Stories

Lesson 3-2

DATE   TIME

Record the fractions you are assigned. For each fraction, write a division number sentence with the fraction as the quotient. Then write a number story to match the number sentence.

Sample answers given.

① Fraction: $\frac{7}{8}$

Division number sentence: $7 \div 8 = \frac{7}{8}$

Number story: There were 7 granola bars in a box. If 8 people split the granola bars equally, how much of a granola bar will each person get?

② Fraction: $\frac{5}{4}$

Division number sentence: $5 \div 4 = \frac{5}{4}$

Number story: Heather has 5 yards of fabric. If she divides it into 4 equal pieces, how many yards will each piece be?

③ Fraction: $\frac{5}{6}$

Division number sentence: $5 \div 6 = \frac{5}{6}$

Number story: Coach Ray brought 5 liters of juice to practice. If he pours the same amount of juice into 6 glasses, how many liters of juice will there be in each glass?

74   5.NF.3

▶ Interpreting Fractions as Division

Math Journal 1, pp. 74–75

| WHOLE CLASS | SMALL GROUP | PARTNER | INDEPENDENT |

Display the fraction $\frac{2}{3}$. Ask: *What division problem would give an answer of $\frac{2}{3}$?* $2 \div 3 = \frac{2}{3}$ *What is a number story that could be represented by this fraction?* Sample answer: Three people share 2 oranges equally. How much orange does each person get? Each person gets $\frac{2}{3}$ orange. Reinforce the connection between the division problem and the fraction by relating the numerator and denominator to the problem context.

Divide students into small groups or partnerships. Assign three fractions to each group, using a mixture of fractions both less than 1 and greater than 1, such as $\frac{7}{8}$, $\frac{5}{4}$, $\frac{5}{6}$, $\frac{8}{6}$, and $\frac{5}{8}$. Have groups record their fractions on journal page 74 and write division number sentences with the fractions as the quotients. Then have students write number stories to match their number sentences. When most groups have completed at least two number stories, invite groups or partnerships to share their favorites.

Ask students to look at their number stories and think about whether they could express the answers to any of their stories in a different way. Remind them that the fair share number stories they solved in Lesson 3-1 resulted in answers that could be written in more than one way. For example, given a problem in which 4 sandwiches were split 6 ways, they found that each student's fair share was $\frac{4}{6}$, or $\frac{2}{3}$, sandwich. Ask: *What number models could summarize this situation?* $4 \div 6 = \frac{4}{6}$, or $4 \div 6 = \frac{2}{3}$ Display both number models. Ask: *Which number model would we expect, given the pattern we have discussed today?* $4 \div 6 = \frac{4}{6}$ Point out that the reason $\frac{2}{3}$ does not seem to follow the pattern is that the numerator and denominator are not the same as the dividend and the divisor.

Have students use their fraction circle pieces to show both $\frac{4}{6}$ and $\frac{2}{3}$. Ask: *What do you notice?* $\frac{4}{6}$ and $\frac{2}{3}$ show the same amount; $\frac{4}{6} = \frac{2}{3}$ Tell students that fractions, like whole numbers, have many equivalent names. Explain that whenever a correct solution to a division problem results in a fraction that *does not* seem to follow the fractions-as-division pattern, there is an equivalent fraction that *does* follow the pattern.

Have students consider another fair share problem in a context where the answer is greater than 1. For example, 6 pieces of posterboard split equally among 4 people results in $1\frac{1}{2}$ pieces of posterboard for each person. Ask: *What number model summarizes this situation?* $6 \div 4 = 1\frac{1}{2}$ *What number model would follow the pattern we discussed?* $6 \div 4 = \frac{6}{4}$ Ask: *Which answer is correct: $1\frac{1}{2}$ or $\frac{6}{4}$? Use your fraction circle pieces to check.* Both answers are correct. The fraction circle pieces show that $1\frac{1}{2} = 1\frac{2}{4} = \frac{6}{4}$. Remind students that a number written using both a whole number and a fraction is called a **mixed number.** Make sure students see that the mixed number and the fraction are equivalent.

Pose additional fair share stories resulting in fractions greater than 1. Have students write a division number model to predict each answer using the pattern of fractions as division. Then have them solve the problem using fraction circle pieces or drawings. GMP4.1 Discuss solutions to each problem, checking that predicted answers are equivalent to the answers found using drawings or fraction circle pieces. *Suggestions:*

- Sam and Kim are having waffles for breakfast. There are 5 waffles, and they decide to share them equally. How much does each person get? Prediction: $5 \div 2 = \frac{5}{2}$. Answer from models: $\frac{5}{2}$, or $2\frac{1}{2}$, waffles. The prediction is correct because $\frac{5}{2} = 2\frac{1}{2}$.

- Ms. Lee has a 10-pound package of clay for her pottery club. If she distributes the clay equally among the 6 students in her club, how many pounds of clay will each student get? Prediction: $10 \div 6 = \frac{10}{6}$. Answer from models: $1\frac{4}{6}$, or $1\frac{2}{3}$, pounds of clay. The prediction is correct because $\frac{10}{6} = 1\frac{4}{6} = 1\frac{2}{3}$.

When students seem comfortable interpreting fractions as division, have them complete journal page 75 independently or in partnerships. Encourage them to model the problems with fraction circle pieces or drawings. Have them label the fractional parts in their drawings and record number models. GMP4.1 Circulate and assist as needed.

More Practice with Fair Shares

Lesson 3-2

DATE   TIME

Solve each number story. You can use fraction circle pieces or drawings to help. Write a number model to show how you solved each problem.

① Davita brought 6 granola bars for herself and the 7 other girls in her camp group for a snack. If they share them equally, what fraction of a granola bar will each girl get?

Solution: $\frac{6}{8}$, or $\frac{3}{4}$ granola bar

Number model: $6 \div 8 = \frac{6}{8}$, or $6 \div 8 = \frac{3}{4}$

② Lucas is making 12 jumbo muffins to sell at his class bake sale. He has 2 bowls full of batter. What fraction of a bowl of batter should Lucas put in each muffin cup?

Solution: $\frac{2}{12}$, or $\frac{1}{6}$ bowl

Number model: $2 \div 12 = \frac{2}{12}$, or $2 \div 12 = \frac{1}{6}$

③ Ms. Cox is combining bottles of hand sanitizer. She has 11 small bottles of sanitizer she wants to divide equally among 3 large containers. How many small bottles should she empty into each large container?

Solution: $\frac{11}{3}$, or $3\frac{2}{3}$ small bottles

Number model: $11 \div 3 = \frac{11}{3}$, or $11 \div 3 = 3\frac{2}{3}$

④ Write a division number story with an answer of $\frac{12}{8}$.

Sample answer: Michelle brought 12 apples to share at her dance practice. There were 7 other girls besides Michelle. The girls decided to split the apples equally. How many apples will each girl get?

Number model: $12 \div 8 = \frac{12}{8}$, or $12 \div 8 = 1\frac{1}{2}$

5.NF.3, SMP4   75

Assessment Check-In   5.NF.3

Math Journal 1, p. 75

Expect most students to find correct solutions and write matching division number models with answers less than 1 for Problems 1 and 2 on journal page 75. Some students may be able to solve Problem 3. Most should be able to use an understanding of fractions as division to write a division number story in Problem 4. To support students who struggle recognizing the connection between fractions and division, provide a series of problems to solve by dividing one whole by divisors they can model with fraction circles (2, 3, 4, 5, 6, 8, 10, or 12). Guide students to make statements like this one: "1 whole divided into 3 parts equals $\frac{1}{3}$." Have them record number models to summarize their solutions and discuss the connections they see between the division problem and the fractional answer.

Evaluation Quick Entry Go online to record students' progress and to see trajectories toward mastery for these standards.

Summarize Have students compare the number models they wrote on journal page 75. Make sure they have interpreted each problem as division. Ask: *Do all of your number models follow the pattern we noticed earlier, where the dividend is the same as the numerator of the answer and the divisor is the same as the denominator of the answer? How do you know?* GMP7.1, GMP8.1 Students should recognize equivalent fractions in Problems 1 and 2 and be able to explain how each number model fits the pattern. Some students may be able to explain how $3\frac{2}{3}$ is equivalent to $\frac{11}{3}$ in Problem 3. Emphasize that although answers to division problems may be written with equivalent names that look different from the dividend and divisor, any fraction can be interpreted as division of the numerator by the denominator.

3 Practice 20–30 min

▶ Playing *Power Up*

Student Reference Book, p. 318; *Math Masters*, p. G11

| WHOLE CLASS | SMALL GROUP | **PARTNER** | INDEPENDENT |

Students practice converting from exponential to standard notation.

Observe

- Which students are able to accurately convert exponential notation into standard notation?
- Which students have developed a strategy for placing the numbers rolled to create the largest possible number?

Discuss

- *How did you decide where to place the numbers you rolled?* **GMP1.2**
- *Is there a rule we could use to create the largest number possible?* **GMP8.1**

| **Differentiate** | **Game Modifications** | **Go Online** | Differentiation Support |

▶ Math Boxes 3-2

Math Journal 1, p. 76

| WHOLE CLASS | SMALL GROUP | PARTNER | INDEPENDENT |

Mixed Practice Math Boxes 3-2 are paired with Math Boxes 3-4.

▶ Home Link 3-2

Math Masters, p. 82

Homework Students solve fair share number stories and write number models summarizing their solutions. **GMP4.1**

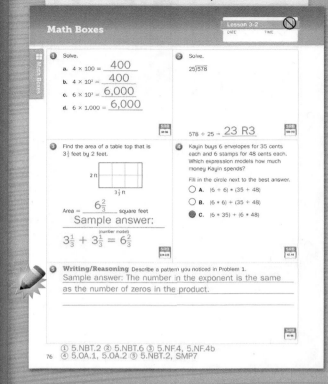

Math Journal 1, p. 76

Math Boxes

Lesson 3-2

① Solve.
a. $4 \times 100 = 400$
b. $4 \times 10^2 = 400$
c. $6 \times 10^3 = 6{,}000$
d. $6 \times 1{,}000 = 6{,}000$

② Solve.
$25\overline{)578}$

$578 \div 25 \to 23 \text{ R3}$

③ Find the area of a table top that is $3\frac{1}{3}$ feet by 2 feet.

Area = $6\frac{2}{3}$ square feet
Sample answer:
(number model)
$3\frac{1}{3} + 3\frac{1}{3} = 6\frac{2}{3}$

④ Kayin buys 6 envelopes for 35 cents each and 6 stamps for 48 cents each. Which expression models how much money Kayin spends?

Fill in the circle next to the best answer.
○ A. $(6 + 6) * (35 + 48)$
○ B. $(6 * 6) + (35 + 48)$
● C. $(6 * 35) + (6 * 48)$

⑤ **Writing/Reasoning** Describe a pattern you noticed in Problem 1.
Sample answer: The number in the exponent is the same as the number of zeros in the product.

① 5.NBT.2 ② 5.NBT.6 ③ 5.NF.4, 5.NF.4b
④ 5.OA.1, 5.OA.2 ⑤ 5.NBT.2, SMP7
76

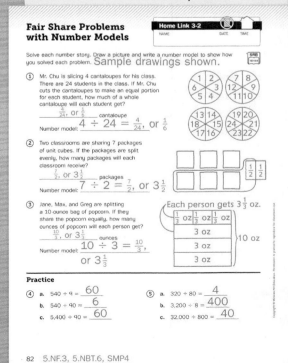

Math Masters, p. 82

Fair Share Problems with Number Models

Home Link 3-2

Solve each number story. Draw a picture and write a number model to show how you solved each problem. Sample drawings shown.

① Mr. Chu is slicing 4 cantaloupes for his class. There are 24 students in the class. If Mr. Chu cuts the cantaloupes to make an equal portion for each student, how much of a whole cantaloupe will each student get?
$\frac{4}{24}$, or $\frac{1}{6}$ cantaloupe
Number model: $4 \div 24 = \frac{4}{24}$, or $\frac{1}{6}$

② Two classrooms are sharing 7 packages of unit cubes. If the packages are split evenly, how many packages will each classroom receive?
$\frac{7}{2}$, or $3\frac{1}{2}$ packages
Number model: $7 \div 2 = \frac{7}{2}$, or $3\frac{1}{2}$

③ Jane, Max, and Greg are splitting a 10-ounce bag of popcorn. If they share the popcorn equally, how many ounces of popcorn will each person get?
$\frac{10}{3}$, or $3\frac{1}{3}$ ounces
Number model: $10 \div 3 = \frac{10}{3}$, or $3\frac{1}{3}$

Each person gets $3\frac{1}{3}$ oz.
$\frac{1}{3}$ oz $\frac{1}{3}$ oz $\frac{1}{3}$ oz
3 oz
3 oz ⎫ 10 oz
3 oz

Practice

④ a. $540 \div 9 = 60$
b. $540 \div 90 = 6$
c. $5{,}400 \div 90 = 60$

⑤ a. $320 \div 80 = 4$
b. $3{,}200 \div 8 = 400$
c. $32{,}000 \div 800 = 40$

82 5.NF.3, 5.NBT.6, SMP4

Application: Interpreting Remainders

Overview Students apply their understanding of fractions as division to report remainders as fractions.

▶ **Before You Begin**
For Part 3, if additional sets of *Prism Pile-Up* cards are needed, copy *Math Masters*, pages G4 and G5. For the optional Readiness activity, make multiple copies of *Math Masters*, page TA12 and cut out whole circles. To work with one small group at a time, you will need 14 whole circles. For the optional Extra Practice activity, students will need Division Story Cards from *Math Masters*, page 83. Cut out the cards ahead of time or plan to have students cut them out before beginning the activity.

Standards

Focus Clusters
- Perform operations with multi-digit whole numbers and with decimals to hundredths.
- Apply and extend previous understandings of multiplication and division.

 Warm Up 5 min

Materials

Mental Math and Fluency Students solve extended division facts.	5.NBT.6

 **Focus** 35–40 min

Math Message Students consider whether they agree with an answer to a division number story.		5.NBT.6
Reporting Remainders as Fractions Students explore when it makes sense to report remainders as fractions and practice finding fractional answers.		5.NBT.6, 5.NF.3 SMP1, SMP6, SMP8
Interpreting Remainders in Division Number Stories Students solve division number stories and decide how to interpret remainders.	*Math Journal 1*, pp. 77–78	5.NBT.6, 5.NF.3 SMP1, SMP6
✓ **Assessment Check-In** See page 236. Expect most students to be able to solve the division number story indicated and interpret the remainder as a fraction.	*Math Journal 1*, pp. 77–78	5.NBT.6, 5.NF.3, SMP1

 Practice 20–30 min

Playing *Prism Pile-Up* **Game** Students practice finding volumes of rectangular prisms.	*Student Reference Book*, p. 319; per partnership: *Math Masters*, p. G6; *Prism Pile-Up* cards; calculator (optional)	5.OA.2, 5.MD.3, 5.MD.3a, 5.MD.3b, 5.MD.4, 5.MD.5, 5.MD.5a, 5.MD.5b, 5.MD.5c SMP1, SMP6
Math Boxes 3-3 Students practice and maintain skills.	*Math Journal 1*, p. 79	See page 237.
Home Link 3-3 **Homework** Students practice solving division number stories and interpreting remainders.	*Math Masters*, p. 86	5.OA.1, 5.NBT.6, 5.NF.3

/// **Go Online** to see how mastery develops for all standards within the grade.

Differentiation Options

Readiness	5–15 min
WHOLE CLASS	**SMALL GROUP**	PARTNER	INDEPENDENT

Enrichment	5–15 min
WHOLE CLASS	SMALL GROUP	**PARTNER**	INDEPENDENT

Extra Practice	10–20 min
WHOLE CLASS	SMALL GROUP	**PARTNER**	INDEPENDENT

Modeling Fractional Remainders

5.NBT.6, 5.NF.3, SMP2

fraction circles; slate;
per group: 14 copies of the red fraction circle from *Math Masters*, p. TA12
(*See Before You Begin.*)

To make sense of fractional remainders, students model division problems using fraction circle pieces. Start by modeling the division problem 13 ÷ 4. Begin distributing 13 whole fraction circles one at a time into 4 equal groups. After 12 circles have been distributed, hold up the last circle and explain that this remainder is 1 whole. Discuss how you could divide the remaining whole into fourths. Exchange the whole for 4 yellow fourths. Distribute 1 yellow fourth piece to each group. Ask: *How many circles are in each equal group?* 3 whole circles and 1 fourth piece, or $3\frac{1}{4}$ circles Summarize how the remainder of 1 was further divided and reported as $\frac{1}{4}$ in the answer. Have students model other division problems using whole circles and additional fraction circle pieces. Pose problems with small numbers so students can distribute all of the whole and fractional pieces into groups. *Suggestions:*
10 ÷ 3 $3\frac{1}{3}$ 14 ÷ 6 $2\frac{2}{6}$ 12 ÷ 8 $1\frac{4}{8}$
Have students write the corresponding number models on slates. **GMP2.2, GMP2.3**

Sharing a Cost

5.NBT.6, 5.NF.3, SMP6

Math Masters, p. 85;
fraction circles (optional)

To explore the connection between fractional remainders and decimals, students solve a number story in which the cost of an item is split four ways. Students think about how dollars are divided in everyday life and develop a strategy to report the answer as dollars and cents.
GMP6.2

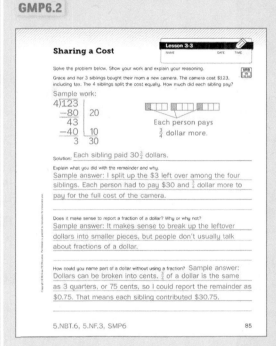

Remainder Tic-Tac-Toe

5.NBT.6, 5.NF.3, SMP6

Activity Card 30;
per partnership: *Math Masters*, pp. 83–84; counters (2 different colors); index cards (optional)

For additional practice recognizing situations that call for fractional remainders, students draw cards showing division number stories (*Math Masters*, page 83). They decide whether to ignore the remainder Cards 2 (14 R0), 4 (8 R4), 8 (5 R1), 11 (7 R2), 12 (2 R2), report the remainder as a fraction Cards 7 (6 R4), 8 (5 R1), 9 (16 R2), 10 (1 R12), 12 (2 R2), or round up Cards 1 (1 R15), 3 (6 R3), 5 (4 R2), 6 (3 R 23). **GMP6.2** They explain their decisions to a partner and record them with counters on a tic-tac-toe board (*Math Masters*, page 84). **GMP6.1**

English Language Learner

Beginning ELL Build on students' understanding of the word *remainder* as what remains so that they make a connection between the terms *remain* and *remainder.* Display a number of objects that does not divide evenly and say: *We are going to share these _____ between the 3 of us: 1 for you, 1 for you, 1 for me ....* Point to the remaining objects at the end, saying: *These are left over. This is what remains. These are the remainder.* Partners do the same using another set of objects, repeating the think-aloud. Ask students to point to the remainder after each round.

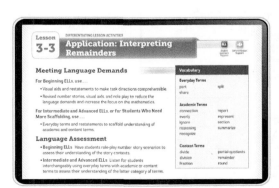

Differentiation Support pages are found in the online Teacher's Center.

Mathematical Process and Practice

SMP1 **Make sense of problems and persevere in solving them.**
GMP1.4 Check whether your answer makes sense.

SMP6 **Attend to precision.**
GMP6.2 Use an appropriate level of precision for your problem.

SMP8 **Look for and express regularity in repeated reasoning.**
GMP8.1 Create and justify rules, shortcuts, and generalizations.

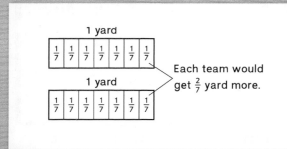

1 yard

Each team would get $\frac{2}{7}$ yard more.

1 yard

Sample drawing for dividing 2 extra yards

---

1 Warm Up   5 min

▶ Mental Math and Fluency

Have students use extended division facts to answer the following questions. *Leveled exercises:*

● ○ ○   How many 3s are in 21? 7
How many 3s are in 210? 70
How many 3s are in 2,100? 700

● ● ○   How many 7s are in 490? 70
How many 70s are in 4,900? 70
How many 700s are in 49,000? 70

● ● ●   How many 80s are in 48,000? 600
How many 800s are in 48,000? 60
How many 80s are in 480,000? 6,000

2 Focus   35–40 min

▶ Math Message

Mr. Burgener has 30 yards of masking tape for a tower-building contest. There are 7 teams and each team needs as much tape as possible to build the tallest tower. Anna says that each team should get 4 yards of tape because $30 \div 7 \rightarrow 4$ R2. Do you agree that the remainder should be ignored? Why or why not?

▶ Reporting Remainders as Fractions

| WHOLE CLASS | SMALL GROUP | PARTNER | INDEPENDENT |

Math Message Follow-Up Ask several students to share their reasoning. Expect some to agree that the remainder should be ignored since there is not enough tape to give each team 5 yards. Others may say that ignoring the remainder wastes 2 yards of tape that could be split up among the teams. If no one mentions this possibility, ask: *Could the 2 extra yards be divided up among the 7 teams so there is no tape left over?* **GMP6.2** Yes. *How can we find out how much more tape each team could receive?* Divide the 2 extra yards of tape among the 7 teams; $2 \div 7$. Display the expression $2 \div 7$, reminding students that they solved fair share problems like this in previous lessons. Have partnerships solve and share their solutions and strategies. *Sample strategies:*

• Draw a picture representing 2 whole yards, each divided into sevenths. Think that $\frac{1}{7}$ from each yard would go to each team. Since $\frac{1}{7} + \frac{1}{7} = \frac{2}{7}$, each team would get $\frac{2}{7}$ yard more tape. (*See margin.*)

• Think about the connection between fractions and division and use the pattern discussed in Lesson 3-2 to determine that $2 \div 7 = \frac{2}{7}$.

Ask: *If the 2 extra yards are divided evenly among 7 teams, how much more tape will each team receive?* $\frac{2}{7}$ yard *How much tape does each team receive in total when the entire roll is used?* $4\frac{2}{7}$ yards Remind students that in previous lessons they solved division stories for which they either ignored the remainder or rounded the quotient up. For some problems, like the one they just solved, it makes sense to divide the remainder and report it as a fraction. Explain to students that today they will apply their understanding of fractions as division to report remainders as fractions.

Pose problem situations for students to consider. For each situation, ask whether it makes sense to report the remainder as a fraction. GMP1.4 Prompt students to explain their reasoning. *Sample situations:*

- Sharing or portioning food Report remainders as fractions because leftovers can be split up.
- Dividing pounds of clay or buckets of sand Report remainders as fractions. These things can be divided into fractional parts.
- Dividing people, animals, phones, or pencils into groups Don't report remainders as fractions. It doesn't make sense to talk about $\frac{1}{2}$ of a person or $\frac{1}{3}$ of a phone.
- Converting measurements in a smaller unit to a larger unit, such as converting 20 inches to feet GMP6.2 Report remainders as fractions. Ignoring the remainder or rounding up would make the measurement imprecise.

Invite students to name other contexts in which it makes sense to report remainders as fractions. Consider displaying a list of these situations for reference when students solve number stories in the future.

Pose several problems in which remainders should be reported as fractions. Encourage students to draw pictures and use partial-quotients division as needed to help them solve. GMP6.2 Have them write number models summarizing their work. *Suggestions:*

- *Ben just adopted a dog. The veterinarian said a bag of dog food should last for 14 days. The bag contains 21 cups of food. If Ben feeds his dog the same amount each day, how many cups of food should he feed the dog?* First divide 21 by 14 to get 1 R7. Then think about dividing the remainder, 7, into 14 equal parts: $7 \div 14 = \frac{7}{14}$. Answer: $1\frac{7}{14}$ cups. Number model: $21 \div 14 = 1\frac{7}{14}$
- *There are 35 mini-bags of pretzels for the 20 students in the after-school program. To avoid wasting the leftovers, the students agree to equally split any remaining bags. How many bags of pretzels will each student receive?* Divide 35 by 20 to get 1 R15. Then think about dividing the remainder, 15, into 20 equal parts: $15 \div 20 = \frac{15}{20}$. Answer: $1\frac{15}{20}$ bags. Number model: $35 \div 20 = 1\frac{15}{20}$
- *Jenna is at an amusement park with her younger brother. One of the rides has a sign saying that the minimum height to ride is 54 inches. Jenna knows that her brother is $3\frac{1}{2}$ feet tall. What is the ride's height requirement in feet?* First divide 54 by 12 to get 4 R6. Then think about dividing the remainder, 6, into 12 equal parts: $6 \div 12 = \frac{6}{12}$. Answer: $4\frac{6}{12}$ feet. Number model: $54 \div 12 = 4\frac{6}{12}$. *Is Jenna's brother tall enough to go on the ride?* No. $3\frac{1}{2}$ feet is less than $4\frac{6}{12}$ feet.

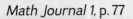

Division Number Stories with Remainders

Lesson 3-3

For each number story:
• Write a number model with a letter for the unknown.
• Solve. Show your work in the space provided. You may draw a picture to help.
• Decide what to do with the remainder. Explain what you did and why.

① Rebecca and her two sisters made pancakes for breakfast. They made 16 pancakes for 5 people. They want to make sure each person gets an equal serving. How many pancakes will each person get? Sample answer

Number model: $16 \div 5 = p$

Sample work:
$3 * 5 = 15$,
so $16 \div 5 = 3$ R1

Quotient: 3   Remainder: 1

Answer: Each person will get $3\frac{1}{5}$ pancakes.

Circle what you did with the remainder.
Ignored it
(Reported it as a fraction)
Rounded the quotient up
Why? Sample answer: Pancakes can be cut into smaller pieces.

② Louis's soccer team is taking a bus to a tournament. They have 32 reusable water bottles. Their water carriers hold 6 bottles each. How many carriers will Louis's team need to bring all of their water bottles on the bus? Sample answer

Number model: $32 \div 6 = c$

Sample work:
| 6 bottles | 6 bottles | 6 bottles |
| 6 bottles | 6 bottles | |

Quotient: 5   Remainder: 2

Answer: Louis's team needs $\frac{2}{6}$ carriers.

Circle what you did with the remainder.
Ignored it
Reported it as a fraction
(Rounded the quotient up)
Why? Sample answer: They needed another carrier for the two bottles left over.

5.NBT.6, 5.NF.3, SMP1, SMP6   77

$21 \div 14 \rightarrow 1$ R7 $21 \div 14 = 1\frac{7}{14}$

$35 \div 20 \rightarrow 1$ R15 $35 \div 20 = 1\frac{15}{20}$

$54 \div 12 \rightarrow 4$ R6 $54 \div 12 = 4\frac{6}{12}$

Division Number Stories with Remainders (continued)

Lesson 3-3

③ Marana saved $80 from her babysitting job. She wants to buy some shirts and pants that are on sale at her favorite store for $17 each. How many items of clothing can she buy? Sample answer

Number model: $80 \div 17 = i$

Sample work:
17)80
−34 2
  46
−34 2
  12 4

Quotient: 4   Remainder: 12

Answer: Mariana can buy 4 items.

Circle what you did with the remainder.
(Ignored it)
Reported it as a fraction
Rounded the quotient up
Why? Sample answer: Mariana did not have enough money left over to buy another item for $17.

④ Jeremy wants to read 100 more books by the end of the school year. There are 36 weeks of school. How many books does Jeremy need to read each week? Sample answer

Number model: $100 \div 36 = b$

Sample work:
36)100
−72 2 $28 \div 36 = \frac{28}{36}$
 28 2

Quotient: 2   Remainder: 28

Answer: Jeremy needs to read $2\frac{28}{36}$ books each week.

Circle what you did with the remainder.
Ignored it
(Reported it as a fraction)
Rounded the quotient up
Why? Sample answer: It makes sense that Jeremy could read part of a book.

78   5.NBT.6, 5.NF.3, SMP1, SMP6

Some students may notice that the fractional parts of mixed numbers can be simplified, or renamed. For example, $1\frac{7}{14}$ can be renamed as $1\frac{1}{2}$. Acknowledge this point and explain that students should feel free to use a simpler equivalent fraction when they are solving a problem and recognize the possibility. However, it is not necessary for them to report quotients as simplified fractions.

Display two number models for each problem, one stating the remainder as a whole number and one reporting the answer as a mixed number with the remainder as the numerator of the fractional part. (*See margin.*)

Ask students to examine the pairs of number models and describe any patterns they notice. Guide a discussion to bring out the following points:

• The whole number remainder is the same as the numerator of the fractional remainder.
• The divisor is the same as the denominator of the fractional remainder.

Encourage students to generalize this pattern. Ask: *How could you report a remainder as a fraction for any division problem?* GMP8.1 Divide the whole number remainder by the divisor for the problem. Use the connection between fractions and division to record the quotient as a fraction.

▶ ## Interpreting Remainders in Division Number Stories

Math Journal 1, pp. 77–78

| WHOLE CLASS | SMALL GROUP | **PARTNER** | **INDEPENDENT** |

Have students solve the number stories on journal pages 77 and 78. If students struggle, suggest that they explain to a partner what the quotient and the remainder represent in each problem. Encourage them to think carefully about what level of precision makes sense when interpreting the remainder. GMP1.4, GMP6.2 Circulate and assist as needed.

✓ **Assessment Check-In**   5.NBT.6, 5.NF.3

Math Journal 1, pp. 77–78

Expect most students to correctly solve Problem 1 on journal page 77 and interpret the remainder as a fraction. Most students should be able to divide the whole numbers in Problems 2–4, and some will be able to interpret the remainders. GMP1.4 If students struggle interpreting remainders, ask questions like these: *What does the remainder mean? Would it make sense to split it into parts?* GMP1.4 If students struggle when working with fractional remainders, suggest that they draw a picture to help them divide the remainder.

Evaluation Quick Entry Go online to record students' progress and to see trajectories toward mastery for these standards.

Summarize Have students share the number stories for which they decided to report the remainder as a fraction. Ask students why a fractional remainder makes sense and how they determined what fraction to report. GMP1.4

 3 Practice 20–30 min

▶ **Playing *Prism Pile-Up***

Student Reference Book, p. 319; *Math Masters*, p. G6

| WHOLE CLASS | SMALL GROUP | **PARTNER** | INDEPENDENT |

Students practice finding volumes of rectangular prisms and figures composed of rectangular prisms.

Observe
- Which students can calculate the volume of their figures and record a corresponding number sentence?
- Which students can explain their reasoning to a partner?

Discuss
- *What was your strategy for finding the volume of the figures?* GMP1.2
- *Is it possible to compare the volumes of two figures without calculating? Explain.* GMP6.4

| Differentiate | **Game Modifications** | Go Online | Differentiation Support |

▶ **Math Boxes 3-3**

Math Journal 1, p. 79

| WHOLE CLASS | SMALL GROUP | PARTNER | INDEPENDENT |

Mixed Practice Math Boxes 3-3 are paired with Math Boxes 3-1.

▶ **Home Link 3-3**

Math Masters, p. 86

Homework Students practice solving division number stories and interpreting remainders.

Math Journal 1, p. 79

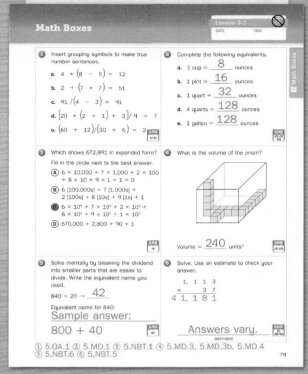

Math Masters, p. 86

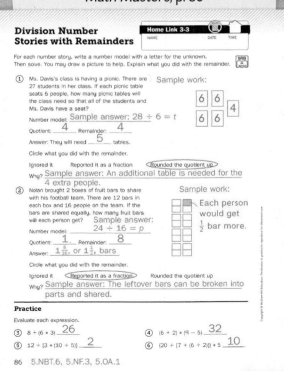

Fractions on a Number Line

Overview Students use number lines to represent, compare, and rename fractions.

▶ **Before You Begin**

Have fraction circle pieces available throughout the lesson and decide how you will display the Fraction Number Lines Poster. For Part 2, plan to have students work in partnerships or small groups so they can share their fraction circle pieces. Consider having copies of *Math Masters*, pages TA12–TA14 available for students who need additional fraction circle pieces. For the optional Readiness activity, consider cutting out fraction strips from *Math Masters*, page 87 in advance.

▶ **Vocabulary**

interval • mixed number • fraction greater than 1 • compare

Standards

Focus Clusters
• Use equivalent fractions as a strategy to add and subtract fractions.
• Apply and extend previous understandings of multiplication and division.

① Warm Up 5 min

Materials

Mental Math and Fluency Students write fractions to model division situations.	slate	5.NF.3

② Focus 35–40 min

Math Message Students represent fractional distances on a number line.	*Math Journal 1*, p. 80	5.NF.3
Representing Fractions on Number Lines Students partition number lines and reason about whole-number benchmarks to locate fractions on number lines.	*Math Journal 1*, p. 80; fraction circles	5.NF.2, 5.NF.3 SMP2, SMP5
Using Number Lines to Solve Problems Students use the Fraction Number Lines Poster to compare and rename fractions and mixed numbers.	*Math Journal 1*, pp. 80–81; *Math Masters*, p. TA15; Fraction Number Lines Poster; fraction circles (optional)	5.NF.2, 5.NF.3 SMP2, SMP5
✓ **Assessment Check-In** See page 244. Expect most students to be able to use the Fraction Number Lines Poster to solve simple fraction comparison and renaming problems.	*Math Journal 1*, pp. 80–81; fraction circles (optional)	5.NF.2, 5.NF.3 SMP5

③ Practice 20–30 min

Playing *Multiplication Top-It: Larger Numbers* **Game** Students practice multiplying multidigit numbers.	*Student Reference Book*, p. 325; per partnership: number cards 0–9 (4 of each)	5.NBT.5 SMP1, SMP6
Math Boxes 3-4 Students practice and maintain skills.	*Math Journal 1*, p. 82	See page 245.
Home Link 3-4 **Homework** Students partition number lines and use them to compare and rename fractions and mixed numbers.	*Math Masters*, p. 89	5.NBT.2, 5.NF.2, 5.NF.3 SMP2, SMP5

 Go Online to see how mastery develops for all standards within the grade.

Differentiation Options

Readiness
10–15 min

WHOLE CLASS | SMALL GROUP | PARTNER | INDEPENDENT

Building Number Lines

5.NF.2, 5.NF.3, SMP2, SMP6

Math Masters, p. 87; large paper (at least 8.5" × 14"); scissors; tape

For experience using length models to represent fractions, students create number lines by folding paper strips. Have students cut out strips from *Math Masters*, page 87, fold two strips in half, and tape them end to end on a large piece of paper. Demonstrate how to draw and label a number line from 0 to 2 beneath the strips, aligning each tick mark with a fold or edge. Label the tick marks $0, \frac{1}{2}, \frac{2}{2}, \frac{3}{2}$, and $\frac{4}{2}$. Note that fractions on number lines are labeled at tick marks. **GMP6.3** Have students run their fingers over the length from 0 to $\frac{1}{2}$, from 0 to $\frac{2}{2}$, from 0 to $\frac{3}{2}$, and so on. Emphasize the way the numbers increase as the lengths increase. Repeat the activity with other denominators. **GMP2.1, GMP2.3**

Paper Strips — Lesson 3-4
NAME   DATE   TIME

Cut on the dashed lines. Write your initials on the back of each strip.

5.NF.2, 5.NF.3, SMP2, SMP6   87

Enrichment
10–15 min

WHOLE CLASS | SMALL GROUP | PARTNER | INDEPENDENT

Exploring Fractions on a Ruler

5.NF.3, SMP2, SMP3

Math Masters, p. 88

To extend their work with fractions on a number line, students partition a ruler into fourths, eighths, and sixteenths of an inch. **GMP2.2** After they mark fractional and mixed number lengths on a ruler, students interpret and explain another person's ing, you may want to remind them that the " symbol means inches.

Exploring Fractions on a Ruler — Lesson 3-4
NAME   DATE   TIME

① Draw tick marks to divide each inch into fourths.
② Draw tick marks to divide each inch into eighths.
③ Draw tick marks to divide each inch into sixteenths.
④ Mark the letter of each location on the ruler below. *Reminder: " means inches.*

A: $\frac{1}{2}$ B: $3\frac{1}{2}$ C: $4\frac{1}{4}$ D: $1\frac{3}{8}$ E: $\frac{1}{4}$ F: $2\frac{1}{16}$
G: $\frac{11}{8}$ H: $5\frac{9}{16}$ I: $2\frac{1}{2}$ J: 3"   K: $4\frac{5}{16}$ L: $5\frac{1}{8}$

⑤ Riccardo said that the stick is $4\frac{3}{4}$ inches long. Explain his mistake.

Sample answer: The stick goes three spaces past 4", but Riccardo mixed up the size of each space. The smallest spaces on a ruler represent $\frac{1}{16}$", not $\frac{1}{4}$".

88   5.NF.3, SMP2, SMP3

Extra Practice
5–15 min

WHOLE CLASS | SMALL GROUP | PARTNER | INDEPENDENT

Renaming and Comparing Fractions and Mixed Numbers

5.NF.2, 5.NF.3

Activity Card 31; *Student Reference Book*, p. 161 or Fraction Number Lines Poster; number cards 1–20 (1 of each); 6-sided die; fraction circles (optional)

For more practice comparing and translating between fractions and mixed numbers, students draw cards and roll a die to generate fractions and mixed numbers. They rename the fractions as mixed numbers and the mixed numbers as fractions greater than 1. Then they compare the values. As an extension, students place the generated numbers on a number line.

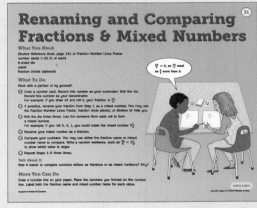

English Language Learner

Beginning ELL To support students' understanding of the terms *locate* and *location*, gesture to numbers on the number line and use terms like *find* and *where* that may be more familiar. For example, ask students: *Where is $\frac{3}{4}$ on the number line? Can you find $\frac{3}{4}$ on the number line? Where is $\frac{3}{4}$ located on the number line? Can you locate $\frac{3}{4}$ on the number line? Show me the location of $\frac{3}{4}$ on the number line.*

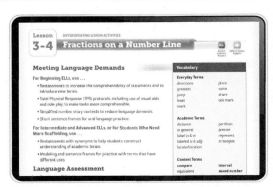

Differentiation Support pages are found in the online Teacher's Center.

Math Journal 1, p. 80

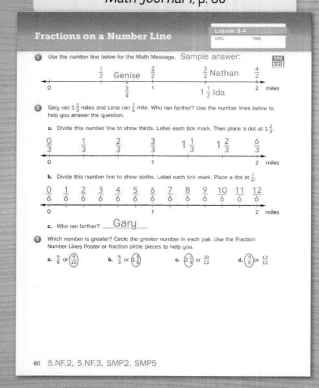

1 Warm Up 5 min

▶ Mental Math and Fluency

Have students write a fraction that models each situation on their slates. *Leveled exercises:*

◉○○ 1 fruit bar shared by 3 people $\frac{1}{3}$
   1 stick of clay shared by 4 people $\frac{1}{4}$
   1 watermelon shared by 12 people $\frac{1}{12}$

◉◉○ 3 containers of paint shared by 10 people $\frac{3}{10}$
   2 cartons of orange juice shared by 9 people $\frac{2}{9}$
   6 apples shared by 8 people $\frac{6}{8}$

◉◉◉ 5 packages of baby carrots shared by 4 people $\frac{5}{4}$
   3 teaspoons of salt used in 2 batches of bread $\frac{3}{2}$
   10 yards of paper for 3 banners $\frac{10}{3}$

2 Focus 35–40 min

▶ Math Message

Math Journal 1, p. 80

Ida, Nathan, and Genise were running at track practice. After 12 minutes Ida had run $1\frac{1}{2}$ miles, Nathan had run $\frac{3}{2}$ miles, and Genise had run $\frac{3}{4}$ mile. Mark the number line in Problem 1 on journal page 80 to show where each runner was after 12 minutes. Be as precise as you can. Label each mark with the runner's name.

▶ Representing Fractions on Number Lines

Math Journal 1, p. 80

| WHOLE CLASS | SMALL GROUP | PARTNER | INDEPENDENT |

Math Message Follow-Up Invite students to show how they marked the number line. Ask: *How did you decide where to place each mark?* Sample answer: I placed a mark for Genise a little before 1 because $\frac{3}{4}$ is less than 1, but more than $\frac{1}{2}$. I placed a mark for Ida halfway between 1 and 2 because $1\frac{1}{2}$ is $\frac{1}{2}$ more than 1. Nathan ran the same distance as Ida. I know because $\frac{3}{2}$ is the same as $\frac{2}{2}$, or 1, and $\frac{1}{2}$ more. So $\frac{3}{2}$ is the same as $1\frac{1}{2}$.

Expect that most students will not have a strategy for marking the number line precisely, but will have roughly estimated where each distance would fall in comparison to the labeled points 0, 1, and 2. Tell students that today they will learn strategies for partitioning number lines to represent fractions as precise locations, as well as estimation strategies for locating fractions greater than 1. They will also use number lines to solve problems.

Discuss some important features of number lines. Emphasize these points:

- Number lines go on forever in both directions. We show this by drawing arrows at each end.
- Tick marks on number lines are evenly spaced. We label tick marks by counting up with a given "jump," or **interval**, such as 1, 10, $\frac{1}{2}$, or $\frac{1}{4}$.
- We decide how to label number lines, including the tick marks on number lines, based on the numbers in a problem.
- Labeled tick marks help us place specific values on the line. Tick marks for whole numbers are sometimes longer or bolder than tick marks for values between whole numbers.
- We use points or dots to represent specific values on a number line.

Encourage students to apply these ideas to the Math Message problem. Ask: *How can we show $\frac{3}{4}$ mile as a precise location on a number line?* Sample answer: We could make tick marks to show $\frac{1}{4}$s so we know where to place $\frac{3}{4}$. Display a number line and make tick marks for 0 and 1. Remind students that the space between 0 and 1 is 1 whole, just as 1 red fraction circle is 1 whole. GMP2.3 Ask: *What is the 1, or the whole, in the Math Message?* 1 mile Label the displayed number line with miles. (*See margin.*)

How could we show fourths on the number line? Use tick marks to divide the whole into 4 equal parts. Think aloud as you demonstrate how to partition the number line into fourths: *If I put a tick mark in the middle, I've split the whole into 2 equal parts. If I put tick marks in the middle of each half, then there are 4 equal parts between 0 and 1.* Have students count the spaces between each tick mark to confirm that the interval between 0 and 1 is divided into 4 equal parts. Ask: *How should we label these tick marks?* $\frac{1}{4}, \frac{2}{4}, \frac{3}{4}$ *Where should we place a point to show the distance $\frac{3}{4}$ mile?* On the tick mark labeled $\frac{3}{4}$ Place a point at $\frac{3}{4}$. Use a different color to trace the line from 0 to $\frac{3}{4}$. Emphasize that the point at $\frac{3}{4}$ is $\frac{3}{4}$ of a unit from 0. If we move from 0 to $\frac{3}{4}$, we have traveled 3 out of the 4 equal parts it takes to travel a whole unit. This point represents $\frac{3}{4}$ mile in the Math Message problem. GMP2.2

Have students look at the mark they made for $1\frac{1}{2}$ and $\frac{3}{2}$ in Problem 1 on journal page 80. Ask students how they could make sure that $1\frac{1}{2}$ miles and $\frac{3}{2}$ miles were labeled as precise locations on the number line. Make sure that both of the following strategies are discussed:

- Think about whole numbers first. Since $1\frac{1}{2}$ is more than 1 whole, jump to the tick mark for 1. Think about the space between 1 and 2 as 1 whole and make a tick mark halfway between 1 and 2 to show halves. Label the mark $1\frac{1}{2}$. Place a dot on the $1\frac{1}{2}$ mark. Count 3 half-unit intervals from 0 to $1\frac{1}{2}$ to check that this is the same as $\frac{3}{2}$.
- Think about the numerator as a counting number. Since $\frac{3}{2}$ is a number of halves, make tick marks for halves. Make a tick mark halfway between 0 and 1 to represent $\frac{1}{2}$. There is already a tick mark for 1. Make another tick mark halfway between 1 and 2 to represent $\frac{3}{2}$. Label the tick marks 0, $\frac{1}{2}, \frac{2}{2}, \frac{3}{2}$, and $\frac{4}{2}$. Then start at 0 and count up 3 halves. Place a dot at $\frac{3}{2}$. Check that this is the same as $1\frac{1}{2}$: $\frac{2}{2}$ is another name for 1, and $\frac{3}{2}$ is $\frac{1}{2}$ more.

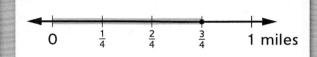

Representing $\frac{3}{4}$ as a precise location

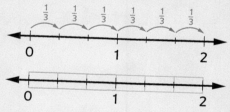

Allow students to use one of these strategies to revise their labels in Problem 1 as needed. Point out that $1\frac{1}{2}$ and $\frac{3}{2}$ are equivalent names for the same number. Remind students that $1\frac{1}{2}$ is called a **mixed number** because it contains a whole number and a fraction. In **fractions greater than 1** (such as $\frac{3}{2}$), the numerator is greater than the denominator, indicating that it represents more than 1 whole.

Return to the Math Message problem. Ask: *What does this number line representation tell us about Genise, Ida, and Nathan? Who ran the farthest in 12 minutes?* GMP2.2 Ida and Nathan ran the same distance. They both ran farther than Genise. *How could you confirm this with fraction circle pieces?* GMP2.3 Sample answer: If I represent $\frac{3}{4}$, $1\frac{1}{2}$, and $\frac{3}{2}$ with fraction circle pieces, I see that $\frac{3}{4}$ is smaller than $1\frac{1}{2}$ and $\frac{3}{2}$, and that $1\frac{1}{2}$ and $\frac{3}{2}$ are the same amount. Point out that the class was just using a number line to **compare** the fractions, or to determine which fractions were greater than, less than, or equal to each other. Ask: *What was useful about representing these distances on number lines?* Sample answer: I could easily see which distance was longer and that two numbers were equivalent.

Invite students to consider how they would represent fractions other than halves and fourths on a number line. Ask: *What if other students ran different distances at track practice? How would you show a distance such as $\frac{1}{3}$ mile?* Divide the space between 0 and 1 into 3 equal parts. Count up 1 hop from 0 and mark $\frac{1}{3}$. $\frac{8}{6}$ *miles?* Divide the space between 1 and 2 into 6 equal parts. $\frac{8}{6}$ is the same as $1\frac{2}{6}$, so count up 2 spaces from 1 to mark $\frac{8}{6}$. *How would you show a distance greater than 2 miles?* Extend the number line past 2.

Pose the following problem: *Gary ran $1\frac{2}{3}$ miles and Lena ran $\frac{7}{6}$ miles. Who ran farther?* Have students partition and label the number lines in Problems 2a and 2b on journal page 80 to help them solve the problem. GMP5.2 As students work, circulate and observe. While it is not crucial for students' partitions to be exactly the same size, students should be attempting to create equal spaces between each pair of whole numbers.

When most students have finished, ask: *Who ran farther, Gary or Lena? How do you know?* Gary ran farther because $1\frac{2}{3}$ is farther to the right on a number line than $\frac{7}{6}$. Explain that the larger a fraction is, the farther that fraction will be from 0 on the number line. Ask: *Does it matter that you represented each distance on a different number line? Why or why not?* No, because the number lines have the same scale. A whole mile is the same length on each number line. Have students use their fraction circle pieces to check the comparison. (*See margin.*)

Point out that it is not always necessary to place fractions in a precise location in order to compare the values. Tell students that thinking about fraction circle pieces can help them place fractions on a number line without completely partitioning the line. Model an example for the class. Display a number line from 0 to 5 and label each whole number. Ask students to represent the fraction $\frac{9}{4}$ with fraction circle pieces. Ask: *Between which two whole numbers does $\frac{9}{4}$ lie?* Between 2 and 3 *How can we place $\frac{9}{4}$ on a number line?* GMP2.3 Sample answer: I can see from my circle pieces that $\frac{9}{4}$ is $\frac{1}{4}$ more than 2. I can move $\frac{1}{4}$ past 2 on the number line and place $\frac{9}{4}$ there.

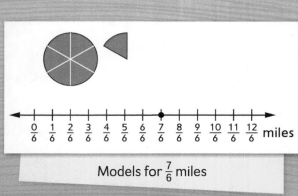

Models for $1\frac{2}{3}$ miles

Models for $\frac{7}{6}$ miles

Make a tick mark and label it $\frac{9}{4}$ on the number line. Give students several more fractions greater than 1 to represent with fraction circle pieces and place on the displayed number line. *Suggestions:* $\frac{10}{3}, \frac{5}{2}, \frac{12}{3}, \frac{9}{2}$

Have students compare the fractions they placed on the number line. Ask: *Which of these fractions is the greatest? How do you know?* Sample answer: $\frac{9}{2}$, because it is farthest to the right on the number line.

Highlight that students now know two strategies for locating fractions on number lines:

- Divide the space between each pair of whole numbers into equal parts and count up to mark each fraction.
- Use fraction circles or think about whole numbers to estimate where fractions fall on the number line.

Explain that both strategies can be important in solving real-world problems.

▶ Using Number Lines to Solve Problems

Math Journal 1, pp. 80–81; *Math Masters*, p. TA15

| WHOLE CLASS | SMALL GROUP | PARTNER | INDEPENDENT |

Display the Fraction Number Lines Poster and distribute *Math Masters*, page TA15, which is identical to the class poster. Explain that much like the blank number lines students just partitioned and labeled, the poster is a tool students can use to compare values and recognize equivalent names for fractions and mixed numbers.

Have students examine the poster and share what they notice. GMP2.2
Be sure to include the following observations in the discussion:

- Each number line shows numbers from 0 to 2. The interval representing 1 whole is the same length for each number line.
- The tick marks on the first line are labeled with whole numbers. The tick marks on the remaining lines are labeled with fractions.
- The poster shows number lines divided into wholes, halves, thirds, fourths, fifths, sixths, eighths, ninths, tenths, and twelfths.

Pose a series of comparison problems for students to solve using the poster. Have volunteers show where they found each fraction on the poster and confirm their answers with fraction circle pieces.
GMP2.2, GMP2.3, GMP5.2 *Suggestions:*

- Which is greater: $\frac{4}{5}$ or $\frac{3}{2}$? $\frac{3}{2}$
- Which is greater: $\frac{8}{6}$ or $\frac{11}{9}$? $\frac{8}{6}$
- Which is greater: $1\frac{3}{4}$ or $\frac{4}{3}$? $1\frac{3}{4}$

Adjusting the Activity

Differentiate To help students interpret the Fraction Number Lines Poster have them identify the fractions on the poster that are equal to 1. Some students may benefit from highlighting fractions equal to 1. Ask: *How do you know these fractions are equal to 1?* Sample answer: The numerator is the same as the denominator, which means we have all of the pieces in 1 whole. *Where will we find fractions greater than 1 on this poster?* To the right of fractions equal to 1 GMP2.2

Go Online   Differentiation Support

Math Masters, p. TA15

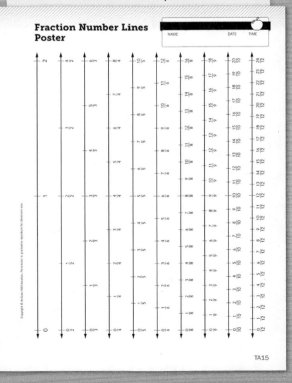

TA15

Fractions on a Number Line (continued)

Lesson 3-4
DATE   TIME

4. Rachel and Dan are growing plants in science class. Rachel reports that her plant is $1\frac{1}{4}$ inches tall. Dan says his plant is $\frac{5}{2}$ inches tall.

   a. Whose plant is taller? __Dan's plant__

   b. How do you know? __Sample answer: $\frac{5}{2}$ is the same as__
   $2\frac{1}{2}$ and $2\frac{1}{2}$ is greater than $1\frac{1}{4}$.

For Problems 5–10, rename each fraction as a mixed number or each mixed number as a fraction greater than 1. You may use the Fraction Number Lines Poster, fraction circle pieces, or division.

5. $\frac{5}{3} = 1\frac{2}{3}$ 6. $1\frac{7}{4} = \frac{16}{9}$

7. $\frac{11}{8} = 1\frac{3}{8}$ 8. $1\frac{5}{6} = \frac{11}{6}$

9. $\frac{16}{5} = 3\frac{1}{5}$ 10. $2\frac{1}{3} = \frac{7}{3}$

Try This

11. a. Rename $\frac{34}{8}$ as a mixed number. __$4\frac{2}{8}$__

    b. Explain your reasoning.
    __Sample answer: $\frac{34}{8} = 34 \div 8$, which is 4 R2,__
    or $4\frac{2}{8}$.

81

NOTE Students are not expected to use an algorithm to translate between fractions and mixed numbers at this time. Some may be able to use their understanding of the relationship between multiplication and division to rename mixed numbers as fractions without using a tool, but this is not a general expectation. Students *should* be able to reason through problems using number lines, fraction circle pieces, and their understanding of fractions as division.

Ask: *How did you locate the mixed number $1\frac{3}{4}$ on the number line?* Sample answer: I used the number line with 4 as the denominator. I found the fraction $\frac{4}{4}$ which is the same as 1. Then I counted 3 more spaces for $\frac{3}{4}$. *How could we rename $1\frac{3}{4}$ as a fraction? Why?* Sample answer: $\frac{7}{4}$ because that is the tick mark I landed on when I counted up $\frac{3}{4}$ from 1. *In general, how could you use the poster to rename a mixed number as a fraction?* Use the number line with the given denominator. Find the fraction that is the same as 1, then count up the spaces for the fraction part. The tick mark you land on is the fraction name for the mixed number. *How could you check your answer with fraction circle pieces?* GMP2.2, GMP2.3 Use whole circles and fraction pieces to represent the mixed number. Exchange the red circle(s) for same-size fraction pieces. Count the total number of fraction pieces.

Point to $\frac{4}{3}$ on the poster and ask: *How can you rename $\frac{4}{3}$ as a mixed number?* $\frac{4}{3}$ is 1 third past 1, so I can call it $1\frac{1}{3}$. *In general, how could you use the Fraction Number Lines Poster to rename fractions greater than 1 as mixed numbers?* GMP5.2 Sample answer: Find the fraction on the poster. Count up to that fraction from 1. Make a mixed number with 1 whole and the fraction for how far the number is from 1. Have students consider other strategies for renaming a fraction greater than 1 as a mixed number. Ask: *How could you use fraction circle pieces to check that $\frac{4}{3} = 1\frac{1}{3}$?* GMP5.2 I could make $\frac{4}{3}$ with 4 orange pieces. I can exchange 3 orange pieces for 1 red circle, or 1 whole, and there is 1 orange piece, or $\frac{1}{3}$, left over. That is $1\frac{1}{3}$ in all. *How could you use division to check that $\frac{4}{3} = 1\frac{1}{3}$?* $\frac{4}{3}$ is the same as 4 ÷ 3. It is 1 R1, or $1\frac{1}{3}$.

Have students complete Problems 3–11 on journal pages 80 and 81. Encourage them to check their answers using fraction circles or division.

Assessment Check-In 5.NF.2, 5.NF.3

Math Journal 1, pp. 80–81

Expect most students to be able to use the Fraction Number Lines Poster to solve simple comparison and renaming problems, such as Problems 3 and 5–8 on journal pages 80 and 81. GMP5.2 Some may use renaming as a comparison strategy in Problem 4 or notice patterns and be able to translate between fractions greater than 2 and mixed numbers in Problems 9–11. If students struggle using the poster effectively, suggest using fraction circle pieces to represent values. Understanding number lines as a visual model for fractions is an important step in developing fraction number sense and in using benchmarks to estimate fraction sums and differences, which students will do in upcoming lessons.

Evaluation Quick Entry Go online to record students' progress and to see trajectories toward mastery for these standards.

Summarize Have students discuss number lines and fraction circle pieces as tools. Ask: *When are fraction circle pieces helpful? When is a number line helpful?* GMP2.3

③ Practice 20–30 min

▶ **Playing *Multiplication Top-It:* Larger Numbers**

Student Reference Book, p. 325

| WHOLE CLASS | SMALL GROUP | **PARTNER** | INDEPENDENT |

Students practice multiplying multidigit numbers. Consider introducing one or more of the variations listed in Lesson 2-8 (Extra Practice) and having students use U.S. traditional multiplication in at least one round.

Observe
- Which students are able to accurately use U.S. traditional multiplication?
- Which students use efficient and accurate multiplication strategies? GMP6.4

Discuss
- *What multiplication strategy do you prefer to use? Why?* GMP1.6
- *How could you use an estimate to determine who won each round?*

▶ **Math Boxes 3-4**

Math Journal 1, p. 82

| WHOLE CLASS | SMALL GROUP | PARTNER | INDEPENDENT |

Mixed Practice Math Boxes 3-4 are paired with Math Boxes 3-2.

▶ **Home Link 3-4**

Math Masters, p. 89

Students partition number lines and use them to compare and rename fractions and mixed numbers. GMP2.2, GMP5.2

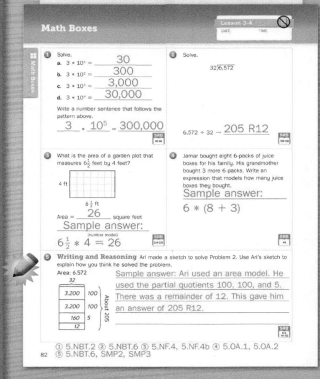

Math Journal 1, p. 82

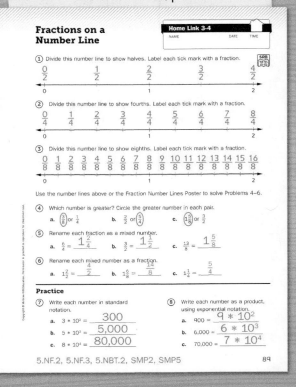

Math Masters, p. 89

Game Strategies

Overview **Day 1:** Students play a version of Top-It with fractions and devise a rule for making the largest possible fraction.
Day 2: Students discuss whether other students' rules work and revise their own rules.

Day 1: Open Response

▶ **Vocabulary**

If possible, schedule time to review students' work and plan for Day 2 of this lesson with your grade-level team.

Standards

Focus Clusters
- Perform operations with multi-digit whole numbers and with decimals to hundredths.
- Apply and extend previous understandings of multiplication and division.

① Warm Up 5 min	**Materials**		
Mental Math and Fluency Students convert from exponential to standard notation.			**5.NBT.2**
②a Focus 55–65 min			
Math Message Students explain how they would place digits in a division problem to get the largest possible quotient.	*Math Journal 1*, p. 83		**5.NBT.6** **SMP7, SMP8**
Sharing Game Strategies Students create and justify a rule for writing a division problem with the largest possible quotient.	*Math Journal 1*, p. 83		**5.NBT.6** **SMP1, SMP7, SMP8**
Solving the Open Response Problem Students play a version of *Top-It* using fractions and devise a rule for making the largest possible fraction.	*Student Reference Book*, pp. 27–29; *Math Masters*, pp. 90, 91, and G14; fraction circles (optional); per partnership: number cards 1–9 (4 of each)		**5.NBT.6, 5.NF.3** **SMP1, SMP7, SMP8**

Getting Ready for Day 2 →

Review students' work and plan discussion for reengagement.

Math Masters, p. TA4, p. TA31 (optional); students' work from Day 1

 Go Online to see how mastery develops for all standards within the grade.

1 Warm Up 5 min

▶ Mental Math and Fluency

Have students convert from exponential to standard notation.
Leveled exercises:

● ○ ○ 10^2 100
     10^3 1,000
     10^4 10,000

● ● ○ 3×10^3 3,000
     8×10^4 80,000
     2×10^6 2,000,000

● ● ● 20×10^2 2,000
     80×10^4 800,000
     900×10^3 900,000

2a Focus 55–65 min

▶ Math Message

Math Journal 1, p. 83

Complete journal page 83. Then explain your thinking to a partner.
GMP7.2, GMP8.1

English Language Learners Support Before the lesson, review
vocabulary associated with division and fractions to help students
understand the problem context. Use Total Physical Response commands
along with written division problems and fractions such as: *Point to the
divisor (dividend) in this problem. Find the quotient of 10 ÷ 2. Show me a
fraction with a numerator of 2. Tell me the fraction with the largest
denominator. Name two fractions with the same numerator.*

Remind students that the way the term *rule* is used in mathematics is
different from its everyday use in games. In mathematics, *rule* is often used
to describe the relationship between two numbers in a pattern.

Standards and Goals for
Mathematical Process and Practice

**SMP1 Make sense of problems and persevere in
solving them.**
  GMP1.2 Reflect on your thinking as you solve
  your problem.

SMP7 Look for and make use of structure.
  GMP7.2 Use structures to solve problems and
  answer questions.

**SMP8 Look for and express regularity in repeated
reasoning.**
  GMP8.1 Create and justify rules, shortcuts,
  and generalizations.

Professional Development

The focus of this lesson is GMP8.1.
Creating and justifying rules means
noticing patterns and relationships
between specific numbers and showing
that these relationships work with any
numbers. An advantage to creating
mathematical rules for the *Top-It* games
in this lesson is that students can use the
rules to choose the largest quotient or
fraction efficiently and therefore have
a consistent strategy for the games.

Go Online ▷ for information about
SMP8 in the *Implementation Guide.*

Math Journal 1, p. 83

Division Top-It   Lesson 3-5
                                   DATE   TIME

Harjit is playing a version of *Division Top-It* with a friend. In this version each player turns
over 3 number cards and places them as the digits in the division problem below. The
player with the larger quotient wins the round.

☐ ☐ ÷ ☐ =

Harjit turns over a 6, a 3, and a 9. How do you think Harjit should place her cards to get the
largest possible quotient? Explain your thinking.
Sample answer: Harjit should use her cards to
make the problem 96 ÷ 3 = 32. 96 is the
largest possible 2-digit number and 3 is the
smallest possible divisor. I know that the larger
the dividend, the more there is to divide up, so
the larger the quotient. I also know that the
smaller the divisor, the larger the quotient. So in
order to get the largest quotient, Harjit should
make the largest dividend and the smallest
divisor.

5.NBT.6, SMP7, SMP8   83

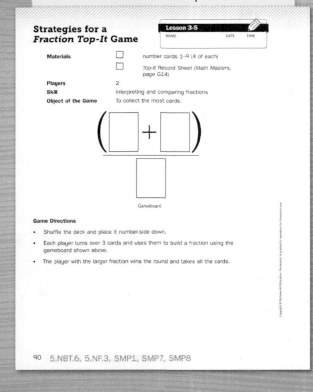

Math Masters, p. 90

Sharing Game Strategies

Math Journal 1, p. 83

| WHOLE CLASS | SMALL GROUP | PARTNER | INDEPENDENT |

Math Message Follow-Up Have students discuss their strategies from the Math Message and consider making a list of strategies that students can refer to during the discussion. Strategies might include solving each possible division problem to compare the quotients or using the largest number as the dividend and the smallest as the divisor.

Tell students to choose one of the strategies the class discussed and use it to find the largest quotient if Harjit draws 2, 2, and 8. **GMP7.2** $82 \div 2 = 41$ Have partners discuss the strategies they used to find the largest quotient and why the strategies work. Sample answers: I created all possible problems with the numbers. I found the quotient for all three problems and got 41 as the largest quotient. I can predict that dividing 82 into 2 equal parts would be more than dividing 28 into 2 equal parts and more than dividing 22 into 8 equal parts. I noticed that 82 was the largest dividend and 2 was the smallest divisor. Have students compare the two sets of cards. *Can you state a rule that Harjit should follow when she plays this version of* Division Top-It? **GMP8.1** Make the largest possible dividend and the smallest possible divisor. Ask: *Why does this rule work?* **GMP1.2, GMP8.1** Sample answer: If you start with the largest possible number and divide it into the fewest number of equal groups, you will end up with the greatest number per group. For example, if we divide a class of 30 students into 2 groups, we would have 15 students in both groups. If we divide the same class into 10 groups, there would be only 3 students in each group. Consider asking a volunteer to use manipulatives or draw a picture to illustrate this rule. **GMP8.1**

Explain that mathematicians are always looking for patterns and rules that will help them solve problems more easily and efficiently. Tell students that today they will create and justify rules while they play a new fraction game. **GMP8.1**

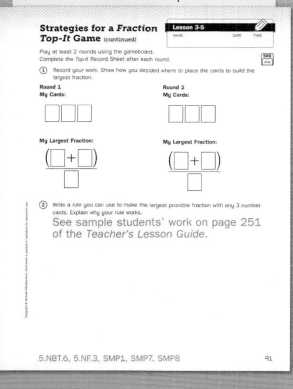

Math Masters, p. 91

▶ Solving the Open Response Problem

Student Reference Book, pp. 27–29; *Math Masters*, pp. 90, 91, and G14

| WHOLE CLASS | SMALL GROUP | PARTNER | INDEPENDENT |

Distribute *Math Masters*, pages 90, 91, and G14. Read and review the directions as a class. Partners or groups should work together to ensure they understand the rules of the game and what they are asked to do. Point out that they will find the sum of two of the cards to create a numerator. They should record their work for at least two rounds of play on the *Top-It* Record Sheet and play as many additional rounds as needed to create and explain a rule. **GMP8.1** Remind students that they can use representations such as fraction circles, number lines, and drawings to help them compare fractions and explain why their rules work.

Students may work together on the problem, but all students should record a solution. Monitor students as they work. Ask: *Why did you place your cards that way? What information did you use to determine how to place your cards?* **GMP1.2, GMP7.2**

Summarize Have students read *Student Reference Book*, pages 27–29 about **GMP8.1**. Ask: *Why is it important for mathematicians to create rules?* Sample answers: Rules can help solve problems more efficiently. Rules help them see connections among different mathematical concepts. Remind students that they will have an in-depth conversation about the problem on Day 2.

Collect students' work so that you can evaluate it and prepare for Day 2.

Adjusting the Activity

Differentiate Students struggling to think of a rule to make the largest fraction can write out all the possible fractions for the rounds. Ask whether they notice a pattern in the numbers that give the largest fraction.

Student Reference Book, p. 27

Create and Explain Rules and Shortcuts

An important part of doing mathematics is looking for shortcuts, rules, and generalizations to make procedures and operations more efficient, so that when you see a similar problem, you can solve it more easily. When you create a shortcut, rule, or generalization, you need to be able to justify it. For example, when you justify a shortcut, you use mathematical properties to explain how and why it works.

Mr. Bates asks his class to figure out an efficient way to multiply powers of 10. He lists these problems:

10,000 * 10   100 * 1,000   1,000 * 100,000

10,000 * 1,000   100 * 10,000

Sophia begins to multiply using her calculator.

10,000 * 10   100 * 1,000   1,000 * 100,000

 Josh, I just noticed that the total number of zeros in the factors is the same as the number of zeros in the product.

 Does that always work? Let's check the last two. You do them by counting zeros, and I'll do them on a calculator.

Sophia   Josh

twenty-seven **27**

Getting Ready for Day 2

Math Masters, p. TA4

Planning a Follow-Up Discussion

Review student work. Use the Reengagement Planning Form (*Math Masters,* page TA4) and the rubric on page 252 to plan ways to help students meet expectations for both the content and process/practice standards. Look for work with rules that apply to a specific group of three numbers rather than to all possible numbers students could draw. Also look for work that shows the strategy of testing all possible combinations of the three cards drawn.

Organize the discussion in one of the ways below or in another way you choose. You may choose to develop a student-friendly rubric using *Math Masters,* page TA31 and facilitate a peer discussion and review as described on *Teacher's Lesson Guide,* Volume 1 page 164. If student work is unclear or if you prefer to show work anonymously, rewrite the work for display.

Go Online for sample students' work that you can use in your discussion.

1. Display work with a rule that involves making all possible fractions and picking the largest. Compare this to work with a rule that involves using larger numbers in the numerator and smaller numbers in the denominator.

    NOTE Consider providing sentence frames for this discussion, such as: "Both rules _____." and "The first rule _____, but the second rule _____."

    Ask: *What do you notice when you compare these two rules?*
    GMP1.2, GMP8.1 Sample answer: Both rules give you the largest fraction. The first rule takes longer because you have to check every possible fraction. The second rule is more efficient because you don't have to create and compare every possible fraction.

2. Show an incorrect rule for Problem 2, such as in Student A's work. Ask: *What do you think is the reasoning behind this rule?* GMP1.2 Sample answer: The student might have confused the terms *numerator* and *denominator* because the student is saying two different things at the same time. Although this student correctly says that if the divisor is smaller then the fraction will be bigger, the student also says that if the denominator is bigger then the fraction will be bigger.

3. Show a correct rule for Problem 2, such as in Student B's work. Ask:
   • *What is the thinking behind this rule?* GMP1.2, GMP7.2, GMP8.1 Sample answer: The student explained that a fraction is like a division problem. The numerator, or dividend, should be large so there is a large number to divide up. This student also said that the denominator, or divisor, should be small so the parts from the division will be bigger.
   • *How could this student improve the explanation?* Sample answer: The student could have explained more clearly why a large numerator and small denominator make a larger fraction.
   • *Can you apply this rule to the* Division Top-It *game that we played in the Math Message? Why or why not?* Sample answer: Yes. If I think about the division problem in *Division Top-It* as a fraction, then this rule will also work. I want to make a larger dividend and smaller divisor in *Division Top-It,* just like I am making a larger numerator and smaller denominator in the fraction game.

4. Show work that has a correct rule for Problem 2, but does not have an explanation for why the rule works. Ask: *What is one thing that this student did well? What could this student do better?* GMP8.1 Answers vary.

Play another round of the game, applying one of the rules students explained in the discussion.

Planning for Revisions

Have copies of *Math Masters,* pages 90, 91, and G14 or extra paper available for students to use in revisions. You might ask students to use colored pencils so you can see what they revised.

Sample student's work, Student A

2. Write a rule that you can use to make the largest possible fraction using any three number cards. Explain why your rule works. My rule is bigger the demominator Smaller the divisor bigger the answer to win the game.

Sample student's work, Student B

2. Write a rule that you can use to make the largest possible fraction using any three number cards. Explain why your rule works.

My rule is you have to make your numorator large so you can get a big sum to ~~divi~~ divided from your small donominator. Also did you know the numerator is the dividend. Second, you have to make your donominator small so it will be bigger parts making the fraction bigger. Also the donominator is the divisor. In conclusion basicly larger the sum of the numerator and the smaller the ~~do~~ donomo-nator the bigger the fraction.

Game Strategies

Day 2: Reengagement

▶ **Before You Begin**

Have extra copies of Math Masters, pages 90, 91, and G14 available to students for revising their work.

Standards

Focus Clusters
- Perform operations with multi-digit whole numbers and with decimals to hundredths.
- Apply and extend previous understandings of multiplication and division.

(2b) Focus   50–55 min

	Materials	
Setting Expectations Students discuss what a complete response to the open response problem would include. They also discuss how to respectfully talk about their own and other students' strategies and explanations.	Standards for Mathematical Process and Practice Poster, Guidelines for Discussions Poster	SMP8
Reengaging in the Problem Students discuss others' rules and why they work.	*selected samples of students' work; Math Masters, p. TA31 (optional)*	5.NBT.6, 5.NF.3 SMP1, SMP7, SMP8
Revising Work Students revise their rules and explanations.	*Math Masters, pp. 90, 91, and G14 (optional); students' work from Day 1; colored pencils (optional)*	5.NBT.6, 5.NF.3 SMP7, SMP8
✓ **Assessment Check-In** See page 254 and the rubric below. Expect most students to be able to interpret the fractions they created in the *Division Top-It* game as division of the numerator by the denominator and to divide correctly.		5.NF.3, SMP8

Goal for Mathematical Process and Practice **GMP8.1** Create and justify rules, shortcuts, and generalizations.	Not Meeting Expectations	Partially Meeting Expectations	Meeting Expectations	Exceeding Expectations
	Does not describe a rule.	Describes a rule that involves testing all possible fractions or that only describes how to place three specific numbers, or provides a correct rule (see Meeting Expectations) without an explanation.	Describes a rule that involves placing the largest two digits in the numerator and the smallest digit in the denominator for any three numbers. Explains that the rule works because the largest numerator has the largest quantity to be divided into equal shares or because a smaller denominator leads to larger equal shares.	Meets expectations and clearly explains why the rule works, using examples of either whole-number division or fraction concepts.

(3) Practice   10–15 min

Math Boxes 3-5 Students practice and maintain skills.	*Math Journal 1, p. 84*	See page 254.
Home Link 3-5 **Homework** Students create a rule for playing another version of a fraction game.	*Math Masters, p. 92*	5.NBT.1, 5.NF.3

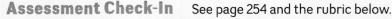

▆▆▆ **Go Online** to see how mastery develops for all standards within the grade.

2b Focus
50–55 min

NOTE These Day 2 activities will ideally take place within a few days of Day 1. Prior to beginning Day 2, see Planning a Follow-Up Discussion from Day 1.

▶ Setting Expectations

| WHOLE CLASS | SMALL GROUP | PARTNER | INDEPENDENT |

Briefly review the open response problem from Day 1. Ask: *What were you asked to do?* Sample answer: Play a version of *Top-It* using fractions, develop a rule to make the largest possible fraction with any three number cards, and explain why the rule works. *What do you think a good response would include?* GMP8.1 Sample answer: It would show which set of cards the student chose for Round 1 and Round 2. It would have a correct rule for making the largest possible fraction using any three number cards. It would explain why the rule works.

After this brief discussion, tell students that they are going to look at other students' work and see whether they thought about the problem in the same way. Refer to GMP8.1 on the Standards for Mathematical Process and Practice Poster. Explain that they will try to figure out how other students created and justified rules.

Explain that mathematicians often make mistakes and learn from them. Remind students that they should feel comfortable analyzing and learning from their own mistakes and the mistakes of others. Refer to your list of discussion guidelines from Units 1 and 2 and encourage students to use sentence frames such as the following:

• I noticed _____.

• I disagree because _____.

▶ Reengaging in the Problem

| WHOLE CLASS | SMALL GROUP | PARTNER | INDEPENDENT |

Students reengage in the problem by analyzing and critiquing other students' work in pairs and in a whole-group discussion. Have students discuss with partners before sharing with the whole group. Guide the discussion based on the decisions you made in Getting Ready for Day 2. GMP1.2, GMP7.2, GMP8.1

▶ Revising Work

WHOLE CLASS   SMALL GROUP   PARTNER   INDEPENDENT

Pass back students' work from Day 1. Before students revise anything, ask them to examine their responses and decide how they could be improved. Ask the following questions one at a time. Have partners discuss their responses and give a thumbs-up or thumbs-down based on their own work.

- *Did you show how you chose to place the cards for each round?*
- *Did you describe a rule to make the largest possible fraction using any three number cards?* **GMP8.1**
- *Did you explain why the rule works?* **GMP7.2, GMP8.1**

Tell students they now have a chance to revise their work. Tell them to add to their earlier work using colored pencils or to use another sheet of paper, instead of erasing their original work. Encourage those who wrote complete and correct explanations on Day 1 to play a few more rounds of the game using their rule. Help students see that the explanations presented during the discussion are not the only correct ones.

Summarize Ask students to reflect on their work and revisions. Ask: *What did you do to improve your rule or explanation?* Answers vary.

✓ Assessment Check-In   5.NF.3

Collect and review students' revised work. For the content standard, expect most students to interpret the fractions they created in the game as division of the numerator by the denominator and to divide correctly. Use the rubric on page 252 to evaluate students' revised work for **GMP8.1**.

Evaluation Quick Entry Go online to record students' progress and to see trajectories toward mastery for these standards.

Go Online for optional generic rubrics in the *Assessment Handbook* that can be used to assess any additional GMPs addressed in the lesson.

Math Journal 1, p. 84

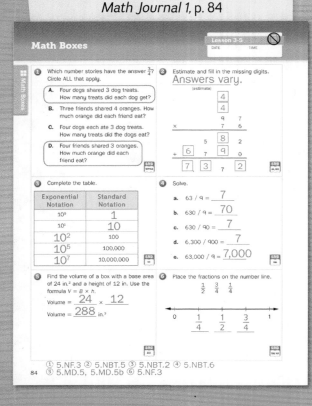

Sample Students' Work—Evaluated

See the sample in the margin. This work meets expectations for the content standard because the student correctly divided the numerators by the denominators in Problem 1 (not shown). With revision, the work meets expectations for the mathematical process and practice standard because the student states a rule and explains why it works. The rule is that the player should place the largest two digits as a sum in the numerator and the smallest digit in the denominator. The student justifies the rule by explaining that having a smaller denominator means fewer groups and thus more in each group, or having more in the numerator means having more to put into each group. **GMP8.1**

Go Online for other samples of evaluated students' work.

3 Practice 10–15 min

▶ Math Boxes 3-5

Math Journal 1, p. 84

WHOLE CLASS	SMALL GROUP	PARTNER	INDEPENDENT

Mixed Practice Math Boxes 3-5 are paired with Math Boxes 3-7.

▶ Home Link 3-5

Math Masters, p. 92

Homework Students create a rule for playing another version of a fraction game.

---

2. Write a rule that you can use to make the largest possible fraction using any three number cards. Explain why your rule works. My rule is to put the biggest two digits in the numerator because it makes the biggest number of any combination plus it leaves the smallest digit and you want the smallest number for the denomiter. This strategy works because the smaller the denominator the less groups you have less groups so you can put more in them instead of spreading them out and having less in each group. You want more in the numerator because you want more to put in each group instead of less in each group.

Math Masters, p. 92

More Fraction Top-It

Home Link 3-5

NAME   DATE   TIME

Eddie and his friend are playing another version of *Fraction Top-It*. Each player turns over 4 number cards and places them as the digits on the gameboard. The player with the larger quotient wins the round.

Eddie's cards are 2, 6, 3, and 4.

① If you were Eddie, how would you place your cards? What is the quotient?

$$\frac{(\boxed{4}+\boxed{6})}{(\boxed{2}+\boxed{3})} = \frac{10}{5} = 2$$

② What rule can Eddie use to create the largest possible fraction? Explain why this rule works.

Sample answer: Eddie should place the largest digits in the boxes in the numerator and the smallest digits in the boxes in the denominator. This will give him the largest possible numerator and the smallest possible denominator, which will also give him the largest fraction because the numerator is the dividend and the denominator is the divisor. A larger dividend is more to divide up, and a smaller divisor means you divide into fewer groups. This gives the largest possible quotient.

Practice

③ Write the value of the 3 in each of the following numbers.

a. 1,322,072 **300,000**   b. 8,236,914 **30,000**

c. 5,703,000 **3,000**   d. 4,091,316 **300**

e. 8,192,038 **30**   f. 7,025,943 **3**

92   5.NF.3, 5.NBT.1

Fraction Estimation with Number Sense

Overview Students use fraction number sense to estimate and assess the reasonableness of answers to fraction addition and subtraction problems.

▶ **Before You Begin**

You may wish to have students cut out the fraction cards (*Math Journal 1*, Activity Sheets 8–13) before the lesson. Decide how you will store them for future use. If you prefer to use fraction cards without visual representations, make copies of *Math Masters*, pages TA16–TA19. For the optional Readiness activity, gather 12 index cards, 25 paper clips, and about 2 feet of string.

▶ **Vocabulary**

conjecture • argument

Standards

Focus Cluster
• Use equivalent fractions as a strategy to add and subtract fractions.

① **Warm Up** 5 min	Materials	
Mental Math and Fluency Students use grouping symbols to make number sentences true.	slate	**5.OA.1**

② **Focus** 35–40 min		
Math Message Students assess the reasonableness of the sum of a fraction addition problem.	fraction circles (optional)	**5.NF.2** SMP1, SMP3
Estimating Fraction Sums and Differences Students assess the reasonableness of answers to fraction addition and subtraction problems.	*Math Journal 1*, p. 85; Fraction Number Lines Poster; fraction circles; slate	**5.NF.2** SMP1, SMP2, SMP3
✓ **Assessment Check-In** See page 260. Expect most students to be able to argue that the answer to the problem identified does not make sense by comparing the sum to the addend $\frac{1}{2}$.	*Math Journal 1*, p. 85; fraction circles	**5.NF.2, SMP3**
Introducing *Build-It* **Game** Students play a game that helps develop fraction number sense.	*Student Reference Book*, p. 293; *Math Masters*, p. G15; scissors; per partnership: fraction cards	**5.NF.2** SMP1

③ **Practice** 20–30 min		
Interpreting Remainders Students practice interpreting remainders in division stories.	*Math Journal 1*, p. 86	**5.NBT.6, 5.NF.3** SMP6
Math Boxes 3-6 Students practice and maintain skills.	*Math Journal 1*, p. 87	See page 261.
Home Link 3-6 **Homework** Students use fraction number sense to assess the reasonableness of answers.	*Math Masters*, p. 96, pp. TA12–TA14 (optional)	**5.OA.1, 5.NF.2** SMP1, SMP3

⫻ **Go Online** to see how mastery develops for all standards within the grade.

Differentiation Options

| WHOLE CLASS | SMALL GROUP | PARTNER | INDEPENDENT | WHOLE CLASS | SMALL GROUP | PARTNER | INDEPENDENT | WHOLE CLASS | SMALL GROUP | PARTNER | INDEPENDENT |

Developing Fraction Number Sense

5.NF.2, SMP6

Student Reference Book,
pp. 174–176; per group: 12 index
cards; 25 paper clips; about 2 feet of string

To help develop fraction number sense, students use a number line to order fractions between 0 and 1. Distribute 8–12 blank index cards and tell students to write on each card a fraction greater than 0 but less than 1. To ensure variety, assign denominators of 2, 3, 4, 5, 8, 9, and tell students not to create duplicate cards. Students order their fractions between 0 and 1 by hanging them on a string with paper clips. Have students explain their thinking as they order the cards.

GMP6.1 Encourage them to refer to *Student Reference Book,* pages 174–176 to review comparing fractions with like numerators or like denominators and comparing fractions to benchmarks, such as 0, $\frac{1}{2}$, or 1. For equivalent fractions, clip the cards together as illustrated.

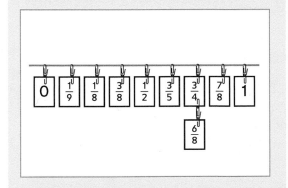

Increase-Decrease

5.NF.2, SMP7, SMP8

Activity Card 32;
Math Masters, p. 93;
per partnership: one spinner from *Math Masters,* p. 94; paper clip

To further develop fraction number sense, students consider whether a given expression increases or decreases in value when the numerator or denominator changes. Students look for patterns, make generalizations, and record their thinking.

GMP7.1, GMP8.1

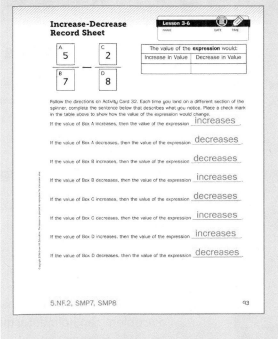

Identifying Unreasonable Answers

5.NF.2, SMP1, SMP3

Math Masters, p. 95;
fraction circles (optional)

To further develop fraction number sense and for more practice assessing the reasonableness of answers, students complete *Math Masters,* page 95. Given two possible answers, they identify the one that doesn't make sense and write an argument defending their reasoning.

GMP1.4, GMP3.1 Encourage students to use fraction circle pieces as well as mental estimation strategies.

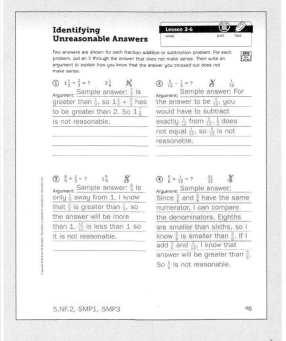

English Language Learner

Beginning ELL Build on students' familiarity with the calculator and its functions to help them understand the term *calculate.* Use think-aloud statements like these: *Let's use the calculator to calculate 8 * 7. Can you calculate 70 ÷ 7 in your head? Do you need to calculate the answer using the calculator? Which way is faster to calculate the sum of 973 + 748, using the calculator or paper and pencil?* Encourage students to use sentence frames like these: "I calculated that _____." and "I can calculate that by _____."

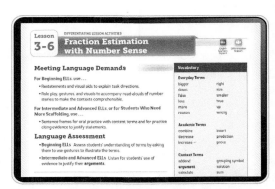

Differentiation Support pages are found in the online Teacher's Center.

Mathematical Process and Practice

SMP1 Make sense of problems and persevere in solving them.

GMP1.4 Check whether your answer makes sense.

SMP2 Reason abstractly and quantitatively.

GMP2.2 Make sense of the representations you and others use.

SMP3 Construct viable arguments and critique the reasoning of others.

GMP3.1 Make mathematical conjectures and arguments.

Academic Language Development

Help students make connections between the thinking process and the terms *reason* (as both a noun and a verb), *reasoning*, and *reasonable* using think-aloud statements to model what they mean. For example: *Fraction circles help me reason, or think, about the size of other fractions for which I only have number models. Seeing representations such as fraction circle pieces helps my reasoning, or my thinking. Seeing representations of fractions helps me determine the reasonableness of an answer, or whether my thinking makes sense. My reason for thinking that _____ is the answer is _____.*

---

1 Warm Up `5 min`

▶ Mental Math and Fluency

Have students rewrite the following number sentences on slates and insert grouping symbols to make them true. *Leveled exercises:*

● ○ ○ $4 + 5 / 3 = 3$ $(4 + 5) / 3 = 3$
$5 * 4 + 2 = 30$ $5 * (4 + 2) = 30$

● ● ○ $3 + 2 * 4 - 2 = 10$ $(3 + 2) * (4 - 2) = 10$
$7 \times 2 + 8 \times 2 = 140$ $7 \times (2 + 8) \times 2 = 140$

● ● ● $5 * 3 + 2 * 4 - 2 = 50$ $5 * \{(3 + 2) * (4 - 2)\} = 50$
$3 * 2 + 6 - 1 - 3 = 12$ $3 * \{2 + (6 - 1) - 3\} = 12$

2 Focus `35–40 min`

▶ Math Message

Christopher solved $\frac{2}{3} + \frac{1}{2}$ and got $\frac{3}{5}$. Sam said, "I think your answer is wrong, and I can prove it without calculating." Show how Sam can prove that Christopher is wrong without calculating a solution. You may use your fraction circles to help you.   `GMP1.4, GMP3.1`

▶ Estimating Fraction Sums and Differences

Math Journal 1, p. 85

| WHOLE CLASS | SMALL GROUP | PARTNER | INDEPENDENT |

Math Message Follow-Up Have students share how they solved the Math Message. Discuss how considering the size of fractions in a problem can help determine whether an answer makes sense. `GMP1.4, GMP2.2, GMP3.1` Be sure the discussion covers the following two strategies:

- Show $\frac{2}{3}$ and $\frac{1}{2}$ with fraction circle pieces. When combined, the pieces make up more than 1 whole. Since $\frac{3}{5}$ is less than 1 whole, the answer can't be $\frac{3}{5}$.

- On the Fraction Number Lines Poster the sum, $\frac{3}{5}$, is just a little more than the addend, $\frac{1}{2}$. Only a little more would need to be added to $\frac{1}{2}$ to get to $\frac{3}{5}$. Because the other addend, $\frac{2}{3}$, is larger than $\frac{1}{2}$, the sum of the two fractions has to be larger than 1 whole. The answer $\frac{3}{5}$ can't be correct.

Tell students that a statement like this, *I think that the sum of $\frac{2}{3}$ and $\frac{1}{2}$ cannot be $\frac{3}{5}$*, is an example of a **conjecture,** or a prediction based on mathematics.

The reasons students gave for why Christopher's answer could not be correct are examples of mathematical **arguments,** or explanations meant to show whether a conjecture is true or false. Remind students how in Unit 2 they used their understanding of whole numbers to make estimates and think about the reasonableness of the answers they got by calculating. Tell them that in this lesson they will use what they know about fractions to make conjectures about whether an answer makes sense. Then they will construct arguments to support their conjectures.

Display the problems below for partnerships to solve, and then discuss as a class. For each problem, students write a number model representing the story on slates. Partners share conjectures about whether the number model is true and make arguments supporting their conjecture. GMP3.1 Stress that students should not try to calculate an exact answer. They may represent numbers with fraction circle pieces, but also encourage them to refer to the Fraction Number Lines Poster or to use mental estimation or their understanding of fractions and operations to reason through the task. GMP1.4, GMP2.2

During the discussion, call attention to the strategies noted in the sample arguments: representing the problem with fraction circle pieces, comparing fractions with the same denominator or the same numerator, comparing two fractions to 1, and comparing the size of the addends (or minuend/subtrahend) to the sum (or difference).

Problem 1: Tiara calculated $\frac{5}{12} - \frac{1}{10}$ and said the difference was $\frac{5}{10}$. Does Tiara's answer make sense? How do you know? GMP3.1 Number model: $\frac{5}{12} - \frac{1}{10} = \frac{5}{10}$

Sample conjectures and arguments:

- Tiara's answer does not make sense. I showed $\frac{5}{12}$ and $\frac{5}{10}$ with fraction circle pieces. $\frac{5}{10}$ is larger than $\frac{5}{12}$. If a number is subtracted from $\frac{5}{12}$, the difference would be smaller than $\frac{5}{12}$, not larger.
- Tiara's answer does not make sense. I noticed that $\frac{5}{12}$ and $\frac{5}{10}$ both have the same numerator. If two fractions have the same numerator, the fraction with the larger denominator is smaller because a larger denominator means the whole is divided into a larger number of smaller pieces. $\frac{5}{10}$ is greater than $\frac{5}{12}$. In a subtraction problem it doesn't make sense that the difference would be larger than the number we started with.

Problem 2: Martin lives $\frac{5}{6}$ mile from school. Maria lives $\frac{1}{12}$ mile farther away from school than Martin. Maria said, "I live $\frac{3}{4}$ mile from school." Does Maria's statement make sense? How do you know? GMP3.1 Number model: $\frac{5}{6} + \frac{1}{12} = \frac{3}{4}$

Sample conjecture and argument:

- Maria's statement does not make sense. Both $\frac{5}{6}$ and $\frac{3}{4}$ are close to 1. $\frac{5}{6}$ is $\frac{1}{6}$ away from 1, and $\frac{3}{4}$ is $\frac{1}{4}$ away from 1. I know that $\frac{1}{6}$ is smaller than $\frac{1}{4}$, so $\frac{5}{6}$ is closer to 1, or greater than $\frac{3}{4}$. Adding $\frac{1}{12}$ to $\frac{5}{6}$ would give a sum greater than $\frac{5}{6}$, not smaller.

Adjusting the Activity

Differentiate Students who struggle assessing the reasonableness of answers may benefit from looking only at the addends (or minuend/subtrahend) and verbalizing a statement about the sum (or difference). For example, suggest that students use fraction circle pieces to model both addends in a problem with the answer covered. Provide additional whole red fraction circles for students to cover as a visual reference and perspective of a whole. Then have students ask themselves: *What do I know about the sum? Will it be less than 1 or more than 1?* This allows students to compare the answer to their estimation statement: *I know the sum has to be more than 1, and the answer, $\frac{7}{8}$, is less than 1, so the answer doesn't make sense.*

Go Online Differentiation Support

Math Journal 1, p. 85

Checking for Reasonable Answers Lesson 3-6 DATE TIME

Christopher solved some fraction problems. Do Christopher's answers make sense? Circle Yes or No. Then write an argument to show how you know.

Name Christopher   Sample arguments given.

Solve.

1. $\frac{2}{7} + \frac{1}{2} = \frac{3}{4}$ Conjecture: Does answer 1 make sense? Yes (No)
   Argument: If you add $\frac{2}{7}$ to $\frac{1}{2}$, the answer has to be more than $\frac{1}{2}$, not less. The answer $\frac{3}{4}$ is less than $\frac{1}{2}$ so it can't be correct.

2. Write >, <, or =. Conjecture: Does answer 2 make sense? (Yes) No
   $\frac{9}{10} > \frac{7}{8}$ Argument: I compared each fraction to 1. $\frac{9}{10}$ is $\frac{1}{10}$ away from 1. $\frac{7}{8}$ is $\frac{1}{8}$ away from 1. $\frac{1}{10}$ is less than $\frac{1}{8}$, so I know that $\frac{9}{10}$ is closer to 1 whole. $\frac{9}{10}$ is greater than $\frac{7}{8}$.

3. $\frac{7}{12} + \frac{1}{4} = \frac{8}{12}$ Conjecture: Does answer 3 make sense? Yes (No)
   Argument: $\frac{7}{12}$ and $\frac{8}{12}$ have the same denominator so they're easy to compare. I know that I would have to add $\frac{1}{12}$ to $\frac{7}{12}$ to get $\frac{8}{12}$. $\frac{1}{4}$ does not equal $\frac{1}{12}$, so the answer doesn't make sense.

4. $\frac{8}{9} + \frac{1}{3} = \frac{8}{12}$ Conjecture: Does answer 4 make sense? Yes (No)
   Argument: I compared the denominators because $\frac{8}{12}$ and $\frac{8}{9}$ have the same numerator. 12ths are smaller than 9ths, so $\frac{8}{12} < \frac{8}{9}$. It doesn't make sense for the sum to be smaller than the addend.

5.NF.2, SMP1, SMP3   85

Student Reference Book, p. 293

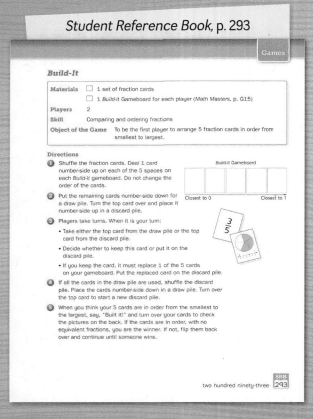

Math Journal 1, p. 86

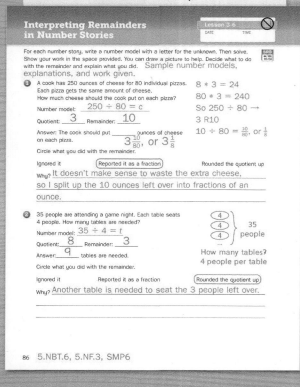

Problem 3: Jason has a bean plant that measures $\frac{7}{8}$ inch tall. A beetle nibbled $\frac{1}{16}$ inch off the top. Jason said, "Now my bean plant is only $\frac{6}{8}$ inch tall." Does Jason's conclusion make sense? How do you know? GMP3.1

Number model: $\frac{7}{8} - \frac{1}{16} = \frac{6}{8}$

Sample conjectures and arguments:

- Jason's conclusion does not make sense. I used the Fraction Number Lines Poster to compare $\frac{7}{8}$ and $\frac{6}{8}$. I can see that $\frac{7}{8}$ is exactly $\frac{1}{8}$ more than $\frac{6}{8}$. So I know that subtracting $\frac{1}{16}$ from $\frac{7}{8}$ would not equal $\frac{6}{8}$.

- It does not make sense to say the bean plant is $\frac{6}{8}$ inch tall. Both $\frac{7}{8}$ and $\frac{6}{8}$ have the same denominator so I can look at the numerators to compare them. I know that $\frac{7}{8}$ is $\frac{1}{8}$ more than $\frac{6}{8}$. So to get an answer of $\frac{6}{8}$, you'd have to subtract $\frac{1}{8}$, not $\frac{1}{16}$, from $\frac{7}{8}$.

Have partnerships complete journal page 85. Explain that, like Sam in the Math Message, they will examine more examples of Christopher's work. Without calculating, they should make a conjecture about whether each answer is reasonable and make an argument supporting their conjecture. GMP3.1 Encourage students to use fraction circles as well as mental estimation strategies.

✓ Assessment Check-In 5.NF.2

Math Journal 1, p. 85

Expect most students to be able to argue that the answer to Problem 1 does not make sense by comparing the sum to the addend $\frac{1}{2}$. Some may be able to write reasonable arguments for Problems 2–4, which require analyzing numerators or denominators or making less familiar comparisons, by using fraction circles or mental strategies. GMP3.1 For students who struggle, refer to the Adjusting the Activity note.

Evaluation Quick Entry Go online to record students' progress and to see trajectories toward mastery for these standards.

▶ Introducing *Build-It*

Math Journal 1, Activity Sheets 8–13; *Student Reference Book,* p. 293; *Math Masters,* p. G15

| WHOLE CLASS | SMALL GROUP | PARTNER | INDEPENDENT |

Build-It gives students an opportunity to develop fraction number sense by comparing and ordering fractions. These skills in turn provide an important foundation for mental estimation with fractions, which students use to assess the reasonableness of answers to fraction computation problems.

Have students cut out the fraction cards (Activity Sheets 8–13) and examine them. Explain that the side with the standard fraction notation is used to play the game, and the representations on the opposite side are used to check their work. Review the directions for *Build-It* on *Student Reference Book,* page 293. Play a sample round with a volunteer and demonstrate how to check the order of the cards using representations. Have partnerships play the game.

Observe

- Which students are following the rules of the game to replace fractions?
- Which students demonstrate an understanding of how to compare fractions with the same numerator? Same denominator?

Discuss

- *How did you check that you were placing fractions in order from smallest to largest?* **GMP1.4**
- *Which fractions are easy to compare? Which are harder to compare? Why?*

Differentiate **Game Modifications** **Go Online** Differentiation Support

---

Summarize Have students explain to a partner how comparing fractions can help them check whether the answers to fraction addition and subtraction problems make sense. **GMP1.4**

---

3 Practice 20–30 min

▶ Interpreting Remainders

Math Journal 1, p. 86

| WHOLE CLASS | SMALL GROUP | **PARTNER** | **INDEPENDENT** |

Students practice solving division number stories and interpreting remainders. They decide whether to ignore the remainder, round the quotient up, or report the remainder as a fraction. **GMP6.2**

▶ Math Boxes 3-6

Math Journal 1, p. 87

| WHOLE CLASS | **SMALL GROUP** | PARTNER | **INDEPENDENT** |

Mixed Practice Math Boxes 3-6 are paired with Math Boxes 3-8.

▶ Home Link 3-6

Math Masters, p. 96

Homework Students use fraction number sense to assess the reasonableness of answers. **GMP1.4** They make arguments to support their reasoning. **GMP3.1** You may wish to send home copies of *Math Masters,* pages TA12–TA14 so students have fraction circles to help them.

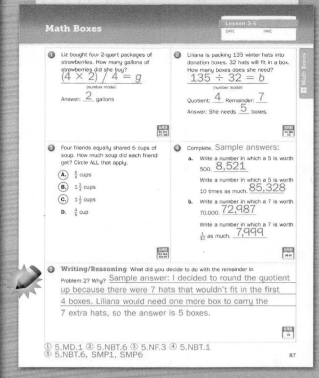

Math Journal 1, p. 87

Math Masters, p. 96

Fraction Estimation with Benchmarks

Overview Students use benchmarks to estimate sums and differences of fractions.

▶ **Before You Begin**
For Part 2, decide how you will display the sample problems for group discussion. For *Fraction Spin* each partnership will need one large paper clip.

▶ **Vocabulary**
benchmark

Standards

Focus Cluster
• Use equivalent fractions as a strategy to add and subtract fractions.

	Materials	
① Warm Up 5 min		
Mental Math and Fluency
Students use estimation to determine whether fraction number sentences are true or false.		5.NF.2
② Focus 45–50 min		
Math Message
Students represent the sum of two fractions on a number line.	*Math Journal 1*, p. 88	5.NF.2
SMP2, SMP3
Using Benchmarks to Estimate Sums and Differences of Fractions
Students use benchmarks to estimate sums and differences of fractions. They represent their estimates on a number line.	*Math Journal 1*, p. 88; fraction circles (optional)	5.NF.2
SMP2, SMP3
✓ **Assessment Check-In** See page 266.
Expect most students to be able to put an X near the correct benchmark and explain their thinking when estimating sums and differences of fractions.	*Math Journal 1*, p. 88; fraction circles (optional)	5.NF.2
Introducing *Fraction Spin*
Game Students practice estimating sums and differences of fractions.	*Student Reference Book*, p. 308; per partnership: *Math Masters*, pp. G16–G17; large paper clip	5.NF.2
SMP3
③ Practice 10–20 min		
Math Boxes 3-7
Students practice and maintain skills.	*Math Journal 1*, p. 89	See page 267.
Home Link 3-7
Homework Students use benchmarks to estimate sums and differences of fractions.	*Math Masters*, p. 99	5.NF.2, 5.MD.5, 5.MD.5a, 5.MD.5b

/// ⟨Go Online⟩ to see how mastery develops for all standards within the grade.

Differentiation Options

Readiness 15–20 min	**Enrichment** 15–20 min	**Extra Practice** 15–20 min
WHOLE CLASS · SMALL GROUP · PARTNER · INDEPENDENT	WHOLE CLASS · SMALL GROUP · **PARTNER** · INDEPENDENT	WHOLE CLASS · **SMALL GROUP** · PARTNER · INDEPENDENT

Readiness

Using Fraction Circles to Place Fractions on a Number Line

5.NF.2, SMP2

Math Masters, p. 97; fraction circles

To prepare for using benchmarks to estimate, students compare fractions to $\frac{1}{2}$ and locate them on a number line. Have students represent fractions between 0 and 1 with fraction circles. **GMP2.1** Ask them whether each fraction is less than, equal to, or greater than $\frac{1}{2}$ and have them explain their thinking. They record the fractions in the appropriate box on *Math Masters,* page 97 and then complete the problems at the bottom of the page. Discuss students' responses.

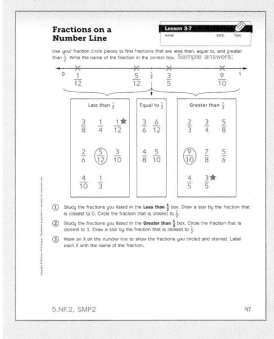

Enrichment

Playing *Fraction Top-It* (Estimation Version)

5.NF.2

Activity Card 33; per partnership: *Math Masters,* p. G14; 30 index cards

To further develop fraction number sense and explore using benchmarks, students play *Fraction Top-It* (Estimation Version). In this version they apply their knowledge of benchmarks to estimate sums and compare them.

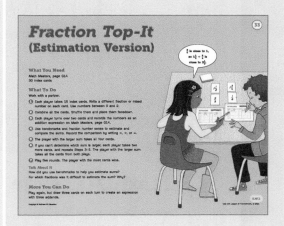

Extra Practice

Using Benchmarks to Estimate Sums and Differences

5.NF.2, SMP1, SMP2

Math Masters, p. 98

For more practice using benchmarks to estimate sums and differences, students complete *Math Masters,* page 98. They make sense of number stories and match the stories to estimates shown on a number line. **GMP1.1, GMP2.2**

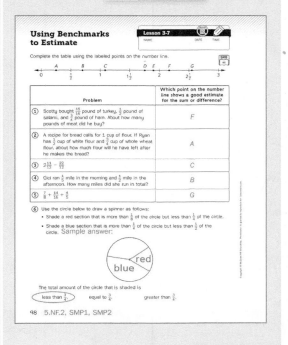

English Language Learner

Beginning ELL Make think-aloud statements using the term *help* to scaffold students' understanding of the term *benchmark* as a point of reference, or a standard according to which things are judged. For example: *I will use the fraction $\frac{1}{2}$ to help me think about the size of $\frac{5}{8}$. I will use $\frac{1}{2}$ as a benchmark to help me think about the size of $\frac{5}{8}$.* Students may benefit from seeing how benchmarks can be useful helpers in the same way helper facts are useful for thinking about nearby facts.

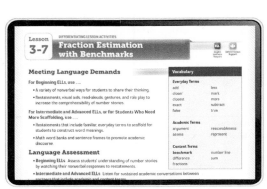

Differentiation Support pages are found in the online Teacher's Center.

Standards and Goals for
Mathematical Process and Practice

SMP2 **Reason abstractly and quantitatively.**
GMP2.1 Create mathematical representations using numbers, words, pictures, symbols, gestures, tables, graphs, and concrete objects.

SMP3 **Construct viable arguments and critique the reasoning of others.**
GMP3.1 Make mathematical conjectures and arguments.
GMP3.2 Make sense of others' mathematical thinking.

1 Warm Up — 5 min

▶ Mental Math and Fluency

Display the following number sentences. Ask students to identify whether they are true or false and then explain their thinking. *Leveled exercises:*

●○○ $\frac{1}{2} + \frac{1}{4} < 1$ True
$\frac{1}{2} - \frac{1}{4} > \frac{1}{2}$ False
$\frac{3}{4} + \frac{1}{2} < 1$ False

●●○ $\frac{3}{4} + \frac{5}{8} > 1$ True
$1 - \frac{2}{9} < \frac{1}{2}$ False
$\frac{1}{4} + \frac{4}{2} > 2$ True

●●● $\frac{6}{10} + \frac{8}{5} + \frac{1}{3} < 1\frac{1}{2}$ False
$\frac{6}{7} + \frac{4}{3} + \frac{1}{2} > 2$ True
$\frac{4}{3} - \frac{1}{3} + \frac{8}{4} > 1\frac{1}{2}$ True

2 Focus — 45–50 min

▶ Math Message

Math Journal 1, p. 88

Place an X to show the sum of $\frac{4}{12}$ and $\frac{3}{12}$ on the number line at the top of journal page 88. GMP2.1 *Be prepared to share your argument for why you placed the X where you did.* GMP3.1

▶ Using Benchmarks to Estimate Sums and Differences of Fractions

Math Journal 1, p. 88

| WHOLE CLASS | SMALL GROUP | PARTNER | INDEPENDENT |

Math Message Follow-Up Display a number line like the one on journal page 88. Invite a volunteer to show where he or she placed the X and to explain his or her reasoning. Ask other students to share their arguments, highlighting how they determined the distance of the sum from the labeled tick marks. GMP3.1 Guide the discussion with questions like these:

• *Is the sum closer to 0 or 1?* 1 *How do you know?* Sample answer: I know the sum is $\frac{7}{12}$. $\frac{7}{12}$ is greater than $\frac{1}{2}$, so it is closer to 1.

• *Is the sum closer to $\frac{1}{2}$ or 1?* $\frac{1}{2}$ *How do you know?* Sample answer: $\frac{6}{12}$ is another name for 1 half. $\frac{7}{12}$ is only $\frac{1}{12}$ away from 1 half but is $\frac{5}{12}$ away from 1 whole. So $\frac{7}{12}$ is closer to $\frac{1}{2}$ than to 1.

Math Journal 1, p. 88

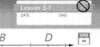

Using Benchmarks to Make Estimates

Lesson 3-7

DATE   TIME

E   A   C   B   D

0    $\frac{1}{2}$ 1   1$\frac{1}{2}$ 2   2$\frac{1}{2}$ 3

Estimate the sum or difference for each fraction number story. Place an X on the number line to represent your estimate. Then circle the best answer.

① Micah bought 1$\frac{1}{3}$ pounds of grapes and 1$\frac{1}{2}$ pounds of bananas. About how many pounds of fruit did Micah buy?

0    $\frac{1}{2}$ 1   1$\frac{1}{2}$ 2   2$\frac{1}{2}$ 3

How much fruit? about 2 pounds   about 2$\frac{1}{2}$ pounds   (about 3 pounds)
Explain your thinking.
Sample answer: 1$\frac{1}{3}$ is close to 1$\frac{1}{2}$, so I added 1$\frac{1}{2}$ + 1$\frac{1}{2}$ in my
head. 1 + 1 = 2, and $\frac{1}{2}$ + $\frac{1}{2}$ is 1, so that's 3 all together.

② Chloe has 2$\frac{1}{2}$ yards of fabric. She will use about $\frac{3}{8}$ yard to make a scarf. How many yards of fabric will she have left?

0    $\frac{1}{2}$ 1   1$\frac{1}{2}$ 2   2$\frac{1}{2}$ 3

How much fabric is left?   about 1$\frac{1}{2}$ yards   (about 2 yards)   about 3 yards
Explain your thinking.
Sample answer: I used a benchmark of $\frac{1}{2}$ for $\frac{3}{8}$. Then I
thought about 2$\frac{1}{2}$ − $\frac{1}{2}$ and knew that would be about 2.

Try This

③ The perimeter of a triangle is 10 inches. One side is 3$\frac{4}{16}$ inches long. Another side is 4$\frac{5}{8}$ inches long. About how many inches long is the third side? Explain how you estimated.
Sample answer: 3$\frac{4}{16}$ is close to 3$\frac{1}{2}$, and 4$\frac{5}{8}$ is close to 4$\frac{1}{2}$.
The length of those two sides is about 8 inches.
So the third side is about 10 − 8, or about 2 inches.

88   5.NF.2, SMP3

Point out how knowing that $\frac{7}{12}$ is just a little more than $\frac{1}{2}$ makes it easier to find the approximate location of $\frac{7}{12}$ on the number line.

Explain that familiar reference points, such as $\frac{1}{2}$, are called **benchmarks**. To estimate sums and differences of fractions, it often helps to think about how the fractions in the problems, as well as their sums and differences, relate to benchmarks like 0, $\frac{1}{2}$, 1, $1\frac{1}{2}$, 2, and $2\frac{1}{2}$.

Point out that it was possible to mentally calculate the exact answer in the Math Message because both fractions had a denominator of 12. Ask: *What are some examples of fraction addition and subtraction problems for which the exact sum or difference would not be as easy to calculate mentally?* Sample answer: Problems in which the denominators are different, such as $\frac{3}{7} + \frac{1}{3}$ Explain that using benchmarks is one way to mentally estimate sums or differences when an exact answer is not obvious. Thinking about benchmarks can also help in assessing the reasonableness of answers. Tell students that today they will focus only on estimating fraction sums and differences. They will not be finding exact answers.

Ask: *Is $\frac{1}{12} + \frac{1}{35}$ closer to 0 or to $\frac{1}{2}$? How do you know?* Guide the class to think through arguments with prompts like these: *Which benchmark is $\frac{1}{12}$ closest to?* 0 *Which benchmark is $\frac{1}{35}$ closest to?* 0 *If we add a number close to 0 to another number close to 0, will the sum be closer to 0 or $\frac{1}{2}$?* 0

Pose the following problems orally. Have students share their arguments and justify their solutions. GMP3.1

• Is $\frac{12}{13} + \frac{7}{8}$ closer to 1 or to 2? Sample answer: The sum is closer to 2 because $\frac{12}{13}$ is close to 1 and $\frac{7}{8}$ is close to 1.

• Is $\frac{1}{2} - \frac{7}{17}$ closer to $\frac{1}{2}$ or to 0? Sample answer: The difference is closer to 0 because $\frac{7}{17}$ is close to $\frac{1}{2}$.

• Is $\frac{18}{19} + \frac{23}{12}$ closer to 2 or to 3? Sample answer: The sum is closer to 3 because $\frac{18}{19}$ is close to 1 and $\frac{23}{12}$ is close to 2.

Have students look again at the number line on the top of journal page 88. Ask them to estimate the sum of $\frac{4}{5}$ and $\frac{1}{2}$ and to draw a dot labeled *A* on the number line at the approximate location of the sum. GMP2.1 Have partners share their thinking, and then discuss as a class. GMP3.1 Ask: *Where did you decide to mark the sum? Why?* Sample answer: The dot for the sum goes a little before the $1\frac{1}{2}$ mark. I know that $\frac{4}{5}$ is close to 1 but a little less than 1. If I add $\frac{1}{2}$ to something a little less than 1, the answer will be a little less than $1\frac{1}{2}$.

Display the fraction number stories on the next page one at a time and have students represent the estimated answer on the same number line, using labeled dots to mark each estimated sum or difference. GMP2.1

Student Reference Book, p. 308

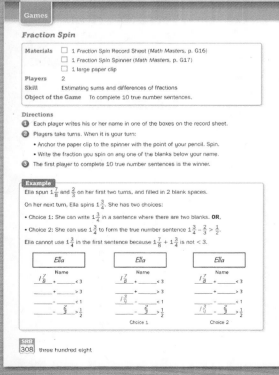

Have partners share their thinking and invite a few students to share their partner's thinking with the whole class. Explain that discussing how they solve problems will help students refine their mathematical communication skills. In addition, when they restate a partner's ideas in their own words, students learn to make sense of others' ideas and expand their understanding of problem-solving strategies. Encourage students to ask questions to make sure they understand their partner's thinking. **GMP3.1, GMP3.2**

- Jake mowed lawns for about $\frac{1}{3}$ hour on Saturday and about $1\frac{3}{4}$ hours on Sunday. About how many hours did Jake mow lawns over the weekend? Label your estimate B. Sample answer: Place the dot a little after 2. I know that $1\frac{3}{4}$ is $\frac{1}{4}$ from 2. Adding $\frac{1}{3}$ makes the total a little more than 2 because $\frac{1}{3}$ is a little more than $\frac{1}{4}$.

- Martin has $2\frac{1}{2}$ yards of fabric. He used $\frac{7}{8}$ yard to make a pillow. How many yards of fabric does he have left? Label your estimate C. Sample answer: Place the dot a little after $1\frac{1}{2}$. I know that if Martin had used exactly 1 yard, he would have $1\frac{1}{2}$ yards left. $\frac{7}{8}$ is little less than 1, so he has a little more than $1\frac{1}{2}$ yards left.

- Emma ran $2\frac{1}{4}$ miles. Stacie ran $\frac{1}{2}$ mile. How many miles did they run all together? Label your estimate D. Sample answer: Place the dot halfway between $2\frac{1}{2}$ and 3. I know that $2\frac{1}{4}$ is between 2 and $2\frac{1}{2}$. Adding $\frac{1}{2}$ makes the total more than $2\frac{1}{2}$ but less than 3, so Emma and Stacie ran between $2\frac{1}{2}$ and 3 miles all together.

- Karsyn has a piece of wood that is 3 feet long. If she uses $2\frac{13}{16}$ feet to make a bookshelf, how many feet of wood will she have left? Label your estimate E. Sample answer: Place the dot very close to 0. I know that $2\frac{13}{16}$ is very close to 3, so if Karsyn uses that much wood, she would have almost none—just $\frac{3}{16}$ foot—left.

Have partnerships complete Problems 1–3 on journal page 88.

 Assessment Check-In 5.NF.2

Math Journal 1, p. 3

Expect most students to be able to put an X near the correct benchmark and explain their thinking for Problems 1 and 2 on journal page 88. Students should consider both the whole number and fraction part of a mixed number when they choose a benchmark. The focus at this time is on making reasonable estimates, so don't worry if some students make estimates that are slightly greater than or less than the exact answer. Those who struggle may benefit from using fraction circle pieces to represent the number stories.

 Evaluation Quick Entry Go online to record students' progress and to see trajectories toward mastery for these standards.

▶ Introducing *Fraction Spin*

Student Reference Book, p. 308; *Math Masters*, pp. G16–G17

| WHOLE CLASS | SMALL GROUP | **PARTNER** | INDEPENDENT |

Fraction Spin provides practice estimating sums and differences of fractions. Students develop fraction number sense by strategically placing fractions to create true number sentences.

Review the directions for *Fraction Spin* on *Student Reference Book*, page 308 and play a sample round with a volunteer. While modeling the game, be sure to share your reasoning about how you used benchmarks, how you decided where to place each fraction, and how you determined whether a sentence was true. Encourage students to do the same as they play. **GMP3.1**

Observe

• Which students are placing fractions strategically?
• Which students can use mental strategies to estimate?

Discuss

• *How do you know your number sentences are true?* **GMP3.1**
• *Do you agree with your partner's reasoning?* **GMP3.2**

| **Differentiate** **Game Modifications** | **Go Online** | Differentiation Support |

Summarize Have students explain to a partner how they used a benchmark to estimate a sum or difference in *Fraction Spin*.

3 Practice 10–20 min

▶ Math Boxes 3-7

Math Journal 1, p. 89

| WHOLE CLASS | SMALL GROUP | PARTNER | INDEPENDENT |

Mixed Practice Math Boxes 3-7 are paired with Math Boxes 3-5.

▶ Home Link 3-7

Math Masters, p. 99

Homework Students use benchmarks to estimate sums and differences of fractions.

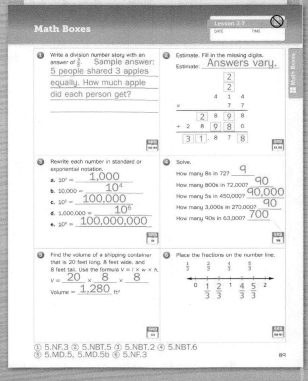

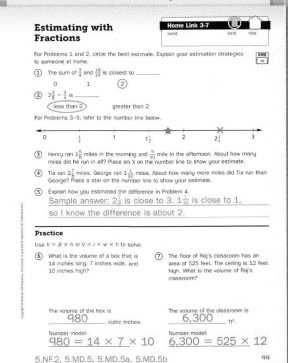

Renaming Fractions and Mixed Numbers

Overview Students rename mixed numbers and fractions greater than 1 by composing and breaking apart wholes.

▶ **Before You Begin**
Make sure fraction circle pieces are available. Students may need to pool their pieces for some problems. You can create extra pieces from *Math Masters,* pages TA12–TA14.

Standards

Focus Cluster
• Apply and extend previous understandings of multiplication and division.

 Warm Up 5 min

	Materials	
Mental Math and Fluency Students compare fractions greater than 1 to whole numbers.		5.NF.3

 Focus 35–40 min

Math Message Students rewrite a fraction as a mixed number.	fraction circles (optional)	5.NF.3 SMP3, SMP5
Renaming by Making and Breaking Apart Wholes Students rename mixed numbers and fractions greater than 1 by composing fractional parts into wholes and breaking apart wholes into fractional parts.	*Math Journal 1,* p. 90; fraction circles	5.NF.3 SMP1, SMP3, SMP5
✓ **Assessment Check-In** See page 272. Expect most students to be able to reason about breaking apart wholes and grouping fractional parts to find correct names for the mixed numbers in problems like those identified.	*Math Journal 1,* p. 90; fraction circles (optional)	5.NF.3 SMP3
Introducing *Rename That Mixed Number* **Game** Students play a game to practice renaming mixed numbers by trading wholes for fractional parts.	*Student Reference Book,* pp. 321–322; per partnership: *Math Masters,* p. G18; number cards 1–9 (4 of each); timer or stopwatch (optional)	5.NF.3 SMP3

③ Practice 20–30 min

Connecting Fractions and Division Students solve division number stories with fractional answers.	*Math Journal 1,* p. 91; fraction circles (optional)	5.OA.2, 5.NF.3 SMP7, SMP8
Math Boxes 3-8 Students practice and maintain skills.	*Math Journal 1,* p. 92	See page 273.
Home Link 3-8 **Homework** Students practice renaming mixed numbers and fractions greater than 1.	*Math Masters,* p. 102	5.NBT.5, 5.NF.3

/// **Go Online** to see how mastery develops for all standards within the grade.

Differentiation Options

Readiness	5–15 min	**Enrichment**	10–15 min	**Extra Practice**	15–20 min
WHOLE CLASS SMALL GROUP PARTNER INDEPENDENT		WHOLE CLASS SMALL GROUP PARTNER INDEPENDENT		WHOLE CLASS SMALL GROUP PARTNER INDEPENDENT	

Renaming Whole Numbers

5.NF.3

base-10 blocks, fraction circles

To prepare for renaming mixed numbers, students review renaming whole numbers. Have students show 24 with base-10 blocks. Expect most to show 2 tens (longs) and 4 ones (cubes). Ask students if they can show 24 in a different way. If students need help, remind them that they could trade 1 ten for 10 ones, and show 24 with 1 ten and 14 ones. Ask: *Is there another trade we can make?* Yes, trade the remaining ten for 10 ones to get 24 ones. Repeat with another number. *Suggestion:* 48 4 tens and 8 ones; 3 tens and 18 ones; 2 tens and 28 ones; 1 ten and 38 ones; 48 ones

Remind students that whole numbers can also be renamed as fractions. Have them show 2 with two red fraction circle pieces. Ask: *Are there any trades we can make to show 2 a different way?* Sample answer: Trade 1 whole for 2 halves. Have students make the trade as you record $2 = 1\frac{2}{2}$. Ask: *Is there another trade we can make?* Sample answer: Trade the remaining whole for 2 halves. Have students trade again as you record $2 = 1\frac{2}{2} = \frac{4}{2}$. Repeat as time permits. *Suggestions:* Show 3 using fourths. $3 = 2\frac{4}{4} = 1\frac{8}{4} = \frac{12}{4}$ Show 4 using sixths. $4 = 3\frac{6}{6} = 2\frac{12}{6} = 1\frac{18}{6} = \frac{24}{6}$ Emphasize that each name represents the same quantity.

Finding a Rule for the Number of Names

5.NF.3, SMP7, SMP8

Math Masters, p. 100

To extend their work with renaming mixed numbers, students rename mixed numbers in as many ways as they can by making and breaking apart wholes. They look for patterns and write a rule to describe how many equivalent names with the same denominator can be found for any given mixed number. **GMP7.1, GMP8.1** They use the rule to determine how many different ways they could rename larger mixed numbers. **GMP7.2**

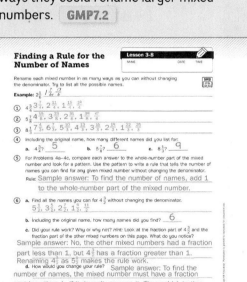

Renaming Mixed Numbers

5.NF.3, SMP1

Activity Card 34; per partnership: *Math Masters,* p. 101; scissors; fraction circles (optional)

For additional practice renaming mixed numbers, students determine which of their cards (from *Math Masters,* page 101) represents another name for a given mixed number or a fraction greater than 1. They compare their strategies. **GMP1.6**

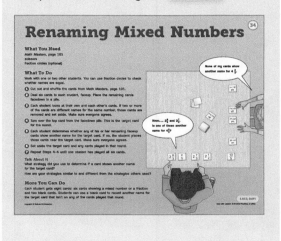

English Language Learner

Beginning ELL To support students' understanding of the term *mixed number,* build on their understanding of *same* and *different.* For example, show a set of red objects and say: *These are all the same.* Then add objects of another color, saying: *Now they are not all the same. Some are different. I have a* mixed *set.* Show a variety of items mixed together and point out that the collection contains different things. Extend this idea to mixed numbers, showing that whole numbers and fractions are different.

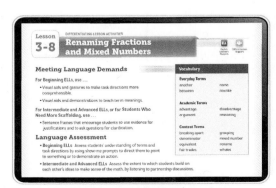

Differentiation Support pages are found in the online Teacher's Center.

Mathematical Process and Practice

SMP1 **Make sense of problems and persevere in solving them.**
GMP1.2 Reflect on your thinking as you solve your problem.

SMP3 **Construct viable arguments and critique the reasoning of others.**
GMP3.1 Make mathematical conjectures and arguments.

SMP5 **Use appropriate tools strategically.**
GMP5.2 Use tools effectively and make sense of your results.

Professional Development

The standards require Grade 5 students to interpret fractions as division (5.NF.3). In Lessons 3-1 and 3-2, students used an equal-sharing interpretation of division to explore connections between fractions and division. For example, they discovered that when 3 objects are shared equally by 4 people, each person's share is $\frac{3}{4}$ of an object. In this lesson, students use an equal-grouping interpretation of division to explore connections between fractions and division. For example, they convert $\frac{21}{5}$ to $4\frac{1}{5}$ by thinking: *How many groups of 5 fifths are in 21 fifths?* Students extend this reasoning to rename mixed numbers in multiple ways, which is a prerequisite skill for mixed-number addition and subtraction.

---

1 Warm Up   5 min

▶ Mental Math and Fluency

Dictate fractions greater than 1. For each fraction, ask: *Between which two whole numbers is this fraction?* Encourage students to explain their reasoning. *Leveled exercises:*

●○○ $\frac{3}{2}$ 1 and 2 $\frac{5}{2}$ 2 and 3 $\frac{7}{3}$ 2 and 3

●●○ $\frac{11}{2}$ 5 and 6 $\frac{17}{3}$ 5 and 6 $\frac{26}{4}$ 6 and 7

●●● $\frac{47}{4}$ 11 and 12 $\frac{89}{7}$ 12 and 13 $\frac{63}{2}$ 31 and 32

2 Focus   35–40 min

▶ Math Message

Write $\frac{21}{5}$ as a mixed number. Explain to a partner how you know that your mixed number is another name for $\frac{21}{5}$. **GMP3.1**   *You can use fraction circles, a number line, or drawings to help you.* **GMP5.2**

▶ Renaming by Making and Breaking Apart Wholes

Math Journal 1, p. 90

| WHOLE CLASS | SMALL GROUP | PARTNER | INDEPENDENT |

Math Message Follow-Up Expect most students to have rewritten $\frac{21}{5}$ as $4\frac{1}{5}$. Invite students to share their arguments for why $\frac{21}{5} = 4\frac{1}{5}$. **GMP3.1** *Sample arguments:*

- I can think of $\frac{21}{5}$ as $21 \div 5$, and I know that $21 \div 5 = 4\frac{1}{5}$. So $\frac{21}{5} = 4\frac{1}{5}$.

- I used 21 dark green fraction circle pieces to show $\frac{21}{5}$. When I put the pieces into groups of 5 to make wholes, there were 4 wholes with one piece left over. So $\frac{21}{5} = 4\frac{1}{5}$. **GMP5.2**

- I counted up 21 fifths from 0 on a number line. I know that each set of 5 fifths makes a whole, so I circled the numbers I landed on after every 5 counts. I know these numbers are the same as 1, 2, 3, and 4. I counted up one more fifth after 4, so that means $\frac{21}{5} = 4\frac{1}{5}$. **GMP5.2**

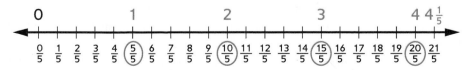

Highlight the strategy of rewriting $\frac{21}{5}$ as a mixed number by putting together groups of 5 fifths to make wholes. Explain that one way to rewrite $\frac{21}{5}$ as a mixed number is to think: *How many groups of 5 fifths are in 21 fifths? There are 4 groups of 5 fifths, or 4 wholes, with 1 fifth left over. That means $\frac{21}{5} = 4\frac{1}{5}$.* Thinking like this works for any fraction.

Ask students if they can think of another mixed number that is equivalent to $\frac{21}{5}$. Some students may suggest replacing the $\frac{1}{5}$ in $4\frac{1}{5}$ with an equivalent fraction, as in $4\frac{2}{10}$. Acknowledge that this is an equivalent name for $\frac{21}{5}$ and $4\frac{1}{5}$, but challenge students to think of a different mixed number with a denominator of 5 that is equivalent to $\frac{21}{5}$.

If no one suggests another mixed number, ask: *Is $1\frac{16}{5}$ another name for $\frac{21}{5}$? Why or why not?* **GMP3.1** Prompt partners to talk about this question, encouraging them to use fraction circle pieces, number lines, or pictures to help them show their thinking. **GMP5.2** *Sample argument:*

- Yes, $1\frac{16}{5}$ is another name for $\frac{21}{5}$. I can show $\frac{21}{5}$ with fraction circles using 21 dark green fifth pieces. If I put 5 fifth pieces together to make 1 whole, and trade those pieces for a red circle, I have 1 whole and 16 fifth pieces left, or $1\frac{16}{5}$.

Point out that students were still thinking about making groups of 5 fifths when they were showing that $\frac{21}{5} = 1\frac{16}{5}$, but this time instead of making as many groups as they could, they stopped after making one group. Ask: *How could we use similar thinking to find other mixed-number names for $\frac{21}{5}$?* **GMP1.2** Sample answer: Make two groups of 5 fifths and trade them for 2 wholes. There are 11 fifths left over, so $2\frac{11}{5}$ is another name for $\frac{21}{5}$. Make three groups of 5 fifths and trade them for 3 wholes. There are 6 fifths left over, so $3\frac{6}{5}$ is another name for $\frac{21}{5}$.

Summarize by noting that $1\frac{16}{5}$, $2\frac{11}{5}$, $3\frac{6}{5}$, and $4\frac{1}{5}$ are all equivalent names for $\frac{21}{5}$. Explain that there are advantages and disadvantages to each way of naming fractions greater than 1 as mixed numbers. For example, naming a mixed number using the biggest whole-number part possible can be useful for deciding which whole number it is closest to. Tell students that in later lessons they will see how other names can be useful when adding and subtracting mixed numbers. Today they will practice finding many names for fractions and mixed numbers without changing the denominator.

Pose another example. Ask: *What is another name for $3\frac{2}{3}$ that has the denominator 3?* Sample answer: $2\frac{5}{3}$ Have volunteers share their arguments for why their fraction or mixed number is another name for $3\frac{2}{3}$. **GMP3.1** *Sample argument:*

- I drew a picture of $3\frac{2}{3}$. Then I divided 1 of the wholes into thirds. Then I had 2 wholes and 5 thirds, so $2\frac{5}{3}$ is another name for $3\frac{2}{3}$.

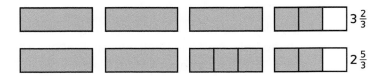

Math Journal 1, p. 90

Renaming Fractions and Mixed Numbers Lesson 3-8

DATE   TIME

Solve by following the steps. You can use fraction circles, number lines, or drawings to help.

① Find another name for $2\frac{3}{4}$.
- Show $2\frac{3}{4}$.
- Break apart 1 whole into $\frac{4}{4}$.

Name: $1\frac{7}{4}$

② Find another name for $\frac{14}{3}$.
- Show $\frac{14}{3}$.
- Make as many groups of 3 thirds as you can.
- Trade each $\frac{3}{3}$ for 1 whole.

Name: $4\frac{2}{3}$

Write another name for each mixed number that has the same denominator. Check that your trades are fair and record them. Sample answers given for Problems 3, 4, and 8.

Example: $3\frac{8}{6}$
Name: $2\frac{14}{6}$
Trade: *1 whole for $\frac{6}{6}$*

③ $2\frac{4}{5}$
Name: $1\frac{9}{5}$
Trade: 1 whole for $\frac{5}{5}$

④ $1\frac{12}{10}$
Name: $2\frac{2}{10}$
Trade: $\frac{10}{10}$ for 1 whole

Fill in the missing whole number or missing numerator.

⑤ $1\frac{4}{3} = 2\frac{\boxed{1}}{3}$ ⑥ $\boxed{3}\frac{1}{6} = 4\frac{7}{6}$ ⑦ $2\frac{4}{2} = 4\frac{\boxed{5}}{2}$

⑧ a. Mojo the monkey has 2 whole bananas and 5 half-bananas. Write a mixed number to show how many bananas Mojo has. $2\frac{5}{2}$ bananas

b. Manny the monkey has 4 whole bananas and 1 half-banana. Do Mojo and Manny have the same amount of banana? Explain how you know. Yes. If Mojo traded 4 of his half-bananas he would have 4 whole bananas and 1 half-banana. That's the same as Manny.

c. Marcus the monkey has the same amount of banana as Mojo. He only has half-bananas. How many half-bananas does he have? Explain your answer. He has 9 half-bananas, because $\frac{9}{2}$ is another name for $2\frac{5}{2}$.

90   5.NF.3, SMP3, SMP5

Games

Rename That Mixed Number

Materials	☐ 1 *Rename That Mixed Number* Record Sheet for each player (*Math Masters*, p. G18)
	☐ number cards 1–9 (4 of each)
	☐ timer or stopwatch (optional)
Players	2
Skill	Renaming mixed numbers
Object of the Game	To find as many possible names for the starting mixed number without changing the denominator.

Directions

1. Shuffle the cards and place them number-side down in a pile.

2. Each player picks 3 cards and uses them to represent the 3 parts of a mixed number: whole number, numerator, and denominator. Players may place the cards in any position they wish.

3. Players write the mixed number in the Starting Mixed Number column for Round 1 on the record sheet. See the record sheet on the following page.

4. Players list as many other names for their starting mixed number as they can by making or breaking apart wholes. All names must have the same denominator as the starting mixed number. Players record each name in the Other Names column of the record sheet for Round 1. Players may choose to set a 2-minute time limit.

5. Players record the number of other names they found for their mixed number in the Points column.

6. Players exchange record sheets and check each other's work. If a player finds another name for their partner's mixed number, they may claim that point and record it in the Points column for their partner's number on their own record sheet.

7. If players disagree on any answer, draw a picture to check the answer. If necessary, cross out any incorrect names and adjust your points.

8. Repeat Steps 1–7 four more times. The player with the most points after 5 rounds wins. See the example on the following page.

three hundred twenty-one SRB 321

Math Journal 1, p. 91

Connecting Fractions and Division

Lesson 3-8
DATE   TIME

Write a division expression to model each story, then solve. You can use fraction circles or draw pictures to help.

1. Olivia is running a 3-mile relay race with 3 friends. If the 4 of them each run the same distance, how many miles will each person run?

   Division number model: _____ 3 ÷ 4 _____

   Fractional answer: _____ $\frac{3}{4}$ mile _____

2. Chris has 3 pints of blueberries for fruit salad. If he splits the blueberries equally into 8 serving bowls, how many pints of blueberries will be in each bowl?

   Division number model: _____ 3 ÷ 8 _____

   Fractional answer: _____ $\frac{3}{8}$ pint _____

3. Four students shared 9 packages of pencils equally. How many packages of pencils did each student get?

   Division number model: _____ 9 ÷ 4 _____

   Fractional answer: $\frac{9}{4}$, or $2\frac{1}{4}$, packages

Use your answers to Problems 1–3 to answer the questions below.

4. Compare the numbers in each division number model with your fractional answer. What do you notice?
   Sample answer: The dividend is the same as the answer's numerator. The divisor is the same as the answer's denominator.

5. Write a rule for finding the fractional answer to a division problem by using the dividend and divisor.
   Sample answer: Make the dividend the numerator and the divisor the denominator.

6. Write and solve your own number story using your rule.
   Sample answer: Six people split 5 apples equally. How much apple did each person get? Each person got $\frac{5}{6}$ apple.

5.OA.2, 5.NF.3, SMP7, SMP8   91

Ask: *When you renamed $3\frac{2}{3}$, did you think about grouping fraction parts into wholes like you did in the last example? Why or why not?* GMP1.2 Sample answer: No, there weren't enough thirds to make a whole. *What did you do instead?* Sample answer: I thought about breaking apart a whole into fractional pieces. *What other mixed-number names for $3\frac{2}{3}$ can we find by breaking apart wholes?* Sample answer: Break apart 2 of the wholes into thirds to get 1 whole and 8 thirds, or $1\frac{8}{3}$. Break all the wholes into thirds to get $\frac{11}{3}$.

Explain that thinking about making wholes from fractional parts and breaking apart wholes into fractional parts are both methods of finding different names for fractions and mixed numbers. Students can think of both of these things as fair trades. Putting 5 fifths together to make a whole is like trading $\frac{5}{5}$ for 1 whole. This is a fair trade because $\frac{5}{5} = 1$. Breaking a whole into thirds is like trading 1 whole for $\frac{3}{3}$. This is a fair trade because $\frac{3}{3} = 1$. As long as the trades students make are fair, they have not changed the value of the number. Checking whether a trade is fair is one way to make sure that two names are equivalent.

Ask: *What trades could we make to find other names for $2\frac{7}{4}$?* Sample answers: Trade 2 wholes for 8 fourths to get $\frac{15}{4}$. Trade 1 whole for 4 fourths to get $1\frac{11}{4}$. Trade $\frac{4}{4}$ for 1 whole to get $3\frac{3}{4}$.

Have partnerships complete journal page 90.

✓ Assessment Check-In 5.NF.3

Math Journal 1, p. 90

Expect most students to be able to reason about breaking apart wholes and grouping fractional parts to find correct names for the mixed numbers in Problems 1–4 on journal page 90. These problems specify how to create a name or allow students to find any name. Some students may be able to solve Problems 5–7, which require them to find a specific name without directions. Some may also be able to write an argument for why two mixed numbers are equivalent in Problem 8. GMP3.1 If students struggle creating names, have them use fraction circle pieces to represent the starting number and trade pieces to make new names.

Evaluation Quick Entry Go online to record students' progress and to see trajectories toward mastery for these standards.

▶ Introducing *Rename That Mixed Number*

Student Reference Book, pp. 321–322; *Math Masters*, p. G18

| WHOLE CLASS | SMALL GROUP | PARTNER | INDEPENDENT |

Tell students that today they will learn a new game to practice renaming mixed numbers in multiple ways and thinking flexibly about fractions and mixed numbers. Distribute a copy of *Math Masters*, page G18 to each partnership. Read *Student Reference Book*, pages 321–322 as a class.

Model a sample turn. Think aloud as you list additional names for the mixed number you created. Detail how you can break apart wholes or make groups of fractional parts, using trading language as appropriate. For example: *I made the mixed number* $2\frac{1}{4}$*. I can break apart 1 whole into 4 fourths to get 1 whole and 5 fourths, or* $1\frac{5}{4}$*. I can also trade the 2 wholes in* $2\frac{1}{4}$ *for 8 fourths, because* $2 = \frac{8}{4}$*. That leaves me with 0 wholes and 9 fourths, or* $\frac{9}{4}$*.*

Have partnerships play the game. Circulate and observe.

Observe
- Which students have a strategy for renaming their mixed numbers?
- Which students can explain why their names are equivalent?

Discuss
- *What strategy did you use to rename your mixed number?*
- *Did the placement of the cards you chose have an effect on the number of names you found? Explain.* **GMP3.1**

| Differentiate | Game Modifications | | Go Online | | Differentiation Support |

Summarize Have students describe a round of *Rename That Mixed Number* and explain how they know one of their names is equivalent to the starting mixed number.

3 Practice   20–30 min

▶ Connecting Fractions and Division

Math Journal 1, p. 91

| WHOLE CLASS | SMALL GROUP | **PARTNER** | **INDEPENDENT** |

Students practice solving division number stories with fractional answers. They describe patterns in division number models and write a rule connecting fractions and division. **GMP7.1, GMP7.2, GMP8.1**

▶ Math Boxes 3-8

Math Journal 1, p. 92

| WHOLE CLASS | **SMALL GROUP** | PARTNER | INDEPENDENT |

Mixed Practice Math Boxes 3-8 are paired with Math Boxes 3-6.

▶ Home Link 3-8

Math Masters, p. 102

Homework Students practice renaming mixed numbers and fractions greater than 1 without changing the denominator.

Math Journal 1, p. 92

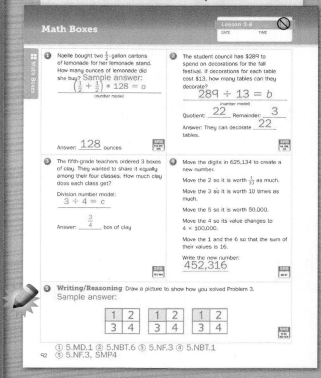

Math Masters, p. 102

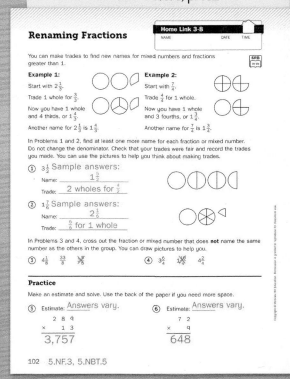

Introduction to Adding and Subtracting Fractions and Mixed Numbers

Overview Students explore strategies and tools for adding and subtracting fractions and mixed numbers.

▶ **Before You Begin**

Throughout the lesson, make sure fraction circle pieces and number lines (*Math Masters*, page TA20) are available for students to use as they solve problems. Partnerships may need to share fraction circle pieces to have enough of each color. If you need extra fraction circle pieces, make copies from *Math Masters*, pages TA12–TA14.

Standards

Focus Cluster
- Use equivalent fractions as a strategy to add and subtract fractions.

---

1 Warm Up   5 min

Materials

Mental Math and Fluency Students identify the value of digits in numbers.		5.NBT.1

2 Focus   35–40 min

	Materials	
Math Message Students solve a fraction addition number story.	fraction circles (optional), Fraction Number Lines Poster (optional)	5.NF.2
Exploring Addition and Subtraction of Fractions and Mixed Numbers Students solve number stories involving fractions and mixed numbers and share their strategies.	*Math Masters*, p. TA20; fraction circles; Fraction Number Lines Poster	5.NF.2 SMP1, SMP4, SMP5
Solving More Fraction Addition and Subtraction Number Stories Students solve problems involving addition and subtraction of fractions and mixed numbers.	*Math Journal 1*, p. 93; *Math Masters*, p. TA20; fraction circles	5.NF.2 SMP1, SMP4, SMP5
✓ **Assessment Check-In** See page 279. Expect most students to make reasonable attempts at solving addition and subtraction number stories that involve fractions and mixed numbers with the same denominator by using drawings, fraction circle pieces, or number lines.	*Math Journal 1*, p. 93; fraction circles (optional)	5.NF.2 SMP4, SMP5

3 Practice   20–30 min

Playing *Division Dash* **Game** Students practice solving division problems.	*Student Reference Book*, p. 301; per partnership: *Math Masters*, p. G13; number cards 1–9 (4 of each)	5.NBT.2, 5.NBT.6 SMP1
Math Boxes 3-9 Students practice and maintain skills.	*Math Journal 1*, p. 94	See page 279.
Home Link 3-9 **Homework** Students solve stories with mixed numbers.	*Math Masters*, p. 104	5.NBT.6, 5.NF.2 SMP1, SMP4

   Go Online ▷   to see how mastery develops for all standards within the grade.

Differentiation Options

Readiness	5–15 min	Enrichment	5–15 min	Extra Practice	5–15 min
WHOLE CLASS · **SMALL GROUP** · PARTNER · INDEPENDENT		WHOLE CLASS · SMALL GROUP · **PARTNER** · INDEPENDENT		WHOLE CLASS · SMALL GROUP · **PARTNER** · INDEPENDENT	

Counting by Unit Fractions

5.NF.2, SMP2

Fraction Number Lines Poster, 3 sets of fraction circles (for demonstration)

To prepare for adding fractions with like denominators, students count by unit fractions as you display fraction circle pieces. Tell students that the red circle is the whole. Hold up an orange piece. Ask: *What fraction does this piece represent?* **GMP2.2** $\frac{1}{3}$ Have students count by $\frac{1}{3}$s as you lay out 3 orange pieces to form a circle. $\frac{1}{3}, \frac{2}{3}, \frac{3}{3}$ Ask: *What is another name for $\frac{3}{3}$?* 1 Lay out 6 more orange pieces one by one as students continue to count by $\frac{1}{3}$s using mixed numbers. $1\frac{1}{3}, 1\frac{2}{3}, 1\frac{3}{3}$ or $2, 2\frac{1}{3}, 2\frac{2}{3}, 2\frac{3}{3}$ or 3 Then have students count back by $\frac{1}{3}$ as you take away orange pieces. **GMP2.2** $3, 2\frac{2}{3}, 2\frac{1}{3}, 2, 1\frac{2}{3}, 1\frac{1}{3}, 1, \frac{4}{4}, \frac{4}{4}$, Next illustrate counting up and back by thirds on the Fraction Number Lines Poster. Point out equivalent names, such as $\frac{3}{3} = 1$, $\frac{4}{3} = 1\frac{1}{3}$, $\frac{5}{3} = 1\frac{2}{3}$, and $\frac{6}{3} = 2$. Repeat the activity with other denominators as needed.

Writing Fraction Stories

5.NF.2, SMP1, SMP4

Math Masters, p. TA11

To extend their work with fraction addition and subtraction number stories, students write their own stories on *Math Masters*, page TA11. Explain that the stories should involve adding or subtracting fractions or mixed numbers with the same denominator. Encourage students to write number models for their stories and to use estimates to check the reasonableness of their answers. **GMP1.4, GMP4.1**

For an extra challenge, you might ask students to write stories that meet requirements like these:

- Write an addition story with an answer that is between 3 and 4.
- Write a subtraction story that requires trading a whole for $\frac{4}{4}$.

Number Stories
Use the space below to write a number story.

Number model: _____

Number Stories
Use the space below to write a number story.

Number model: _____

TA11

Solving Fraction Number Stories

5.NF.2, SMP1, SMP4, SMP5

Activity Card 35; *Math Masters*, p. 103, p. TA20 (optional); 6-sided die; fraction circles

For more practice solving addition and subtraction number stories involving fractions and mixed numbers, students follow the directions on Activity Card 35 to fill in the blanks for the number stories on *Math Masters*, page 103. They write number models for the stories, use tools to solve them, and use estimates to check that their answers make sense. Provide copies of *Math Masters*, page TA20 for students who choose to use number lines. **GMP1.4, GMP4.1, GMP5.2**

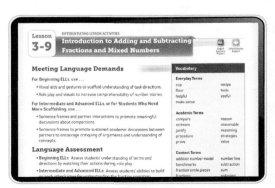

English Language Learner

Beginning ELL To support students with reading fraction terms, provide for student reference a display of unit fraction words, an illustration of the corresponding fraction circle piece, a number line example, and the numerical representation of the fraction. Engage the group in many choral counting opportunities before asking individual students to name fractional numbers.

Differentiation Support pages are found in the online Teacher's Center.

Standards and Goals for
Mathematical Process and Practice

SMP1 Make sense of problems and persevere in solving them.
GMP1.4 Check whether your answer makes sense.

SMP4 Model with mathematics.
GMP4.1 Model real-world situations using graphs, drawings, tables, symbols, numbers, diagrams, and other representations.

SMP5 Use appropriate tools strategically.
GMP5.2 Use tools effectively and make sense of your results.

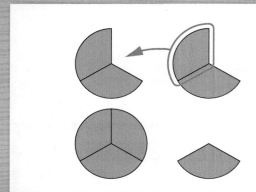

Modeling the Math Message problem with fraction circle pieces

① Warm Up 5 min

▶ Mental Math and Fluency

Display the following numbers and have students identify the value of the specified digit. *Leveled exercises:*

● ○ ○ What is the value of 7 in: 87? 7   78? 70   7,890? 7,000

● ● ○ What is the value of 5 in: 152? 50   25,893? 5,000   962,581? 500

● ● ● What is the value of 2 in: 32,895,476? 2,000,000
813,293,801? 200,000   929,867,135? 20,000,000

② Focus 35–40 min

▶ Math Message

Deena and her brother are making two kinds of bread. Both recipes call for $\frac{2}{3}$ cup of flour. How much flour do they need in all? You may use fraction circles or the Fraction Number Lines Poster to help you.

▶ Exploring Addition and Subtraction of Fractions and Mixed Numbers

Math Masters, p. TA20

| WHOLE CLASS | SMALL GROUP | PARTNER | INDEPENDENT |

Math Message Follow-Up Invite students to suggest a number model for the Math Message problem. GMP4.1 Sample answer: $\frac{2}{3} + \frac{2}{3} = f$ Ask them to estimate the sum using benchmarks. Sample answer: between 1 and $1\frac{1}{2}$ Then have volunteers share their answers and strategies. Be sure to discuss strategies that use fraction circles and number lines. If students are proficient at adding fractions with like denominators, ask them to prove their answer using two different strategies. *Sample strategies:*

• Use fraction circle pieces with the red circle as the whole. GMP5.2
Use 2 orange pieces to show $\frac{2}{3}$ and 2 more orange pieces to show another $\frac{2}{3}$. Put 3 orange third pieces together to make a whole. The total is 1 whole and 1 third, or $1\frac{1}{3}$. (*See margin.*)

• Use the Fraction Number Lines Poster to count up by thirds. GMP5.2
Start at 0. Count up 2 thirds for the flour in the first recipe, then 2 more thirds for the flour in the second recipe, landing at $\frac{4}{3}$. Since $\frac{3}{3}$ is the same as 1, and $\frac{4}{3}$ is 1 third past $\frac{3}{3}$, $\frac{4}{3}$ is the same as $1\frac{1}{3}$.

$\frac{2}{3}$ for first $\frac{2}{3}$ for second
recipe   recipe

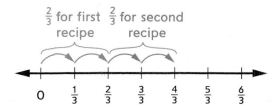

• Think in thirds. 2 thirds and 2 thirds is 4 thirds. It takes 3 thirds to make a whole, so 4 thirds is 1 whole plus 1 more third, or $1\frac{1}{3}$ in all.

Ask: *How many cups of flour do they need?* $1\frac{1}{3}$ cups *Look back at our estimate. Does the answer $1\frac{1}{3}$ cups make sense? How do you know?* GMP1.4 Sample answer: $1\frac{1}{3}$ is between 1 and $1\frac{1}{2}$, so the answer makes sense.

Encourage students to compare strategies using fraction circle pieces to strategies using number lines. Ask: *What did you like about using the fraction circles?* Sample answer: It's easy to see how 3 thirds fit together to make a whole. *What did you like about using a number line?* Sample answer: It was easy to count up by thirds. Tell students that in today's lesson they will solve more problems involving adding and subtracting fractions and mixed numbers. Explain that fraction circle pieces and number lines are both good tools to use when solving these problems.

Distribute a copy of *Math Masters,* page TA20 to each student. Explain that they may use the number lines on this page to help them solve fraction problems. Pose the following story and have students work in partnerships to write a number model, make an estimate, and solve. GMP4.1

Andre bought $1\frac{3}{4}$ pounds of ham and $2\frac{2}{4}$ pounds of beef. How many pounds of meat did he buy in all? Sample number model: $m = 1\frac{3}{4} + 2\frac{2}{4}$; Sample estimate: Between 4 and $4\frac{1}{2}$ pounds

After partnerships have had time to work, have them share strategies. *Sample strategies:*

• First add the whole-number parts: $1 + 2 = 3$. Then add the fraction parts using fraction circle pieces: 3 fourths plus 2 fourths is 5 fourths. The total is 3 wholes and 5 fourths, or $3\frac{5}{4}$. I could trade 4 fourths for 1 whole and rename the answer as $4\frac{1}{4}$.

• Use a number line to think about going up to the next whole number. GMP5.2 Start at $1\frac{3}{4}$. Jump up $\frac{1}{4}$ to get to 2. You still have to add $2\frac{1}{4}$ more. Jump up 2 to get to 4, then jump up $\frac{1}{4}$ to get to $4\frac{1}{4}$.

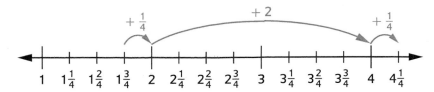

Ask: *How many pounds of meat did Andre buy?* $4\frac{1}{4}$ pounds *Does that answer make sense? How do you know?* GMP1.4 Sample answer: $4\frac{1}{4}$ is between 4 and $4\frac{1}{2}$, so the answer makes sense.

NOTE Be sure to point out that, in this number story, both $3\frac{5}{4}$ and $4\frac{1}{4}$ are correct answers. However, recording the answer as $4\frac{1}{4}$ makes it easier to compare the answer to the estimate and to make sense of the quantity. As students solve problems, encourage them to trade fractional parts for as many wholes as possible when they record their final answer.

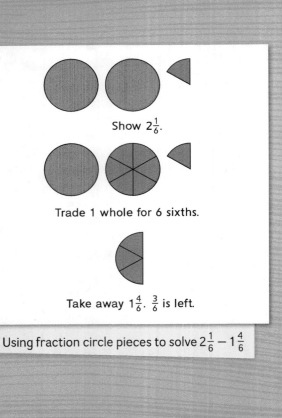

Show $2\frac{1}{6}$.

Trade 1 whole for 6 sixths.

Take away $1\frac{4}{6}$. $\frac{3}{6}$ is left.

Using fraction circle pieces to solve $2\frac{1}{6} - 1\frac{4}{6}$

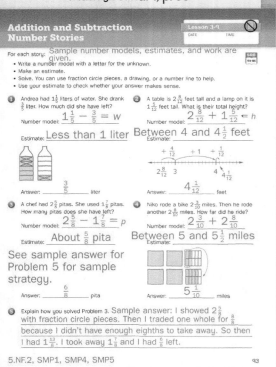

Addition and Subtraction Number Stories

Lesson 3-9

For each story:
- Write a number model with a letter for the unknown.
- Make an estimate.
- Solve. You can use fraction circle pieces, a drawing, or a number line to help.
- Use your estimate to check whether your answer makes sense.

Sample number models, estimates, and work are given.

① Andrea had $1\frac{1}{5}$ liters of water. She drank $\frac{3}{5}$ liter. How much did she have left?

Number model: $1\frac{1}{5} - \frac{3}{5} = w$

Estimate: Less than 1 liter

Answer: $\frac{3}{5}$ liter

② A table is $2\frac{8}{12}$ feet tall and a lamp on it is $1\frac{5}{12}$ feet tall. What is their total height?

Number model: $2\frac{8}{12} + 1\frac{5}{12} = h$

Estimate: Between 4 and $4\frac{1}{2}$ feet

Answer: $4\frac{1}{12}$ feet

③ A chef had $2\frac{5}{8}$ pitas. She used $1\frac{7}{8}$ pitas. How many pitas does she have left?

Number model: $2\frac{5}{8} - 1\frac{7}{8} = p$

Estimate: About $\frac{5}{8}$ pita

See sample answer for Problem 5 for sample strategy.

Answer: $\frac{6}{8}$ pita

④ Niko rode a bike $2\frac{3}{10}$ miles. Then he rode another $2\frac{8}{10}$ miles. How far did he ride?

Number model: $2\frac{3}{10} + 2\frac{8}{10}$

Estimate: Between 5 and $5\frac{1}{2}$ miles

Answer: $5\frac{1}{10}$ miles

⑤ Explain how you solved Problem 3. Sample answer: I showed $2\frac{5}{8}$ with fraction circle pieces. Then I traded one whole for $\frac{8}{8}$ because I didn't have enough eighths to take away. So then I had $1\frac{13}{8}$. I took away $1\frac{7}{8}$ and I had $\frac{6}{8}$ left.

5.NF.2, SMP1, SMP4, SMP5   93

Follow the same procedure for the number story below. Encourage students to think about how pictures, fraction circle pieces, or number lines could be used to solve the problem.

Monique had $2\frac{1}{6}$ graham crackers. She ate $1\frac{4}{6}$ crackers with her lunch. How many crackers did she have left? **GMP4.1** Sample number model: $2\frac{1}{6} - 1\frac{4}{6} = c$; Sample estimate: less than 1 cracker *Sample strategies:*

- Use fraction circle pieces with the red circle as the whole. **GMP5.2** Show $2\frac{1}{6}$. Think: *Can I take away $1\frac{4}{6}$?* No, there are not enough sixths. Trade 1 whole for 6 sixths. Now there is 1 whole and 7 sixths. Take away $1\frac{4}{6}$. $\frac{3}{6}$ are left. (*See margin.*)
- Think: *I can trade 1 whole for $\frac{6}{6}$, so $1\frac{4}{6}$ is the same as $\frac{6}{6} + \frac{4}{6}$, or $\frac{10}{6}$.* Use a number line to count back 10 sixths from $2\frac{1}{6}$. **GMP5.2** Land at $\frac{3}{6}$.

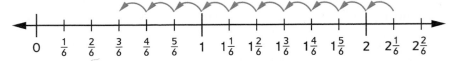

- Draw a picture of $2\frac{1}{6}$ crackers. **GMP4.1** First cross out 1 whole cracker. $\frac{4}{6}$ cracker still needs to be crossed out. Divide the other whole cracker into sixths. Then cross out 4 sixths. $\frac{3}{6}$ cracker is left.

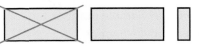

Draw $2\frac{1}{6}$ crackers. Cross out 1 cracker.

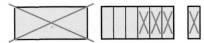

Divide the other whole cracker into sixths.
Cross out 4 sixths. $\frac{3}{6}$ cracker is left.

Ask: *How many graham crackers did Monique have left?* $\frac{3}{6}$ cracker *Does that answer make sense? How do you know?* **GMP1.4** Sample answer: Yes. We estimated that it would be less than 1. $\frac{3}{6}$ is less than 1.

Have students compare strategies. Ask: *How were pictures and fraction circle pieces useful when solving this problem?* Sample answer: They helped me think about how I could trade a whole for 6 sixths. *How was a number line useful?* Sample answer: The number line already shows the wholes divided into sixths. I didn't have to think about a trade.

▶ **Solving More Fraction Addition and Subtraction Number Stories**

Math Journal 1, p. 93; *Math Masters,* p. TA20

| WHOLE CLASS | SMALL GROUP | **PARTNER** | INDEPENDENT |

Have partners complete journal page 93 to practice solving addition and subtraction number stories that involve fractions and mixed numbers with the same denominator. They should write number models with a letter for the unknown and use drawings, fraction circle pieces, or number lines (*Math Masters,* page TA20) as needed to help them solve the problems. **GMP4.1, GMP5.2** Encourage students to use their estimates to check the reasonableness of their answers. **GMP1.4**

Assessment Check-In 5.NF.2

Math Journal 1, p. 93

Expect most students to be able to make reasonable estimates for the problems on journal page 93 and make reasonable attempts at solving them by using drawings, fraction circle pieces, or number lines. **GMP4.1, GMP5.2** Since the problems require making and breaking apart wholes, students may not answer all problems correctly. If students struggle, review how to model the problems with fraction circle pieces and how to use the fraction circle pieces to make the necessary trades.

Evaluation Quick Entry Go online to record students' progress and to see trajectories toward mastery for these standards.

Summarize Have students share their strategies for solving Problem 3 on journal page 93.

3 Practice   20–30 min

▶ Playing *Division Dash*

Student Reference Book, p. 301; *Math Masters,* p. G13

| WHOLE CLASS | SMALL GROUP | PARTNER | INDEPENDENT |

Students practice dividing 3-digit numbers by 2-digit numbers.

Observe

- Which students can find quotients mentally?
- Which students are able to calculate accurately on paper?

Discuss

- *How can you use mental math to divide or check answers?* GMP1.4
- *Explain your division strategy to your partner.*

| Differentiate | Game Modifications | | Go Online | Differentiation Support |

▶ Math Boxes 3-9

Math Journal 1, p. 94

| WHOLE CLASS | SMALL GROUP | PARTNER | INDEPENDENT |

Mixed Practice Math Boxes 3-9 are grouped with Math Boxes 3-12 and 3-14.

▶ Home Link 3-9

Math Masters, p. 104

Homework Students draw pictures and use number lines to solve mixed-number addition and subtraction number stories. GMP4.1 They use estimates to check whether their answers make sense. GMP1.4

Math Journal 1, p. 94

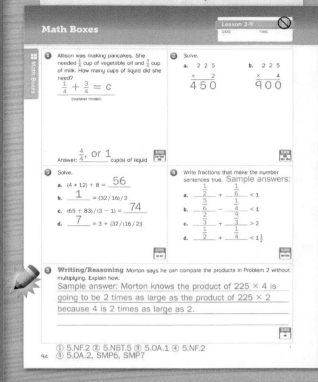

Math Masters, p. 104

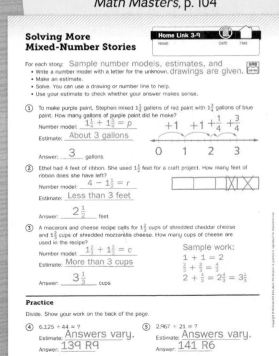

Exploring Addition of Fractions with Unlike Denominators

Overview Students use fraction circle pieces to generate equivalent fractions and add fractions.

▶ **Before You Begin**
Fraction circle pieces should be available to students throughout the lesson. If needed, you can create extra fraction circle pieces from *Math Masters,* pages TA12–TA14. Be sure the Fraction Circles Poster is easily visible to the class. For the optional Enrichment activity, cut apart enough copies of *Math Masters,* page 105 so that each partnership has a half-sheet.

▶ **Vocabulary**
equivalent fraction

Standards

Focus Cluster
• Use equivalent fractions as a strategy to add and subtract fractions.

1 Warm Up 5 min	**Materials**	
Mental Math and Fluency Students compare fractions and sums of fractions.	slate	5.NF.2

2 Focus 35–40 min		
Math Message Students represent a fraction sum with fraction circle pieces.	fraction circles	5.NF.1 SMP5
Thinking in Same-Size Pieces Students practice using same-size pieces to add fractions.	fraction circles	5.NF.1 SMP5
Finding Fraction Sums Students use fraction circle pieces to generate equivalent fractions and find fraction sums.	fraction circles	5.NF.1, 5.NF.2 SMP1, SMP5, SMP6
Practicing Fraction Addition Students practice adding fractions with unlike denominators.	*Math Journal 1,* p. 95; fraction circles	5.NF.1, 5.NF.2 SMP5
✓ **Assessment Check-In** See page 285. Expect most students to be able to use fraction circle pieces to find the sums in fraction addition problems like those identified.	*Math Journal 1,* p. 95	5.NF.1, SMP5

3 Practice 20–30 min		
Explaining Place-Value Patterns Students explain patterns in answers to multiplication problems involving powers of 10.	*Math Journal 1,* pp. 96–97	5.NBT.1, 5.NBT.2 SMP7, SMP8
Math Boxes 3-10: Preview for Unit 4 Students preview skills and concepts for Unit 4.	*Math Journal 1,* p. 98	See page 285.
Home Link 3-10 **Homework** Students solve fraction addition problems.	*Math Masters,* p. 107	5.OA.1, 5.OA.2, 5.NF.1

 Go Online to see how mastery develops for all standards within the grade.

Differentiation Options

Readiness — 5–15 min
WHOLE CLASS | **SMALL GROUP** | PARTNER | INDEPENDENT

Representing Fractions with Fraction Circle Pieces

5.NF.1, SMP2

fraction circles

To prepare for using fraction circle pieces to add fractions, students identify values of various sets of fraction pieces and use fraction circles to show equivalent fractions. **GMP2.2** Tell students that the red circle is the whole. Ask questions like these:

- *Use your pieces to show $\frac{3}{4}$.* 3 yellow pieces *How do you know that they show $\frac{3}{4}$?* Sample answer: 4 yellows make a whole, so 1 yellow is $\frac{1}{4}$ and 3 yellows are $\frac{3}{4}$. *Can you use different pieces to show another name for $\frac{3}{4}$?* Sample answer: I can use 6 dark blue pieces, because each dark blue piece is $\frac{1}{8}$ and $\frac{6}{8}$ is the same as $\frac{3}{4}$.

- *What fraction do 5 light blue pieces show?* $\frac{5}{6}$ *How do you know?* Sample answer: 6 light blue pieces make 1 whole, so 1 light blue piece is $\frac{1}{6}$ and 5 light blue pieces are $\frac{5}{6}$. *Can you use different pieces to show another name for $\frac{5}{6}$?* Sample answer: I can cover each light blue piece with 2 light green pieces. Each light green piece is $\frac{1}{12}$, so that shows that $\frac{5}{6}$ is the same as $\frac{10}{12}$.

Enrichment — 5–15 min
WHOLE CLASS | **SMALL GROUP** | PARTNER | INDEPENDENT

Finding Fractions that Sum to 1

5.NF.1, SMP5

Activity Card 36; *Math Masters*, p. 105; fraction circles

To extend their work with fraction addition, students use fraction circle pieces to find addition problems with a sum of 1. **GMP5.2** They use equivalent fractions to verify that the sum is 1.

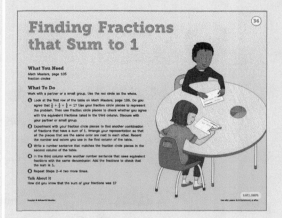

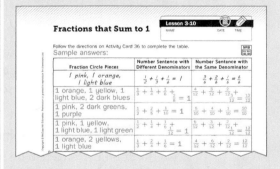

Fractions that Sum to 1

Extra Practice — 5–15 min
WHOLE CLASS | SMALL GROUP | **PARTNER** | INDEPENDENT

Finding Fraction Problems That Do Not Belong

5.NF.1, 5.NF.2, SMP3, SMP5

Math Masters, p. 106; fraction circles

For more practice estimating fraction sums and solving fraction addition problems, students use estimates to eliminate the addition problem in a set that does not have the same sum as the others. Then they find the sum, using fraction circle pieces if they wish. **GMP5.2** For one set, students explain how they knew which addition problem did not share the same sum. **GMP3.1**

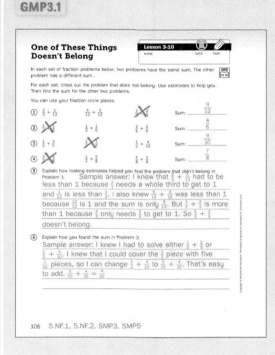

One of These Things Doesn't Belong

English Language Learner

Beginning ELL Use think-alouds as you display a pair of objects that are alike. For example: *These two _____ are like each other. They are alike. They match. This one goes with the other one.* Then replace one of the objects with something that is obviously different. Use statements like these: *This one is not like the other. This one is different. They do not match. They do not go together.* Build on the term *go together* by showing different fraction pairs with like and unlike denominators, asking students to point to the pairs with like denominators.

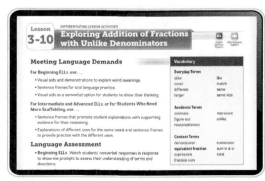

Differentiation Support pages are found in the online Teacher's Center.

Standards and Goals for
Mathematical Process and Practice

SMP1 **Make sense of problems and persevere in solving them.**
   GMP1.3 Keep trying when your problem is hard.

SMP5 **Use appropriate tools strategically.**
   GMP5.2 Use tools effectively and make sense of your results.

SMP6 **Attend to precision.**
   GMP6.3 Use clear labels, units, and mathematical language.

Representing $\frac{1}{4} + \frac{1}{8}$

Covering the yellow piece with two dark blue pieces

Showing that the sum is $\frac{1}{8}$ less than $\frac{1}{2}$

1 Warm Up 5 min

▶ Mental Math and Fluency

Display pairs of expressions and fractions. Have students write the larger value on their slates. Encourage them to explain their thinking. *Leveled exercises:*

● ○ ○ $\frac{3}{4}; \frac{1}{2}$ $\frac{3}{4}$ $\frac{3}{8}; \frac{1}{2}$ $\frac{1}{2}$ $\frac{2}{3}; \frac{1}{2}$ $\frac{2}{3}$

● ● ○ $\frac{1}{4} + \frac{2}{4}; \frac{1}{2}$ $\frac{1}{4} + \frac{2}{4}$ $\frac{2}{8} + \frac{1}{8}; \frac{1}{2}$ $\frac{1}{2}$ $\frac{1}{3} + \frac{1}{3}; \frac{1}{2}$ $\frac{1}{3} + \frac{1}{3}$

● ● ● $\frac{1}{4} + \frac{1}{2}; 1$ 1 $\frac{1}{3} + \frac{1}{2}; 1$ 1 $\frac{2}{3} + \frac{3}{4}; 1$ $\frac{2}{3} + \frac{3}{4}$

2 Focus 35–40 min

▶ Math Message

Use your fraction circle pieces. The red circle is the whole. Place a $\frac{1}{4}$ piece next to a $\frac{1}{8}$ piece. What is $\frac{1}{4} + \frac{1}{8}$? $\frac{3}{8}$ Talk with a partner about how you could use your fraction circle pieces to find the sum. GMP5.2

▶ Thinking in Same-Size Pieces

| WHOLE CLASS | SMALL GROUP | PARTNER | INDEPENDENT |

Math Message Follow-Up Ask: *If the red circle is the whole, what piece represents $\frac{1}{4}$?* The yellow piece *What piece represents $\frac{1}{8}$?* The dark blue piece Display a yellow piece next to a dark blue piece. (*See margin.*) Have students share strategies for finding the sum. GMP5.2 *Sample strategies:*

• Cover the yellow piece with two dark blue pieces. (*See margin.*) This shows that $\frac{1}{4}$ is the same as $\frac{2}{8}$. So $\frac{1}{4} + \frac{1}{8}$ is the same as $\frac{2}{8} + \frac{1}{8}$, or $\frac{3}{8}$ in all.

• Put an extra dark blue piece next to the representation of the sum. (*See margin.*) Then the total is $\frac{1}{2}$. So $\frac{1}{4} + \frac{1}{8}$ is $\frac{1}{8}$ away from $\frac{1}{2}$. $\frac{4}{8}$ is the same as $\frac{1}{2}$, so the sum must be $\frac{1}{8}$ less than that, or $\frac{3}{8}$.

To summarize the Math Message, display 3 dark blue pieces. Explain how students can simply count the pieces to find the total: 1 eighth, 2 eighths, 3 eighths. Display a yellow piece and a dark blue piece. Ask: *Why can't you figure out what fraction this represents by just counting the pieces?* GMP5.2 The pieces aren't the same size.

Explain that there is no fraction name for "a-fourth-and-an-eighth" because individual fractions always represent a number of same-size pieces. Recall how it was necessary when solving the Math Message to think about how many same-size eighth pieces would be the same as $\frac{1}{4} + \frac{1}{8}$. Tell students that in today's lesson they will solve more fraction addition problems by thinking about same-size pieces.

▸ Finding Fraction Sums

WHOLE CLASS | SMALL GROUP | PARTNER | INDEPENDENT

Have partnerships use fraction circle pieces to solve the problem $\frac{1}{2} + \frac{1}{4}$. GMP5.2 Encourage students to think about how to find the sum using same-size pieces. After a few minutes, invite volunteers to share strategies. Ask guiding questions to help them explain their strategies using language that connects the fraction circle pieces to the fractions they represent. GMP6.3 After students explain each step, display corresponding fraction notation, as shown below. Ask:

- *What did you do first?* Sample answer: I put together one pink piece and one yellow piece. *Why did you use those pieces?* The pink piece shows $\frac{1}{2}$ and the yellow piece shows $\frac{1}{4}$, so together they show $\frac{1}{2} + \frac{1}{4}$. Display $\frac{1}{2} + \frac{1}{4}$.
- *What did you do next?* Sample answer: I covered the pink piece with two yellow pieces. *What does that show?* It shows that $\frac{1}{2}$ is the same as $\frac{2}{4}$. Cross out $\frac{1}{2}$ and replace it with $\frac{2}{4}$.
- *How did you figure out the sum?* Sample answer: Now I can see that there are 3 yellow pieces, or 3 fourths. Complete the number sentence by displaying $\frac{2}{4} + \frac{1}{4} = \frac{3}{4}$.

Point out that when students covered the pink piece with two yellow pieces, they changed the fraction $\frac{1}{2}$ to the **equivalent fraction** $\frac{2}{4}$. This gave both fractions in the problem the same denominator. Using equivalent fractions makes fraction addition easier because all fractions represent the same-size piece—in this case, fourths.

Pose the following problems. Have students use fraction circle pieces to make equivalent fractions so both fractions represent the same-size piece.

- $\frac{1}{5} + \frac{1}{10}$ $\frac{2}{10} + \frac{1}{10} = \frac{3}{10}$
- $\frac{1}{2} + \frac{3}{8}$ $\frac{4}{8} + \frac{3}{8} = \frac{7}{8}$

Next ask students to make an estimate for $\frac{1}{2} + \frac{1}{3}$. Sample answer: Between $\frac{1}{2}$ and 1 Then have partnerships use fraction circle pieces to find the sum. GMP5.2 Because $\frac{1}{3}$ pieces do not exactly cover the $\frac{1}{2}$ piece, students may find it challenging to find the right size piece for this problem. Encourage them to continue trying different pieces if their first attempts don't work. GMP1.3

Differentiate **Common Misconception**

Some students may try to cover each addend's piece with one of a different color. For example, to solve $\frac{1}{2} + \frac{1}{3}$, they may cover the pink $\frac{1}{2}$ piece with two yellow $\frac{1}{4}$ pieces and cover the orange $\frac{1}{3}$ piece with four light green $\frac{1}{12}$ pieces. Explain that they have to use one color/size piece to cover the whole sum. Otherwise, they are simply creating a representation of another fraction addition problem with the same sum (for the example above, $\frac{2}{4} + \frac{4}{12}$).

Go Online | Differentiation Support

Professional Development

The purpose of this lesson is to help students see that rewriting fractions with like denominators is an effective strategy for adding fractions. The lesson is meant to be exploratory, so students are **not** expected to use a formal strategy to generate common denominators at this time. Formal procedures for generating common denominators will be introduced in Unit 5. Keep the focus here on student-developed strategies and using fraction circle pieces. If students are already able to generate common denominators and solve the problems mentally or on paper, encourage them to use the fraction circle pieces to show why their strategies work.

Math Journal 1, p. 95

Adding Fractions with Circle Pieces

Lesson 3-10 DATE TIME

Make an estimate. Then use your fraction circle pieces to find the sum. Use the red circle as the whole. Remember to think about using same-size pieces.

SRB 155, 189

Write a number sentence to show how you used equivalent fractions to find the sum.

Example: $\frac{1}{2} + \frac{1}{8} = ?$

Estimate: Between $\frac{1}{2}$ and 1

Show $\frac{1}{2} + \frac{1}{8}$ with fraction pieces.

Cover the $\frac{1}{2}$ piece with four $\frac{1}{8}$ pieces to show that $\frac{1}{2}$ is the same as $\frac{4}{8}$.

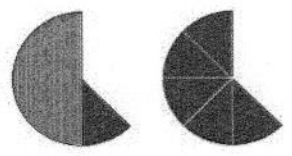

Sum: $\frac{5}{8}$

Number sentence: $\frac{4}{8} + \frac{1}{8} = \frac{5}{8}$

Sample estimates are given.

1. $\frac{2}{3} + \frac{1}{6} = ?$
   Estimate: A little less than 1
   Sum: $\frac{5}{6}$
   Number sentence: $\frac{4}{6} + \frac{1}{6} = \frac{5}{6}$
2. $\frac{2}{5} + \frac{3}{10} = ?$
   Estimate: Between $\frac{1}{2}$ and 1
   Sum: $\frac{7}{10}$
   Number sentence: $\frac{4}{10} + \frac{3}{10} = \frac{7}{10}$
3. $\frac{1}{3} + \frac{1}{12} = ?$
   Estimate: About $\frac{1}{2}$
   Sum: $\frac{5}{12}$
   Number sentence: $\frac{4}{12} + \frac{1}{12} = \frac{5}{12}$
4. $\frac{2}{6} + \frac{1}{4} = ?$
   Estimate: A little more than $\frac{1}{2}$
   Sum: $\frac{7}{12}$
   Number sentence: $\frac{4}{12} + \frac{3}{12} = \frac{7}{12}$
5. $\frac{2}{3} + \frac{1}{4} = ?$
   Estimate: Close to 1
   Sum: $\frac{11}{12}$
   Number sentence: $\frac{8}{12} + \frac{3}{12} = \frac{11}{12}$
6. $\frac{1}{2} + \frac{1}{5} = ?$
   Estimate: Between $\frac{1}{2}$ and 1
   Sum: $\frac{7}{10}$
   Number sentence: $\frac{5}{10} + \frac{2}{10} = \frac{7}{10}$

5.NF.1, 5.NF.2, SMP5 95

Explaining Place-Value Patterns

Lesson 3-10
DATE   TIME

1. Solve.

   a. 45 * 10 = __450__

   b. 45 * 10 * 10 = __4,500__

   c. 45 * 10 * 10 * 10 = __45,000__

   d. 45 * 10 * 10 * 10 = __450,000__

2. Look at your answers to Problem 1.

   a. What pattern do you notice in the number of zeros?
      Sample answer: Every time I multiply by one more 10,
      there is one more zero in the answer.

   b. What pattern do you notice in the value of the products?
      Sample answer: Every time I multiply by one more 10
      the product is 10 times as large.

   c. Do you think the patterns will hold true no matter how many 10s are in the problem?
      Use what you know about place value to explain your answer.
      Sample answer: The pattern will keep going. Every time I
      multiply by 10, I write one more zero. That means the
      product will be 10 times as much because all the digits
      move one place-value position to the left.

96   5.NBT.1, 5.NBT.2, SMP7, SMP8

Explaining Place-Value Patterns
(continued)

Lesson 3-10
DATE   TIME

3. Solve.

   a. $328 * 10^2$ = __32,800__   b. $328 * 10^5$ = __32,800,000__

   c. $328 * 10^7$ = __3,280,000,000__   d. $328 * 10^4$ = __3,280,000__

   e. $328 * 10^3$ = __328,000__   f. $328 * 10^1$ = __3,280__

4. Look at your answers to Problem 3.

   a. What pattern do you notice in the number of zeros?
      Sample answer: The number of zeros at the end of each
      answer is the same as the exponent.

   b. Use the pattern to help you write a rule for how to multiply a whole number by a power
      of 10.
      Sample answer: To multiply a whole number by a power of
      10, first write the whole number. Then write the number
      of zeros shown by the exponent.

   c. Use what you know about place value to explain why your rule will always work.
      Sample answer: The exponent shows how many times you
      are multiplying by 10. Every time you multiply by 10, you
      just write one more zero because multiplying by 10 moves
      all the digits one place to the left.

97

After partnerships have had sufficient time to work, display $\frac{1}{2} + \frac{1}{3}$. Invite volunteers to share strategies. Sample answer: We covered both the $\frac{1}{2}$ piece and the $\frac{1}{3}$ piece with $\frac{1}{6}$ pieces. That changed $\frac{1}{2}$ to $\frac{3}{6}$ and $\frac{1}{3}$ to $\frac{2}{6}$. Then we could add $\frac{3}{6}$ and $\frac{2}{6}$ to get $\frac{5}{6}$. Below the original problem, record the number sentence $\frac{3}{6} + \frac{2}{6} = \frac{5}{6}$.

Representing $\frac{1}{2} + \frac{1}{3}$

Showing that $\frac{1}{2} + \frac{1}{3} = \frac{5}{6}$

Ask: *How was this problem different from the other problems you solved?* We couldn't cover the pink piece with orange pieces. We had to find a smaller piece that would cover both addends. *How did you find the right-size piece?* GMP1.3 Sample answer: I tried different-size smaller circle pieces. I found that three $\frac{1}{6}$ pieces covered the $\frac{1}{2}$ piece, so I tried covering the $\frac{1}{3}$ piece with $\frac{1}{6}$ pieces and it worked. Point out that thinking about equivalent fractions can be helpful in finding the right-size piece. For example, if we know that $\frac{3}{6}$ is the same as $\frac{1}{2}$, the light blue $\frac{1}{6}$ piece might be a good one to try whenever the problem involves $\frac{1}{2}$.

Ask students to recall their estimate for $\frac{1}{2} + \frac{1}{3}$. Ask: *Does our answer of $\frac{5}{6}$ seem reasonable?* Sample answer: Yes. $\frac{5}{6}$ is between $\frac{1}{2}$ and 1. Remind them that they can use estimates to check the reasonableness of the answers they get using fraction circle pieces to make sure they used the pieces correctly.

Pose the following problems and have students make estimates and then solve them using fraction circle pieces. GMP5.2 Invite volunteers to share answers and strategies, encouraging them to clearly explain how they used equivalent fractions in their solutions. GMP6.3

- $\frac{1}{3} + \frac{1}{4}$ $\frac{4}{12} + \frac{3}{12} = \frac{7}{12}$
- $\frac{1}{2} + \frac{2}{5}$ $\frac{5}{10} + \frac{4}{10} = \frac{9}{10}$

▶ **Practicing Fraction Addition**

Math Journal 1, p. 95

| WHOLE CLASS | SMALL GROUP | PARTNER | INDEPENDENT |

Partners practice solving fraction addition problems using fraction circle pieces by completing journal page 95. GMP5.2 Remind students to use estimates to check the reasonableness of their answers.

Academic Language Development To help students share their thinking with a partner, provide a word bank with terms like *benchmark, like, unlike, same, different, numerator, denominator, equivalent,* and *fraction circle piece.* Challenge students to use the fraction circle pieces to illustrate what they say while explaining their solutions.

Assessment Check-In 5.NF.1

Math Journal 1, p. 95

Expect most students to be able to find the sums for Problems 1–3 on journal page 95 using fraction circle pieces. GMP5.2 Some may be able to use their knowledge of equivalent fractions to write appropriate number sentences. Expect students to struggle with finding the sums for Problems 4–6, which require replacing both addends with equivalent fractions. If students struggle solving Problems 1–3, suggest that they start by using their pieces to find several fractions that are equivalent to the first addend and then think about which of those equivalent fractions would be easy to add to the second addend.

 Evaluation Quick Entry Go online to record students' progress and to see trajectories toward mastery for these standards.

Summarize Have students talk with a partner about why it is helpful to think about same-size pieces when they add fractions.

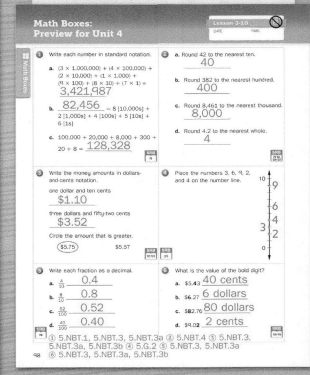

3 Practice 20–30 min

▶ Explaining Place-Value Patterns

Math Journal 1, pp. 96–97

| WHOLE CLASS | SMALL GROUP | **PARTNER** | **INDEPENDENT** |

Students multiply whole numbers by powers of 10. They look for patterns in the answers, use the patterns to write a rule, and explain in terms of place value why the rule works. GMP7.1, GMP8.1

▶ Math Boxes 3-10: Preview for Unit 4

Math Journal 1, p. 98

| WHOLE CLASS | **SMALL GROUP** | PARTNER | **INDEPENDENT** |

Mixed Practice Math Boxes 3-10 are paired with Math Boxes 3-15. These problems focus on skills and understandings that are prerequisite for Unit 4. You may want to use information from these Math Boxes to plan instruction and grouping in Unit 4.

▶ Home Link 3-10

Math Masters, p. 107

Homework Students solve fraction addition problems using fraction circle illustrations.

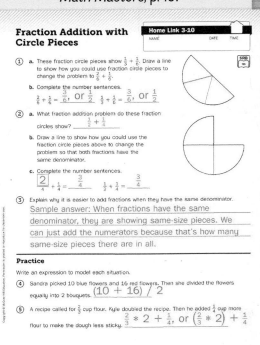

Playing *Fraction Capture*

Overview Students learn a new game to practice breaking apart and adding fractions.

▶ **Before You Begin**

To prepare for *Fraction Capture*, copy and cut apart enough *Fraction Capture* record sheets (*Math Masters*, page G20) for each student to have one. Each partnership will need one gameboard (*Math Masters*, page G19). You may wish to make reusable *Fraction Capture* gameboards by laminating copies of *Math Masters*, page G19 and having students write on them with dry-erase markers.

Standards

Focus Clusters
• Write and interpret numerical expressions.
• Use equivalent fractions as a strategy to add and subtract fractions.

 Warm Up 5 min

	Materials	
Mental Math and Fluency Students add and subtract fractions to solve number stories.		**5.NF.2**

2 Focus 35–40 min

Math Message Students write at least two fraction addition problems with the same sum.		**5.OA.2, 5.NF.1** **SMP1**
Breaking Apart Fractions Students discuss strategies for finding different sets of fractions that have the same sum.	fraction circles (optional)	**5.OA.2, 5.NF.1** **SMP1**
Introducing *Fraction Capture* **Game** Students practice breaking apart fractions and finding fraction sums. They share game strategies.	*Student Reference Book*, p. 305; *Math Masters*, p. TA2 (optional); per partnership: *Math Masters*, pp. G19– G20; two 6-sided dice; fraction circles (optional)	**5.OA.2, 5.NF.1** **SMP1**
✓ **Assessment Check-In** See page 290. Expect most students to be able to find fractions with like denominators that add to the appropriate sum and write a correct addition expression while playing the game *Fraction Capture*.	*Math Masters*, p. G20; fraction circles (optional)	**5.OA.2, 5.NF.1**

3 Practice 20–30 min

Renaming Fractions and Mixed Numbers Students translate between mixed numbers and fractions greater than 1.	*Math Journal 1*, p. 99; *Math Masters*, p. TA20 (optional); fraction circles (optional)	**5.NF.3** **SMP2, SMP5**
Math Boxes 3-11 Students practice and maintain skills.	*Math Journal 1*, p. 100	See page 291.
Home Link 3-11 **Homework** Students complete rounds of *Fraction Capture*.	*Math Masters*, p. 108	**5.OA.2, 5.NF.1, 5.NF.3**

 Go Online ▷ to see how mastery develops for all standards within the grade.

Readiness 5–15 min	**Enrichment** 10–20 min	**Extra Practice** 5–15 min
WHOLE CLASS **SMALL GROUP** PARTNER INDEPENDENT	WHOLE CLASS SMALL GROUP **PARTNER** INDEPENDENT	WHOLE CLASS SMALL GROUP **PARTNER** INDEPENDENT

Writing Fractions as Sums of Unit Fractions

5.NF.1

For experience writing fractions as sums of other fractions, students find sums of unit fractions and then write non-unit fractions as sums of unit fractions. Display the following addition problems one at a time and have students tell you the sums:

- $\frac{1}{4} + \frac{1}{4}$ $\frac{2}{4}$
- $\frac{1}{4} + \frac{1}{4} + \frac{1}{4}$ $\frac{3}{4}$
- $\frac{1}{4} + \frac{1}{4} + \frac{1}{4} + \frac{1}{4} + \frac{1}{4}$ $\frac{5}{4}$

Then reverse the procedure. Display the following fractions and ask students how to write them as sums of unit fractions:

- $\frac{4}{4}$ $\frac{1}{4} + \frac{1}{4} + \frac{1}{4} + \frac{1}{4}$
- $\frac{6}{4}$ $\frac{1}{4} + \frac{1}{4} + \frac{1}{4} + \frac{1}{4} + \frac{1}{4} + \frac{1}{4}$

Repeat using other denominators.

Playing Break It Up!

5.OA.2, 5.NF.1, SMP1

Activity Card 37; per partnership: number cards 1–10 (1 of each), fraction circles

To extend their work breaking apart and adding fractions, students write fractions as sums of other fractions using 1, 2, 3, 4, and 5 different denominators. **GMP1.5** Students check each other's answers. **GMP1.4**

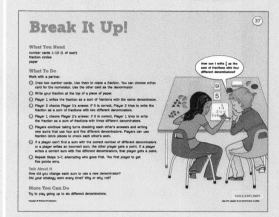

Breaking Apart Fractions

5.OA.2, 5.NF.1, SMP1

Activity Card 38;
Math Masters, p. TA8;
number cards 1–10 (1 of each); fraction circles (optional)

For additional practice breaking apart and adding fractions, students draw number cards to generate fractions and complete name-collection boxes for them using at least three addition expressions. **GMP1.5** They add to check that their answers are correct. **GMP1.4**

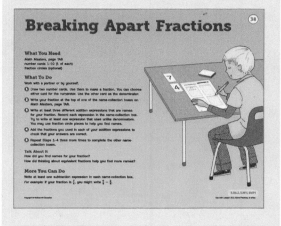

Name-Collection Boxes

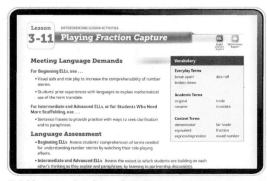

English Language Learner

Beginning ELL To support students' understanding of the term *break apart*, show items that can be broken apart but easily put back together, like a tower of linking cubes. Demonstrate putting the cubes together, then breaking them apart. Think aloud: *I will put all these red cubes together in a tower. Now I am going to break apart my tower to make two smaller towers.*

Differentiation Support pages are found in the online Teacher's Center.

Standards and Goals for
Mathematical Process and Practice

SMP1 **Make sense of problems and persevere in solving them.**

GMP1.3 Keep trying when your problem is hard.

GMP1.4 Check whether your answer makes sense.

GMP1.5 Solve problems in more than one way.

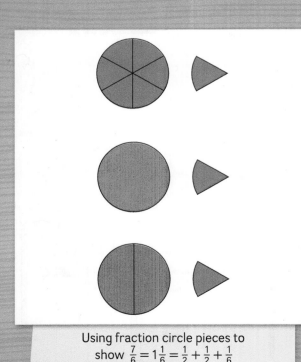

Using fraction circle pieces to
show $\frac{7}{6} = 1\frac{1}{6} = \frac{1}{2} + \frac{1}{2} + \frac{1}{6}$

---

1 Warm Up   5 min

▶ Mental Math and Fluency

Have students solve addition and subtraction number stories involving fractions with like denominators. *Leveled exercises:*

● ○ ○ Charlie used $\frac{1}{4}$ stick of glue to make a card and $\frac{2}{4}$ stick of glue to make a poster. How much glue did he use all together? $\frac{3}{4}$ stick

● ● ○ Jerome ate $\frac{5}{8}$ sub sandwich and Jacki ate $\frac{7}{8}$ sub sandwich. How much sandwich did they both eat? $\frac{12}{8}$, or $1\frac{4}{8}$, sub sandwiches

● ● ● Kristy listened to the radio for $1\frac{3}{4}$ hours on Saturday and $2\frac{1}{4}$ hours on Sunday. How much longer did she listen to the radio on Sunday? $\frac{2}{4}$ hour

2 Focus   35–40 min

▶ Math Message

Write at least two fraction addition problems that have a sum of $\frac{7}{6}$.
GMP1.5 *Be prepared to explain how you know both sums are $\frac{7}{6}$.*

▶ Breaking Apart Fractions

| WHOLE CLASS | SMALL GROUP | PARTNER | INDEPENDENT |

Math Message Follow-Up Ask students to share their answers to the Math Message problem. Sample answers: $\frac{4}{6} + \frac{3}{6}$; $\frac{1}{2} + \frac{1}{2} + \frac{1}{6}$; $\frac{1}{3} + \frac{5}{6}$ If most answers involve adding fractions that all have a denominator of 6, have students think about how they might write a problem involving fractions with different denominators. Encourage them to use fraction circle pieces or drawings to represent their strategies and show that their answers are correct. **GMP1.4** Be sure to include the following strategies in the discussion:

- Change $\frac{7}{6}$ into a mixed number. It takes $\frac{6}{6}$ to make a whole, so $\frac{7}{6}$ is 1 whole plus $\frac{1}{6}$, or $1\frac{1}{6}$. One whole can be broken down into fractions in many ways. For example, $\frac{1}{2} + \frac{1}{2}$. So $\frac{1}{2} + \frac{1}{2} + \frac{1}{6} = \frac{7}{6}$. (*See margin.*)
- Think about equivalent fractions. For example, we know that $\frac{7}{6} = \frac{2}{6} + \frac{5}{6}$. But $\frac{2}{6}$ is the same as $\frac{1}{3}$, so that means $\frac{7}{6} = \frac{1}{3} + \frac{5}{6}$.

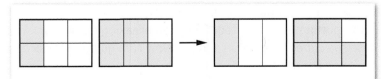

Using a drawing to show that $\frac{7}{6} = \frac{2}{6} + \frac{5}{6} = \frac{1}{3} + \frac{5}{6}$

Display a few more fractions and have students write addition problems with the fraction as the sum. Encourage them to write more than one addition problem for each fraction and to use addends that have unlike denominators. **GMP1.5** Invite students to share answers and strategies and explain how they know their answers are correct. **GMP1.4**
Suggestions:

- $\frac{9}{5}$ Sample answers: $\frac{4}{5} + \frac{1}{3} + \frac{2}{3}$; $\frac{2}{10} + \frac{8}{5}$
- $\frac{5}{8}$ Sample answers: $\frac{1}{2} + \frac{1}{8}$; $\frac{1}{4} + \frac{1}{4} + \frac{1}{8}$
- $\frac{10}{12}$ Sample answers: $\frac{1}{6} + \frac{1}{3} + \frac{4}{12}$; $\frac{1}{2} + \frac{1}{3}$

Differentiate **Adjusting the Activity**

To help students find combinations of fractions that add to the desired sum, suggest that they start by representing the original fraction using fraction circle pieces. Then have them separate the pieces into two or more groups and write a fraction to represent each group. To find fractions with different denominators, students can try covering their groups with different colored pieces.

Go Online Differentiation Support

Tell students that in today's lesson they will learn a new game to practice breaking apart fractions.

▶ Introducing *Fraction Capture*

Student Reference Book, p. 305; *Math Masters*, pp. G19–G20

| WHOLE CLASS | SMALL GROUP | PARTNER | INDEPENDENT |

Read the directions as a class for *Fraction Capture* on *Student Reference Book*, page 305. Display *Math Masters*, page G19 and ask students to imagine that you have rolled a 3 and a 4. Ask:

- *If I formed the fraction* $\frac{3}{4}$, *what are some combinations of fractions I could capture?* Sample answer: $\frac{1}{4}$, $\frac{1}{4}$, and $\frac{1}{4}$; $\frac{1}{2}$ and $\frac{1}{4}$

- *What would I write for my fraction addition expression?* Sample answer: $\frac{1}{4} + \frac{1}{4} + \frac{1}{4}$; $\frac{1}{2} + \frac{1}{4}$

- *If you were my partner, how would you check to make sure my answer is correct?* **GMP1.4** Sample answer: I could use fraction circle pieces to check that the addition expression equals the fraction.

Invite a volunteer to come up and play a few rounds with you. Have students suggest different combinations of fractions that can be captured on each turn.

Academic Language Development

Have students use the 4-Square Graphic Organizer (*Math Masters*, page TA2) to identify four different everyday synonyms for the word *capture*, providing a corresponding illustration for each synonym. For example: *catch, take, grab, seize,* or *bag*. Challenge students working in partnerships to create and role-play interesting sentences using the words they identified and illustrated.

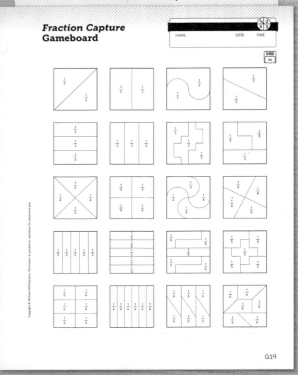

Fraction Capture Gameboard

NAME   DATE   TIME

G19

When students seem comfortable with the rules of the game, distribute one *Fraction Capture* gameboard (*Math Masters*, page G19) and two *Fraction Capture* record sheets (*Math Masters*, page G20) to each partnership. (*See Before You Begin.*) Have students play *Fraction Capture*. Encourage them to find more than one combination of fractions that they could capture on each turn and think about which combination might help them get a better score at the end of the game. **GMP1.5** Allow students to use fraction circles as needed.

Particularly in later rounds of the game, students may struggle finding fractions that add up to the appropriate sum. Encourage them to try several combinations before passing their turn. **GMP1.3** As you observe, look for students who are using interesting strategies and encourage them to share those strategies at the end of the lesson.

Observe

- What strategies are students using to find fractions to capture?
- Do students have strategies for adding fractions to check that their expressions are correct?

Discuss

- *How did you check to make sure your partner's answers were correct?* **GMP1.4**
- *Which denominators make it easier to capture fractions? Which denominators make it harder? Why?*

Differentiate **Game Modifications** **Go Online** Differentiation Support

---

✓ **Assessment Check-In** 5.OA.2, 5.NF.1

Math Masters, p. G20

Expect most students to find fractions with like denominators that add to the appropriate sum and write a correct addition expression for at least one round on their *Fraction Capture* record sheets. Some students may be able to consistently capture appropriate fractions with unlike denominators and clearly explain how they know the sum is correct. If students struggle trying to find combinations of fractions to capture, suggest that they use fraction circle pieces as described in the Adjusting the Activity note.

Evaluation Quick Entry Go online to record students' progress and to see trajectories toward mastery for these standards.

Renaming Fractions and Mixed Numbers

Lesson 3-11

DATE   TIME

Fractions greater than 1 can be expressed as mixed numbers, such as $2\frac{1}{3}$ and $1\frac{4}{5}$, and as fractions with a numerator larger than the denominator, such as $\frac{7}{3}$. You know several ways to rename fractions as mixed numbers and mixed numbers as fractions.

Use fraction circle pieces: Show the original number. Make fair trades between wholes and same-size pieces to rename.

$$2\frac{1}{3} \qquad 1\frac{4}{3} \qquad \frac{7}{3}$$

Use number lines: Use fraction names for whole numbers and count up. The number line on the left shows that $2\frac{1}{3}$ is $\frac{1}{3}$ past 1, so $2\frac{1}{3} = 1\frac{6}{3}$. The number line on the right shows that $2 = \frac{6}{3}$, and $2\frac{1}{3}$ and $\frac{7}{3}$ are both $\frac{1}{3}$ past 2, so $2\frac{1}{3} = \frac{7}{3}$.

Think about making or breaking wholes:

- To rename $2\frac{1}{3}$ as a fraction, think: *How many thirds are in $2\frac{1}{3}$?* There are 2 wholes. I can break each whole into 3 thirds. Two groups of 3 thirds is the same as $2 * 3$ thirds = 6 thirds, or $\frac{6}{3}$. Add one more third to get $\frac{7}{3}$.
- To rename $\frac{7}{3}$ as a mixed number, think: *How many groups of 3 thirds are in 7? What's left over?* There are 2 groups of 3 thirds in 7, with 1 third left over. $\frac{7}{3} = 2\frac{1}{3}$

For Problems 1–3, rename each fraction as a mixed number. Make as many wholes as you can.

1) $\frac{9}{4} = \underline{2\frac{1}{4}}$ 2) $\frac{12}{5} = \underline{2\frac{2}{5}}$ 3) $\frac{15}{8} = \underline{1\frac{7}{8}}$

For Problems 4–6, write at least two other names with the same denominator for each mixed number.

4) $3\frac{4}{5} = \underline{2\frac{9}{5},\ 1\frac{14}{5},\ \frac{19}{5}}$ 5) $2\frac{1}{6} = \underline{1\frac{7}{6},\ \frac{13}{6}}$

6) $4\frac{1}{2} = \underline{3\frac{3}{2},\ 2\frac{5}{2},\ 1\frac{7}{2},\ \frac{9}{2}}$

For Problems 7–9, fill in the missing whole number or missing numerator.

7) $4\frac{2}{5} = \frac{22}{5}$ 8) $2\frac{2}{8} = \frac{18}{8}$ 9) $2\frac{5}{3} = 3\frac{2}{3}$

5.NF.3, SMP2, SMP5   99

Summarize Have students share strategies they used to decide which fractions to capture while playing the game. *Sample strategies:*

- I made the fraction $\frac{2}{3}$. I knew that $\frac{2}{3} = \frac{1}{3} + \frac{1}{3}$, but there was only one $\frac{1}{3}$ section left. So I thought: *How could I use other fractions to make the other $\frac{1}{3}$?* I know that $\frac{1}{3}$ is the same as $\frac{2}{6}$, so I captured two $\frac{1}{6}$ sections and one $\frac{1}{3}$ section to get $\frac{2}{3}$.
- I made the fraction $\frac{6}{4}$. I wanted to capture a $\frac{1}{2}$ to keep my partner from claiming a square. So I thought: *If I capture that $\frac{1}{2}$, how much do I still have to capture?* I know $\frac{1}{2} = \frac{2}{4}$, so I had $\frac{4}{4}$ left to capture. I know that $\frac{4}{4}$ is the same as 1 whole, so I captured three $\frac{1}{3}$ sections along with the $\frac{1}{2}$ section to make $\frac{6}{4}$ in all.

Tell students that they will have the opportunity to play the game again in the future, so they will have a chance to try using their classmates' strategies.

③ Practice 20–30 min

▸ Renaming Fractions and Mixed Numbers

Math Journal 1, p. 99

| WHOLE CLASS | SMALL GROUP | **PARTNER** | INDEPENDENT |

Students use fraction circle pieces, number lines, and multiplication and division to translate between fractions greater than 1 and mixed numbers. GMP2.3, GMP5.2 Have copies of *Math Masters*, page TA20 available for students who want to use number lines.

▸ Math Boxes 3-11

Math Journal 1, p. 100

| WHOLE CLASS | **SMALL GROUP** | PARTNER | INDEPENDENT |

Mixed Practice Math Boxes 3-11 are paired with Math Boxes 3-13.

▸ Home Link 3-11

Math Masters, p. 108

Homework Students add fractions and name fractions they could capture in simulated rounds of *Fraction Capture.*

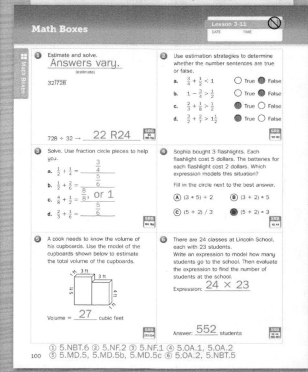

Math Journal 1, p. 100

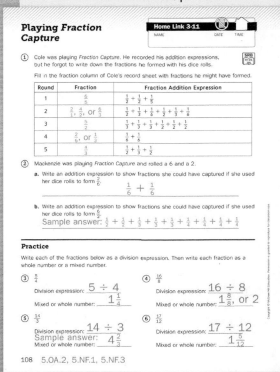

Math Masters, p. 108

Solving Fraction Number Stories

Overview Students identify problem-solving strategies and solve a variety of fraction number stories.

▶ **Before You Begin**

Fraction circle pieces should be available throughout the lesson, although students may choose not to use them for every problem. Partners may share pieces to have enough of each color, or you can create extra fraction circle pieces from *Math Masters*, pages TA12–TA14. Number lines (*Math Masters*, page TA20) should also be available for students to use. Decide how you will display students' work from the Math Message.

Standards

Focus Clusters
- Use equivalent fractions as a strategy to add and subtract fractions.
- Apply and extend previous understandings of multiplication and division.

 Warm Up 5 min   **Materials**

Mental Math and Fluency Students estimate fraction sums using benchmarks.		5.NF.2

 **Focus** 35–40 min

Math Message Students examine a diagram about problem solving and reflect on their own problem-solving habits.	*Student Reference Book*, p. 30	SMP1
Identifying Problem-Solving Strategies Students generate a list of strategies for solving problems and apply them to a fraction number story.	*Student Reference Book*, p. 30; fraction circles (optional)	5.NF.2 SMP1, SMP4
Solving Fraction Number Stories Students use problem-solving questions and strategies to solve a variety of fraction number stories.	*Math Journal 1*, pp. 101–102; *Math Masters*, p. TA20 (optional); fraction circles (optional)	5.NF.1, 5.NF.2, 5.NF.3 SMP1, SMP4
Assessment Check-In See page 296. Expect most students to be able to understand and find correct solutions to fraction number stories like the problems identified.	*Math Journal 1*, pp. 101–102	5.NF.1, 5.NF.2, 5.NF.3

 Practice 20–30 min

Practicing Division Students practice solving division problems and number stories.	*Math Journal 1*, p. 103	5.NBT.6 SMP1, SMP4
Math Boxes 3-12 Students practice and maintain skills.	*Math Journal 1*, p. 104	See page 297.
Home Link 3-12 **Homework** Students make sense of and solve fraction number stories.	*Math Masters*, p. 110	5.NF.1, 5.NF.2, 5.NF.3, 5.MD.5, 5.MD.5b, 5.MD.5c SMP1, SMP4

(Go Online) to see how mastery develops for all standards within the grade.

Differentiation Options

Readiness   5–15 min

| WHOLE CLASS | SMALL GROUP | PARTNER | INDEPENDENT |

Creating a Menu of Fraction Operations

5.NF.1, 5.NF.2, 5.NF.3, SMP1

Math Journal 1

To prepare for identifying what kind of math could help them solve a number story, students look back through their journal pages from Unit 3. Ask: *What math have you done in this unit?* Division with fractional answers, renaming fractions and mixed numbers, estimation, addition and subtraction of fractions and mixed numbers Have students create a menu of these mathematical actions. Ask: *How do you know when you need to divide to solve a real-world problem?* GMP1.1 Sample answer: When I am splitting something into equal groups Have students record this clue next to "divide" on their menus. Continue asking questions about how students know when to rename, estimate, add, and subtract. Students should record clues for recognizing when a situation calls for each action on their menus. Encourage students to refer to their menus as they complete the Focus part of the lesson.

Enrichment   5–15 min

| WHOLE CLASS | SMALL GROUP | PARTNER | INDEPENDENT |

Working Backward to Write Fraction Number Stories

5.NF.1, 5.NF.2, 5.NF.3, SMP1

Activity Card 39;
Math Journal 1, pp. 101–102 (optional);
Math Masters, p. TA11; number cards 1–16
(1 of each); 6-sided die

To extend their work with fraction number stories, students write their own number stories based on fractional answers. Students roll a die to determine whether they will write a story involving division, addition, or subtraction. They discuss what was challenging about writing the stories and the kinds of information they included to help others make sense of their problems. GMP1.1, GMP1.3

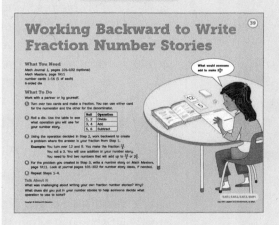

Extra Practice   5–15 min

| WHOLE CLASS | SMALL GROUP | PARTNER | INDEPENDENT |

Making Minestrone

5.NF.1, 5.NF.2, 5.NF.3, SMP1

Math Masters, p. 109, p. TA20 (optional);
fraction circles (optional)

For more practice with fraction number stories, students solve a series of problems in the context of making a pot of minestrone soup. They have to distinguish when to divide, rename, estimate, add, or subtract to solve the problems. Allow students to use fraction circle pieces, number lines, and paper as needed. GMP1.1, GMP1.2

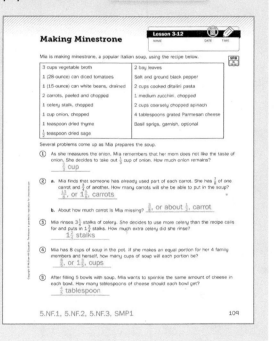

English Language Learner

Beginning ELL As students read and discuss the number stories in this lesson, provide pictures of real-world items and other visual aids related to the objects named in the stories, which will help students attend to the mathematical content. Maintain a display of words with pictures for student reference when reading number stories.

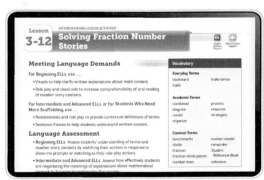

Differentiation Support pages are found in the online Teacher's Center.

Standards and Goals for
Mathematical Process and Practice

SMP1 **Make sense of problems and persevere in solving them.**

　GMP1.1 Make sense of your problem.

　GMP1.3 Keep trying when your problem is hard.

SMP4 **Model with mathematics.**

　GMP4.2 Use mathematical models to solve problems and answer questions.

Student Reference Book, p. 30

A Problem-Solving Diagram

When you solve problems, you work to make sense of the problem. You reflect on your thinking and keep trying when the problem is hard. As a good problem solver, you always check to see whether your answer makes sense by trying to solve the problem in more than one way and by comparing the strategies you use to those others use.

The diagram below can help you think about problem solving. The boxes in the diagram show the type of things you do when you use mathematical practices to solve problems. The arrows show that you don't always do things in the same order.

Organize the information.
• Study the information in the problem.
• Arrange the information into a list, table, graph, or diagram.
• Look for more information if you need it.
• Get rid of information you don't need.

Understand the problem.
• Retell the problem in your own words.
• Figure out what you want to find.
• Figure out what you know.
• Imagine what the answer will look like.
• Make a guess at the answer.

Play with the information.
• Draw a picture, diagram, or another mathematical representation.
• Write a number model.
• Model the problem using objects such as counters or base-10 blocks.

Check your answer as you work.
• Does your answer make sense?
• Compare your answer with a classmate's.
• Does your answer fit the problem?
• Can you solve the problem another way?

Figure out what math can help.
• Can you use addition? Subtraction? Another operation?
• Can you use geometry? Patterns? Other mathematics?
• Try the math. See what happens.
• What units are you using? Label your numbers with units.

30 | thirty

---

1 Warm Up　5 min

▶ Mental Math and Fluency

Display the benchmarks 0, $\frac{1}{2}$, 1, $1\frac{1}{2}$, and 2. Provide addition expressions and have students estimate the sum by naming the closest benchmark. *Leveled exercises:*

●○○　$\frac{2}{3} + \frac{1}{4}$　1　　$\frac{1}{8} + \frac{1}{4}$　$\frac{1}{2}$　　$\frac{1}{12} + \frac{1}{20}$　0

●●○　$\frac{1}{4} + \frac{5}{8}$　1　　$\frac{1}{6} + 1\frac{7}{8}$　2　　$1\frac{1}{8} + \frac{1}{10}$　1

●●●　$\frac{5}{3} + \frac{1}{8}$　2　　$\frac{3}{2} + \frac{3}{4}$　2　　$\frac{1}{10} + \frac{4}{3}$　$1\frac{1}{2}$

2 Focus　35–40 min

▶ Math Message

Student Reference Book, p. 30

Look at the problem-solving diagram on Student Reference Book, *page 30. Think about your own problem-solving habits. What questions do you ask yourself when you begin solving a problem?* **GMP1.1** *What strategies do you use to keep going when a problem is challenging?* **GMP1.3** *Be ready to share your thoughts with the class.*

▶ Identifying Problem-Solving Strategies

Student Reference Book, p. 30

| WHOLE CLASS | SMALL GROUP | PARTNER | INDEPENDENT |

Math Message Follow-Up Invite students to share the questions and strategies they use when solving real-world problems. Expect them to draw some ideas from the problem-solving diagram, but encourage them to add their own questions and strategies. List and display students' contributions. *Sample responses:*

To Get Started **GMP1.1**

• What do I know? What do I need to find out?

• What information is the problem giving me? Do I have all the information I need? Is there information I can ignore?

• Can I draw a picture of the problem? Can I write a number model?

• What math can help? Is this problem like one I've solved before?

• Can I use fraction circle pieces, number lines, or other tools?

• Will using an estimate help?

To Keep Going GMP1.3

- Organize the information in the problem.
- Play with the information. Try solving one part of the problem or working backward.
- Make a written record of what I do.
- Create a model (draw a picture, write a number model, make a graph or diagram).
- Think about how my model is helping me solve the problem. Try another model if it's not working.
- Talk to others working on the problem. Try to make sense of their strategies.
- Look back at what I've tried. Does my work make sense?

Tell students that today they will apply what they have learned about fractions to solve a variety of real-world fraction problems. Point out that the list of problem-solving strategies they just discussed can be an important resource as they work through the problems.

Pose a problem like the following to show how problem-solving strategies can help students get started and keep going when they feel stuck. Use the questions and strategies generated in the Math Message Follow-Up to guide students through the problem-solving process. Emphasize the point that students need to determine what math will help them solve the problem. GMP1.1, GMP1.3

Example: Adeline is mixing fruit juices to make punch. She mixes $1\frac{1}{2}$ gallons of apple juice with $1\frac{1}{2}$ gallons of grape juice. How much fruit punch will she have? *Sample strategies:*

- Let the red fraction circle represent 1 whole gallon of juice. Show $1\frac{1}{2}$ gallons of apple juice using 1 red circle and 1 pink half piece. Show $1\frac{1}{2}$ gallons of grape juice using another red circle and pink half piece. Combine the fraction circle pieces to make wholes. The two red circles and two pink halves together represent 3 gallons of punch in all. GMP4.2
- Model the situation with the expression $1\frac{1}{2} + 1\frac{1}{2}$. Add the wholes mentally, $1 + 1 = 2$. Add the halves mentally, $\frac{1}{2} + \frac{1}{2} = 1$. Combine the sums, $2 + 1 = 3$. Adeline will have 3 gallons of fruit punch. GMP4.2

Ask: *What problem-solving strategies and questions from our list did you use?* Sample references to the class lists might include these:

- I thought about what I knew. Adeline used $1\frac{1}{2}$ gallons of each kind of juice. I thought about what I had to find out: How many gallons of fruit punch is there in all? That's how I knew I had to add the amounts of juice.
- I thought about what math I could use. I knew I had to put together $1\frac{1}{2}$ gallons of one kind of juice and $1\frac{1}{2}$ gallons of the other kind of juice, so I used addition because addition is like putting together.
- I used fraction circle pieces as a tool. They helped me think about how to put halves together to make a whole.

Math Journal 1, p. 101

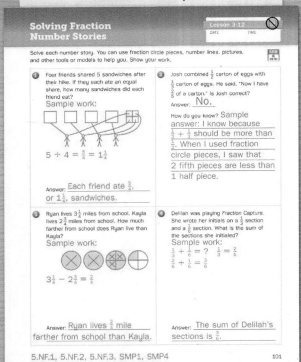

Math Journal 1, p. 102

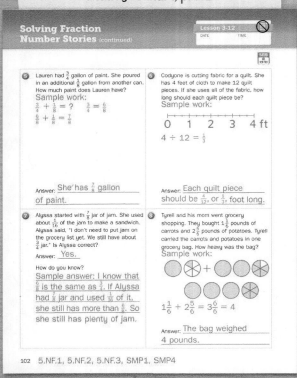

Summarize the discussion by emphasizing the importance of using problem-solving questions and strategies to think through problems and determine what kind of math will help solve the problem.

▶ Solving Fraction Number Stories

Math Journal 1, pp. 101–102

| WHOLE CLASS | SMALL GROUP | PARTNER | INDEPENDENT |

Have partners complete journal pages 101 and 102. After most students have completed Problem 1, have them briefly share how they decided what math (division, estimation, addition, or subtraction) would help them solve the problem. **GMP1.1** Encourage students to think about similar problems they have solved before and to use models to make sense of the situations. **GMP4.2** If they are stuck, encourage them to refer back to the list of problem-solving strategies compiled during the Math Message Follow-Up. **GMP1.3** Circulate and assist as needed.

If students become frustrated while solving problems, encourage them to use sentence frames like this: "I understand _____, but I'm not sure about _____." Encourage students working in partnerships to ask each other questions like this: "I see why you did _____, but why did you _____?" These frames provide a way for students to express uncertainty while remaining responsible for making sense of the problem and persevering to find a solution. **GMP1.3**

 Assessment Check-In 5.NF.1, 5.NF.2, 5.NF.3

Math Journal 1, pp. 101–102

Expect most students to be able to make sense of the problems on journal pages 101 and 102 and find correct solutions to Problems 1–3. Most students will also make reasonable attempts at the remaining problems.
If students struggle identifying what math, models, or tools could help them find a solution, ask questions like these: *What's going on in this problem? What are you trying to find? Could a picture or model help? How have you solved problems similar to this one? What math have you done so far? What else would make sense here?* Encourage students to make a written record of their work so they can reflect on their thinking.

 Evaluation Quick Entry Go online to record students' progress and to see trajectories toward mastery for these standards.

Summarize Invite partners to share the problem-solving strategies they used to keep trying when problems were challenging. **GMP1.3**

3 Practice   20–30 min

▸ Practicing Division

Math Journal 1, p. 103

| WHOLE CLASS | SMALL GROUP | **PARTNER** | **INDEPENDENT** |

Students solve division problems and division number stories. They use estimates to check whether answers make sense and write number models for the number stories. GMP1.4, GMP4.1

▸ Math Boxes 3-12 ✎

Math Journal 1, p. 104

| WHOLE CLASS | **SMALL GROUP** | **PARTNER** | **INDEPENDENT** |

Mixed Practice Math Boxes 3-12 are grouped with Math Boxes 3-9 and 3-14.

▸ Home Link 3-12

Math Masters, p. 110

Homework Students make sense of and solve fraction number stories. GMP1.1, GMP1.3 They can draw pictures or other representations to help them solve the number stories. GMP4.2

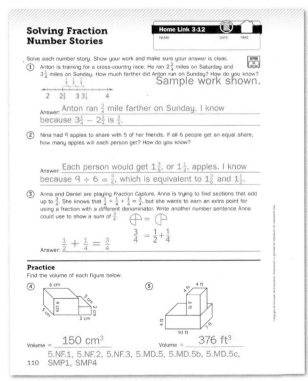

Math Masters, p. 110

Math Journal 1, p. 103

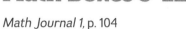

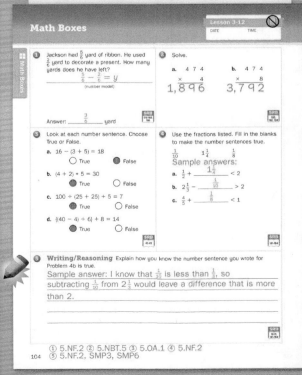

Math Journal 1, p. 104

Fraction-Of Problems, Part 1

Overview Students solve fraction-of problems to build readiness for multiplying fractions by whole numbers.

▶ **Before You Begin**
For Part 2, have counters and copies of *Math Masters,* page TA20 available for students for solving fraction-of problems. Consider whether you want to cut out the *Fraction Of* cards (*Math Journal 1,* Activity Sheets 14 and 15) before the lesson or have students cut them out. Decide how you will store them for future use.

▶ **Vocabulary**
fraction-of problem • representation

Standards

Focus Cluster
• Apply and extend previous understandings of multiplication and division.

	Materials	
1 Warm Up 5 min		
Mental Math and Fluency Students evaluate multiplication expressions involving powers of 10.	slate	**5.NBT.2**
2 Focus 35–40 min		
Math Message Students solve a fraction number story.	*Math Masters,* p. TA20 (optional); counters (optional)	**5.NF.4, 5.NF.4a, 5.NF.6**
Solving Fraction-Of Problems Students discuss the meaning of "$\frac{1}{2}$ of" and "$\frac{1}{3}$ of." They use counters, drawings, or number lines to help them solve fraction-of problems.	*Math Journal 1,* p. 105; *Math Masters,* p. TA20 (optional); counters (optional)	**5.NF.4, 5.NF.4a, 5.NF.6** **SMP2, SMP6**
Introducing *Fraction Of* **Game** Students practice solving fraction-of problems.	*Student Reference Book,* pp. 306–307; *Math Masters,* p. G24; per partnership: *Fraction Of* cards; scissors; counters (optional)	**5.NF.4, 5.NF.4a** **SMP2, SMP6**
✓ Assessment Check-In See page 303. Expect most students to be able to use concrete representations or drawings to help them correctly solve fraction-of problems while playing the game *Fraction Of.*	*Math Masters,* p. G24	**5.NF.4, 5.NF.4a, SMP2**
3 Practice 20–30 min		
Using Models to Estimate Volumes Students estimate volumes of real-world objects.	*Math Journal 1,* p. 106	**5.MD.3, 5.MD.5, 5.MD.5a, 5.MD.5b,** **5.MD.5c, SMP4**
Math Boxes 3-13 Students practice and maintain skills.	*Math Journal 1,* p. 107	See page 303.
Home Link 3-13 **Homework** Students solve fraction-of problems.	*Math Masters,* p. 112	**5.NBT.6, 5.NF.4, 5.NF.4a, 5.NF.6**

 [Go Online] to see how mastery develops for all standards within the grade.

Differentiation Options

Readiness	5–15 min
WHOLE CLASS	**SMALL GROUP**	PARTNER	INDEPENDENT

Reviewing Flexibility of the Whole

5.NF.4, 5.NF.4a, SMP2

fraction circles, counters

To review the idea that the meaning of a fraction depends on the whole, students use their fraction circle pieces to solve problems where they identify fractional parts, given the whole. Pose problems like:

- If the red circle is the whole, then what is $\frac{1}{2}$? **GMP2.2** The pink piece
- If the pink piece is the whole, then what is $\frac{1}{3}$? **GMP2.2** The light blue piece

Continue with similar problems as needed.

Extend this idea to fractions of collections. Have students use counters to help them solve problems like these:

- If these 12 counters are the whole, then what is $\frac{1}{2}$? **GMP2.2** 6 counters
- If these 20 counters are the whole, then what is $\frac{1}{4}$? **GMP2.2** 5 counters

Continue with similar problems as needed.

Enrichment	5–15 min
WHOLE CLASS	SMALL GROUP	**PARTNER**	INDEPENDENT

Interpreting Representations

5.NF.4, 5.NF.4a, SMP2, SMP3

Math Masters, p. 111

To extend their work solving fraction-of problems, students interpret various representations of fraction-of problems on *Math Masters*, page 111. **GMP2.2, GMP3.2**

Interpreting Representations   Lesson 3-13

NAME   DATE   TIME

Each of the problems below shows a representation a student drew while trying to solve a fraction-of problem. Study the representations and figure out what problem the student was solving. Then solve the problem. Fill in the blanks below each representation to show your answer.

① [circles diagram]   ② Total area = 72
                        Area of 1 box = ?

$\frac{1}{6}$ of __18__ = __3__ $\frac{1}{12}$ of __72__ = __6__

③ [number line 0 to 175]

$\frac{1}{7}$ of __175__ __25__

④ [bar diagram]

$\frac{1}{5}$ of __155__ __31__

⑤ Explain how you solved Problem 1.
Sample answer: It looked like the student started by drawing 18 circles, so I thought 18 was the whole. Then the student put a ring around 3 of the circles. There are 6 groups of 3 in 18, so that meant the student was finding $\frac{1}{6}$ of 18, and the answer was 3.

5.NF.4, 5.NF.4a, SMP2, SMP3   111

Extra Practice	5–15 min
WHOLE CLASS	SMALL GROUP	**PARTNER**	INDEPENDENT

Solving Fraction-Of Problems

5.NF.4, 5.NF.4a, SMP7

Activity Card 40; *Fraction Of* Whole Cards; counters

For additional practice solving fraction-of problems, students look for *Fraction Of* Whole Cards for which all three numbers have whole-number answers for a particular fraction. They look for patterns in the whole numbers that produce whole-number answers. **GMP7.1**

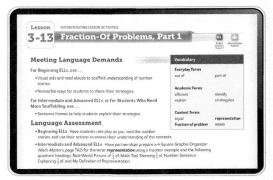

English Language Learner

Beginning ELL Help students understand the meaning of the term *fraction of* by connecting it to *part of.* Display visuals like a whole pizza and a partially eaten pizza and use think-alouds to describe them. For example, say: *A fraction of this pizza was eaten. We ate part of the pizza. This is not the whole pizza. Part of it is not there.*

Differentiation Support pages are found in the online Teacher's Center.

Standards and Goals for
Mathematical Process and Practice

SMP2 **Reason abstractly and quantitatively.**

GMP2.1 Create mathematical representations using numbers, words, pictures, symbols, gestures, tables, graphs, and concrete objects.

GMP2.2 Make sense of the representations you and others use.

SMP6 **Attend to precision.**

GMP6.4 Think about accuracy and efficiency when you count, measure, and calculate.

Common Misconception

Differentiate Some students may try to solve the Math Message by comparing the fractions only. They may reason that Ella eats more carrots than Jackson because $\frac{1}{2}$ is greater than $\frac{1}{3}$. If students do this, explain that $\frac{1}{2}$ of 2, for example, is a different number than $\frac{1}{2}$ of 4. Emphasize the idea that the whole needs to be considered when solving fraction-of problems.

Go Online Differentiation Support

---

1 Warm Up · 5 min

▶ Mental Math and Fluency

Have students write numbers in standard notation on slates. Ask them to explain how they determined the number of zeros to write in their answers. *Leveled exercises:*

● ○ ○ $3 * 10^3$ 3,000
         $5 * 10^2$ 500
         $4 * 10^5$ 400,000

● ● ○ $65 * 10^4$ 650,000
         $39 * 10^5$ 3,900,000
         $72 * 10^6$ 72,000,000

● ● ● $90 * 10^5$ 9,000,000
         $70 * 10^6$ 70,000,000
         $20 * 10^3$ 20,000

---

2 Focus · 35–40 min

▶ Math Message

Ella has 22 baby carrots and her brother Jackson has 36 baby carrots. Ella eats $\frac{1}{2}$ of her carrots and Jackson eats $\frac{1}{3}$ of his. Who ate more carrots? Jackson *Be prepared to explain your answer.*

▶ Solving Fraction-Of Problems

Math Journal 1, p. 105

| WHOLE CLASS | SMALL GROUP | PARTNER | INDEPENDENT |

Math Message Follow-Up Invite students to share their strategies for solving the Math Message problem. *Sample strategies:*

- I used counters to make equal groups. **GMP2.1** When I put 22 counters into 2 equal groups, there were 11 in each group. When I put 36 counters into 3 equal groups, there were 12 in each group. 12 is more than 11, so Jackson ate more carrots. **GMP2.2**

- I drew pictures. First I drew a rectangle with two parts. I kept putting one X in each part until I had 22 Xs. **GMP2.1** There were 11 Xs in each part. Then I drew a rectangle with three parts. I put one X in each part until I had 36 in all. There were 12 Xs in each part. So Jackson ate more because he ate 12. **GMP2.2**

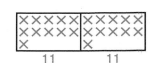

• I thought about division. I know 22 / 2 = 11. I also know that 36 / 3 = 12. That means Jackson ate more.

Use the discussion of strategies to review the idea that "$\frac{1}{2}$ of" means 1 out of 2 equal parts and "$\frac{1}{3}$ of" means 1 out of 3 equal parts. Tell students that problems like this are called **fraction-of problems.** Ask: *Ella and Jackson's neighbor also had baby carrots. He ate $\frac{1}{4}$ of his carrots. Did he eat more or fewer carrots than Ella?* We can't tell. *Why?* We don't know how many carrots he had. Explain that students must identify the whole before they can solve fraction-of problems.

Tell students that in today's lesson they will solve more fraction-of problems and learn a game to play for practice.

Explain that mathematicians often create and use **representations,** such as pictures, number sentences, or groups of counters, to help them solve problems. GMP2.1, GMP2.2 Review how students used representations to solve the Math Message. Have partnerships or small groups solve Problem 1 on journal page 105, encouraging them to create and use representations to solve the problem.

When most groups have solved the problem, invite them to share their representations and strategies. Look for students who used efficient strategies for splitting 56 into equal groups, such as putting 10 beads in each group and then distributing the rest. (*See below.*) Have these students share their strategies and representations and encourage the rest of the class to think about how they might be more efficient when solving fraction-of problems. GMP6.4

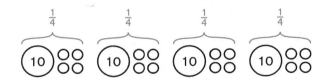

Have groups work together to solve Problems 2 and 3 on journal page 105. If time permits, some groups may also want to try solving Problem 4. If students have trouble getting started, ask guiding questions like these: *What is the whole in this problem? What fraction of the whole do you need to find?* Students may find problems dealing with measurements more challenging than ones concerned with discrete objects because it may be harder to visualize separating the whole into equal groups. Look for students who use number-line or area representations to think through the problems. Invite these students to share their strategies and discuss how the representations were useful. *Sample strategies:*

Problem 2: I drew a number line from 0 to 45 to show the yarn. GMP2.1 I divided the number line into 5 equal parts. Then I could see that the length of each fifth would be 45 / 5, or 9. GMP2.2 Jenna used 9 yards.

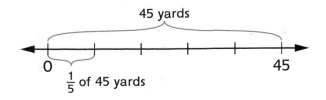

Fraction-Of Problems

Lesson 3-13
DATE   TIME

Work with a partner or a small group to solve the problems. You can use counters, drawings, or number lines to help you. Be prepared to explain how you solved the problems.

① There are 56 beads on a necklace. $\frac{1}{4}$ of the beads are blue. How many beads are blue? See *Teacher's Lesson Guide*, pages 301 and 302 for sample representations for Problems 1–3.

___14___ beads

② Jenna had 45 yards of yarn. She used $\frac{1}{5}$ of it for a knitting project. How much yarn did she use?

___9___ yards

③ Morris has a rectangular garden with an area of 60 square feet. $\frac{1}{10}$ of the garden is planted with bean plants. How many square feet are planted with bean plants?

___6___ square feet

Try This

④ The length of my living room is 24 feet. The width of my living room is $\frac{1}{2}$ the length. What is the area of my living room?

___288___ square feet

5.NF.4, 5.NF.4a, 5.NF.6, SMP2   105

Academic Language Development

Have students work in partnerships to create illustrated "How To" posters to share their strategies for solving fraction-of problems. Encourage them to use sequencing terms like *first, next, then,* and *last.* Provide a word bank of math terms that are used for solving fraction problems like these: *fair share, numerator, denominator, number line,* and *fraction circle pieces.* Allow students time to practice their explanations before presenting to the class.

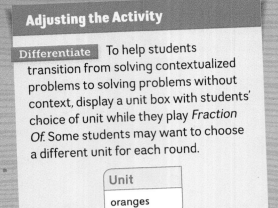
Math Journal 1, p. 106

Using Models to Estimate Volumes

Lesson 3-13
DATE   TIME

A mathematical model of each real-world object is given below.

For each object, use the mathematical model to estimate its volume. Be sure to include a unit when you write the volume.

Then write one or more number sentences to show how you found the volume.

① Volume: 200 cm³
5 * 4 * 10 = 200
(number sentence)

② Volume: 1,872 in.³
12 * 12 * 13 = 1,872
(number sentence)

③ Volume: 36 in.³
12 * 3 * 1 = 36
(number sentence)

④ Volume: 566 cm³
6 * 6 * 6 = 216; 14 * 5 * 5 =
350; 216 + 350 = 566
(number sentences)

⑤ Volume: 56 in.³
Sample answer:
2 * 2 * 2 = 8; 8 * 2 * 2 = 32;
2 * 2 * 4 = 16; 8 + 32 + 16
= 56
(number sentences)

⑥ Volume: 455 cm³
Sample answer: 11 * 6 * 5 =
330; 5 * 5 * 5 = 125; 330 +
125 = 455(number sentences)

106   5.MD.3, 5.MD.5, 5.MD.5a, 5.MD.5b, 5.MD.5c, SMP4

Problem 3: I drew a rectangle to show the garden and divided it into 10 parts. GMP2.1 I can think of an area of 60 square feet as 10 columns of 6 square feet each, so I knew that each $\frac{1}{10}$ was 6 square feet. GMP2.2

Area = 60 square feet
1 column = 6 square feet

To summarize, ask students to name some representations they can use to help them solve fraction-of problems. Sample answers: drawings, counters, number lines Tell students that they will find these representations useful as they learn to play a new game.

▶ **Introducing *Fraction Of***

Math Journal 1, Activity Sheets 14–15; *Student Reference Book,* pp. 306–307; *Math Masters,* p. G24

WHOLE CLASS   SMALL GROUP   PARTNER   INDEPENDENT

Ask: *What is $\frac{1}{3}$ of 14?* Encourage students to use a representation to begin solving the problem. GMP2.1, GMP2.2 After they have had some time to work, ask: *How is this problem different from the other problems you have solved today?* Sample answers: I can't put 14 things into 3 equal groups without splitting them apart. I have to think about what to do with 2 extra pieces. Explain that this problem is different because it does not have a whole-number answer: $\frac{1}{3}$ of 14 is somewhere between 4 and 5. Tell students that they will learn to solve this kind of problem in the next lesson, but for now they just need to be able to tell when a problem does not have a whole-number answer.

Read the directions for *Fraction Of* on *Student Reference Book,* pages 306–307 together and discuss the examples. Direct students to Activity Sheets 14 and 15 in the back of their journals and have partnerships cut out one set of cards. Distribute one copy of *Math Masters,* page G24 to each student and have partnerships play the game. Be sure counters are available to students.

NOTE This game will be played in Unit 5 with additional fraction cards. Save the Set 1 cards for use when the game is played again.

Observe
• Which students recognize, without solving, the problems that have whole-number answers?
• What representations are students creating and using to solve fraction-of problems? GMP2.1, GMP2.2

Discuss
• *How could you check that you solved the problems correctly?*
• *Can you think of a more efficient way to solve the fraction-of problems?* GMP6.4

Differentiate **Game Modifications**   Go Online Differentiation Support

Assessment Check-In   5.NF.4, 5.NF.4a

Math Masters, p. G24

Expect most students to be able to use concrete representations or drawings to help them correctly solve fraction-of problems while playing *Fraction Of.*   GMP2.1, GMP2.2   Some students may be able to solve the fraction-of problems mentally. Do not expect students to immediately know which problems will give them a whole-number answer. They will become better at this with more experience solving fraction-of problems. If students struggle solving the problems using representations or drawings, provide sheets of paper divided into 2, 3, 4, 5, and 10 sections. Have students sort counters into the sections to help them solve the problems. If students need a challenge, have them solve Problem 4 on journal page 105 if they have not already done so.

 Evaluation Quick Entry Go online to record students' progress and to see trajectories toward mastery for these standards.

Summarize Have volunteers share one problem they solved in a round of *Fraction Of* and describe how they found the answer.

3 Practice   20–30 min

▶ **Using Models to Estimate Volume**

Math Journal 1, p. 106

| WHOLE CLASS | SMALL GROUP | **PARTNER** | **INDEPENDENT** |

Students examine mathematical models of real-world objects. They apply volume formulas to the models to estimate the volume of each object.   GMP4.2

▶ **Math Boxes 3-13**

Math Journal 1, p. 107

| WHOLE CLASS | SMALL GROUP | PARTNER | INDEPENDENT |

Mixed Practice Math Boxes 3-13 are paired with Math Boxes 3-11.

▶ **Home Link 3-13**

Math Masters, p. 112

Homework Students solve fraction-of problems.

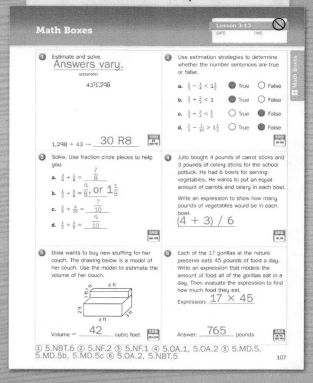

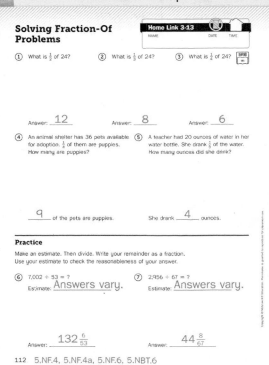

Fraction-Of Problems, Part 2

Overview Students solve fraction-of problems with fractional answers to continue building readiness for multiplying fractions by whole numbers.

▶ **Before You Begin**
Near the Math Message, display a unit box filled in with a unit, such as apples, that students can visualize dividing into fractional parts.

Standards

Focus Cluster
• Apply and extend previous understandings of multiplication and division.

① Warm Up   5 min

	Materials	
Mental Math and Fluency Students solve extended division facts.		5.NBT.6

② Focus   35–40 min

Math Message Students solve sets of related fraction-of problems.		5.NF.4, 5.NF.4a
Solving Fraction-Of Problems with Answers Greater Than 1 Students discuss strategies for solving fraction-of problems with answers that are not whole numbers.	slate, scissors, index cards, stick-on notes (optional)	5.NF.4, 5.NF.4a SMP2, SMP3
Solving Fraction-Of Problems with Answers Less Than 1 Students extend their strategies to solve fraction-of problems with answers less than 1.	Math Journal 1, p. 108; slate; scissors, index cards, stick-on notes (optional)	5.NF.4, 5.NF.4a, 5.NF.6 SMP1, SMP2, SMP3
✓ **Assessment Check-In** See page 308. Expect most students to be able to solve fraction-of problems like those identified that have fractional answers greater than 1.	Math Journal 1, p. 108	5.NF.4, 5.NF.4a, 5.NF.6

③ Practice   20–30 min

Playing Number Top-It **Game** Students use place-value understanding to practice building and comparing multidigit numbers.	Student Reference Book, p. 316; per partnership: Math Masters, pp. G7–G9; number cards 0–9 (4 of each)	5.NBT.1 SMP1, SMP7
Math Boxes 3-14 Students practice and maintain skills.	Math Journal 1, p. 109	See page 309.
Home Link 3-14 **Homework** Students solve fraction-of problems with fractional answers.	Math Masters, p. 114	5.NBT.5, 5.NF.4, 5.NF.4a, 5.NF.6

 (Go Online) to see how mastery develops for all standards within the grade.

Differentiation Options

Readiness 5–15 min	**Enrichment** 5–15 min	**Extra Practice** 5–15 min
WHOLE CLASS · **SMALL GROUP** · PARTNER · INDEPENDENT	WHOLE CLASS · **SMALL GROUP** · PARTNER · INDEPENDENT	WHOLE CLASS · SMALL GROUP · **PARTNER** · INDEPENDENT

Readiness

Solving Fraction-Of Problems with a Concrete Model

5.NF.4, 5.NF.4a, SMP2

stick-on notes, scissors

For experience using a concrete model to solve fraction-of problems with fractional answers, students use stick-on notes to represent and solve problems. Tell the group that you want to find $\frac{1}{2}$ of 3. Display 3 stick-on notes and explain that these notes, taken together, are what they need to find half of. Choose two students and explain that you want to give each of them $\frac{1}{2}$ of the notes. Ask: *Are there enough notes for each student to have 1?* Yes. Give each student 1 note. Ask: *Are there enough notes for each student to have 2?* No. *What can we do with the extra note?* Cut it in half and give $\frac{1}{2}$ to each student. Give each student $\frac{1}{2}$ of a note. **GMP2.1** Ask: *How many notes does each student have?* $1\frac{1}{2}$ notes *So what is $\frac{1}{2}$ of 3?* **GMP2.2** $1\frac{1}{2}$

Repeat the activity with a problem for which there are too few notes for each student to have a whole one. For example, display 2 stick-on notes and choose 3 students. Tell them that you want to give them each $\frac{1}{3}$ of the stick-on notes. To help them get started, ask questions like these: *Are there enough notes for each student to have 1? What else can we do?*

Enrichment

Adjusting Recipes

5.NF.4, 5.NF.4a, 5.NF.6

Math Masters, p. 113; fraction circles (optional)

To extend their work solving fraction-of problems, students find $\frac{1}{2}$ of and $\frac{1}{4}$ of the ingredients in recipes, requiring them to find fractions of fractions and fractions of mixed numbers. Students should be encouraged to use fraction circle pieces or drawings to help them solve the problems.

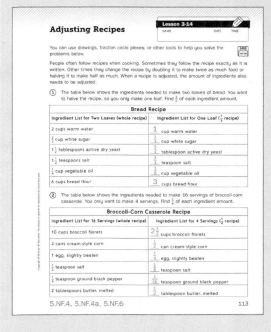

Extra Practice

Solving More Fraction-Of Problems

5.NF.4, 5.NF.4a, SMP1, SMP2

Activity Card 41, number cards 1–20 (1 of each), 6-sided die

For more practice solving fraction-of problems, students use number cards and a die to generate fraction-of problems. They represent the problems with drawings and check whether their answers make sense. **GMP1.4, GMP2.1**

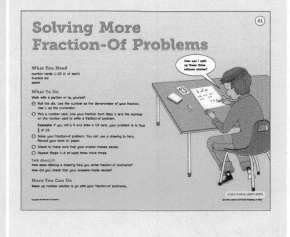

English Language Learner

Beginning ELL To help students understand the everyday term *split* along with the general academic term *separate* and the math content term *divide* as they relate to equal groups, use demonstrations along with think-alouds. For example, demonstrate splitting a set of 20 counters in different ways while making statements like these: *I will split these counters among the four of us. I will divide them equally among us. I will separate them into equal groups. We will each get the same amount when I split them.* Direct students to split/divide/separate varying numbers of objects into groups.

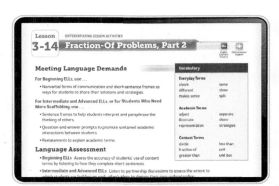

Differentiation Support pages are found in the online Teacher's Center.

SMP1 Make sense of problems and persevere in solving them.
GMP1.4 Check whether your answer makes sense.

SMP2 Reason abstractly and quantitatively.
GMP2.1 Create mathematical representations using numbers, words, pictures, symbols, gestures, tables, graphs, and concrete objects.

SMP3 Construct viable arguments and critique the reasoning of others.
GMP3.2 Make sense of others' mathematical thinking.

Academic Language Development

Encourage students to use sequencing terms, such as *first, next, then,* and *finally* in explanations of their strategies. Provide a word bank with fraction math terms used so far. Be sure to include *fair share, numerator, denominator, split, divide, number line,* and *fraction circle pieces.* Use think-pair-share protocols, allowing students to practice their explanations before presenting to the class.

---

1 Warm Up   5 min

▶ Mental Math and Fluency

Display extended division facts for students to solve. *Leveled exercises:*

● ○ ○ 32 / 4   8
          320 / 4   80
          3,200 / 4   800

● ● ○ 810 / 90   9
          81,000 / 900   90
          810,000 / 9,000   90

● ● ● 42,000 / 700   60
          630,000 / 900   700
          4,800,000 / 6,000   800

2 Focus   35–40 min

▶ Math Message

Display a unit box filled in with a unit like apples. (*See Before You Begin.*)
Solve the following problems:

$\frac{1}{2}$ of 8   4 $\frac{1}{3}$ of 12   4

$\frac{1}{2}$ of 9 $4\frac{1}{2}$ $\frac{1}{3}$ of 13 $4\frac{1}{3}$

                              $\frac{1}{3}$ of 14 $4\frac{2}{3}$

▶ Solving Fraction-Of Problems with Answers Greater Than 1

| WHOLE CLASS | SMALL GROUP | PARTNER | INDEPENDENT |

Math Message Follow-Up Have students share their answers for $\frac{1}{2}$ of 8. Then have them share strategies for $\frac{1}{2}$ of 9. *Sample strategies:*

• I thought about putting 9 apples into two boxes. I knew I could put 4 in each box with 1 left over. Then I had to split the extra apple between the 2 boxes. Splitting 1 apple in 2 equal parts gives $\frac{1}{2}$ apple for each box. So $4\frac{1}{2}$ go in each box, and $\frac{1}{2}$ of 9 is $4\frac{1}{2}$.

• I drew 9 circles in a row and drew a line down the middle.   GMP2.1
There were $4\frac{1}{2}$ circles on each side of the line.

• $\frac{1}{2}$ of 8 is 4. I know that 9 is 1 more than 8, so I just have to find $\frac{1}{2}$ of 1 and add that to $\frac{1}{2}$ of 8. I know $\frac{1}{2}$ of 1 is $\frac{1}{2}$, so $\frac{1}{2}$ of 9 is $4 + \frac{1}{2}$, or $4\frac{1}{2}$.

Repeat the process for the second set of problems. Ask students to share answers for $\frac{1}{3}$ of 12. Then have them share strategies for solving $\frac{1}{3}$ of 13 and $\frac{1}{3}$ of 14. Encourage students to draw pictures illustrating their strategies and to explain their thinking. **GMP2.1** *Sample strategies for $\frac{1}{3}$ of 14:*

- I thought about 14 circles. First I put as many circles as I could into three equal groups. I put 4 in each group and had 2 left over. I split the 2 left over into thirds. That gave me 6 thirds. I put 2 thirds into each group. Each group then had $4\frac{2}{3}$ circles in it. So $\frac{1}{3}$ of 14 is $4\frac{2}{3}$.

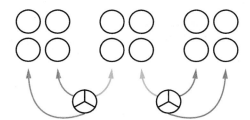

- I drew 14 boxes and divided each one into thirds. I thought about putting 1 third from each box together into a group. Each group would have 14 thirds in it. So $\frac{1}{3}$ of 14 is $\frac{14}{3}$. (*See margin.*)

After each strategy is shared, encourage other students to explain it in their own words. **GMP3.2** Be sure to point out that all the strategies give the same answer. Ask: *How do you know that $\frac{14}{3}$ is the same as $4\frac{2}{3}$?* Sample answer: I know that $\frac{12}{3}$ is the same as 12 ÷ 3, or 4. $\frac{14}{3}$ is two more thirds than that, or $4\frac{2}{3}$.

Pose a few more fraction-of problems with fractional answers greater than 1. Have students work independently or in partnerships to solve the problems and display answers on slates. *Suggestions:*

- $\frac{1}{4}$ of 9 $\frac{9}{4}$, or $2\frac{1}{4}$
- $\frac{1}{5}$ of 7 $\frac{7}{5}$, or $1\frac{2}{5}$
- $\frac{1}{10}$ of 23 $\frac{23}{10}$, or $2\frac{3}{10}$

Tell students that today they are going to use these strategies to solve a different kind of fraction-of problem.

▶ Solving Fraction-Of Problems with Answers Less Than 1

Math Journal 1, p. 108

| WHOLE CLASS | SMALL GROUP | PARTNER | INDEPENDENT |

Ask: *What is $\frac{1}{5}$ of 3?* Have partners work on the problem together. Expect that students may struggle getting started on this problem because they cannot begin by putting a whole number into each fifth.

Adjusting the Activity

Differentiate   Provide scissors and index cards, stick-on notes, or other same-size slips of paper that students can cut apart and manipulate to help them solve the problems in this lesson.

Go Online   Differentiation Support

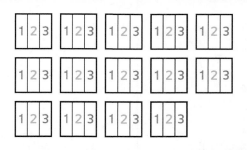

Put all the 1s together, all the 2s together, and all the 3s together. Each group will have 14 thirds.

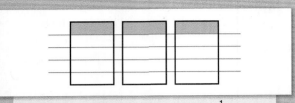

The shaded parts together show $\frac{1}{5}$ of 3.

Encourage students to think through the problem by making a representation and applying some of the strategies they discussed during the previous activity. **GMP2.1, GMP3.2** Look for partnerships using a successful strategy and have them share their strategies and representations. *Sample strategies:*

- I drew 3 rectangles and then drew lines to split them all into fifths. I knew parts above the top line together made $\frac{1}{5}$ of 3. There are 3 fifths above the top line, so $\frac{1}{5}$ of 3 is $\frac{3}{5}$. (*See margin.*)

- I saw a pattern in the earlier problems. The answer was always a fraction where the whole number from the problem was the numerator. The denominator of the answer was always the same as the denominator of the fraction in the problem. So following that pattern, the answer to this problem is $\frac{3}{5}$. I checked to make sure I was right by thinking: *Does $\frac{3}{5} + \frac{3}{5} + \frac{3}{5} + \frac{3}{5} + \frac{3}{5} = 3$?* The answer is yes, because $\frac{3}{5} + \frac{3}{5} + \frac{3}{5} + \frac{3}{5} + \frac{3}{5} = \frac{15}{5}$, or 3. **GMP1.4**

Regardless of which strategies students use, encourage them to use addition, drawings, or other methods to check that their answers make sense. **GMP1.4**

Pose a few more fraction-of problems with answers less than 1 and have students display answers on slates. *Suggestions:*

- $\frac{1}{4}$ of 3 $\quad \frac{3}{4}$
- $\frac{1}{6}$ of 4 $\quad \frac{4}{6}$, or $\frac{2}{3}$
- $\frac{1}{5}$ of 2 $\quad \frac{2}{5}$
- $\frac{1}{10}$ of 9 $\quad \frac{9}{10}$

Have students complete journal page 108. They explain how they can check whether one of their answers makes sense in Problem 7. **GMP1.4**

Math Journal 1, p. 108

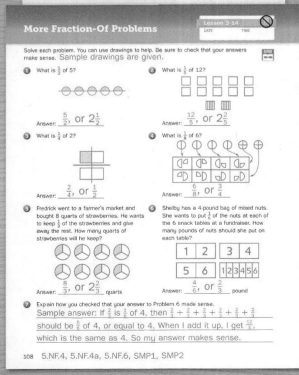

☑ **Assessment Check-In** 5.NF.4, 5.NF.4a, 5.NF.6

Math Journal 1, p. 108

Expect most students to be able to correctly solve Problems 1, 2, and 5 on journal page 108, which have fractional answers greater than 1. Some students may be able to solve Problems 3, 4, and 6, which have answers less than 1. If students struggle trying to solve Problem 1 or 2, suggest representing the problem with slips of paper as suggested in the Adjusting the Activity note. Some students may also benefit from using a unit box to put the problems in context.

✓= **Evaluation Quick Entry** Go online to record students' progress and to see trajectories toward mastery for these standards.

Summarize Ask: *How are the problems in this lesson different from the problems you solve when you play* Fraction Of*?* Sample answer: In *Fraction Of* there are never any leftovers to split up. For these problems I have to think about splitting up leftovers into smaller parts.

3 Practice 20–30 min

▶ Playing *Number Top-It*

Student Reference Book, p. 316; *Math Masters*, pp. G7–G9

| WHOLE CLASS | SMALL GROUP | PARTNER | INDEPENDENT |

Students build their understanding of place-value relationships and expanded form by playing *Number Top-It*. If some of your students have mastered writing whole numbers in expanded form, consider introducing a variation in which they flip over a number card after they have made their 7-digit numbers. Then they look for that digit in their number and determine its value. The player with the higher value wins. Players who do not have the digit in their number cannot win the round. If no players have the digit in their number or there is a tie, they flip over another card.

Observe

- Which students have a strategy for placing their cards on the gameboard to create the largest number?
- Which students are able to fluently read and compare their numbers?

Discuss

- *How does moving a number card one place to the left change its value? To the right?* GMP1.2, GMP7.2
- *If you could choose 7 different number cards to place, which 7 would you choose and in what order would you place them to create the largest number? How would you say that number?* GMP7.2

| Differentiate | **Game Modifications** | Go Online | Differentiation Support |

▶ Math Boxes 3-14

Math Journal 1, p. 109

| WHOLE CLASS | SMALL GROUP | PARTNER | INDEPENDENT |

Mixed Practice Math Boxes 3-14 are grouped with Math Boxes 3-9 and 3-12.

▶ Home Link 3-14

Math Masters, p. 114

Homework Students solve fraction-of problems with fractional answers.

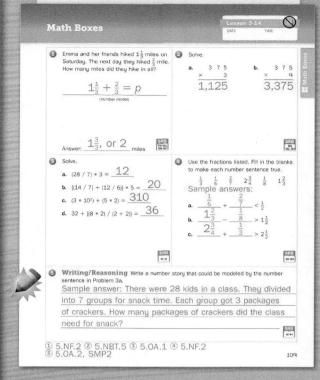

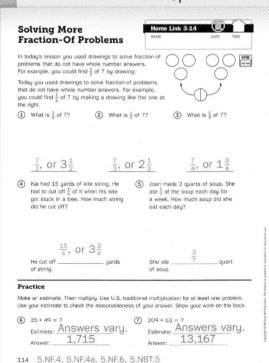

Unit 3 Progress Check

Overview   **Day 1:** Administer the Unit Assessments.
Day 2: Administer the Open Response Assessment.

Day 1: Unit Assessment

 **Quick Entry Evaluation** Record results and track progress toward mastery.

1 Warm Up   5–10 min

Materials

Self Assessment
Students complete the Self Assessment.

Assessment Handbook, p. 23

2a Assess   35–50 min

Unit 3 Assessment
These items reflect mastery expectations to this point.

Assessment Handbook, pp. 24–27

Unit 3 Challenge (Optional)
Students may demonstrate progress beyond expectations.

Assessment Handbook, pp. 28–29

Standards	Goals for Mathematical Content (GMC)	Lessons	Self Assessment	Unit 3 Assessment	Unit 3 Challenge
5.NBT.6	Divide multidigit whole numbers.	3-3, 3-5		2	
	Illustrate and explain solutions to division problems.	3-3, 3-5		2	
5.NF.1	Add and subtract fractions with unlike denominators.	3-10 to 3-12	6	11a, 11b	
5.NF.2	Solve number stories involving fraction addition and subtraction.	3-7, 3-9, 3-12	7	9, 10a, 10b	2b, 3c
	Use estimates to reason about sums and differences of fractions.	3-4, 3-6, 3-7, 3-9, 3-10, 3-12	4	7, 8a–8d	
5.NF.3	Interpret a fraction as division of a numerator by a denominator.	3-2 to 3-5, 3-8, 3-12	2, 3, 5	3–6, 14	1a
	Solve number stories involving whole-number division that leads to fractional answers.	3-1 to 3-3, 3-12	1	1, 2	1a–1c
5.NF.4	Multiply fractions by whole numbers.	3-13, 3-14	8	12a–12d, 13	4a, 4b
5.NF.4, 5.NF.4a	Interpret $\left(\frac{1}{b}\right) \times q$ as 1 part of a partition of q into b equal parts.	3-13, 3-14	8	12a–12d, 13	4a, 4b
5.NF.6	Solve real-world problems involving fraction multiplication.	3-13, 3-14		13	

Standards	Goals for Mathematical Content (GMC)	Lessons	Self Assessment	Unit 3 Assessment	Unit 3 Challenge
SMP1	Make sense of your problem. GMP1.1	3-12		10a	3a, 3b
	Check whether your answer makes sense. GMP1.4	3-3, 3-6, 3-9, 3-11, 3-14		7	4a
SMP2	Create mathematical representations using numbers, words, pictures, symbols, gestures, tables, graphs, and concrete objects. GMP2.1	3-7, 3-13, 3-14		4	
SMP3	Make mathematical conjectures and arguments. GMP3.1	3-6 to 3-8		7	
	Make sense of others' mathematical thinking. GMP3.2	3-1, 3-7, 3-14		10a	3a, 3b
SMP4	Model real-world situations using graphs, drawings, tables, symbols, numbers, diagrams, and other representations. GMP4.1	3-1, 3-2, 3-9		1, 2, 9	2a
SMP5	Use tools effectively and make sense of your results. GMP5.2	3-4, 3-8 to 3-10		11a, 11b	
SMP6	Use an appropriate level of precision for your problem. GMP6.2	3-3		2	1c

 Go Online to see how mastery develops for all standards within the grade.

1 Warm Up 5–10 min

▶ Self Assessment

Assessment Handbook, p. 23

| WHOLE CLASS | SMALL GROUP | PARTNER | **INDEPENDENT** |

Students complete the Self Assessment to reflect on their progress in Unit 3.

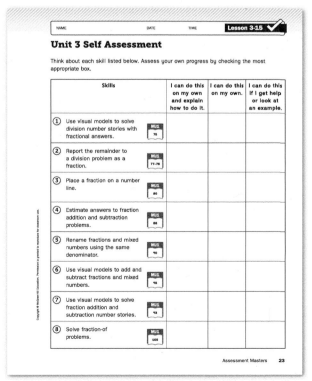

Assessment Handbook, p. 23

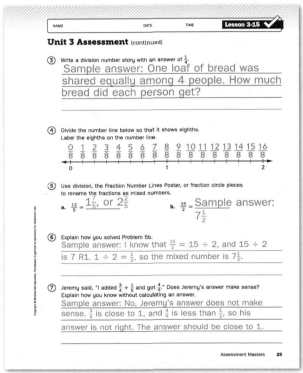

Unit 3 Assessment

Solve the following number stories. Use your fraction circle pieces or a drawing to help you.

① There are three small rounds of cheddar cheese in Grace's refrigerator. She wants to bring a small wedge of cheese in her lunch every day for the next 12 days of school. If she wants to eat the same amount of cheese every day, how much cheese should she bring for lunch each day?

Solution: $\frac{3}{12}$, or $\frac{1}{4}$, round of cheese

Number model: $3 \div 12 = \frac{3}{12}$

② Three families live in the same apartment building. They decided to share a giant 220-ounce container of laundry detergent. If the families split the detergent equally, how many ounces of laundry detergent will each family get?

Solution: Each family will get $73\frac{1}{3}$ ounces.

Number model: $220 \div 3 = 73\frac{1}{3}$

Explain what you did with the remainder and why.   Sample answer: I reported the remainder as a fraction. It did not make sense to throw away the extra detergent. The families could split it up.

24   Assessment Handbook

2a Assess   35–50 min

▶ Unit 3 Assessment

Assessment Handbook, pp. 24–27

| WHOLE CLASS | SMALL GROUP | PARTNER | **INDEPENDENT** |

Students complete the Unit 3 Assessment to demonstrate their progress on the standards covered in this unit.

Generic rubrics in the *Assessment Handbook* can be used to evaluate student progress on the Mathematical Process and Practice Standards.

Unit 3 Assessment (continued)

③ Write a division number story with an answer of $\frac{1}{4}$.

Sample answer: One loaf of bread was shared equally among 4 people. How much bread did each person get?

④ Divide the number line below so that it shows eighths. Label the eighths on the number line.

$\frac{0}{8}$ $\frac{1}{8}$ $\frac{2}{8}$ $\frac{3}{8}$ $\frac{4}{8}$ $\frac{5}{8}$ $\frac{6}{8}$ $\frac{7}{8}$ $\frac{8}{8}$ $\frac{9}{8}$ $\frac{10}{8}$ $\frac{11}{8}$ $\frac{12}{8}$ $\frac{13}{8}$ $\frac{14}{8}$ $\frac{15}{8}$ $\frac{16}{8}$

0   1   2

⑤ Use division, the Fraction Number Lines Poster, or fraction circle pieces to rename the fractions as mixed numbers.

a. $\frac{12}{5} = 1\frac{7}{5}$, or $2\frac{2}{5}$ b. $\frac{15}{2} =$ Sample answer: $7\frac{1}{2}$

⑥ Explain how you solved Problem 5b.
Sample answer: I know that $\frac{15}{2} = 15 \div 2$, and $15 \div 2$ is 7 R1. $1 \div 2 = \frac{1}{2}$, so the mixed number is $7\frac{1}{2}$.

⑦ Jeremy said, "I added $\frac{3}{4} + \frac{1}{2}$ and got $\frac{4}{4}$." Does Jeremy's answer make sense? Explain how you know without calculating an answer.
Sample answer: No, Jeremy's answer does not make sense. $\frac{3}{4}$ is close to 1, and $\frac{4}{4}$ is less than $\frac{1}{2}$, so his answer is not right. The answer should be close to 1.

Assessment Masters   25

Assessment Handbook, p. 25

Differentiate — Adjusting the Assessment

Item(s)	Adjustments
1, 2	To scaffold Items 1 and 2, ask students questions like: *What is the whole? How many ways do we need to split the whole? How could you show that with a drawing or fraction circle pieces?*
3	To extend Item 3, have students write a number story with an answer of $\frac{1}{4}$ that does not use the number 1 or the number 4.
4	To scaffold Item 4, have students first divide the number line so that it shows fourths. Ask how they can use fourths to help them divide the number line into eighths.
5	To extend Item 5, have students write a rule for how to convert any fraction greater than 1 into a mixed number.
6	To scaffold Item 6, have students dictate how they solved the problem as you write their responses for them.
7	To scaffold Item 7, have students model the problem with their fraction circle pieces.
8	To scaffold Item 8, give students a list of fractions they can choose from to complete the number sentences.
9	To extend Item 9, have students write an addition number story with an answer of $1\frac{2}{3}$.
10	To scaffold Item 10, encourage students to refer to *Student Reference Book*, page 30 for strategies they can use to get started and to keep going when solving a problem.
11	To extend Item 11, have students write another addition problem with unlike denominators that gives the same answer for each part of the problem.
12, 13	To scaffold Items 12 and 13, give students counters and encourage them to divide the counters into equal groups.
14	To extend Item 14, have students list all possible fraction and mixed-number names for each mixed number without changing the denominator.

Advice for Differentiation

All of the content included on the Unit 3 Assessment was recently introduced and will be revisited in subsequent units.

Go Online:

 Quick Entry Evaluation Record students' progress and see trajectories toward mastery for these standards.

 Data Review your students' progress reports. Differentiation materials are available online to help you address students' needs.

NOTE See the Unit Organizer on pages 212–213 or the online Spiral Tracker for details on Unit 3 focus topics and the spiral.

Assessment Handbook, p. 26

NAME   DATE   TIME   Lesson 3-15

Unit 3 Assessment (continued)

8. Write a fraction to make each number sentence true. Use your fraction circle pieces or the Fraction Number Lines Poster to help you. Sample answers given.

    a. $\frac{2}{3} + \frac{1}{2} > 1$ b. $2 - \frac{2}{4} > 1$

    c. $1 + \frac{3}{4} > 1\frac{1}{2}$ d. $1 - \frac{1}{4} > \frac{1}{2}$

9. A chef had $2\frac{1}{3}$ heads of lettuce. She used $\frac{2}{3}$ head of lettuce to make a salad. How many heads of lettuce does she have left?

    Number model: $2\frac{1}{3} - \frac{2}{3} = h$

    Estimate: Sample answer: a little less than 2

    Answer: $1\frac{2}{3}$ heads of lettuce

10. Natasha is growing a plant for a science project.
    On Monday she measured the plant and found it was $2\frac{1}{8}$ inches tall.
    On Wednesday she measured the plant and found it was $3\frac{2}{8}$ inches tall.
    Natasha told her teacher that her plant grew $5\frac{3}{8}$ inches from Monday to Wednesday.

    a. What mistake did Natasha make?
    Sample answer: Natasha added the two heights. She should have found the difference between the heights to figure out how much the plant grew.

    b. How much did the plant actually grow from Monday to Wednesday?

    Answer: The plant grew $1\frac{1}{8}$ inches from Monday to Wednesday.

26   Assessment Handbook

Assessment Handbook, p. 27

NAME   DATE   TIME   Lesson 3-15

Unit 3 Assessment (continued)

11. Solve. Use your fraction circle pieces to help.

    a. $\frac{1}{3} + \frac{1}{6} = \frac{3}{6}$

    b. $\frac{3}{4} + \frac{1}{12} = \frac{10}{12}$

12. What is:

    a. $\frac{1}{4}$ of 20? 5   b. $\frac{1}{3}$ of 9? 3

    c. $\frac{1}{2}$ of 9? $\frac{9}{2}$, or $4\frac{1}{2}$ d. $\frac{1}{5}$ of 11? $\frac{11}{5}$, or $2\frac{1}{5}$

13. Jason bought a case of 60 oranges from his school fundraiser. He gave $\frac{1}{5}$ of the oranges to his aunt. How many oranges did his aunt get?

    Answer: 12 oranges

14. Write another name for each mixed number that has the same denominator.

    a. $8\frac{1}{2}$ Sample answer: $7\frac{3}{2}$

    b. $2\frac{2}{5}$ Sample answer: $3\frac{2}{5}$

Assessment Masters   27

Unit 3 Challenge

1. Amelie is setting up a water station for cross-country practice. Ten runners will run by. She estimates that 1 jug of water will serve 3 runners.

   a. How much water does Amelie think each runner will need? $\frac{1}{3}$ jug

   b. How many jugs of water will Amelie need?

      Amelie will need __4__ jugs of water.

   c. Explain how you found your answer. Sample answer: I added $\frac{1}{3}$ ten times because each runner will get $\frac{1}{3}$ jug of water. I got $\frac{10}{3}$, or $3\frac{1}{3}$, jugs of water. You can't buy $\frac{1}{3}$ jug of water, so she will need 4 jugs.

2. There were $5\frac{7}{10}$ liters of water in a bucket. Mollie spilled some of the water. Then she added an additional $1\frac{5}{10}$ liters of water to the bucket. After that, there were $6\frac{4}{10}$ liters of water in the bucket.

   a. Write a number model for the story. Use a letter to represent the amount of water Mollie spilled.

      Number model: Sample answer: $5\frac{7}{10} - w + 1\frac{5}{10} = 6\frac{4}{10}$

   b. How much water did Mollie spill?

      Mollie spilled $\frac{8}{10}$ liter of water.

28 Assessment Handbook

▶ # Unit 3 Challenge (Optional)

Assessment Handbook, pp. 28–29

| WHOLE CLASS | SMALL GROUP | PARTNER | **INDEPENDENT** |

Students can complete the Unit 3 Challenge after they complete the Unit 3 Assessment.

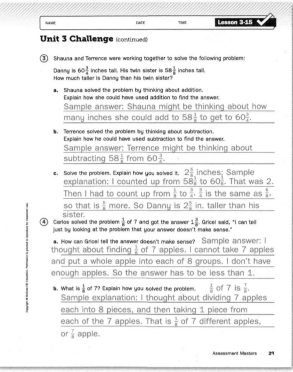

Unit 3 Challenge (continued)

3. Shauna and Terrence were working together to solve the following problem:

   Danny is $60\frac{3}{4}$ inches tall. His twin sister is $58\frac{1}{8}$ inches tall. How much taller is Danny than his twin sister?

   a. Shauna solved the problem by thinking about addition. Explain how she could have used addition to find the answer.
      Sample answer: Shauna might be thinking about how many inches she could add to $58\frac{1}{8}$ to get to $60\frac{3}{4}$.

   b. Terrence solved the problem by thinking about subtraction. Explain how he could have used subtraction to find the answer.
      Sample answer: Terrence might be thinking about subtracting $58\frac{1}{8}$ from $60\frac{3}{4}$.

   c. Solve the problem. Explain how you solved it. $2\frac{5}{8}$ inches; Sample explanation: I counted up from $58\frac{1}{8}$ to $60\frac{1}{8}$. That was 2. Then I had to count up from $\frac{1}{8}$ to $\frac{3}{4}$. $\frac{3}{4}$ is the same as $\frac{6}{8}$, so that is $\frac{5}{8}$ more. So Danny is $2\frac{5}{8}$ in. taller than his sister.

4. Carlos solved the problem $\frac{1}{8}$ of 7 and got the answer $1\frac{3}{8}$. Gricel said, "I can tell just by looking at the problem that your answer doesn't make sense."

   a. How can Gricel tell the answer doesn't make sense? Sample answer: I thought about finding $\frac{1}{8}$ of 7 apples. I cannot take 7 apples and put a whole apple into each of 8 groups. I don't have enough apples. So the answer has to be less than 1.

   b. What is $\frac{1}{8}$ of 7? Explain how you solved the problem. $\frac{1}{8}$ of 7 is $\frac{7}{8}$. Sample explanation: I thought about dividing 7 apples each into 8 pieces, and then taking 1 piece from each of the 7 apples. That is $\frac{1}{8}$ of 7 different apples, or $\frac{7}{8}$ apple.

   Assessment Masters 29

Assessment Handbook, p. 29

Unit 3 Progress Check

Day 2: Open Response Assessment

▶ **Before You Begin**

Have fraction circle pieces available and be sure the Fraction Number Lines Poster is visible to all students.

2b Assess 50–55 min

Materials

Solving the Open Response Problem
Following a brief introduction, students use representations to explain how they know whether the answer to a fraction number story is correct.

Assessment Handbook, p. 30; fraction circles; Fraction Number Lines Poster

Discussing the Problem
After completing the problem, students share their representations and their reasoning.

Assessment Handbook, p. 30

Standards	Goal for Mathematical Content (GMC)	Lessons
5.NF.2	Use estimates to reason about sums and differences of fractions.	3-4, 3-6, 3-7, 3-9, 3-10, 3-12
	Goal for Mathematical Process and Practice (GMP)	
SMP2	Create mathematical representations using numbers, words, pictures, symbols, gestures, tables, graphs, and concrete objects. GMP2.1	3-7, 3-13, 3-14

▶ **Evaluating Students' Responses**

Evaluate students' abilities to use benchmark fractions and number sense to estimate mentally and assess the reasonableness of answers. Use the rubric below to evaluate their work based on **GMP2.1**.

Goal for Mathematical Process and Practice GMP2.1 Create mathematical representations using numbers, words, pictures, symbols, gestures, tables, graphs, and concrete objects.	Not Meeting Expectations	Partially Meeting Expectations	Meeting Expectations	Exceeding Expectations
	Does not attempt to create or use a representation to solve the problem.	Creates a partially correct or incomplete representation that shows $\frac{1}{2} + \frac{3}{8}$ is greater than or not equal to $\frac{2}{5}$, or that $\frac{1}{2}$ is greater than $\frac{2}{5}$.	Creates a correct representation that shows $\frac{1}{2} + \frac{3}{8}$ is greater than or not equal to $\frac{2}{5}$, or that $\frac{1}{2}$ is greater than $\frac{2}{5}$.	Meets expectations and creates more than one correct representation.

3 Look Ahead 10–15 min

Materials

Math Boxes 3-15: Preview for Unit 4
Students preview skills and concepts for Unit 4.

Math Journal 1, p. 110

Home Link 3-15
Students take home the Family Letter that introduces Unit 4.

Math Masters, pp. 115–118

 Go Online to see how mastery develops for all standards within the grade.

NAME   DATE   TIME   **Lesson 3-15** ✓

Unit 3 Open Response Assessment
Running

Clare was training for a running race. She decided to run $\frac{3}{8}$ mile from school to the park. Later she left the park and ran $\frac{1}{5}$ mile home. She told her brother the distances she ran. Her brother said, "You ran a total of $\frac{2}{5}$ mile."

Do you agree with Clare's brother? Use pictures, words, number sentences, or other representations to explain why you agree or disagree.

Answers vary. See sample students' work on page 317 of the *Teacher's Lesson Guide*.

30   Assessment Handbook

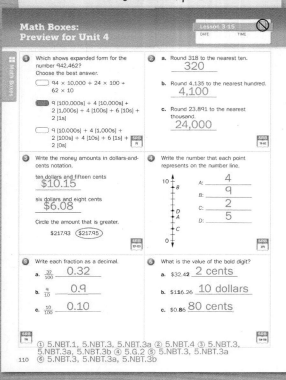

Math Boxes:
Preview for Unit 4   Lesson 3-15

① Which shows expanded form for the number 942,462?
Choose the best answer.

⬭ 94 × 10,000 + 24 × 100 + 62 × 10

⬛ 9 [100,000s] + 4 [10,000s] + 2 [1,000s] + 4 [100s] + 6 [10s] + 2 [1s]

⬭ 9 [10,000s] + 4 [1,000s] + 2 [100s] + 4 [10s] + 6 [1s] + 2 [0s]

② a. Round 318 to the nearest ten.
    320

b. Round 4,135 to the nearest hundred.
    4,100

c. Round 23,891 to the nearest thousand.
    24,000

③ Write the money amounts in dollars-and-cents notation.

ten dollars and fifteen cents
$10.15

six dollars and eight cents
$6.08

Circle the amount that is greater.
$217.93   ($217.95)

④ Write the number that each point represents on the number line.

A: 4
B: 9
C: 2
D: 5

⑤ Write each fraction as a decimal.

a. $\frac{32}{100}$ 0.32

b. $\frac{9}{10}$ 0.9

c. $\frac{10}{100}$ 0.10

⑥ What is the value of the bold digit?

a. $32.42   2 cents

b. $116.26   10 dollars

c. $0.86   80 cents

① 5.NBT.1, 5.NBT.3, 5.NBT.3a ② 5.NBT.4 ③ 5.NBT.3, 5.NBT.3a, 5.NBT.3b ④ 5.G.2 ⑤ 5.NBT.3, 5.NBT.3a
⑥ 5.NBT.3, 5.NBT.3a, 5.NBT.3b

2b Assess   50–55 min

▶ Solving the Open Response Problem

Assessment Handbook, p. 30

| WHOLE CLASS | SMALL GROUP | PARTNER | **INDEPENDENT** |

The open response problem requires students to apply skills and concepts from Unit 3 to decide whether an answer to a fraction number story is correct. Students use estimation and fraction representations to explain their reasoning. The focus of this task is **GMP2.1:** Create mathematical representations using numbers, words, pictures, symbols, gestures, tables, graphs, and concrete objects. Note that representations are not limited to pictorial representations but can also include number models, words, and manipulatives.

Tell students that today they will decide whether an answer to a fraction number story is correct. Distribute *Assessment Handbook,* page 30. Have students read the page and discuss the directions. Ask: *Do you need to find an exact answer for the number story to answer the question? Why or why not?* Sample answer: No. The question asks whether Clare's brother is correct; it doesn't ask for the exact distance. You can use estimation or reasoning to decide whether his answer is correct. Tell students that they can use any mathematical representations they choose to help them solve the problem and explain their answers. Ask: *What types of representations could you use?*   GMP2.1   Sample answers: fraction circle pieces, Fraction Number Lines Poster, number lines, number models, or drawings

Students should complete the problem independently.

Differentiate   **Common Misconception**

Some students will attempt to add the two distances together and get stuck or find an incorrect answer because they haven't yet learned a formal procedure for adding fractions with unlike denominators. Remind these students that they do not need to find an exact answer, and encourage them to use an estimate, fraction representations, benchmarks, or what they know about fractions to explain why they agree or disagree with Clare's brother. Some students may be able to find the correct sum using fraction circle pieces or pictorial representations.

Differentiate   **Adjusting the Activity**

If students struggle getting started, suggest that they represent $\frac{2}{5}$ and use that representation as a reference. They may represent the fraction with a drawing, fraction circle pieces, the Fraction Number Lines Poster, or a number line.   GMP2.1   Ask: *Is $\frac{2}{5}$ greater or less than $\frac{1}{2}$? How do you know?* For students who understand the mathematics but struggle with how to make their arguments, provide sentence frames such as the following: "I _____ with Clare's brother because _____."

▶ Discussing the Problem

Assessment Handbook, p. 30

WHOLE CLASS   SMALL GROUP   **PARTNER**   INDEPENDENT

After students have had a chance to complete their work, invite them to share their representations and explanations with a partner before having a few students share with the class. Discuss both correct and incorrect representations. You may want to begin with an incorrect representation and allow students to suggest how to improve it. A successful strategy can then emerge from this discussion. Consider sharing a few solutions that use different representations but arrive at the same answer. Ask: *How are these representations similar? How are they different?* GMP2.1 Answers vary.

Evaluating Student's Responses   5.NF.2

Collect students' work. For the content standard, expect most students to assess the reasonableness of the brother's answer to the number story and to explain that the brother's sum of $\frac{2}{5}$ is incorrect. Use the rubric on page 315 to evaluate students' work for **GMP2.1.**

See the sample in the margin. This work meets expectations for the content standard because the student correctly explained that $\frac{1}{2} > \frac{2}{5}$, so the brother's sum of $\frac{2}{5}$ is incorrect. The work meets expectations for the mathematical process and practice standards because this student drew and labeled fraction circles to show that $\frac{1}{2} > \frac{2}{5}$, and correctly used words and numbers to explain that $\frac{3}{8} + \frac{1}{2}$ cannot be $\frac{2}{5}$. GMP2.1

 Evaluation Quick Entry Go online to record students' progress and to see trajectories toward mastery for these standards.

NOTE Additional samples of evaluated students' work can be found in the *Assessment Handbook* appendix.

③ Look Ahead   10–15 min

▶ Math Boxes 3-15: Preview for Unit 4

Math Journal 1, p. 110

WHOLE CLASS   SMALL GROUP   PARTNER   **INDEPENDENT**

Mixed Practice Math Boxes 3-15 are paired with Math Boxes 3-10. These problems focus on skills and understandings that are prerequisite for Unit 4. You may want to use information from these Math Boxes to plan instruction and grouping in Unit 4.

▶ Home Link 3-15: Unit 4 Family Letter

Math Masters, pp. 115–118

Home Connection The Unit 4 Family Letter provides information and activities related to the Unit 4 content.

Sample student's work, "Meeting Expectations"

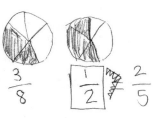

Math Masters, pp. 115–118

Decimal Concepts; Coordinate Grids

Contents

*The standards listed here are addressed in the **Focus** of each lesson. For all the standards in a lesson, see the Lesson Opener.

Focus

In this unit, students read, write, and represent decimals through thousandths in a variety of ways and learn strategies to compare, order, and round decimals. They are also introduced to the first quadrant of the coordinate grid. Finally, they apply whole-number algorithms to add and subtract decimals.

Major Clusters

5.NBT.A Understand the place value system.

5.NBT.B Perform operations with multi-digit whole numbers and with decimals to hundredths.

Supporting Cluster

5.G.A Graph points on the coordinate plane to solve real-world and mathematical problems.

Process and Practice Standards

SMP2 Reason abstractly and quantitatively.

SMP7 Look for and make use of structure.

Coherence

The table below describes how standards addressed in the Focus parts of the lessons link to the mathematics that students have done in the past and will do in the future.

		Links to the Past	Links to the Future
5.NBT.1		In Unit 2, students described place-value relationships in whole numbers, including how the value of a digit changes when it moves to the left or to the right, and applied this understanding to play the game *Number Top-It*. In Grade 4, students worked with place-value concepts in whole numbers through 1,000,000.	In Unit 6, students will apply place-value concepts as they multiply and divide decimals by powers of 10. They will also apply place-value concepts as they compute with decimals. In Grade 6, students will extend their understanding of place value by making sense of the U.S. traditional algorithms for decimal computation.
5.NBT.3 **5.NBT.3a**		In Unit 2, students wrote whole numbers using base-10 numerals, number names, and expanded form. In Grade 4, students used decimal notation to write fractions with denominators of 10 or 100.	In Unit 6, students will write decimals in standard notation to help them find patterns when they multiply and divide decimals by powers of 10. In Grade 6, students will use expanded notation to help them make sense of the U.S. traditional algorithms for decimal computation.
5.NBT.3 **5.NBT.3b**		In Grade 4, students compared decimals to hundredths, recorded those comparisons symbolically, and justified their reasoning.	In future units, students will compare and order decimals in real-world contexts, such as representing data that is collected in decimal notation. In Grade 6, students will compare and order rational numbers.
5.NBT.4		In Grade 4, students used place-value understanding to round multidigit whole numbers to any place.	Throughout Grade 5, students will continue to round decimals as a strategy for making estimates for decimal computation problems. They will also determine when and how to round decimals in real-world problems in Unit 8. In Grade 6, students will continue to round decimals as they make estimates for decimal computation problems.
5.NBT.7		In Grade 4, students solved addition and subtraction problems involving fractions with denominators of 10 and 100, and related these fractions to decimals.	In Unit 6, students will learn strategies for multiplying and dividing decimals. Throughout Grade 5, students will practice decimal computation by solving real-world problems. In Grade 6, students will use U.S. traditional algorithms to add, subtract, multiply, and divide decimals.

Planning for Rich Math Instruction

	4-1 Decimal Place Value	**4-2** Representing Decimals through Thousandths	**4-3** Representing Decimals in Expanded Form	**4-4** Comparing and Ordering Decimals
RIGOR Conceptual Understanding	**Place value in decimals through thousandths** Extending Place-Value Patterns to the Thousandths Place, pp. 332–334	**Decimal representations** Representing Decimals on a Thousandths Grid, pp. 340–342	**Decimal representations** Introducing Expanded Form for Decimals, pp. 346–348	**Decimal comparisons** Exploring Strategies for Comparing Decimals, pp. 352–354
Procedural Skill and Fluency	Journal p. 111, #2, #3	Solving Fraction Number Stories, p. 343 Enrichment, p. 339	Playing Fraction Capture, p. 349	Interpreting Real-World Remainders, p. 355
Applications	Journal p. 111, #3	Solving Fraction Number Stories, p. 343	Journal p. 120, #3 Home Link 4-3, p. 349 Enrichment, p. 345	Interpreting Real-World Remainders, p. 355 Enrichment, p. 351
Rich Tasks and Mathematical Reasoning	Enrichment, p. 331	Collecting Names for Decimals, pp. 342–343	Math Message, p. 346 Representing Decimals in Expanded Form, pp. 348–349 Playing Fraction Capture, p. 349 Journal p. 120: Writing/Reasoning	Introducing Decimal Top-It, pp. 354–355 Home Link 4-4, p. 355
Mathematical Discourse	Extending Place-Value Patterns to the Thousandths Place, pp. 332–334 Extra Practice, p. 331	Representing Decimals on a Thousandths Grid, pp. 340–342 Extra Practice, p. 339	Introducing Expanded Form for Decimals, pp. 346–348 Playing Fraction Capture, p. 349 Extra Practice, p. 345	Extra Practice, p. 351
Distributed Practice	Mental Math and Fluency, p. 332 Playing Fraction Of, pp. 336–337 Math Boxes 4-1, p. 337	Mental Math and Fluency, p. 340 Solving Fraction Number Stories, p. 343 Math Boxes 4-2, p. 343	Mental Math and Fluency, p. 346 Playing Fraction Capture, p. 349 Math Boxes 4-3, p. 349	Mental Math and Fluency, p. 352 Interpreting Real-World Remainders, p. 355 Math Boxes 4-4, p. 355
Differentiation Support	Differentiation Options, p. 331 ELL Support, p. 331 Online Differentiation Support 4-1 Academic Language Development, p. 333 Common Misconception, p. 334	Differentiation Options, p. 339 ELL Support, p. 339 Online Differentiation Support 4-2 Adjusting the Activity, p. 341	Differentiation Options, p. 345 ELL Support, p. 345 Online Differentiation Support 4-3 Adjusting the Activity, p. 347	Differentiation Options, p. 351 ELL Support, p. 351 Online Differentiation Support 4-4 Common Misconception, p. 353 Adjusting the Activity, p. 354

Red text = Game

	4-5 Rounding Decimals	**4-6** Introduction to the Coordinate System	**4-7** Playing *Hidden Treasure*	**4-8** Solving Problems on a Coordinate Grid, Part 1	
	Rounding with decimals Rounding Decimals Using Grids and Number Lines, pp. 358–361	**Understanding the structure of a coordinate grid** Plotting Points on a Coordinate Grid, pp. 366–368	**Understanding the structure of a coordinate grid** Showing Routes on a Coordinate Grid, pp. 372–374 Introducing *Hidden Treasure*, pp. 374–375	**Representing and solving problems on a coordinate grid** Discussing the Results of Operations on Ordered Pairs, pp. 379–381	**Conceptual Understanding**
	Journal p. 126, #1	Journal p. 130, #1, #6	Practicing U.S. Traditional Multiplication, p. 375	Journal p. 137, #1, #6	**Procedural Skill and Fluency**
	Rounding Decimals Using Grids and Number Lines, pp. 358–361 Developing a Rounding Shortcut, pp. 361–363 Home Link 4-5, p. 363	Math Message, p. 366 Plotting Points on a Coordinate Grid, pp. 366–368 Finding Volumes of Soil, p. 369 Home Link 4-6, p. 369	Math Message, p. 372 Showing Routes on a Coordinate Grid, pp. 372–374 Enrichment, p. 371	Journal p. 137, #1, #4, #5	**Applications**
	Developing a Rounding Shortcut, pp. 361–363 Journal p. 126: Writing/Reasoning Enrichment, p. 357	Introducing *Over and Up Squares*, p. 369 Finding Volumes of Soil, p. 369 Enrichment, p. 365	Showing Routes on a Coordinate Grid, pp. 372–374 Introducing *Hidden Treasure*, pp. 374–375 Practicing U.S. Traditional Multiplication, p. 375 Enrichment, p. 371	Discussing the Results of Operations on Ordered Pairs, pp. 379–381	**Rich Tasks and Mathematical Reasoning**
	Developing a Rounding Shortcut, pp. 361–363 Playing *Rename That Mixed Number*, p. 363 Extra Practice, p. 357	Readiness, p. 365	Showing Routes on a Coordinate Grid, pp. 372–374 Enrichment, p. 371	Plotting Ordered Pairs and Transforming Figures, pp. 378–379	**Mathematical Discourse**
	Mental Math and Fluency, p. 358 Playing *Rename That Mixed Number*, p. 363 Math Boxes 4-5, p. 363	Mental Math and Fluency, p. 366 Finding Volumes of Soil, p. 369 Math Boxes 4-6, p. 369	Mental Math and Fluency, p. 372 Practicing U.S. Traditional Multiplication, p. 375 Math Boxes 4-7, p. 375	Mental Math and Fluency, p. 378 Playing *Decimal Top-It*, p. 381 Math Boxes 4-8, p. 381	**Distributed Practice**
	Differentiation Options, p. 357 ELL Support, p. 357 Online Differentiation Support 4-5 Adjusting the Activity, pp. 359, 360 Academic Language Development, p. 359 Common Misconception, p. 360	Differentiation Options, p. 365 ELL Support, p. 365 Online Differentiation Support 4-6 Adjusting the Activity, p. 368	Differentiation Options, p. 371 ELL Support, p. 371 Online Differentiation Support 4-7 Academic Language Development, p. 373 Adjusting the Activity, p. 374	Differentiation Options, p. 377 ELL Support, p. 377 Online Differentiation Support 4-8	**Differentiation Support**

RIGOR

Red text = Game

Planning for Rich Math Instruction

	4-9 **Solving Problems on a Coordinate Grid, Part 2**	**4-10** Open Response **Folder Art** (2-Day Lesson)	**4-11** **Addition and Subtraction of Decimals with Hundredths Grids**	**4-12** **Decimal Addition Algorithms**
RIGOR **Conceptual Understanding**	Representing and solving problems on a coordinate grid Graphing Data as Ordered Pairs, pp. 384–386	Representing and solving problems on a coordinate grid Changing the Logo, pp. 389–390	Addition and subtraction of decimals Adding Decimals with Grids, pp. 400–402 Subtracting Decimals with Grids, pp. 402–405	Addition of decimals Extending Whole-Number Addition Algorithms to Decimals, pp. 408–410
Procedural Skill and Fluency			Adding Decimals with Grids, pp. 400–402 Subtracting Decimals with Grids, pp. 402–405 Home Link 4-11, p. 405	Extending Whole-Number Addition Algorithms to Decimals, pp. 408–410 Practicing Decimal Addition pp. 410–411 Home Link 4-12, p. 411 Differentiation Options, p. 407
Applications	Graphing Data as Ordered Pairs, pp. 384–386 Graphing Ordered Pairs and Interpreting Graphed Data, pp. 386–387 Home Link 4-9, p. 387 Readiness, p. 383	Math Message, p. 389 Solving the Open Response Problem, pp. 391–392 Home Link 4-10, p. 397		Home Link 4-12, p. 411 Enrichment, p. 407
Rich Tasks and Mathematical Reasoning	Math Message, p. 384 Graphing Data as Ordered Pairs, pp. 384–386 Journal p. 140: Writing/Reasoning Home Link 4-9, p. 387 Enrichment, p. 383	Math Message, p. 389 Solving the Open Response Problem, pp. 391–392 Revising Work, p. 396 Home Link 4-10, p. 397	Adding Decimals with Grids, pp. 400–402 Subtracting Decimals with Grids, pp. 402–405 Playing *Over and Up Squares*, p. 405 Journal p. 145: Writing/Reasoning	Math Message, p. 408 Extending Whole-Number Addition Algorithms to Decimals, pp. 408–410 Journal p. 147: Writing/Reasoning
Mathematical Discourse	Graphing Data as Ordered Pairs, pp. 384–386	Changing the Logo, pp. 389–390 Setting Expectations, p. 395 Reengaging in the Problem, p. 395	Adding Decimals with Grids, pp. 400–402 Subtracting Decimals with Grids, pp. 402–405	Extending Whole-Number Addition Algorithms to Decimals, pp. 408–410 Playing *Fraction Of,* p. 411 Enrichment, p. 407
Distributed Practice	Mental Math and Fluency, p. 384 Playing *High-Number Toss,* p. 387 Math Boxes 4-9, p. 387	Mental Math and Fluency, p. 389	Mental Math and Fluency, p. 400 Playing *Over and Up Squares,* p. 405 Math Boxes 4-11, p. 405	Mental Math and Fluency, p. 408 Playing *Fraction Of,* p. 411 Math Boxes 4-12, p. 411
Differentiation Support	Differentiation Options, p. 383 ELL Support, p. 383 Online Differentiation Support 4-9 Adjusting the Activity, p. 385 Academic Language Development, p. 386	ELL Support, p. 390 Adjusting the Activity, p. 391	Differentiation Options, p. 399 ELL Support, p. 399 Online Differentiation Support 4-11 Common Misconception, p. 402 Academic Language Development, p. 402 Adjusting the Activity, p. 403	Differentiation Options, p. 407 ELL Support, p. 407 Online Differentiation Support 4-12 Adjusting the Activity, p. 409

Red text = Game

Notes

4-13
Decimal Subtraction Algorithms

4-14
Addition and Subtraction of Money

4-13	4-14	4-15 Assessment
Subtraction of decimals Extending Whole-Number Subtraction Algorithms to Decimals, pp. 414–416	**Addition and subtraction of decimals** Discussing Decimal Addition and Subtraction Strategies, pp. 420–422	**Lesson 4-15 is an assessment lesson. It includes:** • **Self Assessment** • **Unit Assessment** • **Optional Challenge assessment** • **Cumulative Assessment** • **Suggestions for adjusting the assessments.**
Extending Whole-Number Subtraction Algorithms to Decimals, pp. 414–416 Practicing Decimal Subtraction, pp. 416–417 Home Link 4-13, p. 417 Differentiation Options, p. 413	Math Message, p. 420 Discussing Decimal Addition and Subtraction Strategies, pp. 420–422 Introducing *Spend and Save*, pp. 422–423 Home Link 4-14, p. 423 Enrichment, p. 419 Extra Practice, p. 419	
Math Message, p. 414 Home Link 4-13, p. 417	Math Message, p. 420 Finding Areas of New Floors, p. 423 Home Link 4-14, p. 423	
Extending Whole-Number Subtraction Algorithms to Decimals, pp. 414–416 Playing *Prism Pile-Up*, p. 417 Journal p. 149: Writing/Reasoning Enrichment, p. 413	Introducing *Spend and Save*, pp. 422–423 Enrichment, p. 419	
Extending Whole-Number Subtraction Algorithms to Decimals, pp. 414–416	Discussing Decimal Addition and Subtraction Strategies, pp. 420–422 Introducing *Spend and Save*, pp. 422–423	
Mental Math and Fluency, p. 414 Playing *Prism Pile-Up*, p. 417 Math Boxes 4-13, p. 417	Mental Math and Fluency, p. 420 Finding Areas of New Floors, p. 423 Math Boxes 4-14, p. 423	
Differentiation Options, p. 413 ELL Support, p. 413 Online Differentiation Support 4-13 Adjusting the Activity, p. 416	Differentiation Options, p. 419 ELL Support, p. 419 Online Differentiation Support 4-14 Adjusting the Activity, p. 421 Academic Language Development, p. 422	

Go Online:

Evaluation Quick Entry
Use this tool to record students' performance on assessment tasks.

Data Use the Data Dashboard to view students' progress reports.

Red text = Game

Unit 4 Materials

Lesson	*Math Masters*	Activity Cards	Manipulative Kit	Other Materials
4-1	pp. 119–123; G24	42	counters; base-10 blocks; per partnership: number cards 0–9 (4 of each)	slate; per partnership: *Fraction Of* fraction cards (Set 1), *Fraction Of* whole cards
4-2	pp. 124–125; TA21–TA24	43	fraction circles (optional); base-10 blocks; meterstick; number cards 0–9 (4 of each); counter	slate; colored pencils or highlighters (optional)
4-3	pp. 126–128; TA24; per partnership: TA25, G19–G20	44	number cards 0–9 (4 of each); two 6-sided dice; counter	slate; colored pencils
4-4	pp. 129–130; TA23–TA24; per partnership: TA25–TA26, G14; G15	45	number cards 0–9 (4 of each); 6-sided die	calculator; scissors
4-5	pp. 131–133; 134 (optional); 135; TA2 (optional); TA22–TA23 (optional); TA27; per partnership: G18	46	per partnership: number cards 0–9 (4 of each); stopwatch (optional)	slate; per partnership: timer (optional); paper clip
4-6	pp. 136–139; TA28; per partnership: G25	47	per partnership: two 6-sided dice	slate; straightedge; per partnership: colored pencils, several road maps or political maps
4-7	pp. per partnership: 140–142; 143; G26	48–49	per partnership: number cards 0–10 (8 of each); 6-sided die	slate; per partnership: a world political map or globe showing latitude and longitude; coin; scissors; red pen or crayon; per group: about 15 feet of masking tape, two different colors of paper or masking tape, markers
4-8	pp. 144–146; per partnership: TA23 (optional), TA25; TA28; per partnership: G14	50	per partnership: number cards 0–9 (4 of each); 6-sided die	straightedge (optional)
4-9	pp. 147–150; TA2 (optional); per partnership: G10	51	number cards 0–10 (1 of each); per partnership: 6-sided die	slate; straightedge
4-10	pp. 151–155; TA4; TA31 (optional)		ruler	Guidelines for Discussion Poster; colored pencils (optional); selected samples of students' work; students' work from Day 1
4-11	pp. 156; TA22; per partnership: G25	52–53	number cards 0–9 (4 of each); base-10 blocks; per partnership: two 6-sided dice	slate; colored pencils or crayons
4-12	pp. 157–158; TA7 (optional); TA22–TA23 (optional); G24	54	counters; per partnership: number cards 0–9 (4 of each); per group: stopwatch	slate; per partnership: calculator (optional), *Fraction Of* fraction cards (Set 1), *Fraction Of* whole cards; per group: 1 piece of poster paper
4-13	pp. 159–160; TA7 (optional); TA23 (optional); per partnership: TA25, G6	55	number cards 0–9 (4 of each); 2 counters; per partnership: 6-sided die	slate; calculator (optional); per partnership: *Prism Pile-Up* cards; per group: 1 piece of poster paper`
4-14	pp. 161; TA22 (optional); per partnership: G27	56	per partnership: number cards 0–9 (4 of each), 2 counters	slate; play bills and coins (optional); per partnership: 1 coin
4-15	pp. 162–165; *Assessment Handbook*, pp. 31–41			

Assessment Check-In

These ongoing assessments offer an opportunity to gauge students' performance on one or more of the standards addressed in that lesson.

 Evaluation Quick Entry Record students' performance online.

 Data View reports online to see students' progress towards mastery.

Lesson	Task Description	Content Standards	Processes and Practices
4-1	Apply place-value concepts to write decimals using numerals and words.	5.NBT.1, 5.NBT.3, 5.NBT.3a	
4-2	Generate multiple names for decimals and represent decimals by shading thousandths grids.	5.NBT.3, 5.NBT.3a	SMP2
4-3	Read and write decimals in expanded form.	5.NBT.1, 5.NBT.3, 5.NBT.3a	
4-4	Compare decimals to thousandths using place-value strategies.	5.NBT.3, 5.NBT.3b	SMP2
4-5	Round decimals to a given place.	5.NBT.4	
4-6	Plot and label points on a coordinate grid.	5.G.1	SMP2
4-7	Plot points and use clues to find the location of a hidden point on a coordinate grid.	5.G.1, 5.G.2	
4-8	Plot points on a coordinate grid and apply rules to create new sets of ordered pairs.	5.G.1, 5.G.2	
4-9	Represent problems by graphing points on a coordinate grid and interpret coordinate values in context.	5.OA.3, 5.G.1, 5.G.2	
4-10	Create and apply a rule to enlarge a picture on a coordinate grid.	5.G.2	SMP5
4-11	Find decimal sums and differences by shading grids.	5.NBT.3, 5.NBT.3a, 5.NBT.7	
4-12	Use an algorithm to add decimals.	5.NBT.7	SMP1
4-13	Use an algorithm to subtract decimals.	5.NBT.7	SMP6
4-14	Add and subtract money amounts.	5.NBT.7	SMP6

Virtual Learning Community
vlc.uchicago.edu

While planning your instruction for this unit, visit the *Everyday Mathematics* Virtual Learning Community. You can view videos of lessons in this unit, search for instructional resources shared by teachers, and ask questions of *Everyday Mathematics* authors and other educators. Some of the resources on the VLC related to this unit include:

EM4: Grade 5 Unit 4 Planning Webinar
This webinar provides a preview of the lessons and content in this unit. Watch this video with your grade-level colleagues and plan together under the guidance of an *Everyday Mathematics* author.

Playing *Prism Pile-Up*
Watch clip 3 of 3 to see a teacher play a sample round of the game with a class of fifth graders.

ACI Booklet: Grade 5 Unit 4
This booklet is a collection of PDFs of all the student pages in the unit that include concepts leading up to the ACI. Each page shows you where to find the ACI information in the Teacher's Lesson Guide.

For more resources, go to the VLC Resources page and search for Grade 5.

Spiral Towards Mastery

The *Everyday Mathematics* curriculum is built on the spiral, where standards are introduced, developed, and mastered in multiple exposures across the grade. Go to the Teacher Center at my.mheducation.com to use the Spiral Tracker.

 Spiral Towards Mastery Progress This Spiral Trace outlines instructional trajectories for key standards in Unit 4. For each standard, it highlights opportunities for Focus instruction, Warm Up and Practice activities, as well as formative and summative assessment. It describes the **degree of mastery**— as measured against the entire standard—expected at this point in the year.

Number and Operations in Base Ten

5.NBT.1 — 3-9 Warm Up | 3-10 Practice | 3-14 Practice | 4-1 through 4-5 Warm Up Focus Practice ✓ | 4-8 Practice | 4-9 Practice | 4-11 Warm Up | 4-15 Progress Check | 5-4 Practice | 5-10 Warm Up | 6-1 Focus Practice

 Progress Towards Mastery By the end of Unit 4, expect students to recognize that in multidigit whole numbers, a digit in one place represents 10 times what it represents in the place to its right and $\frac{1}{10}$ of what it represents in the place to its left; recognize that place-value patterns with whole numbers extend to decimals.

Full Mastery of 5.NBT.1 expected by the end of Unit 6.

5.NBT.3a — 4-1 through 4-5 Warm Up Focus Practice ✓ | 4-8 Practice | 4-11 Focus Practice ✓ | 4-12 through 4-14 Warm Up Practice | 4-15 Progress Check ✓ | 5-4 Warm Up Practice | 5-5 Warm Up Practice | 5-10 Warm Up | 6-1 Warm Up Focus Practice | 6-2 Warm Up

 Progress Towards Mastery By the end of Unit 4, expect students to represent decimals through thousandths by shading grids; read and write decimals through thousandths with no placeholder zeros; read and write decimals in expanded form as sums of decimals (e.g., $0.392 = 0.3 + 0.09 + 0.002$).

Full Mastery of 5.NBT.3a expected by the end of Unit 6.

5.NBT.3b — 4-4 Focus Practice ✓ | 4-5 Focus Practice | 4-7 Warm Up | 4-8 Practice | 4-13 Warm Up | 4-15 Progress Check ✓ | 5-1 Practice | 5-3 Practice | 5-4 Practice | 5-8 Warm Up | 6-2 Warm Up Focus Practice | 6-4 Practice

 Progress Towards Mastery By the end of Unit 4, expect students to use grids or place-value charts to compare and order decimals through thousandths when the decimals have the same number of digits after the decimal point; record comparisons using >, =, and < symbols.

Full Mastery of 5.NBT.3b expected by the end of Unit 6.

Key ✓ = Assessment Check-In    = Progress Check Lesson    = Current Unit    = Previous or Upcoming Lessons

5.NBT.4 | 4-5 Focus Practice | 4-12 Focus | 4-13 Focus Practice | 4-14 Warm Up | 4-15 Progress Check | 5-4 Practice | 5-6 Warm Up Practice | 5-9 Warm Up | 6-11 Warm Up | 6-14 Progress Check | 8-5 Focus Practice | 8-11 Focus

Progress Towards Mastery By the end of Unit 4, expect students to use grids, number lines, or a rounding shortcut to round decimals to the nearest tenth or hundredth in cases when rounding only affects one digit.

Full Mastery of 5.NBT.4 expected by the end of Unit 6.

5.NBT.7 | 4-11 through 4-14 Focus Practice | 4-15 Progress Check | 5-1 Practice | 5-3 Practice | 5-9 Practice | 5-12 Practice | 6-4 Practice | 6-6 Practice | 6-8 through 6-13 Focus Practice | 6-14 Progress Check

Progress Towards Mastery By the end of Unit 4, expect students to use grids to add and subtract decimals; use algorithms to add and subtract decimals through tenths with regrouping and through hundredths without regrouping.

Full Mastery of 5.NBT.7 expected by the end of Unit 8.

Geometry

5.G.1 | 4-6 through 4-10 Focus Practice | 4-11 Practice | 4-15 Progress Check | 5-2 Practice | 5-6 Practice | 5-13 Practice | 6-1 Practice | 6-14 Progress Check | 7-10 through 7-13 Focus Practice | 8-2 Practice

Progress Towards Mastery By the end of Unit 4, expect students to understand that an ordered pair of numbers identifies an exact location on a coordinate grid; use coordinates to graph points and to name graphed points in the first quadrant of the coordinate plane.

Full Mastery of 5.G.1 expected by the end of Unit 7.

5.G.2 | 4-7 through 4-10 Focus Practice | 4-15 Progress Check | 5-2 Practice | 5-6 Practice | 5-13 Practice | 6-1 Practice | 6-14 Progress Check | 7-10 through 7-13 Focus Practice | 7-14 Progress Check | 8-2 Practice

Progress Towards Mastery By the end of Unit 4, expect students to understand that information from some real-world and mathematical problems can be represented as ordered pairs and graphed on a coordinate grid; plot points to represent given information.

Full Mastery of 5.G.2 expected by the end of Unit 8.

Key   ✓ = Assessment Check-In   ✦ = Progress Check Lesson   ▱ = Current Unit   ▰ = Previous or Upcoming Lessons

Mathematical Background:
Content

▶ **Extending Place-Value Concepts to Decimals**
(Lessons 4-1 through 4-3)

Decimals, like fractions, are used to represent values between whole numbers. Unlike fractions, however, decimals follow the place-value rules of the base-10 number system. Students first encountered decimal concepts in Grade 4 when they interpreted decimal fractions, or fractions with denominators such as 10 or 100, as parts of wholes partitioned into 10 or 100 equal parts. Students used this fractional meaning to visually represent decimals, translate between fraction and decimal notation, and compare and order decimals to hundredths.

In Grade 5 students situate decimals in the base-10 place-value system and extend their understanding of decimals to the thousandths place. Using visual representations as an anchor for discussion, students review the *10 times as much* and $\frac{1}{10}$ *of* place-value patterns they observed with whole numbers in Unit 2. (*See the Mathematical Background section of the Unit 2 Organizer for more information.*)

Students construct an understanding of the tenths, hundredths, and thousandths places by extending the $\frac{1}{10}$ *of* pattern to the right of the ones place. **5.NBT.1** Knowing that 1 is $\frac{1}{10}$ of 10, students deduce that the next place to the right must be $\frac{1}{10}$ of 1, or $\frac{1}{10}$ (the tenths place). They use the same reasoning to identify hundredths and thousandths, and refer to visual representations to confirm that 10 tenths, 100 hundredths, and 1,000 thousandths each make 1 whole. They then use place-value charts to read and write decimals in words and numerals, as well as to identify the value of individual digits. **5.NBT.3, 5.NBT.3a**

In Lesson 4-2 students represent decimals on thousandths grids. Each thousandth grid is divided into 10 columns, 100 small squares, and 1,000 tiny rectangles. Each column represents 1 tenth, each small square represents 1 hundredth, and each tiny rectangle represents 1 thousandth. To represent the number 0.418, for example, students shade 4 columns, 1 small square, and 8 tiny rectangles. (*See margin.*)

Students reinforce their understanding of decimal place value by working with expanded form in Lesson 4-3. **5.NBT.3, 5.NBT.3a** Each version of expanded form illuminates important aspects of a digit's value. (*See margin.*) For example, representing the value of the 8 in 0.418 as 8 thousandths helps students think of 0.008 as 8 copies of 1 thousandth, which they can then represent as $8 * 0.001$ or $8 * \frac{1}{1,000}$.

Standards and Goals for Mathematical Content

Because the standards within each strand can be broad, *Everyday Mathematics* has unpacked each standard into Goals for Mathematical Content GMC. For a complete list of Standards and Goals, see page EM1.

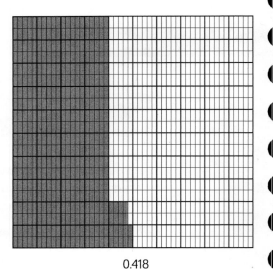

0.418

$0.4 + 0.01 + 0.008$
4 tenths + 1 hundredth + 8 thousandths
$(4 * 0.1) + (1 * 0.01) + (8 * 0.001)$
$(4 * \frac{1}{10}) + (1 * \frac{1}{100}) + (8 * \frac{1}{1,000})$

Versions of expanded form for 0.418

Unit 4 Vocabulary

accuracy	efficiency	round down
addend	expanded form	round up
algorithm	extrapolate	rounding
axes	hundredths	standard notation
balance	interpolate	tenths
column addition	intersect	thousandths
compare	order	thousandths grid
conjecture	ordered pair	trade-first subtraction
coordinate grid	origin	U.S. traditional addition
coordinates	partial-sums addition	U.S. traditional subtraction
counting-up subtraction	perpendicular	*x*-axis
decimal	plot	*x*-coordinate
decimal point	represent	*y*-axis
digit	representation	*y*-coordinate

▶ Comparing, Ordering, and Rounding Decimals
(Lessons 4-4 and 4-5)

Once students extend their place-value understanding to decimals, they can answer questions such as "Which decimal is larger? Which is smaller?" Students may struggle making sense of how a decimal like 0.418 can be less than one with fewer digits, such as 0.8, if they use whole-number reasoning. To address this common misconception, students visually represent the decimals being compared in Lesson 4-4. This encourages them to compare decimals place-by-place starting with the digits of largest value. This is an efficient way to compare decimals, because the digits of largest value have the biggest influence on the value of the decimal. So, when comparing 0.418 and 0.8, students start at the tenths place. Because 4 tenths is less than 8 tenths, 0.8 is greater, and the hundredths and thousandths in 0.418 do not affect the comparison. Students record the results of their comparisons using >, =, and < symbols. **5.NBT.3, 5.NBT.3b**

In Lesson 4-5 students explore strategies for rounding decimals. **5.NBT.4** To extend to decimals what they know about rounding whole numbers, students must recognize decimals as being between other decimals. For example, to round 3.619 to the nearest tenth, it is necessary to see that 3.619 is between 3.6 and 3.7. To help students with these ideas, Lesson 4-5 emphasizes grids and number lines as tools for rounding. Students discuss patterns that they notice to identify a rounding shortcut grounded in the structure of the base-10 place-value system. (*See Lesson 4-5 for more details.*)

▶ Coordinate Grids (Lessons 4-6 through 4-10)

Grade 5 is students' first introduction to coordinate grids. A *coordinate grid* is formed by two number lines, called axes, which intersect at their zero points, forming right angles. This intersection is called the *origin*, and points can be located on the grid using *ordered pairs* of numbers called *coordinates*, indicating a point's distance along each axis from the origin.

NOTE Standards for Grade 5 include defining a *coordinate* system and graphing points on a coordinate plane. Coordinate systems are reference frames in which one or more numbers (coordinates) are used to locate points. Coordinate systems can be 1-, 2-, or 3-dimensional. A coordinate plane is a 2-dimensional coordinate system, as a plane is a 2-dimensional flat surface that extends forever in all directions. In *Everyday Mathematics,* graphing points on a coordinate plane is synonymous with plotting points on a coordinate grid. The axes of a coordinate grid create four different sections, or quadrants, typically numbered I, II, III, and IV, moving counterclockwise from the upper right. Students in Grade 5 work only with the first quadrant of the coordinate plane.

Coordinate Grids *Continued*

By convention, the first number in ordered pairs gives the position of a point along the horizontal axis, or *x-axis*, and is called the *x-coordinate*. The second number is the *y-coordinate* and gives the position along the vertical axis, or *y-axis*. For example, the ordered pair (3, 4) locates a point that is 3 units from the origin in the direction of the *x-axis* and 4 units from the origin in the direction of the *y-axis*. (*See margin.*) Students' work in Grade 5 is limited to points with positive coordinates.

Students are introduced to the features of coordinate grids in Lessons 4-6 and 4-7. They learn two games, *Over and Up Squares* and *Hidden Treasure,* to practice plotting points and identifying coordinates. **5.G.1** In Lessons 4-8 through 4-10 students use their knowledge of coordinate grids to solve mathematical and real-world problems. **5.G.2** They connect points to form figures and study how operations on coordinates transform the figures. **5.NF.5, 5.NF.5a** They also express data sets as lists of ordered pairs, plot the points on a coordinate grid, and use the graph to extend the patterns and solve problems. **5.OA.3** This lays the groundwork for students to use coordinate grids to represent and solve more complex problems in later units and in Grade 6.

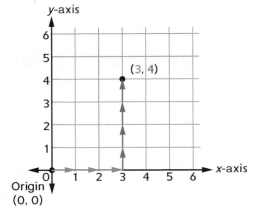

Plotting (3, 4) on a Coordinate Grid

▶ Adding and Subtracting Decimals

(Lessons 4-11 through 4-14)

The final four lessons of Unit 4 focus on adding and subtracting decimals. Students explore these operations by shading hundredths grids to represent sums and differences. (*See examples in margin.*) This initial emphasis on concrete models helps students focus on the meaning of the numbers and the operations and provides a conceptual grounding for strategies based on place value, such as partial-sums addition and counting-up subtraction.

Shading grids to add and subtract also helps address common misconceptions about regrouping. For example, a student adding 0.8 + 0.8 who mistakenly gets a sum of 0.16 will quickly see the error after shading 8 columns of a hundredths grid and then 8 more columns. The sum must be greater than 1 whole, so 16 hundredths does not make sense. Students can use grids both to solve problems and to check that answers found with other methods are reasonable. Students should also be routinely encouraged to make estimates to check the reasonableness of their answers.

Algorithms for decimal addition and subtraction build on the concept of adding and subtracting digits in like place-value positions (adding hundredths to hundredths, tenths to tenths, ones to ones, and so on). This means students can extend strategies for adding and subtracting whole numbers to decimals. In Lessons 4-12 and 4-13, students review whole-number addition and subtraction algorithms from earlier grades and apply algorithms of their choice to compute with decimals. **5.NBT.7**

Note that *Everyday Mathematics* intentionally exposes students to multiple methods of adding and subtracting decimals. The authors expect that students will gravitate toward a method they can use reliably and that makes sense to them; one method is not preferred over another.

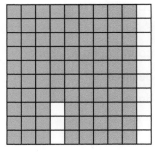

$0.33 + 0.54 = 0.87$

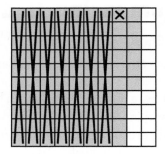

$0.86 - 0.71 = 0.15$

Mathematical Background:
Process and Practice

 See below for some of the ways that students engage in **SMP2 Reason abstractly and quantitatively** and **SMP7 Look for and make use of structure** through **Number and Operations in Base Ten** and the other mathematical content of Unit 4.

> ## Standard for Mathematical Process and Practice 2

Mathematically proficient students reason abstractly and quantitatively. In other words, they make sense of quantities and their relationships in problem situations. **SMP2** To do this in Unit 4, students create, make sense of, and make connections among representations of decimals. **GMP2.1, GMP2.2, GMP2.3** For example, in Lesson 4-1 students consider a picture of a large square (representing 1) and the 10 strips or 100 small squares that compose the large square, using language like *10 times* and $\frac{1}{10}$ *of* to describe the relative sizes of the squares and strips. Abstracting from these ideas, they represent decimals on a place-value chart and with numerals. In Lessons 4-2 and 4-3, students further explore the meaning of decimal quantities by working with two new representations: the thousandths grid and expanded form. Throughout the unit, as students create their own decimal representations and make sense of existing ones, they are challenged to make connections: *How can I represent this decimal on a grid? On a place-value chart? In expanded form?* **GMP2.1** *How are these representations the same? How are they different?* **GMP2.3** *Which representation is most useful to me as I solve this problem?* **GMP2.2**

> ## Standard for Mathematical Process and Practice 7

Students engaged in mathematics notice and make use of patterns, properties, and other mathematical structures—key components of Standard for Mathematical Process and Practice 7. Unit 4 highlights SMP7 as students use place-value patterns to develop decimal concepts and look for patterns to help them interpret graphs.

In Grade 5 students articulate $\frac{1}{10}$ *of* and *10 times as much* place-value patterns in both whole numbers and decimals. **GMP7.1** In Lesson 4-1 they apply these place-value patterns to label the columns to the right of the ones column as $\frac{1}{10}, \frac{1}{100}$, and $\frac{1}{1,000}$ or, using decimal notation, as 0.1, 0.01, and 0.001. **GMP7.2** By making use of place-value patterns, students deepen their understanding of the structure of the base-10 place-value system.

In Lesson 4-9 students look for patterns as they work with data tables, ordered pairs, and coordinate grids. They learn that graphing a data set on a coordinate grid can help them see patterns in the data, allowing them to continue patterns beyond and in between data points to predict other values. **GMP7.1** For example, students are given a table showing the ages of two siblings at different times in their lives. By graphing the data, they can determine the respective ages of the siblings at other times in their lives. **GMP7.2**

Standards and Goals for Mathematical Process and Practic

SMP2 **Reason abstractly and quantitatively.**

GMP2.1 Create mathematical representations using numbers, words, pictures, symbols, gestures, tables, graphs, and concrete objects.

GMP2.2 Make sense of the representations you and others use.

GMP2.3 Make connections between representations.

SMP7 **Look for and make use of structure.**

GMP7.1 Look for mathematical structures such as categories, patterns, and properties.

GMP7.2 Use structures to solve problems and answer questions.

Go Online to the *Implementation Guide* for more information about the Mathematical Process and Practice Standards.

For students' information on the Mathematical Process and Practice Standards, see *Student Reference Book*, pages 1–34.

Decimal Place Value

Overview Students extend place-value patterns to decimals and practice reading and writing decimals to thousandths.

▶ **Before You Begin**

For Part 2, create a 25" by 25" square (Square A) and a 25" by 2.5" rectangle (Strip B) using chart paper. (*See page 332.*) Make one copy of *Math Masters*, page 119 and cut out figures C–F. For best contrast, glue them to colored paper. For the Math Message, display figures A, B, and C. Prepare a large place-value chart with seven columns like the one shown on page 334. Do not label the columns.

▶ **Vocabulary**

tenths • hundredths • thousandths • decimal point • decimal

Standards

Focus Cluster
• Understand the place value system.

① **Warm Up** 5 min	Materials	
Mental Math and Fluency Students identify digits in given places in whole numbers.	slate	**5.NBT.1**

② **Focus** 35–40 min		
Math Message Students determine the value of a 1-by-10 rectangular strip and a 10-by-10 square when a small square is defined as 1.	figures A, B (*See Before You Begin.*); figure C (*Math Masters*, p. 119)	**5.NBT.1** **SMP2**
Extending Place-Value Patterns to the Thousandths Place Students discuss times-10 and $\frac{1}{10}$-of place-value relationships and extend the patterns through the thousandths place.	figures A, B (*See Before You Begin.*); figures C–F (*Math Masters*, p. 119); place-value chart	**5.NBT.1, 5.NBT.3, 5.NBT.3a** **SMP2, SMP6, SMP7**
Reading and Writing Decimals Students use a place-value chart to read and write decimals to thousandths.	*Math Journal 1*, pp. 112–113; *Math Masters*, p. 120 (for display); slate	**5.NBT.1, 5.NBT.3, 5.NBT.3a** **SMP2, SMP6**
✓ **Assessment Check-In** See page 336. Expect most students to be able to write decimals to hundredths using numerals and words.	*Math Journal 1*, p. 112 (optional), p. 113	**5.NBT.1, 5.NBT.3, 5.NBT.3a**

③ **Practice** 20–30 min		
Playing *Fraction Of* **Game** Students practice multiplying fractions and whole numbers.	*Student Reference Book*, pp. 306–307; *Math Masters*, p. G24; per partnership: *Fraction Of* cards; counters (optional)	**5.NF.4, 5.NF.4a** **SMP6**
Math Boxes 4-1 Students practice and maintain skills.	*Math Journal 1*, p. 111	See page 337.
Home Link 4-1 **Homework** Students translate between numerals and words for decimals through thousandths.	*Math Masters*, p. 123	**5.NBT.1, 5.NBT.3, 5.NBT.3a, 5.NBT.5** **SMP6, SMP7**

 Go Online to see how mastery develops for all standards within the grade.

Differentiation Options

Readiness	5–15 min	**Enrichment**	5–15 min	**Extra Practice**	5–15 min
WHOLE CLASS · **SMALL GROUP** · PARTNER · INDEPENDENT		WHOLE CLASS · SMALL GROUP · **PARTNER** · INDEPENDENT		WHOLE CLASS · SMALL GROUP · **PARTNER** · INDEPENDENT	

Readiness · SMALL GROUP · 5–15 min

Representing Times-10 and $\frac{1}{10}$-of Patterns

5.NBT.1, SMP2

base-10 blocks

To build readiness for extending place-value patterns to the thousandths place, students use base-10 blocks to illustrate times-10 and $\frac{1}{10}$-of relationships. Distribute flats, longs, and unit cubes. Hold up 1 unit cube, asking: *How many unit cubes make 1 long?* 10 Have students line up 10 unit cubes next to 1 long to confirm the relationship. Ask: *What fraction of a long is 1 unit cube?* $\frac{1}{10}$ Repeat the process with 10 longs and 1 flat. Ask: *How many unit cubes make 1 flat?* 100 *What fraction of a flat is 1 unit cube?* $\frac{1}{100}$ Summarize by asking: *How are these blocks related?* **GMP2.2** 10 unit cubes make 1 long and 10 longs make 1 flat. A cube is $\frac{1}{10}$ of a long and a long is $\frac{1}{10}$ of a flat. Point out that these relationships match base-10 place-value patterns. Say: *Our whole-number place values start with ones, tens, and hundreds. One times 10 makes 10, and 10 times 10 makes 100. One is $\frac{1}{10}$ of 10, and 10 is $\frac{1}{10}$ of 100.* Ask: *Are these patterns only true for the ones, tens, and hundreds places?* No, they are true for all places. Remind students that in Grade 4 they extended these patterns to the right of the ones place. Explain that in this lesson they will further explore place-value positions that extend even further to the right.

Enrichment · PARTNER · 5–15 min

Writing Many Names for Decimals

5.NBT.1, 5.NBT.3, 5.NBT.3a, SMP2, SMP7

Math Masters, p. 122

To extend their work with decimal place value, students interpret and generate equivalent names for decimals. The equivalent names include fractions, words, equivalent decimals, and expressions. **GMP2.3, GMP7.2**

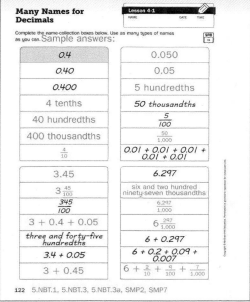

Extra Practice · PARTNER · 5–15 min

Reading and Writing Decimals

5.NBT.1, 5.NBT.3, 5.NBT.3a, SMP2, SMP7

Activity Card 42;
Math Masters, p. 121;
per partnership: number cards 0–9 (4 of each)

For more practice reading and writing decimals using numerals and words, students draw cards to generate decimals to thousandths. Partners take turns forming decimals and reading them aloud. They check each other's work by comparing decimals written in words to numerals recorded in a place-value chart. **GMP2.2, GMP7.2**

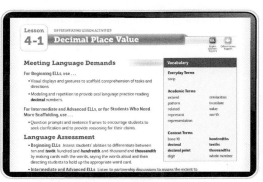

English Language Learner

Beginning ELL To prepare students for describing place-value patterns, introduce the word *pattern.* Use tools like pattern blocks to create both simple patterns and non-examples of patterns. Point to the patterns, identifying each one by saying: *This is a pattern.* Point to non-examples, saying: *This is* not *a pattern.* Assess students' understanding of the term by pointing to examples and non-examples, asking: *Is this a pattern?*

Differentiation Support pages are found in the online Teacher's Center.

Mathematical Process and Practice

SMP2 **Reason abstractly and quantitatively.**
GMP2.2 Make sense of the representations you and others use.

SMP6 **Attend to precision.**
GMP6.3 Use clear labels, units, and mathematical language.

SMP7 **Look for and make use of structure.**
GMP7.2 Use structures to solve problems and answer questions.

Professional Development

Base-10 blocks are used to introduce decimals in *Fourth Grade Everyday Mathematics*. Many students will be familiar with the definition of the base-10 flat as 1, longs as tenths, and unit cubes as hundredths. While base-10 blocks provide a valuable model for demonstrating times-10 and $\frac{1}{10}$-of relationships, it is difficult for students to visualize one-thousandth without redefining the flat as $\frac{1}{10}$ and the big cube as 1. The squares and strips used in this lesson approximate base-10 blocks in shape and size. They are meant to build on students' familiarity with base-10 blocks while providing a single model that can extend from whole-number place value to decimal place value through the thousandths place. Beginning in Lesson 4-2 students will represent decimals through thousandths using hundredths and thousandths grids scaled to the unit square defined in this lesson.

---

① Warm Up 5 min

▶ Mental Math and Fluency

Dictate numbers for students to write on their slates. Then have students identify digits in given places. *Leveled exercises:*

● ○ ○ Thirty-seven thousand, one hundred twenty-four. Circle the thousands digit. Underline the ones digit. 3⑦,124

Fifty-nine thousand, eight hundred forty-six. Circle the tens digit. Underline the hundreds digit. 59,8④6

● ● ○ Six hundred thirty thousand, seven hundred twenty-six. Circle the ten-thousands digit. Underline the hundred-thousands digit. 6③0,726

Seven hundred thousand, fifty-six. Circle the thousands digit. Underline the hundreds digit. 70⓪,056

● ● ● Fifteen million, eight hundred three thousand, twenty-one. Circle the millions digit. Underline the thousands digit. 1⑤,803,021

Two hundred sixty-three million, fourteen thousand, six hundred thirteen. Circle the ten-millions digit. Underline the ten-thousands digit. 2⑥3,014,613

② Focus 35–40 min

▶ Math Message

Display Square A and Strip B as well as Square C from *Math Masters,* page 119. (*See Before You Begin.*)

If the small square represents 1, what does the rectangular strip represent? 10 *What does the large square represent?* 100 *How do you know?* **GMP2.2**

▶ Extending Place-Value Patterns to the Thousandths Place

| WHOLE CLASS | SMALL GROUP | PARTNER | INDEPENDENT |

Math Message Follow-Up Invite students to share answers and discuss their reasoning. They will likely think about the similarities between the figures shown and base-10 blocks as they consider the relative sizes of the figures.

Expect most to recognize that 10 small squares make 1 rectangular strip (so the strip represents 1 ten, or 10) and that 10 strips make one large square (so the large square represents 100). If students seem unsure, move the small square along the strip to show that 10 small squares fit across the strip. Similarly, show that 10 strips would fit across the large square.

Make an explicit connection to place value by attaching the representations to the first three columns in the place-value chart you prepared before the lesson. (*See Before You Begin.*) Have students suggest labels for each column. **GMP2.2, GMP6.3** Hundreds, tens, and ones Tell students they will review place-value patterns in whole numbers and think about how these patterns might extend to places to the right of the ones place.

Remind students of the place-value patterns they identified in whole numbers. Point to the square in the ones column and sweep to the strip in the tens column. As you gesture, ask: *How is 1 related to 10?* Sample answers: 10 times 1 equals 10. 10 is 10 times as much as 1. Repeat the gesture, moving from the tens column to the hundreds column. Ask: *How is 10 related to 100?* Sample answers: 10 times 10 equals 100. 100 is 10 times as much as 10. Repeat the gestures in the opposite direction. Ask: *How is 100 related to 10?* 10 is $\frac{1}{10}$ of 100. *How is 10 related to 1?* **GMP2.2, GMP7.2** 1 is $\frac{1}{10}$ of 10. Have students consider the value of a given digit in different places. Ask: *What would the digit 3 be worth in the ones place?* 3 ones, or 3 *What would it be worth in the tens place?* 3 tens, or 30 *What would it be worth in the hundreds place?* 3 hundreds, or 300 Guide students to generalize these patterns, making sure to cover the following points:

- A digit in a given place is worth 10 times as much as it would be worth in the place to its right.
- A digit in a given place is worth $\frac{1}{10}$ of what it would be worth in the place to its left.

Extend these place-value patterns to introduce decimal place values. Ask: *How might you use these patterns to label the column to the right of the ones column?* **GMP7.2** Sample answer: When we move to the right, the value is $\frac{1}{10}$ of what it is in the place to the left. $\frac{1}{10}$ of 1 is $\frac{1}{10}$, so we could label the column $\frac{1}{10}$s, or tenths. *What might $\frac{1}{10}$ of the small square look like?* **GMP2.2** Sample answer: A rectangle, similar to a base-10 long Show students Strip D from *Math Masters,* page 119. Have them confirm that 10 of the strips would be equivalent to Square C, and attach Strip D to the fourth column of the place-value chart. Label the column **tenths.** **GMP6.3** (*See next page.*)

Academic Language Development

Have partners work to actively build understanding of the terms *tenths, hundredths,* and *thousandths.* Have each partnership prepare a chart with the heading Place Value and four columns labeled: Math Term; What We Think It Means; *Student Reference Book* Definition; and Visual Example and Sentence. Then they list each term in the first column, discussing and recording what they think it means. At the end of the lesson, partnerships complete their posters by writing the definition and creating a visual example and a sentence.

Math Masters, p. 119

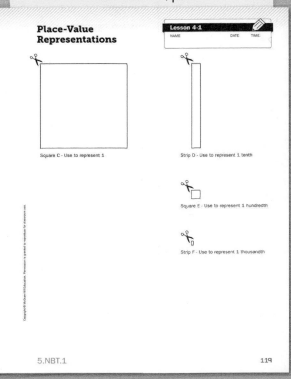

Differentiate Students often struggle interpreting zeros in decimal notation. Some may ignore placeholder zeros, thinking of numbers like 0.078 as 78 hundredths. Some may think that trailing zeros affect the value of a number, thinking, for example, that 0.300 is greater than 0.3. To help students determine whether or not a zero affects the value of a number, encourage them to write the number with and without the zeros and represent the resulting numbers in a place-value chart. Ask:

- *What number do we have if we include the zero(s)?*

- *What number do we have if we leave out the zero(s)?*

Go Online ▸ | Differentiation Support

Math Masters, p. 120

Sample Decimal Representations — Lesson 4-1

120   5.NBT.1, 5.NBT.3, 5.NBT.3a, SMP2

Refer to the next column to the right. Ask: *What pattern do we follow when we move one place to the right?* The value of the place is $\frac{1}{10}$ the value of the place to the left. *What would $\frac{1}{10}$ of the small strip look like?* GMP2.2 A tiny square, similar to a centimeter square Show students Square E from *Math Masters,* page 119. Have them confirm that 10 of Square E are equivalent to Strip D. Attach Square E to the fifth column in the place-value chart. Ask: *How many of those squares would fit into 1 unit square? How do you know?* 100, because there are 10 squares in the strip and 10 strips in the unit square. $10 * 10 = 100$ *What is a name for 1 piece out of 100 equal parts?* GMP6.3 $\frac{1}{100}$ Explain that the square represents $\frac{1}{100}$ and that this column is called the **hundredths** place. Label the place-value chart accordingly. (*See below.*)

Challenge students to extend the place-value chart one more column to the right. Ask: *What would $\frac{1}{10}$ of the hundredth square look like?* GMP2.2 Sample answer: It would be a very small sliver that fits into the square 10 times. Show students Strip F from *Math Masters,* page 119. Have them confirm that 10 of these small rectangles are equivalent to Square E. Attach Strip F to the last column of the place-value chart. Ask: *How many of these rectangles fit into 1 unit square?* 1,000 because $10 * 100 = 1,000$ *What is a name for 1 piece out of 1,000 equal parts?* GMP6.3 $\frac{1}{1,000}$ Tell students that the tiny rectangle represents $\frac{1}{1,000}$ and that this column is called the **thousandths** place. Complete the place-value chart by labeling the final column as thousandths. Display the chart for student reference throughout Unit 4.

Sample place-value chart:

Hundreds	Tens	Ones	.	Tenths	Hundredths	Thousandths

Point to the column between the ones place and the tenths place that contains only a dot. Remind the class that this is called the **decimal point.** A decimal point does not have any value, but it has its own column to emphasize the separation between whole numbers and **decimals,** or numbers written with a decimal point. The values of the digits to the *left* of the decimal point are greater than or equal to 1 (or 0). The values of the digits to the *right* of the decimal are 0 or less than 1. Emphasize that decimal place values follow the $\frac{1}{10}$-of pattern as you move to the right. Tell students that they can use the place-value chart to read, write, and represent decimals.

▶ Reading and Writing Decimals

Math Journal 1, pp. 112–113; *Math Masters,* p. 120

| WHOLE CLASS | SMALL GROUP | PARTNER | INDEPENDENT |

Display the top half of *Math Masters,* page 120 and ask: *How could you represent this number in the place-value chart?* GMP2.2 Sample answer: There are 3 tenths, 4 hundredths, and 5 thousandths. So I can write 3 in the tenths column, 4 in the hundredths column, and 5 in the thousandths column. Explain that when a decimal number has no ones, we often write the number with a 0 before the decimal point. Record 0.345 on the class place-value chart while students record 0.345 in the first row of their own place-value charts on journal page 112. Show students how the place-value chart can help them read decimal numbers. Explain that we read numbers after a decimal point as we would whole numbers and then say the name of the place-value column where the final digit appears. In the number 0.345 the final digit is in the thousandths column, so we read 0.345 as "three hundred forty-five thousandths." Write this number name in words and have students practice saying it aloud. GMP6.3

Next display the bottom half of *Math Masters,* page 120 and have students represent the number on the next row of their place-value charts. GMP2.2 1.078 Ask: *What information does the place-value chart help us see?* Sample answer: The number has 1 one, 7 hundredths, and 8 thousandths.

Explain that when there are digits in both the whole number and decimal places of a number, the decimal point is read as "and." So this number is read: "one and seventy-eight thousandths." Write the number name in words and have students practice saying it aloud. GMP6.3

> **NOTE** When saying numbers aloud, avoid using the word "and" except when it indicates the decimal point. For example, the number 1.307 should be read as "one *and* three hundred seven thousandths," not as "one *and* three hundred *and* seven thousandths."

Display the following decimals: 5.16, 0.8, 2.307, and 7.090. Have students write them in the place-value chart on journal page 112, reminding them to use the decimal point and column labels to carefully place each digit. Students will also use this chart with journal page 113, so make sure that the final four rows on the page remain blank. Watch for students who begin in the right-most column for 1- or 2-place decimals. Ask volunteers to read each decimal aloud as you write the names in words. GMP6.3 Have them identify the value of one of the digits in each of the numbers. *Suggestions:*

- 5.16 Five and sixteen hundredths Value of the 6? 6 hundredths
- 0.8 Eight tenths Value of the 8? 8 tenths
- 2.307 Two and three hundred seven thousandths Value of the 7? 7 thousandths
- 7.090 Seven and nine hundredths, or seven and ninety thousandths Value of the 9? 9 hundredths

> **NOTE** Some students may know that decimals can be read digit by digit with the word *point* to represent the decimal point, such as "one point zero seven eight." Acknowledge this as one way to read decimal numbers, but encourage students to use place-value language when reading decimals in these lessons.

Math Journal 1, p. 112

Reading and Writing Decimals

Lesson 4-1
DATE   TIME

Ones 1s 1s		Tenths 0.1s $\frac{1}{10}$s	Hundredths 0.01s $\frac{1}{100}$s	Thousandths 0.001s $\frac{1}{1,000}$s
	.		□	▫
0	.	3	4	5
1	.	0	7	8
5	.	1	6	
0	.	8		
2	.	3	0	7
7	.	0	9	0
4	.	8		
0	.	4	8	
0	.	0	4	8
6	.	4	0	8

112   5.NBT.1, 5.NBT.3, 5.NBT.3a, SMP2, SMP7

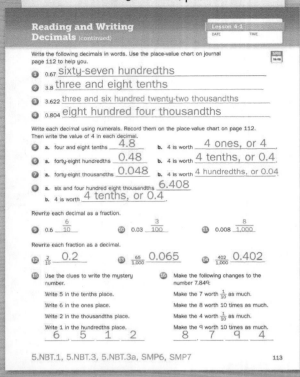

Reading and Writing
Decimals (continued)

Lesson 4-1
DATE   TIME

Write the following decimals in words. Use the place-value chart on journal page 112 to help you.

① 0.67 sixty-seven hundredths

② 3.8 three and eight tenths

③ 3.622 three and six hundred twenty-two thousandths

④ 0.804 eight hundred four thousandths

Write each decimal using numerals. Record them on the place-value chart on page 112.
Then write the value of 4 in each decimal.

⑤ a. four and eight tenths 4.8   b. 4 is worth 4 ones, or 4
⑥ a. forty-eight hundredths 0.48   b. 4 is worth 4 tenths, or 0.4
⑦ a. forty-eight thousandths 0.048   b. 4 is worth 4 hundredths, or 0.04
⑧ a. six and four hundred eight thousandths 6.408
   b. 4 is worth 4 tenths, or 0.4

Rewrite each decimal as a fraction.

⑨ 0.6 6/10   ⑩ 0.03 3/100   ⑪ 0.008 8/1,000

Rewrite each fraction as a decimal.

⑫ 2/10 0.2   ⑬ 65/1,000 0.065   ⑭ 402/1,000 0.402

⑮ Use the clues to write the mystery number.

Write 5 in the tenths place.
Write 6 in the ones place.
Write 2 in the thousandths place.
Write 1 in the hundredths place.

6 5 1 2

⑯ Make the following changes to the number 7,849.

Make the 7 worth 1/10 as much.
Make the 8 worth 10 times as much.
Make the 4 worth 1/10 as much.
Make the 9 worth 10 times as much.

8 7 9 4

5.NBT.1, 5.NBT.3, 5.NBT.3a, SMP6, SMP7   113

Point to 0.8 and ask: *Do you know another way to write eight tenths?* $\frac{8}{10}$
Explain that since digits in the tenths, hundredths, and thousandths places represent $\frac{1}{10}$ s, $\frac{1}{100}$ s, and $\frac{1}{1,000}$ s, it is fairly simple to write decimals as fractions. Point out that we read decimals and fractions in the same way, so reading a decimal aloud can help translate a decimal to fraction notation. Display several decimals and have students write them as fractions or mixed numbers on their slates. Encourage students to read the decimals aloud as they rewrite them. *Suggestions:*

- 0.3 $\frac{3}{10}$
- 0.82 $\frac{82}{100}$
- 0.914 $\frac{914}{1,000}$
- 5.86 $5\frac{86}{100}$
- 2.06 $2\frac{6}{100}$
- 5.804 $5\frac{804}{1,000}$

Have students work in partnerships to complete journal pages 112 and 113. Circulate and assist.

 Assessment Check-In   5.NBT.1, 5NBT.3, 5.NBT.3a

Math Journal 1, p. 113

Expect most students to accurately write decimals to hundredths using numerals and words on journal page 113. This lesson is an introduction to thousandths, so do not expect all students to write and interpret decimals to thousandths accurately. Do expect most to use place-value relationships to write digits in the correct places in Problem 15. Encourage those who struggle to use the place-value chart on journal page 112 to help them write the mystery number.

 Evaluation Quick Entry Go online to record students' progress and to see trajectories toward mastery for these standards.

Summarize Display the number 5.608. Have students read the number to a partner and agree on the value of each digit. GMP6.3

3 Practice 20–30 min

▶ Playing *Fraction Of*

Student Reference Book, pp. 306–307; *Math Masters,* p. G24

| WHOLE CLASS | SMALL GROUP | PARTNER | INDEPENDENT |

Students play *Fraction Of* to practice finding a fraction of a whole number. Have students play the game using the cards that were cut out in Lesson 3-13 (*Math Journal 1,* Activity Sheets 14–15). If additional sets of *Fraction Of* cards are needed, make copies of *Math Masters,* pages G21 and G22. Students should continue to use only Set 1 of the fraction cards until Unit 5. Be sure counters are available.

Observe

- Which students consistently choose a whole that results in a whole-number solution?
- How are students figuring out which wholes will lead to whole-number answers?

Discuss

- *How can you determine which wholes will give you a whole-number solution?*
- *What strategies can you use to score the most points possible?* GMP6.1

| Differentiate | Game Modifications | Go Online | Differentiation Support |

▶ Math Boxes 4-1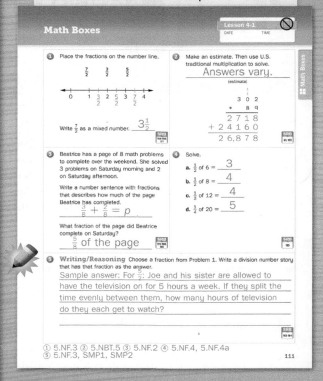

Math Journal 1, p. 111

| WHOLE CLASS | SMALL GROUP | PARTNER | INDEPENDENT |

Mixed Practice Math Boxes 4-1 are paired with Math Boxes 4-3.

▶ Home Link 4-1

Math Masters, p. 123

Homework Students use place-value patterns to translate between numerals and words for decimals to thousandths. GMP6.3, GMP7.2

Representing Decimals through Thousandths

Overview Students represent decimals to the thousandths place using base-10 numerals, number names, fractions, and thousandths grids.

▶ **Before You Begin**

Display the large place-value chart created in Lesson 4-1. Make extra copies of thousandths grids (*Math Masters,* page TA23), and have half-sheet place-value charts (*Math Masters,* page TA24) on hand. Students may find it easier to shade the thousandths grids using colored pencils or highlighters. For the Collecting Names for Decimals activity, use *Math Masters,* pages TA21–TA23 to create a shaded thousandths grid showing 0.300, a shaded hundredths grid showing 0.30, and a shaded tenths grid showing 0.3.

▶ **Vocabulary**

representation • represent • thousandths grid

	Materials	Standards
Warm Up 5 min		**Focus Cluster** • Understand the place value system.
Mental Math and Fluency Students write decimals using numerals.	slate	5.NBT.3, 5.NBT.3a

② **Focus** 35–40 min		
Math Message Students represent a decimal in three ways.	*Math Journal 1,* p. 114	5.NBT.3, 5.NBT.3a SMP2
Representing Decimals on a Thousandths Grid Students discuss how thousandths grids show place-value relationships and use grids to represent decimals.	*Math Journal 1,* p. 114; *Math Masters,* p. TA23, p. TA24 (optional); colored pencils or highlighters (optional)	5.NBT.1, 5.NBT.3, 5.NBT.3a SMP2
Collecting Names for Decimals Students complete name-collection boxes for decimals.	*Math Journal 1,* pp. 114–115; *Math Masters,* pp. TA21–TA23	5.NBT.1, 5.NBT.3, 5.NBT.3a SMP2
✓ **Assessment Check-In** See page 343. Expect most students to be able to give at least two names and shade a thousandths grid to represent decimals like those in the problems identified.	*Math Journal 1,* pp. 114–115; *Math Masters,* p. TA24 (optional)	5.NBT.3, 5.NBT.3a SMP2

③ **Practice** 20–30 min		
Solving Fraction Number Stories Students solve fraction number stories.	*Math Journal 1,* p. 116; fraction circles (optional)	5.NF.2, 5.NF.3 SMP1, SMP4
Math Boxes 4-2 Students practice and maintain skills.	*Math Journal 1,* p. 117	See page 343.
Home Link 4-2 Homework Students represent decimals using base-10 numerals, number names, and thousandths grids.	*Math Masters,* p. 125	5.NBT.1, 5.NBT.3, 5.NBT.3a, 5.NBT.6 SMP2

 Go Online to see how mastery develops for all standards within the grade.

Differentiation Options

Readiness 10–15 min	**Enrichment** 5–15 min	**Extra Practice** 10–15 min
WHOLE CLASS · **SMALL GROUP** · PARTNER · INDEPENDENT	WHOLE CLASS · SMALL GROUP · **PARTNER** · INDEPENDENT	WHOLE CLASS · SMALL GROUP · **PARTNER** · INDEPENDENT

Readiness

Using Place Value to Interpret Decimals To Hundredths

`5.NBT.3, 5.NBT.3a, SMP2`

Math Masters, pp. TA22 and TA24; base-10 blocks

For experience with visual representations of decimals, students interpret and represent decimals using a place-value chart, base-10 blocks, and hundredths grids. Display a flat, a long, and a unit cube, explaining that the flat represents one whole, the long represents one tenth, and the unit cube represents one hundredth. Show how the flat corresponds to the entire hundredths grid, the long to one column, and the cube to one square. Name a decimal like 0.62. Help students record the number in the place-value chart on *Math Masters,* page TA24. `GMP2.1` Ask: *How many tenths are in 0.62?* 6 *How could we show 6 tenths?* 6 longs *How many hundredths are in 0.62?* 2 *How could we show 2 hundredths?* `GMP2.1, GMP2.2` 2 cubes Have students take out 6 longs and 2 cubes and place them next to the hundredths grid on *Math Masters,* page TA22. Ask: *How could you shade the grid to show what you have in base-10 blocks?* `GMP2.3` Shade 6 columns and 2 squares. Repeat as necessary with other decimals to hundredths.

Enrichment

Exploring Decimals with Metric Units

`5.NBT.1, 5.NBT.3, 5.NBT.3a, 5.MD.1, SMP7`

Math Masters, p. 124; meterstick

To extend their work representing decimals to thousandths, students convert between metric units of length. Students measure the lengths of classroom objects and record measurements in millimeters, centimeters, decimeters, and meters. They describe patterns that they notice in their measurements. `GMP7.1, GMP7.2`

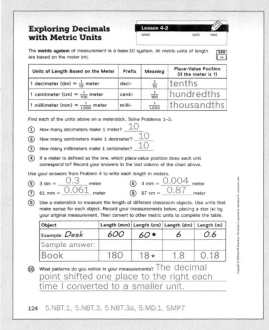

Extra Practice

Representing Decimals with Thousandths Grids

`5.NBT.3, 5.NBT.3a, SMP2, SMP8`

Activity Card 43;
Math Masters, p. TA23;
number cards 0–9 (4 of each); 1 counter

For additional practice representing decimals to thousandths, students draw cards to generate decimals and shade thousandths grids to represent them. `GMP2.1` Students discuss how they decided to shade each grid and develop a rule for representing on a grid any decimal to thousandths. `GMP8.1`

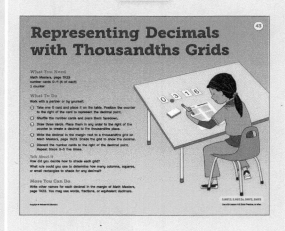

English Language Learner

Beginning ELL Many students have difficulty with the *th* sound. To help students differentiate between and pronounce the terms *ten, tens,* and *tenths; hundred, hundreds,* and *hundredths;* and *thousand, thousands,* and *thousandths,* provide oral language practice with the groups of words. Display the related words and corresponding number forms while pointing to the word endings as you model the pronunciation of individual words. Then have students repeat the words.

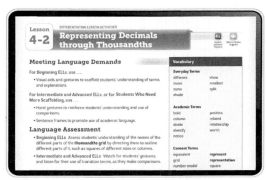

Differentiation Support pages are found in the online Teacher's Center.

Standards and Goals for
Mathematical Process and Practice

SMP2 **Reason abstractly and quantitatively.**

GMP2.1 Create mathematical representations using numbers, words, pictures, symbols, gestures, tables, graphs, and concrete objects.

GMP2.2 Make sense of the representations you and others use.

GMP2.3 Make connections between representations.

Math Journal 1, p. 114

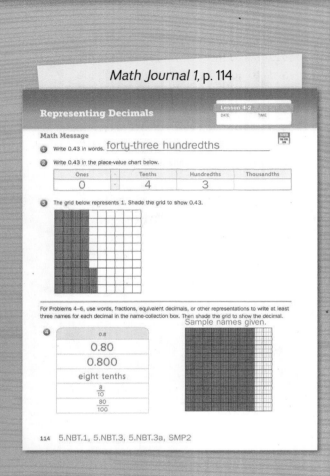

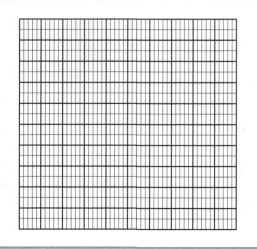

The whole grid represents one.

① Warm Up 5 min

▶ Mental Math and Fluency

Dictate the following decimals for students to write on slates. *Leveled exercises:*

● ○ ○ three and one tenth 3.1
  twenty-seven and eight tenths 27.8

● ● ○ six and seventy-one hundredths 6.71
  seventy-three and sixty-five hundredths 73.65

● ● ● one and nine hundredths 1.09
  forty and two hundred seventy-eight thousandths 40.278

② Focus 35–40 min

▶ Math Message

Math Journal 1, p. 114

Complete Problems 1–3 on journal page 114. GMP2.1 *Be ready to explain how you solved the problems.*

▶ Representing Decimals on a Thousandths Grid

Math Journal 1, p. 114; *Math Masters*, p. TA23

| WHOLE CLASS | SMALL GROUP | PARTNER | INDEPENDENT |

Math Message Follow-Up Have students share their answers. Expect most to recognize that 0.43 can be shown by shading 4 columns of 10 and 3 individual squares. GMP2.1, GMP2.2 If students struggle making sense of the grid, point out that it is the same size and shape as the square used to represent the ones place in Lesson 4-1 and that each square inside the grid is the same size as the tiny square used to represent 1 hundredth. Point out that in the Math Message 0.43 was represented three ways: in words, on a place-value chart, and on a grid. Then ask: *How would each* **representation,** *or way of showing, 0.43 change if we were instead showing 0.432?* Sample answers:

Words: We would say "four hundred thirty-two thousandths" instead of "forty-three hundredths."

Place-value chart: The first three columns would stay the same. We would write a 2 in the thousandths column.

Grid: We would need to split up the squares into smaller parts. We would still shade 4 full columns and 3 squares, but we'd also have to shade thousandths. We would have to divide a hundredth square into 10 tiny pieces to represent thousandths, and shade 2 of them.

Tell students that today they will explore ways to **represent,** or show, decimals to the thousandths. Distribute a few copies of *Math Masters,* page TA23 to each student. Tell them that these grids are called **thousandths grids.** To help students notice the relative sizes of each grid piece, ask questions like these: GMP2.2

- *What do you notice about this grid?* Sample answers: It's the same size as the hundredths grid and as the unit square we used before. It's split into smaller pieces than the hundredths grid.
- *How are the smallest pieces related to the bold squares in the grid?* There are 10 tiny rectangles in each small square created by the bold lines, so each rectangle is $\frac{1}{10}$ of the small square.
- *How are the bold squares related to the columns?* There are 10 small squares in each column, so each small square is $\frac{1}{10}$ of a column.
- *How are the columns related to the whole square?* There are 10 columns in the whole, so each column is $\frac{1}{10}$ of the whole square.

Have students use these relationships to identify the value each grid piece represents. (*See margin.*) GMP2.2, GMP2.3 Ask questions like these:

- *If the whole grid represents 1, what does a column represent? Why?* One tenth because 10 columns make one whole.
- *If the whole grid represents 1, what does a small square represent? Why?* One hundredth because 10 squares make one column and 10 columns make the whole, so there are 10 * 10, or 100, squares in the whole.
- *If the whole grid represents 1, what does the smallest rectangle represent? Why?* One thousandth because 10 small rectangles make one square and 100 squares make the whole, so there are 10 * 100, or 1,000, rectangles in the whole.

Guide a discussion to connect the relative size of each piece to place-value patterns. GMP2.3 Be sure to cover the following points:

- A column is $\frac{1}{10}$ of the whole grid, just like a tenth is $\frac{1}{10}$ of one.
- A small square is $\frac{1}{10}$ of a column, just like a hundredth is $\frac{1}{10}$ of a tenth.
- A tiny rectangle is $\frac{1}{10}$ of a small square, just like a thousandth is $\frac{1}{10}$ of a hundredth.

Return to the Math Message problem. Ask students how they would shade a thousandths grid to show 0.432. GMP2.1 Encourage them to think through the shading digit by digit. For example, there is a 4 in the tenths place, so 4 columns should be shaded, a 3 in the hundredths place, so three small squares should be shaded, and a 2 in the thousandths place, so 2 small rectangles should be shaded. Have students complete the shading on *Math Masters,* page TA23. Acknowledge that the thousandths pieces can be difficult to shade because they are so small.

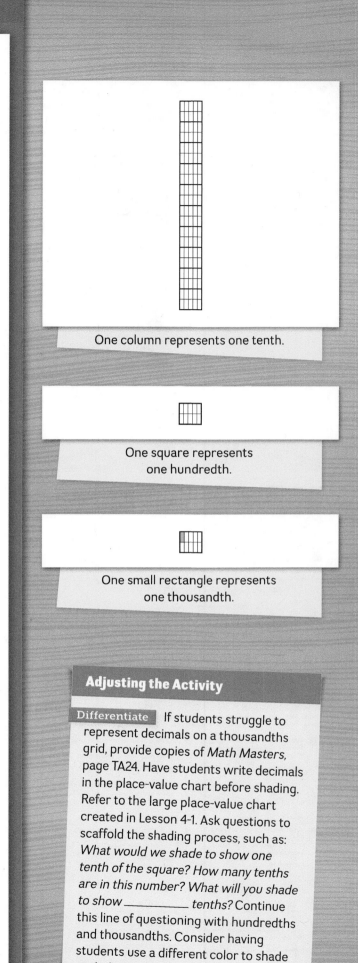

One column represents one tenth.

One square represents one hundredth.

One small rectangle represents one thousandth.

Adjusting the Activity

Differentiate If students struggle to represent decimals on a thousandths grid, provide copies of *Math Masters,* page TA24. Have students write decimals in the place-value chart before shading. Refer to the large place-value chart created in Lesson 4-1. Ask questions to scaffold the shading process, such as: *What would we shade to show one tenth of the square? How many tenths are in this number? What will you shade to show _____ tenths?* Continue this line of questioning with hundredths and thousandths. Consider having students use a different color to shade each digit.

Go Online ⬥ Differentiation Support

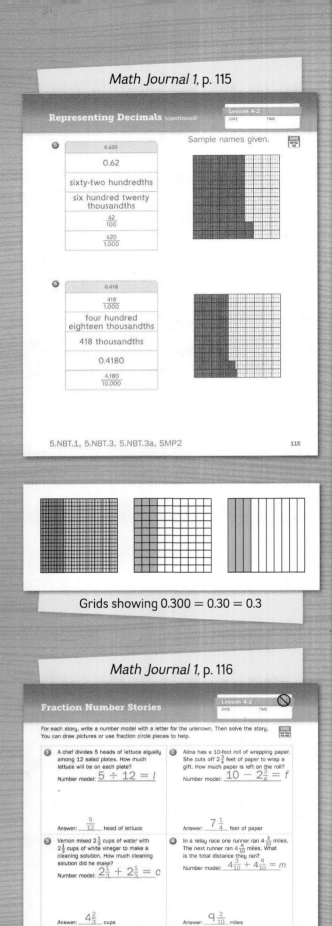

Math Journal 1, p. 115

Grids showing 0.300 = 0.30 = 0.3

Math Journal 1, p. 116

Display several more decimals for students to represent on thousandths grids. Include some decimals with trailing zeros. Have students label each grid with the number they are representing. *Suggestions:*

- **0.285** Shade 2 columns, 8 squares, 5 small rectangles
- **0.067** Shade 6 squares and 7 small rectangles
- **0.504** Shade 5 columns and 4 small rectangles
- **0.250** Shade 2 columns and 5 squares
- **0.300** Shade 3 columns

Have students compare their shaded grids with a partner. Discuss any discrepancies or questions that came up while students were shading.

▶ Collecting Names for Decimals

Math Journal 1, pp. 114–115; *Math Masters,* pp. TA21–TA23

| WHOLE CLASS | SMALL GROUP | PARTNER | INDEPENDENT |

Display a shaded thousandths grid showing 0.300, a shaded hundredths grid showing 0.30, and a shaded tenths grid showing 0.3. *(See Before You Begin.)* Ask students to name the decimal that each grid represents. **GMP2.2** 0.300, 0.30, 0.3 *(See margin.)* Ask: *What do you notice about the grid representations for 0.300, 0.30, and 0.3?* **GMP2.3** The same amount is shaded on each grid. *What does that tell us about the numbers 0.300, 0.30, and 0.3?* They are equivalent. The zeros at the end of each decimal don't affect the value of the number. Reiterate that 0.300, 0.30, and 0.3 are different names for the same number.

Ask: *Are 300, 30, and 3 equivalent whole numbers? Why or why not?* No, the zeros affect the value of the 3. When there are no zeros, the 3 is worth 3 ones. When there's a zero in the ones place, the 3 is shifted to the tens place and is worth 10 times as much. When there are zeros in the ones and tens places, the 3 is shifted to the hundreds place and is worth ten times as much as when it was in the tens place. Emphasize the point that zeros at the end of *whole numbers* affect the value of a number, while zeros at the end of *decimals* do not. With whole numbers, each additional zero causes the digits to move to a different place-value position. With decimals, additional zeros do not cause other digits to move; they remain in the same place-value position.

Tell students that they can use this fact to recognize and write equivalent names for decimals. Remind them that words, fractions, and grids are other ways to name decimals. Have students look at Problem 4 on journal page 114 and suggest names they could write in the name-collection box for 0.8. Have students list their suggestions. Then have them complete journal pages 114 and 115 independently or in partnerships. **GMP2.1, GMP2.3** Circulate and assist.

Assessment Check-In   5.NBT.3, 5.NBT.3a

Math Journal 1, pp. 114–115

Expect most students to give at least two accurate names and correctly shade the thousandths grid for the decimals in Problems 4 and 5 on journal pages 114 and 115. GMP2.1 Also expect most to use words and fractions that directly translate the given decimal, such as writing "eight tenths" for 0.8. Some students may be able to generate more than two names and use equivalent decimals, such as 0.800. For students who struggle representing decimals on a grid, refer to the Adjusting the Activity note on page 341. In addition, suggest using the place-value chart (*Math Masters*, page TA24) to help them write the decimals in words.

Evaluation Quick Entry Go online to record students' progress and to see trajectories toward mastery for these standards.

Summarize Have students work in small groups to compare the names they wrote in the name-collection boxes. Encourage them to explain how they know each is a name for the given decimal. GMP2.2

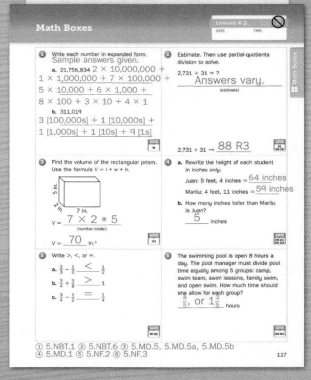

Math Journal 1, p. 117

③ Practice   20–30 min

▶ Solving Fraction Number Stories

Math Journal 1, p. 116

| WHOLE CLASS | SMALL GROUP | **PARTNER** | **INDEPENDENT** |

Students solve number stories involving fractions. They write number models with a letter standing for the unknown to help make sense of the problems and decide what operation to use. GMP1.1 Make sure fraction circle pieces are available. Encourage students to use drawings or the fraction circles to help them solve the problems. GMP4.1

▶ Math Boxes 4-2

Math Journal 1, p. 117

| WHOLE CLASS | **SMALL GROUP** | PARTNER | **INDEPENDENT** |

Mixed Practice Math Boxes 4-2 are paired with Math Boxes 4-4.

▶ Home Link 4-2

Math Masters, p. 125

Homework Students represent decimals to thousandths using base-10 numerals, number names, and thousandths grids. GMP2.1

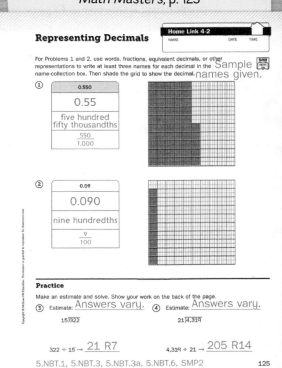

Math Masters, p. 125

Representing Decimals in Expanded Form

Overview Students are introduced to expanded form for decimals.

▶ **Before You Begin**

For Part 2, make sure students have access to at least 3 different colors of pencils, crayons, or markers. Consider preparing place-value charts from *Math Masters*, page TA24 to support students who struggle with expanded form. Continue to display the large place-value chart constructed in Lesson 4-1.

▶ **Vocabulary**

expanded form • digit • standard notation

Standards

Focus Cluster
• Understand the place value system.

① Warm Up 5 min

	Materials	
Mental Math and Fluency Students write whole numbers in expanded form.	slate	**5.NBT.1**

② Focus 35–40 min

Math Message Students shade a thousandths grid to represent digits of a decimal.	*Math Journal 1*, p. 118; colored pencils	**5.NBT.1, 5.NBT.3, 5.NBT.3a** SMP2
Introducing Expanded Form for Decimals Students discuss and translate between different versions of expanded form for decimals.	*Math Journal 1*, p. 118; *Math Masters*, p. TA24 (optional)	**5.NBT.1, 5.NBT.3, 5.NBT.3a** SMP2
✓ **Assessment Check-In** See page 348. Expect most students to be able to translate between standard notation and expanded form for decimals to thousandths.	*Math Journal 1*, p. 118	**5.NBT.1, 5.NBT.3, 5.NBT.3a**
Representing Decimals in Expanded Form Students represent decimals in expanded form and on a thousandths grid.	*Math Journal 1*, p. 119; per partnership: *Math Masters*, p. TA25; number cards 0–9 (4 of each); colored pencils	**5.NBT.1, 5.NBT.3, 5.NBT.3a** SMP2

③ Practice 20–30 min

Playing *Fraction Capture* **Game** Students practice adding fractions.	*Student Reference Book*, p. 305; per partnership: *Math Masters*, pp. G19–G20; two 6-sided dice	**5.OA.2, 5.NF.1** SMP1
Math Boxes 4-3 Students practice and maintain skills.	*Math Journal 1*, p. 120	See page 349.
Home Link 4-3 **Homework** Students write decimals in standard notation and expanded form.	*Math Masters*, p. 128	**5.NBT.1, 5.NBT.3, 5.NBT.3a,** **5.NF.4, 5.NF.4a** SMP2

⫘ **Go Online** to see how mastery develops for all standards within the grade.

Differentiation Options

Readiness	5–15 min	**Enrichment**	15–25 min	**Extra Practice**	5–15 min
WHOLE CLASS · SMALL GROUP · PARTNER · INDEPENDENT		WHOLE CLASS · SMALL GROUP · **PARTNER** · INDEPENDENT		WHOLE CLASS · SMALL GROUP · **PARTNER** · INDEPENDENT	

Identifying the Value of a Digit

5.NBT.1, 5.NBT.3, 5.NBT.3a, SMP2

Math Masters, p. TA24; slate

To explore representing the value of digits in decimals, students discuss ways to determine and express what a digit is worth. Distribute *Math Masters*, page TA24 to each student. Display the decimal 5.634, and have students record the number in their place-value charts. Point to the ones place and ask: *What does this digit mean? How much is a 5 in this place worth?* 5 ones *How do you know?* It is in the ones column on the place-value chart. *How could we write the value of this digit?* **GMP2.1** 5, 5 ones, 5 [1s], or 5 ∗ 1 If no one mentions multiplication, ask: *What do we have 5 of?* Ones *How could we show that we have 5 of something?* Multiply 5 by that thing. Have students record on slates the value of 5 in each of the ways discussed. Continue this questioning with the tenths, hundredths, and thousandths places. Make sure students practice representing decimals with both base-10 numerals and fractions (for example, 0.1 and $\frac{1}{10}$). Repeat with other decimals as needed.

Exploring Decimals through Millionths

5.NBT.1, 5.NBT.2, 5.NBT.3, 5.NBT.3a, SMP2

Math Masters, pp. 126–127

To extend their work with decimals and expanded form, students explore decimals through millionths as they complete *Math Masters*, pages 126 and 127.
GMP2.1, GMP2.2, GMP2.3

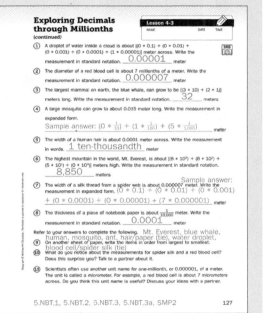

Using Expanded Form

5.NBT.1, 5.NBT.3, 5.NBT.3a, SMP2

Activity Card 44;
number cards 0–9 (4 of each);
1 counter; 6-sided die

For additional practice writing decimals in expanded form, students draw cards to generate decimals and roll a die to determine which version of expanded form to use. **GMP2.1** Students discuss what each version helps them understand about a decimal. **GMP2.3**

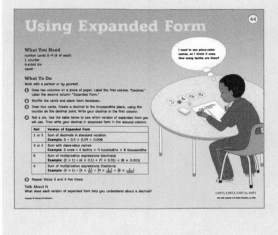

English Language Learner

Beginning ELL To help students understand the meaning of *expand* as in *expanded form*, demonstrate an action like blowing up a balloon or stretching a rubber band. Show that the object is the same object; it is just stretched out. Follow with Total Physical Response prompts that have students clasp their hands and then open their arms wide while repeating the term *expand*.

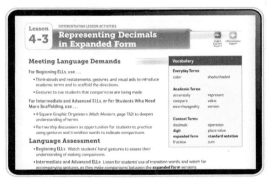

Differentiation Support pages are found in the online Teacher's Center.

SMP2 Reason abstractly and quantitatively.

GMP2.1 Create mathematical representations using numbers, words, pictures, symbols, gestures, tables, graphs, and concrete objects.

GMP2.2 Make sense of the representations you and others use.

GMP2.3 Make connections between representations.

---

1 Warm Up · 5 min

▶ Mental Math and Fluency

Have students write whole numbers in expanded form on slates.

Leveled exercises: Sample answers:

◉○○ **317** $300 + 10 + 7$
   608 $(6 * 100) + (8 * 1)$

◉◉○ **1,956** $1,000 + 900 + 50 + 6$
   2,117 $(2 * 10^3) + (1 * 10^2) + (1 * 10^1) + (7 * 10^0)$

◉◉◉ **10,941** $10,000 + 0 + 900 + 40 + 1$
   318,004 $(3 * 100,000) + (1 * 10,000) + (8 * 1,000) + (0 * 100) + (0 * 10) + (4 * 1)$

---

2 Focus · 35–40 min

▶ Math Message

Math Journal 1, p. 118

Complete the Math Message on journal page 118. GMP2.1 *Then talk with a partner about what you notice.* GMP2.2 *Be ready to share what you and your partner discuss.*

▶ Introducing Expanded Form for Decimals

Math Journal 1, p. 118

| WHOLE CLASS | SMALL GROUP | PARTNER | INDEPENDENT |

Math Message Follow-Up Ask: *What did you notice as you added different colors of shading to your grid?* Sample answer: I didn't have to change the part that I started with. Each time I shaded more, I only shaded an amount to show the new digit. For example, I shaded 1 square in a second color to turn 0.3 into 0.31. I shaded 2 tiny rectangles in a third color to turn 0.31 into 0.312. **Display a shaded grid for the Math Message problem.** Ask: *What number do the first 3 columns of shading represent?* 0.3 *What number does the 1 square in the second color represent?* 0.01 *What number do the 2 tiny rectangles shaded in the third color represent?* GMP2.2 0.002 Point out that the shading shows how 0.3, 0.01, and 0.002 combine to make 0.312. Ask: *What operation is used to show combining or putting together?* GMP2.3 Addition Display the number sentence $0.312 = 0.3 + 0.01 + 0.002$. Remind students that this way of writing a number is called **expanded form.** Highlight how expanded form shows the number broken apart by place value and shows the sum of the value of each **digit.**

Math Journal 1, p. 118

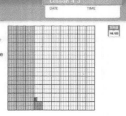

Writing Decimals in Expanded Form   Lesson 4-3
DATE   TIME

Math Message
Shade the grid to represent 0.3. Add shading to the grid in another color so that the grid shows 0.31 in all. Then use a third color to add shading so that the grid shows 0.312.

Use different versions of expanded form to complete the table below.

Standard Notation	Versions of Expanded Form		
	Sum of Decimals in Standard Notation	Sum of Multiplication Expressions (Decimals)	Sum of Multiplication Expressions (Fractions)
Example: 0.568	$0.5 + 0.06 + 0.008$	$(5 * 0.1) + (6 * 0.01) + (8 * 0.001)$	$5 * \frac{1}{10} + 6 * \frac{1}{100} + 8 * \frac{1}{1,000}$
2.473	$2 + 0.4 + 0.07 + 0.003$	$(2 * 1) + (4 * 0.1) + (7 * 0.01) + (3 * 0.001)$	$(2 * 1) + \left(4 * \frac{1}{10}\right) + \left(7 * \frac{1}{100}\right) + \left(3 * \frac{1}{1,000}\right)$
0.094	$0.0 + 0.09 + 0.004$	$(0 * 0.1) + (9 * 0.01) + (4 * 0.001)$	$\left(0 * \frac{1}{10}\right) + \left(9 * \frac{1}{100}\right) + \left(4 * \frac{1}{1,000}\right)$
7.752	$7 + 0.7 + 0.05 + 0.002$	$(7 * 1) + (7 * 0.1) + (5 * 0.01) + (2 * 0.001)$	$(7 * 1) + \left(7 * \frac{1}{10}\right) + \left(5 * \frac{1}{100}\right) + \left(2 * \frac{1}{1,000}\right)$
0.637	$0.6 + 0.03 + 0.007$	$(6 * 0.1) + (3 * 0.01) + (7 * 0.001)$	$\left(6 * \frac{1}{10}\right) + \left(3 * \frac{1}{100}\right) + \left(7 * \frac{1}{1,000}\right)$

118   5.NBT.1, 5.NBT.3, 5.NBT.3a, SMP2

Tell students that in today's lesson they are going to extend their understanding of expanded form to decimal numbers. Remind them that they have already represented whole numbers in expanded form in multiple ways:

- As sums of whole numbers: $400 + 20 + 8$
- As sums with place-value names: 4 hundreds + 2 tens + 8 ones; or 4 [100s] + 2 [10s] + 8 [1s]
- As sums of multiplication expressions: $(4 * 100) + (2 * 10) + (8 * 1)$

Explain that there are also several ways to write decimals in expanded form. Ask students to brainstorm how they could write 0.312 in expanded form, thinking about the examples for whole numbers. Display students' suggestions. Be sure to include each of the expanded forms of 0.312 shown below.

- $0.3 + 0.01 + 0.002$
- 3 tenths + 1 hundredth + 2 thousandths
- 3 [0.1s] + 1 [0.01] + 2 [0.001s]
- $3\left[\frac{1}{10}s\right] + 1\left[\frac{1}{100}\right] + 2\left[\frac{1}{1,000}s\right]$
- $(3 * 0.1) + (1 * 0.01) + (2 * 0.001)$
- $\left(3 * \frac{1}{10}\right) + \left(1 * \frac{1}{100}\right) + \left(2 * \frac{1}{1,000}\right)$

Compare and contrast the different versions of expanded form, highlighting the different ways they represent the value of each digit. **GMP2.2, GMP2.3** Guide the discussion to cover the following points:

- The first example shows the value of each digit in **standard notation.**
- The second example uses words to show place-value names.
- Fraction and decimal names can be used interchangeably to indicate place values.
- All of the different versions represent the same number: 0.312.

Have students consider how to represent decimals greater than 1 in expanded form. Remind them to use what they know about representing whole numbers in expanded form. Ask: *How could we write 7.53 in expanded form?* Sample answers: We could use standard notation and write $7 + 0.5 + 0.03$. We could use multiplication expressions such as $(7 * 1) + (5 * 0.1) + (3 * 0.01)$, or $(7 * 1) + \left(5 * \frac{1}{10}\right) + \left(3 * \frac{1}{100}\right)$. Have students work in partnerships to complete the rest of journal page 118. **GMP2.1, GMP2.2, GMP2.3** Circulate and assist. If students struggle translating decimals into expanded form, provide copies of *Math Masters,* page TA24. Ask questions like these to help students identify the value of each digit: *How many ones? How many tenths? How many hundredths? How many thousandths?* Have students write each digit in the place-value chart before translating to a different notation.

NOTE Some students may remember that they have used exponential notation to represent whole numbers in expanded form. For example, the number 428 can be written as $(4 * 10^2) + (2 * 10^1) + (8 * 10^0)$. Decimals can also be written in expanded form using exponential notation, but this requires the use of negative exponents, which will not be introduced in *Fifth Grade Everyday Mathematics*. If a student asks about exponential notation for decimals, acknowledge that decimals can be written using exponential notation and that students will learn how to do this in a later grade.

Adjusting the Activity

Differentiate Since students have not formally been exposed to fraction or decimal multiplication, they may be unsure of how to interpret versions of expanded form with multiplication expressions. To help students make sense of multiplication expressions, make a connection to repeated addition. For example, say: *We have 3 tenths. We could make 3 tenths by adding 1 tenth + 1 tenth + 1 tenth. How else could we make 3 tenths?* Multiply 1 tenth by 3 Use this same reasoning to explain multiplication expressions representing hundredths and thousandths.

Go Online Differentiation Support

Math Masters, p. TA25

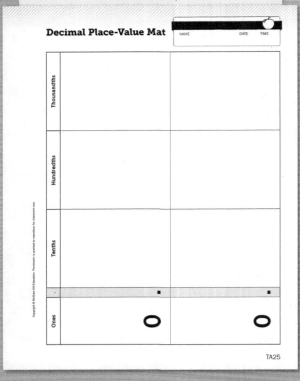

Decimal Place-Value Mat

Math Journal 1, p. 119

Representing Decimals in Expanded Form

Use three number cards to create a decimal on your decimal place-value mat. Record the decimal you created in one of the boxes below. Write the decimal in expanded form. Then shade a thousandths grid using a different color to show the value of each digit. Repeat to complete all four boxes. **Answers vary.**

5.NBT.1, 5.NBT.3, 5.NBT.3a, SMP2

 Assessment Check-In   5.NBT.1, 5.NBT.3, 5.NBT.3a

Math Journal 1, p. 118

Observe students as they complete the table on journal page 118. Expect most to successfully translate between decimals in standard notation and in expanded form as sums of decimals. Some students may be able to represent decimals in expanded form using multiplication expressions. If students struggle interpreting numbers written in expanded form using multiplication expressions, consider using the suggestion in the Adjusting the Activity note.

 Evaluation Quick Entry Go online to record students' progress and to see trajectories toward mastery for these standards.

▶ Representing Decimals in Expanded Form

Math Journal 1, p. 119; *Math Masters,* p. TA25

WHOLE CLASS | **SMALL GROUP** | **PARTNER** | INDEPENDENT

Tell students that they will continue to practice representing decimals in expanded form. Explain that they will draw three number cards and use them to create a decimal on a place-value mat (*Math Masters,* page TA25). After forming the decimal, students write it in expanded form and represent it on a thousandths grid on journal page 119. GMP2.1 They may use any version of expanded form, but they must show the value of each digit in a separate color on the grids. Demonstrate drawing three number cards, creating a decimal, and showing one version of expanded form and a shaded grid. *For example:*

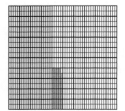

0. 4 3 9

Expanded Form:
0.4 + 0.03 + 0.009

Demonstrate again, but this time show an example that uses a zero card, such as 0.042 or 0.590. Discuss how the placement of the zero affects the value of the other digits. Ask: *Will a zero card affect the look of your representation? If so, how?* Yes. Sample answer: If there is a zero card, we will not use three different colors on the grid.

Provide each partnership with one copy of *Math Masters,* page TA25 and 4 each of number cards 0–9. Partners take turns drawing cards and representing decimals until each partner has had four turns. GMP2.1, GMP2.3 Encourage students to explain their reasoning as they work. Circulate and assist.

Summarize Have partners check each other's work. Ask: *How do you know if a version of expanded form accurately represents a decimal?* **GMP2.2, GMP2.3** Sample answer: Each part of the addition expression in expanded form has to match the value of a digit in the decimal and one color of shading in the grid.

3 Practice 20–30 min

▶ Playing *Fraction Capture*

Student Reference Book, p. 305; *Math Masters*, pp. G19–G20

| WHOLE CLASS | SMALL GROUP | **PARTNER** | INDEPENDENT |

Students practice adding fractions. Encourage students to discuss game-playing strategies before they begin and to try some of the strategies while they play. **GMP1.6**

Observe
- Which students are using an efficient strategy to find fractions to capture?
- Which students are able to check whether their partners' answers are correct?

Discuss
- *Were you able to use a strategy that you learned from a classmate? Explain.* **GMP1.6**
- *Which denominators make it easier to capture fractions? Which denominators make it harder? Why?*

| Differentiate | **Game Modifications** | Go Online | Differentiation Support |

▶ Math Boxes 4-3

Math Journal 1, p. 120

| WHOLE CLASS | SMALL GROUP | PARTNER | INDEPENDENT |

Mixed Practice Math Boxes 4-3 are paired with Math Boxes 4-1.

▶ Home Link 4-3

Math Masters, p. 128

Homework Students translate between decimals in standard notation and decimals in expanded form. **GMP2.1, GMP2.2**

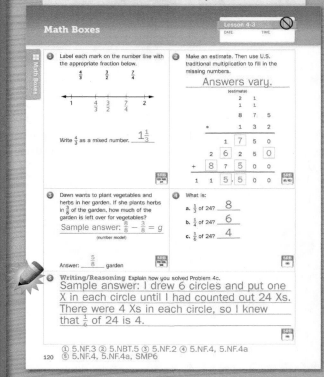

Math Journal 1, p. 120

Math Boxes   Lesson 4-3
                                                    DATE   TIME

① Label each mark on the number line with the appropriate fraction below.

$$\frac{4}{3} \quad \frac{3}{2} \quad \frac{7}{4}$$

1   $\frac{4}{3}$ $\frac{3}{2}$ $\frac{7}{4}$ 2

Write $\frac{4}{3}$ as a mixed number. $1\frac{1}{3}$

② Make an estimate. Then use U.S. traditional multiplication to fill in the missing numbers.

Answers vary.

(estimate)

```
          2   1
          1   1
          8   7   5
    *         1   3   2
        1   7   5   0
    2   6   2   5   0
+   8   7   5   0   0
1   1   5 . 5   0   0
```

③ Dawn wants to plant vegetables and herbs in her garden. If she plants herbs in $\frac{3}{8}$ of the garden, how much of the garden is left over for vegetables?

Sample answer: $\frac{8}{8} - \frac{3}{8} = g$

(number model)

Answer: $\frac{5}{8}$ garden

④ What is:
a. $\frac{1}{3}$ of 24? __8__
b. $\frac{1}{4}$ of 24? __6__
c. $\frac{1}{6}$ of 24? __4__

⑤ Writing/Reasoning Explain how you solved Problem 4c.

Sample answer: I drew 6 circles and put one X in each circle until I had counted out 24 Xs. There were 4 Xs in each circle, so I knew that $\frac{1}{6}$ of 24 is 4.

① 5.NF.3 ② 5.NBT.5 ③ 5.NF.2 ④ 5.NF.4, 5.NF.4a
120 ⑤ 5.NF.4, 5.NF.4a, SMP6

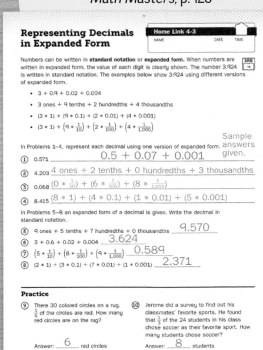

Math Masters, p. 128

Representing Decimals in Expanded Form   Home Link 4-3

NAME   DATE   TIME

Numbers can be written in **standard notation** or **expanded form**. When numbers are written in expanded form, the value of each digit is clearly shown. The number 3,924 is written in standard notation. The examples below show 3,924 using different versions of expanded form.

- $3 + 0.9 + 0.02 + 0.004$
- 3 ones + 9 tenths + 2 hundredths + 4 thousandths
- $(3 * 1) + (9 * 0.1) + (2 * 0.01) + (4 * 0.001)$
- $(3 * 1) + \left(9 * \frac{1}{10}\right) + \left(2 * \frac{1}{100}\right) + \left(4 * \frac{1}{1,000}\right)$

Sample answers given.

In Problems 1–4, represent each decimal using one version of expanded form.

① 0.571 ___$0.5 + 0.07 + 0.001$___

② 4.203 ___4 ones + 2 tenths + 0 hundredths + 3 thousandths___

③ 0.068 ___$\left(0 * \frac{1}{10}\right) + \left(6 * \frac{1}{100}\right) + \left(8 * \frac{1}{1,000}\right)$___

④ 8.415 ___$(8 * 1) + (4 * 0.1) + (1 * 0.01) + (5 * 0.001)$___

In Problems 5–8 an expanded form of a decimal is given. Write the decimal in standard notation.

⑤ 9 ones + 5 tenths + 7 hundredths + 0 thousandths ___9.570___

⑥ $3 + 0.6 + 0.02 + 0.004$ ___3.624___

⑦ $\left(5 * \frac{1}{10}\right) + \left(8 * \frac{1}{100}\right) + \left(9 * \frac{1}{1,000}\right)$ ___0.589___

⑧ $(2 * 1) + (3 * 0.1) + (7 * 0.01) + (1 * 0.001)$ ___2.371___

Practice

⑨ There are 30 colored circles on a rug. $\frac{1}{5}$ of the circles are red. How many red circles are on the rug?

Answer: __6__ red circles

⑩ Jerome did a survey to find out his classmates' favorite sports. He found that $\frac{1}{3}$ of the 24 students in his class chose soccer as their favorite sport. How many students chose soccer?

Answer: __8__ students

128 5.NBT.1, 5.NBT.3, 5.NBT.3a, 5.NF.4, 5.NF.4a, SMP2

Comparing and Ordering Decimals

Overview Students use place-value strategies to compare decimals to thousandths.

▶ **Before You Begin**

Have thousandths grids and decimal place-value charts (*Math Masters*, pages TA23 and TA24) available for students to use. Consider sending home extra copies of these pages to provide support with the Home Link. For the optional Extra Practice activity, you may wish to copy the decimal cards (*Math Masters*, page TA26) on cardstock.

▶ **Vocabulary**

compare • order

Standards

Focus Cluster
- Understand the place value system.

	Materials	
① Warm Up 5 min		
Mental Math and Fluency Students identify place-value positions of digits in decimal numbers.		5.NBT.1
② Focus 35–40 min		
Math Message Students compare two decimals and consider how to explain their thinking.	*Math Masters*, p. TA23 (optional)	5.NBT.3, 5.NBT.3b SMP7
Exploring Strategies for Comparing Decimals Students use thousandths grids, expanded form, and decimal place-value charts to compare decimals.	*Math Masters*, pp. TA23–TA24	5.NBT.1, 5.NBT.3, 5.NBT.3a, 5.NBT.3b SMP1, SMP2, SMP7
Introducing *Decimal Top-It* Game Students practice comparing decimals to thousandths.	*Student Reference Book*, pp. 296–297; per partnership: *Math Masters*, pp. TA25 and G14, p. TA23 (optional); number cards 0–9 (4 of each); 6-sided die	5.NBT.1, 5.NBT.3, 5.NBT.3a, 5.NBT.3b SMP1, SMP2, SMP7
✓ Assessment Check-In See page 355. Expect most students to be able to compare decimals to thousandths using the Decimal Place-Value Mat, expanded notation, or a thousandths grid while playing the game *Decimal Top-It*.	*Math Masters*, pp. TA25 and G14	5.NBT.3, 5.NBT.3b, SMP2
③ Practice 20–30 min		
Interpreting Real-World Remainders Students interpret remainders in division number stories.	*Math Journal 1*, p. 121	5.NBT.6, 5.NF.3, 5.MD.1 SMP6
Math Boxes 4-4 Students practice and maintain skills.	*Math Journal 1*, p. 122	See page 355.
Home Link 4-4 Homework Students compare and order decimals.	*Math Masters*, p. 130	5.NBT.3, 5.NBT.3a, 5.NBT.3b, 5.NF.2

 Go Online to see how mastery develops for all standards within the grade.

Differentiation Options

Readiness 5–15 min

WHOLE CLASS | SMALL GROUP | PARTNER | INDEPENDENT

Testing Ideas about Digits

5.NBT.3, 5.NBT.3b

To address the common misconception that numbers with more digits are always greater than numbers with fewer digits, students compare numbers and count digits. Display a 4-digit number with a digit in the tenths place, such as 158.4. Have students count the digits. Display a greater 3-digit number, such as 412. Have students count the digits. Give a context and unit for the numbers, such as the weight in pounds of two gorillas. Ask: *Which is more, 158.4 pounds or 412 pounds?* 412 *Did the number of digits matter in determining the greater number? Why or why not?* No, because 4 hundreds is more than 1 hundred, so we can tell which number is bigger from the hundreds place. The digits after the hundreds place don't affect the number enough to change which one is greater. Repeat as needed with other numbers, emphasizing the idea that a number with fewer digits can be greater than a number with more digits.

Enrichment 5–15 min

WHOLE CLASS | SMALL GROUP | PARTNER | INDEPENDENT

Exploring Batting Averages

5.NBT.3, 5.NBT.3a, 5.NBT.3b, 5.NF.3, SMP1, SMP6

Math Masters, p. 129; calculator

To extend their work comparing decimals, students calculate and compare baseball players' batting averages. They identify the strongest hitter among the players listed and explain their reasoning.

GMP1.1, GMP6.1 Students may need assistance thinking about how to record repeating decimals as 3-place decimals.

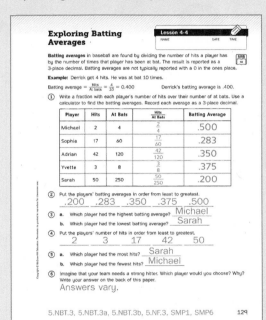

Extra Practice 15–30 min

WHOLE CLASS | SMALL GROUP | PARTNER | INDEPENDENT

Playing *Build-It: Decimal Version*

5.NBT.3, 5.NBT.3a, 5.NBT.3b, SMP1, SMP7

Activity Card 45; *Student Reference Book*, p. 293; *Math Masters*, p. G15; scissors; per partnership: *Math Masters*, p. TA26, pp. TA23–TA24 (optional)

For additional practice comparing and ordering decimals to thousandths, students play a decimal variation of *Build-It*. Have students cut out decimal cards from *Math Masters*, page TA26. Provide thousandths grids and decimal place-value charts (*Math Masters*, pages TA23 and TA24) as needed. Encourage students to use a second comparison strategy to check their work.

GMP1.4, GMP7.2

English Language Learner

Beginning ELL Provide visual input and teacher modeling to scaffold understanding of the terms *greater than, fewer than,* and *less than*. Display the terms *fewer than* and *less than* using small letters and the < symbol. Display the term *greater than* using large letters and the > symbol. Display sets of objects, asking students to point to the set that is *greater than* and the set that is *fewer/less than*. Reword student responses with statements like this one: *Yes, this set has more/fewer _____.* Have students repeat in chorus.

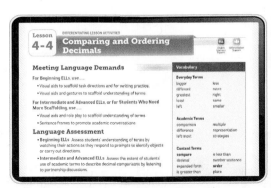

Differentiation Support pages are found in the online Teacher's Center.

SMP1 **Make sense of problems and persevere in solving them.**
   GMP1.4 Check whether your answer makes sense.

SMP2 **Reason abstractly and quantitatively.**
   GMP2.2 Make sense of the representations you and others use.

SMP7 **Look for and make use of structure.**
   GMP7.2 Use structures to solve problems and answer questions.

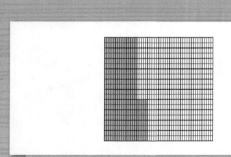

0.34

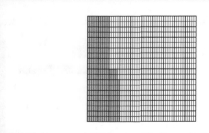

0.248

1 Warm Up   5 min

▶ Mental Math and Fluency

Have students identify the place-value position of the underlined digit in each number. *Leveled exercises:*

●○○ 41.<u>9</u>2 Tenths place
    144.7<u>3</u> Hundredths place

●●○ 8.0<u>9</u> Hundredths place
    0.71<u>2</u> Thousandths place

●●● 6,131.0<u>8</u>1 Hundredths place
    131,096.<u>5</u>4 Tenths place

2 Focus   35–40 min

▶ Math Message

Which is greater, 0.34 or 0.248? You may use a grid or expanded form to help you decide. Be ready to explain how you know. GMP7.2

▶ Exploring Strategies for Comparing Decimals

Math Masters, pp. TA23–TA24

| WHOLE CLASS | SMALL GROUP | PARTNER | INDEPENDENT |

Math Message Follow-Up Invite students to share their answers. Encourage them to use what they know about place value to explain their strategies. Be sure these two strategies are included:

• Shade 3 columns and 4 squares on a thousandths grid to represent 0.34. Shade 2 columns, 4 squares, and 8 tiny rectangles on a second thousandths grid to represent 0.248. (*See margin.*) More of the first grid is shaded, so 0.34 is greater than 0.248. GMP2.2, GMP7.2

• Represent both numbers in expanded form: 0.34 = 0.3 + 0.04; 0.248 = 0.2 + 0.04 + 0.008. Since 0.34 has 3 tenths and 0.248 has only 2 tenths, 0.34 is greater than 0.248. GMP7.2

Summarize by displaying the number sentence 0.34 > 0.248. Remind students that the symbols > (is greater than) and < (is less than) are used to record comparisons. Students should recall that the "open" side of each symbol faces the larger value, in this case, 0.34. The number sentence is read "thirty-four hundredths (0.34) is greater than two hundred forty-eight thousandths (0.248)." Explain that the sentence could also be written 0.248 < 0.34 and read "0.248 is less than 0.34." Establish the idea that the number sentences 0.248 < 0.34 and 0.34 > 0.248 convey the same information.

Explain that to decide which number is greater, we have to **compare** the numbers. Point out that comparing decimals is an important skill since many real-world measurements and statistics, such as race results, the price of gasoline, or baseball batting averages, are expressed as decimals. Today students will use a place-value chart and what they know about place value to learn strategies for comparing and ordering decimals to thousandths.

Display the numbers from the Math Message in a decimal place-value chart (*Math Masters,* page TA24). GMP2.2

Ones 1s	.	Tenths 0.1s	Hundredths 0.01s	Thousandths 0.001s
0	.	3	4	0
0	.	2	4	8

Ask: *What digit should we write in the thousandths place for 0.34?* 0 Record a zero in the thousandths column for 0.34. Explain that when using a place-value chart to compare numbers, it is important to start with the left-most digit and work digit by digit to the right. This is because a digit in each place is worth 10 times as much as it would be in the place to its right, so digits farther to the left have a bigger effect on the value of a number. Point to each place-value column one at a time, starting with the ones place and moving to the right. For each column, ask: *What do you notice about the digit in this place for each number?* GMP2.2 The ones and hundredths are the same. The tenths and thousandths are different. Ask: *Moving from left to right, where is the first difference?* The tenths place *Which is greater, 3 tenths or 2 tenths?* 3 tenths *So which is greater, 0.248 or 0.34?* 0.34 *Does it matter that 0.248 has a larger digit in the thousandths place than 0.34? Why or why not?* GMP7.2 No. 0.34 has more tenths than 0.248 and tenths are worth more than thousandths, so we already know 0.34 is greater.

Remind students that they already showed that 0.34 had one more tenth than 0.248 when they represented the numbers in expanded form. Explain that the place-value chart also makes that difference clear by showing a 3 in the tenths column for 0.34 and a 2 in the tenths column for 0.248. Since 3 tenths > 2 tenths, 0.34 > 0.248. GMP7.2

Next have students use the decimal place-value chart to compare 0.248 and 0.24. GMP2.2 Remind them to compare the numbers digit by digit from left to right. GMP7.2 Ask: *What do you notice about the value of the digits in each place?* The ones, tenths, and hundredths are the same. The thousandths are different. There is an 8 in the thousandths place in 0.248 and a 0 in the thousandths place in 0.24, so 0.248 is greater than 0.24. *How could we record this comparison with a number sentence?* 0.248 > 0.24, or 0.24 < 0.248 *What other representations could we use to check that 0.248 is greater than 0.24?* GMP1.4 Expanded form, thousandths grids Have students use expanded form and thousandths grids (*Math Masters,* page TA23) to compare 0.248 and 0.24 and explain their reasoning in terms of place value. GMP7.2

Next, have students compare 1.2 and 0.25 and share their strategies. Encourage them to use different representations and strategies to check their answer. GMP1.4, GMP2.2, GMP7.2 *Sample strategies:*

- Using a place-value chart, the first difference is in the ones column. 1 one > 0 ones, so 1.2 > 0.25.
- Using expanded form, it's clear that (1 ∗ 1) is greater than (0 ∗ 1), so 1.2 > 0.25.
- Using thousandths grids, it would take a whole grid and 2 columns to show 1.2, and less than one grid to show 0.25. There is more shaded for 1.2, so 1.2 > 0.25.

Ask students to think about all of the decimals they have examined so far. Ask: *Which decimal is the greatest: 1.2; 0.25; 0.248; 0.34; or 0.24? How do you know?* 1.2, because it is the only number with something other than zero in the ones place. Explain that it is possible to compare multiple numbers at once by putting them in the **order** in which they are found on a number line—from least to greatest. Have students order the decimals and share their reasoning. 0.24 < 0.248 < 0.25 < 0.34 < 1.2

▸ Introducing *Decimal Top-It*

Student Reference Book, pp. 296–297; *Math Masters,* pp. TA25 and G14

| WHOLE CLASS | SMALL GROUP | PARTNER | INDEPENDENT |

Tell students they are going to learn a new game to practice comparing decimals. Review the directions on *Student Reference Book,* pages 296–297. Play a round with the class. Ask: *What strategies or representations could you use to compare the two decimals?* GMP2.2 Sample answers: shade thousandths grids; write the numbers in expanded form; use the Decimal Place-Value Mat Demonstrate how to record the comparison on the *Top-It* Record Sheet (*Math Masters,* page G14). Then have students play the game in partnerships.

Observe

- What strategies are students using to compare decimals?
- Which students are comparing decimals accurately? Which students could benefit from shading thousandths grids to check their work? GMP2.2

Discuss

- *What strategy did you use to compare the decimals? What other strategy could you use to check your work?* GMP1.4
- *How did you decide where to place your cards? Did you discover any strategies for creating larger numbers?* GMP7.2

Differentiate Game Modifications **Go Online** ▸ Differentiation Support

---

Math Journal 1, p. 121

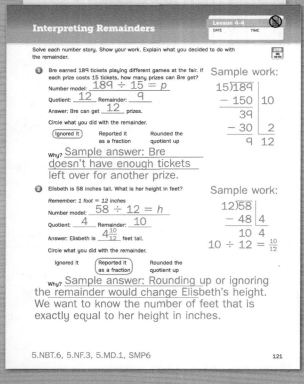

Assessment Check-In 5.NBT.3, 5.NBT.3b

Math Masters, pp. TA25 and G14

Expect most students to correctly compare decimals using the Decimal Place-Value Mat, expanded notation, or a thousandths grid while playing *Decimal Top-It.* GMP2.2 Students should also be able to record decimal comparisons using >, =, or < symbols on the *Top-It* Record Sheet. Some may struggle trying to compare decimals with different numbers of digits or with zero as a digit. Make sure these students are correctly shading thousandths grids or comparing decimals place by place, from left to right, on the Decimal Place-Value Mat. For students ready for a challenge, see the Adjusting the Activity note.

Evaluation Quick Entry Go online to record students' progress and to see trajectories toward mastery for these standards.

Summarize Have students share their favorite method for comparing decimals and explain their preference.

3 Practice 20–30 min

▶ Interpreting Real-World Remainders

Math Journal 1, p. 121

| WHOLE CLASS | SMALL GROUP | PARTNER | INDEPENDENT |

Students practice solving division number stories and decide whether to ignore the remainder, round the quotient up, or report the remainder as a fraction. GMP6.2

▶ Math Boxes 4-4

Math Journal 1, p. 122

| WHOLE CLASS | SMALL GROUP | PARTNER | INDEPENDENT |

Mixed Practice Math Boxes 4-4 are paired with Math Boxes 4-2.

▶ Home Link 4-4

Math Masters, p. 130

Homework Students compare and order decimals to thousandths.

Math Journal 1, p. 122

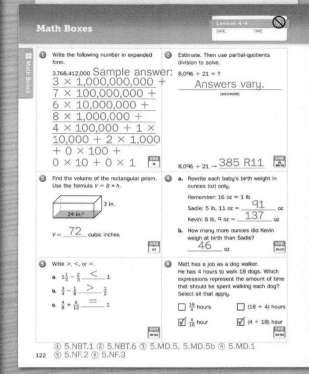

Math Masters, p. 130

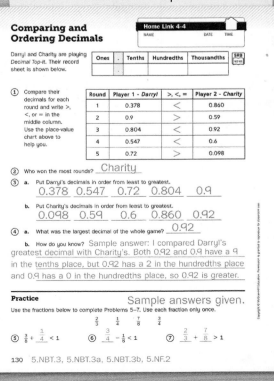

Rounding Decimals

Overview Students use number lines and place-value understanding to round decimals to a given place.

▶ **Before You Begin**
Have copies of hundredths grids, thousandths grids, and blank number lines (*Math Masters,* pages TA22, TA23, and TA27) available for students to use throughout the lesson. To further scaffold rounding with grids, consider using *Math Masters,* page 134. (*See* Adjusting the Activity *on page 359.*)

▶ **Vocabulary**
rounding • round down • round up

Warm Up   5 min

	Materials	**Standards**
		Focus Cluster • Understand the place value system.
Mental Math and Fluency Students write decimals and identify digits in given places.	slate	5.NBT.1, 5.NBT.3, 5.NBT.3a

Focus   35–40 min

Math Message Students use a grid to determine whether 0.28 is closer to 0.2 or 0.3.	*Math Journal 1,* p. 123	5.NBT.1, 5.NBT.3, 5.NBT.3a, 5.NBT.3b SMP2
Rounding Decimals Using Grids and Number Lines Students use grids and number lines to round decimals to the nearest whole, tenth, and hundredth.	*Math Journal 1,* pp. 123–124; *Math Masters,* p. TA27; pp. 134, TA2, TA22, and TA23 (optional)	5.NBT.1, 5.NBT.3, 5.NBT.3a, 5.NBT.3b, 5.NBT.4 SMP2
Developing a Rounding Shortcut Students describe rounding patterns and explain them based on their understanding of place value.	*Math Journal 1,* pp. 124–125; *Math Masters,* pp. 134, TA22, TA23, and TA27 (optional)	5.NBT.1, 5.NBT.3, 5.NBT.3b, 5.NBT.4 SMP2, SMP8
✓ **Assessment Check-In** See page 362. Expect most students to be able to use number lines, grids, or the rounding shortcut to round decimals like those in the problems identified.	*Math Journal 1,* p. 125	5.NBT.4

Practice   20–30 min

Playing *Rename That Mixed Number* **Game** Students practice renaming mixed numbers by trading wholes for fractional parts.	*Student Reference Book,* pp. 321–322; per partnership: *Math Masters,* p. G18; number cards 1–9 (4 of each); timer or stopwatch (optional)	5.NF.3 SMP7
Math Boxes 4-5 Students practice and maintain skills.	*Math Journal 1,* p. 126	See page 363.
Home Link 4-5 **Homework** Students place decimals on a number line and use their understanding of place value to round decimals to the nearest tenth.	*Math Masters,* p. 135	5.NBT.1, 5.NBT.2, 5.NBT.3, 5.NBT.3b, 5.NBT.4 SMP2

Go Online to see how mastery develops for all standards within the grade.

Differentiation Options

Readiness — 5–15 min

WHOLE CLASS · SMALL GROUP · **PARTNER** · INDEPENDENT

Relating Shaded Grids to the Number Line

5.NBT.1, 5.NBT.3, 5.NBT.3a, 5.NBT.3b, SMP2

Math Masters, p. 131

To prepare for identifying values between two decimals, students shade grids to find equivalent names for decimals on *Math Masters*, page 131. Have them shade both a tenths and a hundredths grid to generate equivalent names for tenths. **GMP2.1** Ask them how the hundredths grid helps them think about numbers between each of the given decimals. Sample answer: 0.7 is 0.70 and 0.8 is 0.80. So 0.71 through 0.79 are all in between 0.7 and 0.8. Have students label three values on each number line. **GMP2.3**

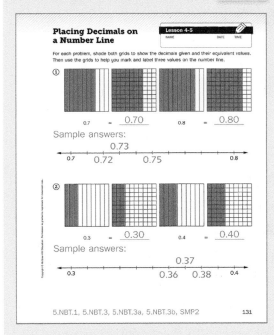

Enrichment — 5–15 min

WHOLE CLASS · SMALL GROUP · **PARTNER** · INDEPENDENT

Rounding Repeating Decimals

5.NBT.1, 5.NBT.4 , SMP7

Math Masters, p. 132

To extend their work rounding decimals, students "unround" rounded calculator displays, identify patterns in repeating decimals, extend those patterns, and re-round the resulting decimal numbers. **GMP7.1**

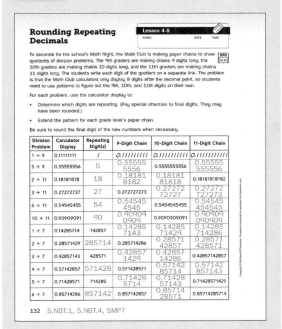

Extra Practice — 5–15 min

WHOLE CLASS · SMALL GROUP · **PARTNER** · INDEPENDENT

Spinning to Round

5.NBT.3, 5.NBT.3a, 5.NBT.4

Activity Card 46;
Math Masters, p. 133;
per partnership: number cards 0–9 (4 of each); paper clip

For more practice rounding decimals, students draw cards to generate a 5-digit number with one digit in the ones place followed by four digits in decimal places. They use a spinner to determine which place value they must round to. Partners check each other's work and read the resulting numbers aloud to each other.

English Language Learner

Beginning ELL To help students understand finding the *closer* or *closest to* number, use think-alouds, role play, and Total Physical Response prompts. For example, stand far away from the door and say: *I am not close to the door. I am far from the door.* Move close to the door and say: *Now I am close to the door.* Move even closer and say: *Now I am closer to the door.* Have students stand at varying distances from the door and make statements like these: _____ *is far from the door.* _____ *is closer to the door.* _____ *is closest to the door.* Scaffold understanding of the term *round* by pairing it with *closer* or *closest*. For example: *Round to the closest tenth.*

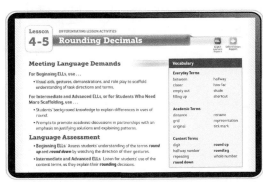

Differentiation Support pages are found in the online Teacher's Center.

Standards and Goals for
Mathematical Process and Practice

SMP2 Reason abstractly and quantitatively.

GMP2.2 Make sense of the representations you and others use.

GMP2.3 Make connections between representations.

SMP8 Look for and express regularity in repeated reasoning.

GMP8.1 Create and justify rules, shortcuts, and generalizations.

Math Journal 1, p. 123

Identifying the Closer Number

Lesson 4-5

DATE   TIME

Math Message

① Shade the grid at the right to show 0.28.

② Is 0.28 closer to 0.2 or 0.3? Use the grid to help you decide. Be ready to explain your reasoning.

0.28 is closer to ___0.3___

③ Label the number line below to show whether 0.28 is closer to 0.2 or 0.3.

0.2 0.21 0.22 0.23 0.24 0.25 0.26 0.27 0.28 0.29 (0.3)

④ Shade the grid at the right to show 0.619. Use the grid to help you solve Problem 5.

⑤ a. Between which two hundredths is 3.619?
   3.619 is between ___3.61___ and ___3.62___

   b. What number is exactly halfway between the numbers you wrote in Problem 5a?
   ___3.615___

⑥ Round 3.619 to the nearest hundredth. Use your answers from Problems 4 and 5 and the number line below to help you.
   ___3.62___

   Round up.

   3.61   3.615   (3.62)
                              3.619

5.NBT.1, 5.NBT.3, 5.NBT.3a, 5.NBT.3b, 5.NBT.4, SMP2   123

---

1 Warm Up   5 min

▶ Mental Math and Fluency

Dictate decimals for students to write on slates. Then have them identify digits in given places. *Leveled exercises:*

● ○ ○ Three and four hundred eighty-two thousandths. Circle the digit in the tenths place. Put an X through the digit in the thousandths place. 3.④8✗

● ● ○ Nine hundred seventeen and two hundred three thousandths. Circle the digit in the hundredths place. Put an X through the digit in the tens place. 9✗72.⓪3

● ● ● Six million three hundred eighty-seven and four hundred twenty thousandths. Circle the digit in the millions place. Put an X through the digit in the hundredths place. ⑥,000,387.4✗0

2 Focus   35–40 min

▶ Math Message

Math Journal 1, p. 123

Solve Problems 1 and 2 on journal page 123. **GMP2.2**

▶ Rounding Decimals Using Grids and Number Lines

Math Journal 1, pp. 123–124; *Math Masters,* p. TA27

| WHOLE CLASS | SMALL GROUP | PARTNER | INDEPENDENT |

Math Message Follow-Up Have students share how they shaded the hundredths grid to show 0.28. Ask: *Is 0.28 closer to 0.2 or 0.3?* 0.3 *How do you know?* Sample answer: I could shade 2 columns to show 0.2 and 3 columns to show 0.3. The shading for 0.28 is closer to 3 full columns than to 2 full columns, so 0.28 is closer to 0.3. Point out that picturing 0.2 and 0.3 on a hundredths grid requires thinking about tenths as equivalent numbers of hundredths. Ask: *How can we rename 0.2 as a number of hundredths?* 0.20 *0.3?* 0.30 Encourage students to use these names to think about the distances between 0.2; 0.28; and 0.3. Ask: *How far is 0.28 from 0.2?* 8 hundredths *How do you know?* Sample answers: I'd have to erase 8 hundredths from 0.28 to show 0.2 on the grid. Thinking in hundredths, 20 hundredths + 8 hundredths = 28 hundredths.

Ask: *How far is 0.28 from 0.3?* 2 hundredths *How do you know?* Sample answers: I'd have to shade 2 more hundredths on the grid to fill in the column and show 0.3. Thinking in hundredths, 28 hundredths + 2 hundredths = 30 hundredths. Summarize by saying that since 0.28 is 2 hundredths from 0.3 and 8 hundredths from 0.2, 0.28 is closer to 0.3 than 0.2. GMP2.2

Tell students that number lines are another representation they can use to find the distance between numbers. Display the number line from Problem 3 on journal page 123. Ask: *How could we show where 0.28 falls on this number line?* Sample answer: Divide the space between 0.2 and 0.3 into 10 equal parts so we can count up by hundredths to 0.28. If students are unsure, remind them that they can think of 0.2 as 0.20, or 20 hundredths, and count up by hundredths until they reach 0.30, or 30 hundredths. Ask: *Could we divide this space into 10 equal parts?* Yes. *How should we label each mark?* 0.21; 0.22; 0.23; and so on Have students draw and label tick marks on the number line in their journals as you do the same on the displayed number line. Once all of the tick marks are labeled, ask students to place a dot at 0.28. GMP2.2, GMP2.3

Ask: *How does this number line show that 0.28 is closer to 0.3 than to 0.2?* Sample answers: The dot for 0.28 is closer to 0.3 than 0.2. There are only 2 hops between 0.28 and 0.3, but there are 8 hops from 0.28 to 0.2. Explain that when they decide which tenth is closest to 0.28, students are **rounding** 0.28 to the nearest tenth. Ask: *Why were we deciding between 0.2 and 0.3? Why didn't we consider whether 0.4 or 0.9 was closer to 0.28?* Sample answer: 0.28 is between 0.2 and 0.3. They are the closest tenths to 0.28. Explain that in real-world contexts, decimals are sometimes shown with more digits than are necessarily useful. Rounding is a way to express decimals with fewer digits and to identify close numbers that are easier to work with or understand. Tell students that today they will use number lines and their understanding of place value to round decimals.

Have students consider how they would round 0.28 to the nearest whole number. Ask: *Between what two whole numbers is 0.28?* 0 and 1 *How do you know?* Sample answer: I thought about a hundredths grid. When we shade 0.28, it's more than 0 but less than 1 whole grid. Draw a number line from 0 to 1. Ask: *How could we use this number line to round 0.28 to the nearest whole number?* Place 0.28 on the number line to see whether it is closer to 0 or 1. If no one mentions it, explain that finding the halfway point between 0 and 1 is a useful way to figure out whether 0.28 is closer to 0 or 1.

Ask: *What number is exactly halfway between 0 and 1?* 0.5 *How do you know?* Sample answer: 1 whole is 10 tenths, or 100 hundredths. 5 tenths, or 50 hundredths, is half of that. Make a tick mark in the middle of the number line and label it 0.5.

Adjusting the Activity

Differentiation This lesson uses number lines to reinforce the meaning of rounding as identifying the closer number. At the same time, labeling a midpoint on number lines helps connect rounding rules to students' understanding of place value. However, some students may reach these same understandings more easily using grids. If you wish for students to have grids available, consider providing copies of *Math Masters,* page 134 for whole-class examples and journal work. You may need to help students choose appropriate grids for specific problems. After choosing a pair of grids, have students shade the decimal they are rounding (the original number) in the grid on the left. They can then use the grid on the right to help them decide how to round.

 Differentiation Support

Academic Language Development

Have partnerships discuss the term *rounding* using the 4-Square Graphic Organizer (*Math Masters,* page TA2) with the following headers: Real-Life Scenario, Number Example, Number Non-Example, and My Definition. Have students explain their organizers to other partnerships.

Have students compare 0.28 to 0.5 using one of the strategies discussed in Lesson 4-4. 0.28 < 0.5 Ask: *Where should we place 0.28 on the number line? Why?* Between 0 and 0.5 because 0.28 is less than 0.5 Place a dot between 0 and 0.5 on the number line and label it 0.28. Note that 0.28 could be placed more accurately if tenths or hundredths were labeled on the number line, but knowing that 0.28 is between 0 and 0.5 is enough information to round to the nearest whole number.

Ask: *Which whole number is closer to 0.28, 0 or 1? Explain.* GMP2.2 0, because it is closer to 0.28 on the number line than 1. Tell students that when they round to a lower number, such as rounding 0.28 to 0, they are **rounding down.** In the Math Message problem, they **rounded up** when they determined that 0.28 was closer to the higher number, 0.3. Reinforce the idea that using a number line and identifying a halfway point make it easier to see whether you should round up or round down.

Present the following problem: *Your parents are at Gus's Gas Station and see that a gallon of gas costs $3.619. They want to estimate how much it will cost to fill up. What is the price of a gallon of gas rounded to the nearest cent, or hundredth of a dollar?*

Ask: *How could a number line help us solve this problem?* Sample answer: If we labeled each end with the hundredths that 3.619 is between and figured out the halfway number, we could see where 3.619 is. If 3.619 is below the halfway number, we round down. If 3.619 is above the halfway number, we round up. Have students shade 0.619 in Problem 4 on journal page 123 to help them identify the nearest hundredths. GMP2.2, GMP2.3 It may help if students think about "emptying out" and "filling up" the partially shaded hundredths square when looking for the nearest hundredths. Ask: *Between which two hundredths is 0.619?* 0.61 and 0.62 *So between which two hundredths is **3**.619?* 3.61 and 3.62 Have students record this information in Problem 5a. Ask: *What number is halfway between 3.61 and 3.62?* 3.615 *How do you know?* Sample answer: If I think of the hundredths as 3.610 and 3.620, I know that 3.615 is in the middle. Tell students to record this information in Problem 5b.

Have students label the number line in Problem 6 and use it to round 3.619. Ask: *What is 3.619 rounded to the nearest hundredth?* 3.62 *Did you round up or round down? Why?* I rounded up because 3.619 is closer to the higher number, 3.62. Return to the problem context, summarizing by saying that to the nearest hundredth of a dollar, the price of a gallon of gas is $3.62.

Next ask students to round the price of gas, $3.619, to the nearest dime, or tenth of a dollar. Distribute blank number lines (*Math Masters*, page TA27) for students to use. Then guide them through the following steps:

- We want to round 3.619 to the nearest tenth. Think: *Between which two tenths is 3.619?* It is between 3.6 and 3.7.
- Label a number line with 3.6 and 3.7 on either end. (*See margin.*)
- Think: *What number is halfway between 3.6 and 3.7?* It's 3.65. Label the tick mark in the middle of the number line as 3.65.
- Compare 3.619 to the halfway point. 3.619 < 3.65, so place 3.619 to the left of 3.65 on the number line. Since 3.619 is closer to 3.6, round down to 3.6. To the nearest tenth of a dollar, the price is $3.60. GMP2.2

Tell students to continue using number lines and identifying halfway points to complete journal page 124. Circulate and observe. If students struggle identifying adjacent whole numbers, tenths, or hundredths, identifying halfway points, or placing numbers on number lines, scaffold by providing hundredths and thousandths grids (*Math Masters,* pages TA22 and TA23) or *Math Masters,* page 134. GMP2.2, GMP2.3

▶ Developing a Rounding Shortcut

Math Journal 1, pp. 124–125

WHOLE CLASS   SMALL GROUP   PARTNER   INDEPENDENT

When most students have completed journal page 124, briefly review answers to Problems 1–6. Once correctly rounded numbers are established, ask: *What do you notice about the halfway numbers?* Sample answer: They all end in 5 and have one more digit than the endpoints. *What do you notice about the numbers we rounded down?* Sample answer: They were less than the halfway number. *What do you notice about the numbers we rounded up?* Sample answer: They were greater than or equal to the halfway number. Explain that when a number is exactly halfway between two other numbers, the convention is to round that number up. Guide a discussion to help students see the following patterns related to the digit to the right of the place they rounded to:

- When the digit to the right is 4 or less, we round down.
- When the digit to the right is 5 or greater, we round up.

Challenge students to explain why these patterns make sense. GMP8.1 Ask: *Why does it make sense to round down if the digit to the right is less than 5?* Sample answer: We can think of the space between the lower and the higher number as 10 equal hops on a number line. 5 is half of 10. If the digit is 4 or less, that's only 4 hops from the lower number, so it's less than halfway. *Why does it make sense to round up if the digit to the right is 5 or greater?* Sample answer: The number being rounded is at least 5 hops from the lower number, so it's halfway or more than halfway to the higher number.

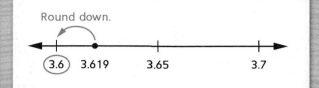

Using a number line to round 3.619 to the nearest tenth

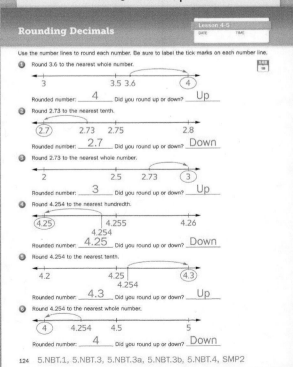

Math Journal 1, p. 124

Rounding Decimals in Real-World Contexts

Lesson 4-5

DATE   TIME

Read about different real-world situations below and round the decimals as directed. Use a number line or grids to help you, if needed.

① At the district track meet each running event is timed to the nearest thousandth of a second using an electronic timer. However, the district's track rules require times to be reported with only 2 decimal places. Round each time to the nearest hundredth of a second.

Note: sec = second(s)

Electronic Timer	Reported Time	Electronic Timer	Reported Time
a. 10.752 sec	10.75 sec	**b.** 55.738 sec	55.74 sec
c. 16.815 sec	16.82 sec	**d.** 43.505 sec	43.51 sec
e. 20.098 sec	20.10 sec	**f.** 52.996 sec	53.00 sec

Explain how you rounded 20.098 to the nearest hundredth. Sample answer: I knew that 8 thousandths was more than halfway to the next hundredth, so I decided to round up. Rounding 9 hundredths up meant having 10 hundredths, which is equal to 1 tenth, so I wrote 20.10.

② Supermarkets often show unit prices for items. This helps customers compare prices to find the best deal. A unit price is found by dividing the price of an item (in cents or dollars and cents) by the quantity of the item (often in ounces or pounds). When the quotient has more decimal places than are needed, some stores round to the nearest tenth of a cent.

Example: A 16 oz container of yogurt costs $3.81.
- $3.81 * 100 cents per dollar = 381¢
- 381¢ ÷ 16 oz = 23.8125¢ per ounce
- 23.8125¢ is rounded down to 23.8¢ per ounce

Round each unit price to the nearest tenth of a cent.

a. 28.374¢   28.4 ¢   b. 19.756¢   19.8 ¢
c. 16.916¢   16.9 ¢   d. 20.641¢   20.6 ¢
e. 18.459¢   18.5 ¢   f. 21.966¢   22.0 ¢

5.NBT.1, 5.NBT.4   125

Say: *It seems that we could use these patterns as a shortcut for rounding. Let's test them with the numbers from our gasoline problem to see whether we get the same result.* Guide students to apply the shortcut as shown below. **GMP8.1**

Round $3.619 to the nearest hundredth of a dollar.

Underline the digit in the place we are rounding to: 1.	3.6<u>1</u>9
We will either round down to 3.61 or up to 3.62.	↓
Look at the digit to the right: 9.	3.62
Think: *Is 9 greater than or less than 5?*	
9 > 5, so round up to 3.62.	
To the nearest hundredth of a dollar, the price is $3.62. This is the result we found using a number line.	

Round $3.619 to the nearest tenth of a dollar.

Underline the digit in the place we are rounding to: 6.	3.<u>6</u>19
We will either round down to 3.6 or up to 3.7.	↓
Look at the digit to the right: 1.	3.6
Think: *Is 1 greater than or less than 5?*	
1 < 5, so round down to 3.6.	
To the nearest tenth of a dollar, the price is $3.60. This is the result we found using a number line.	

Point out that using the patterns resulted in the same rounded value as using the number line. Have students use the strategy of their choice to complete journal page 125. Expect most to gravitate toward the rounding shortcut, but provide additional copies of *Math Masters*, pages 134, TA22, TA23, or TA27 as needed. Circulate and assist.

 Assessment Check-In   5.NBT.4

Math Journal 1, p. 125

Expect most students to be able to round the decimals in Problems 1a–1d on journal page 125 using number lines, grids, or the rounding shortcut. Since this is the first introduction to rounding decimals, students may struggle with Problems 1e and 1f, for which they have to consider how other places change when rounding to the nearest hundredth. Some may be able to clearly explain how number lines can be used to help them round. Help students who struggle by reasoning through the steps involved in rounding with grids and number lines.

 Evaluation Quick Entry   Go online to record students' progress and to see trajectories toward mastery for these standards.

Summarize Have students share how they rounded 20.098 to the nearest hundredth. Ask: *Would it make sense to round to 20.01? Why or why not?* No. 9 hundredths rounds up to 10 hundredths, or 1 tenth, not 1 hundredth. Have volunteers defend their reasoning using a grid or number line. GMP2.2, GMP2.3

3 Practice 20–30 min

▶ Playing *Rename That Mixed Number*

Student Reference Book, pp. 321–322; *Math Masters*, p. G18

| WHOLE CLASS | SMALL GROUP | **PARTNER** | INDEPENDENT |

Students practice renaming mixed numbers in multiple ways.

Observe

• What strategies are students using to rename mixed numbers?
• Which students are able to identify all names (with the same denominator) for their mixed number?

Discuss

• *How did you find multiple names for your mixed number?*
• *Did you notice any patterns in the names you found for each mixed number? Explain.* GMP7.1

| Differentiate | **Game Modifications** | Go Online | Differentiation Support |

▶ Math Boxes 4-5

Math Journal 1, p. 126

| WHOLE CLASS | SMALL GROUP | PARTNER | INDEPENDENT |

Mixed Practice Math Boxes 4-5 are paired with Math Boxes 4-7.

▶ Home Link 4-5

Math Masters, p. 135

Homework Students place decimals on a number line and round decimals to the nearest tenth. GMP2.2

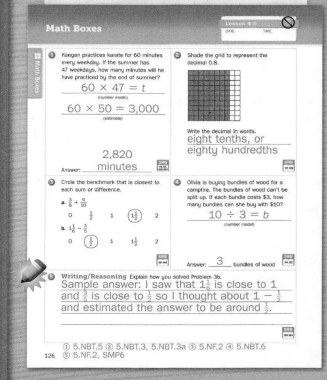

Math Journal 1, p. 126

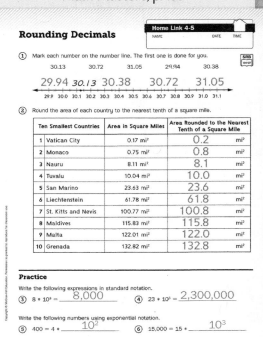

Math Masters, p. 135

Introduction to the Coordinate System

Overview Students are introduced to the coordinate grid and use ordered pairs to plot and identify points.

▶ **Before You Begin**

For the Math Message consider using a map of your city or state with an index of locations in letter-number form. For the optional Readiness activity, gather maps. For the optional Enrichment and Extra Practice activities, have extra copies of blank grids (*Math Masters*, page TA28) available.

▶ **Vocabulary**

perpendicular • coordinate grid • intersect • origin • axes • *x*-axis • *y*-axis • coordinates • plot • *x*-coordinate • *y*-coordinate • ordered pair

Standards

Focus Cluster
• Graph points on the coordinate plane to solve real-world and mathematical problems.

---

① Warm Up   5 min

	Materials	
Mental Math and Fluency Students solve fraction-of problems.	slate	5.NF.4, 5.NF.4a

---

② Focus   35–40 min

Math Message Students locate a city on a map using a letter-number pair.	*Math Journal 1*, p. 127	5.G.1 SMP2
Plotting Points on a Coordinate Grid Students are introduced to the coordinate grid and use ordered pairs to plot points.	*Math Journal 1*, pp. 127–128; *Student Reference Book*, p. 275; *Math Masters*, p. TA28; straightedge	5.G.1 SMP2, SMP6
✓ **Assessment Check-In** See page 368. Expect most students to be able to plot points in the first quadrant of a coordinate grid with occasional reminders about which direction to move first.	*Math Journal 1*, p. 128	5.G.1, SMP2
Introducing *Over and Up Squares* **Game** Students practice using ordered pairs to plot points on a coordinate grid.	*Student Reference Book*, p. 317; per partnership: *Math Masters*, p. G25; colored pencils; two 6-sided dice	5.G.1 SMP2, SMP6

---

③ Practice   20–30 min

Finding Volumes of Soil Students practice finding volumes of rectangular prisms.	*Math Journal 1*, p. 129	5.MD.3, 5.MD.5, 5.MD.5a, 5.MD.5b, 5.MD.5c, SMP1, SMP4
Math Boxes 4-6 Students practice and maintain skills.	*Math Journal 1*, p. 130	See page 369.
Home Link 4-6 **Homework** Students plot points on a coordinate grid.	*Math Masters*, p. 139	5.NBT.1, 5.NBT.3, 5.NBT.3a, 5.G.1 SMP2

 (Go Online) to see how mastery develops for all standards within the grade.

Differentiation Options

Readiness 10–15 min	**Enrichment** 20–30 min	**Extra Practice** 15–30 min
WHOLE CLASS · SMALL GROUP · PARTNER · INDEPENDENT	WHOLE CLASS · SMALL GROUP · PARTNER · INDEPENDENT	WHOLE CLASS · SMALL GROUP · PARTNER · INDEPENDENT

Readiness

Exploring Map Features

5.G.1, SMP2

per partnership: several
road maps or political maps

To prepare for working with coordinate systems, students explore a variety of maps. Provide maps, including ones with a location index in letter-number form (such as, B-3 or F-9). Students examine the maps and discuss interesting features. Call attention to the legend, the scale, the compass rose, the layout of grid lines, the letters and numbers along the sides of the map, and the locations listed with letter-number pairs. Contrast different formats, pointing out that while not all maps have the same features, most are designed to help the reader find useful information. GMP2.2 Have students practice using a location index with letter-number pairs to narrow their search for a specific location.

Enrichment

Creating Designs with Decimal Coordinates

5.G.1, SMP2, SMP6

Math Masters, pp. 137–138,
p. TA28 (optional); straightedge

To further explore coordinate grids, students learn about decimal coordinates on *Math Masters*, page 137. (You may wish to guide students as they study the page.) They then create designs by plotting points with whole-number and decimal coordinates on *Math Masters*, page 138.
GMP2.1, GMP6.4 Provide blank coordinate grids (*Math Masters*, page TA28) for students who wish to complete the Try This activity.

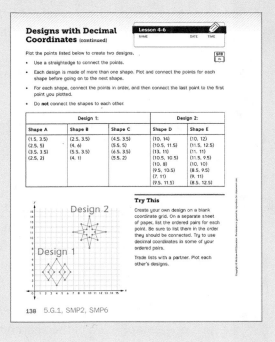

Extra Practice

Plotting Your Initials

5.G.1, 5.G.2, SMP4, SMP6

Activity Card 47;
Math Masters, p. 136, p. TA28 (optional);
straightedge

For more practice plotting points on a coordinate grid, students write ordered pairs and plot points, which they connect to form their initials. GMP4.1, GMP6.4

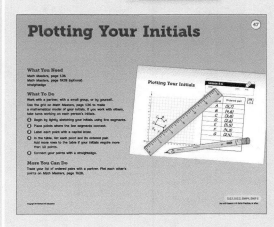

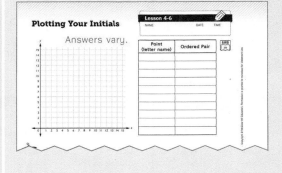

English Language Learner

Beginning ELL Introduce lesson vocabulary to students by preparing an anchor chart titled "Coordinate Grid" and vocabulary cards with the words *coordinate grid, ordered pair, origin, y-axis, x-axis, x-coordinate, y-coordinate,* and *intersect.* Draw a coordinate grid on the chart and use a think-aloud to name different features of the grid as you place the vocabulary cards in the appropriate places. For example, place the vocabulary card for *origin* at (0, 0) as you pronounce the term. Have students repeat it. After modeling all the terms, display another coordinate grid and another set of vocabulary cards. Direct individual students to label the features you name. Encourage them to say the names of the features.

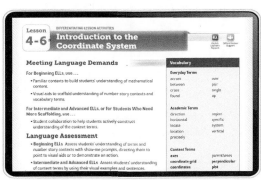

Differentiation Support pages are found in the online Teacher's Center.

Standards and Goals for
Mathematical Process and Practice

SMP2 **Reason abstractly and quantitatively.**

GMP2.1 Create mathematical representations using numbers, words, pictures, symbols, gestures, tables, graphs, and concrete objects.

GMP2.2 Make sense of the representations you and others use.

SMP6 **Attend to precision.**

GMP6.4 Think about accuracy and efficiency when you count, measure, and calculate.

Math Journal 1, p. 127

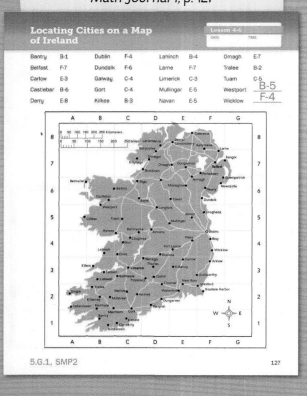

Locating Cities on a Map of Ireland — Lesson 4-6

Bantry	B-1	Dublin	F-4	Lahinch	B-4	Omagh	E-7
Belfast	F-7	Dundalk	F-6	Larne	F-7	Tralee	B-2
Carlow	E-3	Galway	C-4	Limerick	C-3	Tuam	C-5
Castlebar	B-6	Gort	C-4	Mullingar	E-5	Westport	B-5
Derry	E-8	Kilkee	B-3	Navan	E-5	Wicklow	F-4

5.G.1, SMP2   127

---

1 Warm Up   5 min

▶ Mental Math and Fluency

Display fraction-of problems and have students write answers on slates. *Leveled exercises:*

●○○ $\frac{1}{2}$ of 8 4	$\frac{1}{2}$ of 20 10
●●○ $\frac{1}{4}$ of 12 3	$\frac{1}{4}$ of 20 5
●●● $\frac{1}{3}$ of 10 $\frac{10}{3}$, or $3\frac{1}{3}$	$\frac{1}{4}$ of 5 $\frac{5}{4}$, or $1\frac{1}{4}$

2 Focus   35–40 min

▶ Math Message

Math Journal 1, p. 127

Turn to journal page 127. Find the city of Tralee on the map. Be prepared to explain how you found it. **GMP2.2**

> **NOTE** If you are using a map of your city or state instead of the journal page, adjust the Math Message and discussion accordingly. (*See Before You Begin.*)

▶ Plotting Points on a Coordinate Grid

Math Journal 1, pp. 127–128; Student Reference Book, p. 275; Math Masters, p. TA28

WHOLE CLASS	SMALL GROUP	PARTNER	INDEPENDENT

Math Message Follow-Up Point out that the map on journal page 127 shows the island of Ireland. Ask students to share how they located Tralee. Be sure the strategies shared include using the index of locations to find that Tralee is located in square B-2. Point out how column B and row 2 overlap at a square region that includes Tralee.

Ask: *Why is a letter-number pair like B-2 useful for locating places on a map?* It limits the search to a small part of the map. Have students use the index to find other locations on the map. Then point out the locations of Westport (B-5) and Wicklow (F-4) without stating the letter-number pairs. Have students locate the cities and complete the index with the letter-number pair for each city. **GMP2.2**

Point out that in the Math Message students used letter-number pairs to locate *regions* on a map. Tell them that today they will learn how to use ordered number pairs to locate specific *points* on a grid.

Display a number line with points *A*, *B*, and *C* marked at 8, 3, and 6, respectively, as shown below:

Ask: *How could you describe the location of points* A, B, *and* C *on the number line?* **GMP2.2** Point *A* is at 8; point *B* is at 3; point *C* is at 6.

Dictate a few other numbers like 2.5, 4, and $7\frac{1}{2}$ for volunteers to plot and label on the number line. Ask: *Is there a point on the number line for any number you can name? Why or why not?* Yes, because the line goes on forever.

Next label point *D above* the 4 on the number line. (*See margin.*) Ask students to describe the location of point *D*. Expect them to say that the point is above 4, or some number of inches or centimeters above 4. Explain that there is a way to precisely locate *any* point in a 2-dimensional space—even a point that is not on the number line—by drawing two number lines that are **perpendicular,** or at right angles to each other, to form a grid. This kind of grid is called a **coordinate grid.** Display a grid from *Math Masters,* page TA28 and have students examine the grid in Problem 1 on journal page 128. Emphasize how, unlike the map they used in the Math Message, the number lines are labeled *on* the tick marks, not in the space *between* tick marks. The number lines cross, or **intersect,** at the 0 point of each line. This point is called the **origin** of the grid. The number lines in a coordinate grid are called the **axes** of the grid. The horizontal number line is the **x-axis** and the vertical number line is the **y-axis.** Point out the *x* and *y* labels on the axes on the journal page. Have students identify and label the origin on the grid in Problem 1, as you do the same on the display grid.

Explain that locations of points on a coordinate grid are described by pairs of numbers written inside parentheses, such as (3, 4) or (0, 6). The numbers inside the parentheses are called the **coordinates** of the point. Marking a point on a coordinate grid is called **plotting** a point.

Remind students how they used letter-number pairs to locate a map region in the Math Message. Point out the difference between the letter-number pairs used on maps and the ordered pairs used on a coordinate grid, which consist of two numbers. Have students discuss with a partner how they might use the number pair (3, 4) to locate a point on the grid in Problem 1. Invite them to share their reasoning. Then explain and demonstrate how to plot the point, covering the ideas listed on the next page. **GMP2.1**

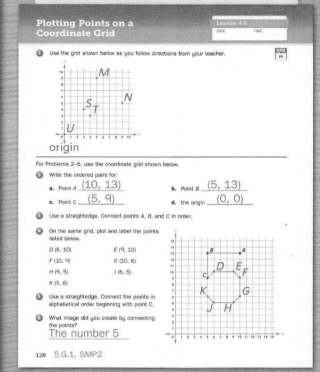

Plotting Points on a Coordinate Grid

128 5.G.1, SMP2

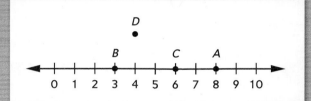

NOTE The phrase *over and up* applies to the first quadrant, but it does not apply to all four quadrants of the coordinate grid. In later grades, when students work with negative numbers in 4-quadrant grids, *over and down* will also apply. In addition to using the terms *over* and *up,* be sure to emphasize the terms *horizontal* and *vertical* when discussing distance from the origin, as these terms will continue to apply in students' work with coordinate grids in later grades.

- The first coordinate is called the **x-coordinate** because it gives the distance traveled from the origin horizontally, or in the direction of the *x*-axis.
- The second coordinate is called the **y-coordinate** because it gives the distance traveled vertically, or in the direction of the *y*-axis.
- Suggest that students think of these directions as *over and up*.
- Model how to plot point (3, 4): Start at 0. Move *over* to the right to locate the 3 on the *x*-axis and then go *up* 4. Draw a dot and label it *S*. Have students plot this point on their grid.
- Emphasize that coordinates describe a precise location—a single *point* on the grid. Contrast this to the work with letter-number pairs on a map, which defined a *region* on the map.

Next ask students to use the same grid to plot a point with coordinates (4, 3) and label it *T*. Ask: *Are points* S *and* T *plotted in the same location?* No. *Does the order of the numbers in parentheses matter?* Yes. *Why?* **GMP2.2** The first and second numbers give different information: the first coordinate tells how far over on the *x*-axis; the second coordinate tells how far up on the *y*-axis.

Explain that pairs of numbers used to locate points on a coordinate grid are called **ordered pairs,** and as the name implies, the order of the numbers is very important. Point out the comma separating the numbers in an ordered pair. Have students practice plotting and labeling a few more points on the grid in Problem 1. *Suggestions: M* (5, 9); *N* (9, 5); *U* (0, 0). **GMP2.1, GMP6.4** When most students seem comfortable plotting ordered pairs, have them look at the grid on the lower half of the journal page and complete Problems 2 and 3 as a class.

Explain that coordinate systems have many real-world applications. For example, many computer graphics programs use a coordinate system to tell the computer exactly what should appear at each point on the screen to create a desired image.

Have students work in partnerships to complete Problems 4–6. Encourage them to refer to *Student Reference Book,* page 275 for more information as they work. Circulate and assist. Ask: *What image did you create by connecting the points?* **GMP2.1** The number 5

✓ Assessment Check-In 5.G.1

Math Journal 1, p. 128

Expect most students to be able to accurately plot the points listed in Problem 4 on journal page 128, with occasional reminders about which direction to move first. **GMP2.1** If students struggle associating each of the coordinates in the ordered pair with the correct axis, consider using the suggestions in the Adjusting the Activity note.

Evaluation Quick Entry Go online to record students' progress and to see trajectories toward mastery for these standards.

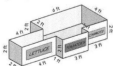

▶ Introducing *Over and Up Squares*

Student Reference Book, p. 317; *Math Masters*, p. G25

| WHOLE CLASS | SMALL GROUP | **PARTNER** | INDEPENDENT |

Demonstrate a sample round of *Over and Up Squares* as you go over the directions on *Student Reference Book*, page 317. Then have students play the game in partnerships.

Observe
- Which students need additional support to play the game?
- Which students can accurately plot points on the grid? **GMP2.1, GMP6.4**

Discuss
- *How did you decide whether the numbers you rolled should be the first or second number in the ordered pair?*
- *Which two numbers do you hope you'll roll on your next turn? Why?*

| **Differentiate** | **Game Modifications** | **Go Online** | 👥 Differentiation Support |

Summarize Have students explain to a partner how they remember which number in an ordered pair corresponds to which axis.

---

③ Practice 20–30 min

▶ Finding Volumes of Soil

Math Journal 1, p. 129

| WHOLE CLASS | SMALL GROUP | **PARTNER** | **INDEPENDENT** |

Students practice finding volumes of real-world rectangular prisms and composite figures. **GMP1.6, GMP4.2**

▶ Math Boxes 4-6

Math Journal 1, p. 130

| WHOLE CLASS | **SMALL GROUP** | **PARTNER** | **INDEPENDENT** |

Mixed Practice Math Boxes 4-6 are paired with Math Boxes 4-8.

▶ Home Link 4-6

Math Masters, p. 139

Homework Students plot points on a coordinate grid to create an outline map of the United States. **GMP2.1**

Math Journal 1, p. 130

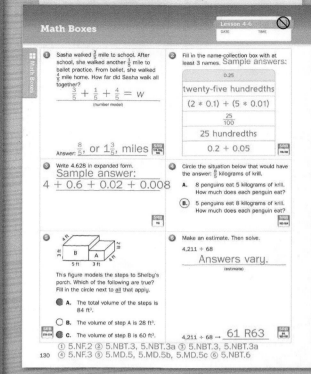

Math Masters, p. 139

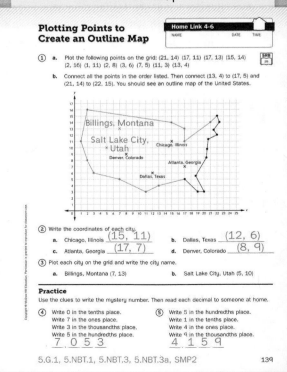

Playing *Hidden Treasure*

Overview Students play a game to practice plotting points on a coordinate grid.

▶ **Before You Begin**
For the optional Readiness activity, you will need about 15 feet of masking tape and additional masking tape in two colors, or colored paper. For the optional Enrichment activity, each small group will need a political map of the world showing latitude and longitude. For the optional Extra Practice activity, have additional copies of *Math Masters*, page 140 on hand.

▶ **Vocabulary**
ordered pair • coordinates

Standards

Focus Cluster
• Graph points on the coordinate plane to solve real-world and mathematical problems.

	Materials	
① Warm Up 5 min		
Mental Math and Fluency Students order decimals from least to greatest.	slate	5.NBT.3, 5.NBT.3b

② Focus 35–40 min		
Math Message Students find the number of blocks between two locations on a map.	*Math Journal 1*, p. 131	5.G.1, 5.G.2 SMP4
Showing Routes on a Coordinate Grid Students use a coordinate grid to find the number of blocks between two locations and use clues to find a location.	*Math Journal 1*, p. 131	5.G.1, 5.G.2 SMP4, SMP6
Introducing *Hidden Treasure* **Game** Students practice plotting points on a coordinate grid and reason about the location of a hidden point.	*Student Reference Book*, p. 311; *Math Masters*, p. G26; per partnership: red pen or crayon	5.G.1, 5.G.2 SMP6
✓ **Assessment Check-In** See page 375. Expect most students to be able to record their guesses as a point labeled with an ordered pair and to report how many square sides the other player's guess is from the hidden point as they play the game *Hidden Treasure*.	*Student Reference Book*, p. 275 (optional); *Math Masters*, p. G26	5.G.1, 5.G.2

③ Practice 20–30 min		
Practicing U.S. Traditional Multiplication Students solve problems using U.S. traditional multiplication.	*Math Journal 1*, p. 132	5.NBT.5 SMP3, SMP6
Math Boxes 4-7 Students practice and maintain skills.	*Math Journal 1*, p. 133	See page 375.
Home Link 4-7 **Homework** Students play a game to practice plotting points and solving problems on a coordinate grid.	*Math Masters*, p. 143	5.NBT.3, 5.NBT.3a, 5.G.1, 5.G.2

 Go Online to see how mastery develops for all standards within the grade.

Differentiation Options

Readiness 10–15 min

WHOLE CLASS | SMALL GROUP | PARTNER | INDEPENDENT

Plotting People and Objects on a Floor Grid

5.G.1, SMP2

per group: about 15 feet of masking tape, two different colors of paper or masking tape, markers

For experience with a coordinate system using a concrete model, students help create a large 6-by-6 coordinate grid on the floor. Use masking tape to make an x-axis and a y-axis on the floor. (If your floor is tile, use the tile edges as grid lines.) Label the axes with whole numbers using different-colored pieces of paper or tape for each axis. Practice plotting and naming points on the grid. **GMP2.1** *Suggestions:*

- Students "plot" themselves on the grid, following directions such as: *Start at the origin. Move along the x-axis to the 4. Then move up 3. What are your coordinates?* (4, 3)
- Students place small objects at different locations on the grid. They ask classmates to report the objects' locations as ordered pairs.

Enrichment 10–20 min

WHOLE CLASS | SMALL GROUP | PARTNER | INDEPENDENT

Using Latitude and Longitude

5.G.1, 5.G.2, SMP4

Activity Card 49; *Student Reference Book*, pp. 276–277; per partnership: *Math Masters*, pp. 141–142; a world political map or globe showing latitude and longitude; coin; scissors

To extend their understanding of coordinate systems, students read *Student Reference Book*, pages 276 and 277 to learn about latitude and longitude. Students record locations on *Math Masters*, page 142 generated from Latitude and Longitude Cards (*Math Masters*, page 141). They use a world political map to find each location. Encourage students to discuss how using latitude and longitude coordinates is similar to their recent work with coordinate grids. **GMP4.2**

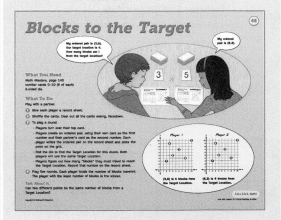

Extra Practice 15–20 min

WHOLE CLASS | SMALL GROUP | PARTNER | INDEPENDENT

Playing *Blocks to the Target*

5.G.1, 5.G.2, SMP2

Activity Card 48; per partnership: *Math Masters*, p. 140; number cards 0–10 (8 of each); 6-sided die

For more practice plotting points on a coordinate grid and comparing the location of two points, students play *Blocks to the Target*. Students record and plot ordered pairs they generate from number cards on *Math Masters*, page 140 and then find the number of blocks it is to a target location. **GMP2.2**

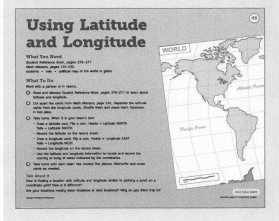

English Language Learner

Beginning ELL To prepare students for playing *Hidden Treasure*, demonstrate the meaning of *hidden* by showing an object and covering it while saying: *Now I see it. Now I don't see it.* Show a picture of a treasure chest to introduce the term *treasure*. Use Total Physical Response routines to let students see and rehearse the action of hiding.

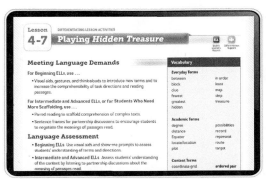

Differentiation Support pages are found in the online Teacher's Center.

Standards and Goals for
Mathematical Process and Practice

SMP4 **Model with mathematics.**
GMP4.1 Model real-world situations using graphs, drawings, tables, symbols, numbers, diagrams, and other representations.
GMP4.2 Use mathematical models to solve problems and answer questions.

SMP6 **Attend to precision.**
GMP6.1 Explain your mathematical thinking clearly and precisely.

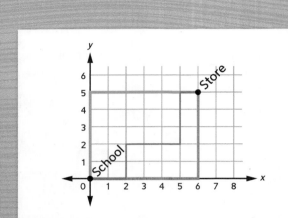

3 possible paths

Math Journal 1, p. 131

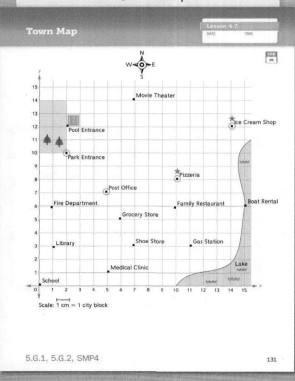

5.G.1, 5.G.2, SMP4   131

1 Warm Up   5 min

▶ Mental Math and Fluency

Have students order the following decimals from least to greatest and record them in order on slates. *Leveled exercises:*

◉○○ 1.83, 0.13, 0.79   0.13, 0.79, 1.83
0.67, 0.24, 0.89   0.24, 0.67, 0.89

◉◉○ 0.3, 0.15, 0.09   0.09, 0.15, 0.3
0.606, 0.64, 0.618   0.606, 0.618, 0.64

◉◉◉ 0.074, 0.05, 0.061   0.05, 0.061, 0.074
0.1, 0.008, 0.03   0.008, 0.03, 0.1

2 Focus   35–40 min

▶ Math Message

Math Journal 1, p. 131

Look at the town map on journal page 131. The grid lines represent the roads in the town. Ben is going to walk from school to the grocery store to buy an after-school snack. What is the fewest number of blocks Ben will have to walk to get to the grocery store? 11 blocks *Be prepared to explain your answer.*   GMP4.2

▶ Showing Routes on a Coordinate Grid

Math Journal 1, p. 131

| WHOLE CLASS | SMALL GROUP | PARTNER | INDEPENDENT |

Math Message Follow-Up Display journal page 131. Make sure students understand that each side of a grid square represents one block. Have students show Ben's route on the map as they explain their thinking.   GMP4.2, GMP6.1   Expect students to describe a variety of different paths. (*See examples in the margin.*) Have them verify that each route described is a total of 11 blocks. Ask students to name the **ordered pairs,** or **coordinates,** that describe the location of the school and the grocery store. School: (0, 0); grocery store: (6, 5) Connect the example of Ben walking a given number of blocks with the ideas about coordinates and distance discussed in Lesson 4-6. Emphasize the fact that the first number in the ordered pair represents how far to travel in the direction of the *x*-axis—in this case, 6 blocks. The second number in the ordered pair represents how far to travel in the direction of the *y*-axis—in this case, 5 blocks.

Ask: *Why do you think all of the routes are 11 blocks in length, even though they are different?* Sample answer: Ben needs to travel 6 blocks in one direction and 5 blocks in another. He needs to go a total of 11 blocks no matter which way he walks. Tell students that today they will find the fewest number of blocks between other locations on the map and learn a new game in which they use distance clues to locate a hidden point on a coordinate grid.

Have students find the fewest number of blocks between other locations on the map. GMP4.2 Students should also record the ordered pair that describes the location of each place. GMP4.1 *Suggestions:*

- What is the fewest number of blocks you would walk from the park entrance (2, 10) to the pizzeria (10, 8)? 10 blocks
- What is the fewest number of blocks you would walk from the library (1, 3) to the ice cream shop (14, 12)? 22 blocks
- What is the fewest number of blocks you would walk from the school (0, 0) to the pool entrance? (2, 12)? 14 blocks
- Which is a longer walk, going from the family restaurant (10, 6) to the boat rental (15, 6), or going from the post office (5, 7) to the fire department (1, 6)? Both walks are the same length: 5 blocks.

Have students find the movie theater and give its location coordinates. (7, 14) Tell them that you are going to give clues to help them find a friend who is somewhere on the map.

Clue #1: *You are at the movie theater. Your friend is at another location on the map that is 9 blocks away. Where could your friend be? Circle the possibilities on your map.* The friend could be at the park entrance, post office, pizzeria, or ice cream shop. Have students explain how they determined the possible locations.

Clue #2: *Your friend's location is also 7 blocks from the boat rental. Put a star next to any location that fits both clues.* Pizzeria, ice cream shop Again have students share why they chose those locations. Ask: *The gas station is 7 blocks from the boat rental. Could your friend be at the gas station? Why or why not?* No. The gas station is 7 blocks from the boat rental, but it is more than 9 blocks from the movie theater.

Clue #3: *Your friend is 6 blocks from the gas station. Where is your friend? What are the coordinates of his or her location?* Pizzeria (10, 8) *How do you know?* GMP4.2, GMP6.1 The ice cream shop is 12 blocks from the gas station, not 6, so it doesn't fit the last clue.

Professional Development

There are two common ways to look at distances in the coordinate plane. One type of distance is *practical distance*. On a coordinate grid, the practical distance between two points is along horizontal and vertical grid lines. Practical distance is sometimes called *taxicab distance* because it measures the distance along a route a taxicab might take.

The other type of distance is *straight-line distance*. On a coordinate plane the straight-line distance between two points is the length of the line segment that connects the points.

In Grade 5 students explore practical distance only. They will explore straight-line distance in later grades.

Academic Language Development

Extend students' understanding of the meaning of the verb and noun forms of the term *coordinate* as bringing together or synchronizing, by using statements like these: *Let's coordinate our plans so that we get there at the same time. Where shall we meet? What are our coordinates?* Lead a discussion about such contexts for students to build their own understanding of the term as bringing multiple things together at one point.

▶ Introducing *Hidden Treasure*

Student Reference Book, p. 311; *Math Masters,* p. G26

| WHOLE CLASS | SMALL GROUP | PARTNER | INDEPENDENT |

Tell students that *Hidden Treasure* is a game similar to the activity they just completed when they used clues to find the location of a friend on a map. Instead of finding a friend, each player will now find the location of a hidden point on another player's coordinate grid.

Go over the rules on *Student Reference Book,* page 311. Display the gameboard on *Math Masters,* page G26 and play a sample round to demonstrate how to complete the grids and how to respond to a player's guesses. Emphasize the importance of saying the ordered pair slowly and carefully, including saying "comma" between the coordinates. Remind students that a gameboard consists of *two* grids.

Have partners play two or more games. Have extra gameboards available. Circulate and assist.

Observe
- Which students accurately record their guesses as a point labeled with the ordered pair?
- Which students use the other player's feedback to make strategic guesses?

Discuss
- *How did you use your grid to decide which ordered pair to guess?*
- *After your partner names an ordered pair, how do you figure out how many "square sides" there are between the guessed point and the hidden point?* GMP6.1

Differentiate **Game Modifications**   **Go Online** Differentiation Support

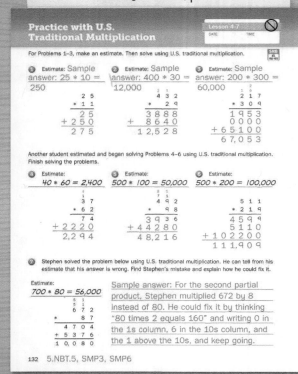

Assessment Check-In 5.G.1, 5.G.2

Math Masters, p. G26

Expect most students to be able to accurately record their guesses on Grid 2 of the gameboard as they play *Hidden Treasure*. Most should also be able to use Grid 1 to report how many square sides the other player must travel to reach the hidden point. Some students may continue to confuse the *x*- and *y*-coordinates when naming the points they plotted. Direct them to *Student Reference Book*, page 275 or to an example of an ordered pair plotted on a classroom display. (See the *Adjusting the Activity* note in Lesson 4-6.)

 Evaluation Quick Entry Go online to record students' progress and to see trajectories toward mastery for these standards.

Summarize Give partnerships 1 minute to find as many ordered pairs as they can that are 10 square sides from the origin, (0, 0). Tell them to use a *Hidden Treasure* gameboard to record their findings. As time permits, students can share their responses and their thinking. GMP6.1 Sample answers: (1, 9); (2, 8); (3, 7); (7, 3)

Math Journal 1, p. 133

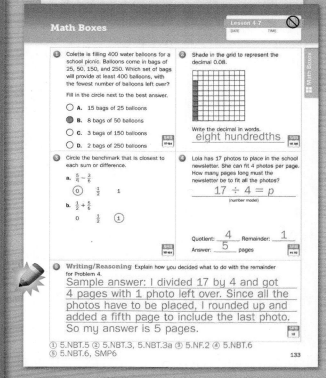

3 Practice 20–30 min

▶ Practicing U.S. Traditional Multiplication

Math Journal 1, p. 132

| WHOLE CLASS | SMALL GROUP | **PARTNER** | **INDEPENDENT** |

Students practice multiplying using U.S. traditional multiplication. They complete work that has already been started and identify a mistake in someone else's work. GMP3.2, GMP6.4

▶ Math Boxes 4-7

Math Journal 1, p. 133

| WHOLE CLASS | **SMALL GROUP** | PARTNER | **INDEPENDENT** |

Mixed Practice Math Boxes 4-7 are paired with Math Boxes 4-5.

▶ Home Link 4-7

Math Masters, p. 143

Homework Students play a game to practice plotting points and solving problems on a coordinate grid.

Math Masters, p. 143

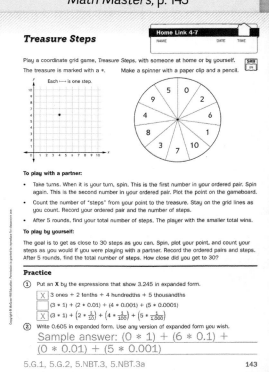

Lesson 4-8

Solving Problems on a Coordinate Grid, Part 1

Overview Students represent mathematical problems on a coordinate grid by plotting points to form pictures and applying rules to ordered pairs.

▶ **Vocabulary**

x-axis • *y*-axis • conjecture

Standards

Focus Clusters
- Apply and extend previous understandings of multiplication and division.
- Graph points on the coordinate plane to solve real-world and mathematical problems.

---

① Warm Up 5 min

	Materials	
Mental Math and Fluency Students estimate fraction sums and differences.		5.NF.2

---

② Focus 35–40 min

	Materials	
Math Message Students plot and connect points to form a picture.	*Math Journal 1*, pp. 134–135; straightedge (optional)	5.G.1, 5.G.2
Plotting Ordered Pairs and Transforming Figures Students use rules to create and plot ordered pairs.	*Math Journal 1*, pp. 134–135; straightedge (optional)	5.NF.5, 5.NF.5a, 5.G.1, 5.G.2 SMP3, SMP6
Discussing the Results of Operations on Ordered Pairs Students discuss the effects on the plotted picture of changing ordered pairs.	*Math Journal 1*, pp. 134–136; straightedge (optional)	5.NF.5, 5.NF.5a, 5.G.1, 5.G.2 SMP3, SMP8
✓ **Assessment Check-In** See page 380. Expect most students to be able to plot points in the first quadrant of a coordinate grid and create new sets of ordered pairs based on rules.	*Math Journal 1*, pp. 134–136	5.G.1, 5.G.2

---

③ Practice 20–30 min

	Materials	
Playing *Decimal Top-It* **Game** Students practice comparing decimals to thousandths.	*Student Reference Book*, p. 296; per partnership: *Math Masters*, pp. TA25 and G14, p. TA23 (optional); number cards 0–9 (4 of each); 6-sided die	5.NBT.1, 5.NBT.3, 5.NBT.3a, 5.NBT.3b, SMP7
Math Boxes 4-8 Students practice and maintain skills.	*Math Journal 1*, p. 137	See page 381.
Home Link 4-8 **Homework** Students practice plotting points and writing coordinates.	*Math Masters*, p. 146	5.NBT.3, 5.NBT.3b, 5.G.1, 5.G.2

 〔 Go Online 〕 to see how mastery develops for all standards within the grade.

my.mheducation.com

Differentiation Options

Readiness	10–15 min
WHOLE CLASS SMALL GROUP **PARTNER** INDEPENDENT	

"What's My Rule?"

5.OA.3, SMP8

Math Masters, p. 144

To prepare for applying rules to ordered pairs, students identify rules and complete "What's My Rule?" tables. Then they create and apply their own rule. **GMP8.1**

"What's My Rule?" — Lesson 4-8

Complete each table below according to the rule.

① Rule: Add 12 to the *in* number.

In (n)	Out (n + 12)
1	13
2	14
4	16
5	17
10	22
17	29

② Rule: Triple the *in* number.

In (n)	Out (3 ∗ n)
1	3
2	6
3	9
6	18
7	21
10	30

③ Rule: Double the *in* number.

In (n)	Out (2 ∗ n)
0	0
2	4
3	6
4	8
4.5	9
10	20

For Problems 4–5, determine the rule and write it in words. Then use the rule to complete the tables. For Problem 6 make up your own rule and complete the table.

④ Rule: Subtract 6 from the *in* number.

In	Out
12	6
15	9
9	3
6	0
7	1
10	4

⑤ Rule: Find half of the *in* number.

In	Out
6	3
2	1
4	2
22	11
10	5
9	4.5

⑥ Rule: Answers vary.

In	Out

144   5.OA.3, SMP8

Enrichment	10–15 min
WHOLE CLASS SMALL GROUP **PARTNER** INDEPENDENT	

"Connect the Dots" Partner Challenge

5.G.1, 5.G.2, SMP2, SMP6

Activity Card 50;
Math Masters, p. TA28; straightedge (optional)

To extend their work plotting points and writing ordered pairs, students create a figure on a coordinate grid and write the coordinates for points that can be connected to make the figure. They exchange their list of coordinates with a partner, and then plot the coordinates and connect them in order to re-create their partner's figure. **GMP2.1, GMP6.4**

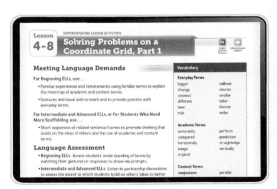

"Connect the Dots" Partner Challenge — 50

I draw a fish. Let's see if my partner's drawing will look like mine.

What You Need
Math Masters, page TA28 paper
straightedge (optional)

What To Do
Work with a partner.
① Draw a simple picture or figure on the top grid on Math Masters, page TA28 using line segments. Do not show your partner. Try to make the line segments end at the intersections of grid lines.
② Draw points at the ends of your line segments.
③ Find the coordinates of the points and list them in order on a blank sheet of paper. List them in order so your partner can draw your figure.
④ When you and your partner have finished Steps 1–3, exchange the list of coordinates you wrote down.
⑤ Plot your partner's coordinates and connect them in order on the bottom grid on Math Masters, page TA28.
⑥ When you are both finished, compare your new figures to the original drawings.

Talk About It
Are the figures the same? If not, what could you or your partner have done differently to make the figures closer to the original drawings?

More You Can Do
Predict what your figure would look like if you divided both the *x*- and *y*-coordinates by 2. Then divide the coordinates by 2 and plot the points to see whether your prediction is correct.

5.G.1, 5.G.2, SMP2, SMP6

Extra Practice	5–15 min
WHOLE CLASS SMALL GROUP **PARTNER** INDEPENDENT	

Plotting a Mystery Word

5.G.1, SMP2

Math Masters, pp. 145 and TA28; straightedge (optional)

For more practice plotting and connecting points on a coordinate grid, students plot three sets of ordered pairs and connect the points to form a three-letter word. **GMP2.1, GMP2.2** They list coordinates they could use to plot another three-letter word.

Plotting a Mystery Word — Lesson 4-8

Plot the 3 sets of coordinates listed below. Connect the points for each set in the order they are listed to form a letter.

Letter 1: (6, 14) (2, 14) (2, 7) (6, 7)
Letter 2: (7, 13) (7, 7) (10, 7) (10, 13)
Letter 3: (11, 7) (11, 13) (15, 13) (15, 10) (11, 10)

What word have you made? **CUP**

Look at a blank coordinate grid (Math Masters, page TA28). Think about how you could sketch another 3-letter word on the coordinate grid.

In the space below, list the ordered pairs that you think you could plot and connect to make another 3-letter word. Check your work by plotting and connecting the ordered pairs you listed on a blank coordinate grid.

Letter 1: Answers vary.
Letter 2: _____
Letter 3: _____

5.G.1, SMP2   145

English Language Learner

Beginning ELL Use a demonstration to help students understand the verb *change*. Show an uninflated balloon, saying: *Here is a balloon. I am going to* change *the size of the balloon by blowing it up.* Blow up the balloon and ask: *Does this balloon look the same? I* changed *its size. I made it look different. It is a different size. It is the same balloon, but now I have filled it with air, so it looks different. I* changed *the size of the balloon. I* changed *the way it looks.*

Lesson 4-8 — Solving Problems on a Coordinate Grid, Part 1

Meeting Language Demands

For Beginning ELLs, use . . .
• Familiar experiences and restatements using familiar terms to explain the meanings of academic and content terms.
• Gestures and visual aids to teach and to provide practice with everyday terms.

For Intermediate and Advanced ELLs, or for Students Who Need More Scaffolding, use . . .
• Short sequences of related sentence frames to promote thinking that builds on the ideas of others and the use of academic and content terms.

Language Assessment
• Beginning ELLs Assess students' understanding of terms by watching their gestures or responses to show-me prompts.
• Intermediate and Advanced ELLs Listen to partnership discussions to assess the extent to which students build on others' ideas to better

Vocabulary

Everyday Terms: bigger, change, connect, different, new, rule, sailboat, shorter, smaller, taller, thinner, wider

Academic Terms: accurately, compared, horizontally, image, original, perform, prediction, straightedge, vertically

Content Terms: conjecture, parallel

Differentiation Support pages are found in the online Teacher's Center.

Standards and Goals for
Mathematical Process and Practice

SMP3 **Construct viable arguments and critique the reasoning of others.**
   GMP3.1 Make mathematical conjectures and arguments.

SMP6 **Attend to precision.**
   GMP6.4 Think about accuracy and efficiency when you count, measure, and calculate.

SMP8 **Look for and express regularity in repeated reasoning.**
   GMP8.1 Create and justify rules, shortcuts, and generalizations.

---

1 Warm Up   5 min

▶ Mental Math and Fluency

Display the benchmarks 0, $\frac{1}{2}$, 1, $1\frac{1}{2}$, and 2. Have students identify the closest benchmark for each sum or difference. *Leveled exercises:*

●○○ $\frac{3}{5} + \frac{1}{4}$ 1 $\frac{7}{8} - \frac{5}{6}$ 0

●●○ $\frac{3}{8} + 1\frac{2}{3}$ 2 $2\frac{1}{4} - \frac{3}{4}$ $1\frac{1}{2}$

●●● $\frac{3}{4} - \frac{2}{10}$ $\frac{1}{2}$ $\frac{1}{9} + \frac{1}{10}$ 0

---

2 Focus   35–40 min

▶ Math Message

Math Journal 1, pp. 134–135

Complete Problem 1 on journal page 134.

> **NOTE** Students may be confused by the fact that some points are listed more than once. Explain that while it is not necessary to plot a point more than once, it will be necessary to connect some points more than once as students create the sailboat image.

▶ Plotting Ordered Pairs and Transforming Figures

Math Journal 1, pp. 134–135

| WHOLE CLASS | SMALL GROUP | PARTNER | INDEPENDENT |

Math Message Follow-Up Display a coordinate grid and invite a volunteer to plot and connect the points from the Math Message. The completed image should look like a sailboat. Discuss any difficulties students had completing the Math Message. Remind them that to plot points accurately, they first move horizontally along the *x*-axis the distance of the first coordinate and then move vertically, parallel to the *y*-axis, the distance of the second coordinate. GMP6.4 Suggest using a straightedge when connecting points. Tell them that in today's lesson they will use rules to create new ordered pairs that will change the sailboat.

Direct students' attention to the rule for New Sailboat 1 on journal page 134. Ask: *What does it mean to double each number of the original ordered pair?* Multiply each coordinate by 2. Point out that the rule has already been applied to the first three ordered pairs. Ask: *Why does it say (16, 2) in the first row?* The original ordered pair was (8, 1). The rule was to double each number: $8 * 2 = 16$ and $1 * 2 = 2$. When it is clear that students understand the rule, have them write the remaining coordinates for New Sailboat 1 on journal page 134.

Math Journal 1, p. 134

Graphing Sailboats   Lesson 4-8
DATE   TIME

1. Find the column labeled Original Sailboat in the table below. Plot the ordered pairs listed in the column on the grid titled Original Sailboat on the next page. Connect the points in the same order that you plot them. You should see the outline of a sailboat.

2. a. Fill in the missing coordinates for New Sailboat 1.

   b. How do you think New Sailboat 1 will be different from the Original Sailboat? Record a conjecture at the top of the column.

   c. Plot the ordered pairs for New Sailboat 1 on the next page. Connect the points in the same order that you plot them.

Original Sailboat	New Sailboat 1	New Sailboat 2	New Sailboat 3
	Rule: Double each number of the original pair.	Rule: Double the first number of the original pair. Leave the second number the same.	Rule: Double the second number of the original pair. Leave the first number the same.
Conjecture:	Sample answer: It will look twice as big.	Sample answer: It will look twice as wide.	Sample answer: It will look twice as tall.
(8, 1)	(16, 2)	(16, 1)	(8, 2)
(5, 1)	(10, 2)	(10, 1)	(5, 2)
(5, 7)	(10, 14)	(10, 7)	(5, 14)
(1, 2)	(2 , 4)	(2 , 2)	(1 , 4)
(5, 1)	(10 , 2)	(10 , 1)	(5 , 2)
(0, 1)	(0 , 2)	(0 , 1)	(0 , 2)
(2, 0)	(4 , 0)	(4 , 0)	(2 , 0)
(7, 0)	(14 , 0)	(14 , 0)	(7 , 0)
(8, 1)	(16 , 2)	(16 , 1)	(8 , 2)

   d. Complete steps 2a–2c for New Sailboat 2.

   e. Complete steps 2a–2c for New Sailboat 3.

   Be sure to apply each rule to the coordinates from the **Original Sailboat**.

134   5.NF.5, 5.NF.5a, 5.G.1, 5.G.2, SMP3, SMP6

Ask: *How do you think New Sailboat 1 will look compared to the Original Sailboat? Why?* **GMP3.1** Sample answer: I think New Sailboat 1 will be a bigger version of the Original Sailboat because the numbers in the ordered pairs are larger. Remind students that in answering this question they have made a **conjecture,** or a prediction based on mathematics, about how the rule will transform the original picture. Have students record their conjectures on journal page 134. Then have them plot and connect the points for New Sailboat 1 on journal page 135. Circulate to ensure students are plotting points accurately. **GMP6.4** When most have finished, have them talk with a partner about whether or not their conjectures were correct.

Refer students to the rule for New Sailboat 2. Ask: *How is this rule different from the rule for New Sailboat 1?* Only the first number in the ordered pair is doubled. *What happens to the second number in the ordered pair?* It stays the same. Apply the New Sailboat 2 rule to one or two ordered pairs from the Original Sailboat to check students' understanding.

Repeat the process for New Sailboat 3. When it is clear that students understand the two rules, have them work in partnerships to complete journal pages 134 and 135. **GMP3.1** Circulate and assist. Remind students as they work that they should be applying each rule to the Original Sailboat ordered pairs to write ordered pairs for New Sailboats 1, 2, and 3. Consider suggesting that students cover the coordinates of the New Sailboats with a strip of paper once they have been written to make sure students are always referring back to the original ordered pairs.

▶ Discussing the Results of Operations on Ordered Pairs

Math Journal 1, pp. 134–136

| WHOLE CLASS | SMALL GROUP | PARTNER | INDEPENDENT |

When most students have finished plotting New Sailboats 2 and 3, guide a discussion about the changes from the Original Sailboat to each of the new sailboats. *Suggestions:*

- Ask: *How is New Sailboat 1 different from the Original Sailboat?* It's bigger. *How much bigger?* Twice as high and twice as wide *Why?* We doubled both the width and the height of the Original Sailboat by doubling both the *x*- and *y*-coordinates.

- Ask: *How is New Sailboat 2 different from the Original Sailboat?* It's wider. *How much wider?* Twice as wide *Why?* We doubled the width by doubling the *x*-coordinates. The height stayed the same as the Original Sailboat because we did not change the *y*-coordinates.

- Ask: *How is New Sailboat 3 different from the Original Sailboat?* It's taller. *How much taller?* Twice as tall *Why?* We doubled the height by doubling the *y*-coordinates. The width stayed the same as the Original Sailboat because we didn't change the *x*-coordinates.

Graphing Sailboats (continued)

Lesson 4-8

Original Sailboat

New Sailboat 1

New Sailboat 2

New Sailboat 3

5.NF.5, 5.NF.5a, 5.G.1, 5.G.2, SMP6   135

Adjusting the Activity

Differentiate   If students seem especially interested in how various rules will change the sailboat image, let them write their own rules for revising the ordered pairs. Suggest that they write a rule, make a conjecture, and then use a blank coordinate grid to plot the new coordinates and create the new image. Encourage them to make generalizations about how different rules change the resulting images on a coordinate grid.

    Differentiation Support

Discussing how the sailboat changes when multiplying and halving coordinates is important prerequisite work for **5.NF.5**, which requires students to understand multiplication as scaling. However, the emphasis on multiplicative rules may lead students to the misconception that if a coordinate gets larger, the image is stretched. While this generalization is true for multiplication rules, this does not hold for rules involving addition. You may wish to assign a small group a rule involving addition or subtraction and use their work to discuss the differences between multiplication/division rules and addition/subtraction rules.

Math Journal 1, p. 136

A New Sailboat Rule

Lesson 4-8
DATE   TIME

① Circle the rule for Sailboat 4 given to you by your teacher.
  • Triple the first number of the ordered pair.
  • Triple the second number of the ordered pair.
  • Double the first number of the ordered pair; halve the second number.
  • Halve the first number of the ordered pair; double the second number.
  • Other: _____

② Make a conjecture about what New Sailboat 4 will look like.
                          Answers vary.

③ Create ordered pairs for New Sailboat 4 based on the rule. Write them in the table below.
④ Plot the new set of ordered pairs and connect the points in the order they were plotted.
⑤ Was your conjecture correct? Explain. Answers vary.

Answers vary.

Original Sailboat	New Sailboat 4
(8, 1)	(,)
(5, 1)	(,)
(5, 7)	(,)
(1, 2)	(,)
(5, 1)	(,)
(0, 1)	(,)
(2, 0)	(,)
(7, 0)	(,)
(8, 1)	(,)

New Sailboat 4

136   5.NF.5, 5.NF.5a, 5.G.1, 5.G.2, SMP3

Divide the class into small groups and assign each group a different rule from the top of journal page 136. Groups should make a conjecture about what the sailboat will look like, apply the rule to the original ordered pairs, and graph New Sailboat 4 on the grid. **GMP3.1** Groups should be prepared to share their rules, conjectures, and graphs with the class when they are finished. *Sample conjectures:*

- **Triple the first number of the ordered pair.** The sailboat should be 3 times as wide but the same height as the original.
- **Triple the second number of the ordered pair.** The sailboat should be 3 times as tall but the same width as the original.
- **Double the first number of the ordered pair; halve the second number.** The sailboat should be twice as wide but only half as tall as the original.
- **Halve the first number of the ordered pair; double the second number.** The sailboat should be only half as wide but twice as tall as the original.

When most groups have finished, invite volunteers to show and describe their results for New Sailboat 4. Ask: *How did the image change when we doubled or tripled the* first *coordinate?* The image got wider. *How did the image change when we halved the* first *coordinate?* It got thinner, or skinnier. *In general, how does an image change if we multiply or divide the first coordinate by a number? Why?* The image is stretched or squished horizontally because the first coordinate tells how far to move horizontally. *How did the image change when we doubled or tripled the second coordinate?* It got taller. *When we halved the second coordinate?* It got shorter. *In general, how does an image change if we multiply or divide the second coordinate by a number? Why?* **GMP8.1** The image is stretched or squished vertically because the second coordinate tells how far to move vertically.

Assessment Check-In   5.G.1, 5.G.2

Math Journal 1, pp. 134–136

Expect most students to be able to accurately plot points on a grid and create new sets of ordered pairs based on rules. Do not expect them to be accurate in their conjectures about how changing the coordinates will change the sailboat image. This is the first time students have considered transforming images on a grid. They will continue to solve mathematical problems on the coordinate grid in later lessons. Students who struggle plotting points accurately may benefit from completing the Extra Practice activity. If students struggle following the rules to create new ordered pairs, suggest changing all the *x*-coordinates first, then all the *y*-coordinates. They may find it easier to keep track of one task at a time.

Evaluation Quick Entry Go online to record students' progress and to see trajectories toward mastery for these standards.

Summarize Have students summarize how to make a figure on the coordinate grid wider, taller, thinner, and shorter. GMP8.1 Encourage them to use mathematical language, such as x-*coordinate*, y-*coordinate*, x-*axis*, y-*axis*, and *point*.

3 Practice   20–30 min

▶ Playing *Decimal Top-It*

Student Reference Book, pp. 296–297; *Math Masters,* pp. TA25 and G14

| WHOLE CLASS | SMALL GROUP | PARTNER | INDEPENDENT |

Students practice comparing decimals to thousandths. Have students record rounds of play on the *Top-It* Record Sheet (*Math Masters,* page G14).

Observe
- Which students have a strategy to make the largest possible decimal?
- When are students having difficulty determining who won a round?

Discuss
- *In which place would you want to put an 8? A 2? Why?* GMP7.2
- *What is your strategy for deciding where to place a 5?*

| Differentiate | **Game Modifications** | Go Online | Differentiation Support |

▶ Math Boxes 4-8

Math Journal 1, p. 137

| WHOLE CLASS | SMALL GROUP | PARTNER | INDEPENDENT |

Mixed Practice Math Boxes 4-8 are paired with Math Boxes 4-6.

▶ Home Link 4-8

Math Masters, p. 146

Homework Students practice writing coordinates for points by drawing shapes on a coordinate grid.

Math Journal 1, p. 137

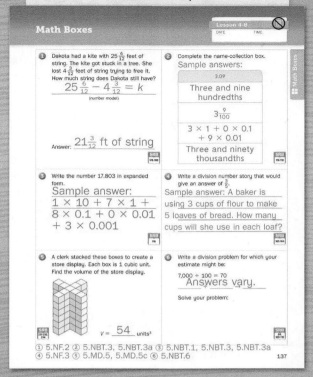

Math Masters, p. 146

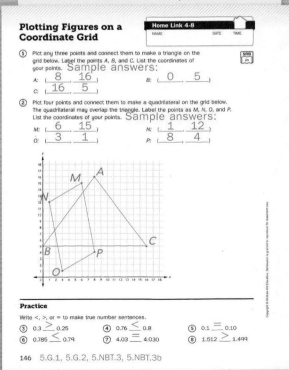

Solving Problems on a Coordinate Grid, Part 2

Overview Students form ordered pairs, graph them, and interpret coordinate values in context.

▶ **Vocabulary**
extrapolate • interpolate

Standards

Focus Clusters
- Analyze patterns and relationships.
- Graph points on the coordinate plane to solve real-world and mathematical problems.

① Warm Up   5 min

	Materials	
Mental Math and Fluency Students find equivalent mixed numbers or fractions with the same denominator.	slate	**5.NF.3**

② Focus   35–40 min

Math Message Students determine whether ordered pairs model a data set.	*Math Journal 1*, p. 138	**5.OA.3, 5.G.1** **SMP4**
Graphing Data as Ordered Pairs Students plot and connect points on a grid and use the resulting graph to answer questions.	*Math Journal 1*, p. 138; straightedge	**5.OA.3, 5.G.1, 5.G.2** **SMP4, SMP7**
Graphing Ordered Pairs and Interpreting Graphed Data Students use a rule to complete a data table, form and graph ordered pairs, and use the graph to answer questions.	*Math Journal 1*, p. 139; *Math Masters*, p. TA2 (optional); straightedge	**5.OA.3, 5.G.1, 5.G.2** **SMP4, SMP7**
✓ **Assessment Check-In** See page 386. Expect most students to be able to use data from a table to write ordered pairs and to plot and connect the points with a line in the first quadrant of a coordinate grid.	*Math Journal 1*, p. 139	**5.OA.3, 5.G.1, 5.G.2**

③ Practice   20–30 min

Playing *High-Number Toss* **Game** Students practice reading, writing, and comparing numbers in standard and exponential notations.	*Student Reference Book*, p. 312; per partnership: *Math Masters*, p. G10; 6-sided die	**5.NBT.1, 5.NBT.2** **SMP8**
Math Boxes 4-9 Students practice and maintain skills.	*Math Journal 1*, p. 140	See page 387.
Home Link 4-9 **Homework** Students plot data and answer questions based on the resulting graph.	*Math Masters*, p. 150	**5.OA.3, 5.NBT.4, 5.G.1, 5.G.2** **SMP4t**

 Go Online to see how mastery develops for all standards within the grade.

Differentiation Options

Readiness · 10–15 min

WHOLE CLASS · **SMALL GROUP** · **PARTNER** · INDEPENDENT

Matching Graphs to Contexts

5.G.2, SMP4

Math Masters, p. 147

For practice interpreting data represented graphically, students match completed graphs to number stories. Read the number stories on *Math Masters,* page 147 together and have students think about what a graph for each story would look like. Ask questions like these: *Where would the graph start? Would the* y*-coordinates increase or decrease as the* x*-coordinates increase? Would the line slant upward or downward?* Have students work in partnerships or small groups to complete the page. **GMP4.2**

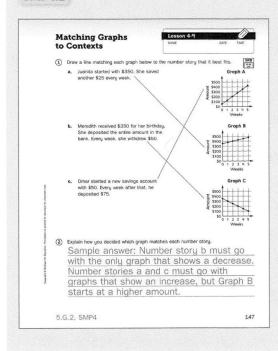

5.G.2, SMP4   147

Enrichment · 5–15 min

WHOLE CLASS · SMALL GROUP · **PARTNER** · INDEPENDENT

Finding Rules for Graphs

5.OA.3, 5.G.1, 5.G.2, SMP4, SMP7

Math Masters, p. 148

To extend their work interpreting data points on graphs, students complete *Math Masters,* page 148.
They examine graphs, identify rules that could have produced the graphs, and create real-world contexts that could be modeled by the graphs. **GMP4.2, GMP7.1**

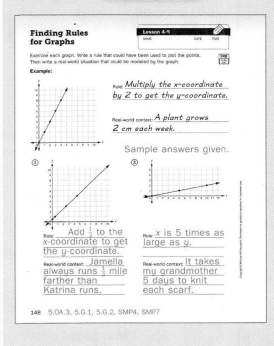

148   5.OA.3, 5.G.1, 5.G.2, SMP4, SMP7

Extra Practice · 5–15 min

WHOLE CLASS · SMALL GROUP · **PARTNER** · INDEPENDENT

Interpreting Data from a Grid

5.OA.3, 5.G.1, 5.G.2, SMP2

Activity Card 51;
Math Masters, p. 149;
number cards 0–10 (1 of each); straightedge

For more practice interpreting points on a line as ordered pairs, students practice plotting points on a grid, connecting them with a line, and using the line to extrapolate and interpolate additional values. **GMP2.1, GMP2.2** Students practice writing coordinates both as ordered pairs and in a table.

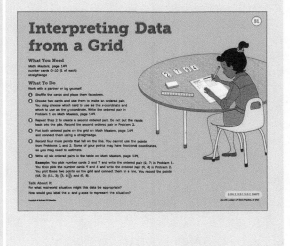

English Language Learner

Beginning ELL Review the term *ordered pair* by focusing on the two words separately. Demonstrate what it means to put things in order by size or height, thinking aloud: *I am putting these _____ in order. They go from shortest to tallest. I have ordered them.* Extend to numerical and alphabetical order. Then show different examples of items that come in pairs, such as socks, shoes, and earrings, emphasizing that pairs come in two. Finally, put the two words together to explain what ordered number pairs are, emphasizing how the numbers come in pairs and are ordered by whether they refer to the *x*- or *y*-axis.

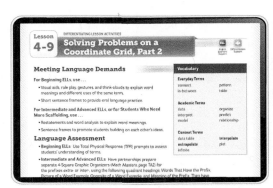

Differentiation Support pages are found in the online Teacher's Center.

SMP4 Model with mathematics.

GMP4.1 Model real-world situations using graphs, drawings, tables, symbols, numbers, diagrams, and other representations.

GMP4.2 Use mathematical models to solve problems and answer questions.

SMP7 Look for and make use of structure.

GMP7.1 Look for mathematical structures such as categories, patterns, and properties.

1 Warm Up 5 min

▶ Mental Math and Fluency

Display the following mixed numbers or fractions. Have students record an equivalent mixed number or fraction with the same denominator on their slates. *Leveled exercises:* Sample answers given for Levels 2 and 3.

● ○ ○ $\frac{3}{2}$ $1\frac{1}{2}$ $\frac{5}{4}$ $1\frac{1}{4}$

● ● ○ $\frac{11}{3}$ $3\frac{2}{3}$ $2\frac{2}{3}$ $1\frac{5}{3}$

● ● ● $1\frac{7}{5}$ $2\frac{2}{5}$ $5\frac{7}{2}$ $4\frac{9}{2}$

2 Focus 35–40 min

▶ Math Message

Math Journal 1, p. 138

Look at the data in the table on journal page 138. Ms. Arazy told her class to use the data to write ordered pairs. Sarah wrote the following ordered pairs: (5, 1)   (3, 7)   (9, 5)   (7, 11)   (12, 8)

Do Sarah's ordered pairs model the data? No. *Why or why not?* GMP4.1 Sample answer: Lilith's age doesn't always come first.

▶ Graphing Data as Ordered Pairs

Math Journal 1, p. 138

| WHOLE CLASS | SMALL GROUP | PARTNER | INDEPENDENT |

Math Message Follow-Up Explain that numerical data is often organized in two-column tables. This allows us to designate one column as *x*-coordinates and the other as *y*-coordinates, which makes it easy to turn the data into ordered pairs and graph them on a coordinate grid.

Discuss students' answers to the Math Message. Help them recognize that although each ordered pair written by Sarah includes both Lilith's and Noah's ages, her pairs do not consistently list one age first. Explain that data tables are often labeled with *x* and *y* to designate which column contains *x*-coordinates and which contains *y*-coordinates. Since Lilith's ages are listed first in the table on the journal page, tell students to use her ages as the *x*-coordinates and Noah's as the *y*-coordinates. Have them write the ages as ordered pairs in Problem 1 on journal page 138. Explain that graphing a data set on a coordinate grid allows us to see patterns in the data. We can use the graph to **extrapolate** and **interpolate**, or continue patterns beyond and in between the data points we have, to predict other values. GMP7.1

Math Journal 1, p. 138

Graphing Data as Ordered Pairs   Lesson 4-9
DATE   TIME

The data in the table show Lilith's and Noah's ages at 5 different times in their lives.

Lilith's Age (years)	Noah's Age (years)
5	1
7	3
9	5
11	7
12	8

Ordered Pairs:
(5 , 1)
(7 , 3)
(9 , 5)
(11 , 7)
(12 , 8)

① Write their ages as ordered pairs. Then plot the points on the grid.

② What do you notice about the points you plotted? Sample answer: They are on a straight line.

③ Connect the points with a line. What other information can we get from this line? Sample answer: ages of Lilith and Noah at other times

④ Use the line to determine the ages of Lilith and Noah at various points in their lives.
   a. When Lilith was 8 years old, Noah was __4__ years old.
   b. When Noah was 6 years old, Lilith was __10__ years old.
   c. How old will Lilith be when Noah is 11? __15__

⑤ Explain how you solved Problem 4c. Sample answer: I followed the y-axis up to 11 since Noah's age is the y-coordinate. Then I followed that grid line over until I hit the line I drew. It was 15 units over, so the x-coordinate would be 15. Lilith is 15.

⑥ Who is older, Lilith or Noah? __Lilith__ How much older? __4 years__

138   5.OA.3, 5.G.1, 5.G.2, SMP4, SMP7

Have students graph the data about Lilith's and Noah's ages to complete Problem 1 on journal page 138. **GMP4.1** As they work, encourage them to look for a pattern. **GMP7.1** They should write what they notice in Problem 2.

Display the journal page with the 5 points plotted. Demonstrate how to draw a line through the points using a straightedge. Ask: *Does it make sense to extend the line past these points? Why or why not?* Sample answer: Yes. Lilith and Noah were once younger than 5 and 1, and someday they will be older than 12 and 8. *How do we indicate that the line should extend beyond the points visible on our graph?* By putting arrows at the ends of the line Some students may point out that the line should not extend below the *x*-axis because Noah cannot be less than 0 years old. Acknowledge this point. Explain that for this problem students could choose to end the line at the *x*-axis instead of including an arrow. For other problems, it may not be possible to determine where the line ends. It is acceptable to include arrows on graphs as long as students keep the context of the problem in mind as they use the graph to answer questions.

Have students use a straightedge to connect the points on their graphs. Ask: *How might this line be useful?* If students have difficulty coming up with ideas, ask: *Have Lilith and Noah only been the five ages shown by the coordinates we plotted?* No, they've been many other ages too. *So what does this line represent?* All the ages they have been at different times in their lives. Students can record their thoughts in Problem 3.

Explain that students will now use the line they have drawn to extrapolate and interpolate from the given data. They will figure out ages for Lilith and Noah for times in their lives that weren't given in the data table. Ask: *How old was Noah when Lilith was 6?* 2 years old *How can you use the graph to answer this question?* Go to 6 on the *x*-axis and trace upward until we reach the line at the point (6, 2). Since the *y*-coordinate represents Noah's age, he was 2. Have a volunteer demonstrate how to find the point on the line representing Lilith at 6 years of age and then trace the grid line to the *y*-axis to find Noah's age. **GMP4.2** Do several more examples as a class, as needed. Then have students complete the journal page. Circulate and assist.

When most students are finished, guide them in discussing their answers. Focus on the "in-between" points on the line: those not originally plotted in the Math Message. Reinforce the idea that even though in the Math Message students plotted only five points, the line represents an infinite number of data points.

Adjusting the Activity

Differentiate If students struggle using the line on the graph to extrapolate or interpolate additional data points, suggest using a straightedge to help locate points on the line. For example, for Problem 4a students should find 8 on the *x*-axis, since Lilith's age is plotted on the *x*-axis. They lay a straightedge along the grid line at 8 on the *x*-axis and make a dot on the line where the straightedge intersects the line. Then they lay the straightedge horizontally across the dot they just drew and identify where the straightedge crosses the *y*-axis. For Problem 4a this will be at 4, which was Noah's age when Lilith was 8.

Go Online Differentiation Support

NOTE Some students may notice that the rule "– 4" relates Noah's age to Lilith's, and they may want to use the rule, rather than the graph, to solve the problems. Keep the focus of the lesson on using the graph to answer questions. Emphasize how the graph can be used to find a missing coordinate of *any* point. If students say they know that Noah was 2 when Lilith was 6 because $6 - 4 = 2$, encourage them to show this using the graph. Students will focus more on finding and using rules in later lessons.

Forming and Graphing Ordered Pairs

Lesson 4-9

DATE   TIME

For each data set, fill in the missing values and write the data as ordered pairs. Plot the points on the grid and connect them using a straightedge. Use the graph to answer the questions.

① Dean is raising money for charity. He earns $2 for each lap he runs around the gym.

Laps Run (x)	$ Earned (y)
1	2
2	4
3	6
4	8

Ordered pairs:
(1 , 2)
(2 , 4)
(3 , 6)
(4 , 8)

a. If Dean has earned $14, how many laps has he run? __7__ laps

b. Put an X on the point on the grid that shows your answer to Part a.

c. What are the coordinates for this point? (7 , 14)

② Sally uses 2 paintbrushes for each paint jar.

Brushes (x)	Paint Jars (y)
2	1
4	2
6	3
8	4

Ordered pairs:
(2 , 1)
(4 , 2)
(6 , 3)
(8 , 4)

a. If Sally uses 6 jars of paint, how many brushes does she need? __12__ brushes

b. Put an X on the point on the grid that shows your answer to Part a.

c. What are the coordinates for the point you marked with an X? (12 , 6)

5.OA.3, 5.G.1, 5.G.2, SMP4, SMP7   139

Some of the points on the line have whole-number coordinates, like the ones students used to answer the questions on the journal page. Other points have coordinates between whole numbers. Ask students to use the graph to determine the following: GMP4.2

• Noah's age when Lilith was $6\frac{1}{2}$ $2\frac{1}{2}$

• Lilith's age when Noah was $1\frac{1}{2}$ $5\frac{1}{2}$

• Noah's age when Lilith was $14\frac{1}{2}$ $10\frac{1}{2}$

Finally, have students share answers to Problem 6. Lilith is 4 years older. Ask: *How do you know?* GMP7.1   When I look at the coordinates of the points on the line, the *x*-coordinate is always 4 more than the *y*-coordinate.

▶ # Graphing Ordered Pairs and Interpreting Graphed Data

Math Journal 1, p. 139

WHOLE CLASS	SMALL GROUP	PARTNER	INDEPENDENT

Have students complete journal page 139 with a partner. They fill in a data table based on a given rule, use the table to write ordered pairs, graph the ordered pairs, and connect them using a straightedge. They use the line on the graph to extrapolate and interpolate additional data points to answer questions. GMP4.1, GMP4.2, GMP7.1

Assessment Check-In   5.OA.3, 5.G.1, 5.G.2

Math Journal 1, p. 139

Expect most students to be able to use data from a table to write ordered pairs and to correctly plot and connect the points with a line on a coordinate grid. Since this is the first time they have been asked to extrapolate and interpolate information from graphed data, students may have difficulty interpreting the data represented by the graph to answer the questions on journal page 139. They will have more practice interpreting data graphed on a coordinate grid in later lessons. If students struggle creating ordered pairs, have them highlight each column of the data table in a different color and use the appropriate color to write each coordinate in the ordered pairs (for example, write the *x*-coordinates in red and the *y*-coordinates in blue). If students struggle plotting points on a coordinate grid, consider using the Readiness or Extra Practice activities in Lessons 4-6 and 4-7.

Evaluation Quick Entry Go online to record students' progress and to see trajectories toward mastery for these standards.

Summarize Invite students to share their solutions to the problems on journal page 139. Ask them to explain how plotting points on a graph and connecting them with a line can help them find answers to questions.

Science Link You may want to draw a clear connection between graphing ordered pairs in mathematics and graphing scientific data. In science the *x*-coordinates are values of the independent variable, or the variable the scientists are manipulating. The *y*-coordinates are values of the dependent variable, or the variable that may change according to a change in the independent variable. For example, if a scientist is investigating the effect of hours of sunlight on plant growth, the independent variable (*x*) will be hours of sunlight (because the scientist can manipulate that) and the growth of the plants would be the dependent variable (*y*).

③ Practice   20–30 min

▶ Playing *High-Number Toss*

Student Reference Book, p. 312; *Math Masters*, p. G10

| WHOLE CLASS | **SMALL GROUP** | **PARTNER** | INDEPENDENT |

Students practice reading, writing, and comparing numbers in standard and exponential notations.

Observe

- Which students have a strategy for building the largest possible numbers?
- Which students are able to translate between exponential and standard notations?

Discuss

- *What's the best card to place in the space for the exponent? Why?* GMP8.1
- *What is the best position to place your smallest card in? Why?*

| **Differentiate** | **Game Modifications** | **Go Online** | 📖 Differentiation Support |

▶ Math Boxes 4-9 ✦

Math Journal 1, p. 140

| WHOLE CLASS | SMALL GROUP | PARTNER | **INDEPENDENT** |

Mixed Practice Math Boxes 4-9 are grouped with Math Boxes 4-12 and 4-14.

▶ Home Link 4-9

Math Masters, p. 150

Homework Students plot and connect points on a coordinate grid and use the graph to answer questions. GMP4.1, GMP4.2

Math Journal 1, p. 140

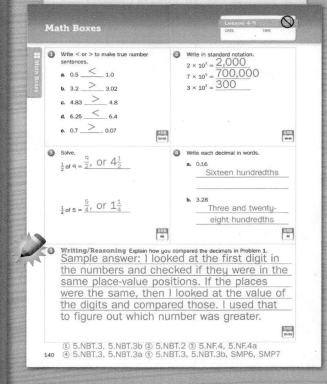

Math Boxes

1 Write < or > to make true number sentences.

a. 0.5 $<$ 1.0

b. 3.2 $>$ 3.02

c. 4.83 $>$ 4.8

d. 6.25 $<$ 6.4

e. 0.7 $>$ 0.07

2 Write in standard notation.

$2 \times 10^3 = $ 2,000

$7 \times 10^5 = $ 700,000

$3 \times 10^2 = $ 300

3 Solve.

$\frac{1}{2}$ of 9 = $\frac{9}{2}$, or $4\frac{1}{2}$

$\frac{1}{4}$ of 5 = $\frac{5}{4}$, or $1\frac{1}{4}$

4 Write each decimal in words.

a. 0.16

Sixteen hundredths

b. 3.28

Three and twenty-eight hundredths

5 Writing/Reasoning Explain how you compared the decimals in Problem 1.

Sample answer: I looked at the first digit in the numbers and checked if they were in the same place-value positions. If the places were the same, then I looked at the value of the digits and compared those. I used that to figure out which number was greater.

① 5.NBT.3, 5.NBT.3b ② 5.NBT.2 ③ 5.NF.4, 5.NF.4a
④ 5.NBT.3, 5.NBT.3a ⑤ 5.NBT.3, 5.NBT.3b, SMP6, SMP7

140

Math Masters, p. 150

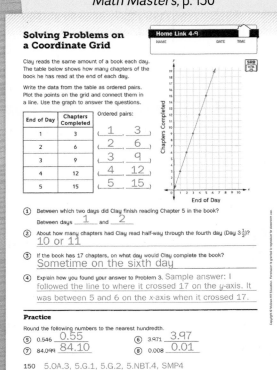

Solving Problems on a Coordinate Grid   Home Link 4-9

Clay reads the same amount of a book each day. The table below shows how many chapters of the book he has read at the end of each day.

Write the data from the table as ordered pairs. Plot the points on the grid and connect them in a line. Use the graph to answer the questions.

End of Day	Chapters Completed
1	3
2	6
3	9
4	12
5	15

Ordered pairs:

(1 , 3)
(2 , 6)
(3 , 9)
(4 , 12)
(5 , 15)

1 Between which two days did Clay finish reading Chapter 5 in the book?

Between days 1 and 2

2 About how many chapters had Clay read half-way through the fourth day (Day $3\frac{1}{2}$)?

10 or 11

3 If the book has 17 chapters, on what day would Clay complete the book?

Sometime on the sixth day

4 Explain how you found your answer to Problem 3. Sample answer: I followed the line to where it crossed 17 on the *y*-axis. It was between 5 and 6 on the *x*-axis when it crossed 17.

Practice

Round the following numbers to the nearest hundredth.

⑤ 0.546 0.55

⑥ 3.971 3.97

⑦ 84.099 84.10

⑧ 0.008 0.01

150   5.OA.3, 5.G.1, 5.G.2, 5.NBT.4, SMP4

Folder Art

Overview **Day 1: Students develop and apply a rule to enlarge a picture on a acoordinate grid.**
Day 2: Students discuss others' rules and pictures on the coordinate grid and revise their work.

Day 1: Open Response

▶ Before You Begin

If possible, schedule time to review students' work and plan for Day 2 of this lesson with your grade-level team.

▶ Vocabulary

x-coordinate • *y*-coordinate

Standards

Focus Cluster
• Graph points on the coordinate plane to solve real-world and mathematical problems.

① Warm Up 5 min	**Materials**	
Mental Math and Fluency Students compare the values of expressions without evaluating them.		5.OA.2

②a Focus 55–65 min		
Math Message Students predict whether a rule for changing the coordinates of an image on a coordinate grid will produce a desired outcome and then check their predictions.	*Math Journal 1*, pp. 141–142; ruler	5.G.1, 5.G.2 SMP2, SMP5
Changing the Logo Students share their predictions and discuss whether their predictions were correct.	*Math Journal 1*, pp. 141–142	5.G.1, 5.G.2 SMP2, SMP5
Solving the Open Response Problem Students create and apply a rule to enlarge a picture on a coordinate grid.	*Math Masters*, pp. 151–153; ruler	5.G.1, 5.G.2 SMP1, SMP2, SMP5, SMP8

Getting Ready for Day 2 →

Review students' work and plan discussion for reengagement.

Math Masters, p. TA4;
Math Masters, p. TA31 (optional);
students' work from Day 1

Go Online to see how mastery develops for all standards within the grade.

1 Warm Up 5 min

▶ Mental Math and Fluency

Display expressions and ask students to answer *yes* or *no* to the related question. Encourage students to share how they know. *Leveled exercises:*

●○○ Is the value of the expression greater than 45×9?

$(45 \times 9) - 50$ No.

$8 + (45 \times 9)$ Yes.

$2 \times (45 \times 9)$ Yes.

●●○ Is the value of the expression less than $798 + 212$?

$(798 + 212) - 125$ Yes.

$(798 + 212) - (798 + 212)$ Yes.

$(798 + 212) / 2$ Yes.

●●● Is the value of the expression exactly half of $5{,}937 \times 31$?

$(5{,}937 \times 31) \div 2$ Yes.

$(5{,}937 \times 31) \times \frac{1}{2}$ Yes.

$2 \times (5{,}937 \times 31)$ No.

2a Focus 55–65 min

▶ Math Message

Math Journal 1, pp. 141–142

Complete journal pages 141 and 142. Be prepared to share your thinking with a partner. GMP2.3, GMP5.2

▶ Changing the Logo

Math Journal 1, pp. 141–142

| WHOLE CLASS | SMALL GROUP | PARTNER | INDEPENDENT |

Math Message Follow-Up Have students share their answers to the Math Message with a partner. Invite partnerships to share their predictions from Problem 1 with the class. Then have a volunteer list the new coordinates from Problem 2 and display the original trapezoid and new trapezoid from Problem 3. GMP2.3, GMP5.2

Standards and Goals for
Mathematical Process and Practice

SMP1 **Make sense of problems and persevere in solving them.**
GMP1.2 Reflect on your thinking as you solve your problem.

SMP2 **Reason abstractly and quantitatively.**
GMP2.3 Make connections between representations.

SMP5 **Use appropriate tools strategically.**
GMP5.2 Use tools effectively and make sense of your results.

SMP8 **Look for and express regularity in repeated reasoning.**
GMP8.1 Create and justify rules, shortcuts, and generalizations.

Professional Development

The focus of this lesson is GMP5.2. Students need to use tools correctly and efficiently, and then understand how their results help them answer a question or solve a problem. In this lesson students use coordinate grids and tables to solve a problem and interpret the information they get from using these tools.

Go Online for more information on SMP5 in the *Implementation Guide*.

Math Journal 1, p. 141

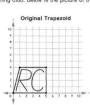

Logo

Lesson 4-10
DATE   TIME

Amy is designing a logo for her school club. She plans to put a trapezoid around the letters RC, which stand for Running Club. Below is the picture of the original trapezoid she drew on a coordinate grid.

Original Trapezoid

Amy decides she wants to include the school's name, so she needs to make the trapezoid wider. She does not want it to be taller. She developed a rule to help her fix the drawing.

Amy's Rule: Double the first coordinate of all the points.

1. If Amy uses her rule, what do you think the new trapezoid will look like? Why? Be specific in your description.
Sample answer: I think the trapezoid will be twice as wide, because I will multiply the x-coordinates by 2, and that will make the x-coordinates twice as large, which will move the points to the right. I do not think the height of the trapezoid will change, because Amy's rule does not change the y-coordinates.

5.G.1, 5.G.2, SMP2, SMP5   141

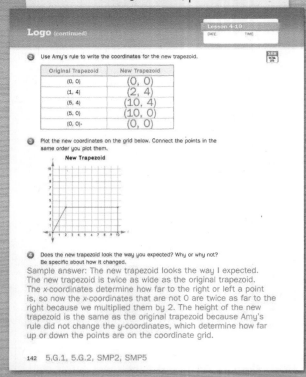
Remind students that the **x-coordinate** is listed first and gives the distance to travel along the *x*-axis horizontally, and the **y-coordinate** is listed second and gives the distance to travel along the *y*-axis vertically. Ask partners to verify that their solutions match. Use gestures in conjunction with modeling to help students visualize how the shape changes when the *x*-coordinate is doubled. Ask: *Did your new trapezoid look the way you expected? How did you answer Problem 4?* GMP5.2 Answers vary. Have students discuss how they could use the image to compare the width and height of the original and new trapezoids. Students might suggest counting units on the grid to determine the width and height or using a ruler to measure. Ask: *How can we use the picture to show that the new trapezoid is two times as wide as the original trapezoid?* GMP2.3, GMP5.2 Sample answer: I can count 5 units between (0, 0) and (5, 0) in the original trapezoid to find the width. I count 10 units between (0, 0) and (10, 0) in the new trapezoid. 10 units is twice as much as 5. This makes sense because Amy's rule said to double the *x*-coordinates.

Ask: *What happened to the point (0, 0) in the new trapezoid when you applied the rule?* Sample answer: When you double the first coordinate of (0, 0), the coordinate does not change because $0 * 2 = 0$. The bottom left corner of both the original and the new trapezoids will share the same point of (0, 0).

Once students can explain how the new trapezoid differs from the original trapezoid, ask: *What tools did you use in the Math Message?* We used a table of coordinates and a coordinate grid. *How did you use the tools?* Sample answer: The table helped us organize the new coordinates when applying the rule. Plotting the coordinates on the grid helped us see how the rule changed the image. Explain that mathematicians often use tools to help them solve problems or answer questions. Today students will have another opportunity to use tools effectively and make sense of the results as they solve a real-world problem using a coordinate grid. GMP5.2

> **NOTE** In this lesson students transform figures on a coordinate grid by applying a rule to the coordinates and plotting the new image on the coordinate grid. The new image is called a transformation image. Students do not need to use the words *transform* and *transformation* when talking about the images, but some might find the terms useful.

▶ Solving the Open Response Problem

Math Masters, pp. 151–153

| WHOLE CLASS | SMALL GROUP | PARTNER | INDEPENDENT |

Distribute *Math Masters,* pages 151–153. Read the problem as a class. Have rulers available. Partners or groups should work together to ensure they understand the directions. Tell students that artists sometimes make smaller sketches of images before making larger images. Coordinates and coordinate grids are tools that can help them make the larger images. GMP5.2

As students make sense of the problem and begin planning, you may want to ask them to think about the size of a standard folder. Consider holding up the original picture on *Math Masters,* page 151 and a student's folder to show the difference in size. The new picture does not have to fill the entire space on the folder, but students should consider the size of a folder as they plan how to use the coordinate grid to enlarge the picture. GMP5.2 Remind them to connect the points in the same order that they plot them so they draw the correct image.

Differentiate **Adjusting the Activity**

If students struggle to get started, suggest that they focus on making the book either taller or wider. Remind them of how they made the trapezoid wider in the Math Message. GMP1.2, GMP2.3, GMP5.2, GMP8.1

To challenge students, ask them to predict how much larger the area of the new book will be than the original book. Have them estimate the area of the original and new book for comparison by counting the squares and partial squares in the picture.

Students can work together on the problem, but everyone should record a solution individually. Monitor students as they work. Ask: *Why did you choose this rule?* GMP1.2, GMP8.1 Sample answer: I knew this rule would make the first coordinate of each ordered pair move to the right (if it wasn't 0), and this makes the book wider. I knew this rule would make the second coordinate of each ordered pair move up (if it wasn't 0), and this makes the book taller. *What specific information can you give when you make your prediction about the new picture?* Sample answer: I can tell how much taller or wider my picture will get.

NOTE It is possible for students to use a rule that results in points in a different quadrant of the coordinate grid. Students will only have experience with Quadrant I in Grade 5. (*See the Mathematical Background section of the Unit 4 Organizer for more information.*) If students use a rule that gives negative numbers, encourage them to think about what this might look like, but ask them to develop another rule that does not involve negative coordinates.

Folder Art — Lesson 4-10

Jake is designing a picture for the cover of his reading folder. He draws a picture of a book on a coordinate grid. After drawing the picture, Jake decides he wants to put a picture of a larger book on his folder.

Jake's Original Picture

① Write a rule that Jake can use to make the picture of the book larger.
Answers vary. See sample students' work on page 397 of the *Teacher's Lesson Guide.*

② Describe what you think the new picture will look like. Be specific about how you think the book will change.
Answers vary. See sample students' work on page 397 of the *Teacher's Lesson Guide.*

5.G.1, 5.G.2, SMP1, SMP2, SMP5, SMP8   151

Folder Art (continued) — Lesson 4-10

③ Use your rule to write the new coordinates.

Original Book	New Book
(2, 0)	
(0, 1)	
(0, 5)	
(1, 5)	
(2, 4)	
(0, 5)	
(2, 4)	
(2, 0)	
(4, 1)	
(4, 5)	
(3, 5)	
(2, 4)	
(4, 5)	

Answers vary. See sample students' work on page 397 of the *Teacher's Lesson Guide.*

④ Plot the coordinates for the new book on *Math Masters,* page 153. Connect the points in the same order that you plot them.

⑤ Compare your prediction in Problem 2 to the new picture. How was your prediction correct? How was your prediction incorrect?
Answers vary. See sample students' work on page 397 of the *Teacher's Lesson Guide.*

152   5.G.1, 5.G.2, SMP1, SMP2, SMP5, SMP8

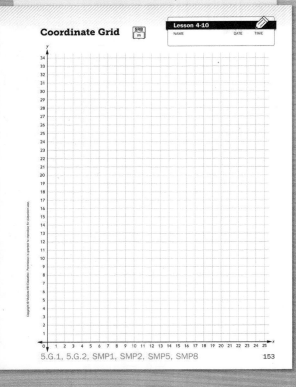

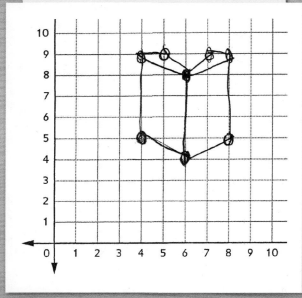

Summarize Ask: *How are the list of coordinates and the plotted points alike? How are they different?* GMP1.2, GMP2.3, GMP5.2 Sample answer: Both the list of coordinates and their plotted points contain the same mathematical information. They are different because the list of coordinates tells you where the points will be, but the plotted coordinates show you how they are spaced. Remind students that they will have a more in-depth conversation about the problem during the reengagement discussion.

Collect students' work so that you can evaluate it and prepare for Day 2.

Getting Ready for Day 2

Math Masters, p. TA4

Planning a Follow-Up Discussion

Review students' work. Use the Reengagement Planning Form (*Math Masters,* page TA4) and the rubric on page 394 to plan ways to help students meet expectations for both the content and process/practice standards. Look for common misconceptions, such as moving the book on the coordinate grid without enlarging it, as well as correct rules and predictions that result in and describe an increase in both the width and height of the image.

Organize the discussion in one of the ways on the next page or in another way you choose. You may wish to develop a student-friendly rubric using *Math Masters,* page TA31 and facilitate a peer review as described in Lesson 2-9 on *Teacher's Lesson Guide,* page 164. If students' work is unclear or if you prefer to show work anonymously, rewrite the work for display.

1. Display an answer to Problem 4, such as Student A's, that shows a rule that did not enlarge the book, but moved it to a new location on the grid. Tell students to compare the image of the original book and the new book. Ask: *Are they the same or different?* GMP1.2, GMP2.3, GMP5.2 Sample answer: The original book and the new book are the same size. The rule didn't make the book larger. The new book was moved over 4 units and moved up 4 units. *What rule did this student use?* GMP8.1 Sample answer: The rule was to add 4 to both of the coordinates in the ordered pairs. *How can this student change the rule to make the book larger?* Sample answer: The student can multiply one or both coordinates instead of adding.

2. Show a student's rule, prediction, new coordinates, and new picture that do not all match, such as Student B's work. Ask:
   - *Look at this student's rule in Problem 1 and prediction in Problem 2. Do you agree that the folder will be taller following that rule? Why or why not?* GMP1.2, GMP8.1 No. If you multiply the x-coordinate by 6, the book will get wider, not taller.
   - *Look at the student's new coordinates in Problem 3 and new picture in Problem 4. Did the student follow the rule written in Problem 1? How do you know?* GMP1.2, GMP2.3, GMP5.2 No. Sample answer: The student's rule said to multiply the x-coordinate by 6, but the student multiplied both the x-coordinate and y-coordinate by 6. *How did multiplying both coordinates by 6 change the picture?* Sample answer: The new picture was 6 times as wide and 6 times as tall. *How could you help this student improve this work?* Sample answer: I would tell the student to either change the rule to multiply both coordinates by 6 or correct the coordinates in Problem 3 to match the rule in Problem 1 and graph the new coordinates.

   Consider choosing and discussing another student's work in which the rule, prediction, new coordinates, and new picture match.

3. Show an explanation for Problem 5 that is incomplete or lacks details, such as in Student C's work. GMP1.2, GMP2.3, GMP5.2 Ask: *What part of the explanation is correct?* Sample answer: This explanation says that the prediction was correct. It says that the new book got taller. *How could this student add details so the explanation is clearer?* Sample answer: The student could tell how much taller the new book is. For example, he or she could say that the new height of the book is 10 units and the height of the original book was 5 units, so the new book is twice as tall as the original book.

Planning for Revisions

Have copies of *Math Masters*, pages 151–153 or extra paper available for students to use in revisions. You might want to ask students to use colored pencils so you can see what they revised.

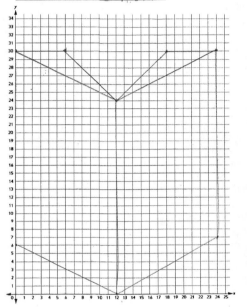

Sample student's work, Student B

1. Write a rule Jake can use to make the picture of the book larger.
Jake can multiply the X coordinate. by 5

2. Describe what you think the new picture will look like. Be specific about how you think the book will change.
The Folder will be taller.

3. Use your rule to write the new coordinates.

Original Book	New Book
(2, 0)	(12, 0)
(0, 1)	(0, 6)
(0, 5)	(0, 30)
(1, 5)	(6, 30)
(2, 4)	(12, 24)
(0, 5)	(0, 30)
(2, 4)	(12, 24)
(2, 0)	(12, 0)
(4, 1)	(24, 6)
(4, 5)	(24, 30)
(3, 5)	(18, 30)
(2, 4)	(12, 24)
(4, 5)	(24, 30)

4.

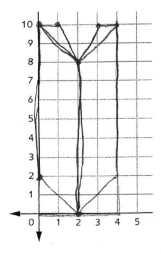

Sample student's work, Student C

5. Compare your prediction in Problem 2 to the new picture. How was your prediction correct? How was your prediction incorrect?
It was correct, it got taller

Folder Art

Day 2: Reengagement

▶ **Before You Begin**
Have extra copies of *Math Masters,* pages 151–153 available for students to revise their work.

Standards

Focus Cluster
• Graph points on the coordinate plane to solve real-world and mathematical problems.

2b Focus   50–55 min

	Materials	
Setting Expectations Students discuss what a complete response to the open response problem would include and how to ask questions about other students' work respectfully.	Standards for Mathematical Process and Practice Poster, Guidelines for Discussions Poster	SMP1, SMP2, SMP5, SMP8
Reengaging in the Problem Students analyze and discuss others' rules, predictions, and pictures.	*Math Masters,* p. TA31 (optional); selected samples of students' work	5.G.1, 5.G.2 SMP1, SMP2, SMP5, SMP8
Revising Work Students revise their work from Day 1.	*Math Masters,* pp. 151–153 (optional); students' work from Day 1; colored pencils (optional)	5.G.1, 5.G.2 SMP1, SMP2, SMP5, SMP8

✓ **Assessment Check-In**   See page 396 and the rubric below.
Expect most students to be able to plot coordinates in the first quadrant of a coordinate grid to represent a picture that has been enlarged.

5.G.2

SMP5

Goal for Mathematical Process and Practice **GMP5.2** Use tools effectively and make sense of your results.	Not Meeting Expectations	Partially Meeting Expectations	Meeting Expectations	Exceeding Expectations
	Incorrectly lists or plots the new coordinates **and** incorrectly describes the changes to the coordinates and plotted image of the original book.	Lists and plots the new coordinates with minimal or no errors **or** correctly describes the changes to the coordinates and plotted image of the original book, specifically referring to how the dimensions (length and width or area) changed.	Lists and plots the new coordinates with minimal or no errors **and** correctly describes the changes to the coordinates and plotted image of the original book, specifically referring to how the dimensions (length and width or area) changed.	Meets expectations and provides a detailed comparison of the prediction to the attributes of the new image, such as: • specifically stating which new dimensions are n times as large as the original • correctly estimating or calculating the relationship between the area of the new and old images

3 Practice   10–15 min

Math Boxes 4-10: Preview for Unit 5 Students preview skills and concepts for Unit 5.	*Math Journal 1,* p. 143	See page 396.
Home Link 4-10 **Homework** Students enlarge a picture of a house on a coordinate grid.	*Math Masters,* pp. 154–155	5.NF.3, 5.G.1, 5.G.2 SMP1, SMP2, SMP5

Go Online ▷ to see how mastery develops for all standards within the grade.

2b Focus

50–55 min

NOTE These Day 2 activities will ideally take place within a few days of Day 1. Prior to beginning Day 2, see Planning a Follow-Up Discussion from Day 1.

▶ ### Setting Expectations

| WHOLE CLASS | SMALL GROUP | PARTNER | INDEPENDENT |

Briefly review the open response problem from Day 1. Ask: *What did the problem ask you to do?* GMP1.2 Sample answer: We had to develop and apply a rule to enlarge a picture on the coordinate grid, and compare our new picture to our prediction. *What do you think a good response would include?* GMP1.2, GMP2.3, GMP5.2, GMP8.1 Sample answers: There should be a rule that makes the picture larger and a prediction of what the new picture will look like. The list of the coordinates and the new picture of the book should follow the rule correctly to make the book larger. There should be a comparison of the prediction with the plotted image of the new book.

After this brief discussion, tell students they are going to look at other students' work and see how they can use others' ideas to make their own work better. Refer to **GMP5.2** on the Standards for Mathematical Process and Practice Poster. Explain that students will analyze how others used the coordinate grid and make sense of their results.

Refer to your list of discussion guidelines from previous units. Remind students that if they are confused during the discussion, they should use sentence frames such as:

- I don't understand _____. • I wonder why _____.
- Could you explain _____?

▶ ### Reengaging in the Problem

| WHOLE CLASS | SMALL GROUP | PARTNER | INDEPENDENT |

Students reengage in the problem by analyzing and critiquing other students' work in pairs and in a whole-group discussion. Have students discuss with partners before sharing with the whole group. Guide this discussion based on the decisions you made in Getting Ready for Day 2. GMP1.2, GMP2.3, GMP5.2, GMP8.1

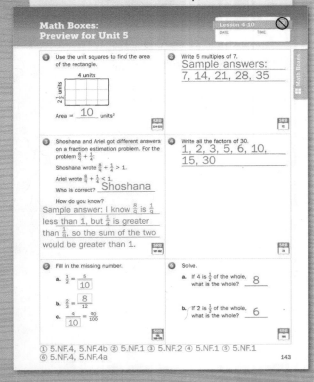

Math Masters, p. 154

▶ Revising Work

| WHOLE CLASS | SMALL GROUP | PARTNER | INDEPENDENT |

Pass back students' work from Day 1. Before students revise anything, ask them to examine their responses and decide how they could be improved. Ask the following questions one at a time. Have partners discuss their responses and give a thumbs-up or thumbs-down based on their own work.

- *Did you write a rule that will make the picture larger?* GMP8.1
- *Did you include specific information in your prediction about how the new picture will change?* GMP1.2, GMP2.3
- *Did you use the table correctly to list the coordinates of the new picture of the book?* GMP5.2
- *Did you use the coordinate grid correctly to plot the image of the new book?* GMP5.2
- *Did you give details when you explained how your prediction was correct or incorrect?* GMP1.2, GMP5.2

Tell students they now have a chance to revise their work. Ask those who wrote complete and correct rules, predictions, and explanations on Day 1 to write a new rule that will change the width and height in different proportions. Help students see that the rules, predictions, and explanations presented during the reengagement discussion are not the only correct ones. Tell students to add to their earlier work using colored pencils or to use another sheet of paper, instead of erasing their original work.

Summarize Have students reflect on their work and revisions. Ask: *Did you change your rule? If so, how did it change your picture?* GMP8.1 Answers vary. *Did you improve your explanation of how your prediction compared to your new picture? If so, how?* GMP1.2 Answers vary.

 Assessment Check-In 5.G.2

Collect and review students' revised work. For the content standard, expect most students to correctly plot the coordinates to represent the new book. You can use the rubric on page 394 to evaluate students' revised work for **GMP5.2**.

 Evaluation Quick Entry Go online to record students' progress and to see trajectories toward mastery for these standards.

Go Online for optional generic rubrics in the *Assessment Handbook* that can be used to assess any additional GMPs addressed in this lesson.

Sample Students' Work—Evaluated

See the sample in the margin. This work meets expectations for the content standard because the student correctly plotted the coordinates to represent the new book. The work meets expectations for the mathematical process and practice standard because the student effectively used the coordinate table and grid to correctly list and plot the new coordinates for the given rule. Furthermore, the student correctly described the changes in the results by saying that "it got longer and wider," specifically pointing to the effect on both the length and width of the figure and indicating the new and original dimensions. If the student had said that the new width and length are 6 times the original width and length, the work would exceed expectations. **GMP5.2**

Go Online for other samples of evaluated students' work.

③ Practice   10–15 min

▶ ## Math Boxes 4-10: Preview for Unit 5

Math Journal 1, p. 143

| WHOLE CLASS | SMALL GROUP | PARTNER | INDEPENDENT |

Mixed Practice Math Boxes 4-10 are paired with Math Boxes 4-15. These problems focus on skills and understandings that are prerequisite for Unit 5. You may want to use information from these Math Boxes to plan instruction and grouping in Unit 5.

▶ ## Home Link 4-10

Math Masters, pp. 154–155

Homework Students enlarge a picture of a house on a coordinate grid. **GMP1.2, GMP2.3, GMP5.2**

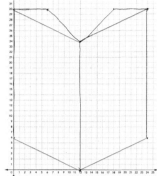

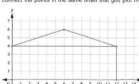

Addition and Subtraction of Decimals with Hundredths Grids

Overview Students shade grids to represent and solve decimal addition and subtraction problems.

▶ **Before You Begin**
For the Math Message, display the shaded grid in the margin of page 400. For Part 2, each student needs at least two copies of *Math Masters*, page TA22 and two different colored pencils. For the optional Extra Practice activity, write problems below the grids on *Math Masters*, page TA22.

▶ **Vocabulary**
tenth • hundredth • addend

Standards

Focus Clusters
- Understand the place value system.
- Perform operations with multi-digit whole numbers and with decimals to hundredths.

① Warm Up 5 min	**Materials**	
Mental Math and Fluency Students use place-value clues to write decimals.	slate	**5.NBT.1**

② Focus 35–40 min		
Math Message Students examine a shaded hundredths grid.		5.NBT.3, 5.NBT.3a, 5.NBT.7 SMP2
Adding Decimals with Grids Students discuss strategies for using grids to solve decimal addition problems.	*Math Masters*, p. TA22; colored pencils	5.NBT.3, 5.NBT.3a, 5.NBT.7 SMP2, SMP6
Subtracting Decimals with Grids Students discuss strategies for using grids to solve decimal subtraction problems. They complete a practice page.	*Math Journal 1*, p. 144; *Math Masters*, p. TA22; colored pencils; base-10 blocks (optional)	5.NBT.3, 5.NBT.3a, 5.NBT.7 SMP2, SMP6
✓ **Assessment Check-In** See page 404. Expect most students to be able to shade grids to represent and solve decimal addition and subtraction problems that involve decimals to hundredths.	*Math Journal 1*, p. 144	5.NBT.3, 5.NBT.3a, 5.NBT.7

③ Practice 20–30 min		
Playing *Over and Up Squares* **Game** Students practice using ordered pairs to plot points on a coordinate grid.	*Student Reference Book*, p. 317; per partnership: *Math Masters*, p. G25; colored pencils; two 6-sided dice	5.G.1 SMP7
Math Boxes 4-11 Students practice and maintain skills.	*Math Journal 1*, p. 145	See page 405.
Home Link 4-11 **Homework** Students shade grids to solve decimal addition and subtraction problems.	*Math Masters*, p. 156	5.NBT.3, 5.NBT.3a, 5.NBT.5, 5.NBT.7 SMP2

Go Online to see how mastery develops for all standards within the grade.

Differentiation Options

Readiness 10–20 min	**Enrichment** 10–20 min	**Extra Practice** 10–20 min
WHOLE CLASS · **SMALL GROUP** · PARTNER · INDEPENDENT	WHOLE CLASS · SMALL GROUP · **PARTNER** · INDEPENDENT	WHOLE CLASS · SMALL GROUP · **PARTNER** · INDEPENDENT

Readiness

Exchanging Base-10 Blocks

5.NBT.3, 5.NBT.3a, 5.NBT.7, SMP2

Activity Card 52;
Math Masters, p. TA22;
per group: 30 longs, 30 cubes,
1 flat; two 6-sided dice

To prepare for adding decimals using grids, students make exchanges with base-10 blocks. Display a flat, telling students that it represents one. Ask: *What does 1 long represent?* 1 tenth *What does 1 cube represent?* 1 hundredth Have small groups follow the directions on Activity Card 52 to practice representing decimals. **GMP2.1, GMP2.2** As you observe, encourage students to use place-value language to describe their exchanges and to draw connections between that base-10 block representations and the grid representations. **GMP2.3**

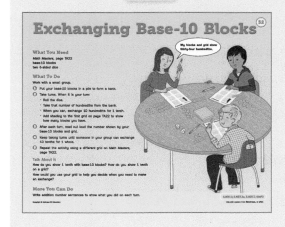

Enrichment

Writing Decimal Addition and Subtraction Fact Families

5.NBT.3, 5.NBT.3a, 5.NBT.7, SMP2

AActivity Card 53; *Math Masters*, p. TA22; number cards 0–9 (4 of each); crayons or colored pencils

To extend their work with addition and subtraction on grids, students shade grids to represent two decimals and write related addition and subtraction number sentences to represent the grids.
GMP2.1, GMP2.2, GMP2.3

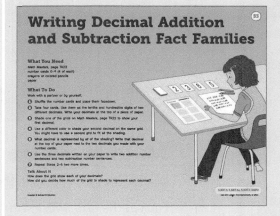

Extra Practice

Solving More Decimal Addition and Subtraction Problems with Grids

5.NBT.3, 5.NBT.3a, 5.NBT.7, SMP2

Math Masters, p. TA22;
crayons or colored pencils

For additional practice solving decimal addition and subtraction problems, students shade grids to represent and solve problems. **GMP2.1, GMP2.2** On *Math Masters*, page TA22, write a decimal addition or subtraction problem under each grid, leaving the grids, sums, and differences blank. Problems can be tailored to the needs of individual students. For example, some may benefit from working only with decimals through tenths, while others may need more practice with problems in which the hundredths digits add up to more than 1 tenth.

English Language Learner

Beginning ELL Introduce or review multiple contexts for using the word *grid* by preparing a pictorial 4-Square Graphic Organizer (*Math Masters,* page TA2) for the term. Show a variety of images, such as a number grid, a multiplicatiton grid, grid paper, and a coordinate grid. Point to each image and then to the term itself in the middle of the organizer. Have students repeat *grid* to help them understand that it is used in many different contexts in mathematics. Use gestures to indicate that in each case a grid is a series of evenly spaced parallel and perpendicular lines.

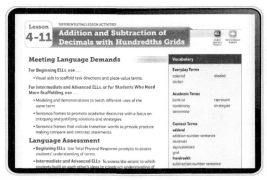

Differentiation Support pages are found in the online Teacher's Center.

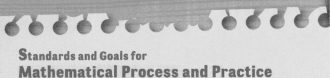

Standards and Goals for
Mathematical Process and Practice

SMP2 **Reason abstractly and quantitatively.**
　　GMP2.1 Create mathematical representations
　　using numbers, words, pictures, symbols,
　　gestures, tables, graphs, and concrete objects.
　　GMP2.2 Make sense of the representations you
　　and others use.

SMP6 **Attend to precision.**
　　GMP6.1 Explain your mathematical thinking
　　clearly and precisely.

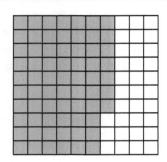

Shaded grid for Math Message

NOTE Although students are
expected to understand decimal
place value to thousandths,
5.NBT.7 only requires them to compute
with decimals to hundredths. Lesson
4-11 focuses on adding and subtracting
decimals to hundredths. Lessons 4-12
and 4-13 extend some computation
methods to thousandths, but students
will not be assessed on computation
with decimals to thousandths.

1 Warm Up　　5 min

▶ Mental Math and Fluency

Read number riddles and have students write numbers on their slates.
Emphasize that there are many correct answers. *Leveled exercises:*
Sample answers given.

◉○○　I have a 7 in the tens place and a 5 in the tenths place.　72.5
　　　I have a 3 in the tenths place and a 9 in the tens place.　290.3

◉◉○　I have a 9 in the hundreds place and a 1 in the hundredths
　　　place.　924.813
　　　I have a 6 in the hundredths place and $\frac{1}{3}$ of that number in
　　　the hundreds place.　214.96

◉◉◉　I have a 4 in the thousands place and a 7 in the thousandths
　　　place.　354,088.267
　　　I have a 4 in the hundredths place and another 4 that's worth
　　　100 times as much.　94.64

2 Focus　　35–40 min

▶ Math Message

Display the grid shown in the margin.

Look at the grid. Work with a partner to answer these questions:

How much is shaded in blue?
How much is shaded in red?
How much is shaded in all?
Write an addition number sentence that represents the grid.　GMP2.2

▶ Adding Decimals with Grids

Math Masters, p. TA22

| WHOLE CLASS | SMALL GROUP | PARTNER | INDEPENDENT |

Math Message Follow-Up Invite volunteers to explain their answers
to the Math Message. Ask: *How much is shaded in blue?* 0.2 *How do you
know?* Sample answer: Each column is a **tenth,** and two columns are
shaded. *How much is shaded in red?* 0.47 *How do you know?* Sample
answer: Columns show tenths and squares show **hundredths.** There are
4 columns plus 7 squares, so that's 0.47 in all. *How much is shaded in all?*
0.67 *How do you know?* Sample answer: Each column is 1 tenth, and there
are 6 columns shaded. There are 7 more small squares shaded and that's
7 hundredths. So 6 tenths and 7 hundredths are shaded, or 0.67 in all.
What number sentence would match what is shown on this grid?
0.2 + 0.47 = 0.67 Record the number sentence below the grid.

Remind the class that decimals are often used in real-world situations that require knowing how to add and subtract them. Explain to students that today they will use grids to help them solve decimal addition and subtraction problems. In the next few lessons they will explore other strategies for adding and subtracting decimals.

Distribute several copies of *Math Masters*, page TA22 and two colored pencils to each student. Have partners work together to solve the decimal addition problems below by shading hundredths grids, using one color to represent the first **addend** and a second color to represent the second addend. To find the sum, they determine the total amount shaded. GMP2.1, GMP2.2 As students work, circulate and make note of the different ways they use the grids to solve the problems. Some students may be able to add decimals without using the grids. Encourage them to check and justify their answers using the grids.

0.33 + 0.54 0.87   0.65 + 0.28 0.93   0.76 + 0.45 1.21

After most partnerships are finished, invite them to share their shaded grids for 0.33 + 0.54. Encourage them to explain as clearly as possible how they used the grids to solve the problems. GMP6.1 Compare different ways of shading and discuss how they show different ways of thinking about combining the decimals. *Sample strategies:*

Strategy 1: I shaded 33 squares for 0.33. Then I started in a new column and shaded 54 squares for 0.54. (*See margin.*) There were 87 squares shaded in all, so the sum is 87 hundredths, or 0.87.

Strategy 2: I shaded 3 columns and then 3 more small squares for 0.33. Then I shaded 5 columns for 0.54. I still needed to shade 4 more small squares for 0.54, so I shaded them in the partially empty column I had for 0.33. (*See margin.*) There are 8 full columns plus 7 more squares shaded. That's 0.87 in all.

If no one shares a strategy similar to Strategy 2, demonstrate it, highlighting how putting hundredths together in the same column makes it easier to see the sum on the grid. For example, when the hundredths are put together, it is easy to see that there are only 3 empty squares in the column, so there are 7 hundredths shown.

Have the class record an addition number sentence under the grid to represent the problem. 0.33 + 0.54 = 0.87 It may be helpful for students to write each addend in the color they used to shade that addend. For example, 0.33 + 0.54 = 0.87.

Next have partnerships share their shaded grids and clearly explain their strategies for solving 0.65 + 0.28. GMP6.1 Be sure to discuss an example similar to Strategy 2, in which a student used the hundredths from the second addend to fill up the partial column from the first addend. (*See margin.*)

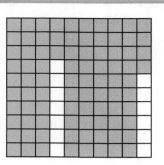

Grid shaded for 0.33 + 0.54
using Strategy 1

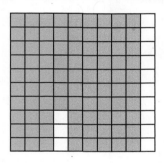

Grid shaded for 0.33 + 0.54
using Strategy 2

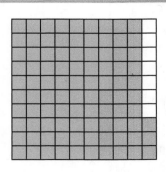

Grid shaded for 0.65 + 0.28

Guide the discussion to help students see that the shaded hundredths fill an additional tenths column. Ask: *How many columns did you have to shade for the tenths?* 6 from 0.65 and 2 from 0.28, or 6 + 2 = 8 in all *How many squares in all did you have to shade for the hundredths?* 5 from 0.65 and 8 from 0.28, or 5 + 8 = 13 in all *What did you notice when you shaded the hundredths?* Together the hundredths filled up a whole column plus 3 more squares. Point out that although it is possible to think about the problem by shading in 8 tenths and 13 hundredths, it can be seen from the finished grid that 10 of the hundredths make an additional column or 1 tenth, so the answer can be expressed as 9 tenths and 3 hundredths, or 0.93. GMP2.2 To summarize, have students record an addition number sentence under the grid. 0.65 + 0.28 = 0.93

Finally, discuss students' strategies for solving 0.76 + 0.45. Students should have found that they needed two grids to show the sum. Have volunteers share their grids and strategies or demonstrate a new strategy they discovered during the discussion of the other problems. GMP6.1
Sample grids and strategy:

- I shaded the tenths first. I needed to shade 7 columns for the 7 tenths in 0.76 and 4 columns for the 4 tenths in 0.45. That was 11 columns in all, so I filled up one grid and started on a second grid. Then I shaded 6 squares for the 6 hundredths in 0.76 and 5 squares for the 5 hundredths in 0.45. That filled up another column, plus one more square. I saw that I had shaded a full grid, and that represents 1. I also shaded 2 more full columns, and that shows 2 tenths. Finally, I had shaded 1 more square, or 1 hundredth. The sum is 1 plus 2 tenths plus 1 hundredth, or 1.21.

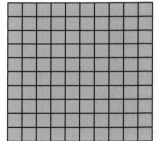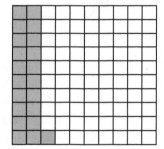

Highlight the idea that 10 tenths make 1 whole, so the 11 tenths in the problem are the same as 1 whole and 1 tenth, or 1.1. Similarly, 10 hundredths make 1 tenth, so the 11 hundredths are the same as 1 tenth and 1 hundredth, or 0.11. Point out that these equivalencies continue the place-value patterns students have used before to add and subtract whole numbers: 1 hundred is the same as 10 tens, 1 ten is the same as 10 ones, 1 one is the same as 10 tenths, and 1 tenth is the same as 10 hundredths. To summarize, have students record an addition number sentence under the grids. 0.76 + 0.45 = 1.21

▶ Subtracting Decimals with Grids

Math Journal 1, p. 144; *Math Masters,* p. TA22

WHOLE CLASS | SMALL GROUP | PARTNER | INDEPENDENT

Have students shade a hundredths grid to represent 0.64. Ask: *How could we use this grid to represent the problem 0.64 – 0.3?* Sample answer: We need to take away 0.3. We can cross out the amount we need to take away or shade the amount we need to take away darker. Have students try a strategy for representing the subtraction and share their shaded grids. GMP2.1 *Sample grids:*

Crossing off 3 columns to represent taking away 3 tenths

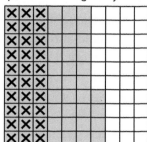

Shading 30 squares darker to represent taking away 30 hundredths

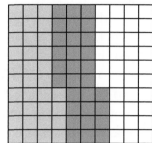

Ask: *How does the grid show 0.64?* It's the total amount shaded. *How does the grid show 0.3?* It's the part I crossed off or shaded darker. *What is the answer to the subtraction problem?* 0.34 *How do you know?* GMP2.2 It's the part that isn't crossed off or shaded darker. Have students record a subtraction number sentence below the grid to summarize the problem. 0.64 – 0.3 = 0.34

Ask: *How is shading a grid to subtract decimals different from shading a grid to add decimals?* Sample answer: Instead of making the shading for the two numbers in the problem separately, you shade to show the bigger number and then shade or cross out to show the number being taken away. *Why would we not want to shade the two numbers in a subtraction problem separately?* GMP2.2 Sample answer: Shading them separately would be like putting them together, and we need to show taking away.

Have partnerships solve the decimal subtraction problems below using hundredths grids. They should shade to show the starting number and then shade darker or cross out to show what is being taken away. Have students summarize their work by writing subtraction number sentences under their grids. If students are able to subtract the decimals without using grids, suggest that they check and justify their answers using the grids. GMP2.1, GMP2.2

0.7 – 0.39 0.31   0.86 – 0.71 0.15   0.52 – 0.14 0.38

Adjusting the Activity

Differentiate To support students working on the subtraction problems, suggest building the two numbers out of base-10 blocks. Have students place the blocks for the smaller number on top of the blocks for the larger number. The uncovered part is the difference. Connect this idea to shading a grid for the larger number and crossing out shading for the smaller number. The shading that is not crossed out represents the difference.

Go Online Differentiation Support

Professional Development

The activities in this lesson build readiness for working with decimal addition and subtraction algorithms. If you see students using strategies that parallel steps of an algorithm, make notes so that you can refer back to their work with grids when discussing the algorithms. *For example:*

- If students shade all the tenths from both addends first and then shade all the hundredths, this is similar to the partial-sums addition algorithm.

- If students shade the hundredths from both addends in the same column and notice that the hundredths combine to make a tenth, this is similar to trading 10 hundredths for 1 tenth using column addition or recording 10 hundredths as 1 tenth using U.S. traditional addition.

- If students talk about "breaking apart" a column so they can cross out more hundredths, this is similar to making a trade in trade-first or U.S. traditional subtraction.

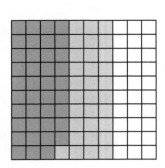

Grid shaded for 0.7 − 0.39 = 0.31

After most have finished, have partnerships share their grids and clearly explain their strategies. **GMP6.1** Highlight strategies in which students thought about "breaking" a tenth (or a column) into 10 hundredths (or 10 squares). *Sample explanations:*

- For 0.7 − 0.39: I shaded 7 columns to represent 7 tenths, or 70 hundredths. Then I double-shaded 3 columns to represent the 3 tenths in 0.39. I still had to take away 9 hundredths, so I double-shaded 9 squares, or 9 hundredths, of the next full column. That left 3 full columns and 1 square that were only shaded once, which means the answer is 0.31. (*See margin.*)

- For 0.86 − 0.71: I shaded 8 columns plus 6 more squares to show 0.86. Then I crossed out 7 columns and 1 square to show the number being subtracted, 0.71. There were 9 squares in one column and 6 squares in another column that were not crossed out. I thought about moving one of the squares over to fill the 9-square column and make 1 tenth. Then I could see that 1 tenth and 5 hundredths were not crossed out, so the answer is 0.15. (*See grid at the left below.*)

- For 0.52 − 0.14: I shaded 5 columns plus 2 squares to show 0.52. Then I double-shaded 1 column and 4 squares to show taking away 0.14. Three columns plus 8 squares were shaded once, so the difference is 3 tenths and 8 hundredths, or 0.38. (*See grid at the right below.*)

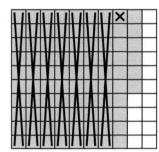
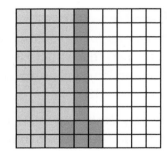

Grid shaded for 0.86 − 0.71 = 0.15   Grid shaded for 0.52 − 0.14 = 0.38

When students seem ready, have them complete journal page 144 independently or in partnerships. They shade grids to represent and solve decimal addition and subtraction problems and explain their thinking. **GMP2.1, GMP2.2, GMP6.1**

Math Journal 1, p. 144

Decimal Addition and Subtraction with Grids

Lesson 4-11

DATE   TIME

For Problems 1 and 2: Sample shading is given.

- Shade the grid in one color to show the first addend.
- Shade more of the grid in a second color to show the second addend.
- Write the sum to complete the number sentence.

①   ②

0.6 + 0.22 = __0.82__   0.18 + 0.35 = __0.53__

For Problems 3 and 4: Sample shading is given.

- Shade the grid to show the starting number.
- Cross out or shade darker to show what is being taken away.
- Write the difference to complete the number sentence.

③   ④

0.47 − 0.20 = __0.27__   0.74 − 0.36 = __0.38__

⑤ Choose one of the problems above. Clearly explain how you solved it. Sample answer: In Problem 2, I shaded 1 column to show the 1 tenth in 0.18 and 3 columns to show the 3 tenths in 0.35. Then I shaded 8 squares for the 8 hundredths in 0.18 and 5 squares for the 5 hundredths in 0.35. There were 5 full columns and 3 more squares shaded, or 0.53 in all.

144   5.NBT.3, 5.NBT.3a, 5.NBT.7, SMP2, SMP6

✔ **Assessment Check-In**   5.NBT.3, 5.NBT.3a, 5.NBT.7

Math Journal 1, p. 144

Expect most students to be able to accurately shade the grids to represent the numbers and find the decimal sums and differences in Problems 1–4 on journal page 144. Some may be able to clearly explain their strategies in Problem 5. If students struggle representing the decimals on the grids, suggest first representing each of the numbers with base-10 blocks and then referring to the blocks to help them fill in the grids.

✔ **Evaluation Quick Entry** Go online to record students' progress and to see trajectories toward mastery for these standards.

Summarize Have partnerships discuss what they found helpful about shading grids to add and subtract decimals.

Math Journal 1, p. 145

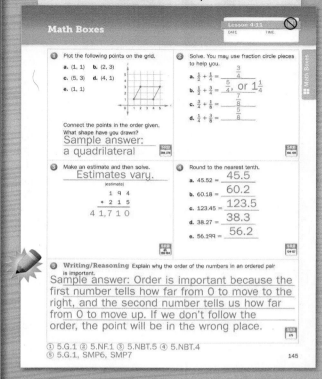

③ Practice 20–30 min

▶ **Playing *Over and Up Squares***

Student Reference Book, p. 317; *Math Masters*, p. G25

| WHOLE CLASS | SMALL GROUP | **PARTNER** | INDEPENDENT |

Students practice plotting points on a coordinate grid.

Observe
- Which students can accurately plot points on the grid?
- Which students have a strategy for choosing which roll to use as the *x*-coordinate and which to use as the *y*-coordinate?

Discuss
- *How did you decide which number to use for each coordinate?* GMP7.2
- *Explain your strategy for connecting as many points as possible.*

| Differentiate | Game Modifications | | Go Online | Differentiation Support |

▶ **Math Boxes 4-11**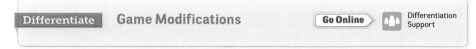

Math Journal 1, p. 145

| WHOLE CLASS | **SMALL GROUP** | PARTNER | INDEPENDENT |

Mixed Practice Math Boxes 4-11 are paired with Math Boxes 4-13.

▶ **Home Link 4-11**

Math Masters, p. 156

Homework Students shade grids to solve decimal addition and subtraction problems. GMP2.1, GMP2.2

Math Masters, p. 156

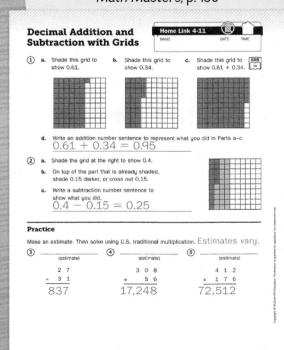

Lesson 4-12

Decimal Addition Algorithms

Overview Students review whole-number addition algorithms and use them to add decimals.

▶ **Before You Begin**
Display the partial-sums work shown on page 408 near the Math Message. For Part 2, each small group needs a piece of poster paper. If students are not familiar with the addition algorithms in this lesson, consider having the class complete the optional Readiness activity. For Part 3, if additional sets of *Fraction Of* cards are needed, make copies of *Math Masters,* pages G21 and G22. For the optional Enrichment activity, each small group will need a stopwatch.

▶ **Vocabulary**
partial-sums addition • algorithm • column addition • U.S. traditional addition • accuracy • efficiency

Standards

Focus Clusters
• Understand the place value system.
• Perform operations with multi-digit whole numbers and with decimals to hundredths.

 Warm Up 5 min

	Materials	
Mental Math and Fluency Students translate between different decimal notations.	slate	5.NBT.3, 5.NBT.3a

②Focus 35–40 min

Math Message Students add decimals using partial-sums addition.		5.NBT.7
Extending Whole-Number Addition Algorithms to Decimals Students review whole-number addition algorithms and explore how they can be used to add decimals.	*Student Reference Book,* pp. 85 and 130; *Math Masters,* p. 157, pp. TA7, TA22–TA23 (optional); poster paper	5.NBT.7 SMP1, SMP6
Practicing Decimal Addition Students practice using decimal addition algorithms and use estimates to check the reasonableness of their answers.	*Math Journal 1,* p. 146	5.NBT.4, 5.NBT.7 SMP1
✓ **Assessment Check-In** See page 410. Expect most students to be able to use an algorithm to add decimals to tenths (with and without regrouping) and decimals to hundredths with no regrouping of tenths or hundredths.	*Math Journal 1,* p. 146	5.NBT.7, SMP1

③Practice 20–30 min

Playing *Fraction Of* **Game** Students practice solving fraction-of problems.	*Student Reference Book,* pp. 306–307; *Math Masters,* p. G24; per partnership: *Fraction Of* cards; counters (optional)	5.NF.4, 5.NF.4a SMP6
Math Boxes 4-12 Students practice and maintain skills.	*Math Journal 1,* p. 147	See page 411.
Home Link 4-12 **Homework** Students solve decimal addition problems.	*Math Masters,* p. 158	5.NBT.7, 5.NF.4, 5.NF.4a SMP1

 Go Online to see how mastery develops for all standards within the grade.

Differentiation Options

Readiness | 15–30 min

WHOLE CLASS | SMALL GROUP | PARTNER | INDEPENDENT

Reviewing Addition Algorithms

5.NBT.7, SMP1

Student Reference Book,
pp. 85–88

For experience with partial-sums addition, column addition, and U.S. traditional addition, students review using these algorithms to add whole numbers. Read the partial-sums addition example on *Student Reference Book,* page 85 together. Then have students help you use partial-sums addition to solve Problem 1 in Check Your Understanding on page 88. As students propose steps, ask guiding questions like these: *What should we do first? Where should we write the 7? How did you get that number?* GMP1.2 Next have partnerships solve Problem 2 using partial-sums addition. Repeat the process for the column addition example on *Student Reference Book,* page 86 and Check Your Understanding Problems 3 and 4. Repeat again with U.S. traditional addition using *Student Reference Book,* page 87 and Check Your Understanding Problems 5 and 6.

Enrichment | 10–15 min

WHOLE CLASS | SMALL GROUP | PARTNER | INDEPENDENT

Adding Times

5.NBT.7

Activity Card 54, per group: stopwatch

To extend their work adding decimals, students time how long it takes each group member to complete a simple task and then add the times together. They compare the sum to how long it takes the group to complete the task in sequence. They discuss why there might be differences in the two times.

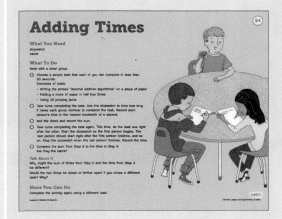

Extra Practice | 10–20 min

WHOLE CLASS | SMALL GROUP | PARTNER | INDEPENDENT

Playing *Decimal Top-It: Addition*

5.NBT.3, 5.NBT.3b, 5.NBT.7, SMP6

Student Reference Book, pp. 298–299; per partnership: number cards 0–9 (4 of each), 4 counters, 1 calculator (optional)

For additional practice adding decimals, students play *Decimal Top-It: Addition.* Have students read the game directions on *Student Reference Book,* pages 298 and 299.

Observe
- Which students can apply algorithms to add decimals?

Discuss
- *What strategy are you using to add? Explain your strategy to your partner.*
GMP6.1

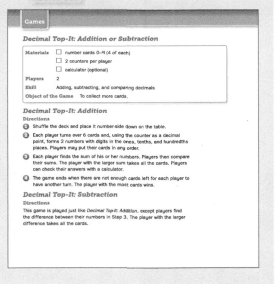

English Language Learner

Beginning ELL Build background knowledge for understanding the term *partial* by demonstrating the meanings of *part* and *partial* using simple jigsaw puzzles. Show puzzle pieces and make statements like these: *This is a part of the puzzle. This is a partial section of the puzzle.* Provide practice with the terms using Total Physical Response commands like these: *Show me one part. Count the parts. Use the parts to make the whole. Put the partial sections together.*

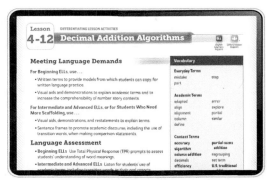

Differentiation Support pages are found in the online Teacher's Center.

Standards and Goals for
Mathematical Process and Practice

SMP1 **Make sense of problems and persevere in solving them.**

GMP1.4 Check whether your answer makes sense.

GMP1.6 Compare the strategies you and others use.

SMP6 **Attend to precision.**

GMP6.4 Think about accuracy and efficiency when you count, measure, and calculate.

		7	2	4
+	1	8	5
700 + 100 →		8	0	0
20 + 80 →		1	0	0
4 + 5 →				9
		9	0	9

724 + 185 solved with partial-sums addition

		2	.	8	4
+	4	.	3	5
2 + 4 →		6	.	0	0
0.8 + 0.3 →		1	.	1	0
0.04 + 0.05 →		0	.	0	9
		7	.	1	9

2.84 + 4.35 correctly solved with partial-sums addition

---

1 Warm Up 5 min

▶ Mental Math and Fluency

Display the following decimals in expanded form. Have students write the decimals in standard notation and in words on slates. *Leveled exercises:*

◉○○ 2.0 + 0.4 + 0.09 2.49; two and forty-nine hundredths
7.0 + 0.3 + 0.01 7.31; seven and thirty-one hundredths

◉◉○ 30.0 + 7.0 + 0.5 + 0.06 + 0.008 37.568; thirty-seven and five hundred sixty-eight thousandths
(6 * 10) + (3 * 1) + (9 * 0.1) + (8 * 0.01) 63.98; sixty-three and ninety-eight hundredths

◉◉◉ 700.0 + 30.0 + 0.07 + 0.001 730.071; seven hundred thirty and seventy-one thousandths
(9 * 100) + (4 * 1) + (8 * 0.001) 904.008; nine hundred four and eight thousandths

2 Focus 35–40 min

▶ Math Message

Display the work for 724 + 185 shown in the margin.

This work shows how to use partial-sums addition to solve 724 + 185.

How could you use partial-sums addition to solve 2.84 + 4.35? Work with a partner to solve the problem. Show your work.

▶ Extending Whole-Number Addition Algorithms to Decimals

Student Reference Book, pp. 85 and 130; *Math Masters,* p. 157

| WHOLE CLASS | SMALL GROUP | PARTNER | INDEPENDENT |

Math Message Follow-Up Ask volunteers to share how they used **partial-sums addition** to solve 2.84 + 4.35. Some partnerships may have produced correct work similar to that shown in the margin. Others' work may contain errors. Be sure to discuss the following common errors:

Alignment error: When recording partial sums, students do not align the sums according to place value, so they add the partial sums incorrectly. Encourage these students to include trailing zeros when recording partial sums.

		2	.	8	4
+	4	.	3	5
2 + 4 →					6
0.8 + 0.3 →			1	.	1
0.04 + 0.05 →		0	.	0	9
		0	.	2	6

Regrouping error: When adding the tenths, students think: *8 tenths plus 3 tenths is 11 tenths* and record the partial sum as 0.11. Suggest to these students that they visualize or shade a hundredths grid (*Math Masters,* page TA22) to show 0.11. Ask: *Does the grid show 11 tenths?* No, it shows 11 hundredths. *Picture or shade a grid showing 11 tenths. What does it look like?* It has 11 columns shaded, which is the same as 1 full grid and another column on a second grid. *How do we write 11 tenths?* 1.1 *Why?* 11 tenths is the same as 1 whole and 1 tenth.

Review the examples of partial-sums addition on *Student Reference Book,* pages 85 and 130 as a class. Ask: *How is using partial-sums addition to add decimals similar to using it to add whole numbers?* Sample answer: I work from left to right, adding the digits in each place separately. *What do you have to remember when you are using partial-sums addition to add decimals?* **GMP1.6** Sample answers: I need to think about place value to make sure I line up the partial sums correctly. Sometimes I may need to regroup ones, tenths, and hundredths.

Explain that partial-sums addition is an example of an **algorithm** for addition, just as partial-products multiplication and U.S. traditional multiplication are algorithms for multiplication. Point out that algorithms are useful because they define a set of steps that can be applied to solve many different problems. Tell students that most whole-number addition algorithms can be adapted to add decimals, and today they will explore two more algorithms they can use to solve decimal addition problems.

Divide the class into groups of three or four students each and distribute a piece of poster paper to each group. Give a copy of *Math Masters,* page 157 to each student. Tell half of the groups that they will explore **column addition** and half of the groups that they will explore **U.S. traditional addition.** Each group should follow the directions on the *Math Masters* page to create a poster about their assigned algorithm.

When most groups have finished, invite them to share their posters. Be sure groups discuss how using each algorithm with whole numbers is similar to using it with decimals. **GMP1.6**

Focus most of the discussion on issues or problems students encountered and on sharing students' advice to others that they recorded on their posters. Expect students to have struggled with the fact that 2.965 and 7.47 do not have the same number of digits. For example, some may have written the addition problem with the right-most digits aligned as shown in the margin, adding thousandths to hundredths, hundredths to tenths, and so on. Make sure students' advice includes considering the values of the digits in the addends. Writing trailing zeros on the addends with fewer digits after the decimal point, as shown in the sample work in the margin of the next page, can help students correctly line up the addends according to place value.

		2	.	8	4
+		4	.	3	5
$2 + 4 \rightarrow$		6	.	0	0
$0.8 + 0.3 \rightarrow$		0	.	1	1
$0.04 + 0.05 \rightarrow$		0	.	0	9
		6	.	2	0

Partial-sums addition with
a regrouping error

Adjusting the Activity

Differentiate Provide hundredths and thousandths grids (*Math Masters,* pages TA22 and TA23) for students to shade to help them make estimates and add decimals. Some students may also benefit from working on computation grids (*Math Masters,* page TA7).

 Go Online Differentiation Support

	2	.	9	6	5
+			7	.	4	7

2.965 + 7.47 with misaligned digits

1s	0.1s	0.01s	0.001s
2 .	9	6	5
+ 7 .	4	7	**0**
9 .	13	13	5
10 .	3	13	5
10 .	4	3	5

2.965 + 7.47 correctly solved
with column addition

$$
\begin{array}{r}
\overset{1}{}\,.\,\overset{1}{} \\
2\,.\,9\ 6\ 5 \\
+\ \ 7\,.\,4\ 7\ \mathbf{0} \\
\hline
1\ 0\,.\,4\ 3\ 5
\end{array}
$$

2.965 + 7.47 correctly solved
with U.S. traditional addition

Math Journal 1, p. 146

Using Algorithms to Add Decimals Lesson 4-12

DATE   TIME

For Problems 1–6, make an estimate. Write a number sentence to show how you estimated. Then solve using partial-sums addition, column addition, or U.S. traditional addition. Show your work. Use your estimates to check that your answers make sense.

① 2.3 + 7.6 = ?
Sample answer:
(estimate)
2 + 8 = 10

② 6.4 + 8.7 = ?
Sample answer:
(estimate)
6 + 9 = 15

③ 7.06 + 14.93 = ?
Sample answer:
(estimate)
7 + 15 = 22

2.3 + 7.6 = _9.9_

6.4 + 8.7 = _15.1_

7.06 + 14.93 = _21.99_

④ 21.47 + 9.68 = ?
Sample answer:
(estimate)
21 + 10 = 31

⑤ 3.514 + 5.282 = ?
Sample answer:
(estimate)
3.5 + 5 = 8.5

⑥ 19.046 + 71.24 = ?
Sample answer:
(estimate)
20 + 70 = 90

21.47 + 9.68 = _31.15_

3.514 + 5.282 = _8.796_

19.046 + 71.24 = _90.286_

⑦ Choose one problem. Answer the questions below.

a. How did you make your estimate? Sample answer: In Problem 5, I knew 3.514 was close to 3.5 and 5.282 was close to 5. I added those easy numbers to get my estimate: 3.5 + 5 = 8.5.

b. How did you use your estimate to check that your answer made sense?
Sample answer: I got 8.796 for my exact answer. That's only a little bit more than my estimate, 8.5, so my answer makes sense.

146   5.NBT.7, SMP1

Tell students that all three algorithms (partial-sums addition, column addition, and U.S. traditional addition) are reliable methods for adding decimals. Explain that when deciding which algorithm to use, they should consider two things: **accuracy** and **efficiency.** Remind the class that accuracy refers to getting the correct answer. Efficiency means getting the answer quickly, or in just a few steps. Ask: *Which algorithm might help you get accurate answers?* Sample answer: I think partial-sums addition would help me get accurate answers because it helps me to think about the partial sums separately. *Which algorithm might be the most efficient one for you to use?* GMP6.4 Sample answer: Column addition seems like the most efficient for me because I can use it the fastest. Emphasize that everyone has different answers to these questions, and it is fine for each student to choose the algorithm that works best for him or her.

▶ **Practicing Decimal Addition**

Math Journal 1, p. 146

WHOLE CLASS	SMALL GROUP	PARTNER	INDEPENDENT

Remind students that when computing, it is important to use estimates to check that their answers are reasonable. GMP1.4 Ask them to make an estimate for the problem 8.35 + 7.24. Then have them share their strategies for making estimates. *Sample strategies:*

- I rounded each decimal to the nearest whole number. 8.35 rounded to 8, and 7.24 rounded to 7. Since 8 + 7 = 15, my estimate is 15.
- I added the whole-number part of each decimal: 8 + 7 = 15. Then I thought about what shaded grids would look like for 0.35 and 0.24. Both grids would be less than half shaded, so the total would be less than 1. My estimate is that the sum is between 15 and 16.

Have students solve 8.35 + 7.24 using the algorithm of their choice and then compare the answer to their estimates to determine whether the answer is reasonable. GMP1.4 15.59; Sample answer: 15.59 is close to 15, so the answer seems reasonable. Then have students complete journal page 146 independently or in partnerships.

 Assessment Check-In 5.NBT.7

Math Journal 1, p. 146

Expect most students to be able to use an algorithm to solve Problems 1–3 on journal page 146, involving decimals to tenths (with and without regrouping) and decimals to hundredths with no regrouping of tenths or hundredths. Some students may be able to solve Problems 4–6, involving decimals to hundredths with regrouping and decimals to thousandths. Some may be able to explain how they used estimates to check the reasonableness of their answers. GMP1.4 If students struggle making estimates and adding decimals, see the suggestions in the Adjusting the Activity note on the previous page.

 Evaluation Quick Entry Go online to record students' progress and to see trajectories toward mastery for these standards.

Summarize Have volunteers share their favorite algorithms for adding decimals and explain their choices.

3 Practice 20–30 min

▶ Playing *Fraction Of*

Student Reference Book, pp. 306–307; *Math Masters*, p. G24

| WHOLE CLASS | SMALL GROUP | PARTNER | INDEPENDENT |

Students practice finding a fraction of a whole number. They should continue to use only Set 1 of the fraction cards until Unit 5 and may use counters to help them solve problems if they wish.

Observe

• Which students are consistently using effective strategies to solve fraction-of problems?

• Which students regularly choose the whole that produces a whole-number answer *and* results in the highest possible score?

Discuss

• *Explain your strategy for picking a number from your whole card.* **GMP6.1**

• *When does a fraction-of problem give a whole-number answer?*

| Differentiate | **Game Modifications** | **Go Online** | Differentiation Support |

▶ Math Boxes 4-12

Math Journal 1, p. 147

| WHOLE CLASS | SMALL GROUP | PARTNER | INDEPENDENT |

Mixed Practice Math Boxes 4-12 are grouped with Math Boxes 4-9 and 4-14.

▶ Home Link 4-12

Math Masters, p. 158

Homework Students practice adding decimals. They use estimates to check the reasonableness of their answers. **GMP1.4**

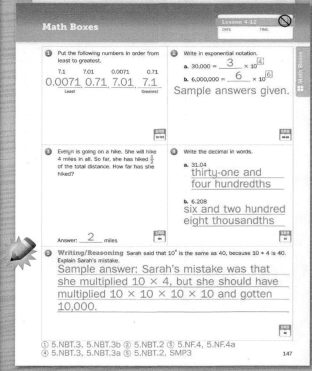

Math Journal 1, p. 147

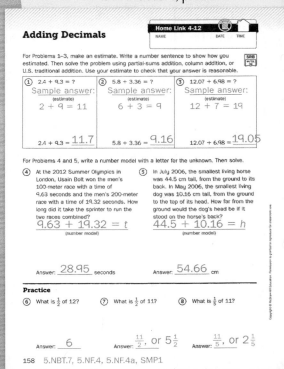

Math Masters, p. 158

Decimal Subtraction Algorithms

Overview Students review whole-number subtraction algorithms and use them to subtract decimals.

▶ **Before You Begin**

For Part 2, each group needs a piece of poster paper. If students are not familiar with the subtraction algorithms in this lesson, consider having the class complete the optional Readiness activity. For Part 3, if additional sets of *Prism Pile-Up* cards are needed, copy *Math Masters,* pages G4 and G5.

▶ **Vocabulary**

algorithm • trade-first subtraction • counting-up subtraction • U.S. traditional subtraction

Standards

Focus Clusters
• Understand the place value system.
• Perform operations with multi-digit whole numbers and with decimals to hundredths.

	Materials	
① Warm Up 5 min		
Mental Math and Fluency Students write decimals that are greater than and less than a given decimal.	slate	5.NBT.3, 5.NBT.3a, 5.NBT.3b
② Focus 35–40 min		
Math Message Students make an estimate for a decimal subtraction number story.		5.NBT.7
Extending Whole-Number Subtraction Algorithms to Decimals Students review whole-number subtraction algorithms and explore how they can be used to subtract decimals.	*Student Reference Book;* *Math Masters,* p. 159; per group: 1 piece of poster paper	5.NBT.4, 5.NBT.7 SMP1, SMP6
Practicing Decimal Subtraction Students practice using decimal subtraction algorithms.	*Math Journal 1,* p. 148; *Math Masters,* p. TA7 (optional)	5.NBT.7 SMP1, SMP6
✓ **Assessment Check-In** See page 417. Expect most students to be able to use an algorithm to subtract decimals to tenths (with and without regrouping) and decimals to hundredths without regrouping.	*Math Journal 1,* p. 148	5.NBT.7, SMP6
③ Practice 20–30 min		
Playing *Prism Pile-Up* Game Students practice finding volumes of prisms.	*Student Reference Book,* p. 319; per partnership: *Math Masters,* p. G6; *Prism Pile-Up* cards; calculator (optional)	5.OA.2, 5.MD.3, 5.MD.3a, 5.MD.3b, 5.MD.4, 5.MD.5, 5.MD.5a, 5.MD.5b, 5.MD.5c SMP1, SMP6
Math Boxes 4-13 Students practice and maintain skills.	*Math Journal 1,* p. 149	See page 417.
Home Link 4-13 Homework Students solve decimal subtraction problems.	*Math Masters,* p. 160	5.NBT.1, 5.NBT.7 SMP1

 **Go Online** to see how mastery develops for all standards within the grade.

Differentiation Options

| Readiness | 15–30 min | Enrichment | 10–20 min | Extra Practice | 10–15 min |

WHOLE CLASS   SMALL GROUP   **PARTNER**   INDEPENDENT   WHOLE CLASS   SMALL GROUP   **PARTNER**   INDEPENDENT   WHOLE CLASS   SMALL GROUP   **PARTNER**   INDEPENDENT

Reviewing Subtraction Algorithms

5.NBT.7, SMP1

Student Reference Book, pp. 89, 91–94

For experience with trade-first subtraction, counting-up subtraction, and U.S. traditional subtraction, students review how to use these algorithms to subtract whole numbers. Read the trade-first subtraction example on *Student Reference Book,* page 89 together. Then have students help you use trade-first subtraction to solve a problem like 75 — 37 = ? 38 As students propose steps, ask guiding questions like these: *What should we do first? Where should we write the 8? Why? How did you get that number?* GMP1.2 Next have partnerships solve a problem like 853 — 471 = ? using trade-first subtraction. 382 Review counting-up subtraction by reading *Student Reference Book,* pages 91 and 92 and posing problems such as 426 — 63 = ? 363 Review U.S. traditional subtraction using *Student Reference Book* pages 93 and 94. Have students solve Problems 3 and 4 in Check Your Understanding. Have students solve the Check Your Understanding problems.

Making a Big Difference

5.NBT.7, SMP3, SMP8

Activity Card 55, number cards 0–9 (4 of each), 2 counters

To extend their work subtracting decimals, students use number cards to create decimals that have the largest possible difference. They write arguments for how they know they made the largest difference and try to find a rule they could use to always get the largest difference. GMP3.1, GMP8.1

Playing *Decimal Top-It* (Decimal Subtraction Variation)

5.NBT.3, 5.NBT.3a, 5.NBT.3b, 5.NBT.7

Student Reference Book, pp. 298–299; per partnership: *Math Masters,* p. TA25, p. TA23 (optional); number cards 0–9 (4 of each); 6-sided die

For additional practice using decimal subtraction algorithms, students play the decimal subtraction variation of *Decimal Top-It.* Have students read about the variation on *Student Reference Book,* pages 298 and 299, and make sure they understand the directions.

Observe
• Which students can apply an algorithm to subtract the decimals?

Discuss
• *Which decimal subtraction algorithm do you like best? Why?*

English Language Learner

Beginning ELL Demonstrate *counting up* by moving your hand up along a number line displayed vertically, with the smaller numbers at the bottom. Use spoken directions and corresponding gestures to direct students to count up. For example, point to 25 on the number line and say: *Count up from 25,* gesturing to 26, 27, 28, and so on as students count. Once students understand counting up when assisted by gestures, point to a number and have students count up without gestures. Then demonstrate using a horizontal number line, so students understand that counting up does not necessarily correspond to physically going up.

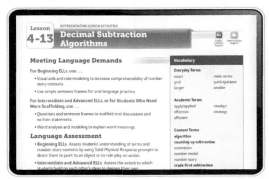

Differentiation Support pages are found in the online Teacher's Center.

Standards and Goals for
Mathematical Process and Practice

SMP1 **Make sense of problems and persevere in solving them.**
GMP1.4 Check whether your answer makes sense.
GMP1.6 Compare the strategies you and others use.

SMP6 **Attend to precision.**
GMP6.4 Think about accuracy and efficiency when you count, measure, and calculate.

Professional Development

Similar to Lesson 4-12, the purpose of this lesson is to expose the class to several different decimal subtraction algorithms so that all students are able to find a method that allows them to solve decimal subtraction problems accurately and efficiently. The most reliable or efficient algorithm may vary from student to student. It is not necessary for students to master every algorithm.

① Warm Up · 5 min

▶ Mental Math and Fluency

Display decimals. For each one, have students write on slates a decimal with the same number of digits that is larger and one with the same number of digits that is smaller. *Leveled exercises:* Sample answers given.

● ○ ○ **0.4** 0.6 is larger; 0.3 is smaller
0.75 0.78 is larger; 0.74 is smaller

● ● ○ **0.53** 0.67 is larger; 0.23 is smaller
0.09 0.12 is larger; 0.03 is smaller

● ● ● **0.219** 0.375 is larger; 0.128 is smaller
0.003 0.999 is larger; 0.001 is smaller

② Focus · 35–40 min

▶ Math Message

Carolyn had 9.48 meters of cloth. She used 7.291 meters to make a dress. How many meters of cloth does she have left?

Write a number model for the number story. Then make an estimate. Do not solve the story.

▶ Extending Whole-Number Subtraction Algorithms to Decimals

Student Reference Book; Math Masters, p. 159

| WHOLE CLASS | SMALL GROUP | PARTNER | INDEPENDENT |

Math Message Follow-Up Ask students to share their number models. $9.48 - 7.291 = c$ Then have volunteers explain their estimation strategies. *Sample strategies:*

• I rounded each decimal to the nearest tenth. 9.48 rounded to 9.5, and 7.291 rounded to 7.3. Since 9.5 − 7.3 = 2.2, my estimate is 2.2.

• I know that 9.48 − 7 = 2.48. I still have to subtract 0.291 more. I know that taking away 0.291 from 2.48 won't get me to 2 because 0.291 is less than 0.48. So my estimate is a little more than 2.

Remind the class that estimates are useful for checking whether exact answers make sense. GMP1.4

Ask: *In an earlier lesson you used grids to subtract decimals. Do you think using grids to solve this problem would be a good strategy? Why or why not?* GMP6.4 Sample answers: I would need a lot of grids to represent 9.48, so that would not be a very efficient way to solve the problem. It would take a long time to shade in that many grids. *Can anyone think of other strategies we could use to solve the problem?* If no one mentions it, remind students that in the last lesson they explored **algorithms** as efficient and effective strategies for solving decimal addition problems. Explain that today they will explore how whole-number subtraction algorithms can be applied to decimals and use algorithms to solve decimal subtraction problems.

Distribute one copy of *Math Masters,* page 159 to each student. Make sure students have their *Student Reference Books.* Divide the class into groups of 3 to 4 students and distribute a piece of poster paper to each group. Assign each group an algorithm: **trade-first subtraction, counting-up subtraction,** or **U.S. traditional subtraction.** Students should follow the directions on the *Math Masters* page and then work with their group to create a poster about their algorithm. See the margin for examples of how to use the algorithms to correctly solve the decimal problem.

When most groups have finished, invite them to share their posters. Discuss observations about how using each algorithm with whole numbers is similar to using it with decimals. GMP1.6

Focus most of the discussion on issues or problems students encountered and on sharing students' advice for others that they recorded on their posters. Be sure to highlight the following common problems and errors:

• When using trade-first subtraction or U.S. traditional subtraction, students may misalign digits because 9.48 does not have a digit in the thousandths place. Even if students align the digits correctly, they may not know what to do with the thousandths column because it appears that there is nothing from which to subtract the 1 thousandth in 7.291. (*See margin.*) Many may simply write 1 as the difference in this column. Ask: *How could we rewrite 9.48 so it has the same number of digits after the decimal point as 7.291?* 9.480 *Why can we write a zero on the end?* Because 48 hundredths is the same as 480 thousandths *Do we know how to take away 1 thousandth from 0 thousandths?* No. *What can we do to get more thousandths?* Trade 1 hundredth for 10 thousandths. It may be helpful for students to visualize this trade by thinking about 0.48 shaded on a hundredths grid and splitting up one of the shaded hundredths squares into 10 thousandths.

9.48 − 7.291 correctly solved with trade-first subtraction

$$0.009 + 0.700 + 1.480 = 2.189$$

9.48 − 7.291 correctly solved with counting-up subtraction

9.48 − 7.291 correctly solved with U.S. traditional subtraction

9.48 − 7.291 with correctly aligned digits, but with no 0 in the thousandths column for 9.48

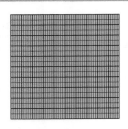

The pink shading shows the decimal part of the smaller number, 0.291. The green shading shows what to add to get to the next hundredth (and tenth). The blue shading shows what to add to get to the next whole number.

• When using counting-up subtraction, students may struggle deciding what to add first. Emphasize how it can be helpful to use counting-up steps to get to the next-highest hundredth or tenth. If necessary, demonstrate how to shade a grid as a way of thinking about counting up to the next-highest hundredth or tenth. Begin by shading 0.291 (the decimal part of the smaller number) on a thousandths grid. Ask: *How much do we need to shade to fill up the partially filled hundredths square?* 9 tiny rectangles, or 9 thousandths, or 0.009 Record that number as the first counting-up step. Ask: *How much is shaded now?* 0.3 *How much do we need to shade to fill up the rest of the square, or get to the next whole number?* 7 columns, or 7 tenths, or 0.7 Add that as the next counting-up step. (*See margin.*)

Summarize by asking: *What is the answer to the Math Message problem?* 2.189 meters *Does that answer make sense according to our estimates?* Sample answer: Yes. We estimated the answer would be a little more than 2.

▶ Practicing Decimal Subtraction

Math Journal 1, p. 148

| WHOLE CLASS | SMALL GROUP | PARTNER | INDEPENDENT |

Tell students that all three algorithms (trade-first subtraction, counting-up subtraction, and U.S. traditional subtraction) are reliable methods for subtracting decimals. Remind them that when choosing which algorithm to use they should consider accuracy and efficiency, just as they did with decimal addition algorithms. Ask: *Which algorithm might help you get accurate, or correct, answers?* Sample answer: I think trade-first subtraction would help me get accurate answers because I can think about making trades before I have to think about subtracting. *Which one might be the most efficient, or the quickest one for you to use?* GMP6.4 Sample answer: U.S. traditional subtraction might be the most efficient because there is the least to write. As with decimal addition algorithms, emphasize that people can have different answers to these questions, and students should choose the algorithm that works best for them.

Have students complete journal page 148 independently or in partnerships to practice using algorithms to subtract decimals. Remind them that when computing it is important to use estimates to check that their answers are reasonable. GMP1.4

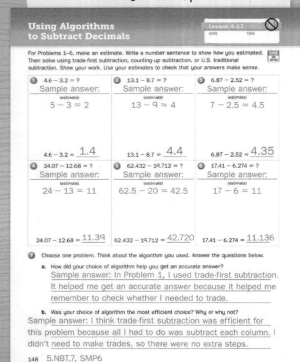

Math Journal 1, p. 148

Assessment Check-In 5.NBT.7

Math Journal 1, p. 148

Expect most students to be able to use an algorithm to solve Problems 1–3 on journal page 148, which involve decimals to tenths (with and without regrouping) and decimals to hundredths without regrouping. Some may be able to solve Problems 4–6, which involve decimals to hundredths and thousandths with regrouping. Some students may be able to explain and evaluate the efficiency of their strategies in Problem 7. **GMP6.4** If students struggle subtracting decimals, suggest focusing on practicing just one algorithm, as suggested in Adjusting the Activity.

Evaluation Quick Entry Go online to record students' progress and to see trajectories toward mastery for these standards.

Summarize Have volunteers share their favorite algorithms for subtracting decimals and explain their choices.

3 Practice 20–30 min

▶ Playing *Prism Pile-Up*

Student Reference Book, p. 319

| WHOLE CLASS | SMALL GROUP | **PARTNER** | INDEPENDENT |

Students practice finding volumes of rectangular prisms and figures composed of rectangular prisms.

Observe
- Which students have a strategy for finding the volumes of figures composed of more than one prism?
- Which students can explain their reasoning to a partner?

Discuss
- *Explain how you would find the volume of the figure on card 16, 17, or 18.* **GMP6.1**
- *What other strategy could you use to find the volumes of these figures?* **GMP1.5**

▶ Math Boxes 4-13

Math Journal 1, p. 149

| WHOLE CLASS | SMALL GROUP | PARTNER | INDEPENDENT |

Mixed Practice Math Boxes 4-13 are paired with Math Boxes 4-11.

▶ Home Link 4-13

Math Masters, p. 160

Homework Students practice subtracting decimals. They use estimates to check that their answers make sense. **GMP1.4**

Math Journal 1, p. 149

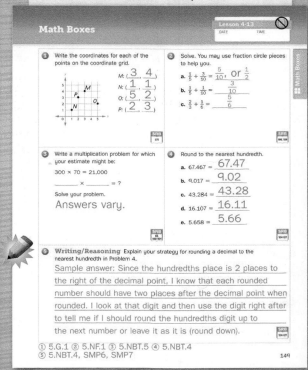

Math Masters, p. 160

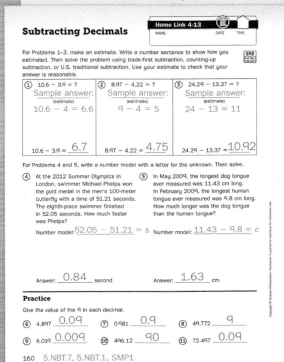

Addition and Subtraction of Money

Overview Students apply decimal addition and subtraction strategies to add and subtract money.

▶ **Before You Begin**

For *Spend and Save*, each partnership will need one counter and one real or play coin showing heads and tails. Cut apart enough copies of *Math Masters*, page G27 for each student to have one record sheet. If possible, make play bills and coins available for students who wish to use them.

▶ **Vocabulary**

balance

Standards

Focus Cluster
• Perform operations with multi-digit whole numbers and with decimals to hundredths.

	Materials	
① Warm Up 5 min		
Mental Math and Fluency Students round decimals.	slate	5.NBT.3, 5.NBT.3a, 5.NBT.4
② Focus 35–40 min		
Math Message Students solve number stories involving money.	*Math Masters*, p. TA22 (optional); play bills and coins (optional)	5.NBT.7
Discussing Addition and Subtraction Strategies Students discuss strategies for adding and subtracting money amounts, and compare them to strategies they used to add and subtract decimals in previous lessons.	play bills and coins (optional)	5.NBT.7 SMP1, SMP4, SMP6
Introducing *Spend and Save* **Game** Students practice adding and subtracting money amounts.	*Student Reference Book*, p. 323; per partnership: *Math Masters*, p. G27; number cards 0–9 (4 of each), 1 coin, 1 counter, play bills and coins (optional)	5.NBT.7 SMP1, SMP6
✓ **Assessment Check-In** See page 423. Expect most students to be able to choose an efficient strategy for adding and subtracting money amounts.	*Math Masters*, p. G27; play bills and coins (optional)	5.NBT.7 SMP6
③ Practice 20–30 min		
Finding Areas of New Floors Students practice finding the area of rectangles with fractional side lengths.	*Math Journal 1*, p. 150	5.NF.4, 5.NF.4b SMP4
Math Boxes 4-14 Students practice and maintain skills.	*Math Journal 1*, p. 151	See page 423.
Home Link 4-14 **Homework** Students solve addition and subtraction number stories involving money.	*Math Masters*, p. 161	5.NBT.6, 5.NBT.7

 Go Online to see how mastery develops for all standards within the grade.

Differentiation Options

| WHOLE CLASS | **SMALL GROUP** | PARTNER | INDEPENDENT | WHOLE CLASS | SMALL GROUP | **PARTNER** | INDEPENDENT | WHOLE CLASS | SMALL GROUP | **PARTNER** | INDEPENDENT |

Connecting Money to Decimals

`5.NBT.1`

To prepare for thinking about amounts of money as decimals, students consider the relationships between dollars, dimes, and pennies. Display the number 2.65, and ask: *What is the value of the 2?* 2 ones *What is the value of the 6?* 6 tenths *What is the value of the 5?* 5 hundredths Remind students that in our place-value system, the value of a digit in a given place is $\frac{1}{10}$ of the value it would have in the place to its left and 10 times the value it would have in the place to its right. Add a dollar sign to the number to show $2.65. Ask: *What is the value of the 2 when we're representing money?* 2 dollars *What is the value of the 6?* 6 dimes, or 60 cents *What is the value of the 5?* 5 pennies, or 5 cents Now ask: *Do the values of digits in amounts of money written in dollars-and-cents notation follow the same pattern as the values of digits in decimal numbers? How do you know?* Sample answer: Yes. There are 10 dimes in a dollar, so that means that 1 dime is $\frac{1}{10}$ of a dollar and 1 dollar is 10 times as much as a dime. There are 10 pennies in a dime, so 1 penny is $\frac{1}{10}$ of a dime and 1 dime is 10 times as much as 1 penny. **Emphasize the idea that it makes sense to think about money amounts as decimal numbers to the hundredths place.**

Playing a Variation of *Spend and Save*

`5.NBT.4, 5.NBT.7`

Student Reference Book, p. 323; per partnership: *Math Masters,* p. G27; number cards 0–9 (4 of each); 1 coin; 1 counter

To extend their work adding and subtracting amounts of money, students play *Spend and Save* but use one of the following variations:

- Players start with $1,000 and draw four cards on each turn. For example, a player who draws 3, 4, 0, 8 could make the money amount $34.08, $84.30, or any other amount that uses the same digits.
- Players start with $10.00, place all three digits after the decimal point, and round each amount to the nearest cent. For example, a player who draws 7, 1, 9 could make the amount $0.719 and round that amount to $0.72 before adding or subtracting.

Adding and Subtracting Money Amounts

`5.NBT.7`

Activity Card 56; per partnership: number cards 0–9 (4 of each), 2 counters

For more practice adding and subtracting amounts of money, partners use number cards to form money amounts and add or subtract the amounts from a running total.

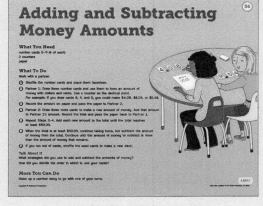

English Language Learner

Beginning ELL Have students role-play the Math Message problems and transactions in *Spend and Save* using play bills and coins and restating *balance* in terms of what is left. After each transaction, ask: *What is the balance? What do you have left? What do you have now? Tell me your balance.* Depending on their English language production level, have students respond by completing short sentence frames, short-phrase answers, or by writing the balance.

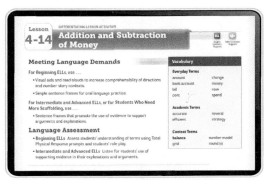

Differentiation Support pages are found in the online Teacher's Center.

Standards and Goals for
Mathematical Process and Practice

SMP1 **Make sense of problems and persevere in solving them.**
GMP1.6 Compare the strategies you and others use.

SMP4 **Model with mathematics.**
GMP4.1 Model real-world situations using graphs, drawings, tables, symbols, numbers, diagrams, and other representations.

SMP6 **Attend to precision.**
GMP6.4 Think about accuracy and efficiency when you count, measure, and calculate.

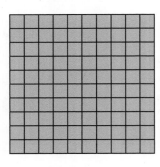

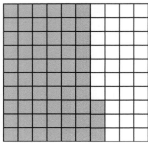

Shaded grids for Math Message Problem 1

① Warm Up 5 min

▶ Mental Math and Fluency

Dictate decimals and have students write them on slates. Have students round the decimals as indicated and record the rounded number on slates. *Leveled exercises:*

●○○ Write 3.7. Round it to the nearest one. 3.7; 4
Write 2.03. Round it to the nearest one. 2.03; 2
Write 7.81. Round it to the nearest tenth. 7.81; 7.8

●●○ Write 0.93. Round it to the nearest tenth. 0.93; 0.9
Write 2.65. Round it to the nearest tenth. 2.65; 2.7
Write 0.681. Round it to the nearest hundredth. 0.681; 0.68

●●● Write 7.2894. Round it to the nearest thousandth. 7.2894; 7.289
Write 8.2938. Round it to the nearest thousandth. 8.2938; 8.294
Write 6.994. Round it to the nearest tenth. 6.994; 7.0

② Focus 35–40 min

▶ Math Message

Solve the following number stories. Explain your strategies to a partner.

1. *You had $1.46 in your piggy bank. You found $0.17 between the couch cushions. How much money do you have now?* $1.63

2. *You bought a vegetable tray for $12.19. You paid with a $20.00 bill. How much change did you get?* $7.81

▶ Discussing Decimal Addition and Subtraction Strategies

| WHOLE CLASS | SMALL GROUP | PARTNER | INDEPENDENT |

Math Message Follow-Up Discuss each Math Message problem. Ask students to suggest a number model and share their answers and strategies. **GMP4.1** As students share, emphasize how their strategies for adding and subtracting amounts of money are similar to the strategies they used in previous lessons to add and subtract decimals. **GMP1.6**
Sample strategies:

Problem 1 $1.46 + $0.17 = t **GMP4.1**

• Think of a dollar as a whole, a dime as a tenth, and a penny as a hundredth. Then use a grid to solve. Shade 1 whole grid, 4 columns, and 6 squares in one color for $1.46. Shade 1 more column and 7 more squares in a second color for $0.17. Now 1 whole grid, 6 full columns, and 3 squares are shaded in all, which is 1.63, or $1.63. (*See margin.*)

- Add the dollars, dimes, and pennies separately. This is like partial-sums addition.

$$
\begin{array}{r r c c c}
 & \$1 & . & 4 & 6 \\
+ & \$0 & . & 1 & 7 \\
\hline
\end{array}
$$

Add the dollars: $1 + $0 = $1, or $1.00 → $1 . 0 0

Add the dimes: 4 dimes + 1 dime = 5 dimes,
or $0.50 → $0 . 5 0

Add the pennies: 6 pennies + 7 pennies =
13 pennies, or $0.13 → $0 . 1 3

Add $1.00 + $0.50 + $0.13 → $1 . 6 3

- Use the U.S. traditional addition algorithm to solve. Make sure to line up the money amounts so that you are adding pennies to pennies, dimes to dimes, and dollars to dollars. (*See margin.*)

Have students compare the strategies they shared. Ask: *Which strategies would help you to be accurate and efficient when adding larger money amounts like $7.42 and $9.15?* GMP1.6, GMP6.4 Sample answers: All of the strategies should give me an accurate answer. Shading grids would not be an efficient strategy because you would have to shade too many grids. Using an algorithm, like partial-sums or U.S. traditional addition, would be more efficient.

Problem 2 $20.00 − $12.19 = c GMP4.1

- Use counting-up subtraction to count up from the cost to the amount paid. Start at $12.19. Count up 1 penny to $12.20. Then count up 80 more cents to $13.00. Then count up $7.00 more to $20.00. In all it took $0.01 + $0.80 + $7.00 = $7.81 to get from the cost to the $20 paid. The change is $7.81.
- Think about $20.00 as $19.00 and 100 cents. This is a little bit like trade-first subtraction because $20 is renamed before doing any subtraction. $19.00 − $12.00 = $7.00. 100 cents − 19 cents = 81 cents. $7.81 is left.
- Use the U.S. traditional subtraction algorithm to solve. (*See margin.*)

Remind students that in previous lessons they used grids to help them solve decimal subtraction problems. Ask: *Would shading grids be an efficient strategy for solving this problem? Why or why not?* GMP1.6, GMP6.4 Sample answer: It would not be efficient. I would have to shade 20 grids and then cross out a bunch of them to subtract.

Ask: *Why do the strategies for adding and subtracting decimals also work for adding and subtracting money?* Sample answer: Money amounts are decimal numbers. Dollars are ones, dimes are tenths, and pennies are hundredths. Since money amounts are decimals, we can use decimal addition and subtraction strategies to solve addition and subtraction problems with money.

$$
\begin{array}{r r c c c}
 & & & 1 & \\
 & \$1 & . & 4 & 6 \\
+ & \$0 & . & 1 & 7 \\
\hline
 & \$1 & . & 6 & 3 \\
\end{array}
$$

$1.46 + $0.17 correctly solved with U.S. traditional addition

$$
\begin{array}{r r c c c}
 & 1 & 9 & 9 & 10 \\
\$\cancel{2} & \cancel{0} & . & \cancel{0} & \cancel{0} \\
- & \$1 & 2 & . & 1 & 9 \\
\hline
 & \$ & 7 & . & 8 & 1 \\
\end{array}
$$

$20.00 − $12.19 correctly solved with U.S. traditional subtraction

Tell students that today they will learn a new game involving adding and subtracting money. Remind them to think about accuracy and efficiency when they choose strategies while playing the game. GMP6.4

▶ Introducing *Spend and Save*

Student Reference Book, p. 323; *Math Masters*, p. G27

| WHOLE CLASS | SMALL GROUP | PARTNER | INDEPENDENT |

Read the directions for *Spend and Save* on *Student Reference Book*, page 323 together. Explain that the **balance** of a bank account is the amount of money that is currently in the account. Ask: *If you spend, will you add or subtract the money amount to find your new balance?* Subtract *Why?* When you spend money, you take it out of your bank account, so the balance will go down. *If you save, will you add or subtract?* Add *Why?* When you save money, you usually put it into your bank account, so the balance will go up.

Play a sample round with a volunteer. Then distribute one *Spend and Save* Record Sheet (*Math Masters*, page G27) to each student and 1 coin and 1 counter to each partnership. Make play bills and coins available to students who might need them to play the game. Have partnerships play the game.

Observe

- Which students have effective strategies for adding and subtracting amounts of money? GMP6.4
- Which students are able to explain their strategies clearly?

Discuss

- *Explain your addition or subtraction strategy to your partner.*
- *How is your strategy similar to your partner's strategy? How is it different?* GMP1.6

| Differentiate | Game Modifications | Go Online ▷ | Differentiation Support |

When most partnerships have finished a game, have each student circle one row on the record sheet. On the back of the sheet, have students show their work and explain their addition or subtraction strategy for that round.

Math Journal 1, p. 150

Assessment Check-In 5.NBT.7

Math Masters, p. G27

Expect most students to be able to choose an efficient strategy for adding and subtracting money amounts. GMP6.4 They should be successful in adding and subtracting some money amounts, but expect that they may struggle adding or subtracting money amounts that require them to regroup multiple times. If students struggle adding and subtracting, consider having them write a number model for the transaction (for example, $96.19 − $4.06 = b$) or act out the problem using play bills and coins.

Evaluation Quick Entry Go online to record students' progress and to see trajectories toward mastery for these standards.

Summarize Have students talk with a partner about which strategies they prefer to use when adding and subtracting decimals that represent money amounts and why.

---

3 Practice 20–30 min

▶ Finding Areas of New Floors

Math Journal 1, p. 150

| WHOLE CLASS | SMALL GROUP | PARTNER | INDEPENDENT |

Students practice finding areas of rectangles with fractional side lengths. GMP4.2

▶ Math Boxes 4-14

Math Journal 1, p. 151

| WHOLE CLASS | SMALL GROUP | PARTNER | INDEPENDENT |

Mixed Practice Math Boxes 4-14 are grouped with Math Boxes 4-9 and 4-12.

▶ Home Link 4-14

Math Masters, p. 161

Homework Students solve number stories involving money.

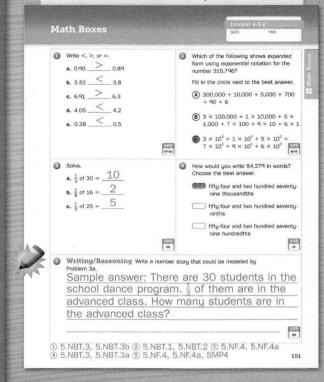

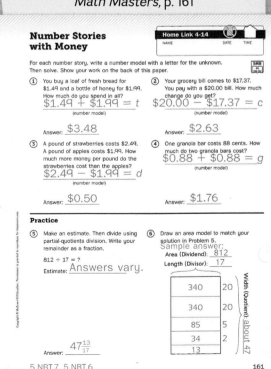

Unit 4 Progress Check

Overview   **Day 1:** Administer the Unit Assessments.
Day 2: Administer the Cumulative Assessment.

Day 1: Unit Assessment

Quick Entry Evaluation Record results and track progress toward mastery.

① Warm Up   5–10 min

Materials

Self Assessment
Students complete the Self Assessment.

Assessment Handbook, p. 31

②ₐ Assess   35–50 min

⭐ **Unit 4 Assessment**
These items reflect mastery expectations to this point.

Assessment Handbook, pp. 32–35

Unit 4 Challenge (Optional)
Students may demonstrate progress beyond expectations.

Assessment Handbook, pp. 36-37

Standards	Goals for Mathematical Content (GMC)	Lessons	Self Assessment	Unit 4 Assessment	Unit 4 Challenge
5.NBT.1	Understand the relationship between the places in multidigit numbers.	4-1 to 4-5		7	
5.NBT.3, 5.NBT.3a	Represent decimals.	4-1 to 4-5, 4-11		12, 13, 17, 18	
	Read and write decimals using numerals.	4-1 to 4-5, 4-11	1	4–7	
	Read and write decimals using number names.	4-1, 4-2	1	1–3, 7	
	Read and write decimals in expanded form.	4-3, 4-4	1	12, 13	
5.NBT.3, 5.NBT.3b	Compare and order decimals.	4-4, 4-5	2	8–11	1
	Record decimal comparisons using >, =, or <.	4-4		8–11	
5.NBT.4	Use place-value understanding to round decimals to any place.	4-5, 4-12, 4-13	3	14, 15	2a
5.NBT.7	Make and use estimates for decimal addition and subtraction problems.	4-12, 4-13		20–23	
	Add and subtract decimals using models or strategies.	4-11 to 4-14	6, 7	17, 18, 20–23	4
5.G.1	Understand and use a Cartesian coordinate grid in two dimensions.	4-6 to 4-10	4	16, 19a, 19b	3a, 3b
5.G.2	Represent problems by graphing points in the first quadrant.	4-7 to 4-10	5	19b	3a, 3b
	Interpret coordinate values of points in context.	4-7, 4-9, 4-10	5		3a, 3b
	Goals for Mathematical Process and Practice (GMP)				
SMP1	Check whether your answer makes sense. **GMP1.4**	4-4, 4-12, 4-13		20–23	
SMP2	Create mathematical representations using numbers, words, pictures, symbols, gestures, tables, graphs, and concrete objects. **GMP2.1**	4-2, 4-3, 4-6, 4-11		12, 13, 17, 18	
SMP3	Make mathematical conjectures and arguments. **GMP3.1**	4-8			2b, 2c
SMP4	Model real-world situations using graphs, drawings, tables, symbols, numbers, diagrams, and other representations. **GMP4.1**	4-7, 4-9, 4-14		19a, 19b	
SMP6	Explain your mathematical thinking clearly and precisely. **GMP6.1**	4-7, 4-11, 4-14		15	

/// **Go Online** to see how mastery develops for all standards within the grade.

1 Warm Up · 5-10 min

▶ Self Assessment

Assessment Handbook, p. 31

| WHOLE CLASS | SMALL GROUP | PARTNER | **INDEPENDENT** |

Students complete the Self Assessment to reflect on their progress in Unit 4.

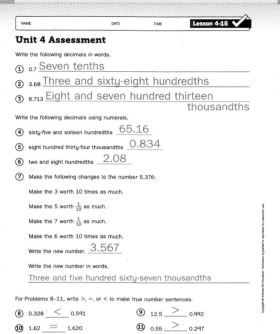

Assessment Handbook, p. 31

2a Assess · 35-50 min

▶ Unit 4 Assessment

Assessment Handbook, pp. 32–35

| WHOLE CLASS | SMALL GROUP | PARTNER | **INDEPENDENT** |

Students complete the Unit 4 Assessment to demonstrate their progress on the standards covered in this unit.

Generic rubrics in the *Assessment Handbook* can be used to evaluate student progress on the Mathematical Process and Practice Standards.

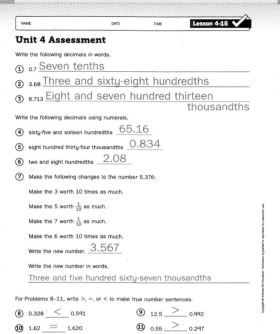

Unit 4 Assessment

Write the following decimals in words.

① 0.7 Seven tenths

② 3.68 Three and sixty-eight hundredths

③ 8.713 Eight and seven hundred thirteen thousandths

Write the following decimals using numerals.

④ sixty-five and sixteen hundredths 65.16

⑤ eight hundred thirty-four thousandths 0.834

⑥ two and eight hundredths 2.08

⑦ Make the following changes to the number 5.376:

Make the 3 worth 10 times as much.

Make the 5 worth 1/10 as much.

Make the 7 worth 1/10 as much.

Make the 6 worth 10 times as much.

Write the new number. 3.567

Write the new number in words.

Three and five hundred sixty-seven thousandths

For Problems 8–11, write >, =, or < to make true number sentences.

⑧ 0.328 $<$ 0.591 ⑨ 12.5 $>$ 0.992

⑩ 1.62 $=$ 1.620 ⑪ 0.55 $>$ 0.297

Unit 4 Assessment (continued)

For Problems 12 and 13, shade the thousandths grid to represent the decimal. Then write the decimal in expanded form.

⑫ 0.215

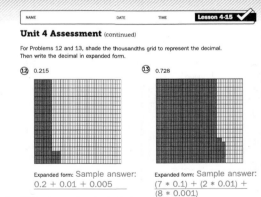

⑬ 0.728

Expanded form: Sample answer: 0.2 + 0.01 + 0.005

Expanded form: Sample answer: (7 * 0.1) + (2 * 0.01) + (8 * 0.001)

⑭ Round each number to the place listed.

Start number	Nearest hundredth	Nearest tenth	Nearest whole number
3.216	3.22	3.2	3
6.076	6.08	6.1	6
0.082	0.08	0.1	0

⑮ Explain how you rounded 6.076 to the nearest hundredth.
Sample answer: 6.076 is between 6.07 and 6.08. There is a 6 in the thousandths place, and that means I need to round the number up. It rounds up to 6.08.

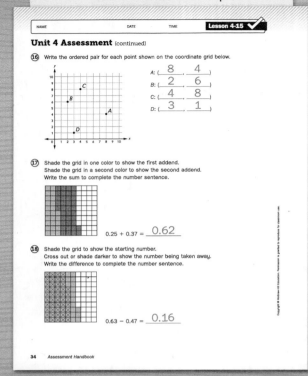

Unit 4 Assessment (continued)

16. Write the ordered pair for each point shown on the coordinate grid below.

A: (__8__ , __4__)
B: (__2__ , __6__)
C: (__4__ , __8__)
D: (__3__ , __1__)

17. Shade the grid in one color to show the first addend.
Shade the grid in a second color to show the second addend.
Write the sum to complete the number sentence.

$0.25 + 0.37 =$ __0.62__

18. Shade the grid to show the starting number.
Cross out or shade darker to show the number being taken away.
Write the difference to complete the number sentence.

$0.63 - 0.47 =$ __0.16__

34   Assessment Handbook

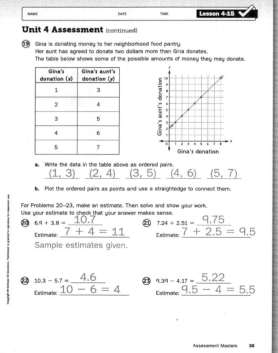

Unit 4 Assessment (continued)

19. Gina is donating money to her neighborhood food pantry.
Her aunt has agreed to donate two dollars more than Gina donates.
The table below shows some of the possible amounts of money they may donate.

Gina's donation (x)	Gina's aunt's donation (y)
1	3
2	4
3	5
4	6
5	7

a. Write the data in the table above as ordered pairs.
  (1, 3)   (2, 4)   (3, 5)   (4, 6)   (5, 7)

b. Plot the ordered pairs as points and use a straightedge to connect them.

For Problems 20–23, make an estimate. Then solve and show your work.
Use your estimate to check that your answer makes sense.

20. $6.9 + 3.8 =$ __10.7__
Estimate: __7 + 4 = 11__
Sample estimates given.

21. $7.24 + 2.51 =$ __9.75__
Estimate: __7 + 2.5 = 9.5__

22. $10.3 - 5.7 =$ __4.6__
Estimate: __10 - 6 = 4__

23. $9.39 - 4.17 =$ __5.22__
Estimate: __9.5 - 4 = 5.5__

Assessment Masters   35

Differentiate   **Adjusting the Assessment**

Item(s)	Adjustments
1–3	To scaffold Items 1–3, have students refer to the class place-value chart created in Lesson 4-1.
4–6	To extend Items 4–6, have students complete a name-collection box for one of the decimals.
7	To scaffold Item 7, have students write the value of each digit in the original number. Have them write the new value of each digit before writing the new number.
8–11	To scaffold Items 8–11, have students compare the whole numbers first. Then, have them shade a thousandths grid to represent each decimal.
12, 13	To extend Items 12 and 13, have students write each decimal in two different types of expanded form.
14, 15	To extend Items 14 and 15, have students explain how they could use a number line to round the numbers.
16	To scaffold Item 16, remind students of the phrase *over and up*.
17, 18	To scaffold Items 17 and 18, have students build the numbers using base-10 blocks before shading the grids.
19	To extend Item 19, ask students to use the line to determine the amount of Gina's aunt's donation if Gina donates $7.
20–23	To scaffold Items 20–23, have students refer to the posters they created in Lessons 4-12 and 4-13.

Advice for Differentiation

All of the content included on the Unit 4 Assessment was recently introduced and will be revisited in subsequent units.

Go Online:

 Quick Entry Evaluation Record students' progress and see trajectories toward mastery for these standards.

 Data Review your students' progress reports. Differentiation materials are available online to help you address students' needs.

NOTE See the Unit Organizer on pages 324 and 325 or the online Spiral Tracker for details on Unit 4 focus topics and the spiral.

▶ Unit 4 Challenge (Optional)

Assessment Handbook, pp. 36–37

WHOLE CLASS	SMALL GROUP	PARTNER	**INDEPENDENT**

Students can complete the Unit 4 Challenge after they complete the Unit 4 Assessment.

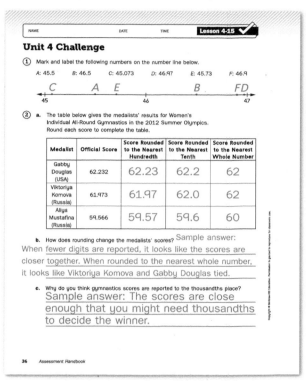

Assessment Handbook, p. 36

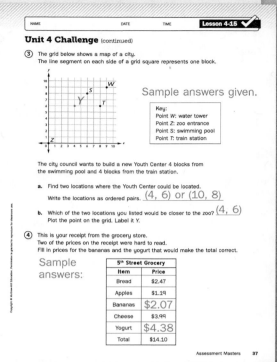

Unit 4 Progress Check

Day 2: Cumulative Assessment

2b Assess   35–45 min

Cumulative Assessment
These items reflect mastery expectations to this point.

Materials

Assessment Handbook, pp. 38–41

Standards	Goals for Mathematical Content (GMC)	Cumulative Assessment
5.OA.1	Write numerical expressions that contain grouping symbols.*	3a
	Evaluate expressions that contain grouping symbols.*	3b
5.OA.2	Model real-world and mathematical situations using simple expressions.*	1a–1c, 3a
	Interpret numerical expressions without evaluating them.	2a, 2b
5.NBT.2	Use whole-number exponents to denote powers of 10.	13a–13c
	Multiply whole numbers by powers of 10; explain the number of zeros in the product.	13a–13c
5.NBT.5	Fluently multiply multidigit whole numbers using the standard algorithm.	6, 7
5.NBT.6	Divide multidigit whole numbers.	9
	Illustrate and explain solutions to division problems.	9
5.NF.2	Solve number stories involving fraction addition and subtraction.	11
5.NF.3	Solve number stories involving whole-number division that leads to fractional answers.	10
5.NF.4	Multiply fractions by whole numbers.	12
5.NF.4, 5.NF.4a	Interpret $\left(\frac{1}{b}\right) \times q$ as 1 part of a partition of q into b equal parts.	12
5.MD.3	Recognize volume as an attribute of solid figures.*	4, 5
5.MD.3, 5.MD.3a	Understand that a unit cube has 1 cubic unit of volume and can measure volume.*	5
5.MD.3, 5.MD.3b	Understand that a solid figure completely filled by n unit cubes has volume n cubic units.*	5
5.MD.4	Measure volumes by counting unit cubes and improvised units.*	5
5.MD.5, 5.MD.5a	Relate packing prisms with cubes to volume formulas.*	5
5.MD.5, 5.MD.5b	Apply formulas to find volumes of rectangular prisms.*	5, 8a
5.MD.5, 5.MD.5c	Find volumes of figures composed of right rectangular prisms.*	8a
	Solve real-world problems involving volumes of figures composed of prisms.*	8a

*Instruction and most practice on this content is complete.

Standards	Goals for Mathematical Process and Practice (GMP)	Cumulative Assessment
SMP1	Make sense of your problem. **GMP1.1**	2b
	Solve problems in more than one way. **GMP1.5**	5
SMP2	Make sense of the representations you and others use. **GMP2.2**	5
SMP4	Model real-world situations using graphs, drawings, tables, symbols, numbers, diagrams, and other representations. **GMP4.1**	3a, 10
	Use mathematical models to solve problems and answer questions. **GMP4.2**	3b, 8a
SMP6	Explain your mathematical thinking clearly and precisely. **GMP6.1**	2b, 4, 8b

3 Look Ahead 10–15 min

Materials

Math Boxes 4-15: Preview for Unit 5
Students preview skills and concepts for Unit 5.

Math Journal 1, p. 152

Home Link 4-15
Students take home the Family Letter that introduces Unit 5.

Math Masters, pp. 162–165

 Go Online to see how mastery develops for all standards within the grade.

2b Assess 35–45 min

▶ Cumulative Assessment

Assessment Handbook, pp. 38–41

| WHOLE CLASS | SMALL GROUP | PARTNER | **INDEPENDENT** |

Students complete the Cumulative Assessment. The problems in the Cumulative Assessment address content from Units 1–3.

Monitor student progress on the standards using the online assessment and reporting tools.

Generic rubrics in the *Assessment Handbook* can be used to evaluate student progress on the Mathematical Process and Practice Standards.

Assessment Handbook, p. 38

NAME   DATE   TIME   **Lesson 4-15** ✔

Unit 4 Cumulative Assessment

① Write the following expressions using numbers.

  a. Subtract fifty-two from one hundred twelve and then multiply by two.
$$(112 - 52) * 2$$

  b. Subtract fifty-two from one hundred twelve and divide the difference by three.
$$(112 - 52) ÷ 3$$

  c. Multiply the difference of one hundred twelve and fifty-two by six.
$$(112 - 52) * 6$$

② **a.** Write the three expressions from Problem 1 in order from least to greatest.
$$(112 - 52) ÷ 3; (112 - 52) * 2; (112 - 52) * 6$$

  b. Did you have to evaluate the expressions in Problem 1 to order them? Explain.
Sample answer: No. Every expression had (112 − 52) in it, and we either multiplied or divided that amount by a different number. Multiplying by 6 would make the largest number, and dividing by 3 would make the smallest number.

③ **a.** Jocelyn is the treasurer of the science club. She collected $2.00 from each of the 30 members and an additional $20 from the principal. Half of this money was used to pay for T-shirts. Write an expression that models the amount of money that was used to pay for T-shirts. Use grouping symbols in your expression.
Sample answer: [2 * 30 + 20] ÷ 2

  b. Evaluate the expression to find the amount of money that was paid for T-shirts.
Jocelyn paid __40__ dollars for T-shirts for the science club.

38   Assessment Handbook

NAME   DATE   TIME   **Lesson 4-15** ✓

Unit 4 Cumulative Assessment (continued)

④ Juan said, "A book doesn't have volume because you cannot pack it with cubes and count the cubes that fill it." Is Juan correct? Explain your answer.

No. Sample explanation: Even though you can't pack a book with cubes, it has volume. Volume is a measure of the amount of space it takes up. You can find the volume of a book by multiplying its length by its width by its height.

⑤ The rectangular prism below is partially packed with centimeter cubes. Find the volume of the prism. Then describe three different strategies you could use to find the volume.

Volume: __72 cm³__ Sample answers:

Strategy 1: I could completely pack the prism with centimeter cubes and count the number of cubes it takes to fill the prism. It takes 72 cubes, so the volume is 72 cm³.

Strategy 2: The picture shows that the area of the base is 24 square centimeters. I can multiply this by the height. 24 * 3 = 72 cm³

Strategy 3: I can use the formula V = l × w × h. The picture shows the length is 4 cm, the width is 6 cm, and the height is 3 cm. 4 × 6 × 3 = 72, so the volume is 72 cm³.

Assessment Masters   39

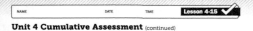

NAME   DATE   TIME   **Lesson 4-15** ✓

Unit 4 Cumulative Assessment (continued)

Make an estimate. Write a number sentence to show how you estimated. Then solve. Use your estimate to see if your answer makes sense. Use U.S. traditional multiplication to solve at least one of the problems.

⑥ 636 * 17 = ?

Estimate: Sample answer:
600 * 20 = 12,000

```
    6 3 6
*      1 7
    4 4 5 2
+   6 3 6 0
  1 0,8 1 2
```

636 * 17 = __10,812__

⑦ 45 * 36 = ?

Estimate: Sample answer:
50 * 40 = 2,000

```
      4 5
*     3 6
    2 7 0
+ 1 3 5 0
  1,6 2 0
```

45 * 36 = __1,620__

⑧ This drawing is a model of Li's laptop computer.

a. What is the approximate volume of Li's computer?

About __2,700 cm³__

b. Explain how you found the volume.

Sample answer: I thought of the computer as 2 prisms and used the formula l × w × h. One prism has a volume of 36 × 25 × 2, or 1,800, and the other has a volume of 36 × 25 × 1, or 900. 1,800 + 900 = 2,700 cm³

40   Assessment Handbook

Item(s)	Adjustments
1	To scaffold Item 1, remind students that the term *difference* refers to the answer to a subtraction problem. Consider dictating these problems to students.
2	To scaffold Item 2, have students highlight which part of each expression is the same.
3	To extend Item 3, have students write another real-world situation that can be modeled by the expression.
4	To extend Item 4, have students list all the different attributes of a book that they could measure.
5	To scaffold Item 5, have students build the prism with centimeter cubes. Remind them of the formulas $V = B \times h$ and $V = l \times w \times h$.
6, 7	To extend Items 6 and 7, have students solve each problem using both partial-products and U.S. traditional multiplication. Ask them to compare the methods.
8	To scaffold Item 8, give students the formula $V = l \times w \times h$ and ask them to label the computer dimensions with l, w, and h.
9	To extend Item 9, have students write a different number story in which they would need to round their answer up.
10	To scaffold Item 10, have students draw a picture to model the problem.
11	To scaffold Item 11, have students use fraction circle pieces.
12	To scaffold Item 12, give students counters and encourage them to divide the counters into equal groups.
13	To scaffold Item 13, have students first write the powers of 10 as products of 10s.

Advice for Differentiation

All instruction and most practice is complete for the content that is marked with an asterisk (*) on page 428.

Go Online:

 Quick Entry Evaluation Record students' progress and see trajectories toward mastery for these standards.

 Data Review your students' progress reports. Differentiation materials are available online to help you address students' needs.

3 Look Ahead 10–15 min

▶ Math Boxes 4-15: Preview for Unit 5

Math Journal 1, p. 152

| WHOLE CLASS | SMALL GROUP | PARTNER | INDEPENDENT |

Mixed Practice Math Boxes 4-15 are paired with Math Boxes 4-10. These problems focus on skills and understandings that are prerequisite for Unit 5. You may want to use information from these Math Boxes to plan instruction and grouping in Unit 5.

▶ Home Link 4-15: Unit 5 Family Letter

Math Masters, pp. 162–165

Home Connection The Unit 5 Family Letter provides information and activities related to Unit 5 content.

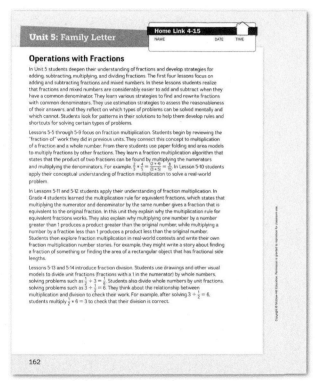

Math Masters, pp. 162–165

Planning Ahead

You may want to administer the Mid-Year Assessment after Unit 4 to check students' mastery of some of the concepts and skills in *Fifth Grade Everyday Mathematics*. See the *Assessment Handbook*.

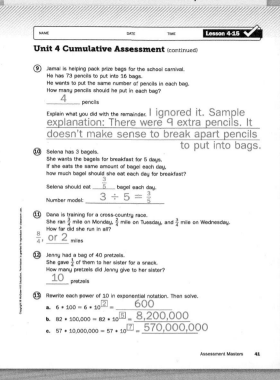

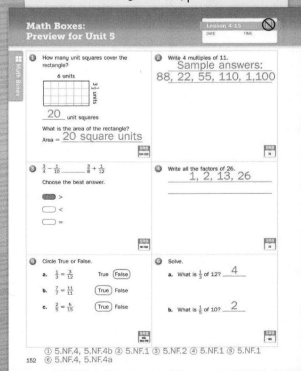

Math Journal 1, p. 152

Everyday Mathematics strives to define terms clearly, especially when those terms can be defined in more than one way.

This glossary focuses on terms and meanings for elementary school mathematics and omits details and complexities that would be required at higher levels. The definitions here are phrased for teachers. Information for explaining terms and concepts to students can be found within the lessons themselves. Additional information is available online. In a definition, most terms in italics are defined elsewhere in this glossary.

0–9

1-dimensional (1-D) (1) Having *length*, but not area or volume; confined to a curve, such as an arc. (2) A figure whose points are all on one *line*. Line segments are 1-dimensional. Compare *2-* and *3-dimensional*.

2-dimensional (2-D) (1) Having *area* but not volume; confined to a *surface*. A 2-dimensional surface can be flat or curved, such as the surface of a sphere. (2) A figure whose points are all in one *plane* but not all on one line. Examples include polygons and circles. Compare *1-* and *3-dimensional*.

3-dimensional (3-D) Having *volume*. Solids such as cubes, cones, and spheres are 3-dimensional. Compare *1-* and *2-dimensional*.

A

accurate (1) As correct as possible for a given context. An answer can be accurate without being very *precise* if the units are large. For example, the driving time from Chicago to New York is about 13 hours. See *approximate*. (2) Of a measurement or other quantity, having a high degree of correctness. A more accurate measurement is closer to the true value. Accurate answers must be reasonably precise.

acute triangle A *triangle* with three acute angles.

An acute triangle

addend Any one of a set of numbers that are added. For example, in $5 + 3 + 1 = 9$, the addends are 5, 3, and 1.

addition/subtraction use classes A category of problem situations that can be solved using addition or subtraction or other methods such as counting or direct modeling. *Everyday Mathematics* distinguishes four addition/subtraction use classes: parts-and-total, change-to-more, change-to-less, and comparison situations. The table below shows how these use classes correspond to those in the standards.

Everyday Mathematics	Standards
change-to-more	add to
change-to-less	take from
parts-and-total	put together/take apart
comparison	compare

Additive Identity The number zero (0). The additive identity is the number that when added to any other number, yields that other number. See *additive inverses*.

additive inverses Two numbers whose sum is 0. Each number is called the additive inverse, or opposite, of the other. For example, 3 and -3 are additive inverses because $3 + (-3) = 0$. Zero is its own additive inverse: $0 + 0 = 0$. See *Additive Identity*.

adjacent angles Two nonoverlapping *angles* with a common *side* and *vertex*.

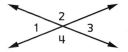 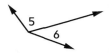

Angles 1 and 2, 2 and 3, 3 and 4, and 4 and 1 are pairs of adjacent angles.   Angle 5 is adjacent to angle 6.

adjacent sides (1) Two sides of a *polygon* with a common *vertex*. (2) Two faces of a *polyhedron* with a common *edge*.

algebra The branch of mathematics that uses letters and other symbols to stand for quantities that are unknown or vary. Algebra is used to describe patterns, express numerical relationships, and model real-world situations.

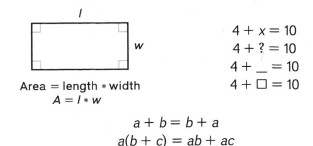

Area = length * width
$A = l * w$

$4 + x = 10$
$4 + ? = 10$
$4 + \_\_ = 10$
$4 + \square = 10$

$a + b = b + a$
$a(b + c) = ab + ac$

Uses of algebra

algorithm A set of step-by-step instructions for doing something, such as carrying out a computation or solving a problem. The most common algorithms are those for basic arithmetic computation, but there are many others. Some mathematicians and many computer scientists spend a great deal of time trying to find more efficient algorithms for solving problems.

anchor chart A classroom display that is co-created by teacher and students and focuses on a central concept or skill.

angle (1) A figure formed by two rays or line segments with a common endpoint called the *vertex* of the angle. The rays or segments are called the sides of the angle. Angles can be named after their vertex point alone as in ∠A; or by three points, one on each side and the vertex in the middle as in ∠BCD. One side of an angle is rotated about the vertex from the other side through a number of *degrees*. (2) The measure of this rotation in degrees.

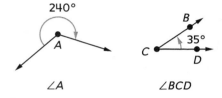

∠A   ∠BCD

-angle A suffix meaning angle, or corner, for example, triangle and rectangle.

apex (1) In a *pyramid*, the *vertex* opposite the *base*. All the nonbase faces meet at the apex. (2) The point at the tip of a *cone*.

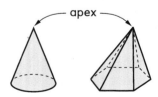

apex

approximate Close to exact. In many situations it is not possible to get an exact answer, but it is important to be close to the exact answer. We might draw an angle that measures approximately 60° with a protractor. In this case, approximate suggests the angle drawn is within a degree or two of 60°. Compare *precise*.

area The amount of *surface* inside a *2-dimensional* figure. The figure might be a triangle or rectangle in a *plane*, the curved surface of a cylinder, or a state or country on Earth's surface. Commonly, the area is measured in *square units*, such as square miles or square centimeters, or other units, such as acres.

A triangle with area 21 square units   A rectangle with area 1.2 cm * 2 cm = 2.4 square centimeters

area model (1) A *model* for multiplication in which the length and width of a *rectangle* represent the *factors*, and the *area* of the rectangle represents the *product*.

Area model for 3 × 5 = 15

(2) A model showing fractions as parts of a *whole*. The whole is a region, such as a circle or a rectangle, representing the unit whole.

Area model for $\frac{2}{3}$

arithmetic facts The addition facts (*whole-number* addends 10 or less); their inverse subtraction facts; multiplication facts (whole-number factors 10 or less); and their inverse division facts, except there is no division by zero. Facts and their corresponding inverses are organized into fact families.

array (1) An arrangement of objects in a regular *pattern*, usually rows and columns. (2) A rectangular array. In *Everyday Mathematics*, an array is a rectangular array unless specified otherwise.

Associative Property of Addition A *property* of addition that for any numbers *a*, *b*, and *c*, (a + b) + c = a + (b + c). The grouping of the three *addends* can be changed without changing the *sum*. For example, (4 + 3) + 7 = 4 + (3 + 7) because 7 + 7 = 4 + 10. Subtraction is not associative. For example, (4 − 3) + 7 ≠ 4 − (3 + 7) because 8 ≠ −6. Compare *Commutative Property of Addition*.

Associative Property of Multiplication   A *property* of multiplication that for any numbers *a*, *b*, and *c*, (*a* × *b*) × *c* = *a* × (*b* × *c*). The grouping of the three *factors* can be changed without changing the *product.* For example, (4 × 3) × 7 = 4 × (3 × 7) because 12 × 7 = 4 × 21. Division is not associative. For example, (21 ÷ 7) ÷ 3 ≠ 21 ÷ (7 ÷ 3) because 1 ≠ 9. Compare *Commutative Property of Multiplication.*

attribute   A characteristic or *property* of an object or a common characteristic of a set of objects. Size, shape, color, and number of sides are attributes.

axis of coordinate grid   Either of the two *number lines* used to form a *coordinate grid.* Plural is axes.

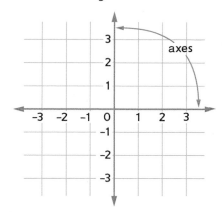

B

bar graph   A graph with horizontal or vertical bars that represent (typically categorical) *data.* The lengths of the bars may be *scaled.*

Fat Content of Foods

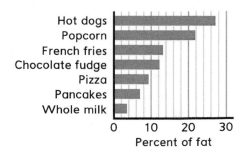

Percent of fat

base   The *side* of a *polygon* or *face* of a polyhedron from which the *height* is measured.

base (in exponential notation)   A number that is raised to a power. For example, the base in 5^3 is 5.

base angles of a trapezoid   Two *angles* that share a *base* of a trapezoid.

base of a number system   The foundation number for a *place-value* based *numeration* system. For example, our usual way of writing numbers uses a base-10 place-value system. In programming computers or other digital devices, bases of 2, 8, 16, or other powers of 2 are more common than *base 10.*

base of a trapezoid   (1) Either of a pair of *parallel sides* in a *trapezoid.* (2) The *length* of this side. The area of a trapezoid is the average of a pair of bases times the corresponding height.

base ten   (1) Related to *powers* of 10. (2) The most common system for writing numbers, which uses 10 symbols 0, 1, 2, 3, 4, 5, 6, 7, 8, and 9, called *digits.* One can write any number using one or more of these 10 digits, and each digit has a value that depends on its place in the number (its *place value*). In the base-10 system, each place has a value 10 times that of the place to its right, and one-tenth the value of the place to its left.

basic facts   Same as *arithmetic facts.*

benchmark   A number or measure used as a reference point. A benchmark can be used to evaluate the reasonableness of counts, measures, or estimates. For example, a benchmark for length is that the width of an adult's thumb is about one inch. The numbers 0, $\frac{1}{2}$, 1, 1$\frac{1}{2}$, and so on may be useful benchmarks for evaluating reasonableness of fraction calculations.

braces   See *grouping symbols* and *General Reference.*

brackets   See *grouping symbols* and *General Reference.*

C

calibrate   (1) To divide or mark a measuring *tool* with graduations, such as the degree marks on a thermometer. (2) To test and adjust the *accuracy* of a measuring tool.

capacity   (1) The amount of space contained by a *3-dimensional* figure. Capacity is often measured in liquid units such as cups, quarts, gallons, or liters. (2) The amount something can hold. For example, a computer hard drive may have a capacity of 64 TB, or a scale may have a capacity of 400 lb.

cardinal number   A number telling how many things are in a *set.* Compare *ordinal number.*

category   A group whose members are defined by a shared *attribute.* For example, triangles are polygons that share the attribute of having three sides.

Celsius A *temperature* scale on which pure water at sea level freezes at 0° and boils at 100°. The Celsius scale is used in the metric system. A less common name for this scale is centigrade because there are 100 degrees between the freezing and boiling points of water. Compare *Fahrenheit*.

cent A penny; $\frac{1}{100}$ of a dollar. From the Latin word *centesimus,* which means "a hundredth part."

center of a circle The point in the *plane* of a *circle* that is equally distant from all points on the circle.

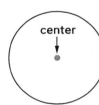
center

center of a sphere The point equally distant from all points on a *sphere.*

center

circle The set of all points in a *plane* that are equally distant from a fixed point in the plane called the center. The distance from the center to the circle is the radius of the circle. The diameter of a circle is twice the radius. A circle together with its interior is called a *disk* or a circular region.

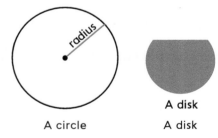
radius

A circle   A disk

close-but-easier numbers Numbers that are close to the original numbers in the problem, but more efficient for mental computation and estimation. For example, to estimate 494 + 78, one might use the close-but-easier numbers 490 and 80.

column addition An addition *algorithm* in which the *addends'* digits are first added in each *place-value* column separately, and then 10-for-1 trades are made until each column has only one digit. Lines are usually drawn to separate the place-value columns.

common denominator A nonzero number that is a *multiple* of the *denominators* of two or more fractions. For example, the fractions $\frac{1}{2}$ and $\frac{2}{3}$ have common denominators 6, 12, 18, and other multiples of 6.

common multiple A number that is a *multiple* of two or more given numbers. For example, common multiples of 6 and 8 include 24, 48, and 72. See *least common multiple (LCM).*

common numerator Same as *like numerator.*

Commutative Property of Addition A *property* of addition that for any numbers *a* and *b*, $a + b = b + a$. Two numbers can be added in either order without changing the *sum.* For example, 5 + 10 = 10 + 5. In *Everyday Mathematics,* this and the *Commutative Property of Multiplication* are called *turn-around rules.* Subtraction is not commutative. For example, $8 - 5 \neq 5 - 8$ because $3 \neq -3$.

Commutative Property of Multiplication A *property* of multiplication that for any numbers *a* and *b*, $a * b = b * a$. Two numbers can be multiplied in either order without changing the *product.* For example, $5 * 10 = 10 * 5$. In *Everyday Mathematics,* this and the *Commutative Property of Addition* are called *turn-around rules.* Division is not commutative. For example, $10 / 5 \neq 5 / 10$ because $2 \neq \frac{1}{2}$.

compose To make up or form a number or shape by putting together smaller numbers or shapes. For example, one can compose a 10 by putting together ten 1s: $1 + 1 + 1 + 1 + 1 + 1 + 1 + 1 + 1 + 1 = 10$. One can compose a pentagon by putting together an equilateral triangle and a square.

composite number A *counting number* that has more than two *factors.* For example, 10 is a composite number because it has four factors: 1, 2, 5, and 10. Compare *prime number.*

composite unit A *unit* of measure made up of multiple copies of a smaller unit. For example, a foot is a composite unit of 12 inches used to measure length, and a row of unit squares can be used to measure area.

3 rows of 5 square units each gives an area of 15 square units.

concave polygon A *polygon* on which there are at least two points that can be connected with a *line segment* that passes outside the polygon. For example, segment *AD* is outside the hexagon between *B* and *C.* Informally, at least one *vertex* appears to be "pushed inward." At least one interior angle has measure greater than 180°. Same as *nonconvex polygon.* Compare *convex polygon.*

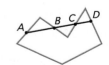

cone A *geometric solid* comprising a circular *base,* an *apex* not in the *plane* of the base, and all line segments with one endpoint at the apex and the other endpoint on the circumference of the base.

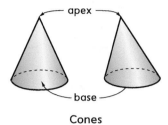

Cones

congruent figures Figures having the same size and shape. Two figures are congruent if they match exactly when one is placed on top of the other after a combination of isometric transformations (slides, flips, and/or turns). In diagrams of congruent figures, the corresponding *congruent sides* may be marked with the same number of hash marks. The symbol ≅ means "is congruent to".

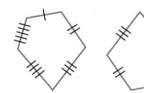

Congruent pentagons

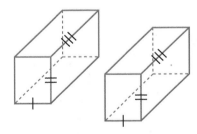

Congruent right rectangular prisms

conjecture A claim that has not been proved, at least by the person making the conjecture.

convex polygon A *polygon* on which no two points can be connected with a *line segment* that passes outside the polygon. Informally, all *vertices* appear to be "pushed outward". Each *angle* in the polygon measures less than 180°. Compare *concave polygon.*

coordinate A number or one of the numbers in a pair or triple used to locate a point on a line, a grid, or in space.

coordinate grid (rectangular coordinate grid) A *reference frame* for locating points in a plane using *ordered pairs* of numbers. A rectangular coordinate grid is formed by two number lines, or axes, that intersect at right angles at their *zero points.*

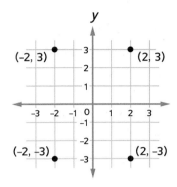

coordinate system (rectangular coordinate system) A *reference frame* in which any point can be located with one or more numbers relative to the *origin* of a number line in one dimension or intersecting perpendicular axes in multiple dimensions. Also called rectangular coordinate system. See *coordinate grid.*

corresponding sides Sides in the same relative position in similar or *congruent figures.*

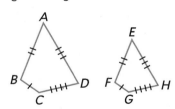

Pairs of corresponding sides are marked with the same number of hash marks.

corresponding terms Terms that occupy the same position within two *sequences.* For example, in the table below, the third term in the "in" column corresponds to the third term in the "out" column. They are corresponding terms.

Rule		in	out
* 3		1	3
		2	6
		3	9
		4	12

counting numbers The numbers used to count things. The set of counting numbers is {1, 2, 3, 4, . . .}. Sometimes 0 is included, but not in *Everyday Mathematics.* Counting numbers are also known as *natural numbers.*

counting-up subtraction A subtraction strategy in which a *difference* is found by counting or adding up from the smaller number to the larger number. For example, to calculate 87 − 49, one could start at 49, add 30 to reach 79, and then add 8 more to reach 87. The difference is 30 + 8 = 38.

cube (1) A *regular polyhedron* with 6 square *faces.* A cube has 8 *vertices* and 12 *edges.*

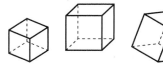

Cubes

(2) In *Everyday Mathematics*, the smaller cube of the base-10 blocks, measuring 1 cm on each edge.

cubic unit A unit, such as cubic centimeters, cubic inches, cubic feet, and cubic meters, used to measure volume or capacity.

cubit An ancient unit of *length,* measured from the point of the elbow to the end of the middle finger. The cubit has been standardized at various times between 18 and 22 inches. The Latin word *cubitum* means "elbow".

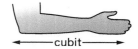

cubit

cup A U.S. customary unit of *volume* or *capacity* equal to 8 fluid ounces or $\frac{1}{2}$ pint.

curved surface A *2-dimensional surface* that does not lie in a *plane.* Spheres, cylinders, and cones have curved surfaces.

customary system of measurement In *Everyday Mathematics*, same as *U.S. customary system of measurement.* See *Tables of Measures.*

cylinder A *geometric solid* with two congruent, *parallel* circular regions for *bases* and a curved *face* formed by all the segments that have an endpoint on each *circle* and that are parallel to a *segment* with endpoints at the *centers of the circles.* Also called a circular cylinder.

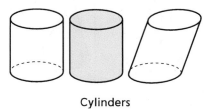

Cylinders

D

data Information that is gathered by counting, measuring, questioning, or observing. Strictly, data is the plural of datum, but data is often used as a singular word.

decagon A 10-sided *polygon.*

decimal (1) A number written in standard *base-10* notation containing a decimal point, such as 2.54. (2) Any number written in standard base-10 notation. See *decimal fraction, repeating decimal*, and *terminating decimal.*

decimal fraction (1) A *fraction* or mixed number written in standard *decimal notation.* (2) A fraction $\frac{a}{b}$ where b is a positive power of 10, such as $\frac{84}{100}$.

decimal notation Same as *standard notation.*

decimal point A mark used to separate the ones and tenths places in *decimals.* A decimal point separates dollars from cents in dollars-and-cents notation. The mark is a dot in the U.S. customary system and a comma in Europe and some other countries.

decompose To separate a number or shape into smaller numbers or shapes. For example, 14 can be decomposed into 1 ten and 4 ones. A square can be decomposed into two isosceles right triangles. Any even number can be decomposed into two equal parts: $2n = n + n.$

degree (°) (1) A unit of measure for *angles* based on dividing a *circle* into 360 equal parts. Latitude and longitude are measured in degrees based on angle measures. (2) A unit for measuring *temperature.* See *Celsius* and *Fahrenheit.* The symbol ° means degrees of any type.

denominator The nonzero *divisor b* in a *fraction* $\frac{a}{b}$. In a *part-whole fraction,* the denominator is the number of equal parts into which the *whole* has been divided. The denominator determines the size of each part. Compare *numerator.*

difference (1) The distance between two numbers on a *number line*. The difference between 5 and 12 is 7. (2) The result of subtracting one number from another. For example, in $12 - 5 = 7$ the difference is 7, and in $5 - 12$ the difference is -7. Compare *minuend* and *subtrahend*.

digit (1) Any of the symbols 0, 1, 2, 3, 4, 5, 6, 7, 8, and 9 in the *base-10 numeration* system. For example, the *numeral* 145 is made up of the digits 1, 4, and 5. (2) Additional symbols in other *place-value* systems, such as, A, B, C, D, E, and F in base-16 notation.

dimension (1) A measurable extent such as *length*, width, or *height*. Having two makes the measured figure 2-dimensional. See *1-, 2-,* and *3-dimensional*. (2) The measures of those extents. For example, the dimensions of a box might be 24 cm by 20 cm by 10 cm. (3) The number of coordinates necessary to locate a point in a geometric space. A plane has two dimensions because an ordered pair of two coordinates uniquely locates any point in the plane.

disk A *circle* and its *interior* region.

displacement A method for measuring the *volume* of an object by submerging it in water and then measuring the volume of water it displaces. The method is especially useful for finding the volume of an irregularly shaped object.

Distributive Property of Multiplication over Addition and Subtraction A *property* that for any numbers a, b, and c:

$$a * (b + c) = (a * b) + (a * c)$$
$$\text{or } a(b + c) = ab + ac$$
$$\text{and } a * (b - c) = (a * b) - (a * c)$$
$$\text{or } a(b - c) = ab - ac$$

This property relates multiplication to a *sum* or *difference* of numbers by distributing a *factor* over the terms in the sum or difference. For example,

$$2 * (5 + 3) = (2 * 5) + (2 * 3) = 10 + 6 = 16.$$

dividend The number in division that is being divided. For example, in $35 / 5 = 7$, the dividend is 35. Compare *divisor* and *quotient*.

$$\text{dividend} \quad \overset{\text{divisor}}{\underset{}{35 / 5 = 7}} \quad \text{quotient}$$

$$\text{dividend} \quad \overset{\text{divisor}}{\underset{}{40 \div 8 = 5}} \quad \text{quotient}$$

$$\text{quotient} \longrightarrow 3$$
$$\text{divisor} \longrightarrow 12\overline{)36} \longleftarrow \text{dividend}$$

divisible by If a *counting number* can be divided by another counting number with no *remainder*, then the first is divisible by the second. For example, 28 is divisible by 7, because $28 / 7 = 4$ with no remainder. If a number n is divisible by a number d, then d is a *factor* of n. Every counting number is divisible by itself.

division symbols The number a divided by the number b is written in a variety of ways. In *Everyday Mathematics*, $a \div b$, a / b, and $\frac{a}{b}$ are the most common notations, while $\overline{)}$ is used to set up some division algorithms. $a:b$ is sometimes used in Europe, $\boxed{\div}$ is common on calculators, and $/$ is common on computer keyboards.

divisor In division, the number that divides another number. For example, in $35 / 7 = 5$, the divisor is 7. See the diagram with dividend. Compare *dividend* and *quotient*.

E

edge (1) Any *side* of a *polyhedron's faces*.

edges

(2) A line segment or curve where two *surfaces* of a *geometric solid* meet.

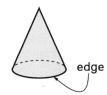
edge

efficient strategy A method that can be applied easily and quickly. For example, adding a group and doubling are usually efficient *strategies* for solving multiplication facts.

endpoint A point at the end of a line segment, ray, arc, or curve. These shapes are usually named using their endpoints. For example, the segment shown is "segment *TL*" or "segment *LT*".

endpoints

enlarge To increase the size of an object or a figure without changing its shape. See *scale factor*.

equal (1) Identical in number or measure; neither more nor less. (2) *Equivalent*.

equal groups Sets with the same number of elements, such as cars with 5 passengers each, rows with 6 chairs each, and boxes containing 100 paper clips each.

equal-groups notation In *Everyday Mathematics*, a way to denote a number of equal-size groups. The size of each group is shown inside square brackets and the number of groups is written in front of the brackets. For example, 3 [6s] means 3 groups with 6 in each group. In general, *n* [*k*s] means *n* groups with *k* in each group.

equal parts *Equivalent* parts of a *whole*. For example, dividing a pizza into 4 equal parts means each part is $\frac{1}{4}$ of the pizza and is equal in size to each of the other 3 parts.

equal share One of several parts of a whole, each of which has the same amount of area, volume, mass, or other measurable or countable quantity. Sometimes called fair share. See *equal parts*.

equation A *number sentence* that contains an equal sign. For example, $5 + 10 = 15$ and $P = 2l + 2w$ are equations.

equilateral polygon A *polygon* in which all sides are the same length.

Equilateral polygons

equilateral triangle A *triangle* with all three sides equal in length. Each angle of an equilateral triangle measures 60°, so it is also called an equiangular triangle. All equilateral triangles are *isosceles triangles*.

equivalent *Equal* in value but possibly in a different form. For example, $\frac{1}{2}$, 0.5, and 50% are all equivalent.

equivalent equations *Equations* with the same *solution*. For example, $2 + x = 4$ and $6 + x = 8$ are equivalent equations with the common solution $x = 2$. Properly applying the rules of *algebra* to any given equation will produce an equivalent equation. For example, adding 4 to both sides of the first equation above will yield the second equation.

equivalent expressions *Expressions* that name the same number or would name the same number if numbers replaced all the *variables*. For example, 17 and $10 + 7$ are equivalent expressions. The expressions $4(a + b)$ and $4a + 4b$ are also *equivalent*.

equivalent fractions *Fractions* that name the same number, such as $\frac{1}{2}$, $\frac{4.5}{9}$, and $\frac{28}{56}$.

equivalent fractions rule A rule stating that if the *numerator* and *denominator* of a *fraction* are each multiplied or divided by the same nonzero number, the result is a fraction *equivalent* to the original fraction.

equivalent names Different ways of naming the same number. For example, $2 + 6$, $4 + 4$, $12 - 4$, $18 - 10$, $100 - 92$, $5 + 1 + 2$, eight, VIII, and ||||| /// are all equivalent names for 8. See *name-collection box*.

equivalent problem In *Everyday Mathematics*, a division problem solved by writing an *equivalent expression*. For example, to solve 35.6 / 0.5, students may solve the equivalent problem 356 / 5.

estimate (1) An answer close to, or *approximating*, an exact answer. (2) To make an estimate.

evaluate a numerical expression To carry out the *operations* in a numerical *expression* to find a single value for the expression.

even number (1) A *counting number* that is divisible by 2 : 2, 4, 6, 8 . . . . (2) An *integer* that is *divisible by* 2. Compare *odd number*.

expanded form Same as *expanded notation*.

expanded notation A way of writing a number as the *sum* of the values of each *digit*. For example, 356 is $300 + 50 + 6$ in expanded notation. Same as *expanded form*. Compare *standard notation* and *number-and-word notation*.

exponent A number used in *exponential notation* to tell how many times the *base* is used as a *factor*. The exponent is typically a superscript or written after a caret. For example, in 5^3 or 5^3, the base is 5, the exponent is 3, and $5^3 = 5 * 5 * 5 = 125$. Same as *power*.

exponential notation A way of representing repeated multiplication by the same *factor*. For example, 2^3 is exponential notation for $2 * 2 * 2$. The *exponent* 3 tells how many times the *base* 2 is used as a factor.

expression (1) A mathematical phrase made up of numbers, *variables*, *operation symbols*, and/or *grouping symbols*. An expression does not contain *relation symbols* such as $=$, $>$, and $\leq$. (2) Either side of an *equation* or *inequality*.

extended facts   Variations of basic *arithmetic facts* involving *multiples* of 10, 100, and so on. For example, 30 + 70 = 100, 40 * 5 = 200, and 560 / 7 = 80 are extended facts. See *fact extensions*.

extrapolate   To *estimate* an unknown value beyond known values. Graphs are useful tools for extrapolation. Compare *interpolate*.

F

face   (1) A flat *surface* on a closed, 3-*dimensional* figure. Some special faces are called *bases*.

a flat face

(2) More generally, any *2-dimensional* surface on a 3-dimensional figure.

a curved face

fact extensions   Calculations with larger numbers using knowledge of basic *arithmetic facts.* For example, knowing the addition fact 5 + 8 = 13 makes it easier to solve problems such as 50 + 80 = ? and 65 + ? = 73. Fact extensions apply to all four basic arithmetic operations. See *extended facts*.

fact family   A set of related *arithmetic facts* linking two inverse operations. For example,

$$5 + 6 = 11 \qquad 6 + 5 = 11$$
$$11 - 5 = 6 \qquad 11 - 6 = 5$$

are an addition/subtraction fact family. Similarly,

$$5 * 7 = 35 \qquad 7 * 5 = 35$$
$$35 / 7 = 5 \qquad 35 / 5 = 7$$

are a multiplication/division fact family. Same as *number family*.

factor   (1) Each of the two or more numbers or *variables* in a *product*. For example, in 6 * 0.5, 6 and 0.5 are factors; in 7*b*, 7 and *b* are factors. Compare *factor of a counting number* n. (2) To represent a number as a product of factors. For example, factor 21 by rewriting as 7 * 3.

factor of a counting number *n*   A *counting number* whose product with another counting number equals *n.* For example, 2 and 3 are *factors* of 6 because 2 * 3 = 6. But 4 is not a factor of 6 because 4 * 1.5 = 6, and 1.5 is not a counting number.

factor pair   Two *factors of a counting number* n whose *product* is *n.* A number may have more than one factor pair. For example, the factor pairs for 18 are 1 and 18; 2 and 9; 3 and 6.

factor tree   A diagram used to get the *prime factorization* of a *counting number.* Write the original number as a product of *factors* not including 1. Then write each of these factors as a product of factors, and continue until the factors are all prime numbers. A factor tree looks like an upside-down tree, with the root (the original number) at the top and the leaves (the factors) beneath it.

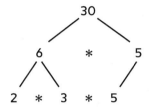
A factor tree for 30

Fahrenheit   A *temperature* scale on which pure water at sea level freezes at 32° and boils at 212°. The Fahrenheit scale is widely used in the United States but in few other places. Compare *Celsius*.

false number sentence   A *number sentence* that is not true. For example, 8 = 5 + 5 is a false number sentence. Compare *true number sentence.*

fathom   A unit of *length* equal to 6 feet, or 2 yards. It is used mainly by people who work with boats and ships to measure depths underwater and lengths of cables. Estimated with arm span.

fathom

figurate numbers   Numbers that can be illustrated by specific geometric *patterns. Square numbers* and *triangular numbers* are figurate numbers.

Square numbers   Triangular numbers

fluid ounce (fl oz)   A U.S. customary unit of *volume* or *capacity* equal to $\frac{1}{16}$ of a pint, or about 29.573730 milliliters. Compare *ounce*. See *Tables of Measures*.

formula   A general rule for finding the value of something, usually an *equation* with quantities represented by letter *variables*. For example, a formula for distance traveled *d* at a rate *r* over a time *t* is $d = r * t$. The area *A* of a triangle with base length *b* and height *h* is shown.

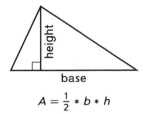

$$A = \tfrac{1}{2} * b * h$$

fraction   (1) A number in the form $\frac{a}{b}$ or *a/b*, where *a* and *b* are *integers* and *b* is not 0. A fraction may be used to name part of an object or part of a collection of objects, to compare two quantities, or to represent division. For example, $\frac{12}{6}$ might mean 12 eggs divided in groups of 6, a ratio of 12 to 6, or 12 divided by 6. Also called a common fraction. (2) A fraction that satisfies the previous definition and includes a unit in both the *numerator* and the *denominator*. For example, the rates $\frac{50 \text{ miles}}{1 \text{ gallon}}$ and $\frac{40 \text{ pages}}{10 \text{ minutes}}$ are fractions. (3) A number written using a fraction bar, where the fraction bar is used to indicate division. For example, $\frac{2.3}{6.5}$, $\frac{1\frac{4}{5}}{12}$, $\frac{\pi}{4}$, and $\frac{\frac{3}{4}}{\frac{5}{8}}$. Compare *decimal*.

Fraction Circle Pieces   In *Third* through *Fifth Grade Everyday Mathematics*, a set of colored circles each divided into equal-size slices, used to represent *fractions*.

fractional part   Part of a whole. *Fractions* represent fractional parts of numbers, sets, or objects.

frequency   (1) The number of times a value occurs in a set of *data*. (2) A number of repetitions per unit of time, such as the vibrations per second in a sound wave.

frequency graph   A graph showing how often each value occurs in a *data* set.

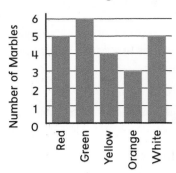

Colors in a Bag of Marbles

frequency table   A table in which *data* are tallied and organized, often as a first step toward making a *frequency graph*.

Color	Number of Marbles
Red	̶H̶ft̶
Green	̶H̶ft̶ /
Yellow	////
Orange	///
White	̶H̶ft̶

front-end estimation   An arithmetic estimation method that retains only the left-most digit in the numbers and substitutes 0s for all others. For example, the front-end *estimate* for 45,600 + 53,450 is 40,000 + 50,000 = 90,000.

function   (1) A set of *ordered pairs* (x, y) in which each value of *x* is paired with exactly one value of *y*. A function is typically represented in a table, by points on a coordinate graph, or by a rule such as an *equation*. (2) A rule that pairs each *input* with exactly one *output*. For example, for a function with the rule "Double," 1 is paired with 2, 2 is paired with 4, 3 is paired with 6, and so on. See *"What's My Rule?"*.

function machine   An imaginary device that receives *inputs* and pairs them with *outputs* using a rule that is a *function*. For example, the function machine below pairs an input number with its double. See *function*.

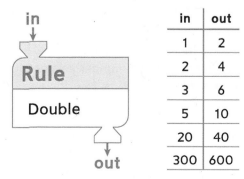

A function machine and function table

G

geometric solid   The *surface* or surfaces that make up a *3-dimensional* figure such as a *prism, pyramid, cylinder, cone,* or *sphere.* Despite its name, a geometric solid is hollow; that is, it does not include the points in its *interior.* Informally, and in some dictionaries, a solid is defined as both the surface and its interior.

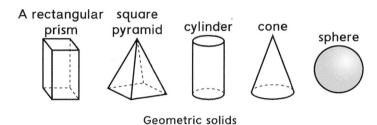

Geometric solids

Geometry Template   An *Everyday Mathematics* tool that includes a millimeter ruler, a ruler with $\frac{1}{16}$-inch intervals, half-circle and full-circle *protractors,* a *Percent Circle,* pattern-block shapes, and other geometric figures.

-gon   A suffix meaning *angle.* For example, a *hexagon* is a plane figure with six angles.

great span   The distance from the tip of the thumb to the tip of the little finger (pinkie), when the hand is stretched as far as possible. The great span averages about 9 inches for adults. Same as hand span. See *Tables of Measures.*

Great span

grouping symbols   Parentheses ( ), brackets [ ], braces { }, and similar symbols that define the order in which operations in an *expression* are to be done. *Nested* grouping symbols are groupings within groupings, and the innermost grouping is done first. A vinculum is a bar or line used to group numbers $\left(\text{as in } \frac{3+5}{2}\right)$ or in conjunction with a radical $\left(\text{as in } \frac{3+5}{\sqrt{1+3}}\right)$ or in a variety of other ways beyond elementary school mathematics.

H

height   (1) The *length* of a *perpendicular* segment from one *side* of a geometric figure to a *parallel* side or from a *vertex* to the opposite side. (2) The line segment itself.

Height of 2-D figures are shown in red.

heptagon   A 7-sided *polygon.*

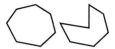

Heptagons

hexagon   A 6-sided *polygon.*

A hexagon

hierarchy of shapes   A classification in which shapes are organized into categories and subcategories. For each category, every defining *attribute* of a shape in that category is a defining attribute of all shapes in its subcategories. A hierarchy is often shown in a diagram with the most general category at the top and arrows or lines connecting categories to their subcategories. See *quadrilateral* and below for examples.

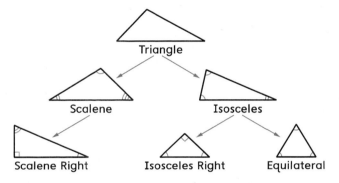

Hierarchy of triangles by angle size

Home Link   In *Everyday Mathematics,* a suggested follow-up or enrichment activity to be done at home.

horizontal   In a left-to-right orientation. Lined up with the horizon.

I

image   A figure that is produced by a transformation of another figure called the *preimage.*

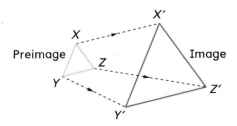

improper fraction   A *fraction* with a numerator that is greater than or equal to its denominator. For example, $\frac{4}{3}$, $\frac{5}{2}$, $\frac{4}{4}$, and $\frac{24}{12}$ are improper fractions. *Everyday Mathematics* does not use this term. Improper fractions are referred to as fractions greater than or equal to 1.

inequality A *number sentence* with a *relation symbol* other than =, such as >, <, ≥, ≤, or ≠. Compare *equation*.

input (1) A number inserted into a *function machine*, which applies a rule to pair the input with an *output*. (2) Numbers or other information entered into a calculator or computer.

integer A number in the set {..., −4, −3, −2, −1, 0, 1, 2, 3, 4,...}. A *whole number* or its opposite, where 0 is its own opposite. Compare *rational numbers, irrational numbers,* and *real numbers*.

interior of a figure (1) The set of all points in a *plane* bounded by a closed *2-dimensional* figure such as a polygon or circle. (2) The set of all points in space bounded by a closed *3-dimensional* figure such as a polyhedron or sphere. The interior is usually not considered to be part of the figure.

interpolate To *estimate* an unknown value between known values. Graphs are often useful tools for interpolation. Compare *extrapolate*.

intersect To share a common point or points.

Intersecting lines and line segments   Intersecting planes

interval (1) The set of all numbers between two numbers *a* and *b*, which may include one or both of *a* and *b*. (2) All points and their coordinates on a *segment* of a *number line*. The interval between 0 and 1 on a number line is the unit interval.

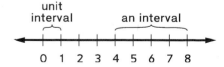

inverse operations Two *operations* that undo the effects of each other. Addition and subtraction are inverse operations, as are multiplication and division.

irrational numbers Numbers that cannot be written as *fractions* where both the numerator and denominator are *integers* and the denominator is not zero. For example, $\sqrt{2}$ and π are irrational numbers. In *standard notation,* an irrational number can only be written as a

nonterminating, nonrepeating decimal. For example, π = 3.141592653 ... continues forever without a repeating pattern. The number 1.10100100010000 ... is irrational because its pattern does not repeat. Compare *rational numbers*.

isosceles trapezoid A *trapezoid* with a pair of *base angles* that have the same measure. See *quadrilateral*.

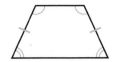

An isosceles trapezoid

isosceles triangle A *triangle* with at least two equal-length *sides. Angles* opposite the equal-length sides are equal in measure.

Isosceles triangles

iterate units To repeat a *unit* without gaps or overlaps in order to measure. To measure *length,* units are placed end-to-end along a path. Unit iteration can also be used to measure *area* (by tiling) or *volume* (by filling).

K

key sequence The order in which calculator keys are pressed to perform a calculation.

kilogram A metric unit of mass equal to 1,000 grams. Though for over a century the kilogram has been defined based on platinum and iridium cylinder kept in Sevres, France, the National Institute of Standards and Technology has been working to define it by unchanging quantum properties of nature. A kilogram is about 2.2 pounds. See *Tables of Measures*.

kite A *quadrilateral* that has two nonoverlapping pairs of *adjacent,* equal-length *sides*. Note that all four sides might be of equal length, so a *rhombus* is a kite.

Kites

L

label (1) A descriptive word or phrase used to put a number or numbers in context. Labels encourage children to associate numbers with real objects. (2) In a spreadsheet, a table, or graph, words or numbers providing information such as the title of the spreadsheet, the heading for a row or column, or the variable on an axis.

lattice multiplication An old *algorithm* for multiplying multidigit numbers that requires only basic multiplication facts and addition of 1-digit numbers in a diagram resembling a lattice.

length The distance between two points on a *1-dimensional* figure. For example, the figure might be a line segment, an arc, or a curve on a map modeling a hiking path. Length is measured in units such as inches, kilometers, and miles.

length of day The elapsed time between sunrise and sunset.

like Equal or the same.

like denominator Same as *common denominator.*

like numerator A number that is the numerator of two or more fractions. For example, the fractions $\frac{3}{11}$ and $\frac{3}{7}$ have a common numerator of 3. Sometimes also called *a common numerator.*

line A *1-dimensional* straight path of points that extends forever in opposite directions. A line is named using two points on it or with a single, italicized lower-case letter such as *l*. In formal Euclidean geometry, line is an undefined geometric term.

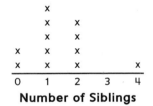

Line *PR* or $\overleftrightarrow{PR}$

line graph A graph in which *data* points are connected by line segments. Also known as broken-line graph.

line of reflection (mirror line) A line halfway between a preimage and its *reflection* image. It is the perpendicular bisector of the line segments connecting points on a preimage with their corresponding points on its reflection image. Compare *line of symmetry.*

line of symmetry A line that divides a figure into two parts that are reflection images of each other. A figure may have zero, one, or more lines of symmetry. For example, the numeral 2 has no lines of symmetry, a square has four lines of symmetry, and a circle has infinitely many lines of symmetry.

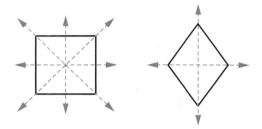

Lines of symmetry are shown in blue.

line plot A sketch of *data* in which check marks, Xs, or other symbols above a labeled line show the *frequency* of each value.

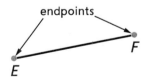

Number of Siblings

A line plot

line segment A part of a *line* between and including two points called *endpoints* of the segment. Same as *segment.* A line segment is often named by its endpoints.

endpoints

E   *F*

Segment *EF* or $\overline{EF}$

line symmetry A figure has line symmetry if a line can be drawn that divides the figure into two parts that are reflection images of each other. See *line of symmetry.*

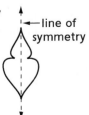

←line of symmetry

liquid volume An amount of liquid measured in units such as liters and gallons. Units of liquid *volume* are frequently used to measure *capacity.*

liter (L) A metric unit of *volume* or *capacity* equal to the volume of a cube with 10-cm-long edges. 1 L = 1,000 mL = 1,000 cm³. A liter is a little larger than a quart. See *Tables of Measures.*

long-term memory *Memory in a calculator* used by keys with an M on them, such as [M−] and [M+]. Numbers in long-term memory are not affected by calculations with keys without an M, which use *short-term memory.*

M

mass (1) A measure of the amount of matter in an object. Mass is not affected by gravity, so it is the same on Earth, the moon, or anywhere else in space. Mass is usually measured in grams, kilograms, and other metric units. Compare *weight.* (2) A standard object used in measuring weight or mass.

Math Boxes In *Everyday Mathematics*, a collection of problems to practice skills. Math Boxes for each lesson are in the *Math Journal.*

Math Message In *Everyday Mathematics*, an introduction to the day's lesson designed for students to complete independently.

mathematical argument An explanation that shows why a claim or *conjecture* is true or false using words, pictures, symbols, or other representations. For example, if a student makes a conjecture that $\frac{1}{2} + \frac{3}{5} = \frac{4}{7}$ is not true, the student might support that conjecture by arguing that $\frac{3}{5}$ is more than $\frac{1}{2}$, so the answer to $\frac{1}{2} + \frac{3}{5}$ is greater than 1. Since $\frac{4}{7}$ is less than 1, $\frac{1}{2} + \frac{3}{5} = \frac{4}{7}$ must not be true.

mathematical process and practice Ways of working with mathematics. Mathematical processes and practices are habits and actions that help people use mathematics to solve problems, such as perseverance, abstract reasoning, and pattern generation.

mathematical structure A relationship among mathematical objects, operations, or relations; a mathematical *pattern*, framework, category, or *property.* For example, the Distributive Property of Multiplication over Addition is a key structure of arithmetic. The number grid illustrates some patterns and structures that exist in our base-ten number system.

measurement scale See *scale of a number line.*

measurement unit The reference unit used when measuring. Examples of basic units include inches for length, grams for mass or weight, cubic inches for volume or capacity, seconds for elapsed time, and degrees Celsius for change of temperature. Compound units include square centimeters for area and miles per hour for speed.

memory in a calculator Where numbers are stored in a calculator for use in later calculations. Most calculators have both a *short-term memory* and a *long-term memory.*

mental arithmetic Computation done by people "in their heads," either in whole or in part. In *Everyday Mathematics*, children learn a variety of mental-calculation *strategies* as they develop automaticity with basic facts and fact power.

Mental Math and Fluency In *Everyday Mathematics*, short, leveled exercises presented at the beginning of lessons. Mental Math and Fluency problems prepare children to think about math, warm up skills they need for the lesson, and build mental-arithmetic skills. They also help teachers assess individual strengths and weaknesses.

meter (m) The basic metric unit of *length* from which other metric units of length are derived. Originally, the meter was defined as $\frac{1}{10,000,000}$ of the distance from the North Pole to the equator along a meridian passing through Paris. From 1960 to 1983, the meter was redefined as 1,630,763.73 wavelengths of orange-red light from the element krypton. Today, the meter is defined as the distance light travels in a vacuum in $\frac{1}{299,792,458}$ second. One meter is equal to 10 decimeters, 100 centimeters, or 1,000 millimeters.

metric system The measurement system used in most countries and by virtually all scientists around the world. *Units* within the metric system are related by powers of 10. Units for *length* include millimeter, centimeter, meter, and kilometer; units for mass and weight include gram and kilogram; units for *volume* and *capacity* include milliliter and liter; and the unit for *temperature* change is degrees Celsius. See *Tables of Measures.*

minuend In subtraction, the number from which another number is subtracted. For example, in 19 − 5 = 14, the minuend is 19. Compare *subtrahend* and *difference.*

mixed number A number that is written using both a *whole number* and a *fraction.* For example, $2\frac{1}{4}$ is a mixed number equal to $2 + \frac{1}{4}$.

model A mathematical representation or description of an object or a situation. For example, 60 × 3 can be a model for how much money is needed to buy 3 items that cost 60 cents each. A circle can be a model for the rim of a wheel. See *represent*.

modified U.S. traditional multiplication A multiplication *algorithm* in which the U.S. traditional algorithm is modified by including 0s that are normally omitted in the partial products. These 0s clarify why the algorithm works and help avoid misalignment of partial products.

multiple of a number _n_ (1) A *product* of n and a *counting number*. For example, the multiples of 7 are 7, 14, 21, 28, . . . and the multiples of $\frac{1}{5}$ are $\frac{1}{5}, \frac{2}{5}, \frac{3}{5}$ . . . . (2) A product of n and an *integer*. For example, the multiples of 7 are . . . , −21, −14, −7, 0, 7, 14, 21, . . . and the multiples of π are . . . −3π, −2π, −π, 0, π, 2π, 3π, ...

multiplication counting principle The principle that one can determine the total number of ways to combine two or more independent possibilities by multiplying. For example, 5 shirts and 3 pairs of pants can be combined 5 × 3 = 15 different ways: (purple shirt, gray pants), (purple shirt, black pants), (purple shirt, tan pants), (green shirt, gray pants), (green shirt, black pants), (green shirt, tan pants), and so on.

multiplication rule for equivalent fractions See *equivalent fractions rule*.

multiplication symbols The number a multiplied by the number b is written in a variety of ways. Many mathematics textbooks and *Second* and *Third Grade Everyday Mathematics* use × as in a × b. Beginning in fourth grade, *Everyday Mathematics* uses ∗ as in a ∗ b. Other common ways to indicate multiplication are by a dot as in a • b and by juxtaposition as in ab, which is common in formulas and in algebra.

multiplication/division diagram A diagram used in *Everyday Mathematics* to *model* situations in which a total number is made up of equal-size groups. The diagram contains a number of groups, a number in each group, and a total number. Also called a multiplication diagram for short. See *situation diagram*.

rows	chairs per row	total chairs
15	25	?

A multiplication/division diagram

multiplication/division use class A category of problem situations that can be solved using multiplication or division or other methods such as counting or direct modeling. In *Everyday Mathematics*, these include equal grouping/sharing, *arrays* and *area*, rates and ratio, *scaling*, and *multiplication counting* situations.

multiplicative identity The number 1. The multiplicative identity is the number that when multiplied by any other number yields that other number. See *multiplicative inverse*.

multiplicative inverses Same as *reciprocals*.

N

name-collection box In *Everyday Mathematics*, a diagram that is used for collecting *equivalent names* for a number.

25
37 − 12   20 + 5
ЖҬ ЖҬ ЖҬ ЖҬ ЖҬ
twenty-five   veinticinco

natural numbers Same as *counting numbers*.

negative numbers Numbers less than 0; the opposites of the *positive numbers,* commonly written as a positive number preceded by a −. Negative numbers are plotted left of 0 on a horizontal *number line* or below 0 on a vertical number line.

negative power of 10 A number that can be written in the form 10^{-b}, which is shorthand for $\frac{1}{10^b}$ where b is a counting number. For example, $10^{-2} = \frac{1}{10^2}$. Negative powers of 10 can be written as fractions or in standard *decimal notation*: $10^{-2} = \frac{1}{10^2} = \frac{1}{100} = 0.01$. Compare *positive power of 10*.

nested parentheses *Parentheses* within parentheses in an *expression*. Expressions are evaluated from within the innermost parentheses outward. For example: 4 ∗ (4 − [2 + 1]) = 4 ∗ (4 − 3) = 4 ∗ 1 = 4.

net A *2-dimensional* figure created to represent a *3-dimensional* figure by cutting and unfolding or separating its faces and sides. A 2-dimensional figure that can be folded to form all the faces of a closed 3-dimensional figure is called a net. For example, if a cereal box is cut along some of its edges and laid out flat, it will form a net for the box.

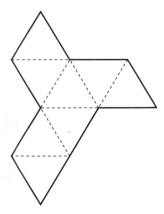

A net of an octahedron   An octahedron

***n*-gon** A *polygon*, where *n* is the number of *sides*. Polygons that do not have special names like squares and pentagons are usually named using *n*-gon notation, such as 13-gon or 100-gon.

nonagon A 9-sided *polygon*.

nonconvex polygon Same as *concave polygon*.

number line A *line* on which points are indicated by *tick marks* that are usually at regularly spaced intervals from a starting point called the *origin*, the zero point, or simply 0. Numbers are associated with the tick marks on a *scale* defined by the unit interval from 0 to 1. Every real number locates a point on the line, and every point corresponds to a *real number*. See *real numbers*.

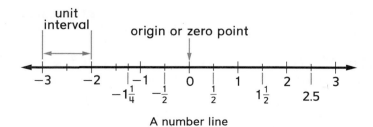

A number line

number model A *number sentence, expression,* or other representation that fits a *number story* or situation. For example, the number story "Sally had $5, and then she earned $8" can be modeled as the number sentence $5 + 8 = 13$, as the expression $5 + 8$, or by

$$\begin{array}{r} 5 \\ +8 \\ \hline 13 \end{array}$$

number sentence Two *expressions* with a *relation symbol,* such as =, <, or >.

$$5 + 5 = 10 \qquad 16 \leq a * b$$
$$2 - ? = 8 \qquad a^2 + b^2 = c^2$$

Number sentences

number sequence A list of numbers, often generated by a rule. In *Everyday Mathematics,* children explore number sequences using Frames-and-Arrows diagrams.

$$1, 2, 3, 4, 5, \ldots \qquad 1, 4, 9, 16, 25, \ldots$$
$$1, 2, 1, 2, 1, \ldots \qquad 1, 3, 5, 7, 9, \ldots$$

Number sequences

number story A story that involves numbers and one or more explicit or implicit questions. For example, "I have 7 crayons in my desk; Carrie gave me 8 more crayons," is a number story.

number-and-word notation A notation consisting of the significant digits of a number and words for the *place value*. For example, 27 billion is number-and-word notation for 27,000,000,000. Compare *standard notation*.

numeral (1) A combination of *base-10 digits* used to express a number. (2) A word, symbol, or figure that represents a number. For example, six, VI, 卌 l, and 6 are all numerals that represent the same number.

numeration A method of numbering or of reading and writing numbers. In *Everyday Mathematics,* numeration activities include counting, writing numbers, identifying equivalent names for numbers in *name-collection boxes,* exchanging coins such as 5 pennies for 1 nickel, and renaming numbers in computation.

numerator The *dividend a* in a *fraction* $\frac{a}{b}$ or *a/b*. In a *part-whole fraction,* the *whole* is divided into a number of equal parts and the numerator is the number of equal parts being considered. Compare *denominator*.

obtuse triangle   A *triangle* with an angle measuring more than 90°.

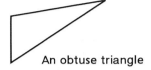

An obtuse triangle

octagon   An 8-sided *polygon*.

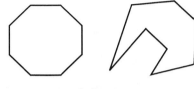

Octagons

octahedron   A *polyhedron* with 8 *faces*. An octahedron with 8 *equilateral triangle* faces is one of the five *regular polyhedrons*.

odd number   (1) A *counting number* that is not divisible by 2. (2) An *integer* that is not *divisible by* 2. Compare *even number*.

open sentence   A *number sentence* with one or more variables that is neither true nor false. For example, $9 + \underline{\quad} = 15$, $? - 24 < 10$, and $7 = x + y$ are open sentences. See *variable* and *unknown*.

operation   An action performed on one or more mathematical objects such as numbers, *variables,* or *expressions* to produce another mathematical object. Addition, subtraction, multiplication, and division are the four basic arithmetic operations. Taking a square root, squaring a number, and multiplying both sides of an *equation* by the same number are also operations. In *Everyday Mathematics*, children learn about many operations along with procedures, or *algorithms,* for carrying them out.

operation symbol   A symbol used in *expressions* and *number sentences* to stand for a particular mathematical operation. Symbols for common arithmetic operations are: addition $+$; subtraction $-$; multiplication $\times$, $*$, $\bullet$; division $\div$, $/$; powering $\wedge$. See *General Reference*.

opposite-change rule for addition   An addition *algorithm* in which a number is added to one *addend* and subtracted from the other addend. Compare *same-change rule for subtraction*.

order   To arrange things according to a specific rule, often from smallest to largest, or from largest to smallest. See *sequence*.

order of operations   A set of rules that tell the order in which operations in an *expression* should be carried out. See *General Reference*.

ordered pair   (1) Two numbers, or coordinates, used to locate a point on a rectangular coordinate grid. The first coordinate x gives the position along the horizontal axis of the grid, and the second coordinate y gives the position along the vertical axis. The pair is written (x, y). (2) Any pair of objects or numbers in a particular order, as in letter-number spreadsheet cell names or map coordinates or functions given as sets of pairs of numbers.

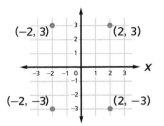

Ordered pairs

ordinal number   A number describing the position or order of something in a *sequence,* such as first, third, or tenth. Ordinal numbers are commonly used in dates, as in "May fifth" instead of "May five". Compare *cardinal number*.

origin   The *zero point* in a *coordinate system*. On a *number line*, the origin is the point at 0. On a coordinate grid, the origin is the point (0, 0) where the two axes intersect.

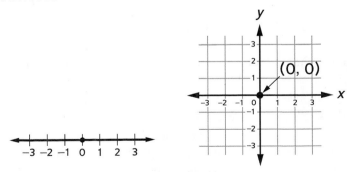

The points at 0 and (0, 0) are origins.

ounce (oz)   A U.S. customary unit equal to $\frac{1}{16}$ of a pound or about 28.35 grams. Compare *fluid ounce*.

outcome   A possible result of a chance experiment or situation. For example, HEADS and TAILS are the two possible outcomes of flipping a coin.

output (1) A number paired to an *input* by a *function machine* applying a rule. (2) Numbers or other information displayed or produced by calculator or computer.

P

parallel *Lines,* line segments, or rays in the same *plane* are parallel if they never cross or meet, no matter how far they are extended. Two planes are parallel if they never cross or meet. A line and a plane are parallel if they never cross or meet. The symbol ‖ means is parallel to.

parallelogram A *trapezoid* that has two pairs of *parallel* sides. See *quadrilateral*.

Parallelograms

parentheses See *grouping symbols*.

partial-products multiplication (1) A multiplication *algorithm* in which partial products are computed by multiplying the value of each digit in one factor by the value of each digit in the other factor. The final *product* is the sum of the partial products. (2) A similar method for multiplying mixed numbers.

partial-quotients division A division *algorithm* in which a partial quotient is computed in each of several steps. The final *quotient* is the sum of the partial quotients.

partial-sums addition An addition *algorithm* in which separate *sums* are computed for each *place value* of the numbers and then added to get a final sum.

part-whole fraction A *fraction* that describes a portion of an object or collection divided into equal parts. In *Everyday Mathematics,* the object or collection is called the *whole* and is the *denominator* of the fraction. The *numerator* is the number of parts of the whole. For example, in the situation Padma ate $\frac{2}{5}$ of the pizza, the whole is 5 pieces of pizza (a whole pizza divided into 5 parts) and Padma ate 2 of the 5 parts.

pattern A repetitive order or arrangement. In *Everyday Mathematics,* children mainly explore visual and number patterns in which elements are arranged so that what comes next can be predicted.

pentagon A 5-sided *polygon*.

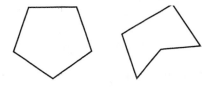

Pentagons

per For each, as in ten chairs per row or six tickets per family.

percent (%) Per hundred, for each hundred, or out of a hundred. 1% = $\frac{1}{100}$ = 0.01. For example, 48% of the students in the school are boys means that, on average, 48 of every 100 students in the school are boys.

Percent Circle A tool on the *Geometry Template* that is used to measure and draw figures that involve percentages, such as circle graphs.

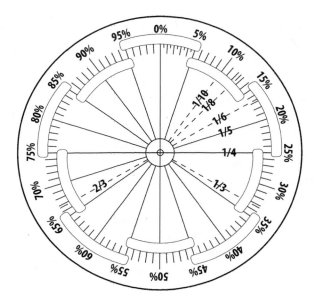

A Percent Circle

perimeter The distance around the boundary of a *2-dimensional* figure. The perimeter of a circle is called its circumference. Perimeter comes from the Greek words for "around measure".

perpendicular   Intersecting at *right angles* or lying on lines that intersect at right angles.

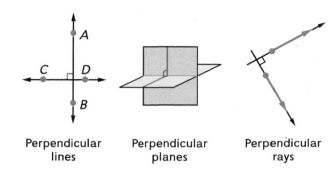

Perpendicular lines   Perpendicular planes   Perpendicular rays

place value   A system that gives a *digit* a value according to its position, or place, in a number. In our standard, *base-ten* (*decimal*) system for writing numbers, each place has a value 10 times that of the place to its right and one-tenth the value of the place to its left.

thousands	hundreds	tens	ones	.	tenths	hundredths

plane   A *2-dimensional* flat *surface* that extends forever in all directions. In formal Euclidean geometry, plane is an undefined geometric term.

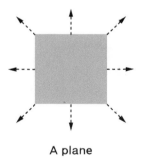

A plane

plane figure   A set of points that is entirely contained in a single plane. For example, squares, pentagons, circles, parabolas, lines, and rays are plane figures; cones, cubes, and prisms are not.

plot   To draw a point or a curve on a number line, coordinate grid, or graph. The points plotted can come from lists, mathematical relationships, or *data.*

point   An exact location in space. Points are usually labeled with capital letters. In formal Euclidean geometry, point is an undefined geometric term.

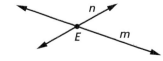

Lines *m* and *n* intersect at point *E*.

poly-   A prefix meaning many. See *General Reference, Prefixes* for specific numerical prefixes.

polygon   A plane figure formed by *line segments* (*sides*) that meet only at their endpoints (*vertices*) to make a closed path. The sides may not cross one another.

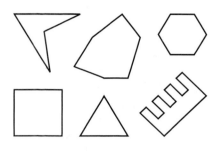

Polygons

polyhedron   A closed *3-dimensional* figure formed by *polygons* with their *interiors* (*faces*) that may meet but do not cross. Plural is polyhedrons or polyhedra.

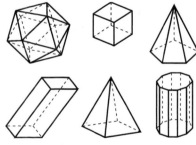

Polyhedrons

positive numbers   Numbers greater than 0; the opposites of the *negative numbers.* Positive numbers are plotted to the right of 0 on a horizontal *number line* or above 0 on a vertical number line.

positive power of 10   A number that can be written in the form 10^b, where *b* is a *counting number.* The positive powers of 10 are 10; 100; 1,000; and so on. Compare *negative power of 10.*

poster   In *Everyday Mathematics,* a page displaying a collection of illustrated numerical *data.* A poster may be used as a source of data for developing *number stories.*

pound (lb)   (1) A U.S. customary unit equal to 16 ounces and defined as 0.45359237 kilograms. See *Tables of Measures.* (2) A unit of measurement for force. Because of gravity, for example, a person who weighs 150 pounds in San Diego weighs about 23 pounds on the moon without gaining or losing mass.

power of 10 (1) A number that can be written in the form 10^b, where *b* is a *counting number*. That is, the numbers $10 = 10^1$, $100 = 10^2$, $1,000 = 10^3$, and so on, that can be written using only 10s as *factors*. Same as *positive power of 10*. (2) More generally, a number that can be written in the form 10^b, where *b* is an integer. That is, all the positive and *negative powers of 10* together, along with $10^0 = 1$.

power of a number *n* The result of n^p for any numbers *n* and *p*. A *whole number* power is a *product* of *factors* that are all the same. For example, $5^3 = 5 * 5 * 5 = 125$ is read "5 to the third power" or "the third power of 5" because 5 is a factor three times. See *exponential notation*.

precise Of a measurement or other quantity, having a high degree of exactness. A measurement to the nearest inch is more precise than a measurement to the nearest foot. A measurement's precision depends on the *unit* scale of the *tool* used to obtain it. The smaller the unit is, the more precise a measure can be. For instance, a ruler with $\frac{1}{8}$ inch markings can give a more precise measurement than a ruler with $\frac{1}{2}$-inch markings. Compare *accurate*.

preimage The original figure in a *transformation*. Compare *image*.

prime factorization A *counting number* written as a product of *prime-number factors*. The Fundamental Theorem of Arithmetic states that every counting number greater than 1 has a unique prime factorization. For example, the prime factorization of 24 is $2 * 2 * 2 * 3$. The prime factorization of a prime number is that number. For example, the prime factorization of 13 is 13.

prime number A *counting number* greater than 1 that has exactly two *whole-number factors*, 1 and itself. For example, 7 is a prime number because its only factors are 1 and 7. The first five prime numbers are 2, 3, 5, 7, and 11. Compare *composite number*.

prism A polyhedron with two *parallel* and congruent *polygonal bases* and lateral *faces* shaped like *parallelograms*. Right prisms have rectangular lateral faces. Prisms get their names from the shape of their bases.

A triangular   A rectangular   A hexagonal
prism   prism   prism

product The result of multiplying two or more numbers, called *factors*. For example, in $4 * 3 = 12$, the product is 12.

property (1) A generalized statement about a mathematical relationship, such as the *Distributive Property of Multiplication over Addition*. (2) Same as *attribute*.

protractor A tool used for measuring or drawing *angles*. A half-circle protractor can be used to measure and draw angles up to 180°. A full-circle protractor can be used to measure and draw angles up to 360°. One of each type is on the *Geometry Template*.

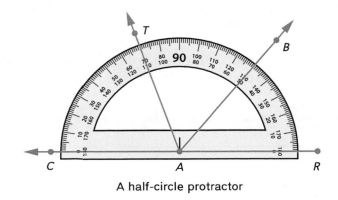

A half-circle protractor

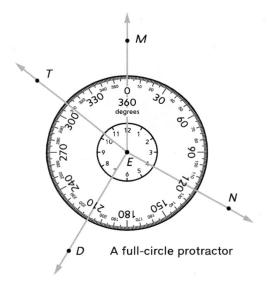

A full-circle protractor

pyramid A *polyhedron* with a polygonal *base* and *triangular* other *faces* that meet at a common *vertex* called the *apex*. Pyramids get their names from the shapes of their bases.

Q

quadrangle Same as *quadrilateral*.

quadrant One of the four sections into which a *rectangular coordinate grid* is divided by the two axes. The quadrants are typically numbered I, II, III, and IV counterclockwise beginning at the upper right.

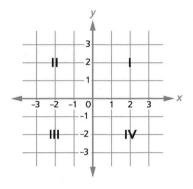

quadrilateral A 4-sided *polygon*. *Squares, rectangles, parallelograms, rhombuses, kites,* and *trapezoids* are organized by defining *attributes* into a hierarchy of shapes.

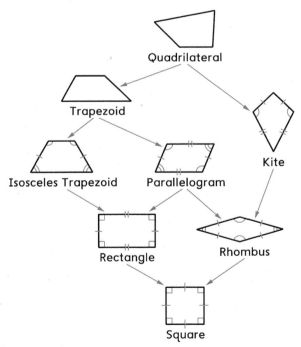

A hierarchy of quadrilaterals

quantity A number with a unit, usually a measurement or count.

quart A *U.S. customary* unit of *volume* or *capacity* equal to 32 fluid ounces, 2 pints, or 4 cups.

quick common denominator (QCD) The *product* of the *denominators* of two or more *fractions*. For example, the quick common denominator of $\frac{3}{4}$ and $\frac{5}{6}$ is $4 * 6 = 24$. In general, the quick common denominator of $\frac{a}{b}$ and $\frac{c}{d}$ is $b * d$. As the name suggests, this is a quick way to get a *common denominator* for a collection of fractions, but it does not necessarily give the *least common denominator*.

quotient The result of dividing one number by another number. For example, in $10 / 5 = 2$, the quotient is 2. Compare *dividend* and *divisor*.

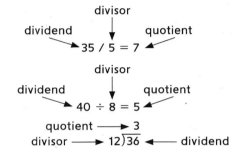

R

rational numbers Numbers that can be written in the form $\frac{a}{b}$, where a and b are *integers* and $b \neq 0$. The *decimal* form of a rational number either terminates or repeats. For example, $\frac{2}{3}$, $-\frac{2}{3}$, 0.5, 20.5, and 0.333 . . . are rational numbers.

ray A part of a line starting at an endpoint and continuing forever in one direction. A ray is often named by its endpoint and another point on it.

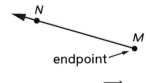

Ray MN or $\overrightarrow{MN}$

real numbers All *rational* and *irrational numbers;* all numbers that can be written as *decimals.* For every real number there is a corresponding point on a *number line,* and for every point on the number line there is a real number.

reciprocals Two numbers whose *product* is 1. For example, 5 and $\frac{1}{5}$, $\frac{3}{5}$ and $\frac{5}{3}$, and 0.2 and 5 are pairs of reciprocals. Same as *multiplicative inverses.*

rectangle A *parallelogram* with four *right angles*. All rectangles are both parallelograms and *isosceles trapezoids*. See *quadrilateral*.

rectangle method A strategy for finding the area of a *polygon* in which one or more rectangles are drawn around all or parts of the polygon through its vertices. The sides of the drawn rectangle(s) together with the sides of the original figure define regions that are either rectangles or triangular halves of rectangles. The areas of these rectangular and triangular regions are added or subtracted to get the area of the original polygon. For example, rectangle *RYSX* was drawn around the original triangle *XYZ* below.

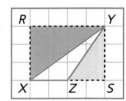

Area of △*XYZ* = area of rectangle *RYSX* −
area of △*XRY* − area of △*YSZ*

rectangular coordinate grid (1) Same as *coordinate grid*. (2) A coordinate grid with perpendicular axes.

rectangular prism A *prism* with rectangular *bases*. The four *faces* that are not bases are formed by either *rectangles* or *parallelograms*. For example, a rectangular prism in which all sides are rectangular models a shoebox.

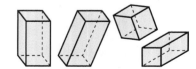

Rectangular prisms

rectangular pyramid A *pyramid* with a rectangular *base*.

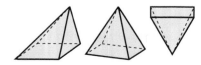

Rectangular pyramids

reference frame A system for locating numbers within a given context, usually with reference to an *origin* or zero point. For example, *number lines,* clocks, calendars, temperature scales, and maps are reference frames.

reflection Point *A′* is a reflection of a point *A* if they are the same distance from a point, a line, or a plane over which *A* has been reflected. If all points on one figure are reflection images of all points on another figure over the same point, line, or plane, the figures are reflection, or mirror images. Figures are most often reflected over points on lines, over lines on planes, or over planes in 3-dimensional space. Reflections, rotations, and translations are types of isometric transformations. Informally called a flip.

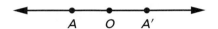

A reflection over a point

A reflection over a line

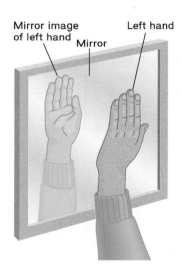

A reflection over a plane

regular polygon A *polygon* in which all sides are the same *length* and all interior *angles* have the same measure.

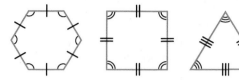

Regular polygons

regular polyhedron A *polyhedron* whose *faces* are all formed by congruent *regular polygons* and in which the same number of faces meet at each *vertex*. There are only five. They are called the Platonic solids, and are shown below.

Tetrahedron   Cube   Octahedron
(4 equilateral triangles)   (6 squares)   (8 equilateral triangles)

Dodecahedron   Icosahedron
(12 regular pentagons)   (20 equilateral triangles)

relation symbol A symbol used to express a relationship between two quantities, figures, or sets, such as ≤, ‖, or ⊂. See *General Reference, Symbols*.

remainder An amount left over when one number is divided by another number. For example, in 16 / 3 → 5 R1, the *quotient* is 5 and the remainder is 1.

repeating decimal A *decimal* in which one digit or block of digits is repeated without end. For example, 0.3333. . . and 0.$\overline{147}$ are repeating decimals. Compare *terminating decimal*.

represent To show, symbolize, or stand for something. For example, numbers can be represented using base-10 blocks, spoken words, or written numerals. See *model*.

rhombus A *parallelogram* with four sides of the same length. All rhombuses are both parallelograms and *kites*. See *quadrilateral*.

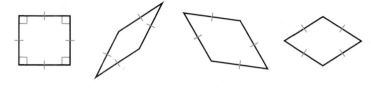

Rhombuses

right angle An *angle* with a measure of 90°.

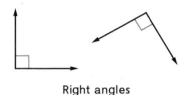

Right angles

right triangle A *triangle* with a right angle.

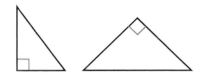

Right triangles

rotation (1) A turn about an axis or point. (2) Point P' is a rotation *image* of point P around a center of rotation C if P' is on the *circle* with center C and radius CP. If all the points in one figure are rotation images of all the points in another figure around the same center of rotation and with the same *angle* of rotation, then the figures are rotation images. The center can be inside or outside of the original image. Reflections, rotations, and translations are types of isometric *transformations*. (3) If all points on the image of a 3-dimensional figure are rotation images through the same angle around a point or a line called the axis of rotation, then the image is a rotation image of the original figure.

round (1) To *approximate* a number to make it easier to use, or to make it better reflect the precision of the data. "Rounding up" means to approximate larger than the actual value. "Rounding down" means to approximate smaller than the actual value. (2) Circular in shape.

rubric A tool used to categorize work based on its quality.

same-change rule for subtraction   A subtraction *algorithm* in which the same number is added to or subtracted from both numbers.

scale   (1) A multiplicative comparison between the relative sizes or numbers of things. (2) Same as *scale factor*. (3) A tool for measuring weight and mass.

scale drawing   A drawing of an object in which all parts are drawn to the same *scale* to the object. For example, architects and builders use scale drawings traditionally called blueprints. Many maps are approximately scale drawings of geographical regions. See *scale factor*.

A woodpecker (8 in.) to $\frac{1}{4}$ scale

scale factor   (1) The ratio of lengths on an *image* and corresponding lengths on a *preimage* in a size change. (2) The ratio of lengths in a *scale drawing* or *scale model* to the corresponding lengths in the object being drawn or modeled. These scales may be represented in a variety of ways including 1 cm = 100 km or 1:10,000.

scale model   A model of an object in which all parts are made to the same *scale* to the object. For example, many model trains or airplanes are scale models of actual vehicles. See *scale factor*.

scale of a number line   The unit interval on a *number line* or measuring device. The scales on this ruler are 1 millimeter on the left side and $\frac{1}{16}$ inch on the right side.

scalene triangle   A *triangle* with *sides* of three different lengths. The three *angles* of a scalene triangle have different measures.

scientific calculator   A calculator that can display numbers using *scientific notation*. Scientific calculators follow the algebraic order of operations and can calculate a power of a number, a square root, and several other functions beyond simple 4-function calculators. Some scientific calculators can do arithmetic with fractions.

scientific notation   A way of writing a number as the product of a *power of 10* and a number that is at least 1 and less than 10. Scientific notation allows one to write large and small numbers with only a few symbols. For example, in scientific notation, 4,300,000 is $4.3 * 10^6$, and 0.00001 is $1 * 10^{-5}$. *Scientific calculators* display numbers in scientific notation. Compare *standard notation* and *expanded notation*.

segment   Same as *line segment*.

sequence   An ordered list of numbers, often with an underlying rule that may be used to generate numbers in the list. Frames-and-Arrows diagrams can be used to represent sequences.

set   A collection or group of objects, numbers, or other items.

short-term memory   *Memory in a calculator* used to store values for immediate calculation. Short-term memory is usually cleared with a [C], [AC], [Clear], or similar key. Compare *long-term memory*.

side   (1) One of the *line segments* that make up a *polygon*. (2) One of the rays or segments that form an *angle*. (3) One of the *faces* of a *polyhedron*.

situation diagram   In *Everyday Mathematics*, a diagram used to organize information in a problem situation in one of the *addition/subtraction* or *multiplication/division use classes*.

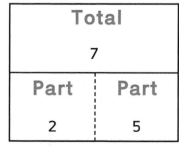

Total	
7	
Part	**Part**
2	5

Susie has 2 pink balloons and 5 yellow balloons.
She has 7 balloons in all.

skew lines Lines in space that do not lie in the same *plane*. Skew lines do not intersect and are not parallel. An east-west line on the floor and a north-south line on the ceiling are skew.

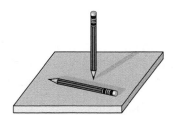

Skew lines can be modeled
with two pencils.

slate In *Everyday Mathematics*, a lap-size (about 8 inches by 11 inches) chalkboard or whiteboard that children use for recording responses during group exercises and informal group assessments.

solid See *geometric solid*.

solution of a problem (1) The answer to a problem. (2) The answer to a problem together with the method by which that answer was obtained.

solution of an open sentence A value or values for the *variable(s)* in an *open sentence* that make the sentence true. For example, 7 is a solution of $5 + n = 12$. Although equations are not necessarily open sentences, the solution of an open sentence is commonly referred to as a solution of an equation.

sphere The set of all points in space that are an equal distance from a fixed point called the center of the sphere. The distance from the center to the sphere is the radius of the sphere. The diameter of a sphere is twice its radius. Points inside a sphere are not part of the sphere.

A sphere

Spiral Snapshot In *Everyday Mathematics*, an overview of nearby lessons that address one of the Goals for Mathematical Content in the Focus part of the lesson. It appears in the Lesson Opener.

Spiral Trace In *Everyday Mathematics*, an overview of work in the current unit and nearby units on selected Standards for Mathematical Content. It appears in the Unit Organizer.

Spiral Tracker In *Everyday Mathematics*, an online database that shows complete details about learning trajectories for all goals and standards.

square A *rectangle* with four sides of equal length. All squares are both rectangles and *rhombuses*. See *quadrilateral*.

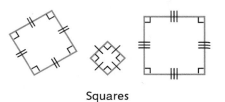

Squares

square corner Same as *right angle*.

square number A *figurate number* that is the *product* of a counting number and itself. For example, 25 is a square number because $25 = 5 * 5$. A square number can be represented by a square array and as a number squared, such as $25 = 5^2$.

square of a number *n* The product of *n* and itself, commonly written n^2. For example, $81 = 9 * 9 = 9^2$ and $3.5^2 = 3.5 * 3.5 = 12.25$.

square pyramid A *pyramid* with a square *base*.

square unit A *unit* to measure *area*. A model of a square unit is a square with each side a related unit of *length*. For example, a square inch is the area of a square with 1-inch sides. Square units are often labeled as the length unit squared. For example, 1 cm² is read "1 square centimeter" or "1 centimeter squared".

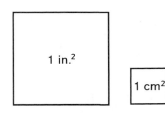

standard form Same as *standard notation*.

standard notation The most common way of representing *whole numbers, integers*, and decimals. Standard notation for real numbers is *base-ten place-value* numeration. For example, standard notation for three hundred fifty-six is 356. Same as *decimal notation*. Compare *number-and-word notation*.

standard unit   A *unit* of measure that has been defined by a recognized authority, such as a government or National Institute of Standards and Technology. For example, inches, meters, miles, seconds, pounds, grams, and acres are all standard units.

straightedge   A tool used to draw *line segments* more accurately than by freehand. Strictly speaking, a straightedge does not have a measuring *scale* on it, so one ignores the marks if using a ruler as a straightedge.

strategy   An approach to a problem that may be general, like "divide and conquer," or more specific, like "adding a group". Compare *algorithm*.

subcategory   A more specific category contained entirely within a given category. Subcategories are usually defined by an attribute shared by some, but not all, of the members of the larger category. For example, right triangles are a subcategory of the larger category of triangles because some, but not all, triangles have a right angle.

subtrahend   The number being taken away in a subtraction problem. For example, in $15 - 5 = 10$, the subtrahend is 5. Compare *difference* and *minuend*.

sum   The result of adding two or more numbers. For example, in $5 + 3 = 8$, the sum is 8.

surface   (1) The boundary of a *3-dimensional* object. (2) Any *2-dimensional* layer, such as a *plane* or a *face* of a polyhedron.

survey   (1) A study that collects *data* by asking people questions. (2) Any study that collects data. Surveys are commonly used to study demographics such as people's characteristics, behaviors, interests, and opinions.

symmetry   The balanced distribution of points with respect to a point, line, or plane in a symmetric figure. See *line symmetry*.

A figure with line symmetry   A figure with rotation symmetry

tally   (1) To keep a record of a count by making a mark for each item as it is counted. (2) The mark used in a count. Also called tally mark.

temperature   How hot or cold something is relative to another object or as measured on a standardized scale, such as degrees *Celsius* or degrees *Fahrenheit*.

term   (1) In an algebraic expression, a number or a product of a number and one or more *variables*. For example, in the equation $5y + 3k = 8$, the terms are $5y$, $3k$, and 8. The 8 is a constant term, or simply a constant, because it has no variable part. (2) An element in a *sequence*. In the sequence of square numbers, the terms are 1, 4, 9, 16, and so on.

terminating decimal   A *decimal* that ends. For example, 0.5 and 0.125 are terminating decimals. Compare *repeating decimal*.

tick marks   (1) Marks showing the scale of a number line or ruler. (2) Marks indicating that two *line segments* have the same length. (3) Same as *tally* (2).

tile (verb)   To cover a *surface* completely with shapes without overlaps or gaps. Tiling with same-size squares is a way to measure area. See *iterate units*.

tool   Anything, physical or abstract, that serves as an instrument for performing a task. Physical tools include hammers for hammering, calculators for calculating, and rulers for measuring. Abstract tools include computational algorithms such as *partial-sums addition*, problem-solving strategies such as "guess and check," and technical drawings such as *situation diagrams*.

Total Physical Response   A teaching technique that facilitates beginning English language learners' acquisition of new English vocabulary through modeling of physical actions or display of visuals as target objects that are named aloud.

trade-first subtraction   A subtraction *algorithm* in which all necessary trades between places in the numbers are completed before any subtractions are carried out. Some people favor this algorithm because they can concentrate on one task at a time.

translation A transformation in which every point in the image of a figure is at the same distance in the same direction from its corresponding point in the figure. Reflections, rotations, and translations are types of isometric transformations. Informally called a slide.

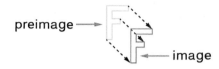

A translation

trapezoid A *quadrilateral* that has at least one pair of *parallel* sides.

Trapezoids with base pairs marked in the same color

trial-and-error method A systematic method for finding the solution of an *equation* by trying a sequence of test numbers.

triangle A 3-sided *polygon*. See *equilateral triangle, isosceles triangle, scalene triangle, acute triangle, right triangle,* and *obtuse triangle.*

Triangles

triangular numbers *Figurate numbers* that can be represented by triangular arrangements of dots. The triangular numbers are {1, 3, 6, 10, 15, 21, 28, 36, 45, . . .}.

Triangular numbers

triangular prism A *prism* whose *bases* are triangular.

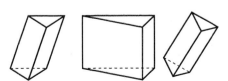

Triangular prisms

triangular pyramid A *pyramid* in which all *faces* are triangular, any one of which is the *base;* also known as a tetrahedron. A regular tetrahedron has four faces formed by *equilateral triangles* and is one of the five *regular polyhedrons.*

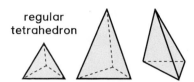

Triangular pyramids

true number sentence A *number sentence* stating a correct fact. For example, 75 = 25 + 50 is a true number sentence. Compare *false number sentence.*

turn-around rule A rule for solving addition and multiplication problems based on the *Commutative Properties of Addition and Multiplication.* For example, if one knows that 6 * 8 = 48, then, by the turn-around rule, one also knows that 8 * 6 = 48.

U

unit A label used to put a number in context. In measuring length, for example, inches and centimeters are units. In a problem about 5 apples, apple is the unit. In *Everyday Mathematics,* children keep track of units in unit boxes.

unit conversion A change from one measurement *unit* to another using a fixed relationship such as 1 yard = 3 feet or 1 inch = 2.54 centimeters. See *Tables of Measures.*

unit cube A *cube* with edge lengths of 1.

unit fraction A *fraction* whose *numerator* is 1. For example, $\frac{1}{2}, \frac{1}{3}, \frac{1}{12}, \frac{1}{8}$ and $\frac{1}{20}$ are unit fractions.

unit square A *square* with side lengths of 1.

unknown A quantity whose value is not known. An unknown is sometimes represented by a _____, a ?, or a letter. See *open sentence* and *variable.*

unlike Unequal or not the same.

U.S. customary system The measuring system used most often in the United States. Units for *length* include inch, foot, yard, and mile; units for weight include ounce and pound; units for *volume* or *capacity* include fluid ounce, cup, pint, quart, gallon, and cubic units; and the unit for *temperature* change is degrees Fahrenheit. See *Tables of Measures.*

U.S. traditional addition algorithm   The *algorithm* for adding multidigit numbers that has traditionally been taught in schools in the United States. This algorithm involves adding digits by place-value columns.

```
    1   1
    3   4   8
+   2   6   3
    6   1   1
```

U.S. traditional long division algorithm   The *algorithm* for dividing multidigit numbers that has traditionally been taught in schools in the United States. This algorithm relies on estimating products of the dividend with the digits in the divisor.

```
        163
    6)978
      - 6
       37
      - 36
        18
      - 18
         0
```

U.S. traditional multiplication algorithm   The *algorithm* for multiplying multidigit numbers that has traditionally been taught in schools in the United States. This algorithm produces partial sums based on multiplying the values of each digit.

```
      1   1
      1   1
      1   2   2
*         7   5
      6   1   0
+ 8   5   4   0
  9,  1   5   0
```

U.S. traditional subtraction algorithm   The *algorithm* for subtracting multidigit numbers that has traditionally been taught in schools in the United States. This algorithm involves subtracting digits by place-value columns, making 10-for-1 trades as needed.

```
    3   14
    4̶   4̶   7
-   1   6   5
    2   8   2
```

use class   In *Everyday Mathematics*, one of several categories of problem situations that one of the basic arithmetic operations can be used to solve. Students use situation diagrams to help model problems from the different use classes. See *addition/subtraction use classes* and *multiplication/division use classes*.

V

variable   (1) A letter or other symbol that can be replaced by any value from a set of possible values. Some values replacing variables in *number sentences* may make them true. For example, to make number sentences true, variables may be replaced by: a single number, as in $5 + n = 9$ where $n = 4$ makes the sentence true; many different numbers, as in $x + 2 < 10$ where any number less than 8 makes the sentence true; or any number as in $a + 3 = 3 + a$ which is true for all numbers. See *open sentence* and *unknown*. (2) A number or data set that can have many values is a variable.

vertex   The point at which the *sides* of an *angle* or polygon, or the *edges* of a *polyhedron* meet. Plural is vertexes or vertices. Informally called a corner.

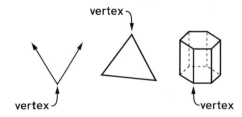

vertical   Upright; perpendicular to the horizon. Compare *horizontal*.

volume   (1) A measure of how much *3-dimensional* space something occupies. Volume is often measured in liquid units such as gallons or liters, or cubic units, such as cm^3 or cubic inches. (2) Less formally, the same as capacity: the amount a container can hold.

W

"What's My Rule?"   A problem in which two of the three parts of a *function* (*input, output,* and *rule*) are known and the third is to be found out. See *function*.

whole   An entire object, collection of objects, or quantity being considered in a problem situation; 100%.

whole numbers   The *counting numbers* and 0. The set of whole numbers is {0, 1, 2, 3, . . .}.

X

x-axis   In a *coordinate grid,* the horizontal number line.

x-coordinate   The first number in a pair of numbers used to locate a point on a *coordinate grid.* The *x*-coordinate gives the position of the point along the horizontal axis. For example, 2 in the point (2, 3) is the *x*-coordinate.

Y

y-axis   In a *coordinate grid,* the vertical number line.

y-coordinate   The second number in a pair of numbers used to locate a point on a *coordinate grid.* The *y*-coordinate gives the position of the point along the vertical axis. For example, 3 in the point (2, 3) is the *y*-coordinate.

Z

zero   (1) The number or numerical digit representing no quantity or value. (2) To adjust a scale or balance before use so that it reads 0 when no object is being weighed. Zeroing is necessary in order to ensure *accurate* measurements.

zero point   Same as *origin.*

General Reference

Symbols

Symbol	Meaning
$+$	plus or positive
$-$	minus or negative
$*, \times$	multiplied by
$\div, /$	divided by
$=$	is equal to
$\neq$	is not equal to
$<$	is less than
$>$	is greater than
$\leq$	is less than or equal to
$\geq$	is greater than or equal to
$\approx$	is approximately equal to
$x^n, x^\wedge n$	nth power of x
$\sqrt{x}$	square root of x
$\%$	percent
$a:b, a/b, \frac{a}{b}$	ratio of a to b or a divided by b or the fraction $\frac{a}{b}$
$a\ [bs]$	a groups, b in each group
$n \div d \longrightarrow a\,\mathrm{R}b$	n divided by d is a with remainder b
$\{\,\}, (\,), [\,]$	grouping symbols
∞	infinity
$n!$	n factorial
$^{\circ}$	degree
(a, b)	ordered pair
$\overleftrightarrow{AS}$	line AS
$\overline{AS}$	line segment AS
$\overrightarrow{AS}$	ray AS
$\ulcorner$	right angle
$\perp$	is perpendicular to
$\parallel$	is parallel to
$\triangle ABC$	triangle ABC
$\angle ABC$	angle ABC
$\angle B$	angle B
$\cong$	is congruent to
$\sim$	is similar to
$\equiv$	is equivalent to
$|n|$	absolute value of n

Prefixes

Prefix	Meaning	Prefix	Meaning
uni-	one	tera-	trillion (10^{12})
bi-	two	giga-	billion (10^{9})
tri-	three	mega-	million (10^{6})
quad-	four	kilo-	thousand (10^{3})
penta-	five	hecto-	hundred (10^{2})
hexa-	six	deca-	ten (10^{1})
hepta-	seven	uni-	one (10^{0})
octa-	eight	deci-	tenth (10^{-1})
nona-	nine	centi-	hundredth (10^{-2})
deca-	ten	milli-	thousandth (10^{-3})
dodeca-	twelve	micro-	millionth (10^{-6})
icosa-	twenty	nano-	billionth (10^{-9})

Constants and Approximations

Constants	Approximations
Pi (π)	3.14159 26535 89793
Golden Ratio (ϕ)	1.61803 39887 49894
Radius of Earth at equator	6,378.388 kilometers 3,963.34 miles
Circumference of Earth at equator	40,076.59 kilometers 24,902.44 miles
Velocity of sound in dry air at 0°C	331.36 m/sec 1,087.1 ft/sec
Velocity of light in a vacuum	2.997925×10^{10} cm/sec

The Order of Operations

1. Do operations inside grouping symbols following Rules 2–4. Work from the innermost set of grouping symbols outward.
2. Calculate all expressions with exponents or roots.
3. Multiply and divide in order from left to right.
4. Add and subtract in order from left to right.

Tables of Measures

Metric System

Units of Length

1 kilometer (km)	= 1,000 meters (m)
1 meter	= 10 decimeters (dm)
	= 100 centimeters (cm)
	= 1,000 millimeters (mm)
1 decimeter	= 10 centimeters
1 centimeter	= 10 millimeters

Units of Area

1 square meter (m^2)	= 100 square decimeters (dm^2)
	= 10,000 square centimeters (cm^2)
1 square decimeter	= 100 square centimeters
1 are (a)	= 100 square meters
1 hectare (ha)	= 100 ares
1 square kilometer (km^2)	= 100 hectares

Units of Volume and Capacity

1 cubic meter (m^3)	= 1,000 cubic decimeters (dm^3)
	= 1,000,000 cubic centimeters (cm^3)
1 cubic centimeter	= 1,000 cubic millimeters (mm^3)
1 kiloliter (kL)	= 1,000 liters (L)
1 liter	= 1,000 milliliters (mL)

Units of Mass and Weight

1 metric ton (t)	= 1,000 kilograms (kg)
1 kilogram	= 1,000 grams (g)
1 gram	= 1,000 milligrams (mg)

U.S. Customary System

Units of Length

1 mile (mi)	= 1,760 yards (yd)
	= 5,280 feet (ft)
1 yard	= 3 feet
	= 36 inches (in.)
1 foot	= 12 inches

Units of Area

1 square yard (yd^2)	= 9 square feet (ft^2)
	= 1,296 square inches (in.^{2})
1 square foot	= 144 square inches
1 acre	= 43,560 square feet
1 square mile (mi^2)	= 640 acres

Units of Volume and Capacity

1 cubic yard (yd^3)	= 27 cubic feet (ft^3)
1 cubic foot	= 1,728 cubic inches (in.^{3})
1 gallon (gal)	= 4 quarts (qt)
1 quart	= 2 pints (pt)
1 pint	= 2 cups (c)
1 cup	= 8 fluid ounces (fl oz)
1 fluid ounce	= 2 tablespoons (tbs)
1 tablespoon	= 3 teaspoons (tsp)

Units of Mass and Weight

1 ton (T)	= 2,000 pounds (lb)
1 pound	= 16 ounces (oz)

System Equivalents (Conversion Factors)

1 inch ≈ 2.5 centimeters (2.54)	1 liter ≈ 1.1 quarts (1.057)
1 kilometer ≈ 0.6 mile (0.621)	1 ounce ≈ 28 grams (28.350)
1 mile ≈ 1.6 kilometers (1.609)	1 kilogram ≈ 2.2 pounds (2.21)
1 meter ≈ 39 inches (39.37)	1 hectare ≈ 2.5 acres (2.47)

Body Measures

1 *digit* is about the width of a finger.

1 *hand* is about the width of the palm and thumb.

1 *span* is about the distance from the tip of the thumb to the tip of the first (index) finger of an outstretched hand.

1 *cubit* is about the length from the elbow to the tip of the extended middle finger.

1 *yard* is about the distance from the center of the chest to the tip of the extended middle finger of an outstretched arm.

1 *fathom* is about the length from fingertip to fingertip of outstretched arms. Also called an arm span.

Units of Time

1 century	= 100 years
1 decade	= 10 years
1 year (yr)	= 12 months
	= 52 weeks (plus one or two days)
	= 365 days (366 days in a leap year)
1 month (mo)	= 28, 29, 30, or 31 days
1 week (wk)	= 7 days
1 day (d)	= 24 hours
1 hour (hr)	= 60 minutes
1 minute (min)	= 60 seconds (s or sec)

Unpacking the Standards

The **Mathematical Standards** include two groups of standards: Standards for Mathematical Content and Standards for Mathematical Process and Practice. The Content Standards define the mathematical content to be mastered at each grade. The Process and Practice Standards define the attributes and habits of mind students need to develop as they learn the content for their grade level.

The Content Standards are organized into **Strands**, large groups of related standards. Within each Strand are **Clusters**, smaller groups of related standards. The strands and clusters themselves define what students should understand and be able to do by the end of the grade. To support daily tracking of the standards, as well as more accurate assessment and effective differentiation, the authors have unpacked the standards into more granular goals.

The summary page for Grade 5 is on the following page. This page summarizes the mathematical content that students should learn in Grade 5.

The chart beginning on page EM3 lists the Standards for Mathematical Content with corresponding *Everyday Mathematics* **Goals for Mathematical Content (GMC)**.

The chart beginning on page EM8 lists the Standards for Mathematical Process and Practice with corresponding *Everyday Mathematics* **Goals for Mathematical Process and Practice (GMP)**.

"… larger groups of related standards. Standards from different strands may … be … related."

"… groups of related standards … Standards from different clusters may … be … related, because mathematics is a connected subject."

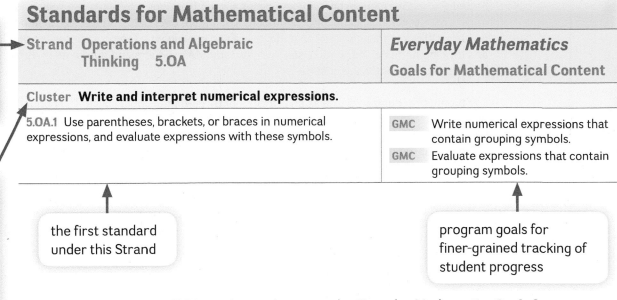

Standards for Mathematical Content

Strand Operations and Algebraic Thinking 5.OA	*Everyday Mathematics* Goals for Mathematical Content
Cluster Write and interpret numerical expressions.	
5.OA.1 Use parentheses, brackets, or braces in numerical expressions, and evaluate expressions with these symbols.	GMC Write numerical expressions that contain grouping symbols.
	GMC Evaluate expressions that contain grouping symbols.

the first standard under this Strand

program goals for finer-grained tracking of student progress

The table on pages EM8–EM11 may be used to trace the *Everyday Mathematics* **Goals for Mathematical Process and Practice** as they unpack the Standards for Mathematical Process and Practice.

The Grade 5 Content Standards are introduced on page 33, as follows:

In Grade 5, instructional time should focus on three critical areas: (1) developing fluency with addition and subtraction of fractions, and developing understanding of the multiplication of fractions and of division of fractions in limited cases (unit fractions divided by whole numbers and whole numbers divided by unit fractions); (2) extending division to 2-digit divisors, integrating decimal fractions into the place value system and developing understanding of operations with decimals to hundredths, and developing fluency with whole number and decimal operations; and (3) developing understanding of volume.

(1) Students apply their understanding of fractions and fraction models to represent the addition and subtraction of fractions with unlike denominators as equivalent calculations with like denominators. They develop fluency in calculating sums and differences of fractions, and make reasonable estimates of them. Students also use the meaning of fractions, of multiplication and division, and the relationship between multiplication and division to understand and explain why the procedures for multiplying and dividing fractions make sense. (Note: this is limited to the case of dividing unit fractions by whole numbers and whole numbers by unit fractions.)

(2) Students develop understanding of why division procedures work based on the meaning of base-ten numerals and properties of operations. They finalize fluency with multi-digit addition, subtraction, multiplication, and division. They apply their understandings of models for decimals, decimal notation, and properties of operations to add and subtract decimals to hundredths. They develop fluency in these computations, and make reasonable estimates of their results. Students use the relationship between decimals and fractions, as well as the relationship between finite decimals and whole numbers (i.e., a finite decimal multiplied by an appropriate power of 10 is a whole number), to understand and explain why the procedures for multiplying and dividing finite decimals make sense. They compute products and quotients of decimals to hundredths efficiently and accurately.

(3) Students recognize volume as an attribute of three-dimensional space. They understand that volume can be measured by finding the total number of same-size units of volume required to fill the space without gaps or overlaps. They understand that a 1-unit by 1-unit by 1-unit cube is the standard unit for measuring volume. They select appropriate units, strategies, and tools for solving problems that involve estimating and measuring volume. They decompose three-dimensional shapes and find volumes of right rectangular prisms by viewing them as decomposed into layers of arrays of cubes. They measure necessary attributes of shapes in order to determine volumes to solve real world and mathematical problems.

Standards for Mathematical Content

Strand Operations and Algebraic Thinking 5.OA	*Everyday Mathematics* Goals for Mathematical Content

Cluster Write and interpret numerical expressions.

5.OA.1 Use parentheses, brackets, or braces in numerical expressions, and evaluate expressions with these symbols.	**GMC** Write numerical expressions that contain grouping symbols.
	GMC Evaluate expressions that contain grouping symbols.
5.OA.2 Write simple expressions that record calculations with numbers, and interpret numerical expressions without evaluating them. *For example, express the calculation "add 8 and 7, then multiply by 2" as 2 × (8 + 7). Recognize that 3 × (18932 + 921) is three times as large as 18932 + 921, without having to calculate the indicated sum or product.*	**GMC** Model real-world and mathematical situations using simple expressions.
	GMC Interpret numerical expressions without evaluating them.

Cluster Analyze patterns and relationships.

5.OA.3 Generate two numerical patterns using two given rules. Identify apparent relationships between corresponding terms. Form ordered pairs consisting of corresponding terms from the two patterns, and graph the ordered pairs on a coordinate plane. *For example, given the rule "Add 3" and the starting number 0, and given the rule "Add 6" and the starting number 0, generate terms in the resulting sequences, and observe that the terms in one sequence are twice the corresponding terms in the other sequence. Explain informally why this is so.*	**GMC** Generate numerical patterns using given rules.
	GMC Identify relationships between corresponding terms of two patterns.
	GMC Form ordered pairs from corresponding terms of patterns and graph them.

Strand Number and Operations in Base Ten 5.NBT	

Cluster Understand the place value system.

5.NBT.1 Recognize that in a multi-digit number, a digit in one place represents 10 times as much as it represents in the place to its right and 1/10 of what it represents in the place to its left.	**GMC** Understand the relationship between the places in multi-digit numbers.
5.NBT.2 Explain patterns in the number of zeros of the product when multiplying a number by powers of 10, and explain patterns in the placement of the decimal point when a decimal is multiplied or divided by a power of 10. Use whole-number exponents to denote powers of 10.	**GMC** Use whole-number exponents to denote powers of 10.
	GMC Multiply whole numbers by powers of 10; explain the number of zeros in the product.
	GMC Multiply or divide decimals by powers of 10; explain the decimal-point placement in the answer.

5.NBT.3 Read, write, and compare decimals to thousandths.	

5.NBT.3a Read and write decimals to thousandths using base-ten numerals, number names, and expanded form, e.g., $347.392 = 3 × 100 + 4 × 10 + 7 × 1 + 3 × (1/10) + 9 × (1/100) + 2 × (1/1000)$.	**GMC** Represent decimals.
	GMC Read and write decimals using numerals.
	GMC Read and write decimals using number names.
	GMC Read and write decimals in expanded form.
5.NBT.3b Compare two decimals to thousandths based on meanings of the digits in each place, using >, =, and < symbols to record the results of comparisons.	**GMC** Compare and order decimals.
	GMC Record decimal comparisons using >, =, or <.
5.NBT.4 Use place value understanding to round decimals to any place.	**GMC** Use place-value understanding to round decimals to any place.

Standards for Mathematical Content

Cluster Perform operations with multi-digit whole numbers and with decimals to hundredths.

5.NBT.5 Fluently multiply multi-digit whole numbers using the standard algorithm.	**GMC** Fluently multiply multi-digit whole numbers using the standard algorithm.
5.NBT.6 Find whole-number quotients of whole numbers with up to four-digit dividends and two-digit divisors, using strategies based on place value, the properties of operations, and/or the relationship between multiplication and division. Illustrate and explain the calculation by using equations, rectangular arrays, and/or area models.	**GMC** Divide multi-digit whole numbers. **GMC** Illustrate and explain solutions to division problems.
5.NBT.7 Add, subtract, multiply, and divide decimals to hundredths, using concrete models or drawings and strategies based on place value, properties of operations, and/or the relationship between addition and subtraction; relate the strategy to a written method and explain the reasoning used.	**GMC** Make and use estimates for decimal addition and subtraction problems. **GMC** Add and subtract decimals using models or strategies. **GMC** Explain decimal addition and subtraction strategies. **GMC** Multiply and divide decimals using models or strategies. **GMC** Explain decimal multiplication and division strategies. **GMC** Make and use estimates for decimal multiplication and division problems.

Strand Number and Operations—Fractions   5.NF

Cluster Use equivalent fractions as a strategy to add and subtract fractions.

5.NF.1 Add and subtract fractions with unlike denominators (including mixed numbers) by replacing given fractions with equivalent fractions in such a way as to produce an equivalent sum or difference of fractions with like denominators. *For example, 2/3 + 5/4 = 8/12 + 15/12 = 23/12. (In general, a/b + c/d = (ad + bc)/bd.)*	**GMC** Add and subtract fractions with unlike denominators. **GMC** Add and subtract mixed numbers with unlike denominators.
5.NF.2 Solve word problems involving addition and subtraction of fractions referring to the same whole, including cases of unlike denominators, e.g., by using visual fraction models or equations to represent the problem. Use benchmark fractions and number sense of fractions to estimate mentally and assess the reasonableness of answers. *For example, recognize an incorrect result 2/5 + 1/2 = 3/7, by observing that 3/7 < 1/2.*	**GMC** Solve number stories involving fraction addition and subtraction. **GMC** Use estimates to reason about sums and differences of fractions.

Cluster Apply and extend previous understandings of multiplication and division.

5.NF.3 Interpret a fraction as division of the numerator by the denominator (a/b = a ÷ b). Solve word problems involving division of whole numbers leading to answers in the form of fractions or mixed numbers, e.g., by using visual fraction models or equations to represent the problem. *For example, interpret 3/4 as the result of dividing 3 by 4, noting that 3/4 multiplied by 4 equals 3, and that when 3 wholes are shared equally among 4 people each person has a share of size 3/4. If 9 people want to share a 50-pound sack of rice equally by weight, how many pounds of rice should each person get? Between what two whole numbers does your answer lie?*	**GMC** Interpret a fraction as division of a numerator by a denominator. **GMC** Solve number stories involving whole number division that leads to fractional answers.

Standards for Mathematical Content

5.NF.4 Apply and extend previous understandings of multiplication to multiply a fraction or whole number by a fraction.	GMC	Multiply fractions by whole numbers.
	GMC	Multiply fractions by fractions.
	GMC	Multiply mixed numbers by whole numbers, fractions, and mixed numbers.
5.NF.4a Interpret the product $(a/b) \times q$ as a parts of a partition of q into b equal parts; equivalently, as the result of a sequence of operations $a \times q \div b$. *For example, use a visual fraction model to show (2/3) × 4 = 8/3, and create a story context for this equation. Do the same with (2/3) × (4/5) = 8/15. (In general, (a/b) × (c/d) = ac/bd.)*	GMC	Interpret $\left(\frac{1}{b}\right) \times q$ as 1 part of a partition of q into b equal parts.
	GMC	Interpret $\left(\frac{a}{b}\right) \times q$ as a parts of a partition of q into b equal parts.
	GMC	Create story contexts for fraction multiplication problems.
5.NF.4b Find the area of a rectangle with fractional side lengths by tiling it with unit squares of the appropriate unit fraction side lengths, and show that the area is the same as would be found by multiplying the side lengths. Multiply fractional side lengths to find areas of rectangles, and represent fraction products as rectangular areas.	GMC	Justify the area formula for a rectangle with fractional side lengths by tiling.
	GMC	Find the areas of rectangles with fractional side lengths by multiplying.
	GMC	Represent fraction products as rectangle areas.
5.NF.5 Interpret multiplication as scaling (resizing), by:		
5.NF.5a Comparing the size of a product to the size of one factor on the basis of the size of the other factor, without performing the indicated multiplication.	GMC	Compare the size of a product to one factor based on the size of the other factor.
5.NF.5b Explaining why multiplying a given number by a fraction greater than 1 results in a product greater than the given number (recognizing multiplication by whole numbers greater than 1 as a familiar case); explaining why multiplying a given number by a fraction less than 1 results in a product smaller than the given number; and relating the principle of fraction equivalence $a/b = (n \times a)/(n \times b)$ to the effect of multiplying a/b by 1.	GMC	Explain the effects of multiplying by fractions greater than 1 or less than 1.
	GMC	Explain the effects of multiplying by fractions equal to 1.
5.NF.6 Solve real world problems involving multiplication of fractions and mixed numbers, e.g., by using visual fraction models or equations to represent the problem.	GMC	Solve real-world problems involving fraction multiplication.
	GMC	Solve real-world problems involving mixed-number multiplication.
5.NF.7 Apply and extend previous understandings of division to divide unit fractions by whole numbers and whole numbers by unit fractions.[1]		
5.NF.7a Interpret division of a unit fraction by a non-zero whole number, and compute such quotients. *For example, create a story context for (1/3) ÷ 4, and use a visual fraction model to show the quotient. Use the relationship between multiplication and division to explain that (1/3) ÷ 4 = 1/12 because (1/12) × 4 = 1/3.*	GMC	Interpret division of a unit fraction by a nonzero whole number and find quotients.
5.NF.7b Interpret division of a whole number by a unit fraction, and compute such quotients. *For example, create a story context for 4 ÷ (1/5), and use a visual fraction model to show the quotient. Use the relationship between multiplication and division to explain that 4 ÷ (1/5) = 20 because 20 × (1/5) = 4.*	GMC	Interpret division of a whole number by a unit fraction and find quotients.

[1] Students able to multiply fractions in general can develop strategies to divide fractions in general, by reasoning about the relationship between multiplication and division. But division of a fraction by a fraction is not a requirement at this grade.

Unpacking the Standards   EM5

Standards for Mathematical Content

5.NF.7c Solve real world problems involving division of unit fractions by non-zero whole numbers and division of whole numbers by unit fractions, e.g., by using visual fraction models and equations to represent the problem. *For example, how much chocolate will each person get if 3 people share 1/2 lb of chocolate equally? How many 1/3-cup servings are in 2 cups of raisins?*

GMC	Solve real-world problems involving division of unit fractions by whole numbers.
GMC	Solve real-world problems involving division of whole numbers by unit fractions.

Strand Measurement and Data   5.MD

Cluster Convert like measurement units within a given measurement system.

5.MD.1 Convert among different-sized standard measurement units within a given measurement system (e.g., convert 5 cm to 0.05 m), and use these conversions in solving multi-step, real world problems.

GMC	Convert among measurement units within the same system.
GMC	Use measurement conversions to solve multi-step, real-world problems.

Cluster Represent and interpret data.

5.MD.2 Make a line plot to display a data set of measurements in fractions of a unit (1/2, 1/4, 1/8). Use operations on fractions for this grade to solve problems involving information presented in line plots. *For example, given different measurements of liquid in identical beakers, find the amount of liquid each beaker would contain if the total amount in all the beakers were redistributed equally.*

GMC	Organize and represent data on line plots.
GMC	Solve problems involving fractional data on line plots.

Cluster Geometric measurement: understand concepts of volume.

5.MD.3 Recognize volume as an attribute of solid figures and understand concepts of volume measurement.

GMC	Recognize volume as an attribute of solid figures.

5.MD.3a A cube with side length 1 unit, called a "unit cube," is said to have "one cubic unit" of volume, and can be used to measure volume.

GMC	Understand that a unit cube has 1 cubic unit of volume and can measure volume.

5.MD.3b A solid figure which can be packed without gaps or overlaps using n unit cubes is said to have a volume of n cubic units.

GMC	Understand that a solid figure completely filled by n unit cubes has volume n cubic units.

5.MD.4 Measure volumes by counting unit cubes, using cubic cm, cubic in, cubic ft, and improvised units.

GMC	Measure volumes by counting unit cubes and improvised units.

5.MD.5 Relate volume to the operations of multiplication and addition and solve real world and mathematical problems involving volume.

5.MD.5a Find the volume of a right rectangular prism with whole-number side lengths by packing it with unit cubes, and show that the volume is the same as would be found by multiplying the edge lengths, equivalently by multiplying the height by the area of the base. Represent threefold whole-number products as volumes, e.g., to represent the associative property of multiplication.

GMC	Relate packing prisms with cubes to volume formulas.
GMC	Represent products of three whole numbers as volumes.

5.MD.5b Apply the formulas $V = l \times w \times h$ and $V = b \times h$ for rectangular prisms to find volumes of right rectangular prisms with whole number edge lengths in the context of solving real world and mathematical problems.

GMC	Apply formulas to find volumes of rectangular prisms.

5.MD.5c Recognize volume as additive. Find volumes of solid figures composed of two non-overlapping right rectangular prisms by adding the volumes of the non-overlapping parts, applying this technique to solve real world problems.

GMC	Find volumes of figures composed of right rectangular prisms.
GMC	Solve real-world problems involving volumes of figures composed of prisms.

Standards for Mathematical Content

Strand Geometry 5.G

Cluster Graph points on the coordinate plane to solve real world and mathematical problems.

5.G.1 Use a pair of perpendicular number lines, called axes, to define a coordinate system, with the intersection of the lines (the origin) arranged to coincide with the 0 on each line and a given point in the plane located by using an ordered pair of numbers, called its coordinates. Understand that the first number indicates how far to travel from the origin in the direction of one axis, and the second number indicates how far to travel in the direction of the second axis, with the convention that the names of the two axes and the coordinates correspond (e.g., *x*-axis and *x*-coordinate, *y*-axis and *y*-coordinate).	GMC Understand and use a Cartesian coordinate grid in two dimensions.
5.G.2 Represent real world and mathematical problems by graphing points in the first quadrant of the coordinate plane, and interpret coordinate values of points in the context of the situation.	GMC Represent problems by graphing points in the first quadrant. GMC Interpret coordinate values of points in context.

Cluster Classify two-dimensional figures into categories based on their properties.

5.G.3 Understand that attributes belonging to a category of two-dimensional figures also belong to all subcategories of that category. *For example, all rectangles have four right angles and squares are rectangles, so all squares have four right angles.*	GMC Understand that shapes in a subcategory have all the attributes of shapes in the parent category.
5.G.4 Classify two-dimensional figures in a hierarchy based on properties.	GMC Classify two-dimenional figures in a hierarchy based on properties.

Standards for Mathematical Process and Practice	*Everyday Mathematics* Goals for Mathematical Process and Practice

1 Make sense of problems and persevere in solving them.

Mathematically proficient students start by explaining to themselves the meaning of a problem and looking for entry points to its solution. They analyze givens, constraints, relationships, and goals. They make conjectures about the form and meaning of the solution and plan a solution pathway rather than simply jumping into a solution attempt. They consider analogous problems, and try special cases and simpler forms of the original problem in order to gain insight into its solution. They monitor and evaluate their progress and change course if necessary. Older students might, depending on the context of the problem, transform algebraic expressions or change the viewing window on their graphing calculator to get the information they need. Mathematically proficient students can explain correspondences between equations, verbal descriptions, tables, and graphs or draw diagrams of important features and relationships, graph data, and search for regularity or trends. Younger students might rely on using concrete objects or pictures to help conceptualize and solve a problem. Mathematically proficient students check their answers to problems using a different method, and they continually ask themselves, "Does this make sense?" They can understand the approaches of others to solving complex problems and identify correspondences between different approaches.	**GMP1.1** Make sense of your problem. **GMP1.2** Reflect on your thinking as you solve your problem. **GMP1.3** Keep trying when your problem is hard. **GMP1.4** Check whether your answer makes sense. **GMP1.5** Solve problems in more than one way. **GMP1.6** Compare the strategies you and others use.

2 Reason abstractly and quantitatively.

Mathematically proficient students make sense of quantities and their relationships in problem situations. They bring two complementary abilities to bear on problems involving quantitative relationships: the ability to *decontextualize*—to abstract a given situation and represent it symbolically and manipulate the representing symbols as if they have a life of their own, without necessarily attending to their referents—and the ability to *contextualize*, to pause as needed during the manipulation process in order to probe into the referents for the symbols involved. Quantitative reasoning entails habits of creating a coherent representation of the problem at hand; considering the units involved; attending to the meaning of quantities, not just how to compute them; and knowing and flexibly using different properties of operations and objects.	**GMP2.1** Create mathematical representations using numbers, words, pictures, symbols, gestures, tables, graphs, and concrete objects. **GMP2.2** Make sense of the representations you and others use. **GMP2.3** Make connections between representations.

Standards for Mathematical Process and Practice	*Everyday Mathematics* **Goals for Mathematical Process and Practice**

3 Construct viable arguments and critique the reasoning of others.

Mathematically proficient students understand and use stated assumptions, definitions, and previously established results in constructing arguments. They make conjectures and build a logical progression of statements to explore the truth of their conjectures. They are able to analyze situations by breaking them into cases, and can recognize and use counterexamples. They justify their conclusions, communicate them to others, and respond to the arguments of others. They reason inductively about data, making plausible arguments that take into account the context from which the data arose. Mathematically proficient students are also able to compare the effectiveness of two plausible arguments, distinguish correct logic or reasoning from that which is flawed, and—if there is a flaw in an argument—explain what it is. Elementary students can construct arguments using concrete referents such as objects, drawings, diagrams, and actions. Such arguments can make sense and be correct, even though they are not generalized or made formal until later grades. Later, students learn to determine strands to which an argument applies. Students at all grades can listen or read the arguments of others, decide whether they make sense, and ask useful questions to clarify or improve the arguments.	**GMP3.1** Make mathematical conjectures and arguments. **GMP3.2** Make sense of others' mathematical thinking.

4 Model with mathematics.

Mathematically proficient students can apply the mathematics they know to solve problems arising in everyday life, society, and the workplace. In early grades, this might be as simple as writing an addition equation to describe a situation. In middle grades, a student might apply proportional reasoning to plan a school event or analyze a problem in the community. By high school, a student might use geometry to solve a design problem or use a function to describe how one quantity of interest depends on another. Mathematically proficient students who can apply what they know are comfortable making assumptions and approximations to simplify a complicated situation, realizing that these may need revision later. They are able to identify important quantities in a practical situation and map their relationships using such tools as diagrams, two-way tables, graphs, flowcharts and formulas. They can analyze those relationships mathematically to draw conclusions. They routinely interpret their mathematical results in the context of the situation and reflect on whether the results make sense, possibly improving the model if it has not served its purpose.	**GMP4.1** Model real-world situations using graphs, drawings, tables, symbols, numbers, diagrams, and other representations. **GMP4.2** Use mathematical models to solve problems and answer questions.

Standards for Mathematical Process and Practice	*Everyday Mathematics* Goals for Mathematical Process and Practice

5 Use appropriate tools strategically.

Mathematically proficient students consider the available tools when solving a mathematical problem. These tools might include pencil and paper, concrete models, a ruler, a protractor, a calculator, a spreadsheet, a computer algebra system, a statistical package, or dynamic geometry software. Proficient students are sufficiently familiar with tools appropriate for their grade or course to make sound decisions about when each of these tools might be helpful, recognizing both the insight to be gained and their limitations. For example, mathematically proficient high school students analyze graphs of functions and solutions generated using a graphing calculator. They detect possible errors by strategically using estimation and other mathematical knowledge. When making mathematical models, they know that technology can enable them to visualize the results of varying assumptions, explore consequences, and compare predictions with data. Mathematically proficient students at various grade levels are able to identify relevant external mathematical resources, such as digital content located on a website, and use them to pose or solve problems. They are able to use technological tools to explore and deepen their understanding of concepts.

GMP5.1 Choose appropriate tools.

GMP5.2 Use tools effectively and make sense of your results.

6 Attend to precision.

Mathematically proficient students try to communicate precisely to others. They try to use clear definitions in discussion with others and in their own reasoning. They state the meaning of the symbols they choose, including using the equal sign consistently and appropriately. They are careful about specifying units of measure, and labeling axes to clarify the correspondence with quantities in a problem. They calculate accurately and efficiently, express numerical answers with a degree of precision appropriate for the problem context. In the elementary grades, students give carefully formulated explanations to each other. By the time they reach high school they have learned to examine claims and make explicit use of definitions.

GMP6.1 Explain your mathematical thinking clearly and precisely.

GMP6.2 Use an appropriate level of precision for your problem.

GMP6.3 Use clear labels, units, and mathematical language.

GMP6.4 Think about accuracy and efficiency when you count, measure, and calculate.

Standards for Mathematical Process and Practice	***Everyday Mathematics*** **Goals for Mathematical Process and Practice**

7 Look for and make use of structure.

Mathematically proficient students look closely to discern a pattern or structure. Young students, for example, might notice that three and seven more is the same amount as seven and three more, or they may sort a collection of shapes according to how many sides the shapes have. Later, students will see 7×8 equals the well remembered $7 \times 5 + 7 \times 3$, in preparation for learning about the distributive property. In the expression $x^2 + 9x + 14$, older students can see the 14 as 2×7 and the 9 as $2 + 7$. They recognize the significance of an existing line in a geometric figure and can use the strategy of drawing an auxiliary line for solving problems. They also can step back for an overview and shift perspective. They can see complicated things, such as some algebraic expressions, as single objects or as being composed of several objects. For example, they can see $5 - 3(x - y)^2$ as 5 minus a positive number times a square and use that to realize that its value cannot be more than 5 for any real numbers x and y.

GMP7.1	Look for mathematical structures such as categories, patterns, and properties.
GMP7.2	Use structures to solve problems and answer questions.

8 Look for and express regularity in repeated reasoning.

Mathematically proficient students notice if calculations are repeated, and look both for general methods and for shortcuts. Upper elementary students might notice when dividing 25 by 11 that they are repeating the same calculations over and over again, and conclude they have a repeating decimal. By paying attention to the calculation of slope as they repeatedly check whether points are on the line through $(1, 2)$ with slope 3, middle school students might abstract the equation $(y - 2)/(x - 1) = 3$. Noticing the regularity in the way terms cancel when expanding $(x - 1)(x + 1)$, $(x - 1)(x^2 + x + 1)$, and $(x - 1)(x^3 + x^2 + x + 1)$ might lead them to the general formula for the sum of a geometric series. As they work to solve a problem, mathematically proficient students maintain oversight of the process, while attending to the details. They continually evaluate the reasonableness of their intermediate results.

GMP8.1	Create and justify rules, shortcuts, and generalizations.

5–6 Games Correlation

Game	Grade 5 Lesson	Grade 6 Lesson	Operations and Algebraic Thinking	Number and Operations in Base Ten	Number and Operations—Fractions	Measurement and Data	Geometry	Ratios and Proportional Relationships	The Number System	Expressions and Equations	Statistics and Probability
Absolute Value Sprint		4-12							●		
Algebra Election		6-6								●	
Baseball Multiplication	1-2, 1-7, 2-5			●							
Build-It	3-6, 5-2	1-3, 1-11		●					●		
Buzz Games (Buzz and Bizz-Buzz)	1-11, 2-12, 5-2			●	●						
Daring Division		6-7							●		
Decimal Domination	6-9, 7-13, 8-7			●							
Decimal Top-It	4-4, 4-8, 4-13, 6-6			●							
Decimal Top-It: Addition or Subtraction	4-12, 5-1, 5-3, 6-4			●							
Divisibility Dash		1-7, 2-3							●		
Division Arrays	2-10			●							
Division Dash	2-10, 3-9			●							
Division Top-It		2-5, 3-9							●		
Division Top-It: Larger Numbers	2-11, 5-7, 6-11			●							
Doggone Decimal	6-8, 7-4	4-9, 6-1		●					●		
Exponent Ball	6-2, 6-7, 7-3, 8-1			●							
Factor Captor		1-1, 1-13, 2-1							●		
First to 100		8-5								●	
Fraction Action, Fraction Friction		3-6								●	
Fraction Capture	3-11, 4-3	1-5, 2-6	●		●				●		
Fraction Of	3-13, 4-1, 4-12, 5-5				●						
Fraction Spin	3-7				●						
Fraction Top-It: Addition	5-11, 7-5				●						
Fraction Top-It (Division)		2-8							●	●	
Fraction Top-It (Multiplication)		2-1							●	●	

Game	Grade 5 Lesson	Grade 6 Lesson	Operations and Algebraic Thinking	Number and Operations in Base Ten	Number and Operations—Fractions	Measurement and Data	Geometry	Ratios and Proportional Relationships	The Number System	Expressions and Equations	Statistics and Probability
Fraction/Whole Number Top-It	5-6, 5-14, 6-13	6-5			●				●	●	
Getting to One		3-1, 5-7, 7-5						●	●	●	
Hidden Treasure	4-7, 5-2	1-14, 2-13, 3-8, 7-1					●		●		
High-Number Toss	2-2, 2-13, 4-9			●							
High-Number Toss (Decimal)		2-8							●		
Landmark Shark		1-11, 2-11, 6-11									●
Mixed-Number Spin		1-8								●	
Multiplication Bull's Eye	2-7	2-10, 6-9		●					●		
Multiplication Top-It: Larger Numbers	2-5, 2-8, 3-4, 5-8			●							
Multiplication Wrestling	2-7			●							
Multiplication Wrestling (Mixed-Numbers)		3-3, 5-4								●	
Name That Number	1-1, 1-5, 2-8	3-11, 4-2, 4-3, 4-10, 5-5, 6-1	●	●						●	
Number Top-It	2-1, 2-4, 3-14			●							
Over and Up Squares	4-6, 4-11						●				
Percent Spin		4-12						●			
Polygon Capture		5-1					●				
Power Up	2-2, 2-12, 3-2			●							
Prism Pile-Up	1-12, 2-6, 3-3, 4-13, 6-6		●			●					
Property Pandemonium	7-7, 7-9, 8-3						●				
Ratio Comparison		5-10, 6-3, 7-8, 8-8						●			
Ratio Dominoes		3-13, 5-1						●			
Ratio Memory Match		4-1						●			
Rename That Mixed Number	3-8, 4-5				●						
Solution Search		4-10, 4-14, 7-1, 7-10, 8-3								●	
Spend and Save	4-14, 5-12, 6-11			●							

Game	Grade 5 Lesson	Grade 6 Lesson	Strand								
			Operations and Algebraic Thinking	Number and Operations in Base Ten	Number and Operations—Fractions	Measurement and Data	Geometry	Ratios and Proportional Relationships	The Number System	Expressions and Equations	Statistics and Probability
Spoon Scramble	**7-2, 7-12, 8-10**	**4-6, 6-7, 7-3**		●	●						
Spoon Scramble for Expressions		**4-6, 6-7**								●	
What's My Attribute Rule?	**7-5**						●				

Sample Work from Student A

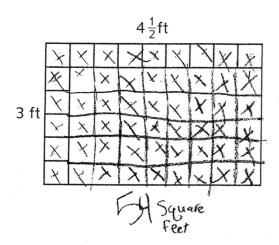

Sample Work from Student B

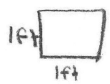

Allyson is going to use 13,5 ft

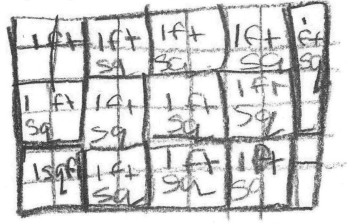

Sample Work from Student C

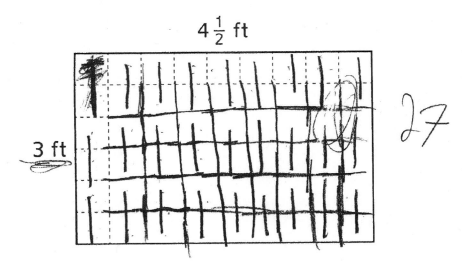

$4\frac{1}{2}$ ft

3 ft

27

1. Why might Justin think they will use 54 square feet of fabric? Do you agree or disagree with Justin's answer? Why? Justin might Think That They Wold need 54 Soyar feet of Fabric Because he Thought That one Square is one Foot and he counted 54 But I agree But I disagree, I disagree Because he Must have Thought That one square is one foot But 2 squares are 1 foot Because There are Six Squares and 3 ft So It most Be 2 Seuars are one Foot I agree Because he have The right awser But I is wrong Because he said It is 54 feet But It is relly 27 feet Because 2 Squares are 1 foot

Sample Work from Student A

4. Yes, it was efficient, it makes sence and gets her a goodestimate. I do belive that she could've done it faster though she could've used divison or done it all in 1 problem. Her way of finding how long it would take her is good though, ste did in the end get a pretty good estimate.

5. $100 \times N = 1,000,000$ $1,000,000/100 = N$

$N = 10,000$

I tapped 100 times in 29 secs.

~ 290000 I used fact extentions

$$10000 \times 29 \over 290000$$

Sample Work from Student B

5. 100 — 2 zeros $\times$ 38

    $\times$ 10,000 $\underset{=}{+}$ 4 zeros   10,000

    1,000,000   6 zeros   380,000

Sample Work from Student C

5. $100 \times 10,000 = 1,000,000$

Sample Work from Student A

2. Write a rule that you can use to make the largest possible fraction using any three number cards. Explain why your rule works. My rule is bigger the demominator smaller the divisor bigger the answer to win the game.

Sample Work from Student B

2. Write a rule that you can use to make the largest possible fraction using any three number cards. Explain why your rule works.

My rule is you have to make your numorator large so you can get a big sum to ~~divet~~ divided from your small donominator. Also did you know the numorator is the dividend. Second, you have to make your donominator small so it will be bigger parts making the fraction bigger. Also the donominator is the divisor. In conclusion basicly larger the sum of the numerator and the smaller the ~~d~~ donomonator the bigger the fraction.

Sample Work from Student A

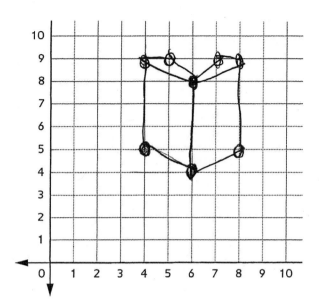

Lesson 4-10

Sample Work from Student B

1. Write a rule Jake can use to make the picture of the book larger.

   Jake can multiply the X coordinate.
                         by 6

2. Describe what you think the new picture will look like. Be specific about how you think the book will change.

   The Folder will be taller.

3. Use your rule to write the new coordinates.

Original Book	New Book
(2, 0)	(12, 0)
(0, 1)	(0, 6)
(0, 5)	(0, 30)
(1, 5)	(6, 30)
(2, 4)	(12, 24)
(0, 5)	(0, 30)
(2, 4)	(12, 24)
(2, 0)	(12, 0)
(4, 1)	(24, 6)
(4, 5)	(24, 30)
(3, 5)	(18, 30)
(2, 4)	(12, 24)
(4, 5)	(24, 30)

4.

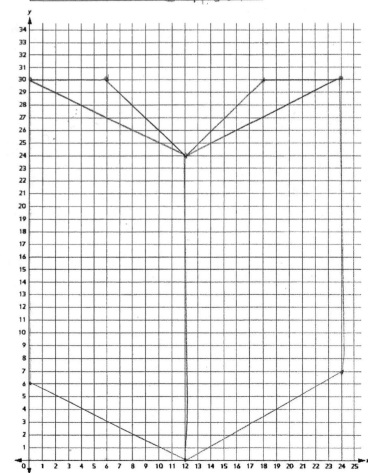

Sample Work from Student C

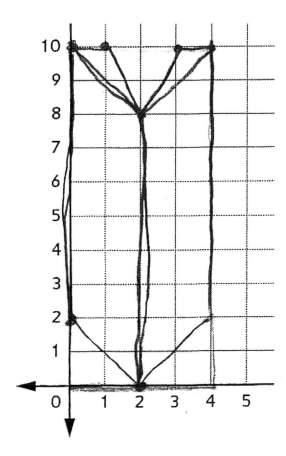

5. Compare your prediction in Problem 2 to the new picture. How was your prediction correct? How was your prediction incorrect?

It was correct, it got taller

5.NF.4, 5.NF.4b

Work Sample #1—
Not Meeting Expectations

This sample work does not meet expectations for the content standards or for the mathematical process and practice standards. This student says that Allyson multiplied 3 by $4\frac{1}{2}$ to get $13\frac{1}{2}$ but agrees with both Allyson and Justin, which indicates a serious misconception about what the multiplication represents. 5.NF.4, 5.NF.4b Although the student multiplied $3 * 4\frac{1}{2}$ correctly, the student goes on to say that $13\frac{1}{2} = 54$ square feet. This lack of clarity shows that the student has not made sense of Allyson's thinking. GMP3.2

2. Why might Allyson think they will use $13\frac{1}{2}$ square feet of fabric? Do you agree or disagree with Allyson's answer? Why?

I think Allyson Said 13 1/2 because she multiplyied 3ft x 4 1/2 and that equals 13 1/2. I agree with both because 13 1/2 = 54 square feet.

Work Sample #2—
Partially Meeting Expectations

This sample work meets expectations for the content standards but only partially meets expectations for the mathematical process and practice standards. This student agrees with Allyson and shows a way to find the area by tiling the rectangle. 5.NF.4, 5.NF.4b The student uses words and a number model to show that Allyson could have multiplied $3 * 4\frac{1}{2}$ to get $13\frac{1}{2}$ but does not reference square feet. The student also draws a picture showing how to count 13 sets of 4 squares and 1 set of 2 squares but does not explain that 4 small squares make up 1 square foot or that 2 small squares make up $\frac{1}{2}$ square foot. GMP3.2

2. Why might Allyson think they will use $13\frac{1}{2}$ square feet of fabric? Do you agree or disagree with Allyson's answer? Why?

Allyson got her answer by multipling $3 \times 4\frac{1}{2} = 13\frac{1}{2}$.

Allyson is right because she multiplied $3 \times 4\frac{1}{2} = 13\frac{1}{2}$ which o exactly the answer.

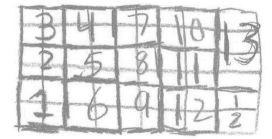

5.NF.4, 5.NF.4b

Work Sample #3—
Partially Meeting Expectations

This sample work meets expectations for the content standards but only partially meets expectations for the mathematical process and practice standards. With revision (the picture and discussion below the line), this student agrees with Allyson and correctly shows a method for finding the area. 5.NF.4, 5.NF.4b The student draws a picture showing how to count 12 big squares made up of 4 little squares and explains grouping 4 of the 6 remaining little squares into a group of 4 little squares and a group of 2 little squares. The student says that there are $13\frac{1}{2}$ squares, which we can infer means big squares. However, the student did not refer to units or what the little and big squares represent. GMP3.2

2. Why might Allyson think they will use $13\frac{1}{2}$ square feet of fabric? Do you agree or disagree with Allyson's answer? Why?

Allyson came up with her answer by doing 4½ x 3=13⅘ So she was wrong because Justin got the right anser

Allyson was right because we were doing sqers on we did 4 little sqers and and then a big one. and I got 12 sq Boxes then There were 6 sqs left So I did 4 little sqes and 2 other little ones and I got 13½ sqiers

□=little sq
▢=big sq

5.NBT.2

Work Sample #1—Meeting Expectations

This sample work meets expectations for the content standard and for the mathematical process and practice standards. This student calculated a reasonable estimate of the time it takes to make 1 million taps using patterns of powers of 10. **5.NBT.2** The student identifies an aspect of Maya's solution that is not efficient by saying Maya could have divided 1,000,000 by 100 to get 10,000. This student shows a more efficient solution using Maya's sample count, division, and extended facts. GMP6.4

4. Did Maya use an efficient strategy? Explain your thinking.

I don't think her way was Efficent because she could of Just divided 1,000,000 ÷ 100 = to get 10,000 with TEN thousand She could multiply that by 22 and get 220,000 Seconds with would be her Answer

5. Estimate the time it would take you to tap your desk 1 million times, without any interruptions. Use the time it took you to make 100 taps in your estimate. Use a strategy that is more efficient than Maya's strategy. Show your strategy on another sheet of paper.

$$1,000,000 \div 100 = 10,000$$

$$10,000 * 22 = 220,000$$

Work Sample #2—Partially Meeting Expectations

This sample work meets expectations for the content standard and partially meets expectations for the mathematical process and practice standards. This student recorded 40 seconds as the time it took to tap the desk 100 times and then made a reasonable estimate of 400,000 seconds for the time it takes to make 1,000,000 taps for Problem 5. **5.NBT.2** For Problem 4 the student explains that Maya's solution was very long and that his or her own solution strategy was faster. However, for Problem 5 the student does not show how to use a more efficient strategy for finding the number of seconds it takes to make 1,000,000 taps. GMP6.4

4. Did Maya use an efficient strategy? Explain your thinking.

No, I dont think Maya's was a good strategie, because she made it very long and confusing. I think it was inificient because she made it very long andconfusing. My way is more efficidnt because it is faster and easier. One way it might be a little efficidnt is because with that many steps, she is probably extra sure she is right.

5. Estimate the time it would take you to tap your desk 1 million times, without any interruptions. Use the time it took you to make 100 taps in your estimate. Use a strategy that is more efficient than Maya's strategy. Show your strategy on another sheet of paper.

$$40 \times 1 = 40$$

$$400,000$$

5.NBT.2

Work Sample #3—
Partially Meeting Expectations

This sample work does not meet expectations for the content standard and partially meets expectations for the mathematical process and practice standards. This student left Problem 5 blank, so he or she did not find a reasonable estimate of the time it takes to make 1,000,000 taps. 5.NBT.2 The student says that Maya's solution is not efficient and that it would be better if she used fact extensions because it would be quicker. However, the student does not show how to find the number of seconds it takes to make 1,000,000 taps. GMP6.4

4. Did Maya use an efficient strategy? Explain your thinking.

I think her way is not so efficient. I think it would be better if she did fact extentions because It would be quicker to do and so it doesn't take up so much space and it's easier for some people to read.

5.NF.3

Work Sample #1—Meeting Expectations

This sample work meets expectations for the content standard and for the mathematical process and practice standards. This student correctly divided the numerators by the denominators in Problem 1 (not shown) and showed the interpretation of a fraction as division by stating that the numerator is the dividend and the denominator is the divisor. 5.NF.3 The student described a generalizable rule and explained why it works in Problem 2. The student's rule is to find the largest numerator and smallest denominator in order to get the largest fraction. The student justified the rule by explaining that having a smaller denominator results in "bigger parts making the fraction bigger." GMP8.1

2. Write a rule that you can use to make the largest possible fraction using any three number cards. Explain why your rule works.

My rule is you have to make your numerator large so you can get a big sum to ~~get~~ divided from your small donominator. Also did you know the numerator is the dividend. Second, you have to make your donominator small so it will be bigger parts making the fraction bigger. Also the donominator is the divisor. In conclusion basicly larger the sum of the numerator and the smaller the ~~d~~ donomo- nator the bigger the fraction.

Work Sample #2— Partially Meeting Expectations

This sample work meets expectations for the content standard and partially meets expectations for the mathematical process and practice standards. The number sentence $\frac{9}{3} = 3$ shows that the student interpreted fractions as division of a numerator by a denominator. 5.NF.3 Although the student's rule of making the largest numerator possible is correct, there is no explanation of why the rule works. GMP8.1

2. Write a rule that you can use to make the largest possible fraction using any three number cards. Explain why your rule works.

The rule is if you have 3 numbers (example) 543, I would pick 5 & 4 the sum is 9. I was trying to get the largest dividend, In a division problem the dividend ~~is~~ is the numartor, So in this case a would be the numarator, So the 3 would be the divisor in a division problem. Also in fraction form the three would the denomin ator.

$$\frac{9}{3} = 3$$

5.NF.3

Work Sample #3—
Exceeding Expectations

This sample work meets expectations for the content standard and exceeds expectations for the mathematical process and practice standards. The work shows an understanding of fractions as division of a numerator by a denominator with correct division in Problem 1 (not shown). 5.NF.3 This student stated a rule and explained why it works, including an example to support his or her reasoning. The student's rule is that the player should place the largest two digits in the numerator and the smallest digit in the denominator. The student explained that this rule works because having a bigger numerator means you start out with more of what you are going to divide up. This student also explained that if the denominator is small, then the amount in each group will be bigger. With revision this work exceeds expectations by using other numbers (14, 7, 2) that support his or her reasoning and showing an understanding of the relationship between whole number division and fractions. GMP8.1

2. Write a rule that you can use to make the largest possible fraction using any three number cards. Explain why your rule works.

To get the biggist possible fraction, put the 2 biggist numbers as the numerator and the smallest number as the denominator. This works because if you put the biggist number on top you start out with the most of whatever you are talking about and if the denominator is small then the groups will be bigger.
For example, $14 \div 7 = 2$
but, $14 \div 2 = 7$
so if there is less groups then there is more in the groups.

5.G.2

Work Sample #1—
Partially Meeting Expectations

This sample work meets expectations for the content standard and partially meets expectations for the mathematical process and practice standards. This student correctly plotted points to represent the new book. 5.G.2 While this student correctly listed and plotted the new coordinates and described the new picture as getting bigger, the student did not explicitly describe which attributes (length, width, area) of the picture got bigger. GMP5.2

1. Write a rule Jake can use to make the picture of the book larger.

Double first and second cordinates

3. Use your rule to write the new coordinates.

Original Book	New Book
(2,0)	4,0
(0,1)	0,2
(0,5)	0,10
(1,5)	2,10
(2,4)	4,8
(0,5)	0,10
(2,4)	4,8
(2,0)	4,0
(4,1)	8,2
(4,5)	8,10
(3,5)	6,10
(2,4)	4,8
(4,5)	8,10

4.

5. Compare your prediction in Problem 2 to the new picture. How was your prediction correct? How was your prediction incorrect?

The picture did get bigger but not big enough.

5.G.2

Work Sample #2—Meeting Expectations

This sample work meets expectations for the content standard and for the mathematical process and practice standards. This student correctly plotted points to represent the new book. 5.G.2 This student listed and plotted the new coordinates without errors. Furthermore, this student specifically indicated the original and new widths and lengths of the books. The student said that the book doubled, which is unclear and might imply that the area doubled. Students are not required to see the relationship between the area of the original book and the area of the new book. However, if this student had clearly stated which dimensions doubled or that the area quadrupled, the work would exceed expectations. GMP5.2

1. Write a rule Jake can use to make the picture of the book larger.

   Double both coordinates

3. Use your rule to write the new coordinates.

Original Book	New Book
(2,0)	4,0
(0,1)	0,2
(0,5)	0,10
(1,5)	2,10
(2,4)	4,8
(0,5)	0,10
(2,4)	4,8
(2,0)	4,0
(4,1)	8,2
(4,5)	8,10
(3,5)	6,10
(2,4)	4,8
(4,5)	8,10

4.

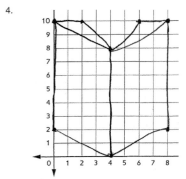

5. Compare your prediction in Problem 2 to the new picture. How was your prediction correct? How was your prediction incorrect? It was correct because, my last width was four, now its 8. My heighth was 5 Now its 10. The Book did double like I Predicted.

5.G.2

Work Sample #3—Exceeding Expectations

This sample work meets expectations for the content standard and exceeds expectations for the mathematical process and practice standards. This student correctly plotted points to represent the new book. **5.G.2** This student listed and plotted the new coordinates without errors. Furthermore, after revision this student correctly described the length and width as doubling, which is a more sophisticated articulation of the multiplicative relationship between the change and the dimensions. **GMP5.2**

1. Write a rule Jake can use to make the picture of the book larger.

Double both coordinates.

3. Use your rule to write the new coordinates.

Original Book	New Book
(2,0)	4,0
(0,1)	0,2
(0,5)	0,10
(1,5)	2,8
(2,4)	4,8
(0,5)	0,10
(2,4)	4,8
(2,0)	4,0
(4,1)	8,2
(4,5)	8,10
(3,5)	6,10
(2,4)	4,8
(4,5)	8,10

4.

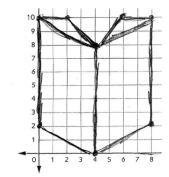

5. Compare your prediction in Problem 2 to the new picture. How was your prediction correct? How was your prediction incorrect?

Yes, my prediction was correct. The book got wider and taller. It got higher because the book started out as 5 units high and I doubled it and got 10 units high and The book started out as 4 units wide and I doubled it and got 8 units wide.

Notes